Field Guide to Appropriate Technology

Field Guide to Appropriate Technology

Edited by

BARRETT HAZELTINE
Brown University

CHRISTOPHER BULL
Brown University

ACADEMIC PRESS

An imprint of Elsevier Science

Amsterdam Boston London New York Oxford Paris San Diego
San Francisco Singapore Sydney Tokyo

ACADEMIC PRESS
An imprint of Elsevier Science
525 B Street, Suite 1900, San Diego, California 92101-4495, USA
http://www.academicpress.com

Academic Press
84 Theobald's Road, London, WC1X 8RR, UK
http://www.academicpress.com

Library of Congress Catalog Card Number: 2001090670

International Standard Book Number: 0-12-335185-5

Transferred to Digital Printing in 2010

Contents

Chapter 2: Energy
Edited by NEWELL THOMAS

Chapter 3:   Food
Edited by TUARIRA ABDUEL MTAITA

Chapter 4:   Health
Edited by STEVEN G. McCLOY

Chapter 5:   Construction
Edited by RICHARD HERBOLD

Chapter 6:   Transportation
Edited by BARRETT HAZELTINE

Chapter 7:   Household Technologies
Edited by BARRETT HAZELTINE

Chapter 8:   Water Supply
Edited by BARRETT HAZELTINE

Contributors

Eric Allemano   International Education and Training Consultant, Long Island City, New York, USA

Elaine L. Bearer   Professor of Biology and Medicine, Brown University, Providence, Rhode Island, USA

David J. Bermon   Latino Health Institute, Boston, Massachusetts, USA

Jonathan Brown   The World Bank, Washington, DC, USA

Thomas J. S. Carlson   Professor and Director of the Center for Health, Ecology, Biodiversity, & Ethnobiology, University of California, Berkeley, Berkeley, California, USA

Philip B. Cedillo   International Rice Research Institute, Makati City, Philippines

Drew Conroy   Professor of Applied Animal Science, University of New Hampshire, Durham, New Hampshire, USA

Andrew Conteh   Professor, Minnesota State University Moorhead, Moorhead, Minnesota, USA

James Crossley   Senior Engineer (retired), Westinghouse Corporation, Wexford, Pennsylvania, USA

Stephen Davis   Clinical Associate Professor of Biology and Medicine, Family Medicine, Brown University, Providence, Rhode Island, USA

Alan Dick   Alaska Native Knowledge Network, University of Alaska Fairbanks, Fairbanks, Alaska, USA

Dimitri Donskoy   Associate Professor and Associate Director of the Davidson Laboratory, Stevens Institute of Technology, Hoboken, New Jersey, USA

Jeff Doff   Monitor Company, Cambridge, Massachusetts, USA

Gwinyai Dziwa   Primary Health Care Trust, Rusape, Zimbabwe

Barbara Earth   Professor of Gender and Development Studies, Asian Institute of Technology, Pathumthani, Thailand

Robert Elliott   Independent Builder, Dighton, Massachusetts, USA

David E. Erikson   New Hampshire Public Schools, Weare, New Hampshire, USA

Katherine Evans   Brown University, Providence, Rhode Island, USA

Majid Ezzati   Post-Doctoral Fellow of Science, Technology, and Environmental Policy (STEP) Program, Princeton University, Princeton, New Jersey, USA

Richard Ford   Professor of Community-Based Development, Clark University, Worcester, Massachusetts, USA

Harry S. H. Gombachika   University of Malawi—The Polytechnic, Chichiri, Blantyre, Malawi

Jill Graat   Brown University, Providence, Rhode Island, USA

Nancie H. Herbold   Professor of Nutrition, Simmons College, Boston, Massachusetts, USA

Richard F. Herbold   Parsons Brinckerhoff, Inc., Boston, Massachusetts, USA

Kwamena Herbstein   Brown University, Providence, Rhode Island, USA

Kylee Hitz   Artisan (Independent), Rehoboth, Massachusetts, USA

Fred V. Jackson   Ecology and Evolutionary Biology, Division of Biology and Medicine, Brown University, Providence, Rhode Island, USA

Daniel M. Kammen   Professor of Energy and Society, University of California, Berkeley, Berkeley California, USA

Float Auma Kidha   Program Officer for AIDS, International Planned Parenthood Foundation, Nairobi, Kenya

Helen King   Rumford, Rhode Island, USA

Etta Kralovec   Vice President for Learning, Training and Development Corporation, Oland, Maine, USA

Ayse Kudat   The World Bank, Washington, DC, USA

Jochen Lohmeier   Organizational Development/Consultants International, Berlin, Germany

Aubrey Ludwig   Urban Environmental Laboratory, Brown University, Providence, Rhode Island, USA

Brian Mangwiro   Faculty of Agriculture and Natural Resources, Africa University, Mutare, Zimbabwe

B. M. Mbinda   Mpala Research Centre, Laikipia, Kenya

Steven G. McCloy   Clinical Assistant Professor of Medicine, Brown University, Providence, Rhode Island, USA, and Director of the San Lucas Health Project, Guatemala

Tuarira Abduel Mtaita   Faculty of Agriculture and Natural Resources, Africa University, Mutare, Zimbabwe

Zephirin Ndikumana   Faculty of Agriculture and Natural Resources, Africa University, Mutare, Zimbabwe

Philip Nelson   Village Development Program, Evangelical Lutheran Church, Central African Republic

Henry Ngaiyaye   University of Malawi—The Polytechnic, Chichiri, Blantyre, Malawi

Martha K. Preus   Organizational Development/Consultants International, Berlin, Germany

Greta Rana   International Centre for Integrated Mountain Development (ICIMOD), Kathmandu, Nepal

Michael A. Rice   Professor, Department of Fisheries, Animal and Veterinary Science, University of Rhode Island, Kingston, Rhode Island, USA

Rachel Salguero   Tufts University School of Medicine, Boston, Massachusetts, USA

Sarah Scheiderich   Department of Anthropology, University of Illinois, Urbana, Illinois, USA

Monique Schumacher   Brown University, Providence, Rhode Island, USA

Burton H. Singer   Office of Population Research, Princeton University, Princeton, New Jersey, USA

Frank Simpson   Professor Earth Sciences, University of Windsor, Windsor, Ontario, Canada

Girish Sohani   Executive Vice President, BAIF Development Research Foundation, Warje, Pune, India

David Sowerwine   International Centre for Integrated Mountain Development (ICIMOD), Kathmandu, Nepal

Fanuel Tagwira   Professor and Dean of Faculty of Agriculture and Natural Resources, Africa University, Mutare, Zimbabwe

Stanslous Thamangani   Faculty of Agriculture and Natural Resources, Africa University, Mutare, Zimbabwe

Newell Thomas   Bright Technologies LLC, Providence, Rhode Island, USA

Francis Vanek   Director, Sustainable Technology and Energy for Vital Economic Needs (STEVEN Foundation), Ithaca, New York, USA

Laura Voorhees   Brown University, Providence, Rhode Island, USA

Kevin J. Weddle   Colonel in the United States Army Corps of Engineers and Professor of Military Strategy, Planning, and Operations, United States Army War College, Carlisle, Pennsylvania, USA

Laura Wells   Brown University, Providence, Rhode Island, USA

Myra Wong   UCLA/RAND Center for Adolescent Health Promotion, Los Angeles, California, USA

Jessica Wurwarg   Brown University, Providence, Rhode Island, USA

Chantelle Wyley   Organizational Development/Consultants International, Berlin, Germany

Acknowledgments

We appreciate the advice and support we received from Professor Robert Armstrong, Africa University, Zimbabwe, and Ralph Rusley, Lutheran Pastor, Retired. We are also very grateful for the professional and sensitive work of the Senior Project Manager from Academic Press, Julio Esperas.

Overview

Edited by BARRETT HAZELTINE
Brown University

INTRODUCTION TO BOOK

The intended audience for this book is a person from the United States or another industrialized country who intends to work in the less industrialized parts of the Third World, hoping to make life better for people in the local community. This book is a collection of short articles about so-called "appropriate technology." The articles are not definitive treatises. They are intended to illustrate the possibilities of the technology described, and also to provide sufficient information for the reader to start implementing that technology. In some cases the best way to start is to do more reading. References are included, for the reader, to both printed material and web sites.

We focus on appropriate technology because we believe it brings special advantages to people in the Third World. It is worth repeating a description of appropriate technology here because not all readers may be familiar with the idea and because repeating a description may focus attention on the reasons for using appropriate technology. One description characterizes appropriate technology as being small scale, energy efficient, environmentally sound, labor intensive, and controlled by the local community. The Intermediate Technology Development Group, an organization that pioneered the use of appropriate technology, uses nearly the same description but adds that the technology must be simple enough to be maintained by the people using it.

This introductory chapter has three articles that we think will be useful to most readers regardless of their particular situation. The first article, by Professor Andrew Conteh, discusses how to implement an appropriate technology project. The article has a particular emphasis on paying attention to the culture of the place one will be working. The second article focuses on how to design things; a person wanting to work in the Third World could design devices or processes that improve the quality of life. Design usually emphasizes things—machines or seeds, for example—but it has a human aspect also. A primary question to answer is "what will the effect of this project/technology be on this individual community?" The third article gives guidelines on how one can be useful in a foreign culture.

We hope the reader will find the articles useful but not use them rigidly. A glance at the table of contents will show the range of technologies discussed. We expect that most readers will pick and choose among the articles, selecting ones that are helpful in their present situation. All technologies change. New ideas are an integral part of small-scale technologies. We urge the reader to be open to change, and to build on what is described in the book.

1

Chapter one deals with planning and implementation. A lesson learned from the history of appropriate technology projects is that the benefits of new technology, no matter how appropriate the technology, do not come by simply creating the technology. The implementation must be planned, and peaple must be convinced to change.

An often-asked question is what good can a person from another culture do in a third world situation. We believe she or he can suggest possibilities by showing how problems similar to ones facing people in a third world village have been approached in other places. The United States and other industrialized conuntries have become stronger when embracing ideas and devices from abroad. The communities of the Third World should have the same opportunity to learn about ideas and devices foreign to them.

A few notes on the presentation are needed. We use the euphemism "Third World" to refer to countries that are significantly less industrialized than the United States. Alternative phrases— underdeveloped, developing, less developed, industrializing, poor, southern, and so forth— seemed either unwieldy or patronizing. A particular objection to "underdeveloped" is that many Third World nations have a culture at least as developed as the United States. We realize "Third World" is a very broad term, used to include countries as different as Brazil and Nepal but hope that the context will make the meaning clear. When we use the unit of dollars we are referring to United States dollars exclusively so we do not use the abbreviation US $. Several of the examples use a local currency. We thought it misleading to translate uniformly into dollars as wage and price scales make such translations misleading for U.S. readers. We have generally used the units of length and weight that fit the example being discussed—feet and pounds in the United States, meters and kilograms elsewhere. A conversion table is given. We have expected, in general, that most of our readers will have United States backgrounds and have usually written from that perspective.

UNITS

1 acre	equals	4.356×10^4	square feet
1 Btu	"	1054.8	joules
1 cubic foot	"	7.481	gallons
1 cubic meter	"	35.31	cubic feet
1 foot	"	30.48	centimeters
1 gram	"	3.527×10^{-2}	ounces
1 hectare	"	2.471	acres
1 inch	"	2.54	centimeters
1 liter	"	61.02	cubic inches
1 liter	"	.2642	gallons

The abbreviation "kilo" means $1000 = 10^3$
The abbreviation "mega" means $1,000,000 = 10^6$

CULTURE AND THE TRANSFER OF TECHNOLOGY

Andrew Conteh

Culture is the newest fad sweeping the literature on international relations, security studies, and international economics.[1] A throng of recent essays and books point to culture as the basic force impelling nation-states (who are the principal actors in international relations), other institutions, and individuals to act and organize themselves as they do. Many of these writings argue that culture's importance is rapidly growing.

The notion that culture affects human behavior is, of course, hardly new. Observations about relative skills and behavior patterns of various societies are as old as human history. In modern times, Max Weber studied the relative economic benefits of Protestant and Catholic cultures; Adda Bozeman and others have focused on culture's role in national decision making; Lucian Rye and Sydney Verba connected national culture to development; and Robert Putnaur studied the relationship between civic culture and democracy. Some scholars contend that with the end of the Cold War, cultural factors have finally emerged as predominant in international relations, the primus inter pares of the engines driving world affairs. One form of cultural expression is through technology, as well as through the discussions of appropriate technology.

The purpose of this short essay is not to outline the details of the scope and contents of the strategies in appropriate technology but rather to explore one of the critical issues in the debate on technology transfer, namely the "appropriateness of technology" within existing cultures. There can be little doubt that cultural attributes play a substantial role in providing human beings with mental, moral, and economic tools for life. It goes beyond any doubts that some cultures are better equipped than others for the successful development of capitalism. Lawrence E. Harrison, a former U.S. Agency for International Development official, makes this argument in "Who Prospers: How Cultural Values Shape Economic and Political Success." "It is values and attitudes—culture," he writes, that differentiate ethnic groups and are mainly responsible for such phenomena as Latin America's persistent instability and inequity, Taiwan's and Korea's economic "miracles," and the achievement of the Japanese. Culture thus plays a critical role in determining the economic fates of nation-states, peoples, and individuals because some cultures underwrite success better than others. Second, cultural perspectives and belief systems strongly influence the way in which national leaders, civil servants, and so forth view policy problems, both individually and collectively over time, and often determine the solutions they choose to deal with them. Third, culture serves as the dominant blueprint for social, economic, and military structures and institutions, thus exercising a strong influence on the behavior, attitudes, and prospects of nation-states in the world community. Finally, culture can also be the chief source of conflict between nation-states, transnationals and nation-states, and individuals—civil servants and foreign experts. Consequently, there appears to be a clash between culture and the process of globalization. In this process culture often emerges as the victor.

And now let's turn our attention to the debate about appropriateness of technology. In recent years the concept of appropriate technology has gained currency in both developing and developed countries. This concept has emerged in response to a recognition that culture and other factors are significant to technology transfer. One of the reasons for this situation has been an overemphasis in the past on the part of a large number of developing countries on capital-intensive "heavy" industrialization and the use of technologies that do not necessarily reflect factor endowments and socioeconomic conditions prevailing in these countries. Because nations wary in their level of cultural resource endowment, the strategies adopted by each nation in the pursuit of technological development would vary from nation to nation depending on each nations particular circumstances.[2]

WHAT IS AN APPROPRIATE TECHNOLOGY?

Appropriate technology is defined as any object, process, ideas, or practice that enhances human fulfillment through satisfaction of human needs. A technology is deemed to be appropriate when it is compatible with local, cultural, and economic conditions (i.e., the human, material and cultural resources of the economy), and utilizes locally available materials and energy resources, with tools and processes maintained and operationally controlled by the local population. Technology is considered thus "appropriate" to the extent that it is consistent with the cultural, social, economic, and political institutions of the society in which it is used. Abubakar N. Abdullalli has suggested that appropriate technology should be self-sustaining, cause little cultural disruption, and should ensure the relevance of technology to the welfare of the local population.[3]

Appropriate technology represents the social and cultural diversions of innovation. The essence of appropriate technology is that the usefulness or value of a technology must be consolidated by the social, cultural, economic, and political milieu in which it is to be used. Most of the groups working in the developing countries tend to view appropriate technology as the main tool in meeting the basic needs of hundreds of millions of people who have been largely left out of the development process.

WHY DO DEVELOPING COUNTRIES NEED APPROPRIATE TECHNOLOGY?

In his book on appropriate technology P. D. Dunn,[4] lists the areas of appropriate technology as follows:

1. The provisions of employment
2. The production of goods for local market
3. The substitution of local goods for those previously imported and that are competitive in quality and cost
4. The use of local resources of labor, materials, and finance
5. The provision of community services including health, water, sanitation, housing, roads, and education

Dunn asserts that appropriate technology should be compatible with the wishes, culture, and traditions of a particular community and not have a socially disruptive effect. While Dunn gives these broad guidelines, other writers have been rather specific on the issue. One reason why developing countries need appropriate technology is the poor state of socioeconomic development of these countries. As Nji Agaga[5] explains it, developing countries are characterized by a predominately agrarian and hence rural population. These countries account for 70 percent of the total world population; of this 75 to 80 percent of them live in rural areas. Adams and Bjork[6] also note that apart from this structural distribution of the population in the developing countries, these countries have low incomes, low literacy rates, poor or inadequate infrastructure (e.g., roads, buildings, equipment, etc.), insufficient health and sanitary facilities, and a chronic shortage of fiscal resources. Yet, Nji points out these countries have plenty of labor, vast rural resources, and a pool of latent untapped agricultural and mining resources. Venkare notes that the transfer of technology to the so-called underdeveloped countries must take into account the following predominant characteristics of these countries.

- The prevalence of a dual economy (urban and rural) with different and often conflicting lifestyles, unequal resource distribution, and a highly intersectional migration propensity: general unemployment and underemployment and consequently inequalities in income distribution; the relative lack of fixed and circulating capital in terms of machines, tools, and money; and a larger population growth rate as compared to the developed countries.
- An awareness of these inequalities (social and economic injustices) and the fact that they can indeed be removed is a good reason to design and implement strategies that can solve or reasonably alleviate these problems. It is to this end that appropriate technology can be used as a means to achieve balanced development in poor countries.

HOW CAN THIS BE DONE?

1. *Clearly identify the problems.* Commonsense knowledge tells us that the problems noted above are obvious. Some writers have thus emphasized the importance of applied research and the use of social indicators to "assess" the ability of a geopolitical unit to provide for the continued enhancement of the human condition in those social domain areas consensually defined as important for social well-being. Thus, applied research is seen as a vital component not only in the implementation of rural development projects but also as a desirable and indispensable prelude to the identification, selection, and development of appropriate technologies to combat destitution and misery in the rural areas.

   Such societal and technological development can be attained by a number of methods.

   a. Identifying existing technologies in the developing countries to select those that are useful from those that are not
   b. Improving the quality and performance of human resources in developing countries considered useful in eradicating poverty in general and in particular rural poverty
   c. Recycling used technology
   d. Adapting imported technology to local needs, materials, and resources
   e. Research and development of appropriate technologies to solve basic human needs

2. *A common definition of the nature and purpose of appropriate technology.* In light of the above-mentioned, it is of paramount importance that the technological analysts, development scientists, and policymakers (local civil servants) all share a common definition of the nature and purpose of appropriate technology. However, the influence from the appropriate technology approach is that the major criteria for the selection of appropriate forms of technologies are country factor endowment, the proportional distribution of labor in the country, land, capital, skills, and natural resources.

3. *Meeting basic human needs requirement.* The technology required for a basic needs strategy in a developing country must concentrate more than the past on meeting the requirements of the small farmer, small-scale rural industry, and the informal sector producer. Such a strategy calls for, and is in turn supported by, a special kind of appropriate technology: a technology that differs from that developed in the industrialized countries and for the industrialized countries.

4. *Technological assessment.* In order to make available to developing countries a greater flow of information, permitting the selection of technologies, the establishment of an "Industrial Technological Information Bank" is proposed, with possible regional or sectoral banks, and of an "International Center for the Exchange of Technological Information," notably for sharing research findings relevant to developing countries: This would enable a survey of technologies available and an assessment of their respective merits, leading to technological choices. Developing countries should improve the transparency of the industrial property market in order to facilitate the technological choices and be able to select an appropriate technology.

5. *Transfer of technologies.* All states should cooperate in evolving an international Code of Conduct for the Transfer of Technology, corresponding in particular to the special needs of the developing countries for a people's oriented development. Nation-states should also adopt and strictly adhere to the Code of Conduct of the Transnational Corporations that is being elaborated by the United Nations.

   To aid developing countries in the transfer and development of technology, the Paris Convention on the protection of industrial property, as well as the other international conventions on patents and trademarks, should be reviewed and revised to meet in particular the special needs of the developing countries.

6. *Information and adaptation.* Developed countries should facilitate access of developing countries on favorable terms and conditions to relevant information and other technologies suited to their needs, as well as on new uses of existing technology, and possibilities of adapting them to local needs.

At present, most products are designed by developed countries for their consumers and are often little adapted to the tastes and cultures of consumers of developing countries.

Adaptation has therefore to be undertaken in the following areas:

a. In the design of the product
b. In the materials used, preferably raw materials of developing countries rather than synthetics
c. In the manufacturing process, particularly on the use of labor and capital
d. On the cost of the product, as the purchasing power of consumers in developing countries is sometimes 1/10 of that in developed countries. Thus, restrictive clauses particularly concerning the adaptations of technologies should be deleted in the transfers of technologies.

In order to protect themselves against the negative effects of inappropriate technology, many developing countries have resorted to adopting "restrictive technology policies," allowing far-reaching public interference in technology transactions such as registration and control of contacts, the regulation of their terms and conditions, the determination of their legal effects, of the applicable law and jurisdiction, of arbitration procedures, and so on. This proliferation and diversity of national or domestic legislation and administrative procedures are surely bound to create a considerable concern on the side of the suppliers of foreign technology. This process demands that foreign technology suppliers acquaint themselves with these domestic norms and work toward establishing suitable local conditions for the efficient absorption of the transferred technology.

An international framework for the transfer of appropriate technology must therefore be based on a series of international legal principles. Among these principles must be the following in particular.

1. Respect for sovereignty and political independence of states
2. Respect for the sovereignty and the laws of the recipient country by enterprises engaged in technology transfer
3. Respect for the receiving country to determine its political and economic foundations
4. Freedom of the parties to negotiate, conclude, and implement technology transfer within the limits of international agreements, domestic laws, and culture
5. Mutual benefit and rewards to both parties
6. Peaceful resolution of all conflicts in accordance with contemporary international legal obligations

Thus, the ultimate goal of the scientific and technological community of the developing countries is the creation of a suitable indigenous technology, whenever this is required, taking into account the specific cultures of these countries. Policymaking capacities should be created or strengthened to elaborate national policies for science and technology and to promote self-reliance and people-oriented development.

CITED LITERATURE

1. See Thomas Sowell, Race and Culture: A World View. New York, N.Y.: Basic Books, 1994, Lawrence E. Harrison, Who Prospers? How Culture Values shape Economic and Political success. New York, N.Y.: Basic Books, 1992; Francis Fukuyama, Trust: The Social Virtues and the Creation of Prosperity. New York, N.Y. The Free Press, 1995 and Samuel P. Huntington, "The Clash of Civilizations?" Foreign Affairs 72 (summer 1993), pp. 22–49.
2. See Merryn Claxton (1994) Culture and Development: A Study. UNESCO, (World Decade For Cultural Development (1988–1997) UNESCO, Paris and A.S. Bhalla, (1979) Towards Global Action for Appropriate Technology, Peramon Press, Oxford.
3. See Abubakar N. Abdullahi "Strategies in Technological take-off in Nigeria: Are we in the Right Direction" Nigerian Journal of Science and Technology, Vol. 1. No. 1. 1983.
4. P.D. Dunn (1978), Appropriate Technology: Technology with a Human Face, Macmillan, London.
5. Nji Agaga, (1992), "The Dialectics Between Appropriate Technology, Public Policy and Rural Development" Implications for Discovery and Innovation in the Third World. Discovery and Innovation, Vol. 4, No. 1, March 1992, pp. 43–44.
6. Adams, Don and Robert M. Bjork, (1969), Education in Developing Areas, David McKary, New York, see also Michael Lipton, (1977), Why Poor People Stay, Harvard University Press, Cambridge Massachusetts.

DESIGN PHILOSOPHIES FOR APPROPRIATE TECHNOLOGY

Francis Vanek

Applications of appropriate technology, as reflected by the range of possibilities included in this handbook, encompass a large number of activities, from construction, pumping, and agriculture to cooking and heating—in short, most of the household and community activities that people require for a comfortable life. It is therefore difficult to cover in brief the philosophies that bind these technologies together. Nevertheless, a brief discussion of some common tenets can help prepare the practitioner to develop and implement a technology with greater likelihood of success.

A successful design stage should help us to approach any project more clearly and hopefully, anticipate any pitfalls, rather than finding out about them after the fact. In some instances this work will focus mainly on structural or mechanical aspects of a device or project, but in the case of appropriate technology, inclusion of societal and economic factors are very important as well.

In this section, we will discuss how design for appropriate technology differs from design for other applications and also the pros and cons of designing at lesser or greater depth before the building/implementation stage. The ideas presented here reflect some 20 years of developing, testing, and disseminating appropriate technology on the part of the STEVEN Foundation and its predecessor, the Ensol Cooperative, working in several developing countries as well as in various parts of the United States. For further reading, the practitioner may wish to refer to the following books. For a longer description of mechanical design, those with some technical background will find Chapter 11 in the *CRC Handbook of Mechanical Engineering* (Kreith 1997) useful. Dickson (1975) lays out the political debate surrounding appropriate technology in a chapter entitled "Intermediate technology and the Third World." Lastly, Weisman's (1998) account of the evolution of the Gaviotas ecological community in Colombia provides a microcosm of the issues faced in the field when introducing these technologies. Lastly, while the focus of this chapter is on physical projects involving either products (e.g., pumps, mills) or fixed infrastructure (buildings, catchment ponds, etc.) the importance of institutional work to support appropriate technology (e.g., community and financial organizing) must be recognized. For further reading on approaches to institutional work, reading on the topic of "project management" in the literature on management of technology (e.g., Slack et al. 1998) may be of interest.

APPROACHES TO THE DESIGN OF APPROPRIATE TECHNOLOGY

Practitioners of design sometimes make the distinction between "hard" and "soft" design. In hard design, a device is broken down into component parts, which are precisely defined in terms of their dimensions and materials. The device requires that each component be fabricated to these standards in order for the device to function. Take the example of an automobile: An engine piston having the wrong circumference or not having a smooth surface could prevent the entire vehicle from moving, even if all other components are built "according to spec."

Appropriate technologies are more often built along the lines of "soft design," in which the design concept is subject to modification in the field so as to better use available materials and knowhow. For example, a given material, such as metal pipe or bar, may not be available in a given size, so that the design must be modified to incorporate a different diameter or size. Alternatively, certain skills or facilities, such as people with welding ability or access to electricity, may not be available, so that an installation may need to be built using hand tools to assemble a device with nuts and bolts or other hardware. This level of flexibility may have advantages beyond the ability to create a working installation under difficult conditions. As the late Jacques Cousteau stated, soft technologies tend to develop from "thousands of free-thinking individuals, taking thousands of small random steps forward, [so as to] make only small mistakes on the way to accumulating a large aggregate success" (Skurka & Naar 1976, p. 9).

Successful design of appropriate technology also requires trading off the benefits of thorough design against the costs in terms of time and effort of undertaking such a design stage. Obviously, if a good design can preempt all problems before materials and labor are used in a project, then the time spent will be well worth the trouble. On the other hand, excessive time spent in the design stage may delay a project unnecessarily, especially in the case of soft design where some amount of trial and error in the latter stages of the project may be inevitable.

Three basic levels of design can be seen to exist at this juncture: mental, picture with dimensions, and picture with supporting calculations. In the first instance, the practitioner does not create any diagram or plan before starting work. An example of this is the case of Gaviotas, where a number of the engineers on site responsible for developing new products worked from memory without producing any type of diagram of their work (Weisman 1998). While many practitioners would be uncomfortable building a device without so much as a paper sketch, these members of Gaviotas were able to develop several highly successful technologies in this way.

For those most designers, however, some sort of sketch will be in order. This may be a copy of a diagram from a publication, or an adaptation of such a diagram; alternatively, it may be a sketch developed from scratch so as to fit into a specific location or meet a specific need. Questions such a diagram will answer are typically the dimensions of the device or installation, or the total material requirements for completing the project.

The last option, involving supporting calculations, is the most difficult, as it will generally require specialized knowledge beyond commonsense use of measurement and arithmetic. These are some questions that we might seek to answer at this level.

1. Will the structure hold the required weight without breaking apart?
2. Will the device deliver adequate heat or power to meet the expected demand?

3. If the financial benefits of the project are known, are they likely to exceed the financial costs over the project lifetime?

Thus, we are trying to find out not only the size and material content of the project at this stage but also to anticipate whether or not the design as given will succeed in terms of the project expectations. In some instances, it will not be possible to complete such calculations, often for lack of knowledge about the input data (e.g., what is the material strength of bamboo, of varying diameter and weight, that one happens to have obtained locally?), so that there is nothing left but to construct the design, knowing that ex post facto modification may be needed. On the other hand, it is good practice to at least consider whether supporting calculations are possible, as they may lead to design changes before beginning construction that either allow completion for a lower financial cost or prevent the project from failing entirely. Take the use of steel angle-iron for the creation of some weight-bearing structure, where a number of different widths and thicknesses will be available. If it can be determined that a narrow, thin cross-section is adequate to support the weight, then the cost of the larger iron will be avoided, and the savings can be diverted to other uses.

GUIDELINES FOR IMPLEMENTATION

Having discussed approaches to appropriate technology design in general, it is now possible to discuss specific guidelines for use in the design process. For instance, the STEVEN Foundation has in the past focused on three specific guidelines for the technologies it promotes: local materials, local knowhow, and local business opportunity, such as for family-run businesses or worker-owned cooperatives. Alternatively, Darrow and Saxenian (1993, p. 7) suggest a more comprehensive list of 11 criteria that encompass these three points but also include several others.

The purpose here is not to definitively choose one set of criteria over another but rather to illustrate the benefit of having a set of criteria and also to show how a given approach will tend to steer the development of a device or project in one direction rather than another. Regarding the first point, the criteria provide guidelines that are not specific to a type of project (e.g., solar, wind, agricultural development, etc.) but rather provide guidance in the use of appropriate technology generally, especially to help the practitioner to be sensitive to the needs of individuals and communities in developing countries. When given due consideration, they will help to answer questions such as "What will the effect of this project/technology be on this individual community?" "Is it compatible with the local situation and needs?" "Are there hidden disbenefits that were not obvious at first glance?"

For the second point, we can continue with the example of the previous three-point approach. The use of local materials will tend to favor fabrication of components from plumbing parts (pipes etc.), hardware (such as angle-iron or reinforcement rod), lumber, or some mixture of the three, depending on availability. Also, local skills usually include carpentry and basic metalworking, including welding, in many countries. One could not, however, in general rely on the availability of precision metal-working such as the use of lathes and milling machines.

By following this approach, a project might evolve in the direction of using local materials in their simplest form in order to fabricate complex structures, regardless of the time required. This would not only encourage the development of local skills but also allow unemployed and underemployed adults to exchange their free time for an opportunity to reduce expenditures on finished subcomponents. However, the resulting installation may require more initial "tweaking" and more input in terms of maintenance than a different version that incorporates some manufactured components.

Solar powered pumping systems illustrate this tradeoff (Figure 1). A typical approach to solar powered pumping is to have a photovoltaic panel produce electric current to power a motor, which then drives a mechanical pump. While the panel and motor are beyond the scope of local production, the pump could be built from local plumbing parts and hardware in the interest of self-reliance. Another approach would be the use of a solar collector to generate steam, which could then turn a steam-engine and power a pump in turn. (Such systems have been produced by the STEVEN Foundation for demonstration purposes since 1984. Historically, it was also available at the end of the 19th century, prior to the introduction of photovoltaics; see Butti & Perlin 1980.) Since this latter system uses no electricity and does not require high-precision manufacturing, it can be produced from mirrors, plumbing parts, and lumber, all of which are likely to be locally available. On the other hand, the photovoltaic system will not place as much demand on a community for initial installation and maintenance and so may

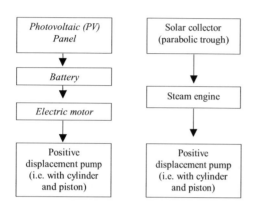

FIGURE 1

Two alternative approaches to solar-powered pumping. The design on the left relies more on high-tech mass production (components shown in italics) but requires less advanced preparation and less learn-as-you-go.

have greater longevity, even if it does not provide the same opportunity for learning as does the steam system.

Beyond the three-point criteria (which focus mainly on educational and economic empowerment), the cultural dimension should be considered as well. Care should be taken to consider the cultural consequences of introducing a new technology. For example, the Gaviotas group developed a pedal-powered cassava grinder for use in rural South American communities, reasoning that this machine would relieve the burden of grinding this vegetable by hand (Weisman 1998). As it transpired, women had traditionally done the hand grinding, which, while tedious, had given them a certain status in the households. When the pedal grinders were introduced, the time required was greatly reduced, and the task was also taken over by men in the household, so that the status of the women was reduced. Some consideration of this cultural attribute at the design stage might have anticipated this negative fallout; perhaps some way could have been found to free the women from the obligation to grind by hand while still supporting the importance of their place in the household.

Understanding of the cultural context for the technology require two-way communication, of course, not just from the outside AT specialist to the community but in the other direction as well. Often the most successful projects are the ones where the community largely shapes the outcome based on information actively requested from the specialist (this approach is advocated by Darrow and Saxenian 1993). In one project in Mexico, volunteers from the STEVEN Foundation arrived planning to build deep-lift pumps for local communities, only to find that the need was for a different design of shallow-lift pump. As the materials and tools had already been assembled, the community members and volunteers set about creating a different pump design that would meet the needs in the particular village. Not only did the villagers largely control the designing and multiple production of the pump, they also continued to improve on the design after the volunteers had returned to the United States. The pumps were used for four or five years before eventually they were made redundant by electrification of the community.

In conclusion, we have seen that appropriate technology is a specific application of technological design separate from the "high-tech" that is used in the industrialized world; that it can be approached by applying specific criteria; and that it is situated in its cultural context. Perhaps the most important lesson that can be drawn in terms of a philosophy is to look before you leap. Careful planning at the design stage can prevent unintended consequences later on, not only in terms of a device or structure which functions as planned, but also in terms of a practice which is truly adopted by the community for the long term.

REFERENCES

Butti, Ken, and John Perlin. (1980). *A Golden Thread: 2500 Years of Solar Architecture and Technology*. Palo Alto, CA: Cheshire Books.

Darrow, Ken, and Mike Saxenian. (1993). *Appropriate Technology Sourcebook*. Stanford, CA: Volunteers in Asia Publications.

Dickson, David. (1975). *The Politics of Alternative Technology*. New York: Universe Books.

Kreith, Frank. (Ed.) (1997). *CRC Handbook of Mechanical Engineering*. Boca Raton, FL: CRC Press.

Skurka, Norma, and Jon Naar. (1976). *Design for a Limited Planet: Living with Natural Energy*. New York: Ballantine Books.

Slack, Nigel et al. (1998). *Operations Management: Second Edition*. London: Pitman.

Weisman, Alan. (1998). *Gaviotas: a Village to Reinvent the World*. White River Junction, VT: Chelsea Green Publishing.

REFERENCES CONSULTED

Arlosoroff, Saul, et al. (1987). *Community Water Supply: The Handpump Option*. Washington, DC: The World Bank.

Ahmed, Kulsum. (1991). *Renewable Energy Technologies: A Review of the Status and Costs of Selected Technologies*. Washington, DC: World Bank.

Aspin, Terry. (1978). *The Backyard Foundry*. Herts, England: Model & Allied Publications.

Butti, Ken, and John Perlin. (1980). *A Golden Thread: 2500 Years of Solar Architecture and Technology*. Palo Alto, CA: Cheshire Books.

Clark, Wilson. (1974). *Energy for Survival: The Alternative to Extinction*. New York: Anchor Press.

Clegg, Peter. (1975). *New Low-Cost Sources of Energy for the Home*. Charlotte, VT: Garden Way Publishing.

Congdon, R. J. (Ed.) (1978). *Introduction to Appropriate Technology: Toward a Simpler Life-Style*. Emmaus, PA: Rodale Press.

Daniels, George. (1976). *Solar Homes and Sun Heating*. New York: Harper and Row.

DeMoll, Lane. (Ed.) (1977). *Rainbook: Resources for Appropriate Technology*. New York: Schocken Books.

Feldman, Stephen, and Robert Wirtshafter. (1980). *On the Economics of Solar Energy*. Lexington, MA: Lexington Books.

Inglis, David. (1978). *Wind Power and Other Energy Options*. Ann Arbor: University of Michigan Press.

Leckie, Jim, Gil Masters, Harry Whitehouse, and Lily Young. (1981). *More Other Homes and Garbage*. San Francisco: Sierra Club Books.

Mazria, Edward. (1979). *The Passive Solar Energy Book: A Complete Guide to Passive Solar Homes, Greenhouse and Building Design*. Emmaus, PA: Rodale Press.

Minan, John, and William Lawrence. (1981). *Legal Aspects of Solar Energy*. Lexington, MA: Lexington Books.

Ritchie, James. (1980). *Successful Alternative Energy Methods*. Farmington, MI: Structures Publishing Company.

Stoner, Carol. (Ed.) (1974). *Producing Your Own Power*. Emmaus, PA: Rodale Press.

Veziroglu, T. Nejat. (Ed.) (1980). *Solar Energy and Conservation: Technology, Commercialization, Utilization*. New York: Pergamon.

BEING A VOLUNTEER OVERSEAS

Adapted with permission from HVO's A Guide to Volunteering Overseas, *published by Health Volunteers Overseas, Washington, DC*

WHO MAKES A GOOD SHORT-TERM VOLUNTEER?

In addition to being well prepared for an overseas assignment, successful short-term volunteers share the following qualities.

- Flexible
- Relaxed
- Innovative
- Organized
- Culturally sensitive
- Committed to sharing their knowledge and skills

In his excellent book *Survival Kit for Overseas Living* L. Robert Kohls suggests that low goal/task orientation, a sense of humor, and the ability to fail are also essential to a successful overseas assignment.

Good listening skills are invaluable to a productive volunteer experience. When you first arrive at your site, listen closely and observe others to develop an understanding of communication patterns, greetings, hierarchy, and protocol. You will need to listen two to three times harder than at home just to begin to understand what is happening in a foreign environment. Some helpful communication techniques include asking open-ended questions and paraphrasing the words of others. These techniques will help to ensure that you understand what is being conveyed and that your message is grasped by your listeners.

You should also be aware of the importance of nonverbal communication—called the "silent language" by many specialists. Nonverbal communication is largely unconscious, spontaneous, and culturally determined. Understanding this form of communication is critical since nonverbal clues often indicate how oral communication should be interpreted: Is the message friendly, sarcastic, or threatening? However, you will quickly discover that body language and the meanings we associate with it are not necessarily culturally transferable. What is a sign of greeting in one culture may well be an obscenity in another cultural context!

Following are other examples of potential miscues in nonverbal communication: While direct eye contact connotes sincerity in our culture, it may be considered rude or disrespectful in another context. Shaking hands may be a sign of professionalism and assertiveness in our culture but may be inappropriate or suggestive between members of the opposite sex in another culture. In addition, intercultural communication experts argue that Americans often need more physical space between them and their listeners than is required in other cultures. Indeed, overseas you may feel crowded in by your colleagues: This is not necessarily a sign of aggression but rather a reflection of different spatial patterns in communication.

CULTURE SHOCK

What is "culture shock"? Culture shock is the term used to describe the more pronounced reactions to the psychological disorientation most people experience when they move into a culture markedly different from their own. Signs of culture shock include homesickness, withdrawal, irritability, stereotyping of and hostility toward host nationals, loss of ability to work effectively, and physical ailments.

Culture shock is a cyclical phenomenon, with the volunteer experiencing at least *two* "low" periods during the course of her or his time overseas. There are also several recognized stages of culture shock, including (1) initial euphoria, (2) irritation and hostility, (3) gradual adjustment and a level of comfort with the culture, and (4) adaptation or biculturalism. How can you cope with culture shock and minimize its negative impact on your work overseas?

Understanding your own culture and its peculiarities may be the first step to combating culture shock. Americans have been described by foreigners as outgoing and friendly and, alternately, as informal and rude. Other qualities ascribed to Americans include hard working, extravagant, wasteful, confident that they have all the answers, disrespectful of authority, and always in a hurry. Many other stereotypes abound about Western culture, and your awareness of them may help you to reject stereotypes of the culture in which you are working. Indeed, being a volunteer means that one must be aware of cultural differences and be able to work effectively in an atmosphere of differing expectations and values. Volunteers, unlike tourists who can choose to remain relatively isolated, must work closely with people within the local cultural context.

Another way to combat culture shock is to learn as much as you can about your host country. Do not be afraid to ask questions, even if they sound silly. This is the only way you can learn about a foreign culture and begin to understand how to function effectively in the environment. Indeed, although you may have gathered copious amounts of information prior to your departure, you will find that there is much more to discover once you arrive in-country.

Above all else, remember that volunteering overseas is as much a learning experience for you as for the people you are training. Having realistic expectations of yourself and others will help you to be flexible and tolerant. Maintain your sense of humor and your sense of adventure, for no matter how well you prepare for your overseas assignment, there will inevitably be problems and challenges. You can, indeed, make a positive impact on community you are working with—just remember that true change is slow and incremental.

Keeping these ideas in mind, how can you prepare yourself for your short-term volunteer assignment?

PREPARING YOURSELF: THE FIRST STEP TO BECOMING AN EFFECTIVE VOLUNTEER

Advance preparation is a critical component of a successful volunteer assignment but one not easy to accomplish since most volunteers are squeezing their trips into already busy schedules. Once you arrive in-country, briefings will usually be short because your host country organizers are busy. So you need to take steps to prepare yourself prior to your departure.

First, you should become familiar with the history and culture of the country you will visit. Good resources for this kind of information include *Country Studies* by the Foreign Area Studies Group. Your local library will have books, articles, or videos that may be useful in your orientation process and the reference section at the end of this article also has a number of excellent reading suggestions. Your hosts will appreciate any efforts you make to familiarize yourself with their country, culture, history and customs.

Second, talk with everyone that you can concerning your assignment to glean as much information as possible. One of the most important sources of information for the short-term volunteer is previous volunteers. They can provide insight into the training program, its goals, your responsibilities, and an understanding of the country and its culture. Try to obtain copies of trip reports from previous volunteers as well as articles and other information about specific countries. Of course, once you arrive in-country, conditions and expectations will undoubtedly differ from those for which you prepared. This is where the qualities of flexibility, cultural sensitivity, innovation, organization, and commitment to shared knowledge and experiences come in!

OTHER TIPS FOR WORKING OVERSEAS

For many in developing countries, the concept of volunteering, in itself, seems strange. In fact, some languages do not even have a word for "volunteer." Volunteers are sometimes thought to have ulterior motives such as practicing experimental medicine or gathering information for the Central Intelligence Agency (CIA).

Volunteers, in turn, may be frustrated by local counterparts who arrive late to work or leave early for other jobs. They may doing professional work in the afternoon and evening, or they may have a second job in an entirely unrelated field such as driving a taxi or acting as a tour guide. You may even feel that counterparts are less concerned and committed than you are. But, remember that these individuals face many daily frustrations and, in fact, often work more than one job just to provide for their families.

You must avoid the overwhelming temptation to demonstrate a method of handling a problem that cannot be done after you leave—for example, using your own computer that you plan to take home with you. Resisting this temptation may be a challenge at times, especially when local leaders ask for help. Sometimes, offering help in such situations may be appropriate in order to cultivate the support and goodwill of the community.

Short-term volunteers must realize that they are not going to change the world in the time that they are working overseas. Be careful not to raise hopes or to make—or even imply—promises that cannot be kept either by you or by the organization that is sponsoring you overseas. Show that you are part of a team that will continue to work side by side with your host counterparts and not just a short-term visitor. Indeed, how you are perceived by your hosts is critical to how effective you will be during your brief stay.

In summary, start your trip with a sensitivity and curiosity to learn from everything you see. Listen, look, and enjoy the differences in cultures. Don't be afraid to ask questions rather than offer expert opinions. Take the time to learn a little bit more about yourself and your own culture through the eyes of your counterparts. Remember that the more empathy and respect you show for the culture and problems of your hosts, the more respect you will command and the more effective a teacher you will become.

OTHER ISSUES TO CONSIDER

Taking Spouses and Families. It may or may not be advisable to take your spouse and family on assignment with you. This decision depends on a variety of factors, including the program site, availability of acceptable activities, suitability of housing, and flexibility and interest of family members. Should your family not find enough to keep busy during your stay, they can become yet another responsibility and burden for the host.

On the other hand, the presence of family can be very reassuring to the volunteer. There are often opportunities at the local health facility or a nearby school or library that can lead to a meaningful experience for family members. Try to get advice from many people about opportunities for family members to become involved in significant work.

Local Politics. Volunteers should remain neutral with respect to expressing their opinions about the national or local government. You may put your hosts in a difficult and potentially embarrassing or dangerous position by commenting on the government or even the administration of the place at which you are working. These are the kinds of observations that you might make in your trip report and to other volunteers.

Equipment. Prior to your volunteer assignment, it is important to determine what equipment and supplies are readily available at the site and, if necessary, to review their usage. The goal of many sponsoring organizations is to provide personnel with the skills and knowledge to enable them to provide better treatment within the framework of existing technology. So while the thought of providing more modern equipment and supplies may be very appealing, remember that maintenance, access to spare parts, and other realities may dictate staying with older, sturdier, and more familiar models and materials already in use.

Licensing Requirements. Some institutions, universities, and hospitals, for example, require documentation about the volunteer's educational and professional background before accepting him or her. Host institutions are concerned with ensuring that all volunteers are appropriately licensed and credentialed before they arrive in the country of their assignment. This process is especially important for those sites that are in an academic setting, since your host counterparts will undoubtedly be well-trained, experienced professionals. Foreign institutions are usually willing to accept certified copies of diplomas and transcripts presented by the volunteer and, in fact, will be surprised if these are not brought by the volunteer.

RETURNING HOME

Returning home can be an unexpectedly stressful experience, especially for the first-time volunteer. Many volunteers have found that the culture shock experienced upon reentry exceeds what they felt upon entering the country of their assignment. In a new situation, people expect things to be different; back home, they expect things to be the same as before they left. The volunteer, however, is often deeply changed by the overseas experience, while family, friends and colleagues at home have continued their lives as usual, unaffected by events in other places. How well volunteers handle reentry depends both on their own preparation and the concern of those around them.

Many volunteers returning home find Western lifestyles to be wasteful and lavish compared to what they have just seen and experienced in the developing country setting. They have been exposed to poverty and have made friends in a different world where much of life is defined by privation and a constant struggle to meet daily needs. Volunteers want to talk about what they have seen and done: they would like their friends and colleagues to share their new knowledge of the world.

However, you may find that many of your friends and colleagues will not be interested in hearing the details of your trip. When they ask about your trip, they will probably want a short answer, not a lengthy discourse on the lack of economic resources, the poor health conditions, the lack of environmental control, the problems of illiteracy and malnutrition, and other everyday conditions you saw.

Individuals who have not worked abroad may have built up defense mechanisms regarding the problems endemic to developing countries. Numbed by the magnitude of the problem described by the media, they cannot share the level of concern felt by the newly returned volunteer. They may ask why the volunteer is not doing more to alleviate problems in this country. Colleagues may become defensive when taken to task for their lack of awareness and action and may in turn question the motivations of the volunteer.

As a result of these communication problems, volunteers may not talk about their experiences unless asked and then may not feel comfortable sharing more than a few impressions. Even when approached enthusiastically, volunteers may find their listeners' attention waning after a brief period. Many volunteers feel lonely and depressed about not being able to share their overseas experiences.

Those who are successful at coping with reentry find ways to integrate their volunteer experience into their lives at home. One of the best ways to do so is to communicate with other volunteers, especially those who are scheduled to go abroad in the near future.

Another successful reentry strategy is to give talks and presentations to interested groups about your overseas experiences. In this way, you can channel your enthusiasm for overseas work to those who share your interests.

GAINS FROM VOLUNTEERING

What can you can expect to gain from your overseas assignment? After all, as a volunteer you are paying your own expenses or at least foregoing earnings and merit something in return for your efforts. Most returned volunteers express that they learned much more than they taught while working overseas. You will undoubtedly observe and learn how professionals in difficult and resource scarce environments use innovative approaches. And, you may be able to contribute to their knowledge by teaching skills and techniques that will make their tasks easier. In this way, you become an integral part of an educational process which is targeted to fostering the independence and professionalism of people in developing countries.

In the process of volunteering overseas, you will also develop and hone your teaching skills and perhaps your listening and learning skills as well. You will be exposed to other cultures, customs and attitudes and will see first-hand their impact upon health care delivery. Finally, you will broaden your awareness of issues such as global health care, international development, and much more.

REFERENCES

Achebe, Chinua. (1959). *Things Fall Apart*. New York: Fawcett Crest.

Hall, Edward T. (1959). *The Silent Language*. Garden City, New York: Doubleday.

Kohls, L. Robert. (1979). *Survival Kit for Overseas Living*. Chicago: Intercultural Press.

Piet-Pelon, Nancy I., and Hornby, Barbara. (1985). *In Another Dimension: A Guide for Women Who Live Overseas*. New York: Intercultural Press.

Samovar, Larry A., and Porter, Richard E. (1988). *Intercultural Communication: A Reader*. 5th Ed. Belmont, CA: Wadsworth Publishing Company.

Planning and Implementation

Edited by BARRETT HAZELTINE
Brown University

Planning and implementation—the management—of a project is important. Good technical ideas are certainly necessary but they are not sufficient to make a sustained improvement in people's lives. As much attention needs to be paid to making the project work as is paid to having an effective technology. This chapter is long because the problem addressed is difficult. The first article, "Development Project Management," is an overview and should definitely be read by anyone who has not managed a project in the Third World. It has become most apparent that neglect of people's concerns, aspirations, and dignity has been a major reason for the failure of many projects. Five articles about the human dimension of planning and implementation are included based on experiences in the Third World. These deal with ensuring community participation, working with people in making change, and dealing with gender unbalances. Because a common purpose is to develop a business that will continue after a volunteer leaves, the next set of articles focus on small businesses. Several deal with how to actually produce goods and services—the usual term is "Operations." Marketing is more than selling—it is the whole process of dealing with potential users. Financial analysis has several aspects: cash flow, costing, discounting future payments. An increasingly common source of funds for small projects is microfinance programs. The mechanics of planning—understanding the business and dealing with administration—are discussed next. The article on project selection looks at particular, important planning problems in detail. Some of this planning material overlaps ideas in the construction chapter and a person doing planning should look there also. Project evaluation is essential and the two articles included complement each other as one is based on the needs of a large organization and the other on the perspective of a grassroots program. Notes follow on managing an ongoing project. Emergency responses to disasters need to be carefully thought of early and some guidance is given. The two following articles deal with training. A measure of the success of a project, especially in the Third World, is its success after the founder departs—the exit strategy article discusses how this departure can be done most effectively. Donors support many Third World projects. The next two articles deal with fund raising. Articles about volunteers' experience with a social service organization in Bolivia and with a Women's Cooperative in Guatemala is included to show what happens, as the expression goes, "on the ground."

DEVELOPMENT PROJECT MANAGEMENT

Chantelle Wyley and Jochen Lohmeier

INTRODUCTION: THE ORGANIZATIONAL CONTEXT OF TECHNOLOGY CHANGE

Appropriate technology innovations are invariably introduced in practice within an organizational context: money, people, and institutions, and equipment. There may be engineer support, maintenance provision, training, or links with a government ministry (e.g., an agricultural extension office). This organizational context is usually structured as a project and, in developing countries specifically, as a development project aiming to change people's lives for the better.

Introducing such change constitutes an intrusion into people's lives, requiring sensitivity and awareness on the part of the change agent. Before we move onto projects and their management, perhaps it is useful to check our understanding of *development*. Here is one definition from the Glossary of the Baobab: *Development Management Networking Web site.*

> Development is the process by which people change a negative (unsatisfactory) situation to an improved one (in which certain pressing problems do not persist). Development involves the satisfaction of basic needs of people living in a state of poverty. Development interventions aim at stimulating the capabilities of people and their societal institutions to (at least) satisfy their material basic needs by using available natural resources (without endangering them), and under changing frame conditions. Good development practice aims to assist poor people to solve their prevailing problems themselves. The development process is always linked to norms and values.

From the same source, here is a definition of a *development project.*

> A set of *temporarily, geographically and sectorally limited measures*, carried out by an agency with predetermined objectives, designed to develop and spread *innovative solutions*. Projects should neither be full-scale implementers of solutions, because of their limited nature in relation to the ongoing development process, nor should they merely provide advisory services because of few possibilities to develop adjusted solutions. A project as the solution-finder is usually a component of a larger or comprehensive or area development programme, which is responsible for providing required supplies and services in order to sustain the solution found.

For an appropriate technology solution to work requires the following.

- It addresses an identified and analyzed problem.
- It is technically sound.
- It is appropriate to the prevailing physical conditions.
- It is culturally appropriate.
- It is introduced in such a manner that it is received favorably.

Thus, technical factors (machine-related) as well as organizational factors (systems- and people-related) are involved in the introduction of technological change. The sensitive interplay between the two needs to be managed.

Development project management requires two sets of skills: project management (the "hardware" of management, such as operational planning and budgeting) and people management (the "software" of management, such as team building, conflict resolution, communication). The latter are particularly pertinent where cultural differences between innovator and user apply.

It is important to emphasize that the two sets of skills are never distinct from one another and responsibility for them cannot be split. Managing any undertaking is an intricate process involving the interplay between the internal self, others (individuals and groups in many constellations: colleagues, work teams, beneficiaries, adversaries, institutions, service providers, etc.), learned skills (e.g., a project planning method), and previous experience. The process is constantly informed by data collected from the complex reality in which the work is situated and constantly adjusted appropriately. Important here are one's observations, formal and informal information systems (e.g., reports, meetings, site visits, conversations with key players). The interpretation and assessment of these data and their use to make decisions, to adjust, to change, and hence move forward constitutes the art of management.

For practical reasons, however, this article limits itself to an orientation to project management and an introduction to some basic tools; people management issues are touched upon briefly but dealt with in more detail elsewhere in this collection.

PROJECT MANAGEMENT

International experience in development projects has resulted in some useful tools and approaches, according to the norms and values of the time and context. Currently prevailing is the notion of *project cycle management* (*PCM*).

PCM is an orientation framework for managing development cooperation and partnership between benefactors and beneficiaries. It is used to assist agencies in developing countries to successfully plan and steer projects. In PCM, the procedure involved in preparing and implementing a project is given a structure: It specifically defines stages in project management from design and planning through to evaluation and back to replanning, as a cycle through which a project may repeatedly move in the various phases of its existence (see Figure 1).

In PCM such stages are connected by structured and linked processes. This avoids the situation of busy and energetic project staff running around spending time and money pursuing well-meaning

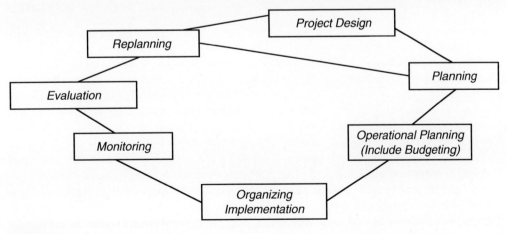

FIGURE 1
Project cycle management (PCM).

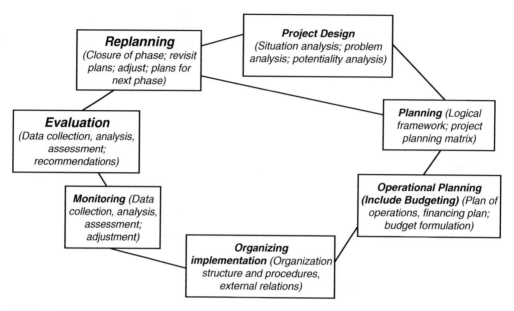

FIGURE 2
Tools for each PCM stage.

goals but finding after a couple of years that impact has been uneven and (given the expenditure) minimal. With PCM, planning can link project activities to impact (to changes in beneficiary behavior and hence to subsequent benefit, the goal), these activities can be detailed (personnel, costs, and equipment) and directly linked to a monitoring system. Indicators of achievement specified at the planning stage can be used for monitoring, and evaluation; all feeding into the next/replanning stage and beginning of the next cycle.

Figure 2 shows the PCM diagram again with some tools related to each stage. This article attempts to give an introductory overview of these issues and related tools.

THE DESIGN OF DEVELOPMENT PROJECTS

The question of which solution and hence intervention is appropriate in a particular situation (i.e., one in which problems exist relating to a lack of satisfying existential needs and a situation of poverty) is sometimes one not given enough attention when solution-driven benefactors interact with developing countries. For example, a motorized transport system to get crops to city markets may be an "obvious" solution to low-income in a potentially productive rural area; more in-depth analysis may show that introducing different types of crops attractive to more local markets is more appropriate—in the face of bad roads and no funds for road maintenance. The tool to use in this case is a *structured situation analysis*.

This is an analysis of the actual situation in which a potential development project is embedded. It takes place prior to project planning and proceeds from the point of view of those directly affected by the problems. The situation analysis consists of an analysis of all participants and stakeholders, an analysis of objectives of the various parties involved, and a problem plus potentialities analysis, as well as an analysis of the project environment in order to define risks or assumptions. It should also consist of an analysis of the target groups of the potential inter-vention. The purpose is to sufficiently understand the context and to gain an in-depth insight into the problems in order to be able to formulate improvements or solutions adjusted to the situation. It demands a communicative approach emphasizing participation and may make use of techniques such as Participatory Rural Appraisal (PRA) and the more recent Participatory Learning and Action (PLA).

To be more specific, a problem analysis looks at existing negative conditions in the lives of people, in a particular instance. It is broken down into an analysis of deficiencies (unsatisfied existential needs) and of constraints (factors that cause the deficiencies). This information is gleaned from professional investigation of the project environment (e.g., sociologists reports, census statistics) and workshops with stakeholders. In the latter, participants may be asked to write deficiencies and constraints on individual cards, and these are then shared and related to one another in a displayed cause-effect "tree" (hence the term *problem tree*). This helps to generate a common understanding of what is at the root of material lack or misery and to make a decision as to where best to intervene to alleviate or arrest it (and this may not be at the root cause).

A potentialities analysis may follow. This explores resources and opportunities that are unutilized or underutilized but may be utilized or developed in order to overcome constraints and hence contribute to objectives. The decisions and choices made (using methods to prioritise) are at the heart of any development intervention: They aim at unlocking potentials in target groups in order to enable them to tackle problems, overcome deficiencies sustainably, and thereby reduce poverty.

The situation analysis may specifically analyse economic, environmental, institutional/political, target group problems and potentials.

As the situation analysis progresses, possible solutions to the problem(s) may arise. An analysis of alternatives is a systematic way of searching for and deciding on problem solutions and hence the work of an impending project. Alternatives analysis follows a problem analysis and is a prerequisite for deciding on action strategies. Alternatives are analyzed using criteria relevant to the situation (a matrix table is useful here). Techniques such as cost-benefit analysis and utilities analysis are used in alternatives analysis. The results—an appropriate intervention in respect of a particular problem—feed directly into the project planning process.

PLANNING

Planning can be defined as a process of communication, clarification, and agreement between individuals and groups who wish to work to change an undesirable situation (e.g., one in which people live in poverty, lacking basic needs). It involves the anticipation of a desirable and realistically achievable vision and scheduling of future actions, together with the utilization of resources, all directed toward achieving defined objectives. Who does what when, how, and why?

In reality plans for development projects are often formulated as the substance of applications for funding and other support—from international development donors, government contracts, or nongovernment partnerships. In some cases the appraisal, design, and planning processes are supported (by funds or the temporary allocation of experienced technical personnel) through potential donors/partners following an initial project idea/application. The institutional requirements of fund applications for projects are often a challenge for development practitioners. Planning formats and procedures are laid down in order to facilitate decision making within funding institutions and may result in plans being drawn up in a bureaucratic manner (so-called "desktop planning"). The intention of planning with beneficiaries and possible service providers is thus compromised; ownership then may remain with the funder/intervener. In addition a funder-orientation may influence the planning of a development intervention in scale, by the amount of funding available, or in orientation by the type of technical support available. This can be dangerous in that the needs of the intended beneficiaries and the appropriateness of the intervention may be sidelined.

There are many different planning methods that assist in structuring thinking and presenting possibilities. The most comprehensive at present is objectives-oriented project planning and its specific variants—for example, Zielorientierte Projektplanung (ZOPP). ZOPP uses the Logical Framework Approach (LFA) to planning which displays objectives according to logically related levels in a matrix format (known as a project planning matrix or PPM). Such strategic planning involves the formulation of an intervention strategy, with detailed operational planning following in the Plan of Operations (PlanOps).

The major advantage of this approach to planning is that it stimulates development interveners to explore and define the (logical) links between their proposed activities and eventual impact, rather than merely hoping that such actions may produce effective and beneficial effects. If carefully and responsibly used, this facility ensures that the voice and needs of the intended beneficiaries are kept to the fore.

Table 1 shows an example of the objectives part of a Project Planning Matrix (PPM) of a rural bus system project.

In addition, the PPM provides for specification of achievements for all these objectives, called "indicators," as well as envisaged external risks on the project's progress. This information for the project in Table 1 is shown in Table 2.

A Project Planning Matrix is usually formulated in a workshop situation (the strategic objectives) with participatory input from stakeholders. Follow-up working groups may work on specifics later (indicators, for example, are best defined by those working in the field rather than stakeholder representatives).

TABLE 1
PPM objectives

Strategic Objectives	Example: Rural Bus System

GOAL
  (impact on beneficiaries)
	Peasants of villages A,B,C increase their
	  income through market production.

PURPOSE
  (utilization of project
  outputs by target groups)
	Peasants arrive at the marketplace
	  safely and in time, and buses
	  operate reliably.

OUTPUTS/RESULTS
  of project (goods and
  services delivered)
	1. Bus maintenance and repair system
	  is introduced and operating.
	2. Standard training courses for bus
	  drivers implemented.
	3. Management system for optimal
	  deployment and flexible bus use
	  established.

ACTIVITIES
  of project
	(For Output 2)
	2.1 Check knowledge, and deficits,
	  of bus drivers.
	2.2 Design appropriate course
	  curriculum.
	2.3 Carry out courses.
	2.4 Evaluate impact of course and
	  revise curriculum.

TABLE 2
PPM specification for achievements

Objectives	Indicators	Risks/Assumptions
Goal: Peasants of villages A,B,C increase their income through market production.	*For the achievement of the development goal:* After year 3 of project implementation income from market sales of more than 70% of peasants is at least stable (or increased).*	
Purpose: Peasants arrive at the marketplace safely and in time, and buses operate reliably.	*For the achievement of the purpose:* 3 years after beginning of the project's implementation phase, more than 50% of the female peasants who produce a marketable surplus are able to transport 80% of their marketable products from their villages to the marketplace, arriving there in the early morning hours.**	*For achieving the development goal:* Competing producers do not receive excessive subsidies from other intervening agents.
Outputs/ Results: 1. Bus maintenance and repair system is introduced and operating in practice. 2. Standard training courses for bus drivers implemented. 3. Management system for optimal deployment and flexible bus use established.	*For the achievement of the results:* 1. After year 2 of project implementation a repair of a serious breakdown of a bus does not take longer than 10 days after the bus reaches the workshop. 2. After year 3 of project implementation accidents caused by bus drivers themselves reduced to below 30% of all accidents. 3. After year 3 of project implementation the bus transport capacities are adjusted to the marketing system.***	*For achieving the purpose* a. Trained bus drivers apply their new knowledge. b. Road improvement measures are implemented (possibly by a project of another donor). c. Ticket prices are in line with the purchasing power of the farmers.
Activities: (For Output 2) 2.1 Check knowledge, and deficits, of bus drivers. 2.2 Design appropriate course curriculum. 2.3 Carry out courses. 2.4 Evaluate impact of course and revise curriculum.	***Specification of inputs/costs of each activity***	*For achieving the outputs/results* 1. . . . 2. Drivers attend courses regularly and are sufficient motivated. 3. . . .

*–*** Project Planning Matrices commonly also include the data sources of the information necessary for indicators (known as the *Means of Verification*, MoVs). In the example above, these could be:
For Goal: * Household income survey at villages A, B, C 3 years after the start of the project
For Purpose: ** Annual survey at market place one month after beginning of the harvesting season
For Outputs: *** 1. Workshop service cards; 2. Police records of bus accidents; and 3. Bus schedules

It is important to emphasize that the PPM is a format. Different to a form to be filled in a bureaucratic procedure, the PPM is only a way to concisely present decisions made during a planning *process*. This process itself is usually a combination of participatory and analytical contributions. Beneficiaries may be asked about their needs, technicians will come in for discussion about appropriate solutions, and workshops may be held to bring all stakeholders together. It is an art to conceptualize the planning process in a way that optimizes (rather than maximizes) necessary contributions from relevant role players in a manageable sequence. The package that can be assembled in the end from all elements elaborated on during the process is the PPM.

OPERATIONAL PLANNING (INCLUDING BUDGETING)

The details of *who* carries out project *activities* as listed in the PPM (the distribution of responsibilities) with *what resources* (the procurement plan for required equipment and materials), at what *cost* (budgets), and by *when* (time schedule) are specified in the operational plan. A tabular format that combines and interlinks all the aspects of such a plan is given here (this is often known as a Plan of Operations or PlanOp). See Table 3.

It is at this point that a realistic picture emerges of the cost of a project. Only with this degree of detail (activities, equipment, materials, staff itemized) can a realistic *budget* be formulated. This stipulates funds for salaries, equipment maintenance, and other running costs, as well as funds for capital expenses such as new equipment and premises.

- At this stage the contribution of various funders can be specified—for example, international donor funding, contributions from within the country (e.g., from the related government ministry coffers), local donations (e.g., community workers' time, land), any credit required (e.g., bank loans).
- It may also be useful to discuss and formulate a *financing plan* listing funding secured and funding still to be sourced, with appropriate strategies to plan for project sustainability.
- It is important to detail training requirements for the project at the PlanOp stage, as these require resources and organization. An extra activity (i.e., training of project staff, community workers, or government extension officers) may need to be added to the activities in the PPM.

This illustrates the important point that *planning is about replanning*: If formulating the PlanOp shows that something is missing from the PPM, this can be amended. If it emerges at this stage that the plan is overambitious and cannot be implemented with the resources at hand, the objectives will have to be made more realistic.

ORGANIZING IMPLEMENTATION

Implementation is all about people having the energy and direction to make things happen. This energy motivates people to work and constitutes the "atmosphere" and "culture" that outsiders sense when they come into an organization. There is no one way to create this in an organization. Having said this, in the development context, being able to see positive impact on the lives of beneficiaries, is a strong motivating factor. Emphasizing the link between daily activities and the organization's purpose and ultimate goal, as outlined in the PPM, is an important element. Alongside this, managers need to ensure the development of smooth, open interaction between individuals and teams with clear systems of communication.

TABLE 3

Format and examples of a plan of operation: a combined work plan and project budget

Project: Rural Buses
Planning Period: WOC

Date of Issue:

Output/Result No.: 2) Standard training courses for bus drivers executed

Activities	Milestone Indicator	a) Assigned to / b) Responsible	I	II	III	IV	Staff Requirement Persons A B C D	Cost	Training Require-ments	Cost	Equipment/ Materials	Cost	Building	Cost	Misc.	Cost	Mainten. & Repair	Cost	Assumptions and Remarks
2.1 Check knowledge and deficits	70% of bus drivers tested by . . .	Operations manager(= op.ma)/ official examiner	--- / ---				A=1, B=-, C=-, D=1	1,000	-- / --	-	Computer / Paper	- / 10	-	-	Travel allow-ances for tests	2,000			Employment records up to date; official examiner cooperates
2.1.1 Devise test	Multiple-choice test agreed on by . . .	op.ma/ official examiner	--- / ---				A=1/4, D=1/4	250 / 250	-	-	Computer	-	-		For exam-iners	50			
2.1.2 Sample testing	Variations in test results are significant	op.ma/ official examiner	---				A=1/4, D=-	250	-	-	Computer	-	-		For exam-iners	50			
2.1.3 Comprehensive testing	Test results available as per schedule	op.ma/ official examiner	--- / ---				A=1/4, D=3/4	250 / 750	-	-	Paper for test printouts	10	-	-	For dri-vers / For exam-iners	1,750 / 150			
2.1.4 Analyze deficits	Deficits are clear	op.ma/ official examiner	---				A=1/4	250	-	-	-	-	-	-	-	-			

Technically, organizing project implementation is a matter of doing the following.

1. Roles and relationships: Defining tasks and responsibilities and developing the organizational structure.
2. Agreeing on *internal decision-making procedures.*
3. Organizing *external relations.*

Distributing tasks and responsibilities (based on operational plans) to individuals is one step; related to this is the formulation of job descriptions for each. A useful guide is to attempt to bring as close together as possible the three elements of task (what needs to be done), competence (what someone is able to do), and responsibility (what someone can be entrusted to do). Individual roles and the cooperation within the project organization go hand in hand.

Another step is to relate the positions to one another by doing the following.

1. Balancing specialization and integration: allocating individuals specialist tasks, and integrating tasks.
2. Grouping different functions and relating these to one another.
3. Defining the lines of command: who can give instructions to whom and who is responsible.

In all these, graphic depictions (e.g., wall charts) of the relationships, building up into the organizational structure, are useful for discussion and elaboration. In the end the type of organizational setup (for example, hierarchical with a strict vertical line of command or a flatter team-driven structure) depends as much on the cultural values of the society in which the project is cited as on what best suits planned project activities.

Once responsibilities are clear, it is useful to outline the mechanics of decision making within the organization. This relates to these two areas.

1. *Short-term steering*: This involves the participation of all staff. Tools for decision making are regular staff meetings with agreed upon and adhered to procedures (notices, agendas, reviews of performance, decision making, planning and replanning, documentation) and action plans (in wall chart format) derived directly from the Plan of Operations and broken down into shorter time periods, teams, and so forth.
2. *Policy matters*: Management and supervisory bodies (e.g., a board of directors/trustees) take responsibility but inform and involve staff. The tool for policy decision making can be a management meeting, the concern of which is specific targets for which management is responsible (defined in the Outputs of the PPM, for example).

Organizing external relations involves the following.

1. Defining the boundaries of the project in relation to its environmental context (societal, institutional, political)
2. Deciding what external factors impact on the project and what is important and needs attention

The process needs to be a conscious one, requiring analysis, with input from staff, adequate space/time for discussion and definition, using visualization and display. Note that this area has a direct relationship to the assumptions/risks column in a PPM and revisiting the planning phase discussion is important.

An overview of organizing and implementing would be incomplete without a mention of leadership. Different leadership styles are appropriate for different circumstances (e.g., authoritarian, military leadership is necessary in the heat of battle, while a consultative, participatory facilitation style is more appropriate if frame conditions are settled). As a leader (of a large organization or a small team) self-awareness, self-management, and sensitivity are key, as is keeping a balance between guiding and being guided. One's own reality is never the same as that of others, and acknowledging and encouraging the sharing of multiple realities build a richer, more productive workspace. Effective conflict resolution—one of the basic skills of management—is based on this.

MONITORING

Monitoring day-to-day project activities often happens spontaneously: Collecting information about what has been done, comparing this to what was intended, and planning further work. The less haphazard, more conscious and systematic this process is, the more effective decisions on project achievements and outputs can be made. Project plans encapsulated in a PPM and PlanOp—with documented indicators of achievements, milestones, resources, and responsibilities—strongly support the development of a useful and effective monitoring system.

Monitoring involves the following.

1. Observation (data collection, of statistics, feelings) of the implementation process, on various levels (namely the use of resources, carrying out of activities, achievement of outputs/results, changes in behavior of target groups, and resulting benefit, as well as risks/assumptions)
2. Assessing these data (in relation to what was planned): interpreting, making meaning of deviations, reflecting, in a process that makes the most of differing realities of the human elements
3. Feeding this back into the project's steering/decision making/learning, in order that adjustments may be made in the allocation of resources (including personnel), energy expended on activities, in order to result in more effective project outputs; or adjustment of the project

The much used and favored management concept of "action learning" depends on a monitoring system integrated into the work life of an organization. Such a system requires a culture of participation and the inclusion of all staff, of transparency (open access to information on which each person bases his/her decisions), and minimized and targeted information collection for managerial decision making.

Monitoring may be done by teams or one person or an external specialist (with clear and accountable terms of reference) and may manifest in reports, meetings, wall charts that assemble information and share it with whoever requires it for decision making (a feedback process).

EVALUATION

In addition to monitoring (= a continuous process), projects need to evaluate their work, to engage in (periodic) events examining in depth what has been done, reflecting on this, in order to take the next step forward into a new phase. Monitoring asks, "How are we doing?" While a phase is running; evaluation asks, "How have we done?" after a phase is completed.

Evaluation can be arranged internally (self-evaluation or self-reflection), involving project staff and management, or externally, as a result of an external impetus or with an external facilitator/evaluator. Evaluations look at performance (resources in relation to activities) and impact (output in relation to change and benefit).

Both internal and external evaluations involve designing the event as a process. In the development context a participatory evaluation is most appropriate, as it involves those who will use the evaluation outcomes in their work. External expertise is included, as well as the experiences of local stakeholders.

In an external evaluation an external person acts to ensure an open and honest process that takes into account appropriate current thinking in the related fields.

A procedure for external evaluation may be as follows.

1. Announcing the evaluation and at the same time affirming the type of evaluation process
2. Deciding on the aim of the evaluation
3. Deciding what is to be evaluated
4. Appointing the evaluation team
5. Formulating a written evaluation plan outlining the process
6. Revisiting the project objectives
7. Redefining (if necessary) indicators to measure these objectives
8. Devising the methods of data collection
9. Preparing and testing the data collection methods chosen
10. Collecting data according to method(s) decided upon
11. Analyzing the data
12. Preparing results for presentation (usually a written report plus verbal presentation)
13. Making recommendations based on the results (included in the report and presentation)
14. Reporting back formally (present report to authorities with summary, conducting workshop reportback(s))
15. Following up on the decision making based on the evaluation results

REPLANNING

Plans are not set in stone. They are not made for set time periods nor for implementation "to the letter." Planning is a management function and a process during which those involved decide on a future vision and devise steps to get there from the present. Plans give an orientation, they direct energy (to overcome confusion and chaos), and they are part and parcel of the process of implementation. Hence planning in most cases takes the form of replanning.

Replanning often concerns the PlanOp: It has proven good practice in a project to organize an annual combined internal evaluation and annual activity plan for the coming budget period. Guiding questions can be What did we intend to do in the last year? What happened in reality? What can we learn from our experiences? How do we assess these? What do we intend to do differently? Then replanning should take place.

Replanning may also involve more comprehensive plans/plans over a longer period of time. A new phase of a project builds on plans from the previous phase. Projects are conceptualized around experiences gained in similar fields and may draw from other plans. Replanning brings closure to one project cycle and focuses attention on the next.

REFERENCES

Baobab: Development Management Networking Web site. (1999).
  http://hagar.up.ac.za/catts/learner/patsy/baobab/index.html.
Chambers, Robert. (1995). *Rural Development Planning: Putting the Last First*. London: Longman.
Commission of the European Communities Evaluation Unit. (1993). *Methods and Instruments for Projects Cycle Management. Project Cycle Management: Integrated Approach and Logical Framework,* Bruxelles, published by The Commission.
Deutsche Gesellschaft fur Technische Zusammenarbeit (GTZ) GmbH. (1997). *ZOPP: Objectives-Oriented Project Planning: A Planning Guide for New and Ongoing Projects and Programmes*. Eschborn, Germany: GTZ.
Deutsche Gesellschaft fur Technische Zusammenarbeit (GTZ) GmbH. (1992). *PFK: Guidelines for Projects Progress Review*. Eschborn, Germany: GTZ.
Deutsche Gesellschaft fur Technische Zusammenarbeit (GTZ) GmbH. (1991). *Methods and Instruments for Project Cycle Management: Outlines*. Eschborn, Germany: GTZ.
Michael Randol, Davine Thaw. (1998). *Project Planning for Development*. Durban, South Africa: Olive Publications.
Mukherjee, Amitava. (Ed.) (1995). *Participatory Rural Appraisal: Methods and Applications in Rural Planning*. New Delhi: Vikas.
Olive Organization Development and Training. (1996). *Contracting: How to Contract Outside Consultants or Providers to Assist Your Organisation*. Durban, South Africa: Olive Publications.
Olive Organization Development and Training. (1996). *Evaluation: Judgment Day or Management Tool? A Manual on Planning for Evaluation in a Non-Profit Organisation*. Durban, South Africa: Olive Publications.
Pfohl, Jacob. (1986). *Participatory Evaluation: A Users Guide*. New York: Private Agencies Collaborating Together (PACT) for United States Agency for International Development/Sri Lanka.
Pratt, Brian, and Jo Boyden. (Eds.) (1985). *The Field Directors' Handbook: An Oxfam Manual for Development Workers*, 4th Ed. Oxford: Oxford University Press for Oxfam.
Pretty, Jules, Irene Guijt, John Thompson, and Ian Scoones. (1995). *Participatory Learning and Action: A Trainers' Guide*. London: International Institute for Environment and Development.
Rauch, Theo, and Jochen Lohmeier. (1997). *Regional Rural Development (RRD)*. Berlin: COMiT.

THE NEW PARTICIPATION: A VIABLE APPROACH FOR SUSTAINABLE DEVELOPMENT PLANNING AND ACTION

Richard Ford

In the last two decades, a revolution has occurred in the role of popular participation in development planning and action. Initially participatory activity focused mostly on local communities in Asia, Africa, and Latin America, maintaining an informal approach. More recently, a number of efforts have created more formal procedures and extended the benefits of participation into important national and even international planning, policy, project design, and action efforts. While there is nothing new about the concept of participation, there are some dramatically new dimensions associated with the recent upturn in participatory methods.

THE RISE OF PARTICIPATORY PRACTICES

At least three circumstances have supported the upsurge in popular participation.

Project and Policy Failures Accelerating Global Poverty

First has been an enormous disaffection, worldwide, with the ability of professional planners and project managers to design projects to solve the basic problems of human existence. By 2000, poverty was, by any measure, as severe and perhaps even more destructive than it was 20 years earlier.

For example, in Kenya—once viewed as a success story in development—absolute rural poverty has remained essentially unchanged from 48 percent in 1982 to 47 percent in 1994. World Bank data show similar stagnation in development and poverty alleviation efforts. For example, in 1997–1998, two billion of the world's people (35 percent) eked out an existence on less than a dollar a day. Because the national governments and international experts have been unsuccessful in achieving adequate water, health, nutrition, livelihoods, employment, resource sustainability, and agricultural production, people are increasingly looking inward to find solutions that they can identify and implement themselves.

Policy reforms to stimulate global privatization offer another example of disaffection. It is increasingly clear that the last two decades of the World Bank and International Monetary Fund market-driven economic reforms, loosely known as structural adjustment programs (SAP), have brought a number of unanticipated and devastating side effects. In 25 of the SAP countries, at least 40 percent of the people live below the poverty line. While privatization and monetary reform have created new economic opportunities and increased efficiency in basic infrastructure services in many of the world's economies, they have also dramatically increased the gap between rich and poor. These policies have driven many small farmers and marginal businesses more deeply into poverty than before the structural adjustment reforms.

Collapse of the State

Second, and related to the first, has been the alarming increase of state collapse in the 1990s: from the USSR to Somalia, from East Timor to the Balkans, and from Afghanistan to Rwanda/Burundi/Zaire (DRC). While many different circumstances have contributed to the collapse of these states, they share the experience that top-down planning and blueprint approaches to solve social and economic problems often do not serve the needs of the people. For sustainable reforms and meaningful alternative state structures that the people will support, much greater levels of participation are required. As an example of an alternative state structure, the people of the Somaliland Republic—the former British Somaliland—have abandoned attempts to replicate a state cast solely in the image of its European colonial interlude. Instead, Somalilanders have created a people-based constitution that blends the wisdom of the traditional clan system with the checks and balances of a European parliamentary state. The "state" of Somaliland has been functioning effectively since 1993. While it continues to be a fragile institution, and while it faces many challenges in the future, Somaliland is maintaining peace and stability at a time when most of the rest of Somalia continues in chaos. It is perhaps ironic that this participation-based people's democracy is yet to be recognized as a legitimate government by any nation in the world.

Changed Agency and Institutional Behavior

A third source of the new participation is courageous leadership from a number of development agencies using new methods and approaches that include heavy elements of participation. International Non-Government Organizations (NGOs) such as ActionAid, Save the Children, Catholic Relief Services, CARE, the American Friends Service Committee, Oxfam, and the Mennonite Central Committee, along with many of the Nordic development organizations, have assumed an aggressive position that participation must be a demonstrated element in project planning, implementation, and monitoring/evaluation that they support. Parallel to the international NGOs have been programs that indigenous organizations have implemented, including groups such as SORRA in Somaliland, SAF in Madagascar, Sarvadoya in Sri Lanka, BRAC in Pakistan, IIDS in Nepal, MYRADA in India, ENDA-Zimbabwe in Zimbabwe, or universities such as Egerton University in Kenya, and many thousands of similar organizations throughout Asia, Africa, and Latin America. These organizations give strong endorsement to the effectiveness of participation, both in word and action. This backing of participatory approaches from local organizations may be the single most important indicator that participatory tools are now a permanent part of development planning and action. As further proof, the dramatic changes that UNDP, UNICEF, and selected units of FAO have achieved in the last decade—moving toward participatory poverty alleviation and resource sustainability—offers additional documentation that participation works.

THE ESSENCE OF THE PARTICIPATORY APPROACH

While many different theories, paradigms, and tool kits are associated with participatory approaches, three fundamental assumptions can be found in most of them.

Local Knowledge

Community and neighborhood residents have knowledge and information, but it needs to be organized. The participatory approach assumes that local resource managers and users have considerable knowledge about their problems and are familiar with locally based ways to solve them. Participatory methods further assume that local residents may not appreciate the enormous power that this information can yield or how systematizing this information can help rank problems, select options to solve the problems, mobilize community groups to take action, and attract external agents to offer assistance. As a first step, participation helps communities to organize and systematize their own information in ways that they will be able to control.

Community Institutions

Local residents have resources, but they need to be mobilized. Neighborhoods and villages can introduce projects, acting primarily on their own resources. Participatory methods help local institutions and leaders to mobilize themselves for effective action. Participation assumes that community institutions are among the most underutilized resources available for development efforts. Participatory approaches therefore do not wait for outside agents to take initiative but instead enable local groups to become the prime movers in development action.

Attracting Outside Help

Outside resources are available but need to be defined in the context of community-based priorities that external agencies can support. While community institutions can take initial steps to solve their own problems, they cannot necessarily do the job alone. External units such as government technical and extension officers, NGOs, and international organizations often can provide critical technical, financial, or managerial assistance that is unavailable to rural communities. Participation creates a setting in which village and outside groups form partnerships with shared goals, inputs, and actions to meet common needs.

THE NATURE OF PARTICIPATORY APPROACHES AND THEIR BENEFITS

Participatory methods use tools for data gathering, analysis, and ranking derived directly from Rapid Rural Appraisal methods developed by Robert Chambers and Gordon Conway and many others engaged in the participatory movement. These tools include village sketch maps, transects, seasonal calendars, livelihood mapping, gender calendars, SWOT analyses, trend lines, time lines, institutional diagrams, resource access ranking, options assessment ranking, community-based monitoring and evaluation, and force-field analysis. The solutions are then organized into community or neighborhood action plans in which specific community groups are identified to carry out particular tasks.

Participatory data gathering and analysis rely almost exclusively on visual data collection instruments. They enable community planners to reach out to many social, ethnic, gender, age, and class elements within a community and collect most data through group discussions. Facilitators use large group meetings to rank information and leave the data with the community for analysis, ranking, and action rather than extract it for external analysis.

These participatory approaches bring at least nine distinct benefits not normally available when centralized planning is practiced.

Community Mobilization

The single greatest advantage of participation is its capacity to mobilize community institutions around issues of sustainable development. Participatory methods have raised awareness of what can be accomplished as well as how local organizations, including women's groups, can do it.

Open Access Derived from Visual Data Gathering Instruments

Participation assumes that local communities function most effectively when data collection tools are visual. The participatory approach utilizes charts, maps, and graphs that residents can understand, comment on, and amend during data collection and analysis.

Interactive Problem Analysis

Use of participation and visual materials enables facilitators to maintain sustained interaction between and among knowledgeable members of the community when defining problems, considering previous successes, and ranking possible solutions.

Local Ownership Resulting from Community-Defined Problems

Ranked solutions, based on local priorities, technical feasibility, ecological sustainability, and cost effectiveness, lead to action that is responsive to local needs. This is accomplished larger through local preparation and implementation of Community Action Plans (CAPS). The local identity leads to community ownership of the action.

Systematized Participation

CAPS provide communities with clearly structured action plans that NGOs, government agencies, international organizations, district and regional development committees, and local groups can support. This process enables development assistance to go directly to the community groups.

Local Leaders Responsible for Follow-up

Inputs in activities designed through participation tend to be items that local communities can manage and control. Implementation, monitoring, and evaluation take place at little cost because locally based leaders have responsibility and capacity for follow-up.

Integrated Sectors

Because community-based approaches focus on problems that a community identifies, they simulate basic sectors of water, agriculture, forestry, micro-enterprise, health, transportation, education, livestock, wildlife, and credit to coordinate with one another and to cooperate rather than compete.

Inexpensive

Participation does not cost a great deal of money. Drawing on skills and experience of local groups and technical officers already in the field, the techniques do not necessarily require high-cost consultants or long-term field visits. While situations may arise in which high tech analysis may be required, it is the exception rather than the rule.

Strengthened Capacities of Local Institutions

Participatory approaches demonstrate that communities do not have to wait for external agencies to come to their assistance. Organizing village information into a systematic plan focuses community attention and mobilizes community groups. Local groups taking initiatives on, for example, water or forestry, have consistently attracted attention of external NGOs, international agencies, or government officers to provide a portion of the needed resources. The growth of these "Community Partnerships" between people and external agencies has been an encouraging development in sustainable resources management and one that offers promise for the future.

OUTGROWTHS AND SPIN-OFFS FROM THE PARTICIPATORY APPROACH

A few brief comments are suggestive of movements that have either started or gained momentum because of the recent expansion of participation. They are offered as a suggestive rather than a definitive list.

New Emphasis on Good Governance and Civil Society

International agencies are now recognizing that democratization is not simply a matter of sponsoring two-party elections. Local organizing, institution-building, and leadership development are critical elements that must run parallel to national legislation and constitution drafting. If checks and balances are to be introduced in societies that have not previously used them, development of local capacities in participatory decision making becomes an essential first step. To be effective, accountability must penetrate from the bottom to the top of a society.

Formal Tools for the Empowerment of Women's Groups

While many groups have adopted and adapted participatory tools for specific target and sector groups, no movement has been more active in this adoption than women. In Somalia women's groups are one of the strongest elements in the peace building process, using participatory tools.

In many parts of Asia, Africa, and Latin America women are using participatory approaches Such as SEGA (Socio-Economic and Gender Analysis)[1] to define problems, pose solutions, manage project activity, and select indicators for community-based monitoring and evaluation. It is likely that participatory tools will serve women's groups well for the next decade as they have in the past.

Introduction of Links Between Local and National Planning

One of the shortcomings of the first decade of participatory planning and action has been a gap between local action and national policy. In the last five years, the World Bank has been experimenting with national participatory poverty assessments, identifying priorities, regions, target groups, and available resources to be enlisted in a nation's campaign to alleviate poverty. While it is much too early to judge how well the process is working, there are reasons to believe that participatory tools can enlist a broad cross section of a society in poverty alleviation. The effort bears watching, especially as organizations such as the World Bank invest more deeply in social funds and human development.

Integrating Economic Analysis with Local Participation

Kenya's Egerton University, working jointly with the University of Arizona and Clark University, has introduced a new approach that blends community-based participation with household surveys. The result is an action plan that participatory tools have helped the community to formulate, with a clear sense of the economic impact on household income from implementing the plan. Known as PAPPA (Policy Analysis for Participatory Poverty Alleviation), the approach has been tested in five villages and is now being expanded to additional communities.

Introducing Local Participation in Regional and National Planning

The greatest weakness of the current participatory movement is its ineffective influence on national planning, priorities, policy, and budget allocations. A few pilot projects exist in which new methods and creative uses of existing tools are linking local and national. For example, in Venezuela, a newly mandated plan for decentralization in planning and budget allocations is using Geographic Information Systems (GIS) to link priorities of local groups with district and regional needs. The GIS enables planners to look at ecological and economic impacts of local plans as well as to build a regional data set, based on information collected through a participatory process.

These spin-off activities are evolving from participatory methods that local people, community residents, and neighborhood leaders have pioneered in the last two decades. While the ultimate outcomes are not yet clear, it is already established that participatory approaches are bringing significant change to the way in which development agencies plan projects and allocate funds. Incorporation of participation into core planning increases local ownership and support as well as sustainability and cost effectiveness. These are qualities that have been sorely lacking in previous development planning, policy, and project design. The revolution of the new participation will surely strengthen the move toward decentralization that has grown in the 1990s and most probably will continue into the next century.

---

[1] A *SEGA Manual* is available from Clark University's Programs in International Development, Community Planning, and Environment, Worcester, Massachusetts 01610, USA.

REFERENCES

Ayako, Aloys B., and Musambayi Katumanga. (1997). *Review of Poverty in Kenya*. Nairobi: Institute of Policy Analysis and Research.
Korten, David. (1995). *When Corporations Rule the World*. Kumarian Press.
World Bank. (1999). *World Development Report 1998/99: Knowledge for Development*. London: Oxford University Press.

PRELUDE TO PARTICIPATORY MANAGEMENT AND EVALUATION OF PROJECT ON CONJUNCTIVE WATER USE

Frank Simpson

Girish Sohani

INTRODUCTION

How does management that is truly participatory begin, where choices for the future have been limited severely for generations by poverty and harsh living conditions? Part of an answer is that the outsiders, entering into a development partnership with the rural poor, should be prepared to *let* things happen rather than *make* things happen.

Recognition of the full range of options for joint action, what things actually *can* happen, arises out of an awareness of both the possibilities and the limitations of the local environment. Ideally, all parties will move toward a common awareness of these, as the partnership matures. Ease of communication between the partners is of major importance to the success of the joint endeavor, right from the earliest days.

The purpose of this account is to describe the main factors contributing to the initiation of participatory management and evaluation of a project, titled *Conjunctive Use of Water Resources in Deccan Trap, India*. The project involved Bharatiya Agro Industries Foundation (BAIF), Pune, India, and University of Windsor (UW) Earth Sciences, Windsor, Ontario, Canada, working in partnership with the tribal and rural people of Akole Taluka, Maharashtra. It ran from 1992 to 1996. The authors were the Canadian (FS) and Indian (GS) project leaders.

A full account of the project research is presented by Sohani, Simpson, et al. (1998). The numerous contributions of the tribal and rural people to different stages of the joint activities are outlined by Simpson and Sohani (2001). A summary of the main technologies for water harvesting and spreading, employed in the project, is presented by Simpson and Sohani (this volume).

THE LAND, THE PEOPLE

Akole Taluka, Ahmednagar District, is in the northwestern part of the State of Maharashtra, in West-Central India. It is located on the eastern flanks of the Western Ghats mountain range. Rugged uplands in the west give way eastwards and southwards across the *taluka* to gently undulating and flat land. A generally thin soil cover, up to several meters thick in the larger valleys, rests on weathered bedrock and the flat-lying basalt lavas that constitute the bedrock. Surface drainage is for the most part southward into the east-flowing Pravara River, by way of numerous, mainly ephemeral streams.

Rainfall varies from about 2,000 mm in the west to 600 mm in the east during the monsoon (June to September), with little or no rain during the remainder of the year. July is the wettest month in the area. Before the project began, the monsoon rains flowed, largely unimpeded, out of the area as surface runoff, laden with eroded soil. In the postmonsoon period (October to January), the flow of streams decrease and many springs show reduced discharge. Drought conditions spread across the area in the premonsoon months (February to May). During April and May, the driest months, temperatures climb into the forties Centigrade.

The tribal and rural people of Akole Taluka are on the lowest rung of the social ladder, with limited access to education, health care, financial support, and to social services in general. Even in the late 1980s, they still had little contact with people from outside of the immediate area and tended to be mistrustful of strangers. They lived in conditions of extreme poverty, farming at a subsistence level of production. Their main crop was rice in the *kharif* growing season (June to September). Mainly cereals were produced during the *rabi* growing season (October to January), but the quality of the crop depended very much on the availability of soil moisture.

FIRST CONTACT

BAIF entered Akole Taluka in the late 1980s to learn firsthand about the needs of the tribal and rural people. This was seen as an important step toward understanding how best to help the rural poor help themselves. At that time, relatively little was known about the villages and outlying areas, mainly because of the remoteness of the location and the underdeveloped status of the people. From the start, BAIF took a flexible approach, with a focus on the needs of the rural family, letting the interaction develop from the responses of the people.

In 1989, dialogues about different options for sustainable livelihoods were initiated with village groups, carefully avoiding requirements specific to particular social groups. As ideas began to emerge from these discussions, BAIF organized "exposure visits" by interested individuals to its program of socio-economic rehabilitation of tribals at Vansda, in the Valsad District of Gujarat.

The villagers saw how other tribal people had gained livelihoods from converting wastelands into productive orchards. They later discussed the applicability of this to their own circumstances. Some families became interested in planting fruit trees and BAIF provided them with assistance, as part of its *wadi* (orchard) project. This led to acceptance of the need to pay attention to the availability of water and improvements to the quality of the soil. In this way, topics in the area of natural resources were opened up for discussion. Families gave high priority to a year-round water supply.

In 1990, BAIF carried out rapid, rural appraisals, dealing with water use and health respectively, in Akole Taluka. These were intended as contributions to the assessment of needs that would serve as a basis for the planning of a project on water supply. The RRAs were carried out by multidisciplinary teams and involved discussions with individuals and groups of villagers.

In June 1991, there was initial contact between the intending partners, when members of the BAIF project team and the UW project leader visited Akole Taluka to assess the feasibility of a joint project. The villages of Ambevangan, Manhere, and Titvi were assessed as representing a significant range of endowment in water resources and, by general agreement, were designated partners in the project.

COMMUNICATION

BAIF's field officers lived in the villages and shared the daily hardships of the tribal and rural people. This regular contact facilitated communication and had the effect of winning the trust of the villagers. Initially, many of the people were suspicious of the motives of the outsiders. However, they were receptive to suggestions that there were benefits to be gained from the exchange of information.

The insights into local knowledge systems, thus obtained, made a significant contribution to the success of the project. Indigenous technical knowledge included important information on botanical indicators of shallow groundwater, the relationship between terrain and spring discharges, and a local classification of soils. Attention to local religious beliefs further heightened the awareness of the outsiders about the views of the villagers, concerning water and gender.

The villagers provided input into design of the project at public meetings, held in meeting halls and schoolhouses throughout its term. These meetings were conducted with considerable formality, in accordance with local, folk, and traditional approaches to communication. At the request of the outsiders, women attended the meetings and expressed their points of view. This was a deviation from custom, which was accepted by the people as a necessity, because of the traditional, water-related hardships, borne by the women.

WATERSHED COMMITTEES

An important step toward participatory management and evaluation of the project was the formation of a watershed committee in each village. The starting concept for this, rooted in tradition, was the *ayojan*, a village planning committee that was responsible for decisions affecting most or all members of a community. Membership of the village watershed committee was determined at public meetings. The number of individuals in a committee was generally between nine and fourteen, with both genders represented.

The village watershed committees were valuable sources of local information on priority concerns, related to the provision of a year-round water supply. They were consulted regularly by the project teams about the appropriateness of particular technologies for water harvesting and spreading, under consideration for possible introduction into the area. Wider approval of these measures was obtained at public meetings. The siting of technologies was finalized on the basis of discussions with the watershed committees and with farmers in their fields.

WOMEN'S GROUPS

Self-help groups also generally arose out of traditional approaches to collective decision making. The women's groups in Akole Taluka came from the tradition of *wavli*, followed by tribal women in Gujarat. This provides protection for the rights of women to have earnings. BAIF institutionalized this concept and introduced income-earning activities for women in Akole Taluka. An important element of this was the synergy that came from shared experiences.

Village women's groups played a major role in the spread of basic concepts on hygiene and sanitation across the area. Isolation of human and animal feces away from water supplies, separation of water uses from the sources of water, purification of water for domestic use, and approaches to personal cleanliness in the home were some of the important issues addressed by these groups.

FIGURE 3
Members of Canadian and Indian teams discuss project outcomes with villagers at public meeting. Village of Titvi, Akole Taluka, Ahmednagar District, Maharashtra State, India.

FIGURE 4
Member of Indian project team (left) at meeting of village women's group. Near village of Manhere, Akole Taluka, Ahmednagar District, Maharashtra State, India.

PARTICIPATORY MANAGEMENT AND AFTER

The people participated in all stages of the project as full partners. They showed considerable interest in the sampling of water, soil, and rock, for laboratory study. Indeed, they frequently assisted in the sampling procedures. The outsiders took pride in gaining new insights into local knowledge systems, as the project continued. This information formed a part of each public discussion that dealt with adoption of new technologies and contributed to the acceptance of innovation by the people.

The villager accepted that economy of water use required attention to water conservation on a watershed scale. They saw that there was potential for improved agricultural production from the resulting improvements to soil moisture. As a result, they invested time and effort in the introduction of technologies to retain monsoon waters in the area. This involved overcoming a traditional bias toward short-term benefits in the practice of agriculture. They assumed ownership of the technologies at the demonstration sites and took responsibility for maintenance, after completion of the project term.

ACKNOWLEDGMENTS

The authors gratefully acknowledge the guidance of the late Manibhai Desai, founder and first president of BAIF. They also thank his successor, Narayan Hegde, for considerable encouragement and helpful discussion.

The project was funded by the International Development Research Centre, Ottawa.

REFERENCES

Simpson, F., and Sohani, G., 2001. "Benefits to villagers in Maharashtra, India, from Conjunctive Use of Water Resources." In: Jeffery, Roger, and Vira, Bhaskar (Editors), *Conflict and Co-operation in Participatory Natural Resource Management: Lessons from Case Studies*. New York, Palgrave, pp. 150–168.

Simpson, F., and Sohani, G., this volume. Water harvesting and spreading for conjunctive use of water resources.

Sohani, G. G., Simpson, F., et al., (1998). *Conjunctive Use of Water Resources in Deccan Trap, India*. Pune: BAIF Development Research Foundation.

SOCIAL ASSESSMENT AND STAKEHOLDER PARTICIPATION

Jonathan Brown and Ayse Kudat

One of the great changes in the decade of the 1990s in project planning and implementation has been the introduction of social assessment—the systematic investigation of social processes and behavior—and of stakeholder participation—the consultation and involvement of key affected groups, especially the poor, the vulnerable, and the excluded. Social assessment (SA) and stakeholder participation (SP) address issues that are especially relevant for intermediate technology: beneficiary commitment, assessing reactions to new ideas and changes in behavior and attitudes, community involvement in project preparation and implementation, and a permanent process of social impact monitoring.

Development projects come in all forms, sites, and sectors: from health care systems in Asia to urban infrastructure in Latin America, from irrigation in the Maghreb to reforestation in Turkey, from education in Africa to reducing environmental pollution in Thailand, from postconflict reconstruction in Bosnia to restructuring the coal sector in Russia. Despite this enormous diversity, some basic common features exist. Every project is a social process, not just a commercial or public investment, and brings into play an array of social actors. Yet for a long time the conventional approach was to treat projects as only economic, financial, or technical interventions. Institutional and environmental analyses were added over the last two decades. Only recently was the need seen to do social analysis, arising from the nature of the development process itself. People's behavior in the development process is generally determined not only by economic rationality, profit seeking, and the market environment but also by a host of cultural variables. This is why adding social analysis to economic analysis can reveal both the potential for development in a specific context and the means for realizing this potential in practice. The integration of economic, financial, technical, institutional, and environmental analysis with social analysis results in a consolidated knowledge that is distinctly more powerful in guiding project planning and implementation than any of its separate parts used in isolation.

Traditionally, development projects have been developed through a top-down approach in which intended beneficiaries and other affected groups have not been involved in planning and execution in a systematic way. This has led to projects being financed that have failed, that have been irrelevant to real needs, and that have been unsustainable, often because the very people who were the intended beneficiaries or other key stakeholders have been excluded from real involvement in decision making. Key stakeholders are the poor, low-income, vulnerable, and excluded social groups. The broader group of stakeholders, who are equally necessary for projects to be successful but are often forgotten, include the private sector, civil society, government and nongovernmental organizations and their members, and others who facilitate or hinder the ability of the poor to have equitable access to the goods and services offered by development projects. The notion of appropriate technology implies that other technology may be inappropriate. Appropriate or inappropriate for whom, by whose definition, and with what consequences? Social assessment and stakeholder participation allow appropriate technology to be applied where it is relevant and useful and with more popular support.

THE RELEVANCE OF THE SOCIAL ASSESSMENT/STAKEHOLDER PARTICIPATION (SA/SP) PROCESS

The SA/SP process would be relevant if it simply ensured that development projects "do no harm" as environmental mitigation plans, for example, do. However, the power of the SA/SP Process lies in its ability to mobilize essential information about people's behaviors and the power of social groups to "do better," to make the chances of success higher, and to improve the quality and quantity of that success. "Doing good" is not enough when "doing better" is easily achievable with the SA/SP Process. First, however, the relevance of the SA/SP process has to be demonstrated in ways that are understandable to development practitioners in areas in which they now need assistance.

Cost Recovery in Azerbaijan and Turkmenistan

The level of tariff increases to recover costs and maintain assets is always a difficult debate between donors and borrowers, a debate in which people are seldom involved. This was the case in a proposed water supply project for Baku, the capital of Azerbaijan. Government claimed the poor could not pay for water. The donors claimed Government could not afford 100 percent subsidy. A social assessment showed that on average, households spent about 17 times more on alternative water supplies, because public water was unavailable, than on their monthly water bills. The poor spent 7 percent of their income on coping strategies for water. The SA/SP process showed that householders would be willing to pay substantially more than their current monthly water charges for better public water service. The poorest elements of the population were prepared to pay 6 percent of their income, a slight decrease from their current payments for coping mechanisms. With debate over cost recovery moved from the ideological to the practical, Government and the donors agreed on a cost recovery system that was both relevant financially and affordable to consumers. An equally dramatic example of the relevance of SA/SP for cost recovery occurred in Turkmenistan where an urban transport project was being prepared. The SA revealed a situation on Ashgabat, the capital city, of poor service in public transport and of tariffs so low that there was no coin small enough for a passenger to pay a single-fare ticket. The result was an unsustainable and deteriorating public transport system in which drivers were not paid and consequently instituted unofficial tariffs. The poorest quarter of households were compelled to spend 12.8 percent of their income on various mechanisms for coping with the lack of public transport or to make "extra" payments for using public transport when it was available. The SA/SP process also revealed that 94 percent of public transport users would accept a 2,000 percent increase in tariffs if there would be a real improvement in service. Tariff increases were immediately put into operation. The quick improvement in public service resulted in the poorest households paying higher fares but incurring lower real costs for transport when compared to the cost of their previous coping mechanisms. Raising tariffs lowered the real cost of transport and cost recovery was never an emotional issue between Government and the donors during project preparation.

Technical Standards in Uzbekistan

The initial technical and financial proposal to provide drinking water in the Karakalpakstan and Khorezm districts of Uzbekistan most affected by the Aral Sea drought conditions was a low cost system of shallow wells and handpumps rather than more expensive surface water solutions. However, an SA with a salinity taste-tolerance survey showed that salinity levels from shallow wells and handpumps of between 5,000 and 7,000 milligrams per liter were considered undrinkable by most consumers. On the other hand, the heavier investments in surface water would have been enormously expensive if it followed Government's objective of only 1,000 milligrams of salt per liter. The SA survey showed that most people would accept 2,000 milligrams per liter, thus permitting smaller investments in surface water investments. Out of this SA/ST process came a definition of appropriate technology that was very different from what was initially suggested.

Institution Building in Russia and Uzbekistan

In the mid-1990s the Russian government was discussing with donors restructuring the coal sector, in particular the institutional mechanism to reduce the level of subsidies and ensure that those subsidies that were provided actually arrived in a timely fashion in the local bank accounts of those who were supposed to benefit from public support. Government, management and labor, and the donors disagreed over which institution should control the planning and implementation of the new subsidy scheme. The SA/SP process showed that the beneficiaries trusted no existing institution—government, management, labor unions, and donors alike were all distrusted. The beneficiaries would trust, however, a new institution with multistakeholder participation including, for the first time, representatives of labor, local communities, and regional officials. Such an institution was created. The SA/SP process also had a major impact on the water sector institution mentioned above in Uzbekistan. The technical solution using groundwater favored by consumers could not be managed by local communities but required a more centralized approach that was distrusted by consumers. This finding caused the water utility to establish a consumer affairs department so that this more centralized institution would be more open and receptive to consumer needs.

Emergency Lending

It is common wisdom that lending in emergency situations, whether the oil spill in the Komi Republic of Russia in 1993 or postconflict reconstruction in the areas surrounding the Nagorno-Karabakh conflict in Azerbaijan in 1996–1997, is too urgent, the resources too small, and the priorities too "evident" to require the SA/SP process. The SA/SP process can, however, be done quickly, and the results are sometimes different from what might be expected. The oil spill in the Komi Republic was three times the size of the Exxon Valdez spill in Alaska, but the SA in Komi showed people's highest priority was not on immediate relief but on prevention of future spills. In Azerbaijan, the SA/SP process revealed a broad and comprehensive pattern of needs of different population groups that required a partnership among donors that had not been initially foreseen by either Government or the donors.

THE ELEMENTS OF THE SOCIAL ASSESSMENT/STAKEHOLDER PARTICIPATION PROCESS

The SA/SP process should be initiated early in the project identification and planning process and should, to different degrees, be continued through project implementation so that there is continual feedback on which to base midcourse corrections if they prove necessary and/or desirable. The SA/SP process is based on the following four elements.

- Identification of key social development and participation issues
- Evaluation of institutional and social organizational issues
- Definition of the participation framework
- Establishment of mechanisms for monitoring and evaluation

Identification of Key Social Development and Participation Issues

Defining information needs and designing an information strategy that identify the key social development and participation issues is a basic step in launching the SA/SP process. This strategy will start with broad sectoral and country-specific information and then focus on project specific data.

Identify Broad Social Development Issues. The preliminary identification may be based on available secondary information such as social development literature relevant to the sector; social impact monitoring studies for similar projects in the same or other countries; existing social development profiles or other background information; country social science studies, area studies, other secondary literature, demographic data, and so forth; and consultations with knowledgeable local and international experts.

Narrow Down the Key Social Development Issues to the Project Context. With the broad social development issues clarified and the key stakeholders identified, the SA/SP process focuses on how the participation of the poor and vulnerable groups may be affected and on how their participation may impact on project ownership and sustainability. The SA/SP proposes specific analyses of potential gains and adverse impacts, with a view toward how specific stakeholder groups may facilitate or hinder the participation of the poor in the development initiative. More general issues, such as social cohesion, equity, social capital, social diversity, social organization, and social exclusion may also be identified, if they pertain to the project.

Identify Stakeholders Whose Participation is of Strategic Importance. The stakeholders will include various social groups, as well as formal and informal agencies in both the public and private sectors, including nongovernmental organizations (NGOs). Which groups and agencies are most directly concerned by the project will emerge from the review of secondary literature. This listing may be complemented by consultations with policy makers, representatives of central and local government, knowledgeable local and international social scientists, and local NGOs. Clearly, this step requires a good understanding of the broader issues in social development and of the technical options that are possible in the project context.

Define a Project-Specific Information Strategy. This strategy will identify the key social actors and their interactions and the social provisions needed to achieve the project's specific economic, technical, and social goals. The information will cover social-economic characteristics of the key stakeholders and other, more broadly concerned groups; information about their problems, constraints, and needs; and ideas for alternative solutions. These data will be used to define eligibility and targeting criteria, to confirm beneficiary identification, and to determine the appropriateness of proposed, alternative solutions for the targeted social groups.

The rigor of the social assessment data-gathering process is important. But the arsenal of social science research techniques used for doing a rigorous SA is broad; often, a mosaic of social science research tools is brought together. Every specific situation demands sociological imagination—a different combination of methods and procedures tailored to the given issues and set of actors. Nevertheless, the initial data gathering must be systematic, quantified, and done by professionals, since the databases will provide an empirical basis for analysis and the baseline for all future monitoring and evaluation. Of particular importance are (1) testing the survey instruments, especially the understanding of the questions by those surveyed; (2) training of survey administration personnel to ensure reliability across the whole sample; and (3) assembling a data analysis capacity that can see the patterns in the data and draw the appropriate conclusions so that the people's answers are faithfully and relevantly reported. Focal groups are an especially good tool for deepening the findings of the systematic surveys.

Define Mitigation Plans. Mitigation measures must be defined where adverse impacts are identified for certain social groups, particularly with regard to national or international norms, especially in the areas of resettlement, impact on indigenous peoples, large-scale unemployment, and so on. Many energy projects, for example those dealing with conventional energy such as dams, adversely affect people unless mitigation plans are carefully prepared and fully implemented.

Evaluation of Institutional and Social Organizational Issues

Identify Blockages to Equitable Access to Project Activities. The poor and vulnerable groups who are intended beneficiaries of project initiatives may encounter difficulties in accessing project resources. The reasons for this can be many and varied: formal and informal institutions, local customs, patterns of social organization, intergroup relations, social institutions (e.g., family, kinship groups, tribal or ethnic affiliations), formal and customary laws and regulations, property rights, subsidy arrangements, central and local government agencies, and information and communications systems. The social assessment determines whether systematic structural blockages exist and, if so, proposes mechanisms to overcome them.

Knowing how to use existing social capital, which may not be immediately obvious, is critical. In 1994 the World Bank helped the government of Tajikistan prepare a preliminary program of reconstruction following severe floods, a pause in the civil strife affecting the area. The social assessment raised the possibility, since new "postcommunist" institutions were very fragile, of mobilizing community-based capital based on what already existed from the communist era or was embedded in traditional structures.

Recommend Strategies for Strengthening Institutional Capacity. Local-level and informal rules—norms, values, and belief systems that shape the attitudes and behavior of social groups—may affect project implementation arrangements. The social assessment institutional analysis will, therefore, not only identify whether structural blockages exist but also will propose modifications to existing arrangements or even entirely new institutional structures to overcome them.

Definition of a Participation Framework

The definition of a participation framework takes the results of the identification of key social development, participation, and institutional issues and makes them a living part of a project to reduce the risks of project failure and enhance the chances of establishing a process of change that is valuable and sustainable for people over the long term.

Formulate the Participation Strategy. The social assessment incorporates two types of participation. First, there is the participation of the poor and vulnerable groups, which is a principle objective of the SA. Second, there is the participation of a broader group of stakeholders—governmental and nongovernmental organizations, donors, and other partners—in project strategy design. The broader stakeholder participation is critical for attaining the participation of the poor and vulnerable groups, and both levels of participation are important in developing support for the specific project proposals and institutional arrangements identified in the first two steps of the SA/SP.

The SA/SP process, therefore, involves the design of an information and communication strategy to ensure stakeholder ownership of these proposals. The strategy usually has the following three elements.

- Mechanisms to share the information from the social surveys and institutional analyses with the broader group of stakeholders and partners (including government, international organizations, and NGOs)
- Mechanisms to ensure the participation of key stakeholders wherever feasible
- Feedback mechanisms that facilitate stakeholder response to the information provided

When direct participation is not feasible (as might occur when the project initiative is very broad), the social surveys and institutional analyses can provide important information themselves about the views of the poor and the vulnerable populations, especially if they are designed with this possibility in mind. In other instances, the specificity of the project may preclude democratic representation in the determination of the mechanisms for key stakeholder participation. In either instance, the SA will still develop specific mechanisms to facilitate the direct involvement of the poor and vulnerable groups in the design, implementation, and monitoring of the project.

Define Implementation Arrangements. On the basis of the stakeholder dialogue aimed at ensuring ownership and commitment to a policy of inclusion, the SA will define the specific responsibilities and monitorable contributions of each stakeholder group. This dialogue will also help determine implementation arrangements, including institutional changes, capacity building, targeting, sequencing, subsidies, and incentives. The implementation plan will also include a joint evaluation of joint social development benefits and risks, including potential conflicts and costs.

Establishment of Mechanisms for Monitoring and Evaluation

The SA/SP process contributes to the monitoring and evaluation (M&E) component of a project by focusing on inputs, processes, outputs, and impacts. Specifically, the SA/SP does the following.

1. Identifies monitoring indicators such as (a) input indicators (benchmarks) to measure and monitor inputs that facilitate participation of the poor or vulnerable groups or meet other social objectives; (b) process indicators for the same purpose; (c) output indicators; and (d) procedures and impact measures to determine whether intended social development changes actually impact people's lives and behaviors.
2. Defines transparent evaluation procedures, including participatory approaches.
3. Ensures that monitoring and evaluation procedures are established for any mitigation plans aimed at avoiding harm.
4. Ensures that all M&E proposed is carefully scheduled, fully budgeted, and properly supervised.

The ideal, of course, is to established a permanent mechanism of social impact monitoring similar to the economic, financial, technical, environmental, and institutional monitoring and evaluation that should exist throughout the life cycle of a project. The lessons of M&E—that it is essential for fiduciary reasons to ensure the appropriate use of funding, that it is essential for development reasons to see that the objectives of helping people are attained, that it is essential for project reasons to permit restructuring when and if needed—similarly require a permanent process of social impact monitoring and evaluation. Successful private sector companies do this as a matter of routine. Public sector institutions should operate similarly, as common practice not as exceptions.

THE LESSONS OF SOCIAL ASSESSMENT/STAKEHOLDER PARTICIPATION

Experience has shown that there are projects where SA/SP is essential: populations who have been historically disadvantaged or excluded from development initiatives; large social and economic inequalities; postconflict and natural disaster situations; acute social problems; large-scale enterprise or sector-wide restructuring; and anticipated major adverse impacts such as involuntary resettlement, indigenous people's contact, and loss of cultural heritage.

There are also additional projects where SA/SP is advisable—for example, where changes in existing patterns of behavior, norms, or values are required; community participation is essential for sustainability and for success; insufficient knowledge exists on local needs, problems, constraints, and solutions; and beneficiary targeting mechanisms or eligibility criteria are unknown.

There are also instances where the SA/SP process may not be necessary or can be reduced in scope—for example, updating medical equipment, expanding the coverage of appropriate educational facilities, introducing improvements in financial management in a banking system. But even in traditional "hardware" projects the SA/SP process can have value if appropriately targeted. In highway projects, for example, the usual Economic justification is the reduction of vehicle operating costs. Yet who benefits from lower-vehicle operating costs and how? And to what standards should highways be improved or maintained? These may sound like technical issues, but they have enormous, and different, impacts on various social groups that can be at the very heart of a highway project's justification, especially when resources are constrained.

It is extremely difficult for the SA/SP process to be carried out effectively if it is not supported by a broad range of groups within a government and the donor community. Mobilizing and maintaining this support among the groups that have money is as important as the quality of the SA/SP itself. Having some donors prepared to go ahead without SA/SP puts a government and the donors who favor SA/SP in a very difficult situation.

SA/SP needs to have adequate funding so that it can do what is required, when it is required. Coming into project preparation when the major decisions have been made undermines the credibility of the process, since SA/SP is then too often seen as delaying, being "always" critical, rather than as the positive force it can be when it is involved from the beginning.

SA/SP needs to be done in a systematic, quantifiable manner by professionals and not through merely qualitative techniques that are open to conclusions that lack substantiation. The process may need, in some cases or in some parts (survey techniques and data analysis, for example), assistance from abroad, but it can, and should, be done in great part by local experts.

SA/SP needs to pass the so-called ASR test—Art, Science and Relevance. The art refers to the need for the process to remain faithful to what the people really say without too much external interpretation that can cloud the basic truths. The science is the ability of the process to understand differences within social groups and to find useful patterns of needs and behaviors to which donors can respond with assistance. In addition, the process needs to be relevant to the requirements of the donors. SA/SP may very well reveal many interesting things of importance to communities and to the government, but it should not neglect the basic objective of securing money from donors to help people.

SA/SP is another instrument in assessment, project preparation, and monitoring and evaluation, which needs to be combined with more traditional economic, financial, technical, environmental, and institutional tools. SA/SP should not and cannot be done in isolation.

Finally, SA/SP represents a way of "sharing" power and responsibility between the disadvantaged people, largely powerless otherwise, and officials in government, the donor community (including international NGOs), and civil society. Empowering people means the people share in the risks and the responsibility of what actually happens.

SOURCES OF FURTHER INFORMATION

The academic literature is rich with information on survey techniques and data analysis, marketing, research methodology, and so on.

For further information on using SA/SP for projects in developing countries and those in transition, see *Social Assessments for Better Development*, Cernea and Kudat, 1997, published by the World Bank. Another compendium of case studies of SA/SP in the rural sector in Turkey and in the former Soviet Union is forthcoming. For other publications and specific examples of social assessments and stakeholder participation by region and by sector, contact the World Bank—e-mail: books@worldbank.org and on the World Wide Web:

http://www.worldbank.org. The Bank Web site has updated information on other resources and on resource people.

Jonathan Brown is operations adviser in the Quality Assurance Group, Office of the Managing Directors of the World Bank. From 1990 to 1997, he managed the unit in the World Bank that provided investment lending and policy advice in infrastructure, energy, and environment for Russia and Central Asia. Mr. Brown also held management positions in the Africa Region of the World Bank from 1981 to 1990 and was the Bank's first resident representative in Senegal from 1978 to 1981. A graduate in history of Yale College, Mr. Brown received an MA in communications from the University of Pennsylvania and a MBA from Harvard Business School.

Ayse Kudat is Social Development Specialist in the World Bank Regional Vice Presidency for Europe and Central Asia. She holds a BA in economics, a graduate degree in social anthropology (Oxford University), and a PhD in social relations (Harvard University) with a major in political science (MIT). Before joining the World Bank in 1989 she served in many positions including director of the Berlin Science Center, Comparative Social Science Institute, coordinator of the 1985 UN Conference on Women, professor of social sciences at the Middle Eastern Technical University, and senior program officer in the UN Center for Human Settlements.

THE HUMAN SIDE OF CHANGE

Martha K. Preus

"THE BEST LAID PLANS GONE AWRY." WHY?

A recent KPMG study indicates that a high percentage of mergers fail. Even if they do "work," most do not meet the planners' expectations. The study indicates that a lack of appreciation for the human side of change is a main reason for failure. Our experience as project managers and organizational development consultants leads us to believe that the same holds true regarding project planning in developing countries. Even when a project is carefully planned and executed, the results and sustainability of the project often falls short of expectations. The main reason for this is that the human side of change is not considered as important as the strategic mission.

Whether private or public, nonprofit or for profit, large or small, international or local, the success and sustainability of a project depends as much on the relationships, values, and beliefs of the stakeholders as on the strategic plan. Without considering the human side of change, most projects will not realize their potential. Unfortunately, time, money, and resources are generally not sufficiently budgeted to address the impact of resistance to almost any proposed change. Resources are usually allocated for tangible things such as technology and hardware, not for intangible feelings and emotions.

"WE HAD SUCH GOOD INTENTIONS, BUT IT ISN'T WORKING."
HOW COME?

A traditional approach has been for donor communities to finance, plan, and implement the project with or without the total commitment of the recipient, believing it was in the recipient's best interest. This is not a "partnership model" and is often met with resistance from various factions within a community. It also sets the stage for dependence on the donor for ongoing support.

Unless a community accepts and commits to a project it will not be successful; furthermore, training, resources, and community support will be required for sustainability.

Increasingly, organizations such as the Deutsche Gesellschaft fur Technische Zusammenarbeit (GTZ) are realizing the importance of partnering with communities. This requires the commitment of resources from both parties. It is no longer a dependent relationship in which the partner with the greater financial resources determines what is best for a community. It is a shared process, as was described earlier in the chapter on project planning and implementation. If the community is unwilling to share the cost and risk (which includes time and human resources), then the project doesn't move forward. All parties must be held accountable for their agreed upon commitments and the responsibility for the outcome. The project is not a gift; rather, it is an improvement for the overall system.

HOW DO YOU MANAGE RESISTANCE TO CHANGE?

Projects change a system. One thing is certain with any change process: There will be people who will aggressively oppose or "silently" sabotage it, regardless of how well it is planned and communicated. Those who are planning the change process (organizational or project leaders and their teams) often underestimate people's ability to resist. There are many books and articles on the process of change, some of which are referenced at the end of the article. Most agree on some key points. For a change to be successful, you need the following.

- Top level commitment
- A shared vision of the future
- Understanding of the need for change
- Critical assessment of present state (SWOT—strengths, weaknesses, opportunities, threats)
- Acknowledgement of the past
- Constant, consistent communication
- Involvement at all levels and input from all stakeholders
- Management of political networks

HOW DO YOU LEAD IN A PARTNERSHIP MODEL?

In his book *Leadership Without Easy Answers* Ronald A. Heifetz states that a leader is someone who helps the community face their problems and find solutions. In this way an external advisor or project manager can provide leadership while supporting the community and developing individuals to take on the responsibilities and accountabilities normally associated with traditional leaders.

We often work with leaders and their "teams" that are coming together on major projects or trying to move from a hierarchical system to a flatter, team-based one. While it is critical to have the support of the present leadership, it is also necessary for that leadership to accept that they cannot have all the answers. They must listen to, understand, and empower others. Leaders must maximize the potential within the group of individuals that make up their team and be willing to share the responsibility, accountability, and rewards. Teamwork is increasingly emphasized as an effective way to manage changes and ensure that the responsibility and accountability for the project are shared throughout the team.

These four concepts and skills are important in a team-based environment.

1. Effective communications
2. Multiple realities
3. Levels of system
4. Dynamics of team evolution

Effective communication requires the following.

- *Active listening*: the art of really hearing what others have to say without trying to frame your response or defend your position. It is a powerful technique for building trust, increasing openness, fostering collaboration, and demonstrating respect.
- *Reflection*: the ability to slow down and take time to consider what was expressed before responding, defending, or challenging.
- *Inquiry*: the skill in asking questions for clarification of a different perspective. Questions reveal the thinking behind the positions or ideas and can help develop trust, understanding, and respect.
- *Advocacy*: refers to the willingness to share your ideas with others and explain why you think as you do. It is important to know when and how to "hang on" or "let go" of your ideas.
- *Constructive feedback*: information given by one person to another, which helps us see ourselves through others' eyes. It is important to develop some norms so that this is done in a caring and effective way.

Multiple realities relates to the awareness that we all experience the same situation through our own filters. This may cause us to perceive the "same thing" in totally different ways—not wrong or right—but resulting in very different perceptions. In any environment this is important, but in a multicultural one it is critical. There will undoubtedly be different norms regarding the following factors.

- Time and space
- Functional background
- Culture and language
- Gender
- Education and learning style
- Age
- Individual personality preference style

Though it takes time to explore individual expectations and understandings, it saves time over the course of working together.

Levels of the system refers to how we look at a situation. There are different vantage points from which we may assess a situation, event, or problem: (1) individual, (2) dyad or small group, (3) organization, and (4) total system or environment.

It is important to consider the impact at all levels when planning and implementing a project. If there is a problem or behavior exhibited on one level of the system, chances are it will be found in other levels of the system. Nothing exists in isolation. Thus, issues or behaviors that are experienced in one level of the system will inevitably impact the others and must be addressed.

Teamwork is the joint effort of a group, which becomes the driving force to bridge the gap between vision and reality. Projects require groups of people working together toward a common vision, sharing a mission with clear goals and objectives, and resources.

An effective organization or team must address the following four areas if they are to realize their potential.

1. Common purpose/goals
2. Roles
3. Procedures
4. Relationships

- *Common purpose/goals*: Project teams come together because they share a common purpose. If they are truly effective, they have compelling vision of the future and a clear definition of mission that is well communicated and agreed upon. Plans and work directly relate to the vision and mission. Specific, measurable goals and objectives are established and published. Priorities are clear, and individuals are committed and enthusiastic. Individual needs are incorporated into the team objectives, and members feel the team's objectives are achievable and reflect their own goals.

- *Group roles*: Effective project teams are made up of members who have clear agreement regarding who is on the team and who is doing what. This relates to how well the organization or team optimizes and leverages the knowledge, skills, and personal traits of each member. All members of high-performing teams participate and are able to influence. Roles and responsibilities are clear to everyone, and functions do not overlap. Members have the required technical skills and knowledge to perform their jobs and are accountable for their performance. Project teams generally deem this important and address it at the outset of any project work.

- *Procedures*: This relates to how the group operates. High-performing teams have clarity and agreement as to operating processes such as problem solving, decision making, conflict management, leadership style, information networks, climate, norms, and quality control. Effective management of the process initiates a review of these procedures from time to time to ensure that important processes and procedures do not become obsolete or dysfunctional. This is an area where teams often take things for granted and do not spend enough time at the beginning of the project work to establish norms. Particularly in a multicultural environment this is very important. In the long run it saves time to work through and establish "ground rules" for expected behaviors and norms.

- *Relationships*: Synergy requires mutual trust, openness, respect, and support. This doesn't "just happen" even if a group shares a common purpose, knows their functional role, and has worked through group processes and procedures. Ongoing constructive feedback and active listening to one another are necessary skills. Interpersonal relationship issues or conflict must be dealt with constructively in a timely manner. High-performing teams experience an atmosphere that is congenial but one in which confluence is not a norm. Relationships usually take time to develop, and time is a resource that is a precious commodity in project work. People have a tendency to jump in and get started with the "real work." However, establishing positive relationships where each member is fully interactive with every other group member is critical to the overall success and timeliness of the project.

Finally, it is critical that a project manager model and communicate the importance of the goals, roles, procedures, and relationships emphasizing the following.

- Effective communication
- Performance measures
- Clarity regarding the vision and mission
- Establishment and review of procedures regarding decision making, problem solving, and conflict management
- Effective team meetings
- Valuing diversity
- An understanding of team dynamics and evolution

Careful attention to these principles will assure that the human side of change is given the attention necessary for successful achievement of the strategic mission.

REFERENCES

Binney, George, Colin, and Williams. (1997). *Leaning into the Future*. London: Nicholas Brealey Publishers.
Charles, Seashore. (1992). *What Did You Say? The Art of Giving and Receiving Feedback*. MA: Douglas Charles Press.
DeGeus, Arie. (1997). *The Living Company*. Boston: Harvard Business School Press.
Goodstein, Leonard, Timothy, Nolan, and J. Pfeiffer. (1993). *Applied Strategic Planning*. New York: McGraw and Hill.
Heifetz, Ronald A. (1994). *Leadership without Easy Answers*. Cambridge, MA: The Belknap Press of Harvard University Press.
Hofsstede, Geert. (1997). *Cultures and Organizations, Software of the Mind*. New York: McGraw and Hill.
Nevis, Edwin. (1996). *Intentional Revolutions*, San Francisco: Jossey-Bass.
Preus, Martha, Shirley, and Wouters. (1998). *Team Development Handbook*. Organizational Development Consultants International Berlin.
Scholtes, Peter. (1998). *The Team Handbook*. Madison WI: Joiner Assoc.
Senge, Peter. (1999). *The Dance of Change*. New York: Doubleday.
Senge, Peter. (1994). *The Fifth Discipline Fieldbook*. Nicholas Brealey Publishing Limited London.
Tanner, Deborah. (1998). *The Argument Culture*. New York: Random House.

GENDER AND TECHNOLOGY IN THE FIELD

Barbara Earth

In this article, I discuss some of the features of gender imbalance in development and present a framework for remaking the relationship. Case studies of three divergent development issues—water, integrated farming, and family planning—reveal the completely different experience of men and women in technological development. Gender dynamics detrimental to women are seen to offset the potential for development in these settings.

International agreements state that gender justice is a necessary component of development. To operationalize gender-just development in the field, a conceptual model for women's empowerment is presented. The model is used to draw out lessons from the case studies and can enhance the gender accountability and thus the overall validity of future development efforts.

INTERNATIONAL MANDATE FOR GENDER JUSTICE IN DEVELOPMENT

International agreements can and should be used as guidelines in development, as they incorporate the values and commitments of the majority of nations. These documents state clearly the rights that all people should be able to enjoy. The purpose of development, in its broadest sense, is to deliver these promises.

Articles contained in international agreements that specifically relate to gender and development include the following.

- "The right to development is a universal and inalienable right and an integral part of fundamental human rights, and the human person is the central subject of development." (Principle 3 of International Conference on Population and Development)
- "The full and equal participation of women in political, civil, economic, social and cultural life, at the national, regional and international levels, and the eradication of all forms of discrimination on grounds of sex are priority objectives of the international community…." (World Conference on Human Rights)
- "States Parties shall take all appropriate measures: a) to modify the social and cultural patterns of conduct of men and women, with a view to achieving the elimination of prejudices and customary and all others practices which are based on the idea of inferiority or the superiority of either of the sexes or on stereotyped roles for men and women…." (The Convention on the Elimination of All forms of Discrimination Against Women)
- "The States Parties to the present Covenant recognize the right of everyone…. b) To enjoy the benefits of scientific progress and its applications…." (The International Covenant on Economic, Social and Cultural Rights)

These promises are lofty, and we are not likely to see them manifested in our lifetime. However, they serve to powerfully legitimate development efforts that strive for gender justice. The ground-level reality may be very far from the ideal, but these promises point the direction for development work that is committed to social change wherein women as well as men will determine the future of their lives and communities.

NECESSITY FOR GENDER ANALYSIS

In order for development initiatives to serve gender justice, the initial analysis at the field level must include the full range of forces differentially affecting the sexes. After diagnosing those differentials, development workers are then challenged to devise means of remaking the relationship between men, women, and development. Implied is that the tools of development—that is, technology[2]—will enhance the lives of both sexes and not privilege one over the other. There are deeply entrenched inherent obstacles to gender justice as described below; however, in the absence of any miracles, the long-term project of remaking the relationship between men, women, and development is the only possible route.

---

[2] Technology includes knowledge and skills as well as hardware.

"GENDER" PRIVILEGES MEN

Gender refers to the socially constructed definitions of men and women. These social definitions bestow power differentially between the sexes. "Power" is a complex variable but contains notions of efficacy in crucial areas of material and social existence, such as control and mastery of resources, and exposure to information, and having a public presence and political voice. These dimensions of power define gender relations. Gender balance of power affects, and in turn can be affected by, the forces of development and technological change.

"Complementarity" of gender roles has been an organizing principle that delineates what the sexes do. Most societies are organized with complementary gender roles. Some are quite strict about what women and men are allowed to do, but most permit some overlap in the gender-assigned occupation of space and the fulfilling of roles and tasks. However, there is a general pattern that privileges men.

The public sphere outside the home where men normally predominate is the hub of economic, cultural, and political aspects of life that define the society. The "productive" economy is located here—that is, where goods and services are exchanged for cash. Though women have entered the productive economy in large numbers, women usually are more closely identified with the "reproductive economy" associated with the home and family. The reproductive role encompasses all the domestic work that enables society to reproduce itself, centering around food, water, shelter, and caring for others.[3]

Reproductive work includes bearing and nurturing the next generation; supplying the household with water and food; supplying fuel and cooking for the family; cleaning the food, the household, the utensils, the clothing, and the children; caring for the sick and elderly; and maintaining the home. The reproductive work performed by women subsidizes the public activities of men. The work must continue on a daily basis for the society to continue. But reproductive work does not afford much status or power to women because it is unpaid and undervalued, and it isolates them from public life.

From a development perspective, gender roles and respective divisions of labor are of critical concern because men and women identify different things as requiring development attention. As men occupy the public sphere, their issues are more visible and their voice is louder. Reproductive work, on the other hand, continues unnoticed because it involves so much maintenance with little product to show for all the labor involved. Women as a group are not prominent in the public sphere, thus their voice is neither practiced nor heard. The case of water described later in this article is a good example of the way that division of labor comprises division of concern. Gender-just development by definition places women's priorities on par with men's. A major obstacle in development is for women's problems to be articulated and recognized fully as society's problems and that the solutions require everyone's effort.

"DEVELOPMENT" PRIVILEGES MEN

Development as a historical process appears also to privilege men. Though women's material condition in some parts of the world may have improved from increased access to goods and services associated with development, their social position vis à vis men has likely deteriorated (Yunxian 1997). I argue now that gender constructs combined with technological development in the age of capitalism have privileged men over women in creating, accessing, and using technology.

---

[3] Fulfilling both productive and reproductive work roles not only increases women's total work hours but also increases the intensity with which they work (Floro 1994).

In previous epochs, men and women likely invented tools and techniques in the course of carrying out their respective tasks; each sex had their realm of responsibility, knowledge, and innovation. Women likely invented tools related to horticulture, methods of making food and clothing, medicines and treatments for the sick (Cockbum 1985). Men were equally inventive in their sphere, including hunting, defense, and traveling long distances to obtain resources.

Social changes associated with metallurgy and mechanical power (lever, wedge, screw, wheel, and inclined plane that made massive construction possible) added weight to the position of men. The forces of the industrial revolution and the production for profit instead of subsistence intensified social stratification by both class and sex (Cockburn 1985).

Besides being occupied by domestic responsibilities, women have been restricted from the realm of innovation because from the beginning of the technological/industrial revolution in Britain, they were limited in owning property, gaining technical education, and registering patents. Thus, women's perspectives and interests may not have been represented in the things that were invented.

Wajcman (1991) suggested that technology itself may be gender biased. As innovations tend to grow on themselves by increments, new ideas and alternative perspectives are seldom introduced. She described innovation as an imaginative process that lies mainly in seeing ways that existing devices can be improved or how techniques successful in one type of application can be applied elsewhere. Thus, even when women do gain access to technological fields, the agenda for invention is already set (Wajcman 1991). The case of modem contraceptive technology is a perfect example of this point of view.

Capitalism as a world system contributes to gender stratification because even in regions and locales where people continue their age-old subsistence patterns, development macro-policies affecting them are driven by the motive of economic growth. Economic growth is maximized by drawing resources as well as labor away from the household and into the productive economy. A direct result is an increase in the burden of women's unpaid work in the reproductive economy.

A FRAMEWORK FOR EMPOWERMENT OF WOMEN

Gender-just development must counter these deeply embedded social and economic biases against women. The following theoretical understanding of "empowerment" is offered as a guide to making development interventions that will counter past trends and enable women to assume their rightful place in community and society.

I have adapted Longwe et al.'s (no date available) framework for empowerment of women. Longwe et al. used categories of entitlements as a guide in planning development projects to ensure that women and men would benefit equally from the project. The entitlements include basic *welfare* (e.g., nutrition, health, etc.); *access to resources* (e.g., education, land, credit, tools, transportation, etc.); *conscientization* (i.e., becoming politically aware); *participation* (gaining strength in numbers); and *control* over development events so as to protect one's interests. These five categories are a way of gauging and correcting gender gaps in development projects. They are the preconditions that development projects must incorporate so as to enable women's voice to develop and be heard in society.

Longwe et al.'s categories are so useful conceptually that it seems appropriate to expand their use. Figure 5 shows empowerment as a process that is gained through the means described by these categories. The process moves from a state of dependency and powerlessness to one of efficacy and empowerment. But the process requires that the preconditions of participation in development are present: access to resources, critical analysis, and group organization.

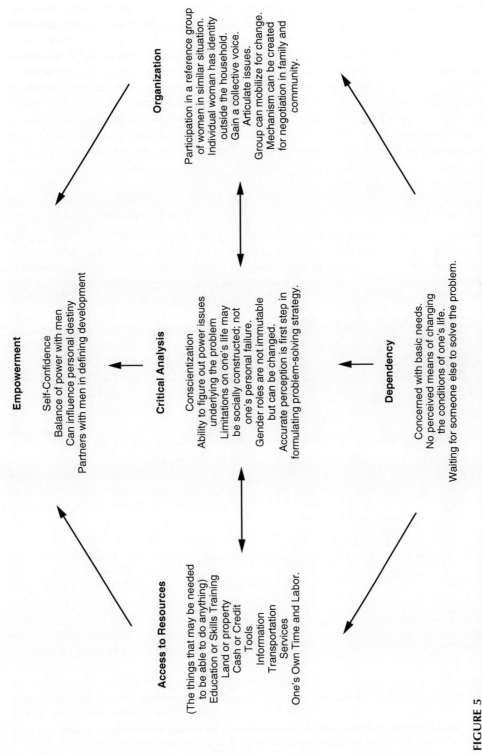

FIGURE 5

The paths to empowerment. Development should open these avenues to women to create and sustain gender balance.

The alchemy of empowerment is mysterious. In some cases, organization and critical analysis can enable women to gain access to resources that were previously denied them. On the other hand, gaining access to resources may enable women's consciousness to move from basic survival to strategic planning. Figure 5 presents the necessary components as equally important and synergistic.[4]

The framework can help identify the crucial variable(s) that are at the root of an existing gender power imbalance. Then appropriate interventions can be initiated that are calculated to change the imbalance. In this way, various existing situations and gender relationships may be mediated by development efforts and renegotiated in a way that is more favorable to women.

I now turn to cases from the field that illustrate gender imbalance in technology and development. The framework for women's empowerment will be used to suggest interventions that would facilitate better terms for women in these settings.

WATER

People need a minimum of 100 liters of water per day for personal health and hygiene (Falkenmark & Widstrand 1992), but this amount is no more than a mirage to increasing millions of people globally. Due to environmental changes associated with "development," dry areas are expanding and becoming drier. The burden falls on women, in terms of both workload and health.

FIGURE 6

Amina collects water from several sources to fill the bucket twice a day. (Sikonge, Tabora Rural, 1994.)

In many areas women carry on their heads or backs the total supply of household water from the source to the household. Men, who are likely responsible for a one-time heavy labor requirement of developing the water source (e.g., digging a well), have little understanding of or empathy for women's daily expenditure of time and energy. They may have no awareness of the health burden borne by women.

Figure 6 shows the serious water situation in one area of central-western Tanzania. As if to mock the inhabitants, relics of previous water projects dot the landscape, including a windmill-driven pump built recently by a foreign NGO and conveniently placed water taps installed during the 1970s by the socialist government. But due to lack of spare parts, oil, and maintenance, none delivers any water (Earth 1996).

During the dry season (April through November) women's days are defined by the search for water. They must go morning and evening to fill buckets from poor sources. As the water holes dry up they must go farther.

[4] This is a political framework in that it aims to transform overall power inequalities between the sexes. Individual capabilities of women will almost certainly be developed in the process of group empowerment.

FIGURE 7
These women are suffering a category of health effects not experienced by men, due to water shortage. They lack not only water but health information and a way of communicating their issue to the men. (Ulyampumba, Tabora Rural, 1994.)

But there is no sentiment that this is a *community* problem; rather, it is seen as women's lot in a difficult land.

In this culture, women are not supposed to bring up problems because they will be seen as "pushing" the men, regarded as unbecoming (Drangert 1993). Women are supposed to rely on the observational powers and goodwill of men. If they were to bring up the problem even tactfully, instead of the problem being addressed, they themselves may be seen as the problem. Even though they are the primary stakeholders, women lack the organization and ability to articulate their position and communicate it to men. Thus, there is no community response, and the situation remains the same.

Following "empowerment" theory (adapted from Longwe et al. no date), group organization is a prerequisite to making change. There must be a meeting of women separate from men to enable the women to formulate their issues. The expressions on the faces of the women and men shown in Figure 7 and 8, respectively, reveal the completely different experience of the sexes, though they occupy approximately the same space. Discussion with the women during this meeting disclosed that they were suffering from menstrual irregularities and severe pelvic pain, a topic that would never have been divulged in men's presence.

Ultimately I learned that the women had cut their own consumption of water to unhealthy levels in order to save their time and labor. Lack of water was indirectly the cause of their symptoms.[5]

[5] Recent international conferences (International Conference on Population and Development; Fourth World Conference on Women) have put women's reproductive health on the development agenda.

FIGURE 8
Men are in charge of cattle and doing business in this society. It is also their duty to provide labor to develop a water source. But if they are not aware of a water problem, there will be no well dug. (Ulyampumba, Tabora Rural, 1994.)

At the time, I was completely engaged with AIDS education and prevention, and could not pursue the water issue. In my mind a plan was forming to try to arrange access to a source of health care and medicine for this community, with specific focus on the women's problem. But upon investigation, it became clear that the underlying cause was lack of water, and until that changed, the medical problems would recur.

The best situation in any field context is for both men and women to be intimately involved with the problem, define it as a community matter, and take joint responsibility. Short of that, there must be a way that women can bring their issues to the attention of men and have the issues taken seriously. This is far more possible if women are organized into a unified voice and have a mechanism for gaining men's/public audience.[6] If there are no organizational capacities or communication mechanisms, they can be facilitated as an aspect of the development project.

Now, it is clear what could have been done in response to the women's water problem. The women's group could have continued to meet on a regular basis to enable the women to achieve solidarity and analyze the problematic situation. As an outsider/health educator, I would have gradually clarified the links between their symptoms and their water situation. Then the focus of the group could have turned to addressing the problem.

---

[6] Callaway (1984) speculated that cross-gender communication may actually be most effective in situations of strict sex segregation wherein women develop their own organization and sense of power separate from men. She found strictly segregated Muslim women to be more clear on their issues and efficacious in pursuing solutions than Christian women who were more integrated in society but lacked their own organization (Nigeria).

Bolt (1994) suggested mapping the women's "dream" village as an exercise to elicit their ideas for improvements. They would be encouraged to envision convenient and ample water sources. They would strategize about how to make their dream village a reality. Communication with the men would be carefully planned.

Ideally, one or more men would be sought as allies to facilitate communication with other men, including village leaders, and begin to set an example for joint action. Drangert (1993) succeeded in getting men to understand the toil involved with water supply by putting them in a situation where they themselves had to do the work. After coming to appreciate the women's perspective, men were more open to water supply improvement.

Drangert (1993) felt that ungendering of tasks is part of a water solution. Just as men can learn to contribute to fetching water for the household, women must begin to have confidence in their ability to initiate and develop new water sources. Both must take responsibility for alleviating the problem. But because ungendering of tasks involves alteration of cultural norms, both men and women need support and assistance to enable them to change.

This case is compelling in that technology probably did not offer a solution. The windmill-driven pump as well as the convenient taps had not survived. Likewise, antibiotics would have offered only temporary relief. A changed gender relation is probably more likely to enhance the overall quality of life as well as the sustainability of water supply in this context.

Though change is not accomplished easily, organization of the women and articulation of their issue is the first step toward resolution of the problem. Communication with men is the second step. Finally, ungendering the tasks associated with water frees the whole community to contribute to the solution. These are crucial tasks of development that require all our sensitivity and ingenuity.

INTEGRATED FARMING

In the dry unyielding northeast province of Thailand, the concept of integrated farming has taken hold with the backing of government and non-government organizations. Integrated farming is a diversified household agricultural economy that includes a fish-stocked pond; field crop (rice); back yard cultivation (vegetables, fruits, herbs); and livestock (cattle, ducks, chicken). Integrated farming supplies all the food needs of the family, as well as surplus that can be sold for cash. The integrated system increases the quality of life of families nutritionally, aesthetically, and economically. Figure 9 shows pond and fields in an integrated farm.

The people of the northeast (Isan) have their roots in Lao culture which is predominantly matrilineal. After marriage, husbands traditionally joined their wife's household to farm the land that was inherited by her. There was no formal marriage certificate nor land title in traditional society. Men's and women's roles were complementary and relatively egalitarian, with decisions taken jointly and the wife controlling the purse.

In 1970, thousands of Isan people were relocated to make way for a large dam project. Each family was given fifteen rai of land in a new location: two rai for a house and thirteen for cultivation. Land ownership became more formalized and state-regulated due to the mass relocation. Old residents were also given legal title to their land.

In a development paradox, the relocation resulted in men gaining title and women losing rights. It was because men's name and signature were used for the land titling. The titling process was "gender blind" in that men were considered to represent the household as "household head" and were therefore called to collect the document. It was just "natural" that men, having more public presence, time and mobility than their wives, would be given, and accept, this bureaucratic responsibility.

FIGURE 9
The couple uses their 13 rai effectively to supply virtually all the family's food needs. This is an integrated farm with a pond, fields and a banana grove. Initially only man were trained. (Personal communication, Ban Kam Pla Lai, November 1997.)

Because the people's tradition is matrilineal, most families continue to share rights and responsibilities as expected by their culture. But serious repercussions for women have resulted. At the worst, it is possible for a woman to completely lose the land she has brought to her marriage; her husband now has the legal right to sell it. Astute women who recognized their vulnerability and tried to get their name added to the title eventually became frustrated with extensive red tape and gave up (personal communication, Ban Kam Pla Lai. Khon Kaen Province, 3 November 1997). See Figure 9 and 10.

Later, when integrated farming was introduced to the area, it was men, as household heads and owners of the land, who were called to attend the training. As has happened frequently in agricultural development, women were initially excluded from accessing technological advances in their own traditional occupation. Again, it was probably not intentional to exclude women; it was just that men were recognized as "head." They probably also had more free time to go to meetings. After several series of trainings someone noticed that women had no access to the new technology, which was becoming very popular, and a concerted effort was made to invite and include them in the training (personal communication, Dr. Tantip, November 1997).

Currently, a process of land allocation is being conducted in some provinces of the Lao People's Democratic Republic, where previously land was free to be used by anyone who cultivated it. Probably for the reasons mentioned above, some villages are registering only men's names on the land titles. Efforts are currently underway to ensure that names of both husband and wife are included in the legal documents (Phengkhay 1999).

This case illustrates how the development process can strip women of rights that they previously enjoyed. In such cases, gender-just development demands that extra measures be taken to ensure women's access to resources such as land and training. Ultimately, ungendering of the reproductive role in the household will balance men's and women's time commitments, thus permitting women the time needed to access training opportunities alongside men. Conscientization of both women

FIGURE 10
This happy couple enjoys a wide variety of food and takes pride in their lovely integrated farm (right). They are partners in every regard except that he alone has title to their land. She tried to get her name added but eventually gave up. (Chansom family, Ban Kam Pla Lai, November 1997.)

and men can assist them to retain important elements of their cultural identity in the midst of development.

FAMILY PLANNING

Modern contraceptive methods (hormonal dosages and intrauterine devices) are probably the most widely dispersed form of modern technology in existence. Access to family planning is nearly universal, perhaps more so than access to electricity or print media.

Arising from dire Malthusian predictions that world population would outstrip resources, the developed North undertook to control population growth rate globally, especially in very populous (poor, underdeveloped) countries. A powerful demographic focus led to the proliferation of family planning programs that would dispense modern contraceptives to the masses. In some cases, the establishment of family planning policy and services has been part of an aid package (Mies in Soutthanome 1999).

The effects of this trend on women have been contradictory and highly controversial. The dominant view in development is that modern contraceptives not only control population growth but also liberate women. For the first time in history, women can control their fertility.

A more critical view, however, reveals a series of gender biases that operate against women. First and foremost, all the hormonal methods (pill, injection, implant) and nearly all the mechanical methods (except for condom and withdrawal) focus only on women as the site of intervention, as if women alone make babies.

In the zeal to control fertility, women's rights have been grossly violated: There have been cases of forced sterilization without informing the women, or against their will, or obtained through bribery; forced abortions; and other intrusive measures. In the push for controlling women's fertility, women have been treated as "baby production machines," where the ends justify the means. Women's real needs, their spirit, and their rights have been neglected by the services, policies, and technologies available through family planning (Souththanome 1999).

Recently, the many shortcomings of family planning have been defined as "unmet need." This concept embraces the desire of millions of couples and individuals to regulate their fertility but for a variety of reasons, do not use modern contraceptives. Their reasons include the following.

1. Side effects of modern contraceptives
   - Physiological changes, such as menstrual irregularities, weight gain or loss, acne, skin color changes
   - Psychological disturbances related to hormonal imbalance
   - Infections
   - Pain
   - Possible injuries
2. Hormonal methods and intrauterine devices offer no protection against sexually transmitted diseases
3. Difficulty in use
4. Lack of information and medical support, causing failure of the method
5. Gender power relations may become more disadvantageous to women
6. Fear
7. Services are inconvenient or not respectful
8. Opposition from husband or others (from Souththanome 1999)

As the sole regulators of human reproduction, many women endure these side effects and unpleasant circumstances so as to bear only the children they want. Many others experience no adverse effects and use them successfully. But the problem of fertility regulation is not solved universally with the mere application of technology. Many women find the methods unacceptable and quit using them. Even in the United States, where access to contraceptives is virtually complete, 48 percent of all pregnancies are unplanned (Souththanome 1999).

Modern contraception is a good example of how technology has evolved in such a way as to privilege men. While men are free to enjoy reproduction, women bear all the risks. Confirming that technology could have developed in another direction, the inventor of "the Pill" has stated that if he could do it over again, he would invent a pill that *brought on* the menses instead of one that prevents them (Souththanome 1999). Most women, like men, prefer the least possible change to their chemistry and physical body.

For those couples who try but quit using modern contraceptives, Souththanome (1999) reported a model for cooperative responsibility that can be used safely by all couples: the natural method without a calendar. It requires only keeping track by marking different symbols for the days with and without menstruation. Even if the people are not literate or do not have a calendar, they can be taught to understand and gauge their sexual activity according to the woman's menstrual cycle, as follows:

Safe   blood   safe

o o o o o o o x x x x o o o o o o o

o o o o o o o o o o o

danger   egg   danger

FIGURE 11
Lao women and their children. Men can share the responsibility of fertility regulation with their wives in a way that is technologically appropriate and does not compromise women's health and well-being.

It is possible for men to become very dedicated to this method, taking responsibility for the daily marking, so as to avoid an unwanted pregnancy or the cost of an abortion. Soutthanome's (1999) field study reported the case of a man who was so upset about his sister's ill health following the use of a hormonal method that, out of love for his wife, he learned the counting method and taught it to her. Several Lao women have confided to me that they have used the counting method successfully for many years.

For couples who reject modern contraceptive methods, it is easier and more respectful to learn the natural cycle of women's fertility than to suffer lack of control over reproduction, unwanted pregnancies, and abortions. Simple, accurate information about reproduction must be made accessible—for example, through existing family planning clinics. Supportive men can be taught to be trainers and thereby set an example for other men. This way incorporates mutual caring and responsibility between partners. In fact, it restores gender balance that is completely distorted by modern contraceptive technology. See Figure 11.

CONCLUSIONS

The three cases I discussed—water, integrated farming, and family planning—are examples of how gender, technology, and development have converged in profound ways to disadvantage women. The changes required are likewise profound.

The coexistence of rapid transition in political economy at the same time as enduring cultural and social expectations is the combination that has left women as a group—that is, half of humanity—relegated to an inferior status relative to men. Now it is the task of development to remake the equation.

International documents leave no doubt that development and the benefits of scientific progress are the right of all people and that sex is not a variable that justifies discrimination. It is also clearly stated that social and cultural patterns of conduct that advantage one sex over the other are to be modified with the intention of eliminating prejudice. These international standards are more than recommendations; they are a mandate for development practice.

Notions of social justice and human rights are often dismissed as northern constructs that cannot be applied to the rest of the world. But international agreements are the result of extensive and painstaking debate, cross-cultural communication, and compromise. They represent a consensus among nations and embody universal standards.

Likewise, professional development workers (from both inside and outside the country) are sometimes criticized for intervening in people's affairs and changing things they know nothing about. It is true that development mistakes abound, usually with women bearing the burden. Thus, it is imperative that development and development practitioners be made to be gender-accountable.

Ultimately, the reconstruction of gender is the work of women and men in the society. However, the framework presented in this chapter clarifies three main areas where development intervention at the field level may be indicated.

1. *Access to resources.* If women are experiencing lack of access to a needed resource that puts them at a disadvantage relative to men, it is a violation of international standards. It is the task of development to address it. This includes renegotiating the gender division of labor such that women gain access to time needed for their own development.

2. *Organization.* If women lack a public presence, group organization must be facilitated so as to enable the women to articulate and communicate their standpoint and their interests. A women's group is the entry point for women to participate in development.

3. *Critical analysis.* Both men and women need critical understanding of social power, gender roles, and gender power imbalance. Men's transformation is as crucial to the empowerment of women as is women's own transformation. Women need support to strive for their own empowerment; men need support to enable them to become allies and partners of women. These two sides of social transformation are the most important tasks of development today.

I have discovered in the field as well as in the classroom that gender issues are burning issues for women. I have also found men to be interested in and thoughtful on these matters. Thus, I do not believe that a development emphasis on gender justice is imposing foreign ideas on unwilling people. Rather, it is responding to the diverse realities and felt needs of people in a myriad of development contexts.

I gratefully acknowledge the villagers in the three case studies for sharing their experiences with me. Thanks also to Dr. Gopal Thapa, associate professor in Natural Resources Management, Asian Institute of Technology, for his critical comments on an earlier draft of this chapter. Finally, thanks to Ms. Kirsten Wilkes for her continuous encouragement as well as technical support. However, any factual or analytical errors are my own.

REFERENCES

Bolt, Eveline. (1994). *Together for Water and Sanitation: Tools to Apply a Gender Approach. The Asian Experience.* The Hague, The Netherlands: IRC International Water and Sanitation Centre.

Callaway, Barbara J. (1984). "Ambiguous Consequences of the Socialization and Seclusion of Hausa Women." *Journal of Modern African Studies* 22(3): 429–450.

Cockburn, Cynthia. (1985). *Machinery of Dominance: Women, Men and Technical Know-How*. London: Pluto Press.

Convention on the Elimination of All Forms of Discrimination Against Women (CEDAW). (1979). http:www.un.org/womenwatch/daw/cedaw/intro.htm.

Drangert, Jan-Olof. (1993). *Who Cares About Water? Household Water Development in Sukumaland, Tanzania*. Linkoping: Linkoping Studies in Arts and Science.

Earth, Barbara. (1996). *Women and Health Education in Rural Tanzania: Lessons in Empowerment*. UMI Number 9619951. Ann Arbor: UMI Dissertation Information Service.

Falkenmark, Malin, and Carl Widstrand. (1992). "Population and Water Resources: A Delicate Balance." *Population Bulletin* 47(3). Washington, DC: Population Reference Bureau, Inc.

Floro, Maria Sagrario. (1994). "Work Intensity and Time Use: What Do Women Do When There Aren't Enough Hours in a Day?" In Gay Young and Bette J. Dickenson (Eds.), *Color, Class and Country: Experiences of Gender*. London: Zed.

International Conference on Population and Development (ICPD) (1994). Gopher://gopher.undp.org/00/ungoph...cpd/conference.

International Covenant on Economic, Social and Cultural Rights (Economic Rights Covenant) (1966). http://www.unhchr.ch/html/menu6/2/fs16.htm.

Keola, Soutthanome. (1999). *Gender Bias: the Basis of Unmet Needs in Family Planning, Nongphaya Village, Vientienne, Lao P.D.R.* Unpublished Masters Thesis, Gender and Development Studies, Asian Institute of Technology.

Longwe, Sarah H., Misrak Elias, Sreelakshmi Gururaja, and Christine Muhingana. (no date). *UNICEF Gender Training Module: Programming for Women's Empowerment*. Draft. New York: UNICEF.

Phengkhay, Chansamone. (1999). *Women and Land Allocation in Lao P.D.R.* Unpublished Masters Thesis, Gender and Development Studies, Asian Institute of Technology.

Wajcman, Judy. (1991). *Feminism Confronts Technology*. University Park, PA: Pennsylvania University Press.

World Conference on Human Rights (Vienna Conference). (1993). http://www.unhchr.ch/html/menu5/d/vienna.htm.

Yunxian, Wang. (1997). *The Household Responsibility System and Women's Position: a Case Study of Two Villages in Zhejiang Province, China*. Unpublished Doctoral Dissertation, Gender and Development Studies, Asian Institute of Technology.

MANAGING A SMALL BUSINESS

OVERVIEW

Readers of this book may not manage their own small business, but they may be called upon to give advice to those who do. Much of the background for managing a small business is discussed in other sections of this chapter. A business consists essentially of producing a good or service and selling it. In order to make this work money has to be available to pay bills before cash comes in. Production of goods and services is discussed in the "Operations" article of this chapter. Selling the good or service is discussed in the "Marketing" article. Some aspects of arranging the money are discussed in the "Financial Viability and Cash Flow Statement" article, and some aspects are discussed in this article. This article starts with a discussion of how to start a business because people starting businesses often need advice. Many of the same ideas used in starting a business apply to improving, even maintaining, a business and may be easier to understand in the startup context.

STARTING A BUSINESS

To start a business one needs an idea for a product or service. Ideas, of course, come from many places. A good place to start is with the entrepreneur's personal experience—what he or she needs may very well be needed by others. If an idea is to be worthwhile it must be for a product or service that people want. Not only must a market exist but the business must be able to supply the market at a profit. The business must have an advantage over its competitors—a valid advantage may simply be location, being the only supplier close to a group of customers.

Another source of competitive advantage is the experience and expertise of the business owner. An important question then is whether the proposed business matches what the owner can do well. Of course, a person thinking of starting a particular business may not have all the expertise needed at first. Knowledge can be gained by talking to people—successful business-people, former teachers, community leaders, and others. Even if one is confident one understands a business opportunity, one should talk to as many people as possible about the opportunity, partly to get many perspectives on the idea, partly to develop a network of people that one can ask again. In some cases, it may be wise not to describe your own plans in detail as the listener might incorporate them in her or his work. Other ways to learn about a business field is simply to observe working businesses in the same field. Other ways are to read about the business or to take courses. Many communities now have agencies, supported by the government or non-governmental organizations, that offer training courses for persons wishing to start businesses.

To start a business successfully one needs to be willing to do certain things, such as take risks, persevere, and seize the initiative. Avoiding situations where the outcome is not clear, giving up easily when faced with difficulties, or waiting for things to happen reduce the chance of making a business work. One also needs the support of one's family—starting a business takes much time and money that might otherwise be used for the family. People start businesses for many reasons, including having lost a job. The people who do best at starting a business seem to be those who want to be their own boss, who seek responsibility, who find the challenge appealing. Some people thrive in situations where success or failure depends almost entirely on one's actions—such people tend to do well in their own businesses. It goes without saying that a person who believes he or she lack the qualities of an entrepreneur can develop them. In addition to the personal reward from creating a new business, a successful entrepreneur will probably have a higher income than someone working for a salary. For this higher income the entrepreneur accepts the risks of doing something new with little external support.

THE BUSINESS PLAN

The next step in starting a business, after one has an idea, is to put together a business plan. Nearly always one will need to raise money to start a business, if only to support oneself until the business produces cash. A business plan will be necessary when raising startup funds as funding agencies or banks will expect to see one. A business plan assists the entrepreneur also by forcing attention on each aspect of the business. Preparing the plan should uncover problems with the proposed idea. It should also help focus the entrepreneur's ideas.

Form of a Business Plan

- Attractive cover
- Executive summary: Include projected starting date and amount of money needed. Summarize the main points in each of the parts below.
- Product or service offered—the business idea.
- Production/operation process: Describe the equipment you will need. Consider spare parts, maintenance, training if any of these may be difficult. Include also ways of sourcing components and raw material if such may be difficult.

- Marketing plan: Describe customers and why they will want your product/service. Give selling price. Describe promotion strategy—advertising, demonstrations, handouts, and so forth. Show distribution plans, including location of outlets, use of wholesalers, assistance to retailers.
- Competition: What your customer would do if you were not there.
- Staff: Probably best to show only staff needed immediately and include a note describing how it will evolve—avoid frightening the reader with a full administration. Identify key people and their expertise.
- Form of business: Sole proprietorship, partnership, corporation (which form), and so forth.
- Forecast of revenues, showing how they build up. Include explanations, even if forecasts were difficult to construct. The selling price stated in the marketing plan is used.
- Costing: Show the calculation of the costs, both direct and indirect, of a single item. Many advisors to small business emphasize the importance of this calculation. It is discussed in the "Financial Viability and Cash Flow Statement" articles.
- Cash flow projection: Monthly for first year; after that at least quarterly until cash flow is positive. The revenue forecast and the cost calculation are used. How to construct a cash flow statement is described in the "Financial Viability and Cash Flow Statement" article.
- Startup capital required: The amount of initial money needed. The amount should come directly from the cash flow projection just made.
- Other financial data: Profit and loss, balance sheet.

Details of the content of each business plan section are described in various sections of this article.

STARTUP FUNDS

The money required for the initial expenses of a new business may come from savings of the owner or her/his family. People outside the business may invest in it in return for a promise to share future profits. In many countries, small business funding agencies exist, supported by the government or NGOs. Microlending organizations, described in a later section, are an example of such funding agencies. In some cases funding agencies give outright grants to assist entrepreneurs. More often they give loans or extend credit, which amounts to the same thing. Some banks lend money for startups, although many do not. Banks nearly always require collateral—a building or a piece of machinery the bank can take and sell if the loan is not repaid. Organizations that support small businesses through courses can often give advice about locating financing. Nearly always a funding agency will require a business plan as part of the request for funding.

The term "equity" is used to describe the money invested in a business. Shares of company stock are equity. The agreement between the owner and the investor should specify the portion of the business that the investor owns; for example, an investment of $1,000 would give the investor a 15 percent claim on the business. In principle, then, the investor would be entitled to 15 percent of the money the business earned. Earnings are paid to the investor in the form of dividends. The amount of the dividend paid usually depends on the earnings for the past year—if earnings are low the dividend will also be low. If the owner needs more money at a later time, then more equity could be sold. The drawback in selling equity, besides the way earnings are distributed, is that a person who controls more than 50 percent of the equity controls the company. If it is not the original owner, then the original owner can lose the company. An advantage of using equity to raise money, compared to borrowing, is that interest must be paid regularly on borrowed money, whether the company is profitable or not. If earnings are low and an interest payment is missed, then the company can be taken over by the lender. Selling equity is less risky than borrowing money, but the owner who used equity must please the stockholders and may have to pay large dividends when the company is successful. An approach to deciding whether to use equity or a loan is to estimate the minimum cash flow expected from the business and borrow as much as possible, keeping the required payments less than this minimum cash flow.

STEPS IN STARTING A BUSINESS

1. Prepare a business plan, indicating the amount of money needed to begin operations.
2. Search for the required funds.
3. Secure the required legal approvals—licenses, incorporation papers, lease agreements, and so forth.
4. Lease or build a structure.
5. Order long lead time items, both production equipment and components.
6. Get utilities connected.
7. Update quotations from suppliers.
8. Hire staff.
9. Install equipment.
10. Commence production.
11. Commence marketing.

Some of these steps can be done in parallel—that is, at the same time. A Gantt chart, described in the "Operations" section, can assist in scheduling.

MANAGING ONGOING OPERATIONS

A few items of advice are often given by owners of small business. One of these concerns members of the extended family. An owner will be asked by such members to extend credit, to give employment, to reduce the selling price, and similar favors. If these requests are not handled prudently, the business can run out of cash and fail. One organization supporting small-scale entrepreneurs in Africa recommends specifically that an entrepreneur not open a business in the same community as her or his family. It appears that such humanitarian issues are especially severe for women. Another piece of advice is the importance of separating business funds and family funds. If money is taken from the business to be used for nonbusiness purposes, it should be recorded and treated as a formal loan or as salary. The risks to the business of repeated but unplanned cash drains should be made apparent to the owner and others. Managing a small business takes much time, so the family of the owner may feel neglected. A good piece of advice is to try to plan for such time problems ahead of time. A final advice item for small business owners is to learn about legal restrictions. One study of small business reported that many owners felt constrained by laws that in fact did not exist.

Small businesses are often constrained by lack of cash—almost never is an entrepreneur able to raise as much money as he or she wants. One way to conserve cash is to hire, rather than purchase, expensive pieces of equipment. In some communities organizations that support small businesses rent machines, either long term or just for a few days or hours—long enough to complete a job. An ongoing business can often borrow money for several weeks or months from a bank. In some countries the term "overdraft" is used for such loans. In order to use an overdraft one should get to know one's bank manager before the money is needed. One needs to convince the banker that the money is going to a sensible, reliable person. In some countries, for some people, credit cards serve the same function as overdrafts. Approval of credit limits on cards is less personal than granting overdrafts, but it is still worthwhile gaining a reputation for being credit-worthy.

Another problem small businesses sometimes have is securing raw materials. Small purchasers may have little influence on suppliers. To gain clout with suppliers several small businesses can get together and make joint purchases. A small business owner may have difficulty finding a place to work. Several small businesses can rent or buy a single structure, saving costs both for the building and for shared equipment. Communities that wish to promote entrepreneurship can establish work spaces with shared facilities, available for nominal rents.

BUSINESS RECORDS

Records can be a nuisance to keep but are useful. They allow a comparison between actual earnings and forecasted earnings. They are the basis of forecasts for future years. They are essential when dealing with banks and other lending agencies. They are a way of checking whether money is being lost from the business through carelessness or dishonesty. The owner of a small business should decide which records are useful to her or him and not keep records that will not be useful. Elaborate accounting schemes suitable for large operations will slow down a small business because too much time and effort are required. Needless to say, records are only useful if they are used—if they are examined regularly to understand how the business is doing and referred to when planning. A sensible procedure to assist in record keeping is to minimize the number of cash transactions—pay by check or credit card as often as possible.

In many businesses cash transactions are essential. For each transaction, such as a customer buying and paying with cash, the amount and source of the money should be recorded, as well as the date. If receipts are routinely issued for cash payments, copies of receipts suffice. If no other written record is made, a note should be made in a cash record book. If cash is received as payment for items bought on credit, the cash inflow is also noted in the book or with a receipt. When cash is taken from the cash box to pay wages or a supplier a written record should also be made. Copies of payslips or receipts from suppliers are suitable records, but if such are not available the cash record book should be used.

The basic record is another book where all transactions are entered, normally at the end of each day. A convenient system is to keep two kinds of records in the same book. One kind of record shows a running balance of the money the business has, including cash kept with the business and money in the bank. The other kind of record classifies the transaction money into categories like labor costs, utilities, and so forth. An example is shown in Table 4. This business does not use a bank and the only fixed expense is utilities.

So when a transaction is made, it is entered in two places. Consider the transaction on 2/7/01. An amount of $1,200 was received for food sold. The amount is listed under Cash In and the Cash Balance is adjusted. Also, $1,200 is entered in the Sales column. The $3,000 paid out to production workers on 2/9/01 is entered in the Cash Out column and in the Labor Costs column. Neither of these transactions involved Material Costs or Utilities, so those columns were not used. At the end of the month, or some other regular time, the numbers in the Sales and various Cost columns are added to give total sales or costs for the month, or period chosen. These costs and sales can be compared to forecasted amounts. The running total given by the Cash Balance

TABLE 4
Record book

Date	Transaction	Cash In	Cash Out	Cash Balance	Sales	Labor Costs	Material Costs	Utilities
2/1/01	Balance			5,000				
2/7/01	Food Sold	1,200		6,200	1,200			
2/9/01	Wages		3,000	3,100		3,000		

column should check with the actual amount in the cash box on the relevant day. Some items, like taxes or loan money received, are not Sales or Costs. They are only entered in the Cash columns.

Another set of records should be kept: a listing of credit transactions. If a sale is made on credit, a sale invoice should be recorded showing the amount of money, the customer, and when cash payment is due. When the payment is made, the payment should be recorded and the entire record kept in case of a dispute in the future. In this way, a credit history is built up for each customer. In the same way, when a purchase is made on credit the transaction is recorded, including the amount, to whom it is owed, and when payment has been promised. Again, after payment has been made the record is marked as paid and retained. It is useful, at the end of a month, or other regular time, to add up all the money owed to the company—its receivables—and the money it owes—its payables.

Reviewing an Ongoing Business

Keeping a business profitable requires regular monitoring. The financial records just discussed—Sales and Costs—should be compared against corresponding numbers from previous months and previous years, as well as against forecasts. Of course, it is not sufficient simply to look at the numbers. Actions should be taken. If sales are down the owner might consider reducing the price, or doing a promotion, or changing the product, or selling at a different place. These alternatives are discussed in the "Marketing" article. In the same way, unexpected increases in costs should be analyzed. Is a new product more difficult to produce? Is an old machine requiring intensive maintenance? When the cause of a problem is understood, remedial steps should be taken. One should not overreact to unexpected numbers. A more common difficulty than overreacting though is not acting at all, hoping that sales will return to their previous level or that costs will go down by themselves.

In addition to financial information, the owner should look at operating data. Probably the essential information is, for each job, the promised date of completion and the projected date of completion—when it is presently thought the job will be completed. A summary, listing jobs that appear to be behind schedule or are close to being behind, should also be prepared. Another very useful piece of information is, for each machine and each person, the spare capacity—the amount of time not yet assigned for the next week, or other period. The Gantt chart described in the "Operations" article is a convenient way of presenting capacity usage information. Estimates of unused capacity assist the owner in scheduling jobs and also in searching out new business. To assess the current health of the business an owner could also record the total production for the previous week and also the fraction of time each machine and each person were actually producing. If some machines are overloaded and some underloaded, then it may make sense to change the marketing emphasis or to purchase new machines.

QUESTIONS OF CONCERN

Whether the organization manufactures a product or supplies a service attention should be paid to how the product or service will actually be made. Some of the questions that need to be considered include the type of machinery to be used, the required capacity of the production facility, its location, sourcing of supplies and energy, scheduling of production and maintenance, quality control, and working with employees. Firm guidelines cannot usually be given for these questions, but some considerations that might be helpful are given here.

CHOICE OF TOOLS

When looking at the type of tools or machinery to be used, one might consider its complexity—usually efficiency and production rate increase with complexity, but so also does the need for skilled operators and maintenance people. Simple machines are usually easier to repair and spare parts easier to obtain. Tight specifications on the finished product often require the use of complex machines—although skilled craftspeople can create very intricate devices. A reasonable guideline is to acquire a machine no more sophisticated than is needed to meet product specifications, with due regard to possible future changes in those specifications. A problem with sophisticated technology is that it often becomes obsolete fairly quickly.

Attention should often be paid to whether the specified equipment, especially sophisticated equipment, is in use elsewhere in the community. Having a unique piece of equipment may give a competitive advantage, but it may mean that advice, and spares, will be difficult to find. The problem is worse if the design is new. If elaborate equipment is purchased, one might insist from the seller that a serviceperson be made available locally. Installing a new machine and making the initial adjustments so it operates at full efficiency is not a trivial task. The problem of learning to use complex equipment also should not be disregarded. When choosing any piece of equipment a strategy should be developed for debugging, for training, for repairs, and for spares-parts and supplies.

Another issue in choosing equipment is poisonous by-products or dangerous processes—safety. Not all unsafe conditions are immediately apparent. Dangers from rotating machinery or unguarded cutting edges are obvious, but fumes or fine dust can do equally serious harm to the respiratory system. One needs to think carefully about the ways a tool might be misused, especially by a careless or inexperienced worker, and prevent as many ways as possible. Equally important is not distributing products that can be dangerous—again, serious attention should be given to how a device or a substance might be used, especially usage different from that intended. Even if the by-product or waste is not toxic, a plan for disposing of it should be made, preferably before much waste has accumulated. Responsible design is concerned with the entire life of the product, including possible aspects of the life not expected or intended.

CAPACITY PLANNING

Capacity planning is choosing the number of tools to be used or the size of the equipment. Of course, tools and equipment need people to operate them, so capacity planning also involves looking at staffing levels. For many small-scale enterprises it is easier to hire more people than purchase new equipment, so capacity planning focuses on equipment. The capacity, of course, should match expected demand, but demand often varies unpredictably over time, so it is hard to know what to expect. General-purpose tools and an adaptable workshop layout make it easier to adjust capacity than specialized tools arranged permanently—the design of the facility is a tradeoff between efficiency and flexibility. An example is a woodworking shop that can make several different pieces of furniture—if the demand for each piece is known, the shop can obtain tools and arrange machinery to make the most profitable piece most efficiently. If the demand is unclear, then the shop should be prepared to make each piece. The more certain one is about the demand the more one can strive for efficiency. Another tradeoff, between the costs of overcapacity and undercapacity, is important in many businesses, especially service businesses. If a restaurant has insufficient capacity and turns customers away, then it will forego the earnings it would have made and may lose customers' goodwill. On the other hand, the cost of a building a restaurant that can serve 100 customers is greater than the cost of one that can serve only 50 customers and must be repaid even if the average number of people served is only 40.

From the start, expansion should be thought about. It may be much more expensive to add capacity after an enterprise is functioning than it would have been to start off with unneeded capacity—putting another service bay in a crowded auto repair shop may be impossible while starting with a larger building might have been possible. When capacity is increased by adding new equipment, the increase may come in a large "chunk"—buying a second drill might double the capacity of a shop, when only a 10 percent increase is really needed. Increasing capacity in this case may pressure the owner to increase marketing. If production consists of several steps, each step done by a different tool, then adding capacity may require that several new tools be obtained. An example is a rest house: If more bedrooms are built, then more bathrooms may be required, as well as more dining space. (A related example, a cookie-baking operation, is given later.)

In some cases the peak demand is much greater than the average demand—highways are much busier during commuting hours than during the rest of the day. An important strategy question is how to meet demand peaks. In some cases—agricultural tools, for example—one can produce ahead of time and store. A maternity clinic cannot effectively store patients, though, and will probably be designed to serve simultaneously the largest number of births expected, so most of the time its capacity is not being used efficiently. The cost of having too little capacity is very high, especially to the mother and new baby, compared to the cost of building an extra room. In a small tailoring shop a demand peak might be met through overtime work or by subcontracting work to neighboring tailors. In some cases prioritizing tasks during demand peaks can satisfy most customers—during rush hours custom orders might not be taken.

Although sophisticated forecasting procedures are available the planner of a small enterprise can probably obtain nearly as satisfactory a forecast using her or his judgment and much discussion with prospective customers. Seldom is sufficient data available to do meaningful statistical studies. The demand for many products, though, varies predictably with the seasons—the demand for new harvesting tools is probably greatest shortly before harvest time. A mathematical technique called "time series smoothing" is helpful when the demand is cyclic.

Estimating demand is only half the problem. Another estimate is needed: the amount of output produced, the capacity. A fruitful way to start is to estimate the time required to make, or process, one item. ("Items" refer to what is worked on—for example, for a clinic a patient is an item.) If a nurse spends an average of 20 minutes per patient, then the nurse's capacity is 3 patients per hour. If a grinding wheel takes 5 minutes to sharpen an axe, then its capacity is 12 axes per hour or 96 axes per 8-hour day. To meet a demand of 150 axes a day, two grinding wheels are needed.

Estimating processing time is more involved when appreciable time is required to set up the tool before use. Before painting a table, one has to get the paint and brushes, clear space, put on protective clothing, and so forth. After painting the brushes must be cleaned. The amount of

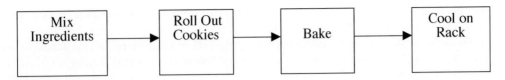

FIGURE 12
Processes in making cookies.

time required for preparation and cleanup is essentially the same if one table is painted or ten, so the time to paint one table is less if ten tables are painted at once rather than just one. If actual painting takes 15 minutes and the preparation and cleanup time is 20 minutes, then the time to paint one table is 35 minutes, while the time to paint ten tables is 170 minutes—that is, 17 minutes a table. If a shop produces several different pieces of furniture, then the capacity depends on the mix of jobs—the capacity is higher if all the pieces of one kind are produced together. Some tradeoffs for this higher capacity are the waiting time for an item to reach production and the necessity for storing the many items produced in a single run.

Many processes require several steps. The capacity is determined by the step with the lowest capacity. Figure 12 shows the steps required in making cookies. The mixing bowl is big enough to hold ingredients for one, two, or three batches of cookies. The Mix Ingredients step takes 5 minutes no matter how many batches are made at once. Rolling out the cookies takes 4 minutes per batch. The oven can cook only one batch at once, and cooking time is 12 minutes. The cookies must cool for 15 minutes. The cooling rack can hold one or two batches. The oven has the lowest capacity—5 batches per hour. The capacity of the rolling step is 15 batches per hour, but only 5 of those batches could be cooked within the next hour. As many as 8 batches can be cooled in an hour, and ingredients for as many as 36 batches can be mixed. Only 5 batches of cookies can be produced in an hour. The so-called bottleneck is the oven.

If another oven is purchased, then the capacity of the baking step is 10 per hour. The bottleneck has moved to the cooling step—only 8 batches per hour can be cooled and, therefore, produced. It was mentioned that capacity sometimes comes in big chunks. A larger oven that could bake 10 batches at once brings that big chunk to this system. The most economical system is often one in which each step has approximately the same capacity—a "balanced line"—because in this case each piece of equipment can be working near capacity. Note that in this example, overall system capacity is not increased by increasing the capacity of the mixing step because excess capacity exists in that stage. Overall capacity is only improved by working on the bottleneck.

This example can also illustrate how scheduling can sometimes affect capacity. In the mixing stage only cookies of the same variety can be mixed together, but three batches can be done at once. If we could always combine three batches, then the capacity would be three times the capacity when only one batch could be done at a time. Such a schedule, of course, might mean that an order would be delayed until two other orders for the same variety was placed. The effort required to use a complicated schedule might also be more than it is worth.

LOCATION

Often a small-scale enterprise is located near the home of the owner. If the question of where to put a facility must be considered, then usually proximity to customers, raw materials, or labor is looked at. Restaurants, hospitals, and other service agencies are usually located near customers. Food processing operations are often placed near where the crops are grown because transporting processed goods is easier than processing raw materials—a bag of cornmeal takes up much less volume than the equivalent cornstalk. Availability of labor is an important factor in locating large factories in labor-intensive industries like automobile production. Projects needing particular kinds of expertise may flourish best in particular locations—innovative software development projects almost always are done near universities.

An entrepreneur must decide whether she or he is better off in an area where many people are doing similar things—such as a business incubation center—or alone. Probably the biggest advantage of being near others doing similar kinds of work relates to keeping up with new developments—the entrepreneur sees what is being done by competitors and talks with them. Also support people—vendors, loan officers, and so forth—are likely to be easily accessible to clusters of entrepreneurs and difficult to find for people outside the network. If one entrepreneur in a cluster cannot meet a customer's request, the request can be passed on to another nearby entrepreneur. Of course, the presence of the Internet changes the meaning of proximity—if communication is normally done electronically, then two people are in close proximity if they know of each other. An important disadvantage of being in a cluster is the converse of the advantage—competitors learn proprietary information. While patents and careful guarding of trade secrets may give a sustained advantage—Coca-Cola has kept its formula secret since the company began—usually new technology can be analyzed and copied fairly easily, even if the competitor is not nearby.

Social factors can be important in selecting a location. In Malawi a small business development group advises entrepreneurs to do their enterprise some distance from members of the extended family so moneys earned in the early stages can be reinvested in the business.

LOGISTICS OF TRANSPORTATION

How best to get the product to the user can be an important matter. Internet-based retailing could not succeed without sophisticated systems for putting orders together, nor without reliable delivery services like Federal Express or United Parcel Service. Projects marketing baskets and ceramic pots produced in villages can be crucially dependent on a vehicle to collect the items. Meat can spoil even during a short trip, especially if the trip is unexpectedly delayed. Road conditions during the rainy season can determine the success of a project. Delivery costs can be especially important for high-bulk, low-unit value products like sugar or fabricated steel door frames, so transportation alternatives need to be carefully considered. Larger producers with economies of scale, such as the ability to consolidate many small loads, may have significant cost advantages. Examples abound of development projects that were unsuccessful because it was not possible to move the goods to market dependably, but no general advice can be given—except to plan thoroughly.

SOURCING

In planning an enterprise, thorough attention needs to be paid to obtaining components parts, energy, and spares. The problem is especially acute if components or spare parts are imported and nearly as acute if transportation is not reliable. Many projects in the Third World have failed because they required parts that became effectively unobtainable. Erratic energy sources may mean unreliable delivery dates—when the electricity goes off, electric motors stop. Petrol (gasoline) engines only work when fuel is available. A major concern when parts are imported in some countries is exchange control—restrictions on payments going outside of the country.

The next article, "Purchasing Policy and Procedures," describes the administration of a purchasing department.

Long lead times for procurement of supplies mean that production planning must be done far in advance, and a change in production level, even of the product made, is difficult. An advantage small-scale enterprises can have over bigger ones is being flexible in meeting customer demands—long lead times reduce that flexibility. The usual strategy for coping with unreliable delivery is to build up a supply of material used and of replacement parts. Keeping a large inventory is costly, as discussed in the next section. Backup energy sources—a diesel-powered generator used when the electric mains fail, for example—are also expensive and need to be serviced and monitored.

A major planning decision is whether to make or buy components. Should a blacksmith making hoes from scrap automobile springs purchase or fabricate the handles? The arguments just given imply that an enterprise planner should be wary of depending on external suppliers, that the effects of a transportation disruption need to be looked at in depth. On the other hand, in some cases using purchased components can reduce costs and result in a higher-quality product. A supplier who concentrates on making a particular part will probably be more efficient in its production and more likely to incorporate new techniques. So an artisan might chose a strategy of buying those parts that he or she cannot make better or cheaper, as long as it appears the supply will be dependable. If the real value of an item made depends on the quality of a particular component—customers might chose one hoe over another because of the shape of the handle—then it may be risky not to make the component. When a component is bought the price should include a fair profit for the seller, so the buyer may be tempted to make the component and retain that profit. The make or buy tradeoffs require careful analysis.

Partnership, informal or formal, with a supplier can be a fruitful strategy. An informal partnership might consist simply of an agreement to supply parts as needed, with a complementary agreement to purchase a minimum number of parts. A supplier-purchaser relationship might go into the design of the component part, how to redesign it for easier production or to improve the final product. The schedule for delivery can also be negotiated, sometimes to the benefit of both parties. The price paid for the component should reflect the necessity for both parties to benefit financially if the relationship is to be sustained. Significant benefits come from treating a supplier as a partner rather than a competitor.

If physical things are being transferred it is usually best when the supplier is near the purchaser. If the enterprise is conducted electronically, as an Internet startup is, then the physical proximity is less important. An ideal place to do small-scale manufacturing may be a location where artisans with different specialties work, some supplying others, each concentrating on a type of production. Another advantage of such a location is that several artisans may share special purpose or expensive tools. An effective way to support the development of small enterprises is to set up these locations where entrepreneurial artisans can work, providing space, utilities, advice as needed, even ways of renting or sharing equipment. Such locations facilitate informal alliances between people working on different parts of a machine or device.

INVENTORY

Inventory is the stock of goods on hand. The basic decision is almost always how large an inventory to keep. Four kinds of inventory need to be considered: finished goods, work-in-progress, raw material, and spare parts. Sourcing decisions—how much to order and how often—affect, of course, the size of the raw material and spares inventories. Inventory choices influence the quality of service and the effort required to manage, and they have significant financial implications. In general, large inventories improve the level of service, reduce strain on a manager, and cost money.

A major reason for keeping a large inventory is that requirements cannot be forecasted precisely. An inventory of finished goods means that an unexpected customer can be served. An inventory of work in progress means that if a machine breaks down subsequent machines in the manufacturing process still have material to process—in the cookie-baking process shown in Figure 10, an inventory of mixed batter would allow the rolling and baking process to continue for a while if the mixer was damaged. The temptation to keep a large work in progress inventory is high—such as inventory reduces the manager's worries. The size of these inventories represents a tradeoff between the cost of keeping items in stock and the cost of a possible lost sale or the cost of a possible production interruption. The costs of possible lost sales or production interruption depends on both the cost itself and the probability that it happens, and estimating these probabilities may be difficult in practice.

A second reason for having a large inventory is that in some cases the cost per item of a component is significantly less if a large number of items is purchased at once—in some cases, suppliers will only sell orders above a certain size. One reason the cost per item may be significantly less is because the internal cost to the organization of processing a purchase may be high. Normally it costs no more to process an order for a thousand pieces than for ten pieces, but, in this case, the processing cost per item will be one hundred times as large if only ten are bought at a time. The size of inventories caused by the cost of purchasing represents, again, a tradeoff between this processing cost and the cost of keeping items in stock. These costs can be estimated fairly accurately. The optimum size of an order can be calculated mathematically. The resulting number is called the "Economic Order Quantity." A related reason for having a large inventory is a big purchase made because one expects a significant price increase to come shortly.

The cost of keeping items in stock can be appreciable for even a small business, but it is often overlooked. Usually the largest part of this cost is the interest that one has to pay on money borrowed to buy the item. Often one starts paying interest as soon as the item is received and continues until the equipment containing the item is sold—actually until the money is collected. If the item is in stock for six months, the interest charges can be significant. If the enterprise does not use borrowed money by paying cash, it is still out the interest the money spent might have earned. Additional costs of keeping items in stock are insurance, rent on the storage space (or the money lost because the storage space cannot be used for another purpose), cost of items damaged or stolen while in inventory, and some cost associated with putting the item into storage and retrieving it. With large inventories a complicated storage system is necessary.

Intangible costs are associated with large inventories. A big risk is that buyers' needs change and the stockpiled item will not be needed. Another is that if six months' supply of an item is purchased at once, then the purchaser cannot benefit from improvements made in the design of the item until six months have passed and the item is reordered. A problem with large work in progress inventories is the other side of reducing the manager's worries. The manager does not have to be as careful to keep equipment functioning if the result of a breakdown will be mitigated by plentiful stock on hand.

Inventories can be monitored in two different ways. One approach is to keep a running record, updated every time an item is taken from the storeroom and every time the stock is replenished. When the inventory record shows that the number in stock is less than a predetermined amount, then an order is placed for more. The predetermined amount depends, of course, on how long after the order is placed before delivery is made and the forecasted usage rate. If computer records are kept in a computer, then a running record is convenient: otherwise it can be cumbersome.

The other approach is periodically to go through the storeroom, or wherever the inventory is kept, and make a count. Reorders are made on the basis of what is still there and estimates of usage. A physical count is more accurate than relying on inventory records. It picks up losses from theft or misplaced items but requires more effort. Another advantage of an actual count is that rarely used items, which are seldom noticed on a running record, become apparent.

A variation of the running record system is called the "two bin" system. Two containers are used. Items are withdrawn from one until it is empty. Then new stock is ordered. Withdrawals are made from the second container until new stock arrives or until the second container is empty. A variation of this system is used in Japan. Folklore is that a completed reorder form sits at the bottom of each container, to be used only when the items are really needed. The size of the bins matches the delivery time and the expected consumption.

In most cases one can reduce inventory costs by focusing attention on the most used items. It is often the case that a very few items in inventory are used extensively and many items are used only infrequently. A third category may exist: items used moderately often. Common practice is to use three categories, labeled A, B, and C, where the items in category A are used extensively, those in category B moderately, and those in C infrequently. In a typical situation, 5 percent of the items are in category A, but these account for 70 percent of the withdrawals. Items in category C comprise 65 percent of the total number of items but only 10 percent of the activity. An effective strategy might be to check the inventory of category A items weekly, of category B items monthly, and of category C items semiannually. The effort not spent on infrequently used items can be usefully spent on the frequently used ones. Nearly always the money tied up in the inventory of high-usage items is, or should be, much greater than that tied up in the inventory of low-usage items, so it makes sense to allocate attention to where the money is.

Focusing attention on the most significant items makes sense in other aspects of managing inventory. When forecasting demand, the planner should devote the most attention to the items representing the largest portion of the inventory. When planning the layout of the storageroom holding the inventory, the planner should put the most used items in the most accessible place.

Computer-based inventory management schemes are available. These are really only effective when all information is entered carefully, so their use may be cumbersome in a small enterprise.

MAINTENANCE

Most pieces of mechanical equipment need regular lubrication. Many need small adjustments as parts wear. Some equipment needs to have filters changed. The manufacturer's instruction sheet nearly always specifies what has to be done and how often. Especially if the equipment is in a location where a breakdown cannot be repaired easily, it is important to do the recommended maintenance. Some maintenance, such as full overhaul, needs to be done by a professional and should be scheduled. The advantage of doing preventive maintenance, rather than waiting for a breakdown, is that the user can do it at a convenient time. Establishing a master schedule for doing maintenance has been worthwhile for many workshops.

Another aspect of maintenance is housekeeping, such as storing tools in appropriate places, keeping cutting tools sharpened, using paint or protective grease to cover corroding surfaces, and removing debris or shavings from work areas. When tools are returned to designated places after use, then missing or damaged items will be evident and the next job can be started expeditiously. Sharp tools are less likely to slip when in use. Falls or other accidents are less likely to happen in a clean workplace. Paintbrushes or spills can be more easily cleaned when fresh.

Most housekeeping chores have to be done eventually, and time is usually saved if they are done at the end of a job rather than at the beginning of the next one.

A final aspect of maintenance is preparing for an emergency failure. A plan is needed, including such matters as whom to call and what to do until repairs are made. If repair service for a particular machine cannot be depended upon, then it may be wise not to buy that machine. The implications of a breakdown need to be considered in depth—will the entire operation be stopped? What kind of backup equipment is reasonable? An advantage of two small machines over a single large one is that if one fails some capacity exists—single-engine planes are more risky than multiple-engine ones. A reliable repair service may be a suitable alternative to a backup machine. The number and kind of spare parts to be held in inventory is affected by how important the pertinent machine is to overall operations and by how quickly spares can be sourced.

SCHEDULING

The owner of a small tailoring business faces a scheduling problem when deciding which order to fill first—several different people have ordered dresses and suits. Different pieces of clothing take different amounts of time to make. Some customers need the clothing very soon. Before doing the actual sewing, thread must be loaded into the sewing machine. Several orders use the same color thread, but others use each use a different color. Certain of her assistants are faster doing dresses than others workers, but some of the assistants that are slower with dresses are the quickest in the shop with trousers.

The chart shown in Figure 13, called a Gantt chart after its originator, is helpful in scheduling. The form of the chart is self-evident. The units of time—days in this case—are listed along the top. The various pieces of equipment to be scheduled are listed in the left-hand column. The times equipment is being used is darkened. In this case three orders are being processed through a shop with 3 sewing machines and 2 pressing machines. On Monday, sewing machine 1 is working on the Dress A order, sewing machine 2 will start the Dress B order at noon, and sewing machine 3 is doing trousers. None of the pressing machines are being used. Pressing machine 1 is only used—this week—on Wednesday.

The chart shows, for example, that if the Dress A order is delayed half a day on Monday, it will still be delivered on time, as subsequent steps will not be delayed. If pressing for Dress A is delayed then it will be necessary to assign Dress B to the other pressing machine. Although pressing machine 2 is only used on Wednesday, without it the three orders could not be completed during the week. Also, if Dress order A were assigned to sewing machine 1 on Thursday, then

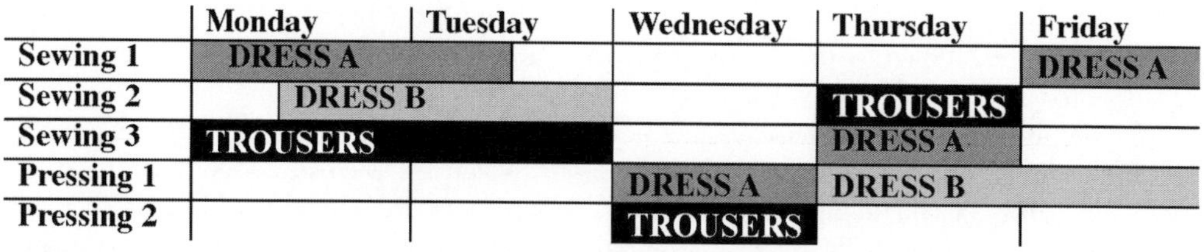

	Monday	Tuesday	Wednesday	Thursday	Friday
Sewing 1	DRESS A				DRESS A
Sewing 2	DRESS B			TROUSERS	
Sewing 3	TROUSERS			DRESS A	
Pressing 1			DRESS A	DRESS B	
Pressing 2			TROUSERS		

FIGURE 13
Gantt or schedule chart.

the job would stay with the same machine for Thursday and Friday, probably increasing efficiency, and the Trousers order would be sewn entirely by sewing machine 3—perhaps a trousers specialist. The chart indicates that this shop could accept more orders. Another Gantt chart is shown in the article "Management of Projects."

Guidelines for scheduling jobs are useful in more complicated situations. Often deadlines for completion of an order are given. One rule is to do the job with the closest completion date first—for example, a lawyer would choose to work on the case coming to trial soonest. Another rule is to do the shortest job first—for example, an automobile service station clears the shop of routine lubrications before tackling big jobs. A commonly used rule in banks or other queues is first come, first served. No rule gives most efficient operation in every case—what "most efficient" means is not even clear.

QUALITY

Quality of an item is one of the things, with price and availability, that customers consider significant when deciding about a purchase. Different customers may want different features in the same object. What makes the quality of a thing high for one user may be of no concern to another. Before making a serious effort to raise quality, an artisan should try to learn what the user really wants. Quality is determined both by the design of an article and the way the article was made. It may also be affected by the service given at the time of purchase or after purchase. A well-balanced axe with a sloppy paint job is of low quality, and even a perfect finish cannot redeem the quality of a badly designed handle.

To obtain quality products a workshop manager needs to assure that artisans in the shop care. The manager needs also to establish a system for chasing down causes of defects and remedying them. An established, highly visible system will send a message that management cares. Frequent discussions about the importance of quality should reinforce the message. An often given piece of advice is that a manager should never accept that some defects are inevitable. The reason for each defect should be explored—the most common defect first—and steps taken to eliminate the cause. In some cases it is practical to examine—inspect—each item produced. When many items are being made, it may be best to take samples—every twentieth item could be inspected. If many items from a sample are found defective, then corrective steps should be taken, as well as more samples. The shop manager should make sure that inspection does not become an excuse for careless work—"it does not matter if this piece is not perfect, it can be corrected by an inspector." The ideal situation is where every artisan cares sufficiently about the quality of the output to look for problems and correct them.

WORK FORCE

Besides choosing equipment and making a schedule, the planner must consider what working in the facility will be like. An argument for using small-scale technologies is that they create appealing jobs—jobs that allow creativity, a sense of accomplishment, autonomy. Appealing jobs are important partly because of the planner's responsibility to other people and partly because as just noted, the quality of what comes out depends on the attitudes of those making the product. For the most part, workers who like their jobs do them carefully and thoroughly. In service industries, especially, customers/clients usually want to interact with a person, not a telephone answering machine or computer, but the value of these interactions depends on the attitude of the company representative. People unhappy with their job usually do not have a positive attitude.

To understand what makes a job appealing one can think about what motivates people. Salary or wages are certainly major motivators, but they are not the only ones. Many people do volunteer work instead of high-paying jobs. Safety at work, job security, relations with a supervisor, administrative policies are a few of the things that influence motivation. A more precise statement is that an unsafe workplace, fear of losing one's job, a discourteous or harassing supervisor, or inflexible company policies can demotivate a worker. Motivation seems to require more than the absence of demotivators though. People are motivated by being part of a group they respect. People are motivated by recognition and encouragement. People are motivated by a sense of accomplishment. The person responsible for planning or operating the enterprise must keep in mind these—and other—factors influencing motivation.

Several specific factors are important when planning a job. It goes without saying that the workplace should be safe. Power saw blades should have guards to prevent loss of fingers. Toxic fumes should be vented. Fire escapes must not be blocked. Beyond safety attention needs to be paid to the specific characteristics of jobs that motivate. Some of these are challenge, responsibility, opportunity to learn or grow, a belief the work is important, opportunity to complete a job, interaction with other workers, interaction with customers/clients, and respect from clients or the general public. These qualities of motivating jobs seem obvious, but it is striking how few jobs seem to be designed with the qualities in mind.

PAY

A worker's pay can be based on either the number of hours—or days—worked, on the amount produced, or on a combination of time and output. A commonly used scheme combining these two methods—time spent and output created—is to pay a base wage for each day worked and an additional bonus for the number of items produced over a standard amount. Probably the most important aspect of the scheme chosen is that it be perceived as fair. Identical work should be seen as earning identical pay. Another aspect of fairness, independent of the basis for calculating the amount, is that the total amount be seen as a reasonable division of the enterprise's earnings between the worker and the owners. Simple schemes for calculating pay are usually more likely to be considered fair. A simple scheme will also offer ease in administration. A scheme should certainly be perceived as giving rewards commensurate with the effort expended.

It would seem that an output-based scheme would be preferable, as it gives the worker some autonomy—if more pay is desired, more work can be done. So-called "piece work," payment strictly on the number of pieces produced, is uncommon in the United States partly because in practice it may violate the minimum wage law. Also, in many situations, the worker does not have complete control of the output—an individual worker may depend on others for inputs. In such cases, incentive bonuses may be given to each member of a group of workers if the group's output exceeds standard. In cases where quality of the output is important, incentive system may be inappropriate because they encourage the worker to rush and the cost of inspecting each piece produced may be excessive. An analogous situation is the risks of alienating customers when sales people are paid on commission. Systems for determining wages need to be thought out carefully—the enterprise tends to get what it pays for.

A survey taken at a site for artisans/entrepreneurs in Zimbabwe showed the following various payment schemes, with their corresponding trade.

Output based	All trades
Production value and commission on sales	Radiator repair
Negotiations per product and what can be afforded	Welding and upholstery
Depending on the level of business	Radiator repair
Sharing of profits	Tailoring and carpentry
Experience (capability and productivity)	All trades
Contractual agreement	All trades
Contract payments for part time workers	Radio and TV repair
Fixed basic and flexible bonus	All trades
Fixed + overtime (permanent workers)	Knitting and sewing
Minimum wage and trainee wage	Timber sales

(From *Survey on Micro- and Small Enterprises on Murahwa Green Market in Mutare*, German Agency for Technical Cooperation Informal Sector Training and Resource Network, P.O. Box 559, Mutare, Zimbabwe and Africa University, P.O. Box 1320, Mutare, Zimbabwe, August 1999)

PURCHASING POLICY AND PROCEDURES

James Crossley

General

This article presents policies and procedures to assist the staff in understanding the responsibilities of the Purchasing Office in acquiring the goods and services necessary for the operation of the organization.

The primary purpose of the Purchasing Operation is to procure all goods and services on the best terms and at the lowest overall cost consistent with an appropriate level of quality. A uniform approach and singular point of contact with suppliers will work toward ensuring that all departments obtain their required goods and services in the specified form, quality, quantity, and timeframe required, and delivered at the most advantageous price. Since these Policies and Procedures serve the interests of the departments as well as policy requirements of the organization, we ask the staff members to give their full support and cooperation. We recognize that emergency or unusual situations will occur. In such instances, departments should discuss the particulars with the Purchasing Department, so whatever assistance is needed may be obtained.

Value Limits

Since this procedure is designed for use with any currency system, the limits shown as ___X___, ___Y___, and soon must be defined by local usage and currency. Limits should be designed to maintain the necessary financial control without creating undue labor and paperwork for minor purchases.

Sole Source Purchasing

As a matter of policy, purchase description should not be written in such a way as to preclude or severely limit competition. Exceptions to this policy are requirements for replacements parts; items available from a single source; or items available from other contracts. Sole source requirements will be accompanied by written justification.

Solicitations

Purchases not exceeding ___X___ value may be accomplished without soliciting competition when prices are considered fair and reasonable. Neither sole source nor price justification is required in these cases.

When purchases are to be in excess of ___X___ but less than ___Y___, reasonable solicitation of qualified suppliers will be made to assure that the procurement is made to the advantage of the organization. Generally, solicitation may be limited to three suppliers. Oral or written quotations may be solicited. The results of all solicitations are required to be attached to departmental requisitions before submission to the Purchasing Department.

When purchases are to be in excess of ___Y___, proposals shall be solicited on a competitive basis from the maximum number of qualified sources consistent with the nature and the requirements of the supplies or services to be procured. The solicitation will be made by issuing Requests for Proposals (RPF) and will be authorized by the Department Head and approved by the Finance Committee. The Chief Operations Officer or equivalent position will coordinate the solicitation and be responsible for issuing the final contract, upon approval of both the Department Head and the Finance Committee. Departments may negotiate the proposals received prior to the finalization of the contract.

Price Justification

Reasonableness of a proposed price should be based on competitive quotations, if at all possible, or on a comparisons with prices found reasonable on previous purchases, current price lists catalogues, advertisements, or personal knowledge of the item being procured.

RESPONSIBILITIES

Department Heads

The Department Head is responsible for the budget of his or her department. No purchase action will be initiated unless authorized by the Department Head or his or her authorized representative. The Department Head must notify the Purchasing Coordinator in writing of the names of any authorized representatives. This official notification should contain a sample of the representative's signature. Such authorizations will be kept on file in the Purchasing Office. Purchase orders will be generated for orders totaling ___Z___ or more. Purchase requests should be made on a Requisition Form and sent to the Purchasing Office. Purchase requests under _____ may be authorized by the department Head without any further approval. Purchases of _____ but less than _____ will require approval by the Chief Operations Officer or equivalent position in addition to the authorization of the Department Head. All requirements of _____ or more will required negotiated fixed-price contracts.

The Department Head will assure that any required prepurchasing approvals have been obtained and that the funds are available and proper. The Department Head or his or her designated representative will be responsible for the proper receiving of the equipment, parts, supplies, or services ordered; the completion of any receiving documentation; and forwarding the documentation to the Purchasing Coordinator.

Purchasing Office

The Purchasing Coordinator will review the Requisition Form for completeness and required authorization and approvals, if applicable. The Purchasing Coordinator will have the Purchase Order prepared, signed, and sent to the vendor, with copies to the requesting department.

For requirements of _____ or more, the Chief Operations Officer or equivalent position will coordinate the solicitation and be responsible for issuing the final contract. The Purchasing Coordinator will be responsible for matching the receiving documentation forwarded from the department with the vendor's invoice and processing the invoice for payment.

The Purchasing Coordinator is available to assist the departments in locating sources for products or services as needed. The Chief Operations Officer or equivalent position responsible for the coordination of major purchases of equipment, furniture, and supplies as well as service contracts.

Methods of Purchasing

There are four types of purchase instruments normally used: (1) Purchase Order, (2) Blanket Purchase Order, (3) Departmental Petty Cash, and (4) Negotiated Fixed Price Contract.

Purchase Order

The Purchase Order is an official instrument that expresses the buyer's part of the contract of sale. The Purchase order will be signed by the Chief Operations Officer or equivalent position. It has the legal force of a binding contract. Except for emergency requirements, all purchase actions for equipment, parts, supplies, and services shall be initiated by a written requisition signed by the Department Head or his or her authorized representative.

The Department will complete a requisition containing the information shown below (under *Requisition*). This will authorize an obligation against the account noted on the requisition. The Purchasing Coordinator will issue a Purchase Order to the vendor with copies of the Purchase Order sent to the requesting department. Upon receipt of the order, the requesting department will complete the receiving documentation and forward it to the purchasing Coordinator. The Purchasing Coordinator will match the receiving documentation with the vendor's invoice and forward the invoice to the Bursar's Office for payment. Any differences discovered during the matching process will be resolved between the Purchasing Coordinator and the Requesting Department.

Emergency Requirements

An emergency purchase will be authorized only when a bonafide emergency exists and the interest of the company would demand such action. In such a case, with the approval of the Faculty or Department Head, the Purchasing Coordinator will issue a Purchase Order number. A confirming requisition (including the Purchase Order Number given) must be forwarded immediately to the Purchasing Office so that a confirming Purchase Order may be issued.

A written justification must be attached.

Requisition

Each requisition will contain the following information.

- Description of other items or services requested in sufficient detail. Use manufacturer's catalogue numbers and description whenever possible.
- Name and address of Vendor. Departments are encouraged to suggest sources of supply with complete addresses, particularly for unusual or nonstandard items. The Purchasing Department will advise the Department Head any time a change of vendors is suggested.
- Quantities desired should be specific (such as box, carton, or package). The unit of issue should relate to the unit cost.

- Unit cost. Either specific or approximate price. If a quotation has been received, it should be attached to the requisition.
- Any terms or conditions of the purchase, such as prepayment (normally required with subscriptions, booklets, and publications); any accounts that are available.
- Freight considerations (such as prepaid, collect, allowed); or if the purchase order should be faxed to the vendor, it should be noted on the requisition.
- Desired delivery date should always be shown. For most stocked items a lead time of 30 working days is normally sufficient.
- The complete account number is to be charged. If more than one account is to be charged, each account number and the amount to be charged should be listed.

Blanket Purchase Order

A blanket or standing purchase order is a simplified method of filling repetitive needs for small quantities of supplies by establishing "charge accounts" with qualified sources of supply. This type of purchase order is designed to reduce administrative costs in small purchasing by eliminating the need for issuing individual purchase orders.

Limits

A blanket purchase order may not be used for individual items costing more than _____.

Procedure

Calls against the blanket purchase order can be made orally or on a requisition as applicable. The receiving department will place the date of receipt, the account number to be charged, and the signature of the individual receiving the items on either the vendor's delivery slip or invoice and forward it to the Purchasing Coordinator *within a day of receipt*. Each delivery slip or invoice received will be matched with the monthly statement by the Purchasing Coordinator and forwarded to the Bursar's Office for payment (see attached flow diagram).

Departmental Petty Cash

Departmental Petty Cash may be used for minor purchases of items from qualified sources with whom there is no blanket purchase order.

Limits: Departmental Petty Cash fund may be used to pay for purchases not in excess of ____.

Procedure: Refer to the Petty Cash procedures received from the Accounting Office.

Fixed Price Contracts

Negotiated Fixed-price contracts will be issued for all requirements in excess of _____. The Purchasing Manager or Agent will coordinate the solicitation of the Requests for Proposals (RFP) and will be responsible for the final contract. The procedures for the solicitation and contracting of both supplies and services shall follow accepted contracting principles.

Ongoing Supervision

It is the responsibility of the Chief Operations Officer or equivalent position to review this program at least annually and to revise as necessary.

MARKETING

Marketing is the whole process of getting the product or service to the user. It includes selling but is much more. In the Third World often the major problem is distribution—establishing a chain of wholesalers and retailers. Promotion, which includes advertising, is ubiquitous in the United States and Europe but less common in the Third World. Setting a reasonable price is a final aspect of marketing. Some comments on exporting are included at the end of the article.

Effective marketing is a strong necessity. Some engineers and aid workers act as if a good product or service will sell itself—that intended users will find out about the product on their own and demand it. Such happens very rarely. Competing products already have the attention of users, and only a few people are eager to experiment on their own with new products. It is unfair to push a new venture onto a group and have it fail because the expected demand did not materialize—because the marketing had not been thought through.

MARKET RESEARCH

Market research is an essential early step in establishing a new enterprise. Probably the most effective method is discussions with potential users, either individual discussions or focus groups. In many societies people tend to be polite, especially with strangers, so answers to direct questions may be what the respondent thinks the questioner wants to hear. If the market research is to be useful, great effort must be made to put the respondent at ease and sympathetic to the needs of the researcher. The question "What do you think your neighbor would be willing to pay for this charcoal stove?" will probably get a more valid answer than "What would you pay for this charcoal stove?," especially in a culture where bargaining is expected. It can be useful to have the artisans who will produce the new product to be part of the research team, so the true needs of the user become apparent. Care must be taken to ensure that the people surveyed are representative of the intended customer group. A common difficulty is to survey people in the most accessible places, the market used by the relatively affluent, when the intended market are those less affluent.

The first step is to compose the questions to be asked. Market research should aim to determine what features the user wants and what price would be reasonable. Where the user would be likely to buy the product is also valuable information, partly to plan where supplies and fuel will be sold. The user's knowledge about the technology employed is worth understanding—can the user change the oil filters in a milling machine? Before actually going out and surveying people it is important to think about what information is needed. After questions are written, try them out on colleagues or a small group similar to the people being surveyed.

Two other issues related to market research are worth noting. In the United States and Europe questionnaires analyzed using statistical techniques are an important aspect of market research. Such can be less effective where questionnaires are unfamiliar and the logistics of distribution are difficult. Another issue is learning about other organizations doing related activities, perhaps producing a similar good or service. In one sense such organizations are competitors and should be understood as such—the article on planning, for example, discusses competitive analysis. In another sense though such organizations can be partners, for example, in working with exporters to build a foreign market. It is certainly worth learning all one can about organizations with related purposes and try to work cooperatively with them.

DISTRIBUTION

Commonly, goods produced in the Third World go through a complicated series of wholesalers before reaching the consumer. Each of these wholesalers adds to the final price paid by the consumer but do not, of course, increase the revenue the producer receives. It appears promising then to try bypassing some of these wholesalers, but such is not always feasible. The worst case for producers, it goes without saying, is when only one buyer is present. The situation is worse if that buyer is the only source of credit—needed to purchase raw materials—for the producer. Extension officers can assist in such cases by finding alternative buyers, perhaps export co-operatives. Extension workers can also assist by educating producers about market conditions so as reduce corruption and discrimination.

If one hopes to bypass wholesalers one must find alternative ways of accomplishing what they do. An important function of wholesalers is bringing together the products of many small producers to create large enough loads for efficient transportation. Exporters, even many urban stores, want to deal in bulk, with a single supplier so aggregating the outputs of small-scale producers is often essential. As just noted, wholesalers may be the most accessible source of cash to people making things. Microfinance organizations (described in another article in this section) are an alternative. Wholesalers may also be in the best position to do packaging. A good wholesaler will advise producers on what customers want, even work with producers in developing new products. Storage, of agricultural goods in particular, needs to be considered. Often economies of scale exist so storage is most efficiently done by a wholesaler.

A marketing project must keep in mind transportation problems in getting goods to the consumer. Villages far from paved roads may be inaccessible in the rainy season. Without reliable transportation to a market food crops may spoil. An essential part of a project to sell handicrafts, like woven baskets, may be a four-wheel drive vehicle to collect from villages. Planning for new roads needs to be done sensitively. Commercial farms may expand to areas made accessible by a new road, draining resources from villages not reached by the road, so the condition of small-scale farmers both on and off the new road may be diminished.

PROMOTION

Promotion is informing and persuading potential customers. In Europe and the United States advertising, an aspect of promotion, is a major activity. Small-scale producers in the Third World tend not to do much promotion, perhaps because opportunities are not common, perhaps because meaningful advertising requires appreciable expense. Cooperatives may do sufficient business to make advertising worthwhile, for example, by printing a brochure. Most small-scale producers rely on the wholesaler to promote their product to the consumer so promotion for such a producer is finding and convincing an appropriate wholesaler.

PRICING

Setting the price is an integral part of marketing. Two factors come into setting the price: cost of making the item and what the customer is willing to spend. The cost of producing will be calculated in the costing article. Estimating what a customer will spend can be difficult. One approach is market research, mentioned previously, but customers may not really know what they will be willing to pay. If the product is intended for an export market, even a market in another part of the country market research will be difficult. Another approach to pricing is to look at competing products, but in some cases it may not be obvious what the customer will perceive as competing products. Advice from experienced marketing people may be helpful here. Of course, most marketing experts will estimate a retail price. From that retail price the wholesaler's markup, or markups, is subtracted to get the price paid to the producer. The retail price may be more than twice the price received by the producer. One of the benefits an aid worker can give is to reduce the markup to raise the revenue to the producer.

Another aspect of pricing is to prevent cutthroat price competition among producers. A buyer can, when several different producers are making a similar product, encourage each to lower its price by threatening to purchase from another supplier. Agreement among producers on a minimum price will prevent a trader from playing one against the other. Producers need each to understand that unilaterally defecting from such agreements—lowering the price below the agreed minimum—may give one major sale but will make it very difficult to maintain the prices in the future.

SALES

When a new enterprise is started it may be necessary to train the sales force. Convincing a potential buyer to purchase requires some skill, as well as practice. In many cultures bargaining at a market is expected. In stores or more other more formal situations customers must be gently encouraged to buy. A salesperson should be expected always to be helpful and courteous. A salesperson should be able to answer questions about the goods being sold and give reasons why a particular item should be purchased. An effective salesperson will suggest, in a nonthreatening way, items the customer might want in addition or as an alternative to what was asked for.

EXPORTING

Two major administrative problems come with exporting to a foreign country: payment and customs. Both can be facilitated by people with experience, so most small-scale producers wanting to export should get help. Most banks are prepared to handle the money problems. Many countries have government agencies that can give information on customs documents and regulations. Shipping agents can do this also. It is best to negotiate a fee early in the transaction. Cooperatives have been formed in some countries to assist small-scale producers in doing the exporting administrative procedures.

The payment question concerns making a guarantee that the producer will be paid when the goods are delivered. Because producers would have difficulty dealing with an unpaid bill in another country, they often want such a guarantee. A *letter of credit* is one such arrangement. When the order is accepted the customer arranges such with its bank. The letter of credit authorizes the producer's bank to pay the producer when the goods are shipped—that is, when shipping documents and invoices are presented. Letters of credit are complicated and hard to modify but are a common mechanism. Another mechanism is a *bill of exchange*. Such is prepared by the producer's bank and accompanies the shipment. When the customer receives the goods, the bill is signed and returned to the producer, who collects the money from the bank.

A major part of the customs problem is learning which forms must be used. At the least a *packing list* is needed, showing exactly what is in each box sent. An *invoice* stating the price of the goods and the terms of the sale should also be included with the shipment. Part of the terms are specifications on who pays various shipping and handling costs. An agent can give advice in this regard. An *airway bill* is prepared by the agent or airline involved. It can be made out to a bank so the customer only receives the goods upon payment. Another part of the customs problem is learning specific regulations for the country of destination—for example, the United States will not allow any goods containing ivory or skins of endangered species to enter. Works of art, especially antiques, may need special certificates from the exporting country.

The paperwork should be done early, so difficulties are solved before the goods are shipped. Airlines and docks charge rent, usually after two days, for goods received but not picked up. It may be worthwhile hiring a reliable agent to clear goods through at the port of destination.

REFERENCES

Pratt, Brian, and Jo Boyden (Eds.). (1988). *The Field Directors' Handbook, An Oxfam Manual for Development Workers*. Oxford: Oxfam Publications.

ILO/IYB Regional Project Office in Harare, Zimbabwe. *Marketing*. ILO/SAMAT, PO Box 210, Harare, Zimbabwe, 1994.

FINANCIAL VIABILITY AND CASH FLOW STATEMENT

A project will not continue, will not be viable, if the amount of money spent is greater than the amount available. In this section the question of how to forecast viability is discussed. Two issues are important. The first is whether the total amount of money brought in by the project—for example, by external grants plus charges to clients—will be greater than the amount of money spent. The second question has to do with timing—whether the money coming in comes in time to make required expenditures. The discussion is terms of cash flow. A planner should estimate how much cash will be required to establish and operate a project. The planner should also estimate how much cash the project will generate—the difference between cash required and cash generated must be raised from project operations. Often projects are started using grants from government or international agencies or from NGOs. An estimate of cash flows is used to determine how large a grant is needed.

Of course, not all costs and benefits from a project are financial. Doing a particular project may give an organization experience valuable for its future. A project may create much good-will for an organization, even if it does not earn much cash. An organization may do a group—portfolio—of projects that are interrelated. A specific project itself may not be financially viable but be essential to the portfolio—for example, a gravel sourcing project supplying raw materials to a construction project. Organizations often have one or several projects that do not earn money but advance the mission of the organization. It is worthwhile understanding the financial implications of such projects, even if they are not intended to be self-supporting.

In principle, it should be easy to determine if a project is financially viable. Income and expenses would each be estimated and compared. If the income—inflow—is greater than expenses—outflow—the project is viable. A problem in determining if financial outflows equal or exceed inflows has to do with what is left at the end of the project. In many cases equipment will have been purchased for the project that is still usable when the project is finished. The financial value of such equipment at the end of the project, the "salvage" value, must be estimated to obtain a valid measure of the outflow. The financial value is the amount of money the equipment could be sold for. If the equipment cannot be sold—for example, because it cannot be removed from the premises—then the salvage value is zero. If the equipment will be used for another project, then the salvage value is what the other project will pay.

CASH FLOW STATEMENT

A cash flow projection is used to determine the amount of cash held by an enterprise at any time. One use is to estimate how much cash is needed to start a project: At least enough initial cash is needed so the cash on hand never goes below zero. It is not sufficient simply to look at the sales, or other forms of revenue, and then look at expenses because for some transactions the cash changes hands some time after the transaction—credit is involved. When customers purchase an item, they may arrange to pay within 30 days, as is done with a credit card. Accountants used the word "receivable" to describe the amount of money owed to an organization for goods purchased but not yet paid for. Similarly, when the organization purchases material used in the products sold, they may not pay cash at that time. The money an organization owes for such supplies is called a "payable." Another reason exists for why financial state of an organization is not simply sales minus expenses that: The organization may have manufactured goods that have not yet been sold. These not yet sold goods are called "inventory." Inventory and receivables are part of the wealth of the company—another part is the machinery and buildings they own. Payables can be thought of as negative wealth. Eventually all this wealth will become cash. Until the wealth does become cash though the organization may have a problem because workers nearly always want to be paid in cash and some suppliers may insist on cash. It is important, therefore, for the organization to forecast how much cash it will have at various times.

Setting up a cash flow forecast—often simply called a cash flow "statement"—is straightforward. A spreadsheet forecasting monthly cash flow is shown in Figure 14. The corresponding enterprise sells snacks to construction workers.

The spreadsheet shows cash from a loan separately from cash generated by operations because such a separation is convenient when estimating the numbers. The enterprise described by the spreadsheet shown received one loan of $5,000 in January. Negative numbers are shown in parentheses: (100).

Cash generated from operations is, of course, cash inflow minus cash outflow. Cash inflow is cash received from sales. Cash outflows are expenses—costs. Costing is a central problem for many entrepreneurs, so it is discussed more extensively in its own article—"Costing"—but is reviewed briefly here. It is convenient to separate cash outflows into two categories: "variable costs" and "fixed expenses." Variable costs depend on the amount of product sold. In this example, the variable costs are the purchases—the meat and bread used in the snacks—and the labor required to put the snacks together. The word "variable" is used because these costs depend on the amount of product sold. Variable costs are sometimes called "direct" costs. Sometimes the variable costs are added together and the sum called "cost of goods sold" or "cost of sales." The difference between the cash inflows from sales and the cash outflows from purchases and labor is called "gross margin." It is a measure of the amount the enterprise is making each month through its business.

Unfortunately, fixed expenses have a claim on some of the cash represented by the gross margin. In this example, the fixed expenses are administration, rent, and utilities. Administration includes such items as office staff, paper, and other supplies. In other businesses licenses, insurance, and custodial staff would need to be included. The word "fixed" is used because the amounts do not depend on how many snacks were sold. The same rent and electricity must be paid for in months when sales are good as when they are poor. Fixed costs are sometimes called "indirect" costs. Earnings are calculated by subtracting the fixed expenses from the gross margin. Earnings represent the amount of cash earned from operations—cash inflow from sales minus all expenses.

Month	Jan.	Feb.	Mar.	Apr.	May	June	July	Aug.	Sept.	Oct.	Nov.	Dec.
Startup Loan	5,000											
Operations												
Food Sales		1,200	1,350	1,500	1,900	2,100	2,900	3,400	3,600	3,800	3,800	3,800
(Purchases)		360	405	450	570	630	870	1,020	1,080	1,140	1,140	1,140
(Labor)		1,000	1,000	1,000	1,000	1,000	1,200	1,200	1,200	1,200	1,200	1,200
Gross Margin		(160)	(55)	50	330	470	830	1,180	1,320	1,460	1,460	1,460
Fixed Expenses												
(Administration)		500	500	500	500	500	500	500	500	500	500	500
(Rent)		450	450	450	450	450	450	450	450	450	450	450
(Utilities)		125	125	125	125	125	125	125	125	125	125	125
Earnings		(1,235)	(1,130)	(1,025)	(745)	(605)	(245)	105	245	385	385	385
Cash Flow	5,000	(1,235)	(1,130)	(1,025)	(745)	(605)	(245)	105	245	385	385	385
Net Cash	5,000	3,765	2,635	1,610	865	260	15	120	365	750	1,135	1,520

FIGURE 14
Spreadsheet for forecasting cash flow.

The entries of the spreadsheet will need to be adapted for different enterprises, but the pattern should apply in nearly all cases. Some projects get periodic cash infusions from government. These cash inputs might be put in the same row as the grants in the spreadsheet shown—perhaps a different label might be needed. Estimating amounts for the various entries requires judgment. The amounts of fixed expenses can usually be estimated fairly straightforwardly—the person running the enterprise has some control over the amounts. In making estimates one looks at the present year's costs and asks suppliers, such as the landlord, whether these costs are expected to change. Sales are harder to forecast and because variable costs depend on sales they can also be difficult to forecast. In situations where it seems hard to make a reasonable estimate, three estimates are sometimes made: corresponding to the most likely situation, an optimistic situation, and a pessimistic situation.

The bottom two rows on the spreadsheet show the net cash flows for each month and cumulative over the year. Cash flow for a month is simply the sum of the cash from grants and the cash from operations. In this simple example cash from grants and from operations came in different months, but such will probably not happen in practice. Net cash, the bottom row, is found by adding the cash generated in a particular month, if it is positive, to the net cash of the previous month. If the cash generated is negative because the enterprise spent more cash than it made during a particular month, then the monthly cash is subtracted from the net cash. Net cash for the first month, of course, is just the cash inflow for that month. Net cash is the cumulative cash flow since the enterprise started—the sum of all the cash in and all the cash out. One can think of this net cash as being our bank balance.

ANALYSIS OF CASH FLOW

What can we learn from the spreadsheet of Figure 14? The cash flow forecast shown indicates that the enterprise does generate cash in August and afterwards that it is economically viable. We see that an initial loan of $5,000 is close to the minimum. A smaller loan would mean that our net cash would become negative. If we started a month with a negative cash balance, we run the risk of not being able to purchase meat or bread that month. A more thorough cash flow forecast would consider ways of coping with negative cash balances—perhaps ask the suppliers for credit. At the least this simple model indicates we have a problem. Perhaps the loaning agency might ask us if they could disburse the loan in two equal parts of $2,500 in January and May. The spreadsheet shown, especially if it were done on a computer, could easily be modified to answer that question.

It may be useful to give some suggestions on what to do in a practical case if the cumulative cash flow is negative. In a real situation businesses buy goods on credit and extend credit to customers—many transactions involve receivables and payables rather than cash. In these cases a business can address its cash flow difficulties by collecting receivables or by seeking additional credit, perhaps requesting a delay before it pays its bills—extending payables. Another way of raising cash in a hurry is to sell items in stock—liquidating inventory, perhaps for less than the usual price. Startup organizations can sometimes get emergency cash from the people who gave the original grant.

The cash flow forecast can be used also when deciding whether an organization should undertake a project. The financial criterion is whether the return—the cash generated—is greater than the cash invested. The spreadsheet only shows one year of operation, and a valid financial analysis should be over the entire life of the project. The spreadsheet suggests the approach though: Look at the cash earned over the life of the project and compare it to the initial grant. If the cash generated is greater than the grant, the project makes financial sense. The problem is slightly more complicated though because if the grant money had not been put into the project, then it might have earned interest in a bank, so the sum of future earnings should be greater than the original investment by the amount of interest that could have been earned. A way of doing this analysis is given in the "Discounted Cash" article in this chapter.

COSTING

REASONS FOR UNDERSTANDING COSTS

Advisors to small-scale business consistently emphasize the importance of understanding what it costs to make an item or provide a service. A major reason for understanding costs is to determine a selling price. Another reason is to gain insight into ways of reducing costs. A knowledge of what is controlling costs will assist in reducing costs, partly by showing where major attention should be paid. Costs are an essential part of a business plan. It is convenient to work with two types of costs: variable costs and fixed costs.

VARIABLE COSTS

Variable costs are costs that vary with the amount produced. If output doubles, then variable costs double. Variable costs are generally the wages of production workers and the costs of the material that goes into the product. In a hair styling business, material costs are the costs of the shampoos, perm lotions, and neutralizers. The labor costs are the money paid to the hair stylists. Not all businesses have variable labor costs—for example, a sidewalk vendor selling fruit does not because her labor costs are the same each day whether she sells much or little. A sidewalk vendor's material costs are the costs of the fruit sold and are a variable cost—the more fruit sold the higher the costs. A cost is variable if it can be directly associated with a transaction: The cost of the orange sold can be associated with the sale of the orange, but the labor cost cannot because the vendor does not have to work more hours as a result of the sale. Variable costs are sometimes called "direct" costs—they arise directly from making the product or doing the service. An example is given below of how a carpenter computes the variable costs in making a table.

FIXED COSTS

Fixed costs are the costs of operating the business. They remain constant—fixed—as the amount of output changes. Rent is a fixed cost, as is insurance and the cost of a license. Usually utilities like electricity and water are fixed costs because hardly any more of either is used when production increases. A bookkeeper's salary or the salary of the shop supervisor is a fixed cost. Small costs such as for lubricants are usually treated as fixed costs for convenience even if they do vary as the amount produced varies. Fixed costs are sometimes called "indirect" costs.

ESTIMATING COSTS

A reason for making the distinction between variable and fixed costs is to simplify the estimation process. Variable labor costs of a product are simply the number of hours required to make the product times the wage rate of the person or persons doing the making. Usually it is not difficult to estimate, even measure, the time required to make an item. Variable material costs are simply the cost of the raw materials that must be purchased to make the product. If some of these raw materials actually are not used in the product but cannot be used elsewhere, then their cost should still be considered part of the variable material cost. An example might be dried paint in the bottom of a can at the end of a job. Variable material costs can be calculated by multiplying the amount of material used by the unit cost of the material. Again, it is usually not difficult to estimate the amount of material needed to make a single item. An estimate for the total variable cost is the sum of the labor and material cost estimates.

Estimating fixed costs is slightly more involved because many more things must be included. Some typical items are rent, electricity, telephone, insurance, stationery, office supplies, and office staff people—all the costs of doing business that are not directly associated with making the product. Usually these can be forecasted without difficulty starting with present costs and asking about increases. Fixed costs are sometimes called "indirect" costs.

Often one wants to know the cost of a single item to determine the price one charges for the item. In this case the amount of the fixed costs must be divided up among each of the items. The simplest way to do this is to divide the total fixed cost by the number of items produced. For example, if the sum of all the fixed costs is $2,000 and 100 items were produced, then the fixed cost associated with each item is $20. The total cost of the item is $20 plus the sum of the variable labor and material costs. To make a profit the sale price must be greater than this cost. Determining how much indirect cost must be associated with each item produced is more difficult if several different items are being produced. One way is to allocate fixed costs proportionally to the variable costs: An item with twice the variable cost of another item will be allocated twice the fixed costs. One reason the choice of how to allocate fixed costs is important is when the relative profit of making two items is compared; an item that carries a high fixed costs may appear unprofitable. Effort needs to be spent to ensure that the system for allocating fixed costs to individual items is perceived as fair by people in the organization.

One fixed cost, which is used extensively by accountants, is "depreciation." Depreciation is the cost of the loss in value of equipment as it gets older. It is not a cash loss—no cash is paid out. A reason for showing depreciation as an expense is to show how the value of the company has changed as the product was made. One way of thinking about depreciation is that money should be set aside regularly to pay for the replacement of the machinery used in production. The total amount of the depreciation equals the original expense of the machinery. The amount that should be set aside each month is the total expense divided by the number of months that the machinery is expected to last. To calculate then the monthly expense for depreciation of a machine, the expected lifetime of the machine is first estimated. Then the initial cost of the machine is divided by this lifetime. The resulting cost is the monthly expense for depreciation. For example, a carpenter buys a table saw for $1,000. The carpenter estimates that the saw will last for 50 months. The monthly depreciation expense then is $20. A cost of $20 needs to be added to the cost of products made during the month. If 10 tables were made that month (and no other products), then a depreciation expense of $2 must be added to the cost of each table. Depreciation is considered a fixed cost because the loss in value is assumed to depend only on the age of the machine, not the amount of use the machine receives. The calculation is not precise because it is difficult to estimate the useful life of a piece of machinery.

A simple example may help in understanding costs. The same carpenter wants to estimate the cost of making a table, based on the following figures.

1. In one month, when 10 tables are made, $120 is spent on wood, $5 on nails, $15 on sandpaper and paint.

2. The total cost for materials is $140, corresponding to $14 per table.

3. The carpenter estimates 8 hours of work are needed to make one table. The carpenter wants to earn $10 an hour. Direct labor costs, then, are $80.

4. Fixed costs for this carpenter are rent—$80, electricity—$10, part-time bookkeeper—$30, all per month. Also a yearly insurance bill of $60 must be paid, plus an annual license fee of $30. In addition, a monthly depreciation of $30—for all tools, including the table saw—has been calculated.

5. The monthly amount for insurance and the license is $7.50 ($90 divided by 12). The other monthly fixed costs are $120 ($80 plus $10 plus $30). Depreciation adds another $30. Total fixed costs for the month are $157.50. Because 10 tables are made in a month, the fixed costs per table are $15.75.

6. Thus, total cost of a table is

Variable labor cost	$80.00
Variable material cost	$14.00
Fixed costs, per table	$15.75
Total cost, per table	$109.75

NOTE ON FIXED AND VARIABLE COSTS

In real life the distinction between fixed and variable costs is not always clear. Also it is harder to make estimates than in the simple example just done. Variable labor costs are not entirely variable because a small business would probably not fire and hire workers each time a job is finished or started, so direct labor costs do not always vary directly proportional to the amount of work done. Fixed costs also change. If the plant is busy, more electricity will be used, and the bookkeeper must work more hours. A more complicated reason why the amount of fixed costs charged to each item—each table in the preceding example—is that the amount charged to each item depends on the number of items made. In the example above, if 15 tables were made in a month, the total fixed cost would probably not change, but the amount charged to each table would be less—$10.50 in this case ($157.50 divided by 15). It has been said that in the short run all costs are fixed and in the long run all costs are variable.

DISCOUNTED CASH FLOW

Money received in the future is generally worth less than money received now because of interest expense. A promise to pay $1,000 next year is worth less than a promise to pay $1,000 tomorrow because if we were to be paid next year, we would have to borrow money for the year. So when the $1,000 was received, we would have to use some of it to pay interest, leaving us with less of the $1,000 available to do what we wanted. Another way of looking at it is to realize that if we were paid tomorrow we could put the money in the bank and earn interest for a year, so the value of $1,000 received tomorrow is greater than the value of $1,000 received next year by the amount of interest earned during the year.

Many management decisions involved spending money now in hopes of receiving earnings in the future—buying a radio station, investing in a new machine tool, or paying for sales promotion are examples. The toy manufacturer thinking about building a factory in Hungary realizes the expenses will be large now, but the future earnings are expected to pay these off. In this article we consider how to account for the value lost because these earnings come in the future. A related problem is how to compare two payment schedules. Is it better to rent an apartment and pay some money each month or buy a house, spending a lot of money up front but having smaller monthly payments? The techniques of this section apply also to these problems.

PRESENT VALUE

First we will calculate the dollar value at the present time of $1,000 to be received in one year if the interest rate being charged is 10 percent. The story is that we will receive $1,000 a year from now; until then we need cash to survive, so we borrow at 10 percent. How much money can we borrow if the $1,000 is to be used to pay back the amount borrowed plus the interest occurred? A little (very little) algebra must be used.

$$PV = \text{Present Value: what the } \$1,000 \text{ is now worth}$$

The amount of interest if PV dollars are borrowed for a year, when the interest rate equals 10 percent is

$$PV \times .1$$

so the total amount that must be paid back is

$$PV + .1PV = PV(1 + .1)$$

But this amount must equal $1,000 because we have $1,000 to pay back the loan, so

$$PV(1 + .1) = 1,000$$

or

$$PV = \$909.09$$

This means that $1,000 to be received in a year when the pertinent interest rate is 10 percent is worth $909.09 right now. The number 909.09 can be checked by calculating the interest we would have to pay if we borrowed this amount for a year at 10 percent—$90.91. If we repaid the original loan and the interest just calculated, the total to be paid back would be

$$90.91 + 909.09 = 1,000$$

The number .90909 is called the discount factor.

Three comments may be useful here. The first is that one should not be afraid of the arithmetic—a spreadsheet program on a computer makes the calculations easy and if a computer is not available a table of discount factors can be used. Table 5 shows such a table. The second comment is that we are not considering the uncertainty of receiving the payment. In the real world not everything always goes as expected, but we assume here that the money is received, or spent, as expected. The final comment is that the phrase "future value" is sometimes used. The future value, in the example, of $909.09 in 1 year, using 10 percent interest, is $1,000.

The calculation is analogous for other interest rates. If

$$r = \text{interest rate—a number, not a percentage, so 10 percent means } r = .10,$$

then interest paid on the amount PV is

$$(PV) (r)$$

so Present Value of $1,000, when the interest rate is r, is

$$PV = \frac{1,000}{1 + r}$$

Money received two years from now is worth less than money received one year from now because we would have to pay two years of interest on the money we borrowed in anticipation. If the value of a dollar received one year from now is about $.91, then the value of money received two years from now is about

$$.91 \times .91 = .8281$$

because the money loses the same fraction of its value each year. The discount factor for n years then is found by multiplying the one-year factor by itself n times.

$$DF = \text{1-year discount factor}$$

$$(DF)^n = \text{n-year discount factor}$$

CHOICE OF DISCOUNT RATE

The question of what rate to use in practice when discounting can be difficult and will be discussed later. If the money will actually be borrowed, then the interest rate charged by the lender makes sense. Companies considering a new investment often use their own funds rather than borrowing externally. These companies usually calculate an equivalent interest rate, often called "cost of capital," to analyze new investments. This cost of capital can be calculated in two ways—either from the interest rate a company would have to pay if it raised new funds or by how much it is making on the funds presently invested. The rationale for using the existing earnings rate is that if the company uses its own funds for a new investment, it is really borrowing from itself and would want to pay an interest rate at least as high as the money would have earned if it had been invested in ongoing operations.

CALCULATION USING DISCOUNT FACTOR TABLE

Probably the easiest way to calculate present values is with a spreadsheet. Probably the hardest is with the formulas, although some hand calculators simplify the calculations by having the formulas programmed in. A way of intermediate convenience is through a table, shown in Table 5. To illustrate the technique, the present value of $5,000 to be received in one year will be calculated assuming 14 percent interest. The table is entered in the top row, corresponding to one year, and the entry under 14 percent—.877 is selected. (Three-place accuracy is nearly always sufficient.) The entry means that the present value of 1 dollar is about 88 cents, after one year @ 14 percent, so $5,000 is worth $4,385. At the same interest rate $10,000 would be worth $5,190 if payment were to be made five years from now.

NET PRESENT VALUE

A common managerial decision is whether to make an investment now to get earnings later. A professional soccer franchise is offered for $20 million. It is expected to generate a cash flow of $1.5 million the first year, $3 million the second, $4.2 million the third, $7 million the fourth, and $8.5 million the fifth. The manager considering this deal expects to sell the franchise the sixth year and hopes to get $10 million for it. (The reason for the low selling prices is related to the high cash flows the last two years—trading off the stars.) The question is whether this is financially worth doing. The manager will think later about how much the fun of owning a professional sports team is worth to her. If the manager did not buy the team, she would put the $20 million back into her razor blade business, which earns 14 percent on money invested. These cash flows are shown in the spreadsheet below.

Year	Cash Flows
1	1.5
2	3
3	4.2
4	7
5	8.2
6	10

TABLE 5
Present value of a dollar received in year n

Year (n)	Anuual Discount Rate									
	1%	2%	3%	4%	5%	6%	7%	8%	9%	10%
1	.9901	.9804	.9709	.9615	.9524	.9434	.9346	.9259	.9174	.9091
2	.9803	.9612	.9426	.9246	.9070	.8900	.8734	.8573	.8417	.8265
3	.9706	.9423	.9151	.8890	.8638	.8396	.8163	.7938	.7722	.7513
4	.9610	.9239	.8885	.8548	.8227	.7921	.7629	.7350	.7084	.6830
5	.9515	.9057	.8626	.8219	.7835	.7473	.7130	.6806	.6499	.6209
6	.9421	.8880	.8375	.7903	.7462	.7050	.6663	.6302	.5963	.5645
7	.9327	.8706	.8131	.7599	.7107	.6651	.6228	.5835	.5470	.5132
8	.9235	.8535	.7894	.6307	.6768	.6274	.5820	.5403	.5019	.4665
9	.9143	.8368	.7664	.7026	.6446	.5919	.5439	.5003	.4604	.4241
10	.9053	.8204	.7441	.6756	.6139	.5584	.5084	.4632	.4224	.3855
11	.8963	.8043	.7224	.6496	.5847	.5268	.4751	.4289	.3875	.3505
12	.8875	.7885	.7014	.6246	.5568	.4970	.4440	.3971	.3555	.3186
13	.8787	.7730	.6810	.6006	.5303	.4688	.4150	.3677	.3262	.2897
14	8700	.7588	.6611	.5775	.5051	.4423	.3878	.3405	.2993	.2633
15	.8614	.7430	.6419	.5553	.4810	.4173	.3625	.3152	.2745	.2394
16	.8528	.7285	.6232	.5339	.4581	.3937	.3387	.2919	.2519	.2176
17	.8444	.7142	.6050	.5134	.4363	.3714	.3166	.2703	.2311	.1978
18	.8360	.7002	.5874	.4936	.4155	.3503	.2959	.2503	.2120	.1799
19	.8277	.6864	.5703	.4746	.3957	.3305	.2765	.2317	.1945	.1635
20	.8195	.6730	.5537	.4564	.3769	.3118	.2584	.2146	.1784	.1486
21	.8114	.6598	.5376	.4388	.3989	.2942	.2415	.1987	.1637	.1351
22	.8034	.6468	.5219	.4220	.3419	.2775	.2257	.1839	.1502	.1229
23	.7954	.6342	.5067	.4057	.3256	.2618	.2110	.1703	.1378	.1117
24	.7876	.6217	.4919	.3901	.3101	.2470	.1972	.1577	.1264	.1015
25	.7798	.6095	.4776	.3751	.2953	.2330	.1843	.1460	.1160	.0923

To see if the investment makes sense, the cash flows should be discounted and compared with the initial investment. This initial investment is shown in the next spreadsheet, as a negative earning at year 0.

Year	Cash Flows
0	−20
1	1.5
2	3
3	4.2
4	7
5	8.2
6	10

Now we can enter the discount factors. We could copy them from the table, but an easier way is to have the spreadsheet calculate them, using the property that the discount factor for every year after the first is the preceding year's divided by the one year factor. She uses a discount rate of 14 percent because that is what she would have earned using her alternative investment.

Year	Cash Flows	Dis. Factors
0	−20	1
1	1.5	0.87719298
2	3	0.76946753
3	4.2	0.67497152
4	7	0.59208028
5	8.2	0.51936866
6	10	0.45558655

Finally, we can multiply the cash flows by the discount factors to get the present values. The sum of the present values is called the "net present value." It is the financial value so to speak, of the investment—in this case negative. It will cost the manager nearly $600,000 over the five years to own this soccer team. Note that if she had not considered the cost of the money—that is, that the $20 million would have earned 14 percent a year invested in her business, the investment would have looked attractive because the undiscounted cash flows add to $13.9 million after the initial investment has been subtracted.

Year	Cash Flows	Dis. Factors	Present Val.
0	−20	1	−20
1	1.5	0.87719298	1.31578947
2	3	0.76946753	2.3084025
3	4.2	0.67497152	2.834880
4	7	0.59208028	4.14456
5	8.2	0.51936866	4.25882
6	10	0.45558655	4.5558655
	Net Present Value		−0.58168

A big reason this investment has a net negative cash flow is that the big earnings come four or five years out. The spreadsheet below shows the same total undiscounted cash flows spread out more evenly in years two through five. The deal looks profitable now.

Year	Cash Flows	Dis. Factors	Present Val.
0	−20	1	−20
1	1.5	0.87719298	1.31578947
2	5.6	0.76946753	4.30901816
3	5.6	0.67497152	3.77984049
4	5.6	0.59208028	3.31564955
5	5.6	0.51936866	2.90846452
6	10	0.45558655	4.55586548
	Net Present Value		0.18462767

Actually, nearly all spreadsheet programs, and many programmable calculators, will compute the net present value directly. One function in EXCEL that does this is called NPV. To use it on the cash flows below one enters the following. (B49–B54 are the locations of the cash flows.)

NPV(14%,B49 − B54).

This function only uses the positive cash flows, so the initial investment must be subtracted off. A more elaborate function which does the same thing and more, is PV.

1	1.5
2	3
3	4.2
4	7
5	8.2
6	10
NPV =	19.4183229
NPV − 20 =	−0.5816771
IRR =	0.13184164

The reason net present value is used is probably obvious. If the net present value is positive, then the investment brings in more cash than is spent, allowing for the money lost on interest because the returns came some time after the initial expenditure.

TIMING OF CASH FLOWS

The preceding calculations amount to assuming that cash is received or spent only once a year. Actually money flows monthly, weekly, or even daily. In principle, if one expected, for example, monthly payments, one could calculate a monthly discount rate by dividing the annual rate by 12 and then discount for each month. If one is using a spreadsheet and one can estimate on a monthly basis when money will flow in or out, these monthly calculations may make sense. In hand calculations, when payments are expected to occur regularly, as taxes and wages do, then one can use correction factors multiplying the payments so the result corresponds to payments discounted over periods shorter than a year. A table of these factors can be found in the Vatter et al. book's (1978).

Let's look at an example: For a 9 percent discount rate, annual amounts paid should be multiplied by 1.043 if the amount is actually disbursed weekly. At the same discount rate annual payments disbursed quarterly should be multiplied by 1.033. These correction factors make the effect of the payments larger, which makes sense—when some of the money flows out earlier in the year, as it would in quarterly or weekly payments, we lose interest on it. Similarly, if we receive money monthly during the year, rather than in a lump sum at the end, then the money is worth more to us because we can invest it as received. These correction factors tend to be small and are often ignored because the actual magnitudes of the cash flows are uncertain. On the other hand, a spreadsheet can often fairly easily be expanded to include the effects of payments at different time.

A related issue has to do with present value calculations using EXCEL or other commercial spreadsheets. Some of these make assumptions about whether payments occur at the beginning or end of the period or require the user to specify when the payments occur. The NPV function in EXCEL assumes end of period payments. The PV function allows the user to specify whether the payments come at the beginning or end.

Internal rate of return is another measure, besides net present value, for evaluating an investment decision. As the name implies, it is basically the interest rate that the project earns. It is useful in comparing investment alternatives—the one with the highest Internal Rate of

Return makes the most money for the company. It is also useful in deciding whether a project is worth doing—if the internal rate of return is less than the cost of the money invested, then the project does not make financial sense. We will return to these issues after we define Internal Rate of Return.

Internal Rate of Return (IRR) is the interest rate that makes the net present value equal to zero. In other words, it is the interest rate that makes the discounted cash flows received exactly equal to the cash invested. The EXCEL function calculated the IRR of the original soccer investment to be about 13 percent. This means that the sum of the discounted cashflows—1.5, 3, 4.2, 7, 8.2, 10—when discounted at about 13 percent sums to 20. A way of thinking about the IRR is to assume the original investment—$20 million in our case—was deposited in a bank. If we wanted to make annual withdrawals of $1.5 million, $3 million, $4.2 million, $7 million, $8.2 million, and $10 million and come out even at the end, the bank would have to be paying an interest rate of 13 percent, more precisely 13.184164 percent. IRRs are nearly always calculated by spreadsheets or programmable calculators. The alternative if one has a table of discount factors is to use trial and error—trying discount factors until the net present value is zero.

PAYBACK PERIOD

Besides npv and IRR, a third measure of the worth of an investment exists. This measure is called payback period and is the length of time in years or months until the initial investment is repaid. A social service agency that is deciding whether to buy a high-speed copier can serve as an example. The copier would cost $12,000. It would save $800 a month. The payback period is 15 months because at the end of 15 months the total savings are $12,000, the original cost. If the monthly savings are constant, as in this example, then the payback period is found by dividing the initial investment by the monthly cash inputs. Otherwise the payback can be calculated by subtracting monthly or yearly payments from the original investment until that investment has been paid back.

Managers find the payback period useful. For one thing, it gives an estimate of how long the organization is at risk. An investment in a factory in another country is at risk of being a loss if the local government closes the factory before the payback period is over. If a company buys a machine to make baseball caps that has a payback period of three years, then it needs to worry if the market for baseball caps will continue for at least three years. A person paying for an automobile through monthly payments wants to make sure the automobile lasts at least until the last payment is made.

American managers are often criticized for expecting too short a payback period, not undertaking projects with payback periods greater than two or three years. It is said that Japanese managers are willing to be more patient and accept longer payback periods if the net present value of the project is large. A reason for accepting these longer paybacks is assumed to be that Japanese investors evaluate companies on their long-term prospects, rather than each quarters' results.

SUMMARY

Whenever the cash flows relevant to a decision come at different times, which is nearly always, then the time value of money must be considered. The present value of a future payment is found by multiplying that payment by a discount factor. Discount factors can be calculated from formulas, looked up in tables, or obtained from spreadsheets. The net present value of a project is the algebraic sum of the cash inflows—often an initial investment—and the outflows—often the cash earned, each discounted appropriately. Internal Rate of Return is analogous to the interest rate that the project investment earns. Payback period is the length of time it takes to recover the original investment.

REFERENCES

Behn Robert D., and James W. Vaupel. (1982). *Quick analysis for Busy Decision Makers.* New York: Basic Books, Inc.

Raiffa, Howard. (1982). *The Art and Science of Negotiation.* Cambridge, MA: Belknap Press, Harvard University.

Vatter, Paul, Stephen P. Bradley, Sherwood C. Frey, Jr., and Barbara B. Jackson. (1978). *Quantitative Methods in Management: Text and Cases.* Homewood, IL: Irwin Publishing Co.

CAPITAL RECOVERY FACTOR

Assume we wanted to compare the costs of two projects; one with a high initial cost but low annual cost, the other with a low initial cost but a high annual cost. An example would be heating a workroom. One alternative would be passive solar: big windows and heat storage. This would cost $1000 to build, but annual costs would be only $50 for cleaning and caulking. The other alternative is a kerosene heater. The initial cost is lower, $100, but the annual cost, mostly kerosene, is much higher, $250. Which of these is chaper if the system must last for 10 years? The problem is made difficult because one probably would borrow the $1000 to build the solar system and must pay back interest as well as the original amount. Without interest charges the passive solar system would cost $1500 (1000 + 50 × 10) and the other $2600 (100 + 250 × 10).

A convenient and valid way to make the comparison uses the *Capital Recovery Factor.* The capital recovery factor gives the amount of money that must be repaid each year if the initial amount is borrowed. Of course, the factor depends on the duration of the loan and the interest rate. The values of the capital recovery factor are shown in Table 6. A complicated formula can be used to find these facors but the table is easier. Most spreadsheet programs give these factors. The table shows, in the upper left hand corner, that if we borrowed $100 for 3 years at 5% interest we would have to pay back $36.70 each year. On the other hand, the lower right hand corner shows that a $200 loan at 20% for 20 years requires annual payments of $41.00.

TABLE 6
Capital recovery factors

Years	Interest Rate						
	5%	8%	10%	12%	14%	15%	20%
3	0.367	0.388	0.402	0.416	0.431	0.438	0.475
5	0.231	0.231	0.264	0.277	0.291	0.298	0.334
7	0.173	0.192	0.205	0.219	0.233	0.240	0.277
9	0.141	0.160	0.174	0.188	0.202	0.210	0.250
10	0.130	0.149	0.163	0.177	0.192	0.199	0.239
15	0.096	0.117	0.131	0.147	0.163	0.171	0.214
20	0.080	0.102	0.117	0.134	0.151	0.160	0.205

To get this annual payment, we multiply the value of the loan by the capital recovery factor, as shown next.

$$200 \cdot 0.205 = \$41.00$$

Banks use similar table to compute mortgage payments for homes or car payments, but theirs have monthly payments.

How does one know which interest rate or which number of years to use in an actual situation? If one really would borrow the money from a bank, then the rate to use, of course, is what the bank charges. If one takes the money out of a savings account, then the interest one would have earned should be used in the calculations. Even if one does not know where the money would come from it is reasonable to use the interest rate charged by a local bank. As far as deciding the number of years to use, one should use the actual number of years if a real loan is taken. If a real loan is not taken then the life of the equipment is a good choice as one often wants to be paying for equipment for about as long as it is useful. (Of course, estimating the life of equipment may not be easy either.)

Now let's compare the two ways mentioned earlir of heating the workroom. Let us assume the relevant interest rate is 12 percent. Let us also assume both our solar system and the kerosene heaters last for 10 years. The capital recovery factor is .177. (See the fifth row and fourth column of Table 6). The annual cost of the solar system has two parts. The first, sometimes called the capital amortization cost, equals the initial cost, $1,000, times the capital recovery factor. The second is the annual maintenance costs, $50. The total annual cost is $227.00. The money spent paying back the large initial expense is the greater part of this annual cost.

Annual Payments	$1000 \times .177 =$	$177.00
Maintenance		50.00
Total annual costs		$227.00

The annual costs for the kerosene heater are calculated next.

Annual Payments	$100 \times .177 =$	$17.70
Fuel & Maintenance		250.00
Total Annual Costs		$267.70

In this case, the solar system is cheaper, although the annual costs are close.

To summarize, then, one compares projects with different initial costs and different annual costs by computing a total annual cost. Part of the annual cost is found from the initial cost by using the capital recovery factor. This part of the annual cost is the principal and interest payments. The rest of the annual cost is out-of-pocket expenses, usually supplies and maintenance.

This financial analysis can be made more accurate in several ways, such as accounting for salvage value. In our example, the salvage values are the worth of the solar system after 10 year and the worth of the 10-year-old kerosene heater. Another refinement is accounting for taxes, which are different for capital improvements like solar greenhouses than for operating costs like fuel. a third refinement is trying to account for inflation—predicting costs and salvage values in the future.

MICROFINANCE

Jill Graat

WHAT IS MICROFINANCE?

Microfinance is a financial operation that provides small loans to struggling businesspeople in order to expand their small enterprises. Informal money lending technologies, usually administered by small scale money lenders, are instituted to give loans that will be quick, convenient, and flexible, required to accommodate their clientele. It is an emerging industry throughout the world. These types of financial institutions were initially created for Third World application but are slowly making their way into the developed world. It is the mission of Micro Financial Institutions (MFI) to alleviate poverty by targeting the poorest sector of the population, the bottom 20 percent, and women especially. A belief in this mission by the staff of a MFI is necessary in order to motivate their clients and inspire the development of their product.

MFIs provide various financial services similar to a commercial bank, but due to the nature of their clientele's portfolio, loan procedures have been adjusted. Profit is not the goal for this type of organization; sustainability is. Self-financing will enable a MFI to expand its operation at a sustainable rate and also reach all those who are in need. However, attaining a status of self-sufficiency won't occur until well into the future until the institution develops into a stable and effective financial resource for the poor. All individuals associated with the MFI must understand that success is attained only when they have reached out to the community and provided its members with the tools they need to improve their lives.

IMPACT ON THE COMMUNITY

- Improvement of livelihood through increased income
- Increased confidence and dignity as a result of loan approval and taking on the responsibility of repaying
- Economic independence
- Empowerment of women within the community and at home
- Employment options, not only for clients but also family and friends
- Increased family welfare—providing options for education for children, increased health
- Increased community access to technologies and market information and resources—benefiting both clients and nonclients

The model used by a MFI is dependent on the need of the clientele. Choosing the correct model seems to be essential for creating a sustainable MFI. Extensive on-site research within a village by the MFI will allow the institution to see the needs of that village and concentrate on them. The research can be most effectively done through the development of a positive relationship with leaders within the community, along with clients. Such relations can lead to gaining information about local market, such as products, services, and the current interest rate charged by other informal financial services.

ORGANIZATION OF THE INSTITUTIONAL STRUCTURE

The initial set up of a MFI infrastructure has various aspects essential for creating a stable operation. These include the type of methodology appropriate for the target community, training procedures, and expectations of the staff.

METHODOLOGY

The type of model used by a MFI can range from individual loans to lending to groups. There are also credit and savings banks that have been established by an MFI within a needy community. The type of model used in a certain area will vary because of the need associated with that specific region. One must recognize that every village is different and that these models are not like cookie cutters. Extensive research on a community must be performed before deciding the type of model that can be implemented in that area. As the program matures it is not uncommon for the MFI to make adjustments in order to fit the need of the clientele. Examples of four common models that MFI can follow in order to get a program set up and running are solidarity group model, Village Banking, Grameen Bank, and the individual lending model.

SOLIDARITY GROUP

The solidarity group model, or peer lending group, is four to five individuals who have come together to take out a loan. Group members are self-selected based on their reputation and the relationship they have with one another. The processes of the group screening and the group pressures imposed upon each member to not default secure loan recovery.

Once a loan is granted, it is the responsibility of the entire group to ensure that all members' payments are on time. If not, the entire group will suffer the consequences. Once the group has repaid their loan and has abided by the established guidelines, they have the opportunity to graduate to a larger loan, if desired.

What drives this model is that a member's reputation can get her into the group, but she must live up to the expectations that have been agreed upon. If she fails, the group will reprimand her, and the MFI and her reputation within the community will be damaged. The members of the group are responsible for the initial formation of the group, for all the administration and organizing of the payment schedule, and for scheduling both the group meetings and the meetings between a borrower and her assigned field officer.

VILLAGE BANKING

Village Banking is a community-based credit and savings association run completely by village members. It is a group of 25 to 50 low-income individuals who join together and take out one large loan. Once the loan is obtained from the MFI, the group selects an administrative committee. The committee decides who will be granted a loan and of what size. The MFI will provide limited administrative assistance, and Village Bank will actually take on some of the operating costs. Village Bank accepts deposits and makes loans from those funds. The loan procedure is the same as for the other models in that the initial loan taken out by the bank will have to be paid back before the group is eligible for another loan. The payback period ranges from 4 to 12 months. What makes this method different is that a savings program is established, which can be used to provide loans for nonmembers, at the discretion of the committee, as well as members.

GRAMEEN BANK

The Grameen Bank model works as follows: A bank is set up in a village with a field officer and some qualified bank workers. It will support 15 to 20 villages—the field officer will already have researched these villages as prospective clientele. Groups of five people are created. Two of the members are eligible to receive a loan. The two members are monitored for a one-month period, and the credibility of the group will then be based on the repayment performance of the two individuals. If the loans are paid back within the 50-week period, then the other members now qualify to receive a loan.

The Grameen Bank model is known globally as the "grassroots" of microfinance models. It originated in Bangladesh by Professor Mohammed Yunus during the early 1970s. It has provided more than $2.1 billion in loans to approximately 2 million people—94 percent have been women. Many of the various models used throughout the world are extensions of the Grameen Bank model. What these other banks have done is taken the basic principles that have made Grameen Bank a success and adjusted them in order to fit the need of their clients.

INDIVIDUAL MODEL

The individual model is a more labor-intensive type of loan program for the MFI. It requires extensive field research on the client's background and high levels of direct monitoring of the client's progress—thus increasing the interaction between the MFI and the client. Personal background research on the clients will come from sources such as referrals by family, friends, and leaders within the community. A cash flow analysis of the household may also be performed, which will provide a guide as to what size of loan is appropriate for this particular client. Unlike the other models, the loan is given directly to the borrower. It is now the sole responsibility of that individual to pay it back. To help secure loan recovery, the client will be assigned to a MFI officer who will schedule frequent meetings to discuss progress and any problems and to answer any questions that may arise. Training in business management skills may be offered with individual loans.

Here are some guidelines for starting a MFI.

ESTABLISHING THE ORGANIZATION

- The lines of authority, job description, institutional rules, mission statement and vision must be defined. It is crucial that entire staff be on the same page with regard to client procedures—understanding and conforming to the fundamental principles of the MFI.
- Staff size should remain small—no more than ten people, depending on the size of clientele. A high officer to client ratio should be sought in order to create a less intimidating and more personable atmosphere for the client.
- Each employee should be skilled in various financial services and should be familiar with administrative practices to lower operating costs.
- A standard manual for client procedures should be developed. It will be used as a reference source for the staff in order to handle any client situations.

- A flexible decision-making structure for the officer is needed, in order to approve or devise the client's loan schedules. The recovery success of the loan will depend on the officer's ability to adapt the loan process to the need of the client.
- Employees should be required to attain a deep knowledge of the culture and customs of the community they will be serving. It is imperative that a field officer does not offend or insult a perspective client with incorrect behavior. Offensive behavior, even if unintended, could tarnish the MFI's reputation—one's reputation is worth more than any amount of cash.

CHOOSING THE CLIENTELE

Choosing clients most likely to repay can be a very difficult task, since many prospective borrowers do not have any form of written financial records or collateral to exchange for a loan. It is key that the officers develop internal village connections and positive relations within the community before interviewing clients can begin. Knowing the community will make it easier to recognize risky clients, gain local market information for products and services, and learn financial information such as prevailing interest rates and the inflation rate. Good information will help reduce unsuccessful loans and decrease loan approval and delivery periods.

LOAN PROCESS

The approval process for a loan must be simple and quick.

- Interview a large number of applicants, 30 to 50, with a variety of products or services.
- To be considered as a client, an applicant must have their product or service already in the local market and it must have expansion capabilities. Innovation is key to having a successful product/service.
- The interview should be informal, with emphasis on the business plan rather than the client's financial history.
- Loan contracts should be finalized informally, either with a handshake or a spoken agreement because many clients are not literate and such formalities as a signature on a piece of paper is not relevant or common within the business experience of the people being dealt with.
- Alternative modes of repayment may be used as loan security, such as collateral (wheel barrow, household tool, etc.). The loan officer must understand that loan eligibility is based not on financial history but on the business plan and the client's character.

LOAN DELIVERY

- Initial loans should be very small sums. Many are approximately $50. The value should depend on the client's situation and the dollar market value.
- The length of the loan repayment period must be determined. Usually the length is no longer than a year.

- The client(s) should be on a weekly or biweekly payment cycle for one year. Frequent payments will decrease the amount that will be due every week, making it easier for the client not to default or miss a payment.
- The interest rate must be assigned. This is described below.
- A month-long grace period before the first repayment is due should be provided.
- Consequences should be immediate if payments are not made as promised—the loan is always due to the assigned field officer on the scheduled day.
- Cash for the loan should be in client's possession within a few days or so—the quicker the better. Many of these clients are used to going to moneylenders who are able to provide the cash very quickly, so their expectations will be the same for a MFI.
- When loan is fully recovered and if there were no problems during the repayment period, the client will now graduate to receive a larger loan if desired.

LOAN RECOVERY

- Each group should assign one member to be the treasurer. It is her responsibility to collect the payments and bring them to the field officer on time. The field officer will then record the payments and report back to the MFI.
- If a loan payment is not made on time, the credit group must take action immediately, reporting it to their field officer. This should be done the following day after the missed payment was due. Any delinquency on loan repayments that is allowed to occur can result in further defaults on loans.
- In case of a missed payment, the field officer should visit the defaulter as soon as possible and discuss why the loan is late and what arrangements can be made. The field officer must remember she is representing the MFI, and thus her actions will be surveyed not only by the client in question but by all within the area. The officer must show respect to the client at all times regardless of the situation.

SUGGESTIONS TO AVOID DEFAULTS

- Match the loan cycles to the economic activities, such as extreme seasonal weather conditions or fluctuating market cycles that are affecting the community.
- Sustain close and strong links between the field officer and the clients.
- Address the borrower's delinquency through a community meeting. The purpose of this meeting is to expose and confront the defaulted. There he/she will be singled out and asked to state why the loan has not been paid.
- If appropriate, increase the number of MFI employees. Such an increase increases MFI's visibility around the village and has been proven to decrease the number of defaults on loan.

INTEREST RATE

The small, short-term, convenient loans provided by MFIs do not come at a small cost to the institution, so the client must pick up the slack. There will be high transaction costs associated with these types of loans. In order to counter these costs, MFI tend to charge relatively high interest rates on their loans.

- Take into consideration the inflation rate affecting the area. If the interest rate assigned to each loan is lower than the inflation rate, money will be lost by the MFI. This is because the inflation rate decreases the value of the loan throughout the year that it is in recovery—for example, if $100 is loaned out when the average inflation rate is 12 percent, that $100 will be losing a $1 a month.

- Set the interest rate equal to that of the inflation rate and then add the necessary percentage to cover operating costs such as administration, staff, defaults, and borrowing. It is common for a MFI to charge 2 to 4 percent above the inflation rate or the prime rate.

- The interest rate charged is usually higher than commercial banks but still lower than that of the moneylenders. That is why it is more appealing for low-income individuals to come to MFI instead of these other options.

- One problem that arises with matching interest rates with inflation rate is that over the course of a year the inflation rate is very unstable. Therefore, the matching interest rate will fluctuate. This can be very complicating and confusing for the client and the MFI, since different amounts of money will be due each time. One idea is to institute a flat rate interest rate plan. This is simple for the client to follow, since there will be one flat rate assigned at the start of the loan cycle that will not change regardless of the inflation rate. A flat rate is also efficient for accounting purposes, since payments will be a consistent amount each time.

- It is possible for a MFI to charge lower interest rates than moneylenders; however, they must counter it with a large number of clients. Since the money collected from the interest of each individual loan is small, having a large clientele makes an appreciable return possible. However, too many clients may cause procedures to become too informal and the loss of the social connection can result in the loss of clients.

ACCOUNTING PRACTICES

Accounting records should be a simple records system in which the various transactions are grouped together by category: outflows or inflows. The results for a specific period—a month or year, for example—should be summarized. Once the accounting system has been set up properly by an accountant, the MFI should hire a bookkeeper to further handle the books. This bookkeeper is then responsible for the daily recording of any financial transactions, both internal and external. These transactions include payment of salaries, loan deposits, withdrawals, invoices, and so on.

Basic accounting statements can be used. Documenting the financial activities enables the MFI to see how their finances are doing and where adjustments may need to take place. Financial records also provide a platform to showcase the activities of the institution to potential investors, clients, donors, or other MFIs. The accounting system should be kept simple and efficient, which is not difficult since the loans being granted are very small and are short term.

The following accounting documents are basic and should be used by any size organization.

Balance Sheet

- Assets, which include what the MFI own.
    —Liquid assets—cash, bank accounts, loans receivable, loans in arrears
    —Fixed assets—furniture, office equipment

* Liabilities, which include what the MFI owe
    —Accounts payable, interest payable, loans payable, reserve for loan loss
* Equity, which is the amount of equity invested or withdrawn directly by the owners
    —Contributed funds, retained surplus, donated equity

Income Statement

* Revenues, which include the inflows of cash during operation
    —Operating fees, interest earned
* Expenses, which include the outflows of cash to initiate operation and other items
    —Rent, utilities, salaries, office supplies, vehicles, depreciation, bad debts

FUNDING

The finances used during the early stages of MFI development will rely heavily on the generosity of donors. Therefore, the process of soliciting donations must already be well underway when an MFI begins operations and will be continued until self-sufficiency is attained. There are various fund-raising avenues that may be pursued, depending on the amount of money needed at the time. Fundraising events such as art shows and concerts are excellent for quick generation of small amounts of cash. These are also an inexpensive way to promote awareness of the MFI and may attract corporation sponsorship. Direct marketing strategies such as phone or mail solicitation are another option but have high costs due to the need for professional help. Direct marketing schemes also require the MFI to have a history or name recognition.

Large donations usually come from wealthy individuals or corporations. Donating to humanitarian causes such as microfinance are attractive to individuals or corporations for various reasons such as a tax break, passion for the cause, a way to improve a public image, or family relevance. The key when soliciting a donor is to identify what that donor wants, including how active or inactive they want to be with the MFI. An MFI, within limits, should attempt to cater to donors' requests. Allowing a donor to become directly involved in the operation enables the development of an intimate relationship between the MFI and the donor. This can strengthen the tie between the two parties, securing the potential for another donation.

As the MFI matures, it should refrain from becoming too dependent on donations and subsidies as a means of continuing operations. Sustainability is one of the goals of a MFI. Continued reliance on outside funding implies that the microfinance industry is not self-sustainable. In any case, eventually donors will become frustrated by repeated requests for donations.

A MFI should be able to finance its activities from generated income. Reducing dependence on donations may require raising interest rates and improving internal efficiency. In order to establish operating efficiency, the MFI may have to institute various cost-reducing measures such training of local people as employees and making adjustments to the program where necessary. An example of this might be to experiment with a different methodology at a smaller scale.

A MFI should not attempt to expand its size or capacity too soon or self-sufficiency will most likely not be attained. Many cases have shown that a MFI will not attain self-sufficiency for 5 to 10 years. This is due to the high cost associated with the initial setup, requiring that during its first few years in operation the MFI slowly work its way out of debt. Established MFI that have reached independence from donations, such as the Grameen Bank and Banco del Sol, have raised money from financial markets. This is risky for immature and struggling MFIs, since they have very little extra cash that they can afford to tie up in the markets.

INTEGRATION OF AN MFI INTO THE COMMUNITY

- The organizers of an MFI must make sure that government leaders are willing to validate and support the institution's actions. Without their approval, regulations that could inhibit the intended development of the MFI program might be enforced—banking regulations often are established to control big banks and do not promote small-scale financial institutions. In order to produce a viable financial service within the community, MFIs must be able to carry out specific procedures within the target area, such as extensive research within the community about local client and market information. Imposed restrictions that inhibit this process will prevent the MFI from providing the necessary services needed for that area.

- The organizers of an MFI must also develop a positive relationship with prominent figures within the community. One strategy is to train such figures to act as or with a field officer. The prominent figures would be instructed about the institutional procedures for handling such issues as visits to client homes and default situations. Involvement of respected village figures will provide added pressures on borrowers from a familiar authority to abide by the rules. Training local leaders creates a social connection by the community toward the MFI, and it also increases the number of field officers at a minimal cost to the institution.

- The local government should be approached to be more flexible with local market regulations. Request should be made to have the local market opened up in order to allow for the clients' products to reach domestic and global markets.

- It is important that officials from the MFI go to the governing figures of that area and provide an in-depth description about the positive impact an MFI would have on their community. Many people in a community will have limited or no knowledge of what an MFI is or does and may be very apprehensive towards the financial institutions presently locate there. The officials should ask for permission to survey the area to see if microfinance is appropriate.

SUMMARY

In summary, when setting up a Micro Financial Institution in a community, there are some keys to remember.

- Have the mission statement established, and make sure that the staff is fully committed to carrying it out in the community.
- Target the community that is in need of microfinance activities.
- Receive permission by local authorities to engage in on-site research. This research will dictate the appropriate methodology to be used, reveal important cultural information about possible clients, and give information about the local market structure.
- Educate staff on the chosen methodology. Provide a standard instruction manual to handle client procedures. Inform staff of the results from the research on the customs/culture and local market of the community.
- Develop an accounting system that has a simple format. Hire a bookkeeper to maintain daily transactions.

- Have a substantial involvement of donors in order to sufficiently sustain the institution during its early development.
- Begin loan approval interviews with perspective clients.
- Assign a field officer to chosen client(s) or groups. Proceed with loan delivery and recovery processes.

PLANNING

The planning discussed here is geared toward determining the overall direction of the organization. In practice, this often results in a choice of markets to be served—people with AIDS, arid region farmers—or products to be offered—small loans to establish new businesses, solar water heaters. The planning might also result in a decision to strengthen an aspect of the organization—expertise sufficient to purify water of any quality to be encountered in the region. Planning is useful because it increases the chance that the outcome of a project will be the desired outcome. Thinking about what you want and where you are beforehand will help get you there. Planning is also useful because it reduces wasted effort—a plan specifies a particular set of steps leading to the goal as directly as possible. The process of putting the plan together can help an organization clarify its mission, improving its effectiveness when it is implementing the plan.

The discussion below draws heavily on Business Planning and Financial Modeling for Microfinance Institutions, A Handbook, Tony Sheldon and Charles Waterfield, Consultative Group to Assist the Poorest, World Bank, Washington DC 20433 USA, 1998. The first article in this chapter also discusses planning.

REASONS FOR PLANNING

A reason for an organization to plan is to decide the most effective way it can achieve its goals. An often quoted remark is "Success comes from doing the right thing, not from doing things right." Planning is selecting the right thing to do. Another part of planning is deciding how to implement the plan, determining the set of specific steps to be done. Planning improves the efficiency of the project once it is implemented. Another important reason for preparing a defensible plan is to gain credibility with donors and other funding groups. A benefit of preparing a plan is that the process builds cohesion within the group—when the plan is being prepared, members of the group have an opportunity to reflect on the group's mission and to influence the strategy. A related benefit is that an agreed upon plan focuses the efforts of everybody in the group on the same objectives and actions.

PROCESS OF STRATEGIC PLANNING

Strategic planning begins with study of three issues: the organization's objectives, its internal strengths and weaknesses, and its external situation, including needs, clients, competition, and government regulations. Possible things the organization can do or directions it might pursue can now be developed, consistent with the study just completed. The next step is evaluating and comparing these alternative directions or steps so as to select the most promising. From the chosen strategy then operational plans—detailed plans for implementing the strategy—are designed.

To outline the planning process, we use a fictitious microfinance organization.

Objectives

An organization's objectives can be broken down into missions and goals. Missions specify the problems the organization addresses. Goals are the specific things the organization intends to accomplish. The mission of a microfinance organization might be to give small loans in order to assist poor people in a particular community wishing to start businesses. A corresponding goal is to be able to give 75 loans within the first year and to increase this number to 230 within three years. The mission should specify the organization's intended clients—members of a particular community in our case. A full mission statement would state the organization's values—in this case, perhaps, to enhance clients' self-determination and increase the availability of goods and services in the community. Missions specify the issues the organization is addressing and the organization's core values—what it wants to do and how. Goals are specific: to assist the members of the organization in realizing what actions have to be taken when and to allow members to evaluate progress.

Time spent thinking about missions and goals is worthwhile. Without an agreed sense of mission members of an organization may attempt to take the organization into too many different areas—for example, a microfinance organization might try to offer literacy classes in its community. The risk when an organization tries too many things is that the original purpose does not get done. Almost always a community will have many needs, but an organization is more likely to improve conditions if it focuses its efforts. Of course, the members of an organization can agree to change the mission—perhaps literacy is needed more than small loans—but the choice should be made deliberately. Goals should be realistic but challenging, so they should be set carefully. Not attaining a goal after working hard will discourage members of the organization, but people tend to respond to challenging goals by working harder. Vague goals such as "Next year we want to help as many people as possible" are less challenging because an argument can be made that any sort of performance did satisfy the goal. Of course, a risk exists that focus on a specific goal will divert attention from the organization's real mission. Satisfying the goal of 98 percent loan repayment may mean that loans are not being given for risky projects that could benefit the community more than those that have an assured payback.

Internal Strengths and Weaknesses

Knowledge of an organization's internal strengths and weaknesses—its capacity—is essential when deciding what an organization should do. Considering internal strengths and weaknesses is sometimes referred to as doing a "resource assessment" or an "institutional assessment." Knowledge of where an organization is weak is a guide to what it should do to strengthen itself. In looking at internal strengths and weaknesses a microfinance organization might look at the capabilities of its loan officers: how well they have performed, their experiences, their training. The organization might also look at its human resource management—how well they recruit and train. Is staff turnover high? Is an effective incentive system in place? Equipment needs should be considered: Are appropriate computers available? Are the offices conducive to serious discussions with clients? Financial capabilities are important: Does the microfinance organization have access to sufficient funds to meet its goals? Does it have a suitable mix of funding sources, without too much dependence on external grants? Internal strengths and weaknesses can often be broken down, as was just done, into questions about people, about facilities, and about finances.

External Threats and Opportunities

Usually the external factors most important when planning include client needs, competing organizations, and government regulations. In assessing client needs, one looks at how many possible clients have a particular need. One also looks at trends over time. Is it likely the number of people with a particular need will increase? A microfinance organization would look at how many entrepreneurs operate in the area the organization serves, whether this number will increase or decrease in the future, and how many of these entrepreneurs are likely to become clients. The organization might also look at trends in the businesses these entrepreneurs do. Competition needs to be studied, both from other microfinance institutions and from other lenders—for example, informal credit schemes and the clients' suppliers. Government policies need to be examined closely because they can destroy a business or create an opportunity for a successful enterprise. A low interest rate ceiling may prevent a microfinance organization from recovering its full costs. Technological change—such as the wide availability of the Internet—may create significant opportunities and needs to be monitored regularly.

The first part of strategic planning then consists of examining an organizations objectives and then looking at its internal strengths and weaknesses and its external threats and opportunities. The acronym SWOT is sometimes used to describe the process of looking at Strengths, Weaknesses, Opportunities, and Threats.

Choice of Strategy

With objectives in mind and a SWOT analysis completed, the organization can choose a strategy. New strategies often consist of new markets or new products. If a microfinance bank notes that its clients are using another institution for foreign exchange transactions, it might decide to establish its own forex department. If the SWOT analysis uncovered an ethnic group in its geographical area unserved by any microfinance organization, it might consider ways of reaching that group. Some strategic possibilities are internally directed. Offering a new product may require that the organization gain new expertise. The strategy in this case can focus on the expertise: "to be the most knowledgeable microbank in the agricultural area." A valid strategy can be simply to reduce costs by instituting internal efficiencies.

In practice, often a single strategy emerges clearly from the analysis. When it does not, several alternative strategies can be formulated and one chosen. The founder of Amazon.com is supposed to have considered over 20 different products before deciding to sell books. Developing alternative strategies requires creativity. It also requires understanding of the important characteristics of an enterprise—realizing what clients/users will support. The choice of which alternative to implement is made basically on a benefit/cost basis. In the armaments business the question is which alternative will deliver the "most bang for the buck." Does a microfinance organization help the poor more by opening another office in a different part of town or by offering classes in accounting for beginning entrepreneurs? In selecting an alternative one balances the possibilities of external opportunities with the likelihood that the organization has the internal capabilities to exploit these opportunities. In making the selection one balances numerical estimates of revenues and costs with judgments about users' choices, unforeseen expenses, possible future changes in technology, and other risks. A microfinance organization would be significantly affected by all the uncertainties just listed and also by governmental policy changes.

OPERATIONAL PLANS

A strategic plan must be translated into operational plans. What specific actions must be taken? What people are needed to do these things? When should they be done? In the case of a microfinance organization, a schedule and a budget must be developed for every product and branch office. When will the product be offered? What must be in place before the product can be offered? When will the branch be opened? What has to be done to open the branch?

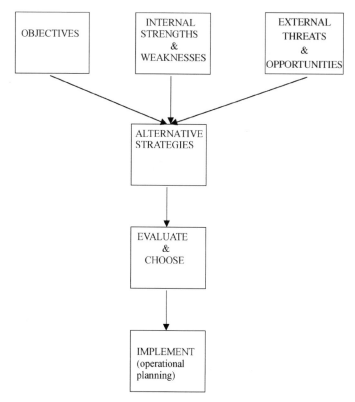

FIGURE 15
Strategic planning.

What people must be hired or assigned? What facilities must be opened or sourced? What legal permits are necessary? How much money will be required to start? When is the money needed?

Specific plans are also needed for day-to-day operations. These plans may have to be modified once operations start, but unless they are made carefully beforehand it is possible that essential steps will be overlooked. One thing that must be set up initially is a way of knowing if operations are proceeding as expected—a so-called control system. Are costs and revenues on budget? For a microfinance organization the outstanding balance of overdue loans might be calculated each week, as well as the number of loans made and total amount of money disbursed. A manufacturing project might monitor number of items produced, number sold, and labor costs each week. The planning process starts with the formulation of overall objectives and results in a way of measuring if specific objectives were reached.

The chart in Figure 15 summarizes the planning process.

PROJECT SELECTION

Project selection starts by looking at community needs—doing a needs assessment. A needs assessment will nearly always uncover many candidate projects: How should an organization decide which needs to address? The organization must also look at its and the community's capabilities—do a resource assessment. Project selection is done by going back and forth between what is needed and what is feasible for the organization. Success comes to an organization by picking its shots: Is a project feasible? Will doing the project make an impact? Can the organization do the project better than other organizations? Is a comparative advantage possible? Can the project contribute to the organization's skills and reputation? Risk and reward must be considered—nearly always the projects that can offer major benefits are risky. Local people, of course, have special expertise and must be involved in selecting projects for implementation.

NEEDS ASSESSMENT

Selecting a project that addresses the needs of a community involves defining the need and involving community members in the process. The success of a project depends on an accurate needs assessment that involves the participation of many different groups in the community. Conducting a needs assessment is not only a way to identify what is missing in the community; it is a source of community buy-in as well. An accurate picture of a community comes from soliciting opinions in formal interviews and collecting the information that emerges in casual conversation. Each setting provides unique opportunities for community commentary. Consequently, neither source of information is complete without the other. Surveying the entire community may not capture their attitude toward the project nor does it provide a relaxed opportunity during which to make suggestions. On the other hand, judging a community's need solely from informally conversations may exclude critical groups like government officials and may not be based on a representative sampling. This section discusses how to balance both methods as well as how to incorporate participation in each step.

FORMAL NEEDS ASSESSMENT

The goal of a needs assessment is to gather an accurate picture of the resources available and unavailable to a community by soliciting information from a representative sample of the population. In assessing the needs of a community, it is important to balance the breadth and depth of the information gathered. This goal requires talking to as many randomly selected people as possible and interviewing each person as extensively as possible. In order to balance breadth and depth, formal needs assessments often take the form of short surveys of many people and long focus groups with selected community members.

EXPANDING THE BREADTH OF INFORMATION: SELECTING A REPRESENTATIVE SAMPLE

By selecting a representative sample of the community, one safeguards against soliciting opinions from one group—those people involved in neighborhood organizations, for example—while overlooking the agenda and interests of another group-subsistence farmers located farther from town, for example.

The less time you have to devote to conducting surveys, the more critical it is that the sample is representative. For example, by talking to every other name in the phone book, you will probably talk to representatives from every group in the community, since you will have spoken to a large portion of the population. However, talking to every other person is time consuming, especially if the phone book is for Manhattan, so talking to every fiftieth name may be more plausible. The drawback of decreasing the pool of people you contact is that there are more opportunities to miss a certain perspective of the issue at hand. To ensure the selection of a representative sample, one must be careful to talk to as many different people as possible. For example, talk to people in different areas of the market buying items of all kinds and price so that the people interviewed represent a cross section of income levels, occupations, and interests. More simply, talk to people leaving a church at different doors or at five-minute increments so that one family or social group does not dominate your results.

Selecting a representative sample is tied closely to the method of solicitation. Most obviously, in the developing world, using a phone book to select names would omit a significant portion of the population, while only representing the opinions and status of people above a certain income level. Door-to-door canvassing, or selecting people randomly at a street corner, in a market, or at a soccer game may be a better way to find a representative sample. However, each of these methods of selection has the potential to skew the result. Sampling from one neighborhood, surveying market-goers instead of farmers or churchgoers instead of a secular community, can create a distorted picture of a community's need. There may not be an ideal method to select a sample, but by identifying how the mode of selection may skew your sample, you can try to counterbalance by canvassing in several neighborhoods or by visiting churches and schools.

ADDING DEPTH TO THE INFORMATION: CONDUCTING FOCUS GROUPS

One of the drawbacks to surveys is that they do not offer detailed information about a community. Because you are stopping people on the street or visiting their homes unannounced, the survey must be short so as to impose as little as possible. In order to avoid sacrificing an understanding of the complexity of the situation or its history, it is helpful to bring people together for a more probing discussion about a need.

Focus groups often build on the findings of the surveys. Once a need is identified through the surveys, or trends emerge from the responses, focus groups provide a forum to follow-up on these trends. Focus groups should have a facilitator that guides the discussion around the topic of nutrition and food availability, and someone should record the discussion. Again, focus groups should be conducted with as many different groups as possible, representing a wide range of interests. These groups should be divided in a way that is appropriate for the issue. For example, on the topic of nutrition, you may want to talk to a group of farmers and a group of market-goers, or you may want to structure your groups around ethnicity or income level. The combination of in-depth focus groups with a few strategically selected segments of the population balanced with general surveys of a larger portion of the community will allow you to get a sense of the overall need as well as some of the intricacies surrounding it.

INFORMAL NEEDS ASSESSMENT

Surveys and focus groups can reveal a substantial amount of information about a community's resources, but surveys and focus groups cannot always capture the attitudes and opinions of community members about a project or a need. The formality of interviews and the fact that the assessor is an outsider can lead to reticence in community members. Both of these factors are inherent to the situation and therefore hard to overcome. One barrier to performing an accurate assessment is that community members in developing countries may tell outsiders what they think the outsider wants to hear. People are not necessarily maliciously lying; they simply want to please the outsider. As a result, informal conversations can often lead to discovering information or discovering the "real situation" that would never surface through formal assessment. As opposed to direct questioning, in informal conversations, community members are not put in a situation in which they feel the need to please the outsider.

More fundamentally, community members will undoubtedly have strong opinions and stereotypes about Westerners. Whether or not you fit the stereotype, it will be imposed upon you by community members. Stereotypically, Westerners are patronizing, dominating, and often destructive. Simultaneously, there is often an obsession and admiration with the West that makes for a tricky love-hate relationship. An outsider cannot escape playing a role in this relationship regardless of whether or not the outsider is aware of this dynamic. In any situation it is critical to consider how you may be perceived separately from your intentions, but when the subtext of the relationship is extremely polarized, as it is in developing countries, it is essential to analyze how you are perceived. By addressing this perception through personal relationships with community members, it is possible to move through the stereotypes so that the community's perceptions mirror your intentions.

RESOURCE ASSESSMENT

Resources are the capabilities of the organization—the internal resources—and of the community—the external resources. In general resources are people, technology, and money. Money is discussed in the "Financial Assessment" section. People resources basically refer to what the people in the organization can do well, but it is convenient to also include the organization's administration. Technology resources include technological expertise of the people in the organization and also the equipment—tools, computers, and machines—possessed by the organization or available to it. Resource assessment for a single entrepreneur thinking of starting a new business, perhaps tailoring, is captured by the Five Finger approach.

1. Do I know the technology?
2. Where will I sell?
3. Will I make a profit?
4. Do I have money to start?
5. Will I be able to contribute some money to my family?

(The last three questions are, of course, financial.)

People resources are evaluated by looking at the skills and experience of people in the organization, if it already exists, and people likely to join the organization. An exciting new project may very likely attract new people into an organization. In fact, an exciting new project might be essential to retain good people in an existing organization. A related question is whether an existing organization is too stressed presently to take on a new project—new people will not contribute at first. A new project, especially if it is larger than what the organization had done previously, may require significant accounting, computer, or administrative skills. Does the organization have these? If not, how will they get them? If an organization is considering several alternative new projects, it may be fruitful to consider briefly the administrative requirements of each and then match those requirements to the experiences of people in the organization. If an organization is just starting, it is especially important that administrative requirements be kept in mind. More organizations probably failed because they could not handle the business aspects than because the technology—machinery, medical test, seed, and so forth—being pro-moted did not work.

Administrative procedures—such as cost accounting, hiring and training practices, salary structures—need to be assessed. Able people can establish or improve these but not overnight and not without significant effort. The "culture" of the organization should be considered. An organization that rewarded innovation and flexibility in meeting community needs may need to change significantly to handle a large external grant requiring accountability. The analogue is a mature organization that must deal with an entirely new clientele. Just as administrative systems can be instituted so the culture can be changed, but such a change requires understanding.

Technology resource assessment consists partly of looking at what people in the organization can do, after training if necessary. The other part is looking at what technology is available locally. To create a sustainable project, one should be selected that does not depend on imported material or expertise. A project that requires expensive material or imported supplies is unlikely to be maintained after an expatriate designer leaves. The same is true for technical knowledge. If local people do not already have skills required for maintenance and upgrading, then training must be provided. The more a project uses local material and knowledge, the more likely it is to be maintained and used by the community after the founder has left. The assessment issue is understanding what expertise is present in the organization and what hiring and training can acquire. Local people can usually learn a great deal rapidly if trained sensibly but if the need for training is not attended to—assessed—the project will probably have difficulties.

Other organizations in the community will probably be involved in similar technologies. Hardly ever does only one person learn about a new technology. It is in everyone's best interest to form collaborations that facilitate beginning the project and build capacity in local organizations. Using the resources available from other organizations and community groups involves more people in the project, increasing their investment in the project's success. The assessment issue is knowing what is going on in the community.

OFFICIAL APPROVAL

Attention must be paid to the attitudes of government officials. Ways of gaining support of regulators vary depending on the situation—the same is true in the United States. At the least, one should inform officials of plans. Respect and integrity will usually go a long way toward gaining approval. The issue can be very important in many cases, even though general advice is not possible.

FINANCIAL ASSESSMENT

The basic question in a financial assessment of a project is whether the benefits from the project—earnings or community improvements—are greater than project costs. If the costs are less than the benefits, then the project does not make financial sense. The question is complicated because (1) some intangible benefits—maternal health, for example—cannot easily be given a monetary value, (2) some costs are paid only at startup—tools, for example—and others reoccur—workers' wages, for example, and (3) benefits usually come some time after the costs and the delay makes the benefit less valuable—a community is better off if a new health clinic is available next month rather than next year. A second financial question is how much cash is required to start the project. Even if a project offers a huge payoff, it cannot be started unless the startup money can be raised. These two financial questions correspond to 3. and 4. of the Five Finger approach. 3. Will I make a profit? 4. Do I have money to start?

Basic techniques of financial analysis are described in the "Financial Viability" and "Discounted Cash Flow" sections. An essential idea is that money spent later costs us less than the same amount of cash spent now. If we spend money two years from now, we can earn interest until we spend it. Because two years' interest will be earned, the amount we have to set aside now is less than if we have to spend the money now. The same argument applies to money received in the future—we are losing the interest we might have collected if we had the money now. Even if the cash involved would not have really earned interest, we are still worse off until the money is actually received or, in the case of money paid in the future, better off to be able to use the money until we pay. The result is that future cash payments and receipts must be multiplied by a "discount factor" to make them comparable to cash paid or received now. How to estimate the numerical value of the discount factor is described in the "Discounted Cash Flow" section.

The financial assessment process can be straightforward and easily explained with an example. A community that is not connected to electric power utility is presently using automobile storage batteries and dry cells to operate the electrical appliances in the village—radios, cassette players, and small lights. The amount of cash being spent each year on dry cells is $100 (We will do this calculation using US$). A storage battery is purchased each year for $50. It is recharged by putting it in a farmer's truck and run up and down the highway for an hour. The farmer charges $15, and recharging must be done four times a year. Storage batteries must be replaced after five years. A photovoltaic system is proposed using rechargeable dry cells. The photovoltaic system will cost $600 and will last ten years. The only maintenance required is rebuilding the wooden support structure after five years—a cost of $75. The annual discount factor has been estimated at 0.909—corresponding to an interest rate of 10 percent. We need to do a financial assessment of the photovoltaic system.

We start by calculating the benefits and costs that we can put a cash value on. Benefits in this case are the money not spent on dry cells and the storage battery. These costs are shown in the chart below. The annual costs for storage batteries is $4 \times \$15$ plus $50 in year 1 and 6.

Year	1	2	3	4	5	6	7	8	9	10
Dry Cells	100	100	100	100	100	100	100	100	100	100
Storage Battery	110	60	60	60	60	110	60	60	60	60
TOTAL COST	210	160	160	160	160	210	160	160	160	160
Discount Factor	1	0.909	0.826	0.751	0.683	0.621	0.564	0.513	0.467	0.424
Discounted Costs	210	145.5	132.2	120.2	109.3	130.4	90.32	82.11	74.64	67.86
Total cost of dry cells and storage battery					$1,162					

The total cost of these dry cells and storage battery for ten years, after discounting, is $1,162. This cost would not have to be paid if the PV system were installed, so the benefit of the PV system is $1,162.

The costs of the photovoltaic system are shown in this chart.

Year	1	2	3	4	5	6	7	8	9	10
Photovoltaic System	600	0	0	0	0	50	0	0	0	0
Maintenance	0	0	0	0	0	75	0	0	0	0
TOTAL COST	600	0	0	0	0	125	0	0	0	0
Discount Factor	1	0.909	0.826	0.751	0.683	0.621	0.564	0.513	0.467	0.424
Discounted Costs	600	0	0	0	0	77.62	0	0	0	0
Total cost of photovoltaic system						$677.6				

We see the cash benefits—$1,162—are significantly higher than the cash costs—$676.60. This project makes financial sense. If the cash benefits came out smaller than the costs, then a project should be looked at carefully before proceeding, although intangible benefits may make a money-losing project worthwhile.

One might now look at intangible costs and benefits of the photovoltaic system. A person in the village points out that trips to the nearby town to buy batteries are an opportunity to visit relatives. If the batteries are recharged in the village, this opportunity will be lost. Another person says she is worried about chemicals leaking from discarded dry cells and would be grateful for rechargeables. The farmer whose truck had been used in recharging the storage battery may be disappointed to lose a source of earnings. Recharging batteries locally may be more dependable, since stores sometimes run out. People involved—those living in the village—will have to weigh these intangibles with the cost saving and make a decision.

In many cases, aid agencies need to compare different projects to choose those for funding. The costs of these projects, as well as the benefits, may be different. For example, the same village might also be considering a wind-powered system to grind corn. A cash benefit-cost analysis, similar to the one for the photovoltaic system, shows benefits of $1,000 and costs equal to $400. Which project should the agency fund? A common approach is to use the benefit/cost ratio. For the PV project the ratio is 1162/676.60 = 1.72. For the wind-powered project the benefit/cost ratio is 1,000/400 = 2.5. Because the latter project has a higher benefit/cost ratio it looks more promising. People at the aid agency would argue that $1.00 put into the PV system gives $1.72 benefit, but $1.00 put into the wind system gives $2.50 benefit. Of course, a benefit/cost ratio is only useful when intangible benefits and costs are also considered.

Another aspect of a financial analysis is determining the initial—startup—costs. Even if the photovoltaic system in the example has benefits greater than costs, it may not be feasible for the village if the $600 required to buy the panels and the rechargeable batteries cannot be raised. An advantage of using dry cells is that the initial costs are low. In many cases projects will be funded by outside agencies that will scrutinize the startup costs carefully—some agencies only give startup money. Organizations should not be discouraged by high startup costs because many times external agencies will donate funds for initiating projects.

A final aspect of a financial assessment is consideration of risk. How likely is the photovoltaic system to fail completely after seven years? To be blown over in a windstorm? Will the cost of dry cells increase or decrease significantly in the next ten years? It is risky to make a major commitment to a technology not yet used in a region. It can also be a risk not to embrace new technologies. A way to understand the implications of various happenings is simply to work out the costs and benefits for likely scenarios. It might be reasonable to expect the cost of dry cells will be reduced greatly so annual costs for using dry cells would be only $25 rather than $100. In this case the first chart becomes the following.

Year	1	2	3	4	5	6	7	8	9	10
Dry Cells	25	25	25	25	25	25	25	25	25	25
Storage Battery	110	60	60	60	60	110	60	60	60	60
TOTAL COST	135	85	85	85	85	135	85	85	85	85
Discount Factor	1	0.909	0.826	0.751	0.683	0.621	0.564	0.513	0.467	0.424
Discounted Costs	135	77.27	70.21	63.84	58.06	83.84	47.94	43.61	39.7	36.04
Total cost of dry cells and storage battery					$655.50					

Now the photovoltaic system looks much less attractive because its cost is greater than the benefit. Those responsible for project selection need to decide which scenarios are plausible and analyze these in depth. Accounting for risks in most cases requires much good judgment.

STRATEGIC ASSESSMENT

The long-term implications of selecting a particular project need also to be considered. Some implications relate to people. If a village installs a photovoltaic system, will the knowledge gained by people in the village permit further improvements? If a photovoltaic system is installed, will certain people in the village—those deciding whose batteries are recharged first—gain power, reducing equality within the community? Some implications are technological. Will a new project bring equipment into a village that can be used for something else? A clinic that does tuberculosis screening could probably easily extend its work to include AIDS.

Some implications relate to long-term financing. Initiating a project may make an organization uniquely prepared for future developments and thus likely to receive support in the long term. If an organization promotes PV systems, it may create a niche for itself, doing something that no other organization is doing. Some projects may have negative long-term financial implications, as when the project requires continued infusions of money or foreign investment. To be successful in the software development business—often proposed for India—a company must be prepared to invest regularly in computer hardware. Earnings cannot be invested in the community if they must be reinvested in the business. The last finger of the Five Finger approach gets to the difficulty: "Will I be able to contribute some money to my family?" Strategic assessment requires much foresight. A pattern for going about the analysis is to think about what a needs and resource assessment would be like five years from now if the project is done.

THE CHOICE

Final project selection is based, of course, on the assessments just described. At the least these assessments should indicate which projects should not be considered further. One issue that should be taken into account is consistency. Projects that are inconsistent with what the organization has done in the past or with the organization's sense of itself will probably be hard to implement. Efficient implementation requires both that the new technology work and that the new organization function effectively. A major change in the organization's operations may be necessary if the organization's environment has altered greatly—a school may need to respond to the demand for computer training—but such a change will still not be easy. A project is also inconsistent if it requires so much resources that other worthwhile projects cannot be done.

Another issue is how other organizations will change, perhaps in response to our change. Long-term success for an organization probably requires that it do at least one service or business better than its competitors. Unless a community organization is the only one doing a particular service—working with teen-age girls, for example—or is recognized for doing the service best, it is vulnerable to losing clients or financial support. In considering an opportunity for a project, one needs to consider if another organization will address the same opportunity. If so, the opportunity is less attractive—perhaps a collaborative effort should be considered.

A major concern for an outside advisor is imposing her or his views when making a selection. Although an assessment may highlight a particular need, it is not enough that a project is a response to that need. The project must also be a response to a request from the community to meet the need. Determining a community need can be an imposition of the interviewer's values. For example, assessing a community and determining that there is a need to provide educational opportunities for war veterans may reflect your values more than a need expressed by the community. This is not to say that the need does not exist or that war veterans should not have educational opportunities, but a project based on the assessor's values rather than a community request may not garner the support or participation of many community members.

VALUE CHAIN

When evaluating alternatives as part of strategic planning, it is important to comprehend how the enterprise functions, what aspects are essential for its operations. Understanding the things that are actually done—the activities performed—gives insight into improving the business, into recognizing what is important in the business. A value chain shows these various things that the business does. The value chain for an organization that supplies basic agricultural tools for a farmers' cooperative is shown below.

DESIGN OF TOOL	PROCUREMENT OF MATERIAL	DESIGN OF WORKSHOP	MANUFACTURE	DELIVERY

FIGURE 16
Value chain.

The value chain can be used in cost reduction studies or in initial planning. Using the value chain in Figure 16 to analyze how costs might be reduced, the workshop director would look at each step individually: Is it possible to redesign the tools to reduce costs? Can metal be bought more cheaply? When the director is trying to decide if a new enterprise is worth doing—for example, should the organization manufacture cooking utensils—an important question is whether the organization can do it better than competitors. If the enterprise does not excel in one of the activities in the value chain, then it is unlikely the enterprise is making a significant contribution. The value chain suggests possible ways the organization could excel. The workshop described by Figure 16 might be better than any other in the logistics of delivering tools to farmers. When attempting to improve operations one must keep in mind the steps that make the organization better than others and not diminish them.

PROJECT EVALUATION

This section is based heavily on the A.I.D. Evaluation Handbook, *April 1987, Agency for International Development, Washington, DC.*

PURPOSES OF EVALUATION

The purpose of evaluating projects is to assist managers in making well-informed decisions. Both ongoing and completed projects should be evaluated. The information gained is used for somewhat different purposes. An evaluation of an ongoing project may suggest ways of improving the remainder of the project. An evaluation of a completed project should be used to improve future projects, even decide which future projects should be undertaken. If the results of an organization's activities are not both analyzed and documented, the organization will have difficulty learning and will not improve its performance.

The reason for evaluating ongoing projects is to plan the next set of steps. When a project is proposed not all possibilities and problems can be forecasted. The present relevance of the original project objectives should be reassessed, as well as assumptions made when the project is planned. In planning an AIDS awareness project, the project director may have believed that urban and rural populations needed to be served differently, but in the initial stages of the project it may become clear that sufficient movement between cities and the country takes place so the populations can be treated identically. Another reason for evaluating is to ascertain whether any aspect of a project is not progressing as expected—those that are not may need extra attention or may be abandoned. Evaluating an ongoing project can encourage discussion about how the project can be improved

The reason for making a final evaluation is to make future projects more effective. The evaluation should include both a set of lessons learned and recommendations about future projects. A lesson learned from a drought relief program is that timely and well-documented warnings of famine are not always heeded by government officials and donors. A recommendation is that effective systems be set up to communicate warnings and mobilize appropriate action. Another reason for doing a final evaluation is to promote dialogue—in the drought relief case, among aid agencies, NGOs, and national governments—about the problem addressed by the project. Final evaluations are especially important when follow-up projects are planned.

It goes without saying that for an evaluation to be of benefit it must be read. It is the responsibility of the writer to make it clear and functional. It is also someone's responsibility to make the evaluation available to those who will benefit from it—the project director in the case of an ongoing project and the project planner in the case of a new project. It is equally the responsibility of the project director to make use of the evaluation. People planning new projects, especially, should search out evaluations of related projects and keep the lessons given in mind.

WHAT TO EVALUATE

An evaluation of an ongoing project should give the manager information about how much money was spent and how much has been accomplished to date. Is the project on schedule? The evaluation should also consider whether the external environment has changed since the project was initiated—do the same needs and constraints exist? Privatization of the government ministry that operated the telephone service will affect a project offering wireless telephony to villages.

Short-term effects, intended or not, of the project should be monitored—for example, has a skills training project taken men from their homes and thus decreased child care? Progress toward sustainability should be considered—what can be done now to improve the likelihood that the project will continue after external funding ceases?

The evaluation of a completed project should include the same kind of information. Was the project effective? Did it accomplish its objectives? Was the project efficient? Were the benefits produced at a reasonable cost? Are the project objectives still relevant to what the organization wants to do? What consequences, both positive and negative, did the project have beyond its stated objectives? A project intended to strengthen an engineering program at a university may have the unintended consequence of a significant increase in the demand for telephone service as students learn to use e-mail. A final aspect to be evaluated, again, is sustainability. What will be left when project funding has stopped? The evaluator should note that the questions being asked are deeper than merely a listing of the project outputs—things built or number of people trained.

OUTPUT FROM EVALUATION

The two most important outputs from an evaluation are the lessons learned and the recommendations. Lessons learned can relate specifically to the project itself or can relate to broader questions about the usefulness of a particular approach. A project to improve an arid region cereal could show the advantages of doing the research at a local university. It could also demonstrate that emphasis should be on distribution of existing seeds rather than development of new ones. The list of lessons is an opportunity to give advice to colleagues. Recommendations are actions that will improve performance, either of an ongoing project or of a new project similar to the one evaluated.

A large agency of the United States government, U.S. A.I.D, uses the following structure for reports from evaluations.

1. Executive summary
2. Purpose of project
3. Purpose of evaluation
4. Findings—history and outcomes
5. Conclusions—implications of outcomes
6. Recommendations for changes in current project or design of future projects
7. Lessons learned—about project design and broad actions. Some conclusions from no. 5 may be repeated here, but the emphasis is on generally applicable insights.

PROCESS OF EVALUATION

The results of an evaluation will be more useful if the evaluation process is considered from the beginning, when the project is initially planned. The data useful in making an evaluation should be identified before the project begins. For example, an evaluation may require comparison of two situations—for example, the number of calories in a typical child's diet before and after a new cereal was introduced. The baseline data must be collected early in the project. If a purpose of the evaluation is to allow a midcourse correction, then thought must be given early as to what data will indicate a need for project modification. In designing the project the question "What data will I need to know to determine whether the project is successful?" needs to be kept in mind

These are the major steps in preparing an evaluation.

1. Incorporate an information component in the design of the activity as part of original proposal. Some important questions are the needs addressed by the evaluation, the variables to be tracked, and the data sources.
2. Decide when to evaluate. An important determinant is when the information is needed.
3. Plan the evaluation. The objectives/uses of the evaluation need to be determined. Specific questions to be addressed in the evaluation need to be listed—questions getting to the issues above of effectiveness, efficiency, relevance of project to present needs, consequences, and sustainability. The plan also needs to specify a schedule and who will do the project.
4. Write a short document, stating the plan succinctly. This document is to be used by the evaluator, by people involved in the project, those who funded the project, and others interested.

A question that needs to be decided is whether to use external or internal evaluators. An external examiner can bring fresh perspectives and state-of-the art knowledge. He or she does not have allegiances to the people who managed the project and thus may be more objective and able to serve as an arbitrator among parties. An internal examiner knows the organization and so can get up to speed quickly, understands the constraints on recommendation, and has a better chance of following up on recommendations. In many situations it is easier and less costly to assign an internal evaluator than to hire an external one

SUMMARY

Evaluation is important if an organization is to improve its performance. Evaluations should consider as relevant effectiveness, efficiency, relevance of project to present needs, consequences, and sustainability. The most valuable part of an evaluation for a user should be the lessons learned and the recommendations. The process of making an evaluation should be planned logically from the beginning of the project.

HOW TO EVALUATE

Katherine Evans

After a project has been planned and implemented, an evaluation allows project designers to take stock of the project's success, get credit for that success, and learn how to improve the project. The degree to which an evaluation can do any or all of these things depends on including community members in the evaluation and asking the right questions.

WHY EVALUATE?

Most fundamentally, evaluation is a means for demonstrating project effectiveness to donors, supervisors, staff, and community members. Evaluations answer the question "To what extent were stated project objectives attained?" In this way, evaluation measures intended impact.

Evaluation also identifies unintended impact. By asking, "What results did the program produce?" project staff can note all of the effects of a project and make adjustments if the project has negative side effects.

Evaluation examines the characteristics of a project that worked and didn't work. Evaluations establish a feedback loop to designers who can improve the project's performance as a result. An organization cannot learn without data on its impact, provided by an evaluation, or the capacity to adapt.

Evaluations can also point to potentially productive directions for the future. Through identifying a project's strengths and weaknesses, evaluations can inform strategic planning discussions.

Finally, evaluations increase the number of people concerned with the project's success. Participative assessment allows community members to take part in determining a project's goals and the method to achieving those goals. Effective performance gives stakeholders reason to maintain an organization or a project.

QUESTIONS TO CONSIDER BEFORE BEGINNING AN EVALUATION

For Whom is the Evaluation?

Evaluations differ in their emphasis and foci depending on the intended audience. An evaluation performed for a donor will show how the money was spent and examine the organizational costs compared to the total benefit created for community members by its efforts. This type of evaluation focuses on a cost-benefit analysis and measuring impact, often for comparative purposes.

Evaluations intended for internal staff serve a troubleshooting function. They identify areas to strengthen, potential problems, and effective characteristics. Internal evaluations often allow for more contextualization and more qualitative data. This article will focus on internal evaluations, aiming to improve project effectiveness.

Who Performs the Evaluation?

There is an active debate about whether evaluations are more effective and accurate if they are performed by an external or internal evaluator. Each type of evaluator offers different benefits and detractions. An external evaluator presumably is an unbiased party who can evaluate a project more objectively than an internal evaluator. An external evaluator may receive more honest responses from staff and community members whom they interview. However, external evaluators often have limited time for an evaluation that can result in a superficial assessment.

Conversely, an internal evaluator can understand the context, activities, and objectives of a project better than an external evaluator. An internal evaluator may have a better sense of the actual situation and the dynamics of a project. Advocates of internal evaluators generally believe that an evaluator cannot know a project without participating in it. In the end, the decision of who will perform the evaluation may be determined by a supervisor or by resources.

How Will the Evaluation be Used?

Evaluations promote accountability to donors and to beneficiaries because they document how allocated money was spent and they demonstrate progress toward intended outcomes. Evaluations describe the materials and activities used by a project to reach its goals and to what extent it was successful.

Evaluations also create a lasting description of a project so that other organizations can avoid reinventing the wheel. An evaluation should be detailed in its description of the site, activities, and materials so that other programs can use it as a model.

Evaluations assess implementation for project improvement. In this way, the evaluator assumes a role in a project's planning, development, and refinement in addition to monitoring implementation and reporting progress. An evaluation can prompt project designers to reexamine their thinking about why or how the project will accomplish its goals.

When?

Ideally, evaluation is a continuous process in the life of a project. Donors generally require an evaluation annually or with any request for funding renewal. Annual evaluations provide valuable information to donors and staff, but they do not institutionalize evaluations. An organization that learns at every step is a result of more frequent formal and informal assessment.

As discussed in the sections on project selection and implementation, evaluations are more effective if they are part of the initial planning stages of the project. Designers who create and evaluation schedule, identify goals and indicators, and include stakeholders in that process, lay the foundation for organizational learning and responsiveness.

What Are the Resource Implications?

An external evaluator, lengthy interviews, and sophisticated data processing take time and money away from project activities. However, evaluations justify their cost because they improve project efficiency and identify weaknesses. In the end, the shape of an evaluation inevitably must take into account available resources and opportunity costs.

HOW TO PERFORM AN EVALUATION

Answers to question about a project's implementation and impact can come from the project proposal, participant surveys or interviews, and observation made by the evaluator. The project proposal provides information about the critical features/activities of a project to an external evaluator. Self-reports from participants provide information about the implementation of the critical activities and their effectiveness. Identifying all possible groups of stakeholders and talking to representatives from each group provide a more objective picture of the project because different groups may have different uses of the program, and therefore their perceptions of its effectiveness vary. Personal observations serve to balance some of the distortion in answers from participants who may tailor their answers for an outsider. However, the presence of an observer may alter what takes place; it is not possible to be a fly on the wall.

The first step in evaluating a project is deciding what to look for. This question was answered in the planning stages of the project when designers determined indicators that reflected the goals of the project and measured the baseline for each of those indicators.

For the purposes of creating an accurate description of a project as a reference for other organizations, the next step is placing the indicators in political, cultural, and geographic context. In the case of water purity, contextualizing that indicator means explaining larger influences on the indicator. For example, have recent economic events left more people homeless, or has a recent election led to an increase in resources allocated for public sanitation projects.

After describing the setting of the project and the external factors that affect the success of a project, an evaluation describes the project's activities: How is a project implemented?

This part of the evaluation also involves interviewing participants and/or making observations. Questions include: In what activities were community members intended to participate? Did they participate in those activities? Who makes policy decisions? What informs those decisions? What is the theory behind a project's activities? (Why do the designers think the means of the project will achieve project goals?) For example, the goal of the project may be family planning, and the activities try to achieve this indirectly by offering classes to women in nutrition, household finance, and weaving on the theory that empowering women through relevant education will empower them to make reproductive decisions. Identifying the theory behind a project's activities can provide insights into whether that theory is justified in the context of this project.

Surveys and interviews are the primary means of collecting information from community members. An evaluator may choose to rely on interviews because he/she can ask follow-up questions, however the presence of the interviewer may affect the answers, in which case, a survey that is distributed to participants and collected anonymously may be preferable. Interviews and surveys can be closed- or open-response. A survey that is closed-response asks multiple-choice questions; an open-response survey provides space for participants to fill in their own answers. Closed responses are easier to tally and process, but open responses may evoke opinions that would not otherwise be disclosed. The basic question is "What impact has this project made in your life (positive or negative)?" Clearly, the questions should be specific to the goals and indicators of the project. Examples of questions to community members include "What is the goal of this project? Is the project successful? What activities do you participate in? What has changed in your life as a result of this project? What do you like about it? What would you change?" as well as questions that indirectly examine to what extent the project has reached its goals in relation to the community member. Another consideration to keep in mind is capturing externalities in your questioning. For example, classes in nutrition may not only attain the goal of decreasing malnutrition but may also affect prevailing attitudes about educating women. Understanding how a project impacts a community beyond its specific goals is important to understanding the dynamics of the community and the dynamics of development. A project becomes truly sustainable when there is the achievement of multiple benefits from any one cost and more stakeholders are created from the perpetuation of the activity. Asking questions about broader changes that may have occurred as a result of the project can capture these externalities.

Just as with a needs assessment, collecting information only through formal avenues omits critical feedback that may be revealed only indirectly and offhand. For this reason, it is good practice for the evaluator or project designer to have an open-door policy, making it clear to participants that comments and questions are welcomed.

DIFFICULTIES IN EVALUATION

A recurring concern in evaluations is ensuring their authenticity. If an evaluation is not a faithful representation of the actual situation, then it is not useful in documenting progress or improving performance. Authenticity is determined by the validity of the assessment—is your method of measurement relevant and complete—and the reliability of the assessment—do you get consistent results from the method of measurement?

Controlling and attributing causality is another obstacle to performing effective evaluations. Projects with intangible goals are especially difficult to evaluate. An evaluator cannot isolate the causes of empowerment or political participation, nor can an evaluator attribute an increase in these to a project. Projects that provide a direct service are not as difficult to evaluate, but even in our example of water purity it must be determined if improvements can be attributed to new latrines installed by the project or to a change in government, which invests more money in sanitation.

Other difficulties arise if donors, supervisors, staff, and community members use different definitions of impact and performance. Does an evaluator use the donor's standard of success or community members'?

Including multiple stakeholders in the evaluation can mitigate each of these difficulties. By consulting representatives from donor, project directors, staff, and community groups, an evaluator increases his or her ability to represent the actual situation in the evaluation. Community members who use the project and those who do not may have disparate opinions. The same is true for project directors and project staff. Consulting many stakeholder groups can also help clarify issues around causality because each group can contextualize a project differently, so an evaluator can identify causes attributable to the project by identifying common responses from each group. Including multiple stakeholders in initial discussions about the goals of the project and indicators of success avoids problems that arise if donors, staff, and community members disagree on those definitions. In addition to avoiding pitfalls in evaluation, including multiple stakeholders in assessment is cost-effective because it combines capacity building and performance monitoring.

DISSEMINATION

A major criticism of NGOs is that internal evaluations are rarely available. Consequently there is little documentation of their impact. Making your findings available to stakeholders and other groups avoids this criticism and makes the project more transparent. By distributing the evaluation to stakeholder groups, policy changes that occur as a result of the evaluation can be understood by community members. Disseminating the evaluation widely is an important step in gaining legitimacy in the eyes of stakeholders because it clarifies the motivation behind policy decisions. Distributing the evaluation to other organizations allows outside groups to use the project as a model, through learning from its mistakes and successes. In this way, evaluations can double as a publicity tool as well as a learning tool.

MANAGEMENT OF PROJECTS

James Crossley

WHAT IS IT?

Project management is the management or administration of an activity that has a definite starting point and a definite ending point. Examples could be the installation of a village well, the installation of an automatic money machine at a bank, the implementation of a new purchasing system at a company. Project management therefore differs from operations management, such as the daily supervision of a fast food store, the continuous management of airline operations, or the daily supervision of a farm.

Project managers, sometimes called administrators or supervisors, must also have management skills in other disciplines. They must manage people, they must have knowledge of financial controls, and they may be involved is areas such as personnel health and safety, the environment, and other concerns.

THE BASIC PRINCIPLES

Project management is concerned with the three W's.

- What are we going to do?
- When are we going to do it?
- Who is going to do it? (The resources needed)

A project will have three phases

- The planning phase
- The construction or installation phase
- The implementation or startup phase

THE 3 W'S APPLY TO ALL PHASES OF A PROJECT

What Are We Going to Do?

What we are going to do is defined by the *scope document*. This can also be called the *job specifications*. Individual steps in the project are known as activities. The greatest cause of failure of a project to be completed on time and within budget is a poor scope document or failure to follow the scope document. "Scope creep" is common in many projects—the addition of items to the job without the necessary budget increase. The scope document should also identify possible items in the project that are *not* within the scope and should clearly state this. Let us consider the scope document for a simple job—the installation of a wind-powered water system for a town.

WIND-POWERED WATER SYSTEM FOR TOWN OF XYZ—SCOPE DOCUMENT

1. Hydraulic studies show that the aquifer will support the withdrawal of 20,000 liters of water per day. (Reference study document)
2. The system shall be designed to remove 10,000 liters of water per day, based on a wind velocity of 15 Km/hr to 40 Km/hr.
3. There shall be storage capacity for 100,000 liters to allow for down periods due to wind conditions and/or equipment maintenance.
4. The project will be responsible for procuring all items shown on bill of materials WM-XYZ.
5. The project will not be responsible for the following.
   a. Obtaining the land (1/2 HA minimum)
   b. The access road to the site
   c. Water piping from the storage tank to the use connection
   d. Maintenance costs after the system is turned over for use, except that any equipment warrantees will be honored
6. The project will provide 24-hour security for the site until the system is turned over for use. At that time, security is the responsibility of the owner.

Bill of material for wind powered water supply system for XYZ

Item No	Description	Quantity	Specification
1	Package wind tower with power blades and drive shaft to pump. Height 12 meters	1	Quotation #99-7 from Tower Africa
2	Spare parts package of blades, bearings	1	Quotation #99-8 from Tower Africa
3	Pump-10,000 litres/day flow at 50 meters hydraulic head	1	Quotation #11-98-7 from Goulds Pump Africa
4	Spare parts package of bearings.	1	Quotation #11-98-8 from Goulds Pump Africa
5	Sand for concrete	2 cu m	
6	Bag cementm-40 kg ea	20	
7	2 Cm max aggregate	2 cu m	
8	Water supply for concrete & misc use	500 1/day	
9	Chain link fence 2 m high with 6 strand upper barbed wire	60 m	
10	Gate 6 m high with locking facility-i m wide	1	
11	Storage tank-10,0000 I capacity-high density polyproplene 10 mm wall. 4 cm pipe size bottom and top nozzles. Sealable inspection port in top. 2 cm top vent line with insect screen.	1	Quotation # 99-12-7 Plastitank PLC
12	PVC sanitary grade plastic pipe, 4 cm nominal diameter 3 mm wall thickness.	30 m	
13	PVC sanitary grade plastic elbows, 4 cm nominal diameter weld fitting	12	
14	PVC sanitary grade plastic pipe valves 4 cm nominal diameter ball type-weld fitting	3	
15	PVC pipe glue, solvent type	1/2 liter	
16	Pressure gauge- 0 to 20 M range- 4 cm fitting	1	

It can be seen that the scope document can be extensive even for a small job. For a large project such as a power plant or a large office building, the scope document can and should be several hundred pages. Money spent in preparing a good scope document is probably the best money spent on a project.

One problem in project administration is determining an approximate cost before design work begins. This is usually necessary to obtain budget approval. A budget estimate should be prepared. This can be done by investigating similar installations, obtaining data from technical or business journals, discussions with vendors, and so on. Any estimate obtained this way must be adjusted for inflation—not just past inflation but inflation projected up to the time that funds will be spent on the project. Care must also be taken that comparable scopes are used for the projects. A good budget estimate—after all considerations such as inflation, location, and scope differences are factored in—are considered to be accurate plus or minus 30 percent of the estimated value.

When Are We Going to Do It?

Determining what will be done when in the project is called the *project schedule*. Project schedules are displayed in a graph form. The earliest and simplest of these is called a Gantt Chart. A portion of a Gantt chart for the wind-powered water system described in the scope above would look like Figure 17.

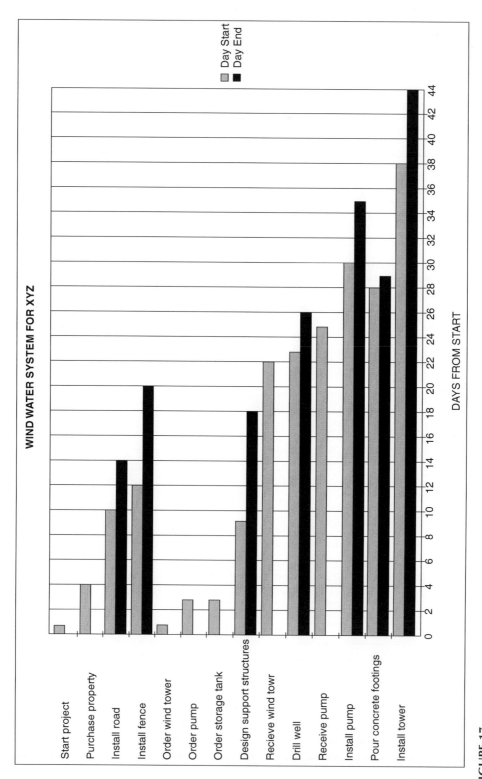

FIGURE 17
Gantt chart.

A more advanced scheduling tool is the *critical path method* (CPM). Here the various activities are laid out to show precursors or dependencies. For example, pouring concrete depends on both the completion of the design of the supports and the completion of the road. If either of these are delayed, the pouring of concrete will be delayed and the tower installation will be delayed. The critical path can be determined by adding up the required times for all activities where there are dependencies. Delay in any activity on the critical path will result in an equivalent delay in the final completion date of the project. There may also be activities where there is no action, such as the time required for concrete to cure before it has adequate strength. This author was involved in a project to install a water system in a Central American village. No CPM chart was prepared. A volunteer team went to the village and poured the concrete base for a storage tank. They were unaware that concrete must cure for several days and had to leave without installing the tank. It was necessary to arrange for another group of volunteers to go down and install the tank much later. Much time was lost and extra money spent.

Who Is Going to Do It?

The other concern that must be addressed in the successful completion of a project is the resources necessary. These resources can be people, such as engineers, secretaries, purchasing agents, craftsmen, and so forth, or can be equipment (a crane may be necessary to unload the tower and the pump).

One of the problems of a project manager is the temporary nature of the job and people assigned to work for that manager. How does a project manager motivate an engineer who is assigned to the project when such a project manager is probably not the permanent boss and might not influence possible promotions or raises for that engineer? The project manager may also be required to accept personnel that are "surplus" in their current positions because they do not have the required skills or because of poor work performance. *Beware*!

To get an estimate of labor and overhead costs for the project, the manager should make a list before the project starts of all personnel and equipment resources that will be required, along with an estimate of how long each resource will be needed.

There are computer programs that can assist in managing projects, from very simple ones such as Project Manager Pro, which could be used for small projects, to Primavera, which is used for large projects such as power plants. All of these computer programs takes dedicated time to learn how to use them effectively, from 40 to 80 hours for a small program to hundreds of hours for the larger. Do not attempt to use a computer program unless you have this much time to learn the process *before the project starts*.

Remember the three W's!

DESIGN AND CONSTRUCTION OVERSIGHT PROGRAM

James Crossley

INTRODUCTION

Design, construction, and beneficial use of capital installations require a systematic and disciplined program to assure that architects, designers, and contractors meet the specified needs of the owner (client). The following program is designed to do this. It is assumed that the Director of Planning or other named officer with equivalent responsibility will be the staff officer responsible for design and construction oversight.

1. Approval of Design

- The Director of Planning must review construction drawings with the architect engineer on a regular basis during the design phase. The costs of these meetings (at least monthly) must be budgeted. Errors and oversights caught early in the design process will save significant money.
- At least 30 days prior to going out for bids (tenders) on construction costs, the architect will supply a complete set of construction drawings for the clients review.
- Within 15 days of receipt of the drawings, the client will review the drawings and will sign off the final approval. Minimum sign-offs are the Director of Planning and the senior staff officer of the facility being designed.
- Based on past experience, construction budgets should contain a minimum 5 percent contingency for design omissions plus an appropriate contingency for inflation.

2. Awarding of Bids (Tenders)

- Obviously bid price is a major factor in the award, but contractors must also be qualified based on previous experience. The Director of Planning must review the qualifications and must interview previous clients for contractor performance. The director of Planning will prepare a written report of this review with recommendations for the appropriate Board committee.

3. Oversight of Construction

- The Director of Planning will produce weekly reports to appropriate management on the status of the construction. Adequate clerical/secretarial help must be provided for these reports. These reports shall contain the following.
  - —The architect/engineer's original estimate of percent completion and the budget expenditures by the report date
  - —The Director of Planning's estimate of percent completion and the budget expenditures by the report date
  - —Any deficiencies noted in construction
  - —Any predicted labor or materials problems that could impact the schedule or cost
  - —Reports of meetings with the architect/engineer
- Construction oversight must be the top priority of Director of Planning. Other duties in the organization will be secondary.
- A simple project management computer program should be used to track projects.

4. Progress Payments for Construction

- Construction contracts must be written to provide payment only for work adequately completed. The Director of Planning is responsible for determining the percentage of any task completed and only then for approving a progress payment.
- Construction contracts must be written to provide for a 10 percent holdback of the final payment until all work is completed and systems adequately tested.

5. Construction Deficiencies

- The Director of Planning will bring to the contractor immediately any noted deficiency in construction. Verbal notices will be followed within 24 hours by a written notice with a copy sent to the architect/engineer.
- The Director of Planning will maintain a notebook containing notes of all conversations with contractors and with the architect/engineer.

6. Change Notices

- Only the Director of Planning may issue change notices for construction work. The Director of Planning will review with the finance committee if the sum total of the changes approach the contingency percentage allowed.
- Change notices will be issued to the architect/engineer, not to the contractor.
- Change notices of major significance will be sent to the architect/engineer by certified mail or will be hand delivered.

RESPONSIBILITY FOR GETTING THE JOB DONE PROPERLY

James Crossley

In different countries, and often in different organizations in the same country, various titles can be used for the person responsible for getting the job done (sometimes called reaching the objectives). This person may be titled manager, supervisor, coordinator, administrator, director, or even commissioner. Although there may be fine differences in an organization between such titles, the importance is the responsibility to get the job done—safely, on time, and within budget.

An important item in defining responsibilities in an organization is the job description. A successful project or operation will have job descriptions for all persons in the organization, from the chief officer to the sweeper. Writing a good job description is not easy, since it must well define the daily routine responsibilities for the position but must have sufficient flexibility to handle unexpected conditions that will certainly arise.

We highly recommend that as soon as the staffing plan for an operation or project starts to develop, job descriptions should follow immediately. Job descriptions are not a fixed document. They should be reviewed and revised at least annually, more often if necessary.

EMERGENCY RESPONSE PROGRAM

James Crossley

In the last few years there has been increased emphasis in developed countries on the advance preparation of emergency response programs for industrial establishments, hospitals, universities, government agencies, and large commercial enterprises. These programs address all or some of the threats to life and property from fire, accidents, natural disasters, chemical releases, terrorism, or civil strife. It is recommended that an Emergency Response Program include the following sections.

RESPONSIBILITIES

- Who directs the response to an emergency during normal working hours (8 AM–4:30 PM)?
- Who directs the response to an emergency during off hours?
- How will the responsibility be delegated if either of the above persons are away?
- Who is responsible for purchasing and maintaining emergency response equipment?
- Who is responsible for training?

TRAINING

- What training is needed for fire response?
- What training is needed for medical emergencies?
- What training is needed for chemical spill response?
- Who should be trained to respond with minimum risk of personnel injury?
- What training and retraining records should be maintained?
- How should drills be conducted and how often?

EQUIPMENT

- What equipment is needed for fire response?
- What equipment is needed for medical emergencies?
- What equipment is needed for chemical spill response?
- What reserve of water is needed for adequate fire protection, and how can this reserve be maintained?

INFORMATION TRANSMITTAL

- What method should be used to sound an alarm in case of fire, medical emergency, or chemical spill?
- When should the fire department be called?
- What training or instructions must be given to nonresponding personnel to assure their safety?
- Which off-site persons must be called in an emergency?
- What is the backup capability if the electric power or telephone system is out of service? Is a backup radio practical and desirable?

EXAMPLE OF AN EMERGENCY RESPONSE PROGRAM

This is the form that would be used by management.

EMERGENCY RESPONSE PROGRAM

Program Content

A. *Emergency Response Procedures.* Actions to be taken by staff when a specific emergency occurs. Written procedures will be prepared for training purposes in the following areas. Emergency notification instructions will be posted prominently on walls and at all telephones.

A - 1 Fire Response Procedures

A - 2 Medical Response Procedures

A - 3 Chemical Spill Response Procedures

A - 4 Civil Unrest Response Procedures

B. *Training Requirements.* Training required for responders

B - 1 Training required for fire response

B - 2 Training required for medical response

B - 3 Training required for chemical spills

C. *Responsibilities.* Who of the staff are responsible for what activity?

C - 0 Overall responsibility for program

C - 1 Responsibility for fire response training and procedures

C - 2 Responsibility for medical response training and procedures

C - 3 Responsibility for chemical spill response training and procedures

D. *Equipment Inspection and Maintenance.* Frequency of inspections and record keeping

D - 1 Fire response equipment lists with frequency of inspection records

D - 2 Medical response equipment lists with frequency of inspection records

D - 3 Chemical response equipment lists with frequency of inspection records

PROGRAM REVIEW

The program will be reviewed and updated annually by the Director of Physical Planning or any person designated by the management and a report made to management by January 1 of each year.

DISASTER RELIEF

Probably it is impossible to find nontrivial guidelines that will speak for all disaster relief situations, but lessons learned from various ones may be suggestive. This article is based on reports from several relief efforts.

Appropriate Technology would appear to offer much in disaster relief because it is flexible—able to be adapted to local conditions—and simple—easy to deploy. Further, it is controllable by users, inexpensive, uses local materials, and addresses basic needs: food and shelter. An influx of refugees shares many problems, for the host country, with disasters caused by natural disturbances. Refugee camps have made good use of owner built shelters, ventilated pit latrines, intensive agriculture, emergency water supplies, and small-scale industries.

MANAGEMENT OF LARGE PROGRAMS

These suggestions are based on the experiences of the U.S. Agency for International Development with the drought in southern Africa in the early 1990s. They focus on large—regional and national—programs.

The first lesson concerns the difficulty in making a timely response. Large organizations—governments and international aid agencies—tend to take a long time to react, possibly because a large number of people must be involved. The first response then to a disaster or threat of a disaster needs to be mobilizing donor support. Alerting governments is usually difficult. Local governments often act as if they do not want to know, and potential donor governments see many requests. Gaining political visibility for the disaster can be an important action for a small assistance group.

Because arranging for the transport and distribution of food from abroad takes much time, it may be best to provide for commercial purchase of food for the first few weeks after the disaster. Even when food from international aid agencies has reached the disaster area, it may be best to use existing commercial distribution channels—supplemented with vouchers for people who cannot pay. Decentralized distribution programs—managed by local organizations—seem to work best, possibly because planning many small programs in parallel is more feasible than planning a single massive one. The U.S. Agency for International Development advises that free food distribution should be avoided. It is cumbersome, and it disrupts the free market.

Not only should existing food distribution channels be used, but, if possible, existing NGOs in the area should be utilized. International agencies can serve by supporting the local NGOs with training and administrative help. Mechanisms for sharing of expertise among NGOs need to be put in place, especially for highly decentralized programs. Building the capacity of local NGOs is valuable because they are likely to remain in the area after the disaster's effects have ceased to attract international attention.

Many national governments are relatively new and have not had experience with natural disasters. Advice given sensitively can be helpful. The outcome has been better when national planning is in place for responding to natural disasters. Experienced people to manage such plans should be identified before the disaster, and they should be deployed early.

A problem for aid agencies is to decide how much the need for assistance stems from the disaster and how much stems from preexisting conditions in the area—so-called structural problems. One reason the question is important is responsibility to donors; it is dishonest to use money for a different purpose than the giver expected. A related question is assessing the need for relief food. In emergencies, especially, it is important that criterion for deciding who receives assistance be straightforward and transparent. Related to the issue of who receives free food is the strategy of requiring work in exchange for the food. A problem with "work for food" schemes is that it is usually women—mothers especially—who need the food, and the schemes have the result of taking mothers away from child care.

REFERENCE

Appropriate Technology, Vol. 19, No. 3, December 1992.

EXIT STRATEGY

"A.I.D. (U.S. Agency for International Development) requires that evaluations examine several broad concerns that are applicable to virtually any type of development assistance....
Sustainability. Are the effects of the project likely to become sustainable development impacts—that is, will they continue after A.I.D. funding has stopped?"
A.I.D. Evaluation Handbook, April 1987

SUSTAINING A PROJECT

A major problem with many projects intended to assist people in the Third World is that the project does not continue after the initial funding stops or after the expatriate who established the project goes home. People responsible for assistance projects need to create a strategy for continuing the project after they exit. The record is not encouraging—studies of projects supported by professionally managed international organizations indicate that less than 20 percent of these projects continue after initial funding stops. Of course, in the United States not all new social service projects endure, but the survival rate seems worse for internationally funded projects.

One piece of advice for a project planner is to plan for sustainability from the beginning of the project. In the initial plans, budgets should be estimated for the years after the initial funding has terminated. The planner has to be realistic about the likelihood that new external grants will be received. Funding agencies often do not want to continue funding a project indefinitely. It is usually harder to get grants for continuing operations than for startups. For one thing, donor agencies are worried about a "dependency syndrome" developing. Many projects are expected to sustain themselves eventually from fees charged users. A common difficulty is for a project to charge a low rate for products/services while it is externally funded. When the external funds are no longer available it is usually difficult to raise the price charged continuing customers/clients. Sources of income in the long term, and the extent to which these can be depended on, needed to be identified when the project begins.

The ideal situation, of course, is that local people will take over the project when the initial funding terminates. For such to occur local participation should be part of the project from the beginning. Local participation includes training successors for each expatriate position. The psychological aspects of turning a project over to local people should not be minimized. The person establishing a project may understand rationally reasons for having the people who will benefit from the project manage it but will still be emotionally bound to it. Strong involvement, from the start, of those who will eventually operate the project may make it easier for the foreign advisor to let go.

Another aspect of sustainability, besides money and people, is equipment. If the project originators are really to make an exit, then the equipment left behind must be usable and maintainable in the long term. Spare parts, consumable items, access to expertise for maintenance, and upgrading must be made available locally. A graceful exit will only have been made if both the hardware and the software to keep the project going are in place.

TRAINING

Based on notes by Eric Allemano

Training people to operate a project is as important as establishing the technology and setting up the management systems. People who will run a project will undoubtedly be able and committed but unless they know how to do what is needed they cannot succeed. Training can be done in many ways, which is why this article cannot give a specific technique. Whatever way is chosen, the essential objective is to give the person or persons responsible for the project confidence that he, she, or they can do the required tasks. Able people with confidence and understanding of the basics can usually solve problems—or know how to get help, if necessary.

Training courses should be useful and efficient. They are useful if they impart material that is relevant and understandable. A course is not useful if it trains for jobs that exist in limited numbers. Most villages can only support one bakery, so training a dozen bakers from a single village is not useful, or fair. Needless to say, the course should deal with the technology the graduates will use—no need to explain sophisticated chemical instruments for quality control if they will not be available on the job. A course is efficient if it is as short as possible, directed to actual needs. As much as possible, training should be tied to the reality of the job being trained for and the graduates' future positions as independent artisans. Lecturing at person for several days is unlikely to develop her or his sense of independent control. In many cases it is important to teach management skills along with technical skills.

TRAINERS

Technical skill is important for a trainer but usually not sufficient. Nearly always a trainer is needed with experience both in the subject matter and in a situation similar to the project's site. Team teaching may make sense—an expatriate and a local person. If a team is used it is important the expatriate allow the local person to be a full member of the team. A trainer must be a leader—able to motivate and empathize. Political factors—relationships with key village personnel, for example—may impinge on the selection of trainers, but the success of the project depends on an appropriate choice, so judgment and diplomacy must be used. Trainers should be given much emotional and technical support. Some will not have been in a similar role before and be nervous, both about the responsibility and about teaching techniques. Training sessions, if only to reassure the prospective trainers, are necessary. These training sessions need to be ongoing, partly to support the trainers, partly to gain feedback on the training and allow ideas to be shared.

DOING THE TRAINING

Training can be done in many ways. Two quite different approaches are a formal class or a participatory discussion. Variations in between these extremes are certainly possible, as is a mixture. The ultimate objective is usually ensuring that graduates can do something on their own, rather than pass a written test. Many people learn best by doing, and involvement in a discussion can be a first step for doing management things. As much as possible, hands-on teaching should be done with actual devices or at least models. People other than the instructor should get a chance to handle the devices and models. An important objective for many projects is to establish sustainable operations. Such need confident, committed owners. Confidence and commitment can be nurtured by giving as much responsibility as possible to the class. The trainer is there to assist the students in their learning, not to lecture.

Training sessions must be planned. A set of learning objectives should be established beforehand. A schedule should be established—the first 5 minutes for "ice-breakers," the second 10 minutes to show the usefulness of the material, then a presentation of the material for 20 minutes, then a discussion of possible problems, and so forth. (Even simply going around the room and asking each participant to give her or his name and home can be an effective ice breaker.) Often, giving students exercises to do in class is appropriate. In addition, longer assignments done away from class make often sense. These longer assignments should be discussed, perhaps by the entire group, at the next session. In many cases, each session, or the entire course, should end with the students preparing an action plan: "The Following Things Are Needed to Begin a Welding Business."

CLASS ADMINISTRATION

Some administrative matters should be considered. One question is where the classes should be held. If they are held close to where the participants live, attendance will be more convenient, but distractions are more likely. A gender issue is present here, and elsewhere: Classes away from the participants' homes are more likely to attract men. It is important to be sensitive to similar gender issues.

The size of the class is significant. Serious discussions may be difficult with more than five or six participants. If group learning is important, sufficient trainers should be employed so groups are small. Even classes taught traditionally should be limited to perhaps a dozen participants—to ensure opportunities for questions and involvement.

It is recommended that participants sit in a circle, with the trainer one of the circle members. Such sends the message that the participants are as important as the trainer. If the trainer wishes to use a chalkboard, then a semicircle makes sense. Without a chalkboard a large piece of paper can be used, or diagrams can be made in the dirt in the middle of the circle. Complicated presentation mechanisms often draw attention away from the message being presented. It goes without saying that the trainer should learn names as soon as possible.

REFERENCES

ILO/IYB Regional Project Office in Harare, Zimbabwe. (1997). *Marketing.* ILO/SAMAT, PO Box 210, Harare, Zimbabwe.
Pratt, Brian, and Jo Boyden (Ed). (1988). *The Field Directors' Handbook, An Oxfam Manual for Development Workers.* Oxford: Oxfam Publications.

THINKING AND LEARNING TOGETHER: FACILITATING EDUCATIONAL CHANGE

Etta Kralovec

Educational change can be facilitated by allowing people to "name the world," allowing educators to learn how people view the world and to learn the obstacles in working with them. The following report is based on experience in Swaziland focused on democratic principles in the classroom.

In 1996, Swaziland, a kingdom surrounded by South Africa, was in the process of writing a new constitution. The king had overridden the constitution years before, so the country had ostensibly been without a constitution for ten years. The government, international aid agencies, community organizations, and non-governments organizations (NGOs) all had a huge stake in the outcome of discussions about the new constitution. There was civil unrest over the issue of who could be at the tables where discussions about the new constitution were being held. In addition, the United States government is very interested in helping African countries reconstruct themselves as democracies. The United States Information Agency was working closely with NGOs in Swaziland to help articulate a set of principles that could be agreed upon that were central to a more democratic political structure. Forward-thinking educators had realized that a democracy demanded an education system that not only taught democratic principles but demonstrated them as well. As a specialist in education and a Fulbright Fellow in neighboring Zimbabwe, I was invited to help in that effort. I was to speak to a national meeting of NGO leaders and to work with the Ministry of Education to design professional development workshops for headmasters, teachers, and ministry personnel. My ministry workshops were intended to help Swazi educators begin thinking about how democratic principles might be put in place in the education system and in daily classroom practice. My work with NGOs was intended to help them learn how to educate the rural folks about democracy.

PREPARING FOR THE WORKSHOPS

My first trip to the country allowed me to meet with key Ministry of Education personnel and NGO leaders to establish the parameters for my work in the country. Our first set of meetings were in the Ministry of Education offices in the Capitol. I had been in Sub-Sahara Africa long enough to have become familiar with the ritual involved in the kind of meetings held at the Ministry offices. Lots of greetings, lots of tea, lots of jokes, and little of substance. I went from office to office being introduced to folks as someone from Zimbabwe, but everyone knew I was really from the United States, as the Minister of Education made clear to me in our short meeting. Our meeting with the minister included myself, my host from the American Cultural Center, and the Ministry sponsor of my project. The minister, in traditional dress—a draped cloth, jewelry, and sandals—got right to his point. He wanted me to know that they did not want Swazi children to be like American children, and if democracy turned out rude, ill-mannered children, he wanted nothing to do with it. I assured him that while democracy might be a little messier than other systems of social organization, democratic classrooms did not in and of themselves turn out rude children. My ministry sponsor assured me that while the Minister of Education had some strong feelings about my work in the country, he and the educators I would be working with were all very excited about our project. I spent time with university professors and school administrators who helped me to understand how their educational system was organized and where the levers for change might be.

In my meetings with NGO officials, we discussed their hopes for using my workshops with educators to spark a deepened commitment to the principles of democracy as they might apply to education. Beyond that, their key interest was in my helping them to develop techniques for educating the rural people about the constitutional process and about democracy in general. They believed that in order for democracy to take hold, the process of writing the constitution must be democratic, and in order for that to happen, the silenced, rural folks must be brought into the discussion. The NGOs were definitely at odds with the government over the issue of the constitution, and they made very clear that while the Minister of Education might not want democracy in the schools, the NGOs certainly did. They were organizing a two-day conference for all the NGOs in the country, and I was to give an address to the whole group and run workshops for folks interested in educating rural communities.

I had two months to construct a program that included a day-long session for administrators who were predisposed to think that all Americans were rude and pushy and a two-day session for teachers who were not paid enough, worked too hard, and were required to be at my sessions. My last two days in the country would be spent working with folks whom I felt deeply aligned with and who had high hopes that I could help them construct educational programs for use with rural people.

THE WORKSHOPS

My workshop for the administrators began with introductory remarks about democracy being a social fabric and the ways in which schools must develop and nurture a love of democracy among the young. I had the participants share their goals and concerns about our work at the very beginning of a session. As they are doing some small group work on what democracy means to them, I listed the goals and concerns for all see, without giving the names of who expressed the thoughts. There was a lot of concern expressed about the nature of Swazi culture being at odds with democracy. I suggested that we not talk about democracy in the abstract but rather talk about it concretely by focusing on decision making and how it might become more democratic. All agreed that democratic decision making in the Ministry of Education would not put Swazi culture at risk. By listing all major decisions and who was responsible for them, we started to identify those decisions that could be delegated to those closer to the point of delivery. At the end of our day together, we were all much more comfortable with democracy being a way of doing business rather than a child-rearing system that created unruly kids.

I delivered the same message during my two days with the teachers and worked in much the same way. In this group, we looked carefully at two aspects of their work. We examined classroom practices, seeking to find ways that we could introduce group decision making and voting into daily classroom practice. I shared with them activities in U.S. schools, like electing milk monitors, which help even the youngest students learn the meaning of voting.

We also looked carefully at the curriculum, looking for openings where we could add lessons on democracy. I was struck by a key issue that the teachers raised the second day and that we spent much of that day working on. One teacher asked how we taught people to be good losers. To him, that was the key issue in democracy and one that needed to be addressed before any changes could be made.

My work with the NGO conference began with the same message about democracy. In addition, I shared with them my methodology for working with folks. I ran the participants in the workshop through the process that I was teaching them to use with the rural folks. After discussing the principles of the method, developed by Paulo Freire, we sat in a circle and listed on a board examples from their lives of democracy in action. Participants listed events from relationships that they thought were democratic, meetings that were run democratically, and times when they had been consulted in decisions. We split into small groups and each group took a "democratic event" and listed the components that made it democratic. We got back together and shared our findings. In this way, they learned the method by doing it.

CONCLUSION

My work as an educator is based on the philosophy of Paulo Freire. Freire believes that oppressed people need to "name their world," and in the process they become empowered. This principle has been used effectively in adult education all over the world. His book *Pedagogy of the Oppressed* includes the method he developed for teaching adults to read.

The method has been used in a variety of settings. By allowing people to "name the world," we learn how they view the world and what the obstacles are in our work with them. In Swaziland, it was important for me to know the attitudes some people had toward democracy so that we could address those conceptions before we started our real work. There are many Web sites devoted to the thought and works of Freire, and for anyone preparing to work in difficult and culturally sensitive areas, his insights are well worth pondering. For those preparing to teach or work with educators in the developing world, *Pedagogy of the Oppressed* should be required reading.

REFERENCE

Friere, Paulo. (1981). *Pedagogy of the Oppressed*. Translated by Myra Bergman Ramos. New York: Continusum.

GRANT WRITING

OUTLINE OF PROPOSAL

The organization to which the proposal is to be sent will have guidelines. These should be followed, including special requirements of the particular agency. These requirements may not seem germane to the project being proposed, but they can be based on legal constraints of the funding organization. Some funding programs, for example, require that all proposals include a discussion about the impact of the project on the natural environment and also a discussion of the implications of the project on the status of women. If the special requirements of the donor agency are not met, the proposal will not be considered for funding. Some organization's guidelines are specific about the form of the proposal. Others are quite general. Suggestions are given next about the structure of a proposal when the writer is given much freedom.

A general outline for a proposal is given next.

    Executive summary
    Narrative
        Goals of project and specific objectives
        Activities
        Intended results
        Measures to verify results furthered program goals
    Budget
    List of participants

The executive summary includes a summary of the main points in each of the parts below. It should also include projected starting date and amount of money needed.

The narrative section should describe goals that the project is working toward and the specific objective of the project. For example, a goal is to give women financial independence, or a specific objective is to train seamstresses to market the wall hangings and pillowcases they have made. It is useful to give assumptions linking objectives to goals—an assumption here is that money earned will be available to the person earning it.

The activities part of the narrative describes what will take place when the project is implemented. A monthly schedule is helpful. Here also assumptions might be listed. For this project to be successful, it is assumed women can find time to attend classes. The intended results of the project should be listed: twenty-five seamstresses will be able to form a cooperative to market the fabrics they have embroidered. Ways of verifying if the intended results were obtained should be given, as well as a discussion showing how the results made progress toward the overall goal. For the project used as an example, one might point to the potential cash earnings of the cooperatives and the confidence gained by its members.

The purpose of the budget is to show inputs needed—generally mostly cash but possibly also other things, such as technical assistance. The proposal should show that these inputs are appropriate—sufficient to complete to complete the project, not so much that some will be wasted. Many funding agencies like to fund only parts of projects so the budget should show other sources of funding expected. Contributions by the organization making the proposal are especially compelling. In making the budget, attention should be paid to particular requirements of the agency to whom the proposal will be spent. Some agencies will not fund particular activities, such as salaries for research professionals. Showing assumptions about need for inputs is useful here as well.

Besides cash, the other main input is people. Those who will work on the project should be listed with their experience and training. Some assurance should also be given that the people listed will actually work on the project, if it is funded.

FUNDRAISING

Kwamena Herbstein

Many Non-Governmental Organizations are supported in part by donations. The following article addresses three matters: how to ask people for money, planning a fundraising campaign, and where to get information about foundations that support charitable efforts.

ASKING FOR MONEY

The following paragraphs are based on advice culled from Kim Klein, of the Alliance for Nonprofit Management.

Asking people for money is both the most difficult and the most important part of fundraising. Every community-sponsored organization uses a variety of methods to ask for money, such as direct mail appeals, special events, pledge programs, products for sale, and so on. But the hardest way for an organization to raise money is for board, staff, and volunteers to ask people directly for donations. It is almost impossible to have a major gifts program without face-to-face solicitation of the prospective donors.

An organization raising funds should be aware of the sources of the fears people have in asking for money and should channel a significant amount of time into overcoming such fears in its fundraising staff. The organization should spend a good amount of time recruiting employees who were not only passionate about the purpose of the organization but who were capable of developing a thick skin, a prerequisite for the task of asking for donor support.

One should take note of the fact that four topics are taboo in polite conversation: politics, money, religion, and sex. Additionally, asking people what their salary is or how much they paid for their house or their car is considered rude. In many families, the men take care of all financial decisions, and it is not unusual, even today, for wives not to know how much their husbands earn, for children not to know how much their parents earn, and for close friends not to know each another's income. Many people don't know anything about the stock market, what the difference is between a " bear" and " bull" market, or what the rising or falling of the Dow Jones means for the economy.

TRAINING STAFF

A lot of fear often surrounds the idea of asking for money from others. To aid in dispelling these fears, two easy exercises should be performed during training for staff members. The first exercise would involve having an individual group member be a facilitator and have members of the group share their feelings about their fears about asking for money. Using a chalkboard, the group would look objectively at their collective fears about asking for money. Fear from hearing a strange noise in the night in one's home would be presented as similar to the fear of asking for money. The group would be presented with two common responses to fear. The first would be a reaction associated with hiding in a safe place and imaging all the worst reasons for the noise. The second, more sensible, and more difficult reaction would be associated with getting up and turning on all the lights often and hopefully discovering that the noise was caused by something as benign as a dog in the house, a dripping pipe, the wind, or nothing at all.

This exercise would result in staff members realizing that their fears were really irrational. Group members would have it explained to them that the likelihood of the feared outcome—namely, a refusal to give money—would tend to be much more unlikely than they thought likely in reality. Additionally, staff members would have it explained to them that given the odds of a refusal, it was worth taking the risks of the feared refusal outcome of a request for money.

To begin the first exercise, each group member would be asked to imagine asking someone for money and then to think about all the possible negative outcomes that could result from this. Each participant would contribute her or his ideas on this subject, likely ranging from an insulting reply to a physical assault or to public humiliation. The facilitator would then proceed to examine each feared outcome individually. After a while a possible list of responses might also include the person saying "no," the person shouting at the requester, the person giving the money but not really wanting to and then consequently resenting the asker, the person thinking that the only reason the asker was nice to them was to get money, the person saying "yes" and then asking for money in return for their own cause, or the person asking questions about the requesting organization or individual that the asker was not able to answer properly.

After this brainstorming exercise, the group members would be lead into realizing that all the possible fears could be lumped into three categories. First, fears associated with extremely unlikely responses such as being punched, being sued, or even having a heart attack. Second, fears of things that might happen but that could be dealt with if they did, such as the person asking for money in return or the person asking questions the requester could not answer about what the money was specifically to be used for. Third, fears of things that could certainly happen sometimes, such as an outright refusal.

Examining the third type of fear first, it would be worth noting that the most common fear associated with asking for money is an outright refusal. However, group members scared of this outcome would have it explained to them that fundraising by definition often involves receiving a "no" for an answer. Furthermore, it would be indicated that just as it is a privilege to be able to make a request for money from a possible giver of money, it is also the giver's right and privilege to reject the request. The individual being asked may have just spent a lot of their money, or may have just been asked to contribute to several other organizations, or may simply have other current priorities. Group members would be urged to not take rejections personally in this regard.

Furthermore, in the second fear category, if the individual or organization to which the request was made turned around and made a request back to the original asker, the members would be urged to realize that they don't necessarily owe their donor any reciprocal favors. To be supportive of their cause, one might want to contribute some money back in return; however, one is never obligated to do so. Additionally, in response to the fear of not being able to answer questions posed by the potential granter, the answer of "I don't know" or " I'll find out and let you know" should be presented as perfectly appropriate temporary responses.

Additional fears could be dealt with one at a time by looking at the specific fear category into which each fear falls. In some cases, it could be inappropriate to ask for money, but this is far less often than one would think. Group members would be made to realize that when they considered asking a person for money but hesitated for some reason, that they should always make an effort to ask themselves about whether their hesitation was based on some assumptions being made about the organization or individual they were making the request to. Some of these assumptions would probably be unfounded.

REASONS FOR GIVING

To begin the second exercise, each group member would be asked to consider their own personal reasons for giving or not giving philanthropic money to requesting individuals or organizations and to realize that people in general have very similar reasons. This approach would help fundraisers get an idea of what motivates typical organizations or individuals to donate money.

In this exercise, group members would be asked to envision a scenario where someone they knew had come to ask them for a gift to aid in some worthwhile project. Group members would be asked to further imagine that the requested gift was an affordable one but not something that they would give to anyone who asked. Group members would then write down on a sheet of paper all the reasons they would say "yes" to this request after which they would list all the reason they would reject the request.

Following this the facilitator would then write down on a blackboard all the "yes" reasons and all the "no" reasons. Common "yes" reasons might include the following: because the giver liked the person asking, the giver believed in the cause to which the money would be directed toward, the giver felt they would get something back for their money, the giver believed they would receive a tax deduction, the giver felt generous, the giver felt they knew that their money would be well used, the giver just got paid, the giver wanted to help provide support to their friend, the giver felt guilty about saying "no," the giver knew people in the group making the request, or the giver liked the approach planned in general.

Common "no" responses might be because the potential giver disliked the person making the request, the potential giver didn't believe in the cause that the request was being made for, the potential giver didn't have the money to give, the potential giver was in a bad mood that day, the requesting organization had a bad reputation, the potential giver felt they would rather give their money toward other purposes, the potential giver had already been asked several times that week for money, the potential giver did not know or understand what their donation would be used for, or the potential giver felt that the person or organization making the request was being naive.

The group would then spend some time discussing the two sets of reasons. Taking note of the "no" list, it would be explained to the group that the answers given fell into categories. The first category involves reasons that were not the fundraiser's fault and that could not have been known beforehand. The group would be lead to realize that the asker generally would have no way of knowing that the potential giver did not have money at the time they were asked for it or that a potential donor was in a bad mood, or had been already asked on numerous occasions that week for money. If these were the reason for the rejection, the asker could only thank the prospect for their time and go on to the next potential donor.

The second category would involve reasons that appear to be "no" but are really "maybe." In this case, if the potential donor knew more about the organization, knew how the money was to be used, or knew that the reasons for the bad reputation had been cleared up, then she or he might still be willing to give. In this case, the "no" answers would be really "maybe" answers in a sense that the giver would change his or her mind if they thought that the money would be well spent. They simply needed some convincing that this was the case. It is important to be willing to listen to the reasons for a rejection of the request and be willing to persuade the rejecters to change their minds.

The overall aim of the above exercises would be to impress upon participants that people have varied reasons for saying "yes" or "no" to requests but generally have many more reasons for saying "yes." Fundraisers should realize that there is no reason to be scared to ask for money, since the worst that can happen in response to a request is that the request be declined. When this happens, the decline is often for reasons beyond the control of the fundraiser.

PLANNING A FUNDRAISING CAMPAIGN

Clearly, the most important step in any fundraising effort is to ask for support. Few people make a contribution without being asked for it. The bigger questions though are who, how, and when to ask. Should family members be asked? Should requests be made near the end of the year when people are beginning to think about tax deductions? Should letters be sent, should calls be made, or are requests best made face-to-face? Many organizations should begin by contacting its board of directors who support its goals and objectives and should be able to express this support through a possibly significant financial contribution. Following this, letters would be written to a wide variety of possible donors of money. These would include CEOs of large and successful companies and generally wealthy individuals who show a high probability of being potential donors of money. In these letters would be explained the reason for the establishment of the organization and why it would be important to the global community and to its clients.

Taking into account the fact that face-to-face requests are almost always a necessity when requesting large contributions (usually referred to as major gifts), the organization's leaders must be prepared to meet with potential donors to make their requests for funds. For major gifts, that is, gifts for amounts defined within a certain range such as above $500, members of the board of directors should make some of the donation requests. These directors, when asking for donations would mention that they had made "major gift" contributions themselves. Request letters are often sent out near the end of the year to take advantage of the combination of holiday spirit and imminent tax bills.

Most nonprofit organizations devote major efforts in raising money from foundations. A foundation is a nongovernmental, nonprofit organization that has a fund or endowment; is managed by its own trustees and directors and maintains or aids charitable, educational, religious, or other activities serving the public good; and makes grants, primarily to other nonprofit organizations.

There are more than 51,000 private and community foundations in the United States today, and the number grows weekly. There are four types of foundations: independent foundations established by a person or family of wealth (this comprises the largest group); company-sponsored foundations (also called corporate foundations), created and funded by business corporations, operating foundations established to operate research, social welfare, or other charitable programs deemed worthwhile by the donor or governing body; and community foundations supported by and operated for the benefit of a specific community or region. The first three types are private foundations. Technically, community foundations are public charities, but because their primary activity is grant making, they can be included in the foundation universe. In addition, various religious charities such as the United Church of Christ have a history of looking favorably upon charitable projects.

The wealthier countries of both the East and West and most OPEC countries have a government aid budget. These funds are given (or occasionally lent at subsidized interest rates) to governments of poor countries. The aid may be given directly (bilateral) or via an international or "multilateral" agency such as the World Bank or the UN agencies. A "small proportion of this amount though, may be channeled through voluntary agencies such as Oxfam or through international research institutes."[7] The agency of the United States government that gives aid is the Agency for International Development, abbreviated USAID.

SOURCES OF FUNDRAISING INFORMATION

fdnCenter.org

A prominent Web resource is fdnCenter.org, created by the Foundation Center. This site provides a comprehensive list of foundations and advice on raising funds from them. This could be used as a first point of call to get information on starting a nonprofit organization in addition to gaining funding. For instance, in order to procure and file the appropriate forms required to incorporate as a nonprofit and apply for tax-exempt status, an organization contacts its state's charity registration office. Procedures vary from state to state, and it is necessary to consult with an attorney or a technical assistance agency whose staff had experience in the area of fundraising. One can inquire at a Foundation Center library or Cooperating Collection for the name of a technical assistance group that assists new and emerging nonprofits. Additionally local bar associations might be able to suggest pro bono or reduced cost legal services as well.

The mission of the Foundation Center is to foster public understanding of the foundation field by collecting, organizing, analyzing, and disseminating information on foundations, corporate giving, and related subjects. Their audiences include grant seekers, grant makers, researchers, policymakers, the media, and the general public. The organization does not actually offer the service of preparing and presenting proposals.

guidestar.org

GuideStar is a searchable database of more than 620,000 nonprofit organizations in the United States. GuideStar is used by typing in a name in a Charity Search box to find a charity or using an Advanced Search to find a charity by subject, state, zip code, or other criteria. Users can not only scroll through a list of current projects, but can also use search engines to search for specific charities or projects of interest to them.

---

[7] *For Richer For Poorer,* UK: Oxfam Press, 1986, p. 11.

A SOCIAL SERVICE ORGANIZATION IN BOLIVIA

Katherine Evans

The Instituto Feminina de Formación Integral, which translates to the Feminine Institution for Integral Formation, was established by five women in 1981. When asked what IFFI stands for, the directors and staff laugh with embarrassment and explain that they had to have an innocent name because they were formed a during a dictatorship. They needed a name that was acceptable.

Their initial objective was to work in semi-urban environments with women in two ways. They wanted to enhance the economic participation of women through self-sufficient centers of production and to facilitate their political participation by incorporating women into community organizations like the juntas vecinales. They quickly realized that working with women meant working with children, so they added educational programs. This period, more than others, was characterized by its collective nature. The founding members of IFFI and the women involved in the factories made communal decisions and tried to embody a socialist model.

The year 1987 marked the beginning of neoliberal policies in Bolivia. In response to the political and economic changes, IFFI changed their programs as well, moving into social services more directly. They began to construct children's dining halls, working with issues of nutrition. Their objective was to become self-sustaining by organizing the parents to produce something that would finance the cost of the dining room. They worked primarily with alternative products, but they realized that they were not going to be able to reach their goal of sustainability. Although the financiers had supported their efforts for three years beyond the initial projection, IFFI decided to terminate the programs and either return the money or enter a period of reassessment.

The donors agreed to support the reassessment, and so in 1993, IFFI entered what they call an institutional crisis. They reevaluated their programs and eliminated most of them. There had been a smaller project focusing on gender, begun in 1991, that was conducted at the dining halls. The project researched the needs of these women and began to consider capacity building of those women involved in other organizations. The phase of reflection and readjustment ended in 1994 when one of the founders, who had been particularly vocal and visible, died. Her death engendered solidarity among the donors and called the staff to action. At the end of 1994 and beginning of 1995, IFFI conducted surveys throughout the neighborhoods of Cochabamba for six months. Their surveys generated the three axes in which they now work: gender, popular education, and political rights.

They now have four separate programs, each with their own team of staff that works largely independently of one another. Mujeres y Capacitación (MYC), which translates to Women and Capacity Building, is the program in which I participated. Their focus is gender and civil rights. They conduct workshops on political capacity building for women in positions of political or organizational leadership. These workshops had 30 to 50 participants. They also had workshops that trained the participants to be counselors for abused women in their neighborhoods. These workshops had 20 to 30 participants, many of whom were part of the other workshops as well. Another program's aim is the economic participation of women through training in nontraditional areas based on research about job availability and salaries in Cochabamba. Their third program addresses issues of nutrition and reproductive health with pregnant women and new mothers. Their most recent program, which is not as stable as the others, involves the establishment of libraries and preschools with activities emphasizing literacy and educational support.

Each of the members of the team I was a part of, along with the other director, repeatedly referred to the diagnóstico that was done in 1992 when I asked about who participated in the program planning. This was a six-month process performed at a time of institutional crisis, when the staff realized that the previous programs involving self-sustaining dining rooms was not feasible. They faced the option of closing and returning the remainder of the funds to Canadian International Development Agency (CIDA) or regaining their bearings and reassessing the needs and possibilities existing in Cochabamba. CIDA agreed to sponsor the diagnóstico, which took the staff members into the neighborhoods of Cochabamba to talk to women involved in politics and women's organizations.

The six-month diagnóstico may have begun the culture of evaluation that pervades IFFI now, or the evaluative culture may be the result of pressure from donors and trends in development work. Regardless, evaluations are omnipresent in the institutional life of IFFI. Every program undergoes an evaluation at the end of each year, and, depending on the program, they have one or two other evaluations throughout the year. These are described as internal evaluations. They involve a week of preparation by the program team who present a report to the entire staff who, in turn, critique the results. The report summarizes the goals and activities of the program and compiles evaluations completed by participants and staff.

In addition to the internal evaluations there are external ones. These are much less frequent and occur at the close of each project at the request of donors. In these evaluations, someone is hired from La Paz to come and interview clients and staff and then report her impressions to the donor. I was present for an internal and external evaluation that took place back-to-back, creating an altogether frenzied atmosphere for two weeks.

The evaluations completed by the women who attend the workshops were overwhelmingly positive. The objectives of the program were listed on the forms. These included contributing to the exercise of the rights and citizenship of women in the neighborhoods; supporting women their positions as leaders of the juntas vecinales and members of the Supervisory Committee; creating a space for recreational activities and mobilizations; and finally, supporting the incorporation of the demands of women in the Operative Plans in the districts. The women were asked to comment on whether each objective was met and why. They were also asked to make suggestions about how to improve IFFI's work. There were many calls for IFFI to come to the barrios and conduct workshops with women there, instead of only with those who are politically involved. Two other common responses were that the women had learned to exercise their rights and had become more self-aware through the workshops. They agreed that the workshops supported and increased the visibility of the work of women who were part of the municipal government. One woman commented that they should advise women more frequently about how to confront domestic violence. I interviewed ten women separately from the evaluation, and their responses were similar to those given to IFFI. The women were given many options for commentary and suggestions in the evaluations, which they seemed to use candidly.

When I spoke with each of the members of MYC about their indicators of success, they mentioned the attainment of each of the objectives listed above as well as a few other indicators. Many of the team members said that they view the growing number of women who are leaders in their juntas vecinales as proof that IFFI's work is effective. Apparently a large portion of the newly elected women have completed IFFI's workshops. They consider the number of participants in their workshops as an indicator of their effectiveness. People mentioned that the fact that IFFI is growing is a result of the demand for their services. Others pointed to more subjective indicators including the participants' increased sense of security, confidence, and self-esteem, and their comments that they have found their voice through the workshops

I was told, however, they are beginning to run into problems with their donors regarding the indicators used in that the donors are no longer content with qualitative ones. Donors have started to push for proof of efficiency through numbers—that is, how many workshops, how many women, how much impact. People are concerned that they will start to sacrifice emphasis on the process of capacity building in order to obtain numbers about results.

The results of the evaluations were plastered to the walls of MYC's office on large pieces of newsprint (the preferred medium for all of IFFI's operations). Their conclusions were the following.

- They did not have time to get to the neighborhoods.
- There was not enough time for communication within the team.
- The radio program was isolated.
- There was a need for autonomous initiatives among staff.
- They needed to define the policies that they want to influence.

On another piece of paper, was written "ambitious plans: go to the rural areas in order to diffuse knowledge of the rights of citizens; what's urgent: workshop attendance (saturation/ exhaustion); personal conflicts: the process for analysis." The responses on the newsprint seem to reflect the concerns expressed by the participants and the concerns expressed by the staff accurately. On the whole, the participants called for IFFI to go to the neighborhoods and the staff asked for more independence.

Several authors make the distinction between vision, mission, and objectives reflecting three different levels of purpose. IFFI makes these distinctions as well. One director describes the vision of IFFI as the existence of a "just society where there is equity, democracy, and solidarity." Another director articulates the mission saying that the objective of IFFI in their four programs is "that the population can contribute and construct a different type of development based in the three axes: a more egalitarian society, participation of women in elections, and influencing and generating debate about what it means to introduce the demands of women in the district plans." There are certainly different characterizations of the purpose of IFFI among the directors and staff, but there is also a sense of communal understanding of why they are there and what they are trying to achieve. A program manager demonstrates this when she says, "I am not familiar with the specific objectives of IFFI because I haven't read the premise, but in my program it is capacity building, so that women exercise their citizenship, their right to a life without violence, and the rights of women."

The argument could be made that the mission is not as clear as it should be, which holds some truth, but it is clear that the organizational culture derived from the mission and the leadership has permeated the personnel's sense of collective purpose. Describing the sources and function of NGO culture, Alan Fowler, who has studied NGOs writes

"Individuals implicitly create unique organizations in terms of the ways of thinking using language, beliefs in myths and symbols and ways of doing things so that [NGO culture] co-defines its identity in addition to formal written statements about governance vision and mission"

One does not need to read the official mission statement to know that, at IFFI, the staff is struggling for equity and empowerment through their work with women.

The leaders, as mentioned above, largely influence the organizational culture. In IFFI, the culture is derived primarily from the theory and practice of decision making, which is determined by the tenet of democracy and decentralized authority in IFFI. I was told that in the first five years of its existence "the question of self-management developed a lot; the theory was profoundly influenced by socialism—in that time we wanted a new society, a revolution, so everything was collective." That ideal was incorporated into the decision-making structure of IFFI. All staff members who have been there for three months are eligible to be part of the assembly, which makes a majority of the decisions. This group elects the directors, approves or disapproves the reports and evaluations presented by each of the programs, determine statutes, policies, internal rules, institutional functions and strategies. The assembly gives mandate to the direction, which consists of the two official leaders, but "[the assembly] gives [the directors] a lot of room to function. There are two directors so that power is not concentrated in the hands of one person, and the directors rotate after four years so that leadership is generated within the organization. There are also two consultative bodies, an external and internal one. The external one consists of three people from outside the organization and the two directors. The directors and heads of programs constitute the internal consultative body. The directors emphasized the absence of hierarchy in their descriptions of this structure: "We try to establish a philosophy in the population of democratic relations and participation, so we try to do the same within IFFI."

In many ways, this seems to be the ideal implementation of many authors' normative view of decision making. However, it should be noted that there is a significant amount of discontent among staff members who see a substantial gap between the democratic rhetoric and the reality of power dynamics between management and staff. When I asked one staff member what she would like to change, she responded that things are not democratic and she would like them to be more so. She paralleled the organizational assembly to the national congress (not a flattering image), saying, "It is only there to ratify things and to register your presence. Another responded similarly remarking, "We have a way of functioning like the state. You have to pass a process— live the process institutionally to receive recognition. It cost me years to have the place I have, but I am not the head of a program yet, but at least I'm listened to." It is not very likely that the two groups exist in complete harmony in any organization, but it is important to remember that to some extent, the gap between theory and practice exists here as well.

The staff members of IFFI expressed another common criticism that illustrates the way the culture of IFFI is changing as a result of its growth. In the first week I was there, I observed a meeting of the MYC team in which the new regulations that were going to be put in place were discussed. One regulation involved checking in with the receptionist when you entered and exited the building. Another regulation involved submitting a personal plan of tasks you wanted to complete for the week, when you were going to complete them, whether they were completed, and the reason. A director explained these changes as a way to be more responsible with the financing. (I think these changes were engineered by the two directors, not requested by donors.) The new regulations evoked resentment from staff members who had been there for several years and were accustomed to an atmosphere of trust based on the commitment of the staff to their cause. They complained about the amount of time it took to fill out the forms and the forms were often completed half-heartedly at the end of the week instead of the beginning.

One of the founders and a former director has a very critical perspective on the culture of IFFI and the changes it has undergone. She said that she thought IFFI was losing sight of its objectives and that the sense of community among the workers is diminishing, explaining that there has been a big turnover in the old guard. She criticized IFFI for spending more time becoming a bureaucracy than doing things. She contrasted the initial period of collective decision making without directors to the present situation where there is "more authority, power and manipulation." Although her opinions are probably more extreme than those of the rest of the staff, it is also likely that her opinions represent some of the staff's sentiments.

Decision making also reflects the influence and participation of external stakeholders. Operational and strategic decisions are different. The operational decisions involve the day-to-day implementation of the projects and policies determined through strategic decisions. In the case of IFFI, the operational decisions are made primarily by staff within each team. The women with whom IFFI works and their donors do not play an instrumental role in these decisions. The women do, however, influence the strategic ones through evaluations, and through less systematic consultation, which will not be discussed here. The influence of the community and the donors has to do with the relationship IFFI has with these constituencies and their attitudes toward their participation.

Both directors discussed the equity that exists between IFFI and their stakeholders. In an interview, I used the word "beneficiaries" to ask about their involvement in decisions; the director corrected me, saying that they do not use the term *beneficiary* "because it implies a vertical relationship. IFFI has a horizontal relationship with them and believes that they are protagonists in the programs that develop. If the population does not construct their own proposal, then it is not sustainable." This clearly implies the effect the women have on planning. Another director, discussing the workshops on domestic violence, said, "It is a proposal that they [the women] were seeking, and with them we have the idea with the theory and the practice of what violence is."

However, the relationships do not entirely embody the ideals of equity. The fact remains that the staff members are leading the workshops, and regardless of the degree of participation, a distinction between the staff and clients will inevitably exist. When I was interviewing a few of the women about their experiences with IFFI, several of them mentioned that the IFFI staff members are professionals but expressed appreciation that even though the staff members were educated, the women could speak to them as equals. There is a dependence on IFFI for skills, information and methodology that betrays some of the equality that is said to exist, but the influence of the women on the staff is evident in how they are talked about by the MYC team members.

In describing their relationship with the donors, I was told that they negotiate how to spend the money with the financiers, but IFFI decides the content of the proposal. The donors have a voice in decisions about how much to spend on each of the things that IFFI has proposed, but their decision-making power does not extend beyond that.

> "We have the luck of having good relationships with our donors—it is horizontal, not one giving, the other receiving; we are not interested in donors who only care about giving money. … We've had the same ones for many years and they are advancing with us in the process of reflection."

She went on to say that they have declined donors whose idea of the process was that they were donors and IFFI was an acceptor.

A SUMMER INTERNSHIP IN GUATEMALA

Monique Schumacher

I worked with a group of mothers, who had taken the initiative to form a women's cooperative, La Cooperativa de Mujeres Xelabaj. My initial project proposal outlined a summer of putting the business's products on the Web and teaching about e-commerce. Upon arrival I found that the cooperative, having just formed in April of this year, had many obstacles to overcome before tapping the e-commerce market.

The women were at a low morale. They were making blender covers and aprons and attempting to sell their merchandise to the students at the Spanish language school, Casa Xelaju.

They found little success, as the product design did not fit the tourist market's buying patterns. As my first task I worked with the cooperative to design purses, wallets, shirts, hand towels, and placemats, products more likely to sell, taking care not to lose the items' Mayan characteristics. I struggled to preserve the beauty of their indigenous cultural designs, while not alienating potential customers by offering items with only colors so bright as to be out of the bounds of consumer taste. At first the women were skeptical of these changes, but I persuaded them to let me, at least, test my ideas.

With the diversification and redesign of product line, students from the foreign language schools in the city began buying, sparking motivation within the cooperative. For though the initial designs were largely mine, the women, after seeing the selling possibilities, showed interest in new product creation. I had gained their trust. I brought in fashion magazines from the United States to illustrate American consumer preferences. A response to my efforts came rapidly. One women said that we should pay more attention to the students' dress, while another promised a hand towel design with an embroidered Mayan symbol. Soon many women were coming up with new styles and other possibilities for merchandise.

As there was an overwhelming acceptance of mediocre work, quality remained a large impediment to future exportation and even local sales. "You can't charge even three quetzals (40 cents) for a wallet whose zipper cannot close," I told the women time and time again. This concept was difficult for the women to understand, as they did not in grow up in a service-oriented, "the customer is always right" society. In order to rectify this lack of attention to well-finished products, we created a quality check system. On a rotating basis one member was put in charge of approving all the finished products; to get credit, the item had to meet a series of standards.

With all this discussion about quality, we still, however, remained without the financial resources to purchase high-end materials. Faced with this lack of capital and to zero in on the varying tastes of Casa Xelaju's international student body, we developed the option of custom-made products. Those interested could come to the basement of the language school where the women worked, meet the members of the cooperative, and from samples choose their material of choice, making a down payment, to cover costs.

Quetzaltenango has 15 Spanish language schools, providing a population of students and tourists, potential consumers for the cooperative's product line. Capitalizing on the student desire to connect with the Mayans and their general altruism towards cooperatives, I began advertising the group heavily in local American gathering spots. With the introduction of special order products and distribution of information to the Americans and Europeans about the existence of the group, the women successfully marketed their products at a local level.

If not working in the cooperative, the women of Xelabaj raise money for their families by washing clothes in the city. Though some had an innate entrepreneurialism, for many the cooperative was their first business endeavor. Thus, they lacked a foundation of commerce knowledge. In response I began teaching a weekly business class with discussion/lectures on topics ranging from presentation and selling skills, to the time value of money, to basic ledger keeping. The mathematically orientated classes were the most difficult as only 4 out of the 15 women had literacy. Though I tried to use physical objects such as marbles in order to illustrate concepts, not everyone was able to grasp these more arithmetic concepts. I ended up working longer hours with the cooperative's treasurer and others who were interested to ensure that some members knew this material well.

With these classes and local level sales, we began the process of looking for export markets. My Internet research showed that the export market for Guatemalan crafts was fairly saturated. Our product line, however, offered more than handicrafts, as I had worked to transform the pants and skirts into something an everyday American would wear. Though confident that the cooperative's merchandise could one day attract international buyers, the time demands on this project went beyond my three-month constraint. I could, however, accomplish the groundwork.

With an official descriptive document to create, legalization papers to work out, and export contracts to be negotiated, we had a lot to do. We began by meeting to discuss the cooperative's history, objectives, and their mission statement. I thoroughly enjoyed these meetings as each woman put forth her own interpretation as to what we had been doing, each expressing their own fears and goals for the group. Using their words, I put together a bilingual pamphlet. The steps toward legalizing the cooperative were long and complicated. I translated the laws of cooperatives into a more basic Spanish and found a lawyer through a Guatemalan professional association. Finally I wrote alternative trade organizations about the existence of Xelabaj, and to those who were interested, we sent sample products and more formal information.

Sadly, the summer ended to quickly with these final projects only half completed. Fortunately, my advisor was able to locate another American willing to take my place as intern and continue with the process of guiding the cooperative through its development.

CHAPTER 2

Energy

Edited by NEWELL THOMAS
Bright Technologies LLC

Inexpensive, reliable energy brings amenities to rural villages. Energy availability also increases possibilities for rural industry. Some common uses of energy in villages are grinding grain, illuminating homes, and operating radios or related electronic devices. Grinding grain using machine-powered equipment saves women appreciable energy and frees up time. The alternative is hand pounding, often done with a large mortar and pestle. Grinding mills can be powered by electricity, a diesel engine, a water wheel, or in other ways. A typical small mill has capacity to grind sufficient corn for domestic use for a small village. The amount of energy per user is small. Similarly, a small amount of electricity in a household, sufficient for three or four electric lights and a radio, can improve living conditions significantly. The national utility in South Africa provides an inexpensive, minimal service to low-income homes—simply three or four outlets in one room, no fuse box, meter, or house wiring. The expense of a meter is avoided by charging a flat rate, independent of usage. Essential issues when considering how to increase energy availability in a village are the type of energy—electrical, mechanical, or other—needed and cost.

For some applications, several different approaches are possible—water can be pumped by a wind driven machine or by an electric motor powered by photovoltaics or by a diesel pump. In deciding the approach, cost is important. So is reliability—photovoltaics have a reputation of working with long periods with minimal maintenance; hydroelectric systems need to be watched and adjusted. Kerosene lamps, an alternative to electric lights, will not work without kerosene, and in many cases, kerosene is an imported fuel. The dependability of the supply will depend on both vagaries of foreign exchange and transportation logistics. Some energy systems— a grinding mill or a hydropowered electric generator—may require village-level ownership and thus building an organization. A final concern on choosing an approach is the resources available: Hydropower needs a stream with sufficient flow and drop, windpower needs wind, and solar power needs sun.

Renewable energy sources—solar, hydro, wind, and so forth—are attractive because they are nonpolluting and do not require imported fuel. They have made less progress in the Third World than might be expected, perhaps because the technologies are not entirely developed, perhaps because the technologies are not well known. In many cases the alternative to a renewable energy source is a diesel engine that does emit pollutants and requires imported—in most cases—fuel. The advantage of the diesel engine is its reliability—the design has been perfected over many years. It is unfair and unwise to promote a technology, including renewables, that is not as reliable, when installed, as a diesel engine. Renewable energy can be less expensive and socially more beneficial in many Third World situations, and its use should be considered very seriously, but it should not be installed unless it be useful.

A very common energy source, which is somewhat renewable, is wood fuel in particular or biomass in general. In much of the world, wood is still the cheapest fuel, but this is not true everywhere. In Nepal, where deforestation has taken a toll, alternatives to wood for cooking and space heating make sense. In general, though, promoting an alternative to wood fires will not be easy, although certainly important. Although it seems very evident to outsiders that wood fires should be avoided because they produce noxious smoke and require much time for fuel gathering, local people may feel differently. Wood smoke may be seen as useful in eliminating mice in thatched roofs; the health implications are long term and not compelling. Promoting alternatives to wood fires may very well be a useful thing to do, but it should be tempered with an understanding of the perspective of local people. The lesson—take very seriously the concerns of the projected user—applies generally, not only in regard to wood fuel. A related issue is making decisions about what local people need: Is it more important to use electric power for small industry or for home lighting? Unless the user decides and sets priorities, the system probably will not sustain itself.

The titles of the articles for this section are probably self-explanatory. The energy-planning article deals with choosing between forms of energy. Because electricity is particularly important and can be generated in several ways, the basics of designing an electrical system are discussed in a separate article. Water power and wind power are widely used. Both can produce either mechanical or electrical power. Solar energy can be used to heat a building or heat water—the technology is discussed in the "Solar Heating" article. Solar energy can also be used to produce electric energy through photovoltaics, the subject of an article. As noted above, much of the energy used in the Third World goes to cooking on wood stoves, discussed in an article. Charcoal is a convenient alternative form of wood—its production is described in an article. An alternative to wood fires is a solar oven. A construction of one of these is also described. Biomass—animal manure or plant residues mainly—can be converted into methane fuel in a methane digestor. Fuel cells are an alternative way of generating electricity through the combination of hydrogen gas and oxygen in the air. Fuel cells are useful because they allow energy to be stored as hydrogen gas. Another, more common storage mode is a battery, the subject of an article. Diesel oil is the fuel of choice in many places but an alternative is ethanol, which can be produced from corn and other grains. Ethanol is the only widely used renewable fuel for motorized transportation. An abundant energy source near oceans would appear to be tidal power. Economical systems for extracting energy from tides have not been implemented—some notes are given in an article.

ENERGY PLANNING

This article is based heavily on "Decentralised Rural Electrification: Technical and Economic State of the Art and Prospects," Bernard Chabot, Senior Advisor on Renewable Energy, ADEME, 500 route des Lucioles, 06560 VALBONNE—FRANCE, 1995

Supplying electricity to an isolated home or community is the major problem addressed in this article. A user able to get power from a national or regional grid through an inexpensive tie-in will nearly always save money by doing so. Extending the power grid one kilometer

(0.6 miles) or further to serve low-consumption users is generally uneconomical. In some cases it may make economic sense for a homeowner connected to the grid to generate power herself or himself and sell surplus power to the grid, but the more common problem is to get at least a small amount of power into a village to improve life there. In many cases, providing power for lighting several rooms and for operating a radio, television, or cassette player will make a significant improvement.

Energy is used in a household in other ways than lighting and electronics. In rural homes in the Third World much energy is used in cooking. Efficient cook stoves and solar cookers are discussed in a separate article. Space heating and provision of hot water can be done using solar energy, also discussed in an article. Actually in much of the Third World the fire used for cooking also warms the house, and hot water is cooked on the same fire. In many areas firewood for cooking is becoming scarce, causing deforestation or economic hardship. Electric cookers would conserve firewood, but in most cases they cost more to use than wood stoves. Kerosene—also called "paraffin"—is a common substitute for expensive firewood. Refrigeration retards food spoilage. The most common refrigerators are electric powered, but refrigeration can be done using propane gas or kerosene. Household energy needs in the Third World at present seem to be most effectively met by providing a small amount of electricity and by managing fuel wood.

Basic characteristics of electrical power are discussed in the next section. Implications of that discussion are (1) for some important applications no substitute for electricity exists, (2) losses in transporting electricity can be appreciable, and (3) a major use for electricity in low-income homes is lighting. It is clear that an effective energy plan must provide electricity to homes. To avoid transmission losses in simple electrical systems the generator should be physically close to the load. Complex power systems use fairly elaborate "transformers" to reduce transmission losses. It is worthwhile using high-efficiency fluorescent lighting fixtures in small-scale electrical systems.

Electricity can be generated using small-scale technologies in several ways. Photovoltaic devices—solar cells—produce electric power directly from sunlight. Small scale hydro and wind systems use the energy in moving water or air to rotate a generator, which produces electric power. A diesel-powered generator, common in the Third World and as a backup in the United States, burns petroleum-based fuel in a motor that drives an electric generator. Of these only the diesel generator uses fuel. Renewable energy sources—hydro, solar, wind—usually have higher construction costs and lower operating costs than a diesel generator. The initial cost of renewable energy systems may be so high that government loans or subsidies may be required, even though the lifetime costs are less than for a diesel generator. The generators used by commercial power utilities are basically the same, but larger, as diesel-powered generators. Coal, oil, natural gas, or nuclear energy, as well as hydro, powers these generators.

Moving power, "distribution," from generator to user can be a significant problem. Commercial utilities distribute power from large generating installations through complex power grids, complex in part to minimize transmission losses. Village-level electric systems often use a diesel generator and distribute power through a small grid. If each household generated its own electricity, distribution would be simpler. Household-size diesel powered generators—as well as photovoltaic, hydro, and wind electric systems—are feasible, even sold commercially in the United States. Larger generators, though, are more economical per unit of electricity produced. The planning tradeoff is between generation and distribution efficiency. The generation costs of electric power are generally much lower for a power utility than for a household- or village-size system and a modern large commercial generator emits fewer pollutants, per unit of electricity produced, than a village size diesel powered system. The initial costs, though, of connecting an isolated house to the national power grid are often prohibitive.

One approach to distribution is use of battery-powered lights and appliances and a central battery charging facility. The author visited a village where automobile storage batteries were used to power a sound system—the batteries were recharged by an automobile making its usual trips. A photovoltaic-powered charging facility serving a village where household electric needs are met through rechargeable batteries can make economic sense.

Planning an energy system for a village or a single household is more than choosing hardware. Information has to be made available to users so an informed choice is made. Part of the information needed is about the technology—its capacity and cost. Another part needed is about system inputs, for example—how much sunlight can be expected? How reliable are the winds? Is the supply of diesel fuel dependable? Another "software" consideration is the skills needed for operation and maintenance. How can technical support be made available? Decentralized electric systems usually require more knowledgeable users than customers of the national utility. A final planning consideration should be "upward compatibility"—how easily can the initial power system be expanded to serve more users, to allow more things to be done with electricity?

Important factors in energy planning are sometimes overlooked. Inexpensive and reliable electricity in a village may make small manufacturing—carpentry, food processing, spinning and weaving—viable and thus improve the overall economic situation, even reduce the flight of young people to cities. Water pumping can increase agricultural production and make cleaner water available. Electric power can improve health care and education. Productive time is increased by allowing people to work in the evenings and by the availability of telephones. Hydro, photovoltaic, solar, and wind energy do not cause global warming.

"Co-Generation" consists of using the same power plant to generate both electricity and heat. The heat usually is in the form of steam. It is generally used to power machinery or for space heating. Mid-sized users, a school or factory, in the United States find the cost and efficiency of such a system greater than that of two separate systems basically because the heat lost in a conventional steam driven electric generator is put to work in a co-generation system. Cogeneration only makes sense, of course, when the heat energy can be put to use. An example where cogeneration might be practical is a village where both a small-holder tea factory and community electrification are being planned.

BASICS OF ELECTRICITY

NECESSITY FOR ELECTRICITY

Although electricity is somewhat cumbersome to generate, it adds significantly to people's lives. Radios, televisions, and computers can only be powered by electricity. Interior lighting in a home is done much more conveniently using electricity than by alternatives. In a hospital, alternatives are even less useful. Electric motors can make small-scale manufacturing feasible, giving a village a way to raise cash. Although a water wheel or steam engine can be used to power small machines, the apparatus is complicated and dangerous—pictures of 19th-century factories show the overhead shafts and belts required before electricity was available. Electrical supplies, however, can be complicated and expensive, and the difficulties and cost increase as the amount of power produced increases. In most cases it is probably not sensible to use electricity to cook a meal or warm a room when a simpler kind of energy would do, although electric cookers have been proposed as a way to reduce deforestation.

Electricity can be supplied to a village usually in five ways: extension of the national power grid, hydropower, photovoltaics, wind power, and a diesel-powered generator. Combinations of sources may make sense—a diesel generator can back up a hydro system during dry periods. Photovoltaics and wind power often complement each other—when the wind blows the sun does not shine. A user who is connected to the utility company does not have to worry about storing electricity—when unneeded power is generated from, for example, a wind generator that power can be sent to the utility. Power sent into a utility earns money—in a sense power is being stored in the utility. In planning an electric system the costs of each alternative need to be weighed, including storage costs and, in the case of diesel generation, fuel costs. One of the purposes of this article is to learn enough about electricity to plan the capacity of the electrical supply. We start with the basics.

VOLTAGE, CURRENT, RESISTANCE

The fundamental aspects of electricity will now be discussed. *Voltage* is analogous to pressure. It is the amount of force or push that the electricity has. The units measuring voltage are *volts*. A flashlight battery pushes with 1.5 volts and an automobile battery with 12 volts (when new!). An electric outlet in the United States pushes with the equivalent of 120 volts. The word "equivalent" is used because actually the voltage is changing continually, as discussed below.

Current is the amount of electricity that flows. Electric current actually consists of the motion of small charges, called electrons. Current is measured in *amperes*. An ampere is the flow of a certain number of electrons through a conductor in a second. Fuses or circuit breakers in a home will blow or trip when more than the rated current flows through them. A 15-ampere fuse will protect a circuit designed for 15 amperes by melting when more than 15 amperes flows through it. A 20-ampere fuse can carry more current without melting and so does not protect a circuit designed to carry 15 amperes. Electric current always flows in a complete loop or circuit. In some cases the return path can be through the ground or the metal frame of the structure, as in an automobile, but usually the return path is a wire.

The amount of current that flows in a wire depends on the voltage. It also depends on the size and material of the wire or the *resistance* of the wire—the greater the resistance, the smaller the current. Resistance is measured in ohms. The abbreviation for ohm is capital omega, Ω. A 4Ω loudspeaker will allow twice as much current to flow as an 8Ω speaker, which is why it is risky to replace an 8Ω speaker by a 4Ω one. (Amplifiers can be damaged if too much current is drawn from them.)

Current, voltage, and resistance are related as shown in the following equation, which says that current increases as voltage (or pressure) increases but decreases as resistance increases. The letter I represents current, V represents voltage, and R represents resistance.

$$I = \frac{V}{R}$$

Often the preceding equation is written as follows. This relationship is known as Ohm's Law.

$$V = IR$$

EXAMPLE OF OHM'S LAW

Commercial electric wire comes in various thicknesses—the thicker the wire, the more cross-sectional area for the current to flow through so the smaller the resistance. A so-called #4 wire 100 feet long has a resistance of 0.025Ω, while the same length of #10 wire, which is thinner, has a resistance of 0.1Ω. If 80 amperes flows through the #4 wire then the voltage lost in the wire—that is, the pressure used up pushing the current through the wire is 2 volts.

$$V = 80 \times .025 = 2 \text{ volts}$$

The voltage loss in a #10 wire (resistance of 0.1Ω) would be 4 times as large.

$$V = 80 \times .1 = 8 \text{ volts}$$

One reason the lower-resistance wire is not always used is cost. One hundred feet of insulated #4 wire would cost about $120, while 100 feet of insulated #10 wire would cost about $23.

POWER

A fourth quantity exists, besides voltage, current, and resistance. It is called *power*. Power is the amount of work done per second. Power is measured in *watts*. A 100-watt amplifier puts out twice as much power as a 50-watt one. By law in the United States, an appliance manufacturer must indicate on an appliance how much power the appliance uses. (In some cases current rather than power needs are given.) A typical toaster, for example, uses 1,200 watts, while a vacuum cleaner uses 520 watts. An electric company charges for the total amount of work done, or energy used, so an electric bill shows charges for "watt-hours" or, more usually, "kilowatt-hours." A kilowatt is 1,000 watts.

Power is the product of voltage and current.

$$P = V \times I$$

This equation makes sense. One can increase work done or power consumed by pushing electrons harder or by pushing more of them—in other words, by increasing voltage or by increasing current.

The preceding formula for power can be used to calculate the amount of current an appliance uses. The reason one cares about the amount of current that will flow is that not caring may result in a blown fuse. Appliances sold in the United States are generally meant to be used with 120 volts. If an appliance is rated at 1,200 watts, the following equation shows it uses 10 amperes from a 120-volt circuit.

$$P = V \times I$$

$$1200 = 120 \times I$$

$$\text{or } I = 10 \text{ Amperes}$$

If a home has 15-ampere fuses, then the total current in one circuit must be less than 15 amperes so other appliances plugged into the same circuit as this 10-ampere appliance must use less than 5 amperes. A 60-watt light bulb takes 0.5 amperes, so the number of 60-watt bulbs that can be powered from a 15-ampere circuit is 30.

Table 1 is a summary of the fundamental electric quantities.

TABLE 1
Fundamental electric quantities

Quantity	Description	Unit	Abbreviation	Symbol
Voltage	Pressure	Volt	V	V
Current	Amount of flow	Ampere	A	I
Resistance	Resistance	Ohm	Ω	R
Power	Voltage × current	Watt	W	P

HOW MUCH ELECTRIC POWER DOES A FAMILY NEED?

In order to design an electric system, one must estimate needs. The power needs of some appliances are given in Table 2.

Two remarks are in order. The stereo, which uses 100 watts, probably only puts out 50 watts of sound—the rest of the input power is given off as heat. In estimating electric power needs one must consider the input power required, not the desired output power. The other remark concerns the refrigerator. The power usage shown applies when the motor is on. Most of the time the refrigerator motor is off because only when the temperature rises above the thermostat setting does the motor turns on. The refrigerator takes power, therefore, only about a quarter of the time it is plugged into an electric outlet. The length of time it is actually drawing power depends on how well it is insulated and how often the door is opened. Also, most motors, including refrigerator ones, draw much more current when starting than when running—just as a car requires more gas when accelerating than when going at a constant speed. The starting current, which lasts for only a short time, is about four times the rated current. Because the starting current is large, time delay fuses are used in homes. These fuses will only blow if the current exceeds the rated value for five seconds or so—long enough to start the motor but not long enough to overheat the house wiring. It takes time for the house wires to get hot enough to do damage, just as it takes time for a toaster to get hot.

Now let us estimate minimal electric energy needs for a small family, using Table 2. The question is not so much what we would like but what is really needed for a reasonable life style. Of course, each person's conception of what is reasonable differs, which is why technology planning is difficult—but let us give it a try. We might start by deciding that we need power from about 6:00 P.M. to 11:00 P.M.—that is, five hours. During this time we need four 75-watt lightbulbs on at once, plus a stereo. We might also agree that either a sewing machine or an electric drill would be used, but not both at once, again for five hours. We certainly will need a refrigerator, but if we are careful, it will draw power less than five hours a day. If we need a water pump, then again we have to estimate how many hours it will be on. Can it fill a tank when no other appliances are on? For now, let us ignore the water pump. Our minimal power needs are shown in Table 3.

TABLE 2
Power needs of appliances

Transistor Radio	4W
Small Stereo	100W
Black & White TV	45W
3/8" Portable Drill	260W
Refrigerator	300W
Sewing Machine	100W
Water Pump	335W

TABLE 3
Minimal electric needs

4 75-Watt Lightbulbs	300 W	5 hours
Stereo	100 W	3 hours
Sewing Machine or Electric Drill	260 W	4 hours
1 Refrigerator	300 W	5 hours
Total	960 W	5 hours (worst case)

Our daily energy needs are approximately 960 watts for five hours, or 4,800 watt-hours. Thus, we must design our system to supply 960 watts at any one time and to supply 4,800 watt-hours (4.8 kilowatt-hours) over a day. If we use a 12-volt system, then we will need 80 amperes for those five hours—dividing the power requirement by the voltage gives the current.

It is worth noting how much of the power is going into incandescent lightbulbs. In our design we really should have specified fluorescent bulbs, which use much less power to produce the same amount of light—these bulbs do cost more. Fluorescent bulbs use about 25 percent of the power of incandescents for the same illumination, so the 300 W for light can be reduced to 75 W. The total power needed, then, is 735 watts rather than 960. In the next few sections we will see how to design the system.

PRACTICAL GENERATORS

The principle of electric power generation is simple. The basic physical principle is that a changing magnetic field through a coil of wire produces a voltage across the ends of that coil. The construction of a generator is straightforward. A generator is shown in Figure 1. A magnet, usually an electromagnet, is rotated inside a stationary coil of wire. In Figure 1, the stationary coil is really a set of coils connected together. These are labeled "Stator" in the figure, where the rotating electromagnet is labeled "Rotor." A small current is fed into the rotor through the slip rings, producing a magnetic field that points in a direction perpendicular to the rotor shaft. As the rotor rotates inside the stator, the magnetic field in the stator coils changes. This changing magnetic field produces voltage in the coils. The coils are connected in series, and the two ends lead to the electric load—that is, these ends of the coil go to the generator terminals. An electric company's generator terminals are connected to transformers, switchgear, and transmission lines but ultimately go to the metal pieces inside the electric outlet in a home. When an appliance is plugged into the outlet, the circuit is completed—current flows out of one end of the coil, through the connecting wires to the load, and back from the load to the other end of the stator coil. The voltage produced by the rotating electromagnet is the force that drives the current through the circuit.

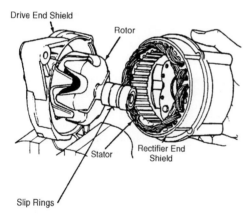

FIGURE 1
Schematic of electric generator.

The circuit is completed slightly differently in an automobile and some other applications. In an automobile one end of the generator coil is attached to the metal frame of the car and the other to the load. The load must then be attached between a wire from the generator and the metal frame. Because only one wire is used, it looks as if no return path for the current exists. Current has to flow, however, in a complete circuit—from the generator to the load and back because the generator does not produce electrons. It only pushes them along the circuit.

While the principle of a generator is simple, construction is not easy. Because the generator turns rapidly, balance and good bearings are necessary. Also, winding coils is tedious. The result is that it does not make sense to try and build one's own electric generator unless one has access to a machine shop, a coil winder, and also to some experience.

Every automobile contains a generator like the one shown, which recharges the battery and powers the lights, spark plugs, radio, and so forth. One possibility, then, in a homemade electric power system is to use scrap automobile components. Another possibility is to buy a commercial generator.

An intermediate possibility between using a scrap automobile generator and a commercial generator is to use an induction motor as a generator. Induction motors are very common—most electrical motors in use are induction motors. They are rugged and probably more reliable than an electric generator. In normal use electricity is applied to a motor, causing it to rotate, producing mechanical power. If, however, the shaft of the motor is turned by, for example, a water turbine, then electricity will be produced. The motor does have to be modified slightly, by the addition of capacitors, to become a generator. Choosing the correct value for the capacitors is not simple. Some technical training is probably required to design and operate an electrical system using an induction motor, so the approach will not be considered further here but is described in detail in Nigel Smith's book, *Motors as Generators for Micro-Hydro Power.*

In an automobile a device called a "voltage regulator" is placed between the generator and the battery. Voltage regulators are needed because the voltage from the generator depends on the engine speed, which varies. Let us see why. The magnitude of the voltage from a generator depends on how fast the magnetic field is changing. When the generator is rotating rapidly, the voltage produced will be large, and when the generator is going slowly, the voltage out will be small. The generator is driven by the car's engine, so when the automobile is going fast, as on an interstate, the battery is charged rapidly. When the engine is idling at a light, the voltage produced is small. Why do we care about the variation in voltage? One reason is that a high voltage might damage the battery. Another is that when the generator voltage is less than the battery voltage, current will flow back from the battery into the generator. The voltage regulator prevents this battery discharge at low generator speeds and also reduces the output voltage at high speeds. A regulator should be used in a homebuilt system, although an operator could monitor the output and make adjustments manually. The voltage regulator is built into most modern automobile alternators, so it is hard to avoid getting one if an alternator for a recent car is purchased. A commercial electric generator will usually come with a built-in voltage regulator, and, as one would expect, the utility companies have regulator mechanisms for their generators.

Because spare auto parts are so easily available, it is tempting to design a home-built system around them. Before we settle on a 12-volt system, however, let us go back and see the implications of some of the calculations we did earlier. If we need 960 watts and will use 12 volts, then according to our power equation, we would need 80 amperes.

$$960 = 12 \times I$$

$$\text{or } I = 80A$$

If we want to live some distance from the generator—to reduce the noise, for example—we would have to run the electric current through connecting wires. If our house was 50 feet (16 meters) from the generator, then we would need 100 feet (32 meters) of wire to get a complete circuit. We saw in previous equations that if we used #4 wire, our voltage loss along the 100 feet of wire would be 2 volts, so of the 12 volts generated, only 10 volts would be available to do anything useful. If we tried to save money and used #10 wire, which has a higher resistance, we would lose an intolerable 8 volts. Actually, #4 wire is very thick and difficult to handle, and even #10 is thicker than used in most homes. Also, 20 feet (7 meters) of wire is probably used up internally just going from outside of the house to the outlets, so with 100 feet of wire, the house and the generator could be only 30 feet (9 meters) apart.

It appears, then, that a 12-volt system has a real problem; to get sufficient power in the load, one needs large currents, which mean large voltage losses in the wiring. The usual solution is to use higher voltages, which is why the standard voltage in U.S. homes is 120V and in most of the rest of the world, 240V. Small 120-volt or 240-volt generators are not nearly as common as automobile generators and are much more expensive.

DC AND AC POWER

Another difference exists between the voltage used in an automobile and that in a home—dc vs. ac. The abbreviation "dc" stands for direct current, meaning that current always flows in the same direction—one generator terminal is always positive and the other is always negative. Batteries give dc voltage and must be recharged using dc. In fact, the reason dc is used in automobiles is just that—to recharge the battery. The reason 12 volts are used is that the chemical process inside a single battery cell produces only 2 volts. To get higher voltages, cells must be connected in series, and 6 cells are as many as are practical, otherwise the battery would be too big to fit into the automobile conveniently. Also, the connections between cells are costly, so the manufacturer does not want to use more cells than necessary. See Figure 2.

AC voltage is actually much more common than dc. One reason is that the generator in Figure 1, and most generators, actually produce ac, as one might expect from a rotating device. As the magnet rotates the magnetic field changes direction. Other devices, called "rectifiers," are used to convert the ac to dc. In most present-day automobiles the generator is actually called an "alternator." The rectifiers are an integral part of the alternator so that the voltage into the battery is dc. Rectifiers are discussed in the Appendix to the "Photovoltaics" article.

A major advantage of ac is that transformers can be used to change the voltage level. Step-up transformers increase the voltage but decrease the current; the power is not changed from the input to the output of a transformer. Step-down transformers decrease the voltage and increase the current. Utilities generate at medium voltage, transform to high voltage for transmission, and transform down to low voltage for the user. High-voltage transmission means low current and thus small losses in the lines. Low voltages, though, are safer, which is why the user is given 120/240 volts. The reader may have noticed large, metal, cylindrically shaped objects on

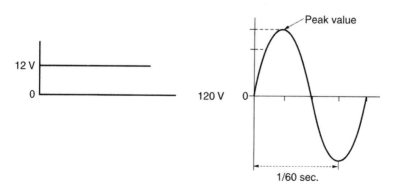

DC voltage as in an Automobile   AC voltage as in a U.S. Home

FIGURE 2
DC and AC voltage.

electric utility poles. These are transformers, which step down the voltage from 440 volts, used for local transmission, to household voltage. An arc welding machine needs high currents but does not need high voltage once the arc is created, so a step-down transformer is used to increase the current and decrease the voltage.

A final advantage of ac is that it can drive motors at a constant speed, corresponding to the number of times the voltage goes up and down in one second. In the United States the voltage alternates 60 times a second. The number of alternations per second is called frequency. In most of the rest of the world the power company frequency is 50 cycles per second. The reason we care about the frequency is that we must make sure it is correct; otherwise electric clocks would not run at the proper speed.

A problem with small-scale hydro- and wind-powered generators is that the generator speed, and thus the voltage frequency, may vary depending on the amount of wind or water. These frequency variations will speed up or slow down electric motors, as noted above. If the frequency of the voltage applied to a motor is significantly lower—10 percent or so—than the rated value, the motor may be draw excessive current and be damaged. The magnitude of the generated voltage also depends on the generator speed. Appliances are not usually damaged if the voltage applied is less than the rated value. Incandescent lamps last longer if the voltage is low but give off less light. At voltages higher than the rated value incandescent lights have an appreciably shorter life. Fluorescent bulbs start less reliably at low voltages but are not damaged by moderate, of the order of 20 percent, overvoltages.

As mentioned before, rectifiers convert ac to dc. Equipment called "inverters" convert dc to ac. Inverters are more complicated than rectifiers, partly because they must produce the correct frequency. A typical price for a 1,000-watt inverter is about $450. Inverters also waste some power, so only about 80 percent of the dc input power is converted to ac output power—the rest is given off as heat. Inverters are basically switches that turn the dc voltage on and off. We would use an inverter if we had a dc power source, such as a battery, and wanted ac power.

STORAGE

Some generators, such as those powered by wind or solar, do not produce power continuously, so the electricity must be stored. The easiest way to store electricity is with batteries, but, as mentioned before, only dc can be stored in batteries. Other methods of storage are cumbersome but possible. Utilities sometimes store electricity by pumping water into a reservoir on top of a mountain when excess power is available and then using that water to drive turbines when power is needed. In this way, the peak power from a steam-powered generator can be greater than the capacity of the generator. As both the pumps and turbines are ac equipment, ac electricity is being stored.

A scheme analogous to storage is to sell excess electricity back to the utility company. This is straightforward in practice, since when current flows through the meter in the opposite direction—from house to utility—it drives the meter backwards, reducing the bill. When one generates more electricity than one is using, then electricity is being sold to the utility company. When more electricity is being used than is being generated, then electricity is being bought from the utility. This process of buying and selling amounts to storing electricity with the utility. By law a utility must buy electric power from homeowners, although the price paid may be different from what the price charged. The utility insists that the frequency of electricity sent in be exactly right so the generator or regulator in the homeowner's system must be accurate. Of course, such storage with the utility is not possible if one is not connected to the power grid.

Batteries are a relatively mature technology; in other words, most of the complications have already been worked out. Automobiles use so-called lead-acid batteries—the plates are lead and the fluid is sulfuric acid. When charging, a chemical reaction takes place at the plates. At discharge the reverse reaction takes place, letting current flow out. Nickel-cadmium batteries are similar, somewhat smaller and more reliable, but more expensive. Lead-acid batteries are probably the most practical for a home-built electrical system because they are readily available. Automobile batteries actually are designed to give a large amount of current for a few seconds to start the engine, which is a different application than giving sustained current for several hours or even days, so in a home system a so-called "Deep Cycle" battery, designed for golf carts, recreational vehicles, or small boats is used. A typical deep-cycle battery might supply 9 amperes for 10 hours or 15 amperes for 5 hours before needing recharging. Such a battery costs about $80. Batteries do need care, mostly checking the fluid level. They work best when warm and need protection from freezing.

In our family example, 80 amperes were needed for 5 hours. As each battery can give 15 amperes we will need 6, which gives some spare capacity.

OVERALL SYSTEM

Now we will design the overall system. We have several choices. Let us opt for a simple one that could be built from scrap automobile parts. The system is shown in Figure 3.

Let us review what is in each block. The "mechanical driver" turns the generator. It could be a water wheel, a wind-driven propeller, or a diesel motor. The automobile generator produces 12 volts dc. The generator is probably really an alternator with built-in rectifiers. Such alternators/rectifiers are available at auto supply stores or junkyards. Because so many are around, the price is low—perhaps $50—although the operating life may be only three or four years. The voltage regulator shown is the same as used in automobiles and is purchased from the same places. It controls the amplitude of the voltage produced as the generator speed varies. The batteries store energy during the times when the generator is not moving. "Deep-cycle" batteries should be used because the batteries must produce current for five hours or so. The number of batteries to be used was calculated above. The fuses protect the system if something goes wrong and a harmful amount of current would be drawn from the batteries. The "dc appliances" are special appliances that use 12 volts dc. Many of these are made to be used with recreational vehicles. The inverter converts 12 volts dc to 120 volts ac, so ordinary household appliances can be used. An advantage of 120 volts over 12 volts is that the same power can be transmitted at a lower current, which means less loss in the connecting lines. Actually, one would probably not use both ac and dc in

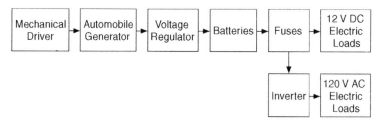

FIGURE 3
Economical homebuilt electric system.

TABLE 4
Prices of components of electric
system

Used Automobile Generator	$50
Used Voltage Regulator	$15
6 Deep-Cycle Batteries	$480
Fuses and Connections	$10
Inverter (500 Watts)	$300
Total	$855

the same system, especially since the inverter is one of the most expensive parts of the system. Here though we are assuming we need some of our output power to be ac.

COST OF THE SYSTEM

Representative prices for each component are shown in Table 4. A reasonable assumption is that each of these would last three years. (Actually the fuses and connections should last nearly indefinitely.)

If we assume the capital recovery factor described in the Planning and Implementation chapter, is .41 (3 years, 12 percent), the annual cost is

$$.41 \times 855 = \$350/\text{year}$$

The reader might compare this amount to her or his annual electric bill. Note that the cost associated with the mechanical driver (water wheel or windmill) has been omitted.

One might decide to use 12-volt dc loads only and spare the cost of the inverter. While most appliances are designed for 120 volts, one can buy refrigerators, vacuum cleaners, and other appliances for recreational vehicles that use 12 volts dc. Of course, automobile lights and radios use 12 volts dc. A 12-volt refrigerator costs $520, a vacuum cleaner costs $25, and a fan is $24. Use of a 12-volt system does mean one must be wary of losing significant voltage along the connecting wires, so the generator must be close to the load.

An alternative to the homemade system of Figure 3 would be to buy a commercial ac generator. These are sold for various applications and some are discussed in the "Wind" article. Others are meant to be used with diesel engines for emergency power. They cost a good deal more than our system—$3,000 or so—but would probably last longer.

REFERENCES

Leckie, Jim, Gil Masters, Harry Whitehouse and Lily Young. (1981). *More Other Homes and Garbage*. San Francisco: Sierra Club Books.

New Alchemy Institute. *The Journal of the New Alchemists*. Woods Hole, Falmouth, MA 02543. Available from the Green Center, 237 Hatchville Rd., E. Falmouth MA, 02536.

Real, Goods. *Alternative Energy Sourcebook 2001*. Real Goods, 966 Mazzoni Street, Ukiah, CA 95482. http://www.real-goods.com.

Salm, Walter. (1965). *Home Electrical Repairs Handbook*. Fawcett Book #736, Fawcett Publications Inc., Greenwich, CT 06830.

Smith, Nigel. (1999). *Motors as Generators for Micro-Hydro Power*. London: Intermediate Technology Publications.

HYDROPOWER

Small-scale water power is a reliable and often economical source of energy where sufficient water flow exists. Successful installations have been made in Nepal and Pakistan, Malawi and Zimbabwe, Panama and Peru, and other countries. Water power was used extensively in early New England. The technology is well proven and needs little maintenance. It can be integrated with an irrigation scheme. Local people can do much of the construction, saving money and imparting ownership. The total cost is usually much less than connecting to the national electric grid—in present-day rural New England, for example, the cost of installing electric lines is about $20,000 a mile. Either electrical or mechanical power may be obtained. Two common uses of mechanical power are grinding grain and sawing wood.

AMOUNT OF POWER AVAILABLE

The equation for the power which can be generated from a stream is

$$P = \frac{HQe}{11.8}$$

where

    H = drop in feet
    Q = amount of water in cubic feet per second (ft^3/s)
    e = efficiency of wheel and generator
    P = output power in kilowatts

This formula makes sense. It shows that the power increases when distance dropped increases. (Incidentally, this drop is usually called the "Head.") It also shows that the more water, Q, that drops, the more power available. Power is wasted both in the water flow and the generator, so the actual power obtained is less than the power available. This wastage is accounted for by the efficiency, e, which is less than one. The factor 11.8 is used to convert units, so the power comes out in kilowatts, when H and Q are measured using feet and cubic feet per second.

    The simplest way to use the energy in the water is a water wheel. These can be built economically with simple tools and care. The alternative is to purchase a turbine—a 1-kilowatt turbine costs about $5,000. Turbines require little operating attention. Installation consists chiefly of building the channels that bring water to the turbine and take the waste water away. A skilled machinist with a basic machine shop could build a turbine; for plans see Inversin (1986) and *Mother Earth News* (1983). Water wheels and turbines are discussed below.

    A picture of an overshot wheel, probably the easiest to build, is given in Figure 4. The principle of operation is probably clear. The water flows from the "flume" into the buckets. The weight of the water turns the wheel. The "Sluice Gate" controls the amount of water that flows. The head, H, is measured from where the water leaves the flume to where it falls out of the buckets. Because some of the water falls out early, and because energy is lost in bearing friction, the efficiency of the wheel is about 75 percent. In most cases a water wheel would drive a generator through belts or gears because the water wheel itself rotates much slower than the rated speeds of most generators (think of how fast an automobile generator rotates). These belts or gears use up energy and their efficiency is also about 75 percent, so the overall efficiency of the wheel-turbine system is between 50 percent and 60 percent.

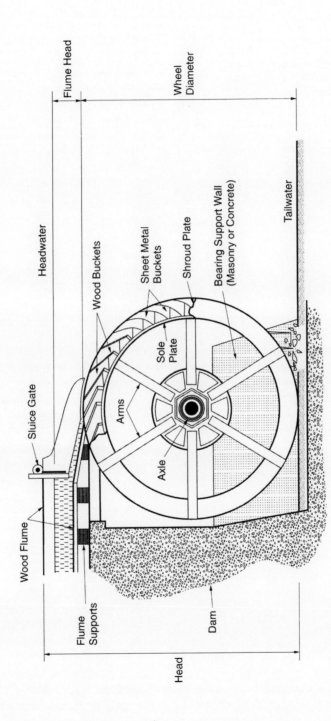

FIGURE 4
Overshot water wheel (Reproduced from Mother Earth News. (1983). *Mother's Energy Efficiency Book*. Hendersonville, NC: Mother Earth News).

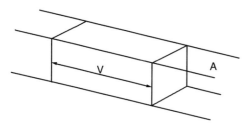

FIGURE 5
Water flowing in flume.

Now let us calculate the power available from a small stream in terms of water velocity and steam size. First we will see how we would calculate Q, the quantity of water that flows through our wheel in a second. A sketch of the flume with an opened sluice gate is given in Figure 5. The cross-sectional area of the flume is A square feet.

The velocity of the water is v feet/second. The amount of water that flows past the opened sluice gate in one second is all the water that was within a distance v (times one second) from the sluice because water closer will have time to get to the gate and water further away will not. So the amount of water reaching the wheel in a second is

$$Q = A \times v$$

so, using our previous equation

$$P = \frac{HAve}{11.8}$$

Now we can estimate the power from a typical water wheel. Assume the head is 9 feet, corresponding to the flume being about 10 feet above the tail water, because clearance is needed above and below the wheel. Assume the cross sectional area of the flume is 2.0 feet by 0.66 feet, that is, 24 inches by 8 inches. Assume the water is moving at 2 feet per second. We will estimate the overall efficiency at 55 percent. The power is calculated in following equation

$$P = \frac{9 \times 2 \times 0.66 \times 2 \times 0.55}{11.8} = 1.13 \text{ kW}$$

This amount of power is more than enough to meet the needs of the family of 4 discussed in the article, "Basics of Electricity." Note that the flow, Q, is 2.64 cubic feet per second, as shown in following equation

$$Q = Av = 2 \times 0.66 \times 2 = 2.64 \text{ ft}^2/\text{s}$$

In other words, a stream carrying 2.64 ft^3 sec and dropping 10 ft would have sufficient power to meet the minimal needs of a family.

SYSTEM DESIGN

Now that we understand the basic physics of water power let us see how we would design a whole hydroelectric system. A block diagram in Figure 6 shows the steps required in designing a small hydroelectric system for a given stream. We will go through these steps one by one.

VERIFYING THAT SUFFICIENT POWER EXISTS

If we are to build a hydroelectric system, we must certainly have a stream flowing through our property. We need to estimate both the head, H, and the quantity of water, Q. The head is the distance the water drops within the length of stream to be used. It can be estimated using a string and tape measure (see Figure 7). To do this one would tie one end of the string to a stake driven close to the water level, move downstream as far as practical, pull the string taut and horizontal (a line level would aid in getting the string horizontal), and measure the vertical distance from the string to the water level. One would repeat this process until the useful length of the stream is covered. Another possibility is to use a surveyor's level.

The quantity of water in a flume, of course, cannot be greater than the quantity of water in the stream. To estimate the water in the stream we need the cross-sectional area of the stream and the velocity. Estimating the cross-sectional area is basically simple, but water flows more slowly near the banks and the bottom, so in estimating the area one disregards the three inches or so near the banks and bottom. An easy way to estimate the velocity is to put a twig, or something else that floats, in the water and time it as it goes a fixed distance. More sophisticated ways of measuring the stream velocity are given in Inversin (1986), Leckie et al. (1981) and the *Jade Mountain* catalog.

Water flow is seasonal, so if one is serious about using the power, one should look at the stream at different times of the year. In New England, August tends to be the low water month. Streams will usually flow under ice in the winter, so winter operation is possible. Actually the most dangerous time for small hydropowered installations is probably in the spring, when the whole thing could be lost in a flood. It is prudent to check the stream at each of these times—spring, winter, and fall. It is also prudent to be wary of even careful estimates, especially if they are close

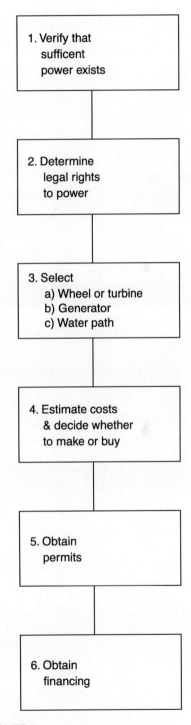

FIGURE 6
Steps in system design.

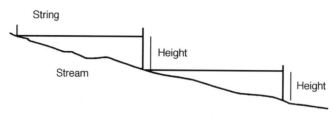

FIGURE 7
Side view of a stream.

to the limits of what is needed or can be handled. Besides the errors in making the estimates, one must realize that nature is not dependable. If one needs 960 watts, and the estimate of the output power is only 1,200 watts, it is probably best to redesign, since 1,200 watts is really too close to 960 watts to be reliable. One should increase the power, if possible, or decrease the load required. A sophisticated way of estimating the annual power available. Using rainfall data for the entire catchment area is given in Inversin (1986). Such an estimate is important when a fairly large system, involving a community, is being designed. At the least the estimate will indicate how often the system will not produce rated power because of lack of water.

Once the head and quantity of water in the stream are known, then we can use our equation to calculate the maximum power available from the stream. Now we will go to the second box in Figure 6, our design flowchart. The next step shown is to select a waterwheel or turbine.

TYPES OF WATERWHEELS AND TURBINES

The expression "waterwheel" usually refers to a large-diameter—10 feet or so—wooden wheel that turns relatively slowly, perhaps one revolution per minute. The word "turbine," on the other hand, refers to a smaller metal wheel, 1.5 feet or so in diameter, that turns much faster—several thousand revolutions per minute. Waterwheels are easier to make, although they are large and heavy and therefore difficult to move from the shop to the site. To make a turbine, one needs access to a machine shop. Again, detailed plans are given in Inversin (1986) and *Mother Earth News*. Turbines for small-scale hydro projects are commercially available.

Three types of waterwheels are shown in Figure 8. As was mentioned above, the overshot one is the easiest to build and is the most efficient: 65 to 85 percent. It needs a head of at least 10 feet. In principle, one could use as large a head as available, but as the diameter of the wheel and the head are approximately equal, heads over 30 feet are not really feasible for an overshot wheel. The publication "D.I.Y. plan 7-timber Waterwheel," *Mother Earth News*, and Leckie et al. (1981) all give plans for a wooden waterwheel.

The breast wheel is used for heads between 6 feet and 10 feet. The wheel shown is a low breast wheel because the water enters below the center of the wheel. Such a wheel has an efficiency of 65 percent. The real problem with such wheels is the breastwork, the close-fitting wooden cover that keeps the water in the buckets. These need to be made carefully to reduce friction. Debris floats down the river and can get caught between the wheel and the breastwork, so a screen is needed. Breast wheels can be made with large diameters, independent of the head, and large diameters create high torque. High torque is useful if the wheel powers machinery or tools or grinding stones but is not important for electric generators, so breast wheels are not used often for hydroelectric projects.

Breast wheels work better than overshot wheels in flood conditions because the wheel turns in a direction that pushes the spent water away from the wheel. In an overshot wheel, in flood conditions, the wheel pushes water into the space behind the wheel. This water acts as a drag on the wheel. Breast wheels were used more often than overshot wheels in colonial New England textile mills because of this ability to get rid of the used water.

The undershot wheel works with the lowest head. It is also the least efficient: 25 to 45 percent. Here the kinetic energy of the water pushing on the blades creates the force.

Turbines

Two turbines are shown in Figure 9. Turbines are the most efficient way to generate electricity. Commercial systems all use turbines. Many small-scale systems also use turbines. The efficiencies range from 75 to 85 percent.

The Pelton turbine uses a nozzle to direct the water onto the blades at a high force. These turbines are used at high head situations, especially when the total amount of flow is small. A small commercial turbine and generator producing 1.4 kilowatts costs about $5,000 (see *Real Goods* publication). Other commercial turbines are described in the *Jade Mountain* catalog.

A crossflow turbine works on another principle. The water flows through the blades on the top, pushing them through the central cylinder, and, finally, out through the bottom, pushing on those blades also. The word "push" is misleading. In the "Wind Power" article, "lift" and "drag" devices are discussed—the crossflow turbine actually uses lift. Such a turbine was successfully built by an inexperienced group in Zaire. Mother Earth Plans sell drawings. If these plans are used, the cost will be cost about $50. A crossflow turbine will work with heads as low as 3 feet.

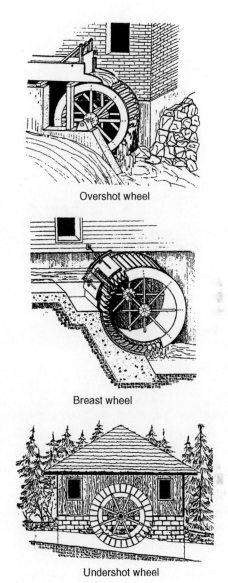

Overshot wheel

Breast wheel

Undershot wheel

FIGURE 8
Types of water wheel.

A final type of turbine, which we will not consider further because they are only used in bigger installations, is the reaction type turbine, or Francis turbine. They usually are totally immersed in the water, with the rotating axis vertical. The weight of the water pushes on the blades.

An alternative to building or buying a turbine is to use a rotary water pump. A pump operated as a turbine will not be as efficient as a turbine of the same size. Pumps though are more common and thus less expensive. Their availability may make them a good choice when electric power must be obtained quickly.

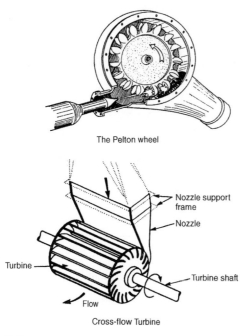

The Pelton wheel

Nozzle support frame

Nozzle

Turbine

Turbine shaft

Flow

Cross-flow Turbine

FIGURE 9
Turbines.

Speed Control

Both waterwheels and turbines will slow down when the load, mechanical or electrical, on them increases. This speed reduction may cause problems—grain will be ground less finely or electric appliances will not work well. To increase speed, more water can be sent to the wheel or turbine by opening a gate or valve on the sluice or penstock. Probably the simplest approach is to do this manually. Automatic, feedback systems are used with large installation. The decrease in speed is sensed by a flyball, and the signal actuates a valve, sending additional water to the turbine or wheel. Such a system is complex (see the *Jade Mountain* catalog).

An alternative speed control for an electric system keeps a constant load on the generator at all times but switches between loads as required. In a typical situation, an electric water heater is used as a default load. When power is needed for lighting, the water heater is switched off and the lights switched on. When the lights are not needed, the electric power goes into the heater. Such a system probably only makes sense when a genuine need for the hot water exists.

Waterworks

In planning channels for bringing water to the wheel and back to the stream, (step 2c in Figure 6) a major decision is whether a dam will be built. Dams are needed if a pond is required to store water. If one needs electricity only 6 hours a day, then one could store water for the other 18 hours. In this case the power available when used is 4 times the continuously available power. Actually, ponds are probably used most often to store energy over longer periods, such as during the summer, so the spring flood water is available in the dry months, like August. As we will see later, a dam can also reduce the length of the channel used to bring water to the wheel.

On the other hand, a dam is not a trivial thing to build, as we will see below, and a dam failure can be a major disaster for people downstream. The process of obtaining permits certainly will be easier if a dam is not used so advantages exist in simply diverting water from a stream and returning it—so-called "run of the river." According to Inversin (1986), most small-scale hydropower systems in the Third World are run of the river.

Figure 10 shows a side view and a top view of a hydroelectric site. The head race is the channel dug out to carry the water to the flume, and the tail race transports the water away from the wheel. In order to get the water to flow with sufficient speed, these must have a downward slope, perhaps 2:1,000, which means a drop of 2 feet in a 1,000-foot channel. If the water does not flow rapidly, a wider race must be used. Races reduce the head, so one has to trade off the additional drop in the stream gained from longer races with the loss in head in the channel.

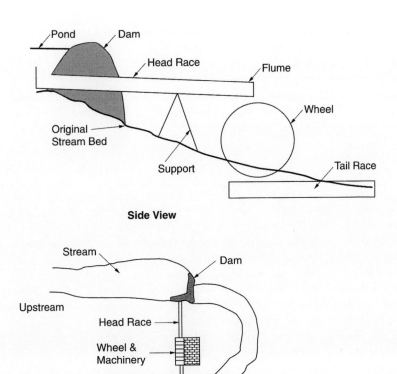

FIGURE 10
Hydroelectric site.

In colonial America some races were a quarter-mile long. The total head available includes the height of the water in the pond above the outlet, as well as the drop through the wheel, but the usable head is measured from where the water drops on the wheel. The inlet location to the head race needs to be chosen carefully so silt does not accumulate in front of the inlet and so the race is not damaged by rocks pushed down the stream at high water.

If a turbine were used in the same installation as Figure 10, it would be located at the lowest point, where the water goes back into the stream. Instead of using an open channel for the head race, a PVC (polyvinylchloride) or metal pipe is used to enclose the water going from the stream to the turbine. This forces the water to flow through the nozzle in a Pelton turbine, or directly into the crossflow wheel. This pipe is called a "penstock." PVC pipes of a suitable size, 4 inches or so, cost about $.50 per foot.

SIZE OF POND

As noted, building a dam to obtain a pond complicates greatly a small-scale project. A pond is needed though to store water when the stream does not always carry enough water to give required power.

Let us estimate how large a pond must be to store sufficient water to give us power for a week. In a previous equation we saw that if the water flow is 2.64 cubic feet/second, the power will be 1.13 kilowatts, sufficient for our needs. To simplify the analysis we will assume no water flows into the pond while we are drawing water out and that we draw water out 24 hours a day for the entire week. How much water must be stored to allow 2.64 cubic feet/second for 1 week? We convert seconds to minutes to hours to days to weeks and multiply by 2.64. We need a pond of about 1.6 million cubic feet. How big is that?

$$60 \times 60 \times 24 \times 7 \times 2.64 = 1,596,672 \text{ ft}^3$$

To make this number meaningful, we note that a square pond 5 feet deep would need to have sides of length 565 feet. The pond would be the size of six football fields—a large but not unrealistic size. Our actual pond could be smaller because we probably would use less power after midnight and because our stream would add water during the week. If power were used only 4 hours a day, the pond could be about the size of one football field.

Actually what we mean by 5 feet deep is that the level of the pond would drop by 5 feet when water was drawn out. We would probably not want to empty the pond entirely, for the sake of the wildlife, if nothing else, so the pond should be deeper than 5 feet.

The pond is used for storage. Another mode of storage is batteries. We could compare the cost of making the pond with that of buying the batteries, but we must factor in alternative use of the land and also the desirability of having a pond for recreation or similar purpose.

DAMS

In picking a site for a dam one wants to maximize the head. One also wants to arrange the waterworks so that the generator, and thus the wheel, is close to the house to minimize electrical transmission losses. On the other hand, one probably does not want to live too close to either a wooden waterwheel or a turbine-generator combination, both of which can be noisy and vibrating. If one has a choice, one should avoid streams with silt, which can damage a turbine. Finally, as will be described next, leaks can be a problem, so one needs to select a site for the pond and the dam that will have low leakage.

Some dams, which can be built by hand or with only simple earth-moving equipment, are shown in Figure 11. Dams can fail in two ways: (1) seepage underneath, which undermines the dam and (2) pressure that pushes the whole dam downstream. Seepage is minimized by using clay for the entire dam, for a core, or for a seal on the sides and bottom. The earthfill dam in Figure 11 has a clay core. Pig manure is another sealant, as is plastic sheeting. Heavy dams are less likely to slide away than light ones, so rocks improve safety. In building a dam it may be necessary to build a temporary dam upstream to hold the water back during construction.

Figure 11 is probably self-explanatory. In an earthfill dam the slope of the upstream side should be about 1:3, so the water pushes down more than sideways and large rocks should be used on the surface facing the water to avoid erosion. The slope of the downstream side can be steeper, perhaps 1:2. Rockfill dams can have a deeper slope than earthfill ones—perhaps 1:1.5. In the rockfill dam of Figure 11 concrete is laid over the rocks to prevent leakage. The crib dam is made of logs stacked a few feet apart with gravel in between. Boards covering the face of the dam reduce leakage. Crib dams are best for low heights—5 feet or less. The gravity dam is made of concrete and is the most expensive.

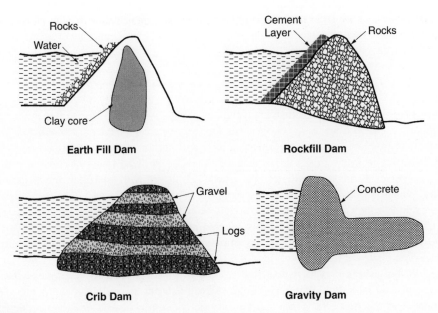

FIGURE 11
Types of dams.

In any case, all dams need to have a spillway, where overflow water can safely get over the dam. It is obvious that a way must exist to get the water from the pond to the head race, or the penstock. This water cannot be taken from the top of the pond, or else the pond would only be useful when it was full. The water outlet, then, should probably go through the dam. It will be easier to repair the dam if the pond can be completely emptied, so a drain at the bottom of the dam is usually also included.

Step 4 in Figure 6, dealing with financing, is done using techniques in the Planning and Implementation Chapter.

REGULATORY REQUIREMENTS

Step 5 in Figure 6 refers to legal requirements. In the United states several agencies have jurisdiction over hydroelectric installations and the laws differ state by state and possibly county by county. Noncompliance will probably mean removal of the installation and possibly a fine. The state agency dealing with environment will probably be interested in your project. So will the fisheries department. The Federal Energy Regulatory Commission licenses hydroelectric projects. Normally obtaining such a license requires an experienced lawyer; however, an exemption is usually granted if the project is less than 100 kilowatts, the typical case for a small-scale project. The local building inspector should pass on construction plans. Regulatory agencies will be much more concerned about a hydropowered project if it contains a new dam, so if the stream has an abondoned dam site, one should try to reconstruct it rather than building on a new site.

An analogous issue in the Third World is alternative use of the water, often irrigation. No government agency may be responsible for adjudicating conflicting use of water, so a group planning a small-scale hydro project should deal directly with the local group managing the irrigation project. In the ideal case the irrigation and power projects can be integrated but such may require significant planning.

ENVIRONMENTAL CONSIDERATIONS

Ponds change streams and thus change the ecology. They give habitats for some varieties of fish and reptiles and store water in the dry seasons for mammals such as deer or otters. On the other hand, they destroy areas of running water that trout and salmon need. Dams are a major obstacle to fish returning from the ocean to spawning grounds at the headwaters of streams. State government agencies in the United States concerned with fishing will nearly always insist on "fish ladders," which are basically small pools arranged in steps so fish can jump from one pool to another. Ladders, however, may not be enough because the pond behind a dam may not be a proper spawning place for a fish used to spawning in running water.

The wheel or turbine itself can do environmental harm. Live fish, and other creatures, can be hurt in the blades of a turbine or, less likely, in the buckets of a waterwheel. They can cause damage to the turbine or waterwheel, too, so the trash racks used to keep sticks out of the penstock or head race should be fine enough to keep fish out as well.

Other environmental effects need to be considered. A major, and obvious, concern is the integrity of the dam. Noise and the visual impact can be important. A wheel creaks and a generator vibrates. The noises are not objectionable if one is 30 feet or farther away, but if living quarters are closer, soundproofing may be needed. One has to use one's judgement about the visual impact. Waterwheels have had a long history in the United States and are often an attraction for visitors.

IN-STREAM TURBINE

A promising approach to obtaining power from running water is to submerge a turbine in a river and use the force of the current to drive the turbine. An analogous device is a turbine dragged through the water behind a moving boat. The river turbine has been used to drive a small irrigation pump. The turbine behind the boat has been used to recharge a storage battery on a sailboat. The amount of energy collected is usually small but useful in particular applications. More information is given in Reference by Inversin.

REFERENCES

Center for Alternative Technology. (1978). "D.I.Y. Plan 7-Timber Waterwheel." Mackynlletts, Powys, Wales.

Inversin, Allen. (1986). *Micro-Hydropower Sourcebook: A Practical Guide to Design and Implementation in Developing Countries*. NRECA International Foundation, Arlington, VA.

Jade Mountain, Appropriate Technology for Sustainable Living. (1999–2000). Vol. XIII, No. 1. PO Box 4616, Boulder, CO 80306-4616. Telephone 800-422-1972.

Leckie, Jim, Gil Masters, Harry Whitehouse, and Lily Young. (1981). *More Other Homes and Garbage*. San Francisco: Sierra Club Books.

Mother Earth News. (1983). *Mother's Energy Efficiency Book*. 105 Stony Mountain Road, Hendersonville, NC 28791.

Mother Earth News. "Mother's Crossflow Turbine Plans." See address above. $15.

National Center for Appropriate Technology. "Micro-Hydropowered Power." U.S. Department of Energy, DOE/ET/01752-1.

Real Goods. Alternative Energy Sourcebook. Real Goods. 966 Mazzoni Street, Ukiah, CA 95482. Telephone 707-468-9214, $10.

WIND POWER

Wind power makes economic sense—compared to diesel generators, power grid extension, or photovoltaics—at sites with adequate wind. Average wind speeds above 9 miles per hour (4 meters per sec)—see Table 5 to estimate wind speeds—are needed. Probably the primary use of wind power on farms has been pumping water for livestock or crops, an application that matches the capabilities of wind machines well. Wind-powered electric systems are effective, and large commercial installations are supplying electric utilities in California and the U.S. Midwest, as well as in Denmark and the Netherlands. Wind generators producing 500 watts at rated wind speed cost about $1,000 (see *Jade Mountain* and *Real Goods* catalogs). Wind powered generators complement photovoltaic devices well; often the sun is strongest during the times the wind is slowest. A combined system may be appropriate in a specific situation. Before the Rural Electrification Act in the early 1930s, nearly all farms in the Midwest got their power from the wind, although the present technology is different.

THE PHYSICS OF WIND POWER

Wind power originates in the kinetic energy of the wind—that is, the energy the wind has because of its motion. Nearly all wind power generators use a rotating propeller that captures the kinetic energy when the wind pushes on the propeller blade. The reader may recall that kinetic energy of an object equals one half of its mass times the square of its velocity. In this case the object

TABLE 5
Beaufort scale of wind force

	Observations	Speed (mph)
Calm	Calm. Smoke rises vertically.	0–1
Light Air	Direction of wind shown by smoke drift but not by wind vanes.	1–3
Light Breeze	Wind felt on face. Leaves rustle. Ordinary vane moved on wind.	4–7
Gentle Breeze	Leaves and small twigs in constant motion. Wind extends light flags	8–12
Moderate Breeze	Raises dust and loose paper. Small branches are moved.	13–18
Fresh Breeze	Small trees in leaf begin to sway. Crested wavelets form on inland waters	18–24
Strong Breeze	Large branches in motion. Whistling in telephone wires. Umbrellas used with difficulty.	25–31
Near Gale	Whole trees in motion. Inconvenience is felt when walking against the wind.	32–38
Gale	Breaks twigs off trees. Generally impedes progress.	39–46
Strong Gale	Slight structural damage occurs (chimneys and roofs)	47–54
Storm	Seldom experienced inland. Trees uprooted. Considerable structural damage occurs	59–63

of interest is a block of air that goes through the blades. Power is energy per second, and the amount of the air reaching the blades per second depends on the velocity of the air, so the power depends on the velocity cubed. The power also depends on the size of the block of air captured by the propeller—that is, on the area covered by the propeller.

$$P = V^3 \times A \times e \times 5.3 \times 10^{-6} \ (\text{watts/mph}^3 \ \text{ft}^2)$$

where
   P = power out in kilowatts
   V = wind velocity in miles/hour (mph)
   A = area swept out by blades in ft^2
   e = efficiency of propeller

The factor 5.3×10^{-6} arises because of the units chosen. Actually, this factor depends somewhat on the density of the air—the number of pounds a cubic foot of air weighs—as the mass of a cubic foot of air depends on its density. We will, however, ignore the small change in density caused by changes in temperature and altitude—hot air is less dense than cold air, as is air at high altitude. Efficiencies for various kinds of propellers are provided later.

One implication of the equation is that power increases rapidly as the wind speed increases. When the wind speed doubles, the power goes up 8 times, or a 25 percent increase in wind speed doubles the output power. This variation of power with wind speed is important because wind often blows in gusts, as shown in Figure 12.

Most of the power comes at high velocities rather than at the average velocity, so use of the average wind speed in Equation 1 will give a wrong estimate of the power—two winds may have the same average speed, but one could have gusts at higher speed and thus have higher power. It is much easier, however, to measure the average wind speed than the variations in speed, and the National Weather Service reports average wind speeds. A correction factor is, therefore, needed in the equation to get an accurate estimate of the power when the value for the wind velocity, V, is the average wind speed. This correction factor equals 2.5—that is, the power calculated from the equation needs to be multiplied by 2.5 if the speed, V, is the average wind speed.

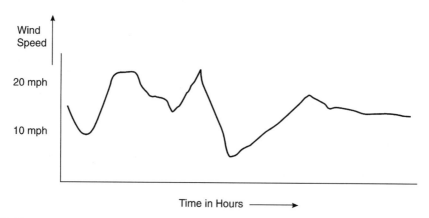

FIGURE 12
Variation in wind speed.

Wind speed increases with height, which makes sense, since the wind speed just at the surface of the ground level is 0. The increase is described approximately by the next equation, which is based on measurements. (Actually the proper exponent in the following equation depends on the terrain, but we ignore this dependence.)

$$\frac{V_1}{V_2} = \left(\frac{h_1}{h_2}\right)^{0.2}$$

By substituting values into this equation, we can see that increasing the height of a tower from 5 feet to 30 feet will increase the velocity by a factor of 1.43. This would approximately triple the power output.

Estimating the wind speed is an important matter in designing a wind generator. The National Weather Service gives average wind velocity for each of their stations—usually at airports. *The Wind Power Book* (Park 1981) gives these averages by month. Wind, though, is a very local thing—sites even 100 feet apart may have significantly different winds, so one has to be careful in using published values. One approach to estimating wind speed at a specific site is to compare that site with that of the Weather Service's and estimate how similar wind conditions are. Other suggestions are given in Thomas (1993). Another approach is simply to measure the wind. Commercial equipment is available, or the *More Other Homes and Garbage* book describes how to make such measuring equipment. Of course, careful measurements, even over a long time, will give only estimates, not perfect forecasts, as the wind is not reliable. The descriptions in Table 5 can be helpful but are not really accurate.

TYPES OF WIND MACHINES

Two types of wind machines exist—the "drag" and the "lift" types. A typical example of each is shown in Figure 13.

The drag type rotates because the wind pushes against it. Square rigged sailing ships moved because of drag. The rotor in Figure 13 is propelled because the wind is caught by one surface and flows over the other. It is called Savonius, after its inventor. It rotates about a vertical axis, and it has the advantage of high torque, especially when starting–because the wind has a large area to push against. It has a disadvantage at high speeds because the wind pattern is badly disrupted ("turbulence" is produced), so the efficiency falls off. Also, the curved surfaces cannot move faster than the wind because they are being pushed. The details of a Savonius rotor will be discussed below.

The lift type, the other rotor of Figure 14, is more efficient at high speeds, because it produces less turbulence, but has a low torque, and thus can be hard to start under load. The principle of operation, which is similar to that of a sailboat or airplane wing, is illustrated in Figure 14.

When the wind blows by a blade, the wind "tries" to keep going in the same direction. The result is that on the bottom of the blade, the wind is concentrated near the blade, while on the top it is away from the blade. Therefore, the region below the blade is at high pressure, and the region above the blade at low pressure, so the blade is pulled, or lifted, as shown. Sailboats, unless they are going down wind, move by the same principle–the wind is deflected to produce low pressure on one side of the sail and high pressure on the other side. The pressure difference produces a force.

An essential consideration in dealing with wind machines is what happens when the wind speed is very high. A way of preventing the rotor from going too fast is needed. A mechanical

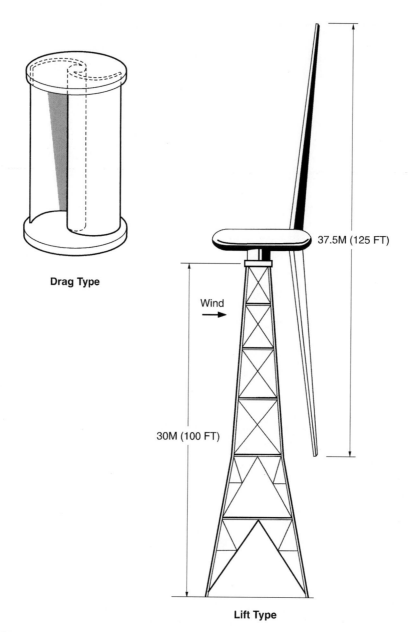

Drag Type

37.5M (125 FT)

Wind

30M (100 FT)

Lift Type

FIGURE 13
Typical wind machines.

brake can always be used but is cumbersome. The blade angle of a lift type device can be adjusted, either manually or automatically, to eliminate the lift when overspeeding occurs. A horizontal axis machine can be rotated so it does not face the wind. Usually this rotation is sideways so the wind blows on the end of the blade rather than the front, but it can be vertical, tipping the rotor so it faces up. Propellers used in commercial wind generators are designed to lose lift, "stall," at wind speeds higher than rated.

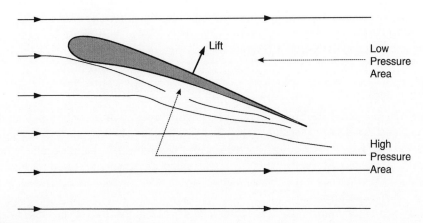

FIGURE 14
How lift is created.

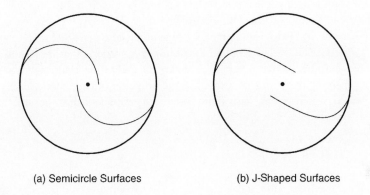

(a) Semicircle Surfaces   (b) J-Shaped Surfaces

FIGURE 15
Savonius rotor.

Savonius Rotor

Top views of two Savonius rotors are shown in Figure 15. The model shown in Figure 15a is simple to make. It can be constructed by slicing an oil drum down the central axis and welding the two halves to form an S-shaped rotor. The second model, Figure 15b is somewhat more efficient because the shape gives some lift when the blade faces the wind. Most of the torque produced, however, comes from the wind pushing on the concave surface. Of course, the wind exerts the most force when it flows directly into one of the pieces, so the torque produced varies as the rotor spins. This variation in torque can be reduced if three or four Savonius-type rotors are stacked on top of each other, offset in angular position.

Savonius rotors are easy to build. They are often used for pumping water. As indicated above they produce high torque even when starting, so they are also useful for grinding mills and other machinery. On the other hand, they are not efficient—about 15 percent when coupled to a load. It is not necessary to face them into the wind, since they always are. If the wind direction is changing rapidly, not having to move the rotor to face the wind is a big advantage. The vertical axis may mean that gears are not needed to bring the power down from the tower top. One does not have to worry about overspeed conditions, since high winds push against the other surface and keep the speed under control. A disadvantage of Savonius rotors for electrical power generation, though, is that the low speed probably means that gears or pulleys must be used to match the rotor to a generator, since most electric generators work at high speeds.

Propeller-Type Rotors

Propeller-type rotors, the other type in Figure 15, are high-rpm (revolutions per minute) devices, so they are ideal for electric generators. The efficiency is high—40 percent for the entire system at high wind speeds but lower at lower wind speed. One problem is that the optimum angle of attack of the blade depends on the wind speed. Propellers are usually designed for high speeds, which reduce the efficiency at low speed, although some have intricate mechanisms to control the angle of attack or pitch.

Some design problems do exist. If the rotor is upwind from the tower then some mechanism is needed to aim the rotor into the wind. Usually a tail is used. An advantage of a tail is that a spring mechanism can be added to move the tail so as to face the rotor out of the wind when the velocity is too high. If the rotor is downwind from the tower, as in Figure 13, a tail is not needed, but the rotor blades get less wind when behind the tower in the "shadow" of the tower. This shadowing can create harmful vibrations.

The exact shape of the blades is important for high efficiency, so a good set of plans is required (see *Mother's Plans*). The design problem for a propeller blade is that the actual linear speed, when the blade is turning, increases along the blade. The end of the blade is moving much faster through the air than the parts near the center, so the propeller must be cut with a varying pitch. Moderately skilled woodworkers have built successful propellers, but the task is not simple, especially if a large blade is needed.

Sailwing Rotors

A variation on the propeller rotor is the sailwing, shown in Figure 16. The sailwing can be built cheaply by an amateur. The blades are made of cloth sails. They can be connected to the hub with springs so that in high winds they automatically feather (face the wind directly, thus losing lift.) The National Center for Appropriate Technology plans give details. Either cotton or Dacron can be used for the sails and should last three or four years (compared with 80 to 100 years for a wooden propeller). Wide sails give high torque, and narrow sails give high speed. Sailwings, therefore, can be used for both pumping and electricity but not both at once. They have about 15 percent efficiency. Even with springs to feather the blades at high wind velocities, one has to be careful when storms come. At velocities over 30 mph, it is necessary to furl the sails to avoid damage.

Siting

If a wind machine is used for water pumping or for other mechanical devices, then the choice of the site may be constrained. An equation above shows that the higher the location, the better. In any case the rotor should be at least 10 yards above wind obstacle standing within 100 yards.

Tops of gently sloped hills, as on the left in Figure 17 are suitable. The wind climbs the hill smoothly and is concentrated at the top. Steep hills, on the right in Figure 17, are less suitable, since turbulence occurs at the corners. Tops of buildings tend to be undesirable for the same reason, plus the difficulty in building a stable mounting for the tower. A valley aligned parallel to the direction of the wind can channel and concentrate it.

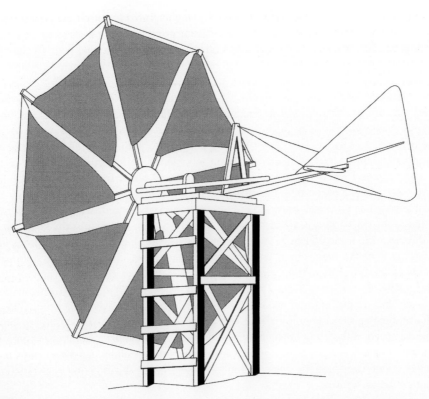

FIGURE 16
Sailwing rotor.

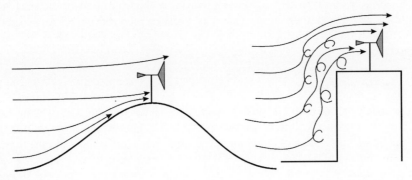

FIGURE 17
Suitable and unsuitable wind machine sites.

TOWER AND TRANSMISSION

Towers can be homemade, but unless one has some structural engineering skills, it may be best to purchase one. The cost is between $500 and $1,000. Tilting towers, hinged at the base, simplify maintenance and repairs. Towers can be made from steel lattice work or a single tube. The former is cheaper but the latter considered more attractive. Guy wires add strength but require a larger "footprint." Generators producing the most energy, as you might expect, need the strongest towers, since they slow down the most wind. Towers for high-speed purchased generators, therefore, should probably be purchased. Towers are especially stressed when the wind changes either speed or direction because of the twisting, so a site with erratic winds needs a particularly strong tower. Towers deserve serious thought because the consequences of a falling wind machine can be disastrous.

"Transmission" refers to the means for getting power down from the rotor to the ground. If the wind machine powers an electric generator, one solution is to put the generator up on top of the tower connected directly or through gears to the rotor, as is done in many commercial wind generators. In this case the tower must support the additional weight, and servicing the generator can be difficult. If the generator is on top, electric wires must, of course, be brought down to the ground, so it is necessary to prevent these wires from becoming tangled and breaking. An elegant solution is to use slip rings and brushes. Slip rings are two rings of brass, each encircling the turntable on which the generator is mounted. The generator output is connected to the rings. Stationary carbon bars, called brushes, rub against the rings. No matter which direction the rotor is facing, current can flow through the rings to the brushes and then down through wires to the ground. The expense of slip rings can be avoided if the operator is able to check the tower regularly.

If the generator is on the ground or if the windmill is being used to drive a mechanical device, then a mechanical transmission is needed. Often these transmissions serve the additional purpose of increasing the speed of rotation above that produced by the rotor. It is useful to be able to increase the speed of drag type machines, which tend to rotate much slower (100 rpm) than the rated speed of an electric generator (3600 rpm). If the wind machine has a horizontal axis, like a propeller or sail wing, then the transmission must also change the direction of the shaft rotation from horizontal to vertical.

A scrap automobile differential can be used on a tower top to change the direction of the shaft rotation. In a rear wheel drive automobile the differential serves the same function, causing the drive shaft, which extends back from the engine, to drive the rear axle. An automobile differential will also compensate for the motion of the rotor housing as the wind direction changes. A properly installed differential will, further, increase the shaft rotation speed above the rotor speed—actually the differential is installed here opposite to the way it is in an automobile, where the engine goes faster than the wheel.

Transmission System Design

To see how much the rotor speed must be increased by the transmission system, let us first estimate how fast the rotor will turn. We use the tip/speed ratio, which is the ratio between the speed of the tip of the blade and the wind speed. This has been worked out theoretically and measured experimentally. For a sail wing it is about 5, for a propeller as high as 8, and for a Savonius rotor about .9.

Assume we have a sailwing and want to know how fast the rotor revolves. If the wind velocity is 10 miles per hour and the tip/speed ratio is 5, then the tip is moving at 50 miles an hour, or 4,400 feet per minute. If the length of a sail is 10 feet, then a revolution is $2\pi \times 10$ feet, so the number of revolutions per minute is

$$\text{Revolutions/minute} = \frac{4400}{20\pi} = 70.1 \text{ rpm}$$

If we want to drive an automobile generator at about 1,000 rpm, we need to gear up about 14 times. Automobile gearboxes can be modified to give this increase in speed, but the modification is somewhat difficult. Commercial gearboxes, which can also be purchased in the United States in farm equipment stores for about $100, are lighter and probably more reliable than scrap automobile parts.

Rather than a vertical shaft and gearbox one could use a chain drive (like a bicycle) or even a belt drive. Chain drives are heavy and require regular oiling and tightening. Belts and pulleys are the cheapest solution but waste some power because they slip. Timing belts have grooves and do not slip but cost more and need matching pulleys.

Electric Generation and Storage

Three wind-driven electric generating systems are shown in Figure 18.

In the "Basics of Electricity" article storage of electricity was considered. Batteries are the best choice in a standalone system. If we are to use batteries we need direct current, and, since the batteries are 12 volts, we should generate at 12 volts. Actually, higher voltages are possible because batteries can be connected in series, so we could store 24 volts, 36 volts, or any multiple of 12. Appliances normally run on 120 volts ac; however, some appliances are available for 12 volts dc, so a 12-volt system makes sense. The top system in Figure 18 shows a 12-volt generator charging batteries, which feed 12V dc appliances.

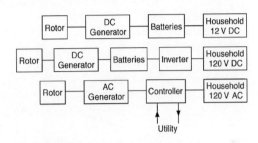

FIGURE 18
Some alternative wind electric systems.

If we prefer to use 120V, or 240V, ac appliances, which are more common and thus easier to obtain, we could use an inverter to go from 12 volts dc to 120/240 volts ac, as in the middle system of Figure 18. An advantage of 120/240 volts over 12 volts is that for the same power a 120/240 volt system uses one-tenth the current, so losses in the connecting wires are less. Another way to get 120/240 volts ac is to buy an ac generator. Many ac generators have speed with controls, so the voltage produced is exactly 60 cycles per second. The output from these generators can be interchanged with the utility—the national power grid. This is the bottom system in Figure 18. The grid here is always connected to the controller. When the wind is blowing strongly, the ac generator produces excess power and sends some to the grid and the rest to the household loads. When the wind is not blowing strongly, the household gets its electricity from the grid. In this system, the wind generator's function is to reduce the household electric bill.

Design

We will design the simplest system in Figure 18. We will use a sailwing rotor. The dc generator could be an automobile alternator with rectifiers and a voltage regulator. We assume that we need enough storage for four days without wind—that is, we want to be able to charge our batteries in 24 hours with enough energy to supply the family in case wind does not blow for the next three days. The desired capacity is sufficient power to supply 960 watts for five hours a day—the same example as in the "Basics of Electricity" article.

First, let us calculate how big a rotor is needed to charge the batteries during one day to supply our home for the next four days. In other words, during 24 hours we want to recharge the batteries so they will give 960 watts for the next five hours and for five hours during each of the next three days.

Let us assume the average wind velocity reported by the weather service is 10 miles per hour, and let us use a factor of 2.5 to account for the ratio between the average power and the power calculated from the average velocity.

$$P_{needed} = V^3 \times A \times e \times 5.3 \times 10^{-6}$$

where

P_{needed} = power needed from the generator during the 24 hours the wind is blowing
V = wind velocity = 10 mph
A = area swept by rotor
e = efficiency for a sail wing rotor = .15

The energy needed to get 960 watts for five hours a day for four days is

$$\text{Energy needed} = 960 \times 5 \times 4 = 19,200 \text{ watts hours}$$

The generator must supply this amount of energy in 24 hours. The power, P_{needed}, required from the generator is

$$P_{needed} \times 24 = 19,200 \text{ watt hours}$$

$$P_{needed} = 800 \text{ watts}$$

Now that we know P_{needed} we can solve our equation for the area swept out by the blades of the rotor. We find that area to be 402.5 ft^2. The corresponding blade length is 11.3 feet—A = πr^2, where r is the blade length.

An 11-foot blade is somewhat long, so we might decide to use two rotors, towers and generators. In this case the blades of each would be 8 feet long. To avoid power loss wind machines should be separated from each other by at least a distance of 15 times the rotor diameter, or about 120 feet in our case.

We also ought to check how much current the generator must supply to the batteries when they are recharging—power equals voltage times current.

$$P_{needed} = 12 \times I$$

$$I = 66.67 \text{ amperes}$$

This large an automobile generator is uncommon, but we could use a small truck alternator. If we cannot find a big enough alternator, we could drive two smaller alternators with the same rotor.

We saw in our previous equation that the propeller rotates at about 70 revolutions per minute. An automobile generator needs to be going at least 500 revolutions per minute to produce rated current, although 1,000 revolutions would be better. So we need to increase the speed by a factor of at least 8. Pulleys and a belt are the cheapest way although gears are more reliable. We will mount the pulleys, belt, and generator on top of the tower and use slip rings and brushes to connect its output to the wires supplying the house.

Blades 11 feet long would require a tower at least 21 feet tall, even taller would be better. Let us choose 30 feet—roughly the height of a three-story building. We certainly would not want to build higher without professional help. The design of the rotor, generator, transmission, and tower is now complete.

Now let us estimate the number of batteries needed. We will use deep-cycle batteries costing $80 each and draw 9 amperes from each. According to the manufacturer, the batteries in this case will last 10 hours before needing a recharge. The energy stored in each 12-volt battery is the power times the hours used.

$$\text{Energy/battery} = 12 \text{ volts} \times 9 \text{ amperes} \times 10 \text{ hours} = 1080 \text{ watt-hours}$$

The energy needed for 960 watts for 5 hours a day for 4 days is 19,200 watt-hours. The number of batteries needed, then, is 18.

Maintenance

Maintenance consists chiefly of checking the blades periodically and lubricating moving parts. Propeller blades can be hit by airborne debris, birds, or insects when moving rapidly—they should be examined at least once a year. The leading edge of the blade may need to be replaced from time to time. The high speeds and unprotected environment that wind machines operate in stress bearings and other components. Manufacturer's guidelines should be followed in lubricating and adjusting generators and other moving parts. Lightning can damage both the electrical system and the tower—good grounding is important.

Water Pumping

Water pumping for irrigation, cattle feeding, or even a village water supply is usually done with a traditional wind pump, shown in Figure 19. Displacement pumps, which operate at low speeds, are used. These wind machines use drag. They have high starting torques and operate at low wind speeds. Such windpumps have been used for 100 years or so. Commercial ones are generally obtainable at farming supply stores. A pump with a 4-foot diameter rotor in a good situation will pump about 1,000 gallons a day, sufficient for 50 cows or 10 families or 3 small farms.

FIGURE 19

Windpump. Reprinted with permission from *Wind Catchers: American Windmills of Yesterday and Tomorrow,* by Volta Torrey. Copyright © 1976 by the Stephen Greene Press.

Environmental Concerns

The major environmental degradation comes from noise. Windmills tend to hum or whine, and several together can be annoying. Modern wind generators are much less annoying. Another annoyance is that metal blades interfere with television reception. The television signals bounce off the blades and into the TV intermittently, so a disturbing pattern is superimposed on the picture. Blades made of wood, plastic, or cloth do not cause TV interference but do not last as long as metal ones. Incidents of birds hitting blades have been reported. Whistles emitting sound at frequencies above human hearing are expected to keep birds away. In the United States objections have been raised to the appearance of wind generators—especially to the towers. While "visual pollution" is a matter of taste, some people find a wind generator an attractive thing.

REFERENCES

Newell, Thomas. (1993). *Introduction to Small Wind Systems.* American Wind Energy Association, 777 N. Capitol Street, NE Suite 805, Washington, DC 20002-4239. Telephone 202-408-8988.

Eldridge, Frank. (1975). *Wind Machines.* National Science Report. Available from Superintendent of Documents, U.S. Government Printing Office, Washington, DC 20402, Stock Number 038-000-00272-4.

Jade Mountain, Appropriate Technology for Sustainable Living. (1999–2000). Vol. XIII, No. 1. PO Box 4616, Boulder, CO 80306-4616. Telephone 800-422-1972.

Leckie, Jim, Gil Masters, Harry Whitehouse, and Lily Young. (1981). *More Other Homes and Garbage.* San Francisco: Sierra Club Books.

Mother's Plans, PO Box 70, Hendersonville, NC, 28793.

National Center for Appropriate Technology. (1977). "D.I.Y. Plan 5—Sail Windmill." Machynlleth, Powgs, Wales.

New Alchemy Institute. *The Journal of the New Alchemists.* (1974). Woods Hole, Falmouth, MA, 02543. Available from the Green Center, 237 Hatchville Rd. E. Falmouth, MA, 02536.

Park, Jack. (1981). *The Wind Power Book.* Palo Alto, CA: Chishere Books.

Real Goods, Alternative Energy Sourcebook. Real Goods, 966 Mazzoni Street, Ukiah, CA 95482.

PHOTOVOLTAICS

Photovoltaic devices, also known as solar cells, produce electric current when light shines on them. They have no moving parts, so reliability is high—the lifetime is 20 years or more. No maintenance of the cell is required, although the surface on which the light shines must be kept clean. Installation is simpler than that of a hydropower or wind power system. No fuel needs to be transported, in contrast to a diesel or propane-powered generator. Because the systems are easy to transport and set up, they can be located near the load, reducing electrical transmission costs. Photovoltaic devices complement wind-powered generators well; often the sun is strongest during the times the wind is slowest. Photovoltaic power is almost always more expensive than power from the utility grid, so solar cells tend to be used when the alternatives are even more costly or inconvenient—the usual application is in remote locations. In Africa they are used in rural mission hospitals. In many parts of the world, including the United States, they are used extensively to power inaccessible microwave relay station. They have been used in parks to avoid unsightly, obtrusive power lines.

PRINCIPLE OF OPERATION

Detailed understanding of what goes on inside a solar cell requires much physics, but basic insight can be gained from simple ideas (see the Appendix to this article). Most solar cells now in use are made of silicon—a very common element. (Sand is silicon dioxide.) One reason that solar cells are expensive is that the silicon must be very pure—less than 1 impurity atom per 1,000,000 silicon atoms. Silicon, like every material, consists of atoms, and atoms contain electrons—small charged particles. When light shines on silicon, the electrons gain energy from the light and become free of the atoms. These free electrons can move. Their motion is current flow. We say silicon is a semiconductor because it only conducts current when illuminated or when electrons are liberated from the atoms in other ways, such as heat.

The real problem with making a solar cell is devising a way to collect all these free electrons so they will flow in the same direction in an external circuit. The method for accomplishing the collection employs small quantities of impurities. Usually these impurities are added to separate layers of silicon. Different kinds of impurities are mixed with the silicon. The effect of one type of impurity is to give one layer a small positive charge. The other type gives the other layer a slight negative charge, so a small voltage is produced between one layer and the other. Any electrons produced in the more negative material are attracted to the other, more positive, layer. Thin metal strips are put on this layer to collect the electrons so they can flow through a load and back to the first layer. These strips are visible on photovoltaic devices used in calculators. The number of impurity atoms introduced is very small, again, about 1 impurity atom to 1,000,000 silicon atoms.

Photovoltaic devices are expensive because they are made of thin layers of closely controlled material, deposited over large areas. Maintaining tolerances, both on the purity of the base material and the placement of impurity atoms, requires highly trained people and complex equipment. The cost of devices will go down as manufacturing experience accumulates, as more photovoltaic devices are put in service.

Photovoltaics work well with little maintenance because nothing really can go wrong in a functioning cell. Of course, in fabrication a connection may not be made well or the layers may not come out right, but such defects appear immediately. A working cell will work a very long time—on the order of 20 years. The only problem is heat. When a cell is hot, extra electrons are produced, and these electrons can overcome the voltage between the layers, reducing the efficiency of the cells. Cell temperatures should be kept below about 140°F. The cell is not, however, permanently damaged by overheating unless the material actually melts, at temperatures well over 300°F.

A typical solar cell puts out .56 volts. This value depends on the material used and is not adjustable. The amount of current produced depends on the area of the cell and the brightness of the sun. In full sun, a solar cell will produce about .28 amperes per square inch of cell; however, 20 percent of the voltage and current is lost internally, so one gets .45 volts and .224 amperes from a square inch cell in maximum sunlight. Maximum sunlight means no clouds and also that the cell faces the sun directly so the sun's rays are perpendicular to the cell's surface.

PANELS OF SOLAR CELLS

Solar cells can be purchased individually or mounted on panels, also called modules. On the panels the cells can be connected in either series or parallel or both—see Figure 20. If the cells are connected in series, the voltages add. In a parallel connection, the currents add—a parallel connection of several cells is exactly the same as one cell with a larger area.

A typical single cell sold commercially is close to circular in shape, about 4 inches in diameter. The output current in full and direct sunlight, then, is about 2.8 amperes, and the output power is about 1.26 watts. The values for output current shown in Figure 20 assume that each cell is this standard size—circular with a 2-inch radius.

Commercial panels come in many different sizes and configurations. A typical power module that nominally produces 40 watts and sells for $245 from Astro Power (30 Lovett Avenue, Newark, Delaware 19711) is shown in Figure 21. This module produces a maximum of 17 volts and 2.80 amperes. Modules from other vendors are shown in the *Jade Mountain* catalog.

Solar cells are somewhat fragile. They are manufactured very thin to reduce the amount of material required. It is therefore best not to handle them unnecessarily, so unless one has an application where a single cell is needed, purchase of an entire panel makes sense. One interesting application where single cells were deployed heat from the cells—produced by the solar energy, not converted to electricity—is heating a home. In this case, each cell was mounted on its own heat exchanger, conducting heat from the cell to the heating system (Loferski et al. 1988).

Another electronic device, actually very similar to a solar cell, should be mentioned: a diode. Diodes allow current to flow in only one direction. A diode is shown in the system diagram in

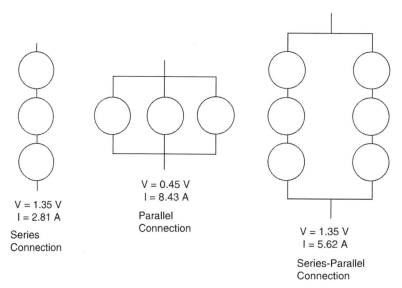

FIGURE 20
Connections of solar cells.

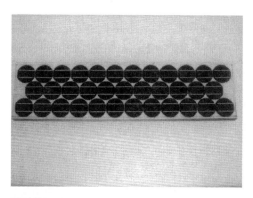

FIGURE 21
Commercial photovoltaic panel.

Figure 22. As one might expect from looking at the picture, current flows, from left to right in the diode shown but not from right to left. The purpose of the diode will be discussed shortly. (Diodes are also explained in the Appendix.)

ELECTRIC SYSTEMS USING SOLAR CELLS

A diagram for an entire system is shown in Figure 22. The operation of the system is probably obvious—current produced by sunlight on the cell array goes through the diode and both charges the battery and powers the 12-volt dc load. A diode is needed because when the cell is not illuminated—at night, for example—no voltage is produced, so current would flow back into the array from the battery.

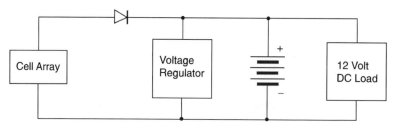

FIGURE 22
Solar panel with battery.

The diode in Figure 22 is to prevent this battery drain. A simple voltage regulator is also included in the circuit to prevent overcharging the battery.

If electric power is needed when the sun is not shining, storage is required. Batteries are the usual storage device. In some applications, such as refrigeration or irrigation, no storage is needed. If the user is connected to the power grid, then electricity can be sent into the grid, as described in the "Basics of Electricity" article. In this case, an inverter is needed—to convert dc to ac—but a battery is not used.

As described in the "Basics of Electricity" article, a choice of voltage levels and of ac versus dc must be made. The voltage available from a solar cell is dc, and if we are using automobile-type batteries, we would probably use 12 volts as our output voltage. In this case, of course, we could only use 12-volt dc appliances. If we prefer to use 120 volts to reduce line losses, we could connect many (267 to be precise) solar cells in series. The 120-volt dc could be stored by connecting ten 12-volt batteries in a series. Such a system would have lower line losses than a 12-volt system, but many motors only work on ac, so an inverter would probably still be necessary. Most commercially available inverters contain transformers and are designed for 12-volt dc input and give 120-volt ac output. The *Jade Mountain* catalog describes inverters—a typical price is $500.

A photovoltaic system is ready to operate once it is connected together. For effective operation, though, it is necessary to consider cooling. Although the silicon will not be damaged unless the temperature is very high, the adhesive holding the cells on the panel will melt at a lower temperature, as will the solder. The effective maximum temperature is about 250°F, so the design must ensure that the cell temperature is less than that.

We can see then that the cells themselves are only part of the cost of a whole installation. The additional costs come from the panel and its mounting, storage, and the inverter. A very approximate estimate is that these so-called "balance of system costs"—the additional costs beyond the cells themselves—are equal to the cell costs. Incidentally, another term that one sometimes sees is "power conditioner," referring to a sophisticated inverter that puts the voltage and current in a condition—proper magnitude, frequency, and wave shape—to be sold to the utility.

The array must be positioned properly. Maximum energy is received by a cell array when it faces the sun directly. On the equator at noon, an array lying horizontally on the earth's surface would face the sun directly—actually this is only true on the equinoxes, but seasonal effects will be ignored. As an array is moved away from the equator it must be tilted up to remain facing the sun directly. The angle of the tilt is equal to the latitude, as a simple sketch will show. An array at the proper tilt will only face the sun directly at noon. The total amount of energy received over the course of a day is increased by about 30 percent if the array "tracks" the sun—moves during the day to face the sun directly. Usually the additional cost of a tracking mount is more than the gain from the additional energy, so tracking mounts are seldom used for small arrays, although some less elaborate systems are designed to be redirected manually once or twice a day. Reflectors and lenses can be used to concentrate the sun's rays. A Fresnel lens, which is thin, is particularly useful here as it absorbs very little of the sun's energy.

SYSTEM DESIGN

Usually a photovoltaic system is put together from modules containing several dozen cells, although one can buy individual cells and construct a module. The design below assumes one is starting with individual cells. It is easy to simplify the approach when modules containing several cells are used. The most important question in designing a system is to estimate the area of cells needed—that is, the number of cells or size of module. The example from the "Basics of Electricity" article will be continued. We need 960 watts for 5 hours. A typical commercial cell 4 inches in diameter has an area of 12.56 square inches. How many cells are needed? We know we need 4,800 watt hours, and we know the power from a 1-inch cell (when the sun is shining directly on it) at solar noon is 1.26 watts.

What makes the calculation difficult is that if we direct the panel to get maximum energy at noontime, the rest of the time the panel will receive radiation at an angle, which is less effective. The effect of glancing radiation and lack of radiation at night has been calculated. In a representative situation the average effective radiation power received during a day is 0.2 times the power of the peak radiation. Peak radiation is the radiation received at the optimum time of day—solar noon. Energy is power received multiplied by the length of time it was received. So if we multiply the maximum radiation received by 0.2 and by 24 hours, we get the same total energy as we would get by adding the energies received at each instant of time.

$$E = P_{peak} \times .2 \times 24 \text{ watt-hrs}$$

where E = energy received per day.

$$P_{peak} = \text{peak power from the cell—power at noon on a cloudless day}$$

Peak power from 1 cell is 1.26 watts, and the energy desired from N cells is 4,800 watt-hours, so

$$4800 = 1.26 \times N \times .2 \times 24$$

$$N = 794 \text{ (number of cells needed)}$$

Typical cells are circular, four inches in diameter, so 9 cells fit into 1 square foot. The area of cells, then, is 89 square feet, perhaps 6×15—smaller than most roofs.

In the case where 40 watt modules were to be used, rather than single cells, the previous equation can be used where N_{40} is the number of 40-watt modules

$$4800 = 40 \times N_{40} \times .2 \times 24$$

$$N_{40} = 25 \text{ (number of 25 watt modules needed)}$$

A typical size for a 40-watt module is 3.5 feet by 1.5 feet (about 1 meter by .5 meters), and a typical cost is $200 per module.

The cost of photovoltaic devices is usually given in dollars per peak watt. In this case, the number of peak watts from the system is the peak watt from an individual cell (1.26 watts) times the number of cells, or

$$\text{Peak watts} = 1.26 \times N = 1000.4$$

A representative cost for cells in 2000 is $3 a peak watt. (In 1978 it was $9.) So our cost for the cells would be about $3,000. If the balance of systems cost were equal to the cost of the cells, then the total cost would be $6,000. For a 20-year system using money borrowed at 12 percent, the annual costs would be $803.28. If we used 40-watt arrays, the modules would cost $5,000, and we would have to add the cost of mounting racks, batteries, and a voltage regulator, so the total cost would be about the same.

WHEN ARE PHOTOVOLTAICS APPROPRIATE?

Photovoltaics electric systems differ from wind power and hydro systems because they cannot be constructed by an amateur or even a good machinist. High-technology equipment is needed to purify the silicon and get the impurities in the correct places. On the other hand, photovoltaic systems are easy to put together from purchased panels and are environmentally benign. In many ways, photovoltaics seem to be an ideal energy source—reliable, quiet, pollution free, land conserving, and locally controllable. Their widespread use appears to have significant advantages to society.

In countries without an extensive power grid, photovoltaic power generation makes much sense. Water pumps in remote regions, electric fences to keep elephants in national parks, and power for village centers are all applications, besides hospitals and communication equipment, where photovoltaics are being used now. Solar electric systems will work unattended for long periods of time; the biggest problem is usually the battery. A small amount of electric power can often make a significant difference in a remote application, and photovoltaics can produce that small amount of power. The cost of alternatives is often much higher than that of photovoltaics in those applications.

REFERENCES

Carts-Powell, Yvonne. (1997). "Solar Energy Closes in on Cost-Effective Electricity." *Laser Focus World,* December, 67–75.

Home Power: The Hands-on Journal of Home-Made Power. Ashland, Oregon. PO Box 520, Ashland, Oregon, 9752.

Jade Mountain, Appropriate Technology for Sustainable Living. (1999–2000). Vol. XIII, No. 1. PO Box 4616, Boulder, CO 80306-4616. Telephone 800-422-1972.

Loferski, J. J., J. M. Ahmad, and A. Pandey. (1988). "Performance of Photovoltaic Cells Incorporated into Unique Hybrid/Photovoltaic/Thermal Panels of a 2.8 KW Residential Solar Energy Conversion System." 1988 Annual Meeting, American Solar Energy Inc., Cambridge, MA.

INTRODUCTION TO PHYSICS OF SEMICONDUCTORS

This section is based heavily on Chapter 2 of Electronic Circuits and Applications, *by Barrett Hazeltine. Kendall-Hunt Publishing Company, 1978.*

THE FLOW OF CURRENTS IN METALS

The reader undoubtedly knows that all substances are composed of atoms and that atoms in turn are made up of electrons, protons, and neutrons. The electrons can be thought of as being small particles with a negative charge that surround the nucleus, which is made up of protons that have a positive charge and neutrons that have no charge. There are just as many protons

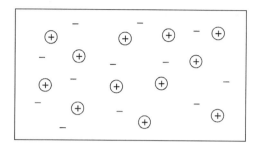

FIGURE 23
Atomic structure of a metal.

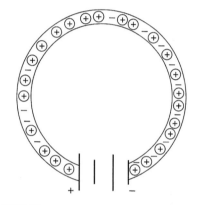

FIGURE 24
A conductor connected to a battery.

as electrons in an atom, and because an atom has just as many electrons as protons, it is electrically neutral. The protons and neutrons making up the nucleus are much heavier than the electrons and do not move around.

The picture one has of a piece of metal, then, is indicated in Figure 23. The circles represent the nuclei (the plural of "nucleus.")

The short lines represent electrons. The atoms making up metals have one or two electrons that are located relatively far from the nucleus. These electrons are called "valence" electrons. The electrons that are not valence electrons are closely associated with the nucleus and neutralize some of the protons of the nucleus. The valence electrons are not closely associated with the nucleus and can drift away from it. It is convenient, therefore, to consider a piece of metal as a collection of nuclei, each with a positive charge, with electrons floating in the space between the nuclei. The floating electrons are valence electrons that have drifted away from the nuclei and are, therefore, not associated with any particular atom. When an electron drifts away from an atom, the nucleus of that atom is left with an unneutralized proton; thus, that nucleus has a net positive charge. Of course, the piece of metal as a whole has no net charge because for every unneutralized proton in a nucleus, there is an electron floating somewhere in the metal.

Using this picture of the interior of a piece of metal, it is easy to see how electric current is conducted. If we attach a conductor between the terminals of a battery, as in Figure 24, we see that electrons near the positive battery terminal will be attracted to the terminal. This will mean that the positive nuclei in the region where the electrons were will not be neutralized.

Thus, the region near the positive battery terminal will become slightly positive. Electrons from other parts of the conductor, farther away from the positive battery terminal, will be attracted to this positive region. The region these electrons leave will become positive, which will permit other electrons still farther away to flow toward the positive terminal of the battery. The result is that there is a continual flow of electrons through the entire conductor, away from the negative battery terminal, toward the positive terminal. At the positive terminal these electrons go into the battery, which pushes them out of its negative terminal. According to our model the battery energy is spent pushing the electrons out. (The reader should recall that according to the usual definition, current flows out of the positive terminal of the battery into the negative terminal. The usual definition of current was created before people realized current was carried by negative charges.)

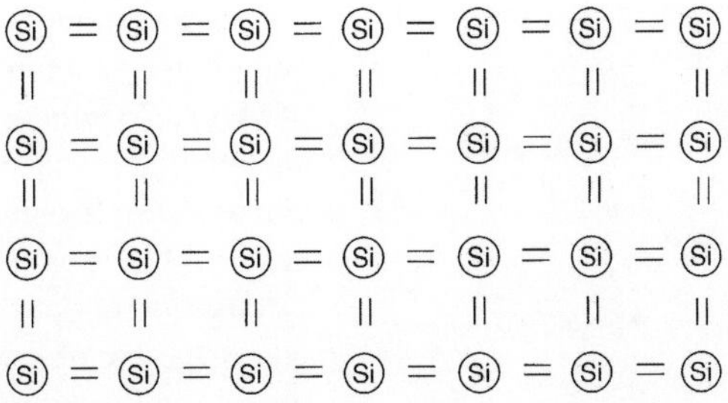

FIGURE 25
Internal structure of semiconductor material.

SOLAR CELLS

Solar cells are usually made out of crystalline germanium or crystalline silicon. (In our discussion, we will consider silicon, but the results apply equally well to germanium.) Because the material is crystalline, the nuclei of the atoms are arranged in a regular pattern, as shown in Figure 25.

In silicon (as well as in germanium) each nucleus has a place for 8 valence electrons, but each atom has only 4 valence electrons. Thus, by sharing valence electrons between adjacent nuclei, each nucleus can fill all its places. Normally, at room temperature nearly all the valence electrons are shared between neighboring nuclei and tend to stay at the same location.

There are a few valence electrons, however, that have enough energy to get away from these places of sharing, and these energetic electrons float through the interior of the semiconductor just as the valence electrons do in a metal. Because there are so few electrons that are mobile at room temperature, only a small amount of current will flow through the semiconductor at room temperature. (This is the reason the name "semiconductor" is used.) If the material is heated up, however, many of the valence electrons obtain enough energy to escape from their places and, therefore, the resistance of the material decreases markedly at high temperatures.

p-TYPE AND n-TYPE IMPURITIES

Adding impurities to the silicon allows appreciable current to flow at room temperature. As was mentioned, the reason there are only a few electrons available at room temperature to carry current is that each silicon atom has four valence electrons and has places for eight valence electrons. By sharing valence electrons each atom fills its eight places. Now let us assume a small amount of arsenic is mixed with the silicon, when the silicon is molten. The arsenic atom is basically similar to the silicon atom and, therefore, fits right into the regular crystal structure. The arsenic atom differs from the silicon atom in one respect, however; it has five valence electrons. Of course, it can only share four of these valence electrons with neighboring silicon atoms; the remaining electron is free to wander through the crystal. An impurity like arsenic that adds free electrons to the material is called an "n-type impurity"—the "n" referring to the negative charge of the electron.

There are also "p-type impurities" which add positive "holes" to the material. Let us see exactly what a hole is. A p-type impurity such as gallium has only three valence electrons, one less than silicon has. Therefore, when gallium is mixed in with silicon, the gallium atoms will

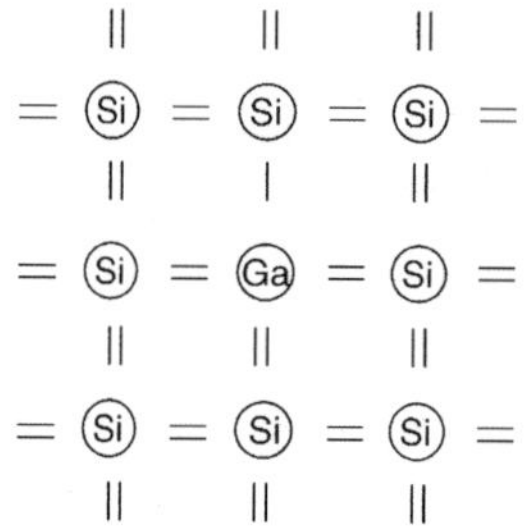

```
     ||        ||        ||
  =  (Si)  =  (Si)  =  (Si)  =
     ||        |         ||
  =  (Si)  =  (Ga)  =  (Si)  =
     ||        ||        ||
  =  (Si)  =  (Si)  =  (Si)  =
     ||        ||        ||
```

FIGURE 26
Silicon with gallium impurity added.

not have enough electrons to share with the neighboring silicon atoms. This is illustrated in Figure 26, where a gallium atom is shown surrounded by silicon atoms. (The chemical abbreviation for silicon is Si and for gallium is Ga.) The point of Figure 26 is that an electron is missing from the otherwise periodic structure of the crystal. The place where the missing electron should be is called a hole.

(The hole in Figure 26 is directly above the gallium atom.) In a material containing p-type impurities, it is possible for an electron associated with another silicon atom to slip into a hole, thus filling it.

If this happens, another hole is left at the place from which the electron came. Not only is a hole produced in the crystal structure but one of the positive charges on the nucleus of an atom adjacent to this new hole is no longer neutralized. It is convenient to pretend that this positive charge is associated with the hole itself. When an electron from another atom slips into this new hole, it leaves a net positive charge somewhere else; thus, in effect, the hole migrates through the material, acting like a particle carrying a positive charge.

BOUND CHARGES

Not only do the moving holes and electrons carry a charge, but the region from which the holes and electrons came have a charge. Let us study this more closely. The nucleus of the gallium atom contains just enough protons to neutralize its own electrons. Therefore, when a hole leaves a gallium nuclei—that is, when an electron slips into the hole adjacent to the gallium nucleus—it adds a negative charge to the region near the gallium atom. The negative charge is associated with the nucleus and it is not free to move through the conductor. It is therefore called a "bound" charge. The entire crystal is still electrically neutral because, as we have seen, the hole produced when the electron slipped into the hole near the gallium nucleus is positive.

We have been talking about material containing a p-type impurity. In a material containing an n-type impurity, the impurity adds an extra electron to the crystal. When this extra electron leaves the neighborhood of the arsenic atom it came from, that region takes on a positive charge because one of the protons in the arsenic nucleus is no longer neutralized by a valence electron. We will see soon that these charges, bound in a region by the presence of impurities, are the basis of the operation of semiconductor diodes and solar cells.

THE p-n JUNCTION

Silicon with an n-type impurity added is called n-type silicon. Similarly, silicon with a p-type impurity is called p-type silicon. Because holes predominate in p-type material, they are called "majority carriers." Electrons are "minority carriers" in p-type materials. In n-type material the designations are reversed; electrons predominate so they are the majority carriers. There are only a few holes in p-type material, so they are minority carriers.

Let us see what happens when a piece of p-type silicon is joined to a piece of n-type silicon forming a so-called "p-n junction," as illustrated in Figure 27a. The p-type material is shown as being full of positive holes, while the n-type material contains negative electrons. The holes in the p-type material and the electrons in the n-type material are, of course, moving around. When the p-type and the n-type materials are joined, therefore, some holes will go from the p-type silicon to the n-type and some electrons from the n-type to the p-type. An electron, when it gets to the p-type material, will probably meet a hole fairly soon because there are so many holes in the p-type material. It is apparent that when an electron meets a hole, it will slip into the hole. When this happens both the hole and the electron disappear, in the sense that neither is available to carry current. The result is that shortly after the two materials are joined, in the p-type material there are fewer holes near the junction than elsewhere. Similarly, in the n-type material there are fewer electrons near the junction than there are elsewhere in the mate-

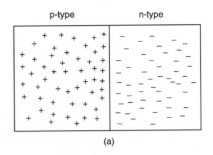

(a)

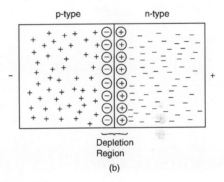

Depletion
Region

(b)

FIGURE 27

(a) p-n junction when first formed. (b) p-n junction when steady state is reached.

rial because the electrons near the junction have combined with holes from the p-type material and vanished. The region near the junction where there are few majority carriers is called the "depletion region." This region is indicated in Figure 27b.

When the electrons in the n-type material near the junction disappear by combining with holes, then the depletion region in the n-type material takes on a positive charge. The origin of this positive charge is the fifth proton in the nuclei of the impurity atoms—the bound charge mentioned above. Originally these protons were neutralized by electrons wandering through the crystal. When the free electrons disappear, the region takes on a positive charge. The bound positive charges are circled in Figure 27b.

The result, then, of putting a piece of n-type material together with a piece of p-type material is that the p-type material near the junction becomes negatively charged and the n-type material near the junction becomes positively charged. In other words, a voltage is produced across the junction with the n-type material at the higher voltage. The polarity of the p-type material and of the n-type is indicated in Figure 27b. An easy way to remember which material becomes positive and which negative is to realize that the n-type loses electrons to the p-type and therefore becomes positive. The p-type gains electrons and therefore becomes negative.

Such a voltage is produced whenever two semiconductors are joined together. In fact, even when two different metals are joined together, a voltage is produced because the electrons in one metal will be more energetic than the electrons in the other metal, and therefore there will be a net flow of electrons from one piece of metal to the other. The piece of metal that acquires more electrons than it loses will be at the lower voltage. This potential difference produced when two dissimilar metals are joined together is called the "contact potential."

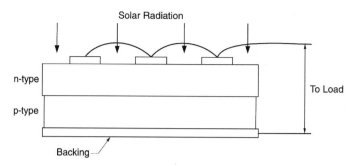

FIGURE 28
Photovoltaic device.

The voltage that is established between the n-type and the p-type materials prevents further flow of electrons and holes because the n-type material, having become negative, repels the further flow of electrons. Thus, a very short time after the junction has been formed, no more holes or electrons will flow across the junction.

PHOTOVOLTAIC DEVICES

A photovoltaic device is just a pn junction—see Figure 28. Electrical power is created because energy from the sun, or any light source, gives electrons sufficient energy to leave their bonds.

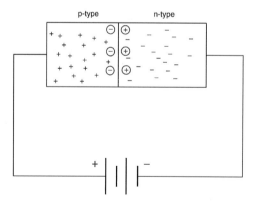

FIGURE 29
p-n junction and battery—current flowing.

The free electrons in the p-type material, which are minority carriers, go to the n-type material and then to the load. The n-type material collects the electrons because it is positive.

Commercial solar cells can be made in different ways. A typical way is to deposit a layer of silicon on a metal backing. Impurities are introduced in two stages, so a p layer is produced adjacent to the backing and an n layer is on top. Energy from the sun goes through the n layer and liberates electrons within the p layer. These electrons go to the n layer, where they are collected by thin metal strips on the surface. The external load is connected to these strips and the backing.

SEMICONDUCTOR DIODE

A diode allows current to flow in only one direction. It is basically a pn junction. Assume a battery is connected across the junction with the positive terminal of the battery connected to the p-type material and the negative terminal connected to the n-type region, as in Figure 29. When the battery is connected in this direction, its voltage will counteract the voltage developed

across the junction and thereby allow electrons from the n-type region to go into the p-
type region. A small voltage—just sufficient
to overcome the voltage at the junction produced by the layers of bound charge—will
allow a fairly large current to flow across the
junction because many electrons are in the n-
type material.

   If the polarity of the battery shown in Figure
29 is reversed—that is, if the positive terminal
is connected to the n-type region and the negative terminal to the p-type region—then the
battery voltage will add to the voltage produced across the junction, as in Figure 30.
(Recall that the p-type material became negative and the n-type became positive when the
junction was formed.) This added voltage will
repel the flow of charges across the junction.

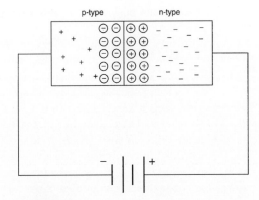

FIGURE 30
p-n Junction and battery connected—no current
flow.

COOK STOVES

Improved cookstoves for village use have received much attention, both research into the
technology and dissemination efforts. Improved cookstoves offer several advantages: decreased
fuel consumption—reducing deforestation and time spent gathering firewood, reduction in
smoke—leading to better health, faster cooking—allowing women more time for other activities,
such as child care. An appreciable amount of the energy consumed in Third World countries
goes into firewood, so an increase in cookstove efficiency is worthwhile. At least as important
is that an appreciable portion of many rural women's time is spent gathering firewood and
cooking. New developments in cookstoves have not spread as quickly as one might have
expected. Perhaps the cultural factors associated with cooking have been underestimated. Perhaps
an inefficient cookstove is desired because it warms a family space. Perhaps the time spent
gathering firewood is not perceived as wasted because it is social time. Stoves can be built in
many ways. To understand construction, it is helpful to understand combustion.

STOVE TECHNOLOGY

For combustion to take place, three things are needed: fuel, air (oxygen) to react with the fuel,
and an ignition source to get the reaction started. A match is a good example of a combination
of these components. The heat for ignition is generated by friction when the match is rubbed
across the striker. This heat ignites the chemicals in the matchhead, which in turn lights the
body of the match. Once combustion is started, it will continue until the fuel (the wood) or the
air is consumed.

   The task facing stove designers is how to control the combustion and how to use the heat
generated most efficiently. We will consider control first. An open fire is controlled by the
loading of fuel and the whim of the wind. In a strong breeze a fire will consume its fuel
more quickly and give off more heat in a given amount of time. The reason the rate of
combustion depends on the wind is that hot air rises, so a fire creates an updraft, pulling air

in from the side and pushing it out above the flame. Wind increases the amount of air flowing into the fire.

A pot supported by three stones with a fire beneath is the most prevalent open fire construction. If the fire is built on the ground, the fuel closest to the ground is starved for air and does not burn completely. Because insufficient air creates incomplete combustion, these fireplaces tend to be smoky, creating health and hygiene problems, although smoke does keep insects away. The efficiency of an open fire (heat that ends up in the pot divided by heat potential of the fuel $\times$ 100) is 10 to 15 percent. Three-stone fireplaces cost nothing and can be built nearly anywhere. They provide heat, light, and a social place, so they do have advantages.

The first tack for improving efficiency is to control the flow of air around the fire. One method is to raise the fire off the ground by building it on a grate. Then the air will flow from beneath the fire and assure complete combustion of the fuel. To gain greater control, one can build the fire in a box with a damper to control the air intake and a flue to vent the exhaust gases out of the box.

By controlling air flow you assure complete combustion and control the rate of fuel consumption—less draft will slow the fire, more draft will speed it up. When things burn in air, they do so at a fairly constant, fuel-dependent temperature. What can be changed is the rate of consumption and therefore the amount of heat liberated in a given length of time. Fires do not so much burn hotter or colder but faster or slower. The faster a fire burns, the "hotter" it appears to be.

A typical burn demonstrates the different ways a fire is controlled. A fire begins in a fuel-limited condition until complete ignition is achieved. The fire is then draft limited—that is, controlled by the air until the fuel is depleted enough to affect the rate. The fire continues to be fuel limited until it exhausts the fuel supply.

Heat is transferred by three mechanisms: conduction, convection, and radiation. The problem in designing a stove is to maximize the energy transferred to the cooking pot and minimize that wasted. Heat can be radiated directly from the flames to the pot. It can also be carried—convection—by the hot gases to the pot. In both cases, the designer should do all that is possible to increase the effectiveness of the heat transfer—for example, by directly exposing the pots to the flame or by ensuring the hot gases cover the bottom and sides of the pots. Unburned fuel between the fire and the pot reduces the amount of heat transferred. So does direct passage of hot gases up the chimney.

Efficiency, defined above as the ratio between useful heat produced and potential heat in the fuel, increases if losses are reduced. Losses come from incomplete fuel combustion and continued burning after cooking is completed. They also come from heat escaping through stove walls, cooking vessels, the chimney, and uncovered spaces. Of course, some of this lost heat is useful in warming the room.

As we said, an open fire smokes. The smoke can be unhealthy for the cook and, especially in a closed room, for other members of the family, producing both eye irritation and lung difficulties. On the other hand, the smoke can keep insects away and even small animals—lizards in a thatch roof, for example. The smoke might also be useful for curing meat hung high in a room.

STOVE CONSTRUCTION

Let's get back to stoves. We have three components that we can arrange to make a stove; the firebox, the damper, and the flue/chimney. The other parameter we control is the material out of which the stove is constructed. Typically stoves are categorized as low mass or high mass. This refers not to their weight but to their thermal mass. A high-mass stove will heat slowly and hold the heat for a long time. High-mass stoves are constructed of mud clay, sand, bricks, or rocks—materials usually present in rural areas.

Low-mass stoves heat and cool quickly. They can be made of metal or fired clay. Metal stoves are often made from scrap metal such as discarded oil drums and paint cans.

It is easy to see that each type of stove has its strong points and drawbacks. The high-mass stove is probably the simplest to construct, it retains its heat well, and it is helpful for space heating as well as cooking. Most of the high-mass materials are fairly brittle, however, and tend to crumble under use.

Low-mass stoves are simple and portable. A low-mass stove, made of sheet metal, in the courtyard of a Zambian house is shown in Figure 31. They are used primarily in urban areas for outdoor cooking. Their drawbacks include being unstable under heavy loads, requiring fuel to be cut into small pieces, and losing much heat through radiation.

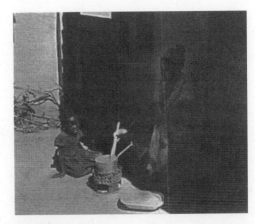

FIGURE 31
Low-mass stove.

SAND/CLAY STOVES

The designers of the Lorena stove, shown in Figure 32, had the considerations above in mind. Its high mass makes it useful for space heating in addition to cooking. It has an efficiency of 20 to 25 percent. Although this is better than an open fire, it may not be sufficient to justify the cost, $20, mostly for the chimney. One advantage of the Lorena Stove is that it is large enough to serve as a work area, as well as for cooking.

The operation of the Lorena stove in Figure 32 is straightforward. The two dampers control the flow of air. The sharp bends in the tunnel create turbulence under the pots, which ensures even heating. The dampers are simply small pieces of sheet metal, with handles, fitting tightly into grooves that hold them in position. The chimney is rolled metal and is the most expensive part.

The Lorena system involves building a solid sand/clay block, then carving out a firebox and flue tunnels. The block is an integral sand/clay mixture that, upon drying, has the strength of a weak concrete (without the cost). The stove is made by molding wet clay, building the stove in stages. Construction takes three or four days. Homemade tools can be used, such as a sifting screen, a shovel, a bucket, and mason's trowel.

The mixture contains two to five parts of sand to one part of clay, though the proportions can differ widely. Pure clay stoves crack badly because the clay shrinks as it dries and expands when it is heated. Sand/clay stoves are predominantly sand, with merely enough clay to glue the sand together. The mix should contain enough clay to bind the sand grains tightly together but enough sand to prevent the clay from shrinking as a mass.

CERAMIC STOVES

A contrasting example to the Lorena is the Damru Chulha, designed at the Agricultural Tools Research Center in Bardoli, India, shown in Figure 33. Combustion is improved by the iron grate onto which the fuel is loaded. A clay ring, the "heat shield," is used to reduce heat loss from the cooking vessel. The Damru Chulha is about 30 percent efficient. It is simple enough to be built

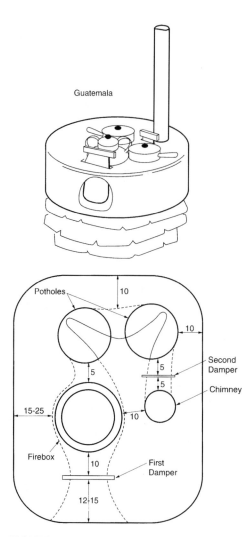

FIGURE 32
Lorena stove.

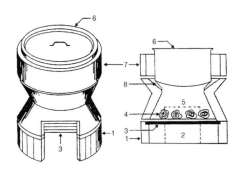

FIGURE 33
Damro chulha stove.

by a local potter at a cost of about $4. Of course, this stove does not warm the room effectively and cannot be built by the homeowner. It is heavier than a metal stove.

The Kenya Ceramic Jiko, shown in Figure 34, is a portable, improved charcoal-burning stove consisting of an hourglass-shaped metal cladding with a n-interior ceramic liner that is perforated to permit the ash to fall to the collection box at the base. A thin layer of vermiculite or cement is placed between the cladding and the liner. A single pot is placed on the rests at the top of the stove.

If used and maintained properly, the Kenya Ceramic Jiko can reduce fuel use by 30 to 50 percent, although not surprisingly there is considerable variation based on the extent of training and outreach efforts, stove quality, and cooking practices. The stove also reduces emissions of products of incomplete combustion (carbon monoxide, nitrogen and sulphur oxides, and various organic compounds), as well as particulate matter, the latter of which contributes to acute respiratory infection. The degree of emissions reduction is estimated to be 20 percent (Kammen 1998).

FUTURE DEVELOPMENT

The set of examples given here is not exhaustive—much opportunity for creative design remains. Stove technology has shown rapid change over the last 20 years—more change is possible. Efficiencies of 35 to 40 percent are possible, but it is necessary to recognize that the stoves must be in harmony with local custom to gain acceptance. Improved designs can serve as a valuable tool for teaching users what makes a stove efficient and thus allowing them to create the next generation of stoves.

REFERENCES

Campbell, John R., and Shewaynesh Bezuayenae. (1991). "Improved Stove in Urban Ethiopia." *Appropriate Technology,* Vol. 17, No. 4, 29–31.

Clarke, Robin. (1985). *Wood-Stove Dissemination.* London: IT Publications.

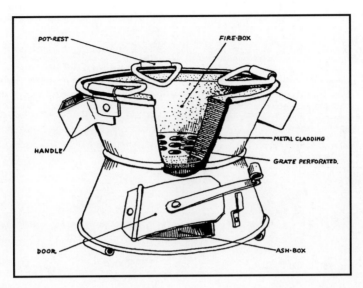

FIGURE 34
Kenya ceramic jiko stove. Reproduced by permission from *More Innovations for Development,* IT publications, London, 1991.

Evans, Ianto, and Michael Boutette. (1981). "Lorena Stoves." The Appropriate Technology Project of Volunteers in Asia. Stanford.

Kammen, Daniel M. (1998). "Research, Development and Commercialisation of the Kenya Ceramic Jiko and other Improved Biomass Stoves in Africa." University of California, Berkeley.

SOLAR COOKERS

Solar cookers use energy from the sun to cook food. Many are used in China and India. The promise seems high, but worldwide usage has not been as large as one would expect. The usual explanation is that their use is significantly inconsistent with people's lifestyles. The advantages are many: protection of forests, no smoke, no fuel cost, no time spent gathering firewood, slow cooking that conserves nutrients, little danger of burning food. Inherent problems are the daylight and good weather requirement and the long cooking time. The necessity that the person doing the cooking must be out in the sun is also given as a reason for the low acceptance.

Two types of solar cookers are in use. The first is a solar cooking box. Plans and construction for such are given a separate article. Solar cooking boxes are well-insulated boxes with a cover that reflects sunlight into the box. In typical use the vegetable or meat to be cooked is placed in the box early in the day and is ready to serve after a few hours. No attention is needed from the cook. Solar cooking boxes keep the food warm in the afternoon and evening. The second type is a reflector cooker. These concentrate the sun's radiation, using a large reflector shaped like a microwave antenna, onto the cooking vessel. Reflector cookers achieve higher heats but must be reaimed frequently, since the sun moves. The reflector can be made of metal or of metal foil on a wooden frame. A large reflector is somewhat cumbersome to handle and is fragile. Perhaps because a cooking box is easier to use it is more common. Figure 35 shows a solar cooker.

FIGURE 35
Reflector cooker.

SOLAR OVEN CONSTRUCTION MANUAL

Daniel M. Kammen
Shea Van Boskirk
Monique Nditu

ACKNOWLEDGMENT

This solar oven construction manual would have been impossible without the assistance and input of dedicated solar oven carpenters and chefs in numerous communities in Kenya, and without the work of EarthCorps volunteers. In particular Francis Muthoka of Ambassadors Development Agency, Josephine Mwota, Christine Mwende, Daniel and Francisca Musyoka, and Patrick Ngalla have all tirelessly worked to improve the design and to spread the solar oven technology throughout Kenya. We would like to give particular thanks to Professors Odhiambo,

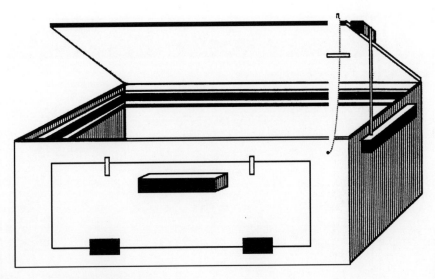

FIGURE 36   Completed solar oven

The solar oven is a well-insulated wooden box with two panes of glass as a top. Inside the wooden box is an inner cardboard box. Insulation is placed between the set of wooden and cardboard walls. The cardboard walls are covered with aluminum foil, or they can be painted black. Also show is a cover, connected by hinges to the box, with its inner surface covered with aluminium foil so it can also be used as an additional reflector.

Munavu, and Genga, and Drs. Kola and Rabah for their support of the renewable energy project in East Africa, and to Professor Lankford of George Mason University. This research is supported by the Department of Physics at the University of Nairobi and the Kenya National Academy of Sciences, the African Academy of Sciences, the United Nations Cultural, Educational and Scientific Organization, and Earthwatch and its Research Corps.

GENERAL COMMENTS ON CONSTRUCTING A SOLAR OVEN

The plans follow after these notes.

1. All dimensions listed in the plans are in inches—for example, 6.5 = 6.5" = 6.5 inches (1" = 2.54 cm).

2. The carpentry involved in constructing a solar oven is very simple and straightforward. Plan ahead and work carefully and you will have an excellent oven.

3. Good carpentry is a tradeoff between the dimensions you calculate, the accuracy of your work, and the reality of the wood. Measure everything over and over again. This will minimize your errors and give you a recourse to fix problems that arise.

4. The goal of the construction is to produce a well-insulated (well-sealed) box. To accomplish this you should do the following.
   • Check and recheck your measurements.
   • Test the fit of all pieces, then assemble them with glue and nails.
   • Minimize the number of nails used.

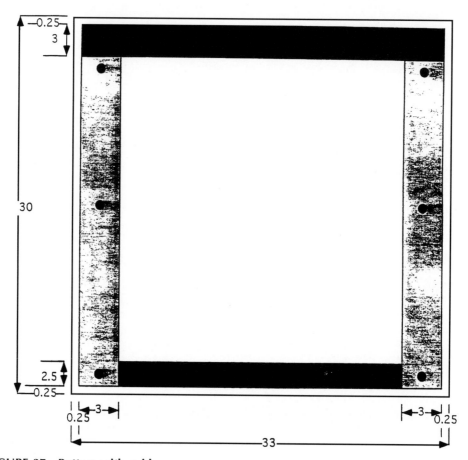

FIGURE 37   Bottom with guides

The bottom of the oven is a large plywood sheet with thin strips of plywood nailed and glued to it. The thin sheets leave a quarter-inch gap all around the edge. This serves as a guide for the oven walls.

- Use lots of glue to seal the oven.
- Use lots of paint to seal the oven, particularly along the seams.

5. The following are some common problems and recommendations.
   - Warped plywood—keep it dry, and place weights on it overnight.
   - The materials available are different from those listed—there are no "secrets" to the oven. A well-insulated box of almost anything will work.
   - Irregular wood cuts lead to gaps along the walk—match "good" sides of wood together, and be aware of what surfaces need to be flat and which are unimportant.

6. Ovens can be built of any size.

TOOLS AND MATERIALS NECESSARY TO CONSTRUCT A SOLAR OVEN

Tools

Wood saw

Coping saw

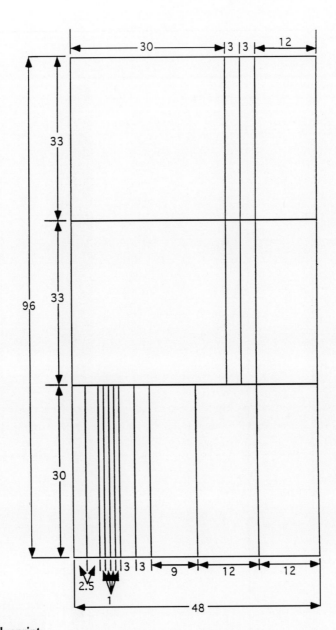

FIGURE 38   Blueprint
This is the layout of the plywood and dimensions of how it should be cut.

TOOLS (Continued)

Metal saw
Regular and star screwdrivers
Hand drill
Drill bits
Measuring stick/square
Hammer

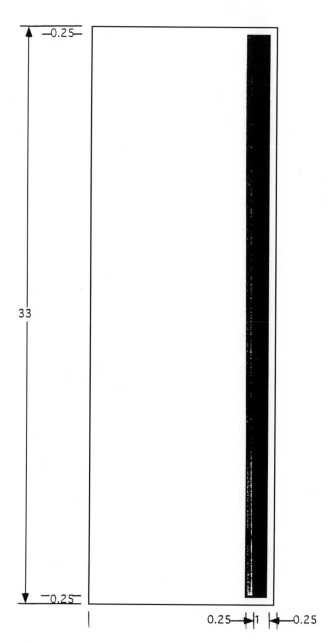

FIGURE 39   Back wall and shelf
A 1/4" is left on both the length and width of the shelf. The space is where the side wall (also 1/4" thick will fit). A top sheet of glass fits over the shelf. These are shelves upon which the two sheets of glass rest. The upper shelf is 1" by 1", and the lower shelf is 3 inches wide and 1/4" thick. The top sheet of glass rests on the upper shelf, while the seconed sheet of glass rests on the lower shelf.

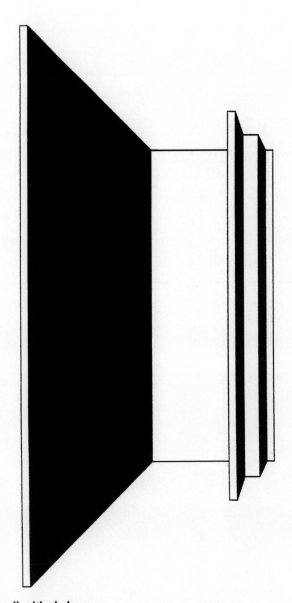

FIGURE 40   Back wall with shelves
The plywood back with shelves is then attached to the bottom piece of wood.

TOOLS (Continued)

Silicone gun
Utility knife
Wood file
Pencil
Paintbrush
Measuring tape
C-clamps (optional)

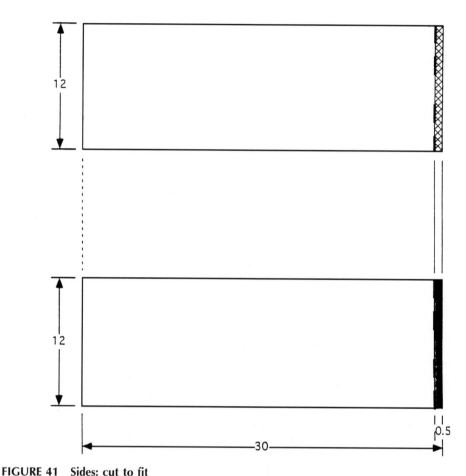

FIGURE 41   Sides: cut to fit
A 1/4" is cut from each side of both plywood walls so that they fit between the back and front walls.

MATERIALS

48" × 96" (4' × 8') sheet of 1/4" thick plywood

~216" of 1" × 1" wooden rods

~3 square yards of cardboard or hardboard

1" nails (with heads, and some headless)

1/4" tacks

23" square hinges

24" T-hinges

1 liter of wood glue

Half tube of high-temperature silicone sealant

Assorted wood screws (0.5" to 1.5" in length)

21" × 24" (approx.) 16–24 gauge metal sheet

1 wide roll of heavy aluminum foil

1/4 liter flat black paint

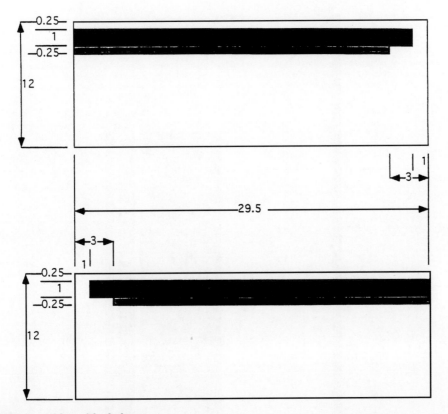

FIGURE 42   Sides with shelves
This is how the side walls appear once the two shelves are attached. The dark-shaded area is the upper shelf, which is a 1" × 1" upon which the top sheet of glass rests, while the light-shaded area is 3 inches wide and 1/4" thick and forms the lower shelf upon which the second sheet of glass lies.

MATERIALS (Continued)

1 liter of oil-based exterior paint

2–3 yards of twine, nylon string, or thin rope

2 sheets sandpaper (80 grade)

2 sheets of 4-mm-thick clear glass:

Lower sheet: 27" × 30"

Upper sheet: 29" × 32"

One sack of insulating material (wood shavings, rice husks, paper, etc.)

Symbols used in plans
  ~ approximately
  " Inches
  ' Feet
The materials cost about Ksh 1,600, or U.S. $32.

There are a wide variety of solar oven designs that all work very well. Further, each time we hold a solar oven workshop, new modifications emerge. We ask that you please send us comments on this design, changes, and whole new designs that you have constructed.

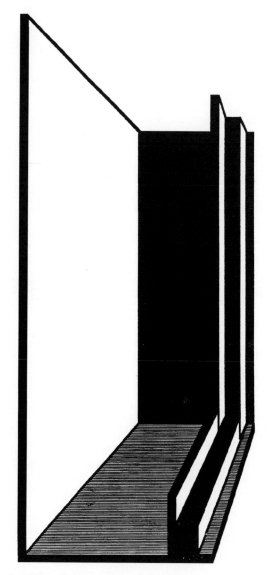

FIGURE 43 A perspective view of the bottom of the oven
The back wall has two shelves fixed on it. The upper shelf is made from the 1" × 1" wood. It supports the top sheet of glass and measure 29" × 32". The lower shelf is 3 inches wide and 1/4" thick. It forms the shelf for the small sheet of glass (27" × 30"). The shelves of the back and side walls fit tightly together—forming a level surface for the glass. The side wall is attached to the 1/4" space on the bottom guide. It has two shelves fixed on the upper side.

Nairobi Contact:
  Monique Nditu. Project Director
  African Academy of Sciences
  PO Box 14798
  Nairobi, Kenya
  Tel: (254-2) 884401/2/3/4/5
  Fax: (254-2) 884406
  add.ass@elci.gn.apc.org

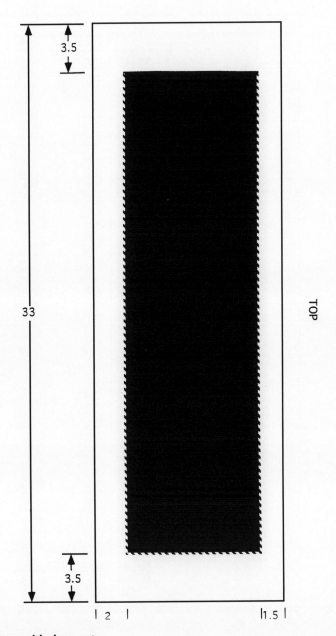

FIGURE 44   Front with door cuts

This is the front part of the plywood from which the door of the cooker is cut. The shaded part is the door, which will be cut out (and saved to be used later). 1.5 inches are left above the door for shelves that will support the glass.

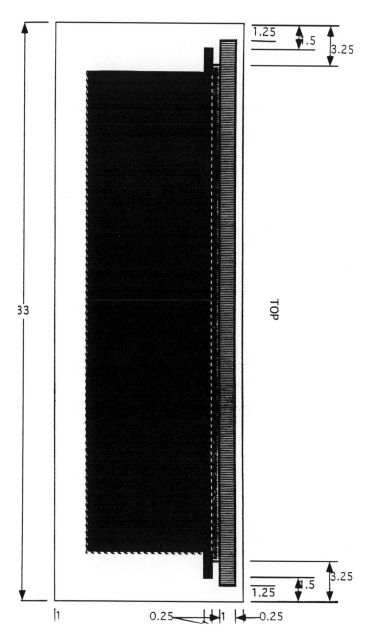

FIGURE 45
Inside front with top shelf.

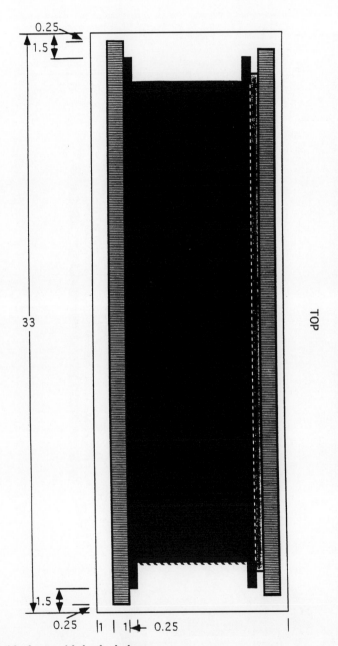

FIGURE 46   Inside front with both shelves
1/4" U-shaped end gap between 1" × 1" wooden rod and lower 1/4" piece will "hold" the similar shelf pieces from the side walls. This is the front part, which has shelves fixed on.

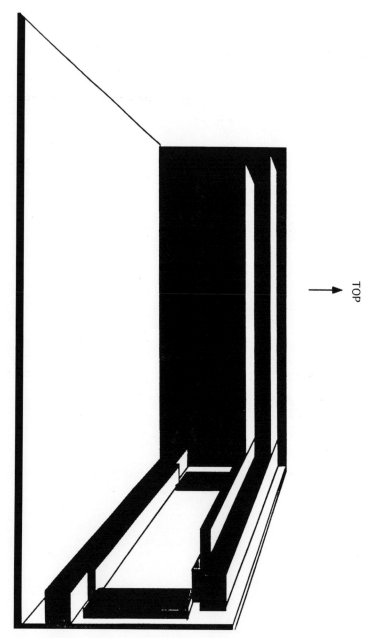

FIGURE 47
Perspective view of oven with one wall and the inside of the front door completed.

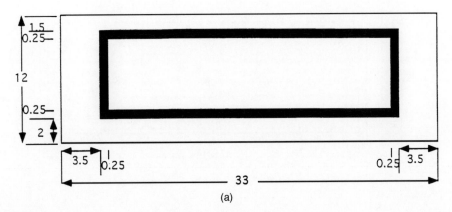

(a)

FIGURE 48a
Front: outer view.

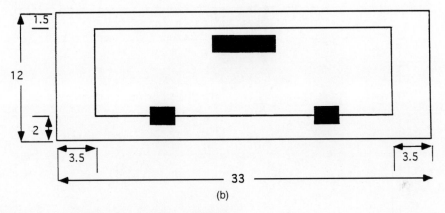

(b)

FIGURE 48b   Front with door, handle, and hinges
The door is fixed to the front part and attached with hinges (shown). A handle (dark bar) made of scrap wood is also attached (glue and, if needed, two nails).

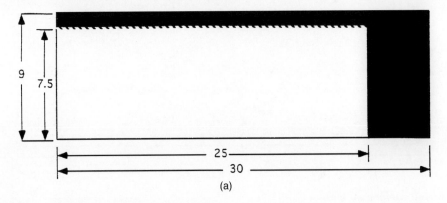

(a)

FIGURE 49a   Inner panel for door: cut to fit
A piece of the "scrap" wood cut from the front to make the door hole is used to add a second panel to the door. It is cut and then fixed onto the door to strengthen it.

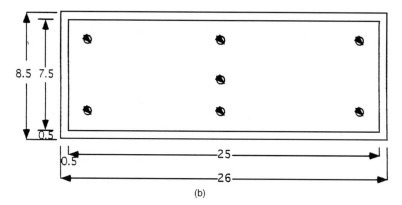

(b)

FIGURE 49b   Door with inner panel
The panel is attached to the door with screws and glue.

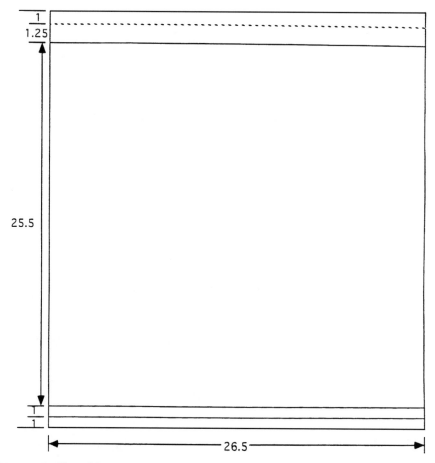

FIGURE 50   Cardboard bottom
1" is folded and lies on the bottom. 1.25" is then folded vertically. These right angle folds make a lip to be attached to the oven floor. The bottom cardboard forms the floor of the cooking area . 25.5" is the length of the inner box. 1" is folded downward to cover the 1" piece that is placed on the bottom. 26.5 inches is the width of the cardboard inner box.

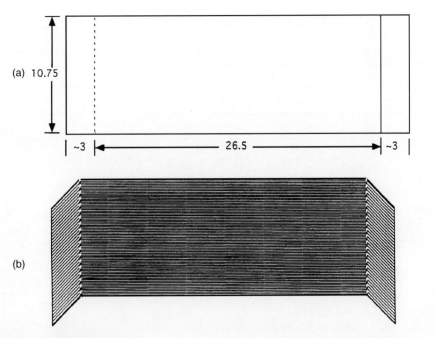

FIGURE 51   Cardboard back
(a) Cardboard forms the inner wall: the "cooking area." This piece forms the back of the inner wall and fits in the inside of the bottom guides. ~3" extend along the sides of the inner wall and hold the back inner wall firmly. (b) Once the cardboard is measured, it is folded and fitted in the inside of the bottom guides.

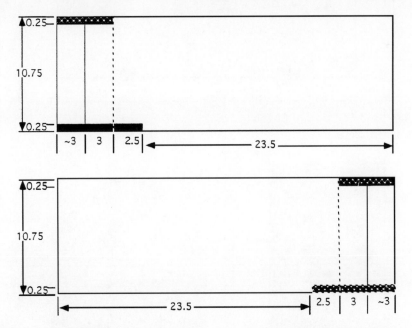

FIGURE 52   Cardboard sides
This is one of the side pieces of cardboard thet forms the inner box of the oven, The side length of the oven is 23.5". It is folded several times to "cap" the end, holding in the sawdust or other insulation. The dark strips are cut off to make room for the floor pieces and shelves (1/4" thick). The 3" fold covers the space between the outer and inner box (this is the space that is filled with sawdust). The next 3" fold bends backward along the side of the oven wall.

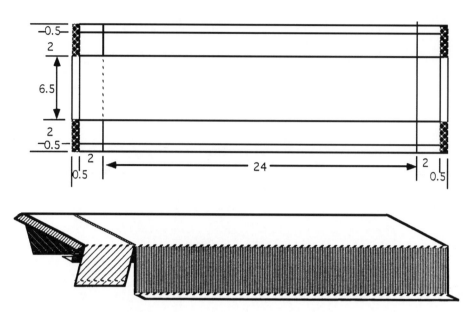

FIGURE 53   Cardboard door

The pattern is then covered with aluminum foil (outside only) and stuffed with sawdust (which itself can be held in a folded piece of aluminum foil). This is the folded pattern.

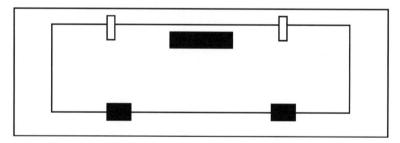

FIGURE 54

Front with door latches.

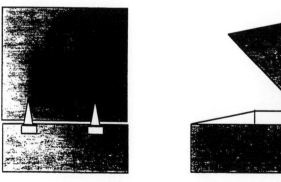

FIGURE 55

Lid with hinges.

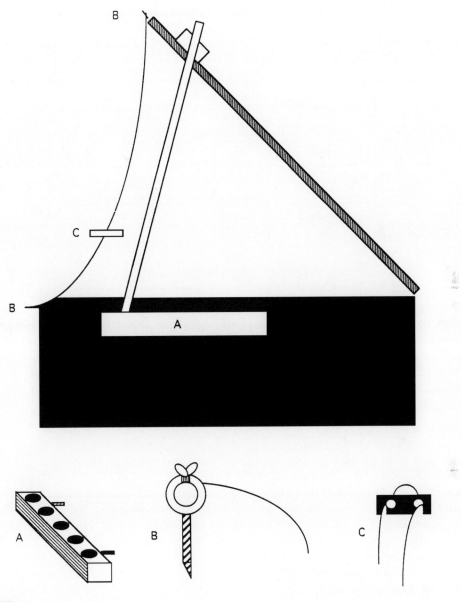

FIGURE 56
LID prop and string.

SOLAR HEATING

Providing hot water and keeping a building warm are important needs in Europe and the United States. In the Third World such needs have been less compelling—the small amount of hot water used can be heated over an open fire; the same fire provides heat for a room. Hospitals and, to a lesser extent, schools use appreciable amounts of hot water, and solar hot water heating is appropriate. Accommodations serving tourists in national parks often aim to be environmentally

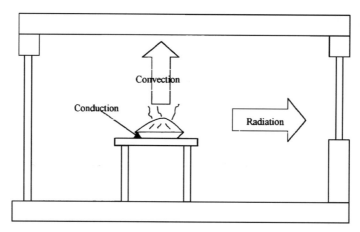

FIGURE 57
Heat transfer from a pie.

benign but comfortable—a good match for solar heating. Amenities in rural areas tend to reduce crowding in urban areas. Solar heating does have a place in the Third World, but because the major use is presently in temperate areas, this discussion is written from the United States perspective. In the United States 1 in 50 homes has some type of flat plate collector installed for hot water. In Greece and Israel solar hot water systems are common. Solar heating is certainly a technology that a user can understand, maintain, and—at least in part—design and construct. We start with some theory.

SOLAR SPACE HEATING

Let's look at the problem of keeping a home warm in the winter. Two processes determine indoor temperature: (1) the heat gained from the sun or the burning of fuel and (2) the heat lost to the outside. Our goal is to maximize the heat gained and to minimize the heat lost. We maximize heat gained by efficiently capturing sunlight or burning fuel. We minimize the heat lost by reducing the ways and speed at which it travels to the outside.

To reduce our heat losses we need to understand how heat travels. Suppose we bake a pie. When it is done, we take it out of the oven and put it on a table to cool. After an hour we know that the pie will be cool enough to serve. The question is, What's happened to the heat? It is gone, but where and how? See Figure 57.

Fact: It is the nature of things that heat flows from hot to cold.

It is a difference in temperature that drives heat from one place to another. Nature strives to have everything at the same temperature. So the heat from the pie warmed the table, the air surrounding it, and the walls of the kitchen. If there are windows in the room, some of the heat has gone through them. The amount of heat in the pie was small, and the heat went to so many places that we may not notice much change in temperature in many of the objects that received the heat, although the table under the pie would be warm to the touch. To measure the increase in temperature of the walls or the air in the room would call for an extremely sensitive thermometer. Even if the temperature difference is very small, any difference in temperature will cause heat to flow from warm to cool.

Fact: Heat travels three ways: by conduction, convection, and radiation.

Our pie warmed the table by conduction. Conduction requires some medium or material through which the heat can travel. There is no perceptible motion of the medium itself. Convection took some of the heat from our pie and warmed the air in the kitchen. For convection to take place we need a fluid (gas or liquid). Unlike conduction, the motion of the medium carries the heat. As the pie warmed the air around it, that air, being lighter, rose, and cooler air took its place. This is called natural convection. You can see it in steam rising from a cup of coffee or smoke rising from a fire. If we serve the pie too hot, our guests may have to use forced convection (blowing) to cool it.

Radiation took heat from the pie out the window. In contrast to conduction and convection, radiation needs no medium, only a difference in temperature. It's heat traveling by radiation that warms our hands by a fire or burns us at the beach.

But what does this pie knowledge have to do with our house? In our house the process is reversed—we don't want to lose heat, we want to keep it. Let's consider each of the ways that heat is transferred and see what we can do to minimize the loss.

HEAT LOSS THROUGH CONDUCTION

Conductive heat loss occurs through the exterior walls, floors, and the roof—anywhere the exterior temperature is lower than the interior. Not only that, the rate of heat loss depends on the difference in temperature. The greater the difference, the faster the heat travels. The rate also depends on the material. Copper conducts heat rapidly, making it a good material for pots and pans but a poor one for our walls. Dead (unmoving) air is a poor conductor. If we incorporate some dead air space in our walls, we reduce our heat loss. Insulation (fiberglass batting, cellulose, etc.) works using this principle.

To evaluate building materials people developed a measure of how quickly they transfer heat. This measure is called the heat transfer coefficient, abbreviated k. Table 6 gives the heat transfer coefficients of some common materials. The coefficient k is also called the thermal conductivity of the material.

If we stop and think about what will affect the heat loss through a wall, in addition to the material and the temperature difference, we uncover two more factors—the area and the thickness. A large wall area will conduct more heat than a small area. It takes heat longer to travel through a thick wall. When we calculate the heat lost through a wall during a day, we must account for the thickness of the wall and its area.

Now let's see if we can come up with a compact (read that mathematical) expression for the relation between these factors. First some abbreviations:

Q = the heat loss per second

A = area

t = thickness

T_{in} = indoor temperature

T_{out} = outside temperature

We've said that as the area and the temperature difference increase the heat loss also increases. This can be written as

$$Q \propto A(T_{in} - T_{out})$$

TABLE 6
Heat transfer coefficients

Material	Conductive Heat Transfer Coefficient ((Btu × in)/(hr × sq ft × F))
Aluminium	1400
Brass	715
Brick	5
Cellulose	1.66
Clay, dry	4
Concrete	12
Cotton	0.39
Glass wool	0.27
Glass	5.5
Granite	15.4
Gypsum	3
Ice	15.6
Iron	326
Limestone	10.8
Paper	0.9
Plaster, Cement	8
Sand	2.28
Soil	12
Water	4.1
Wood	0.9
Wool	0.264

$\propto$ means "is proportional to." As we increase the thickness, the heat loss decreases. Therefore,

$$Q \propto \frac{A}{t}(T_{in} - T_{out})$$

What we really want is Q to be equal to something, not just proportional. We also have not put in k, the heat transfer coefficient of the material. Fortunately, the heat transfer coefficient takes care of both these problems. The equation used to evaluate conductive heat transfer is

$$Q \propto \frac{A \cdot k}{t}(T_{in} - T_{out})$$

To check this equation we'll look at the units attached to our variables. Heat is a form of energy so it should be measured in joules. Q is the rate of heat loss so its units are joules per second. Actually joules per seconds is equal to another unit—watts, abbreviated "w." We measure temperature in degrees Celsius (°C), area in square meters (m^2) and thickness in meters (m).

When we add the units our equation looks like this

$$Q(w) = (A(m^2)k/t(m))(T_{in} - T_{out})(°C)$$

To make the units on the right of the "=" match watts, the heat loss coefficient had better be measured in watts per meter per °C.

When you consult a handbook to find values for heat transfer coefficients, you will see they are tabulated in w/(m°C) or equivalently (Btu/hr/(ft°F)) units. The units of k show why it is also called thermal conductivity. The quantity k in English units is equal to the amount of heat conducted in one hour through a piece of material one foot thick and one square foot in area, when the temperature difference between the two sides of the material is 1° Fahrenheit.

As an example, consider house insulation. A cool but reasonably comfortable interior temperature is 20° Celsius. On a chilly day the exterior temperature might be 2°C. How much heat is lost in a 1 meter × 1 meter block of fiberglass insulation 0.1 meters thick? The value of k is 0.033. The equation just worked out shows that we would lose 5.94 watts through the insulation. A block of steel of the same size, with a thermal conductivity of 54, would lose 9,720 watts. These examples show how much the rate of heat transfer is controlled by the material.

Recall that our original mission was to minimize the heat loss. If we look at the heat loss equation, we can see that decreasing the area, decreasing the heat transfer coefficient, and/or increasing the wall thickness will decrease the heat loss. So the solution is obvious, make the wall very thick, the house very small, and from a material that has a small heat transfer coefficient. Unfortunately, here is where we come hard against reality. We must also weigh material cost and availability, construction cost, and reasonable building size into our solution. We will come back to estimating our losses when we work out a practical example later in the chapter. Now let's explore another mechanism for heat loss.

HEAT LOSS THROUGH CONVECTION

Convection is the transfer of heat from solid surfaces by moving liquids and gases—our pie cooled faster because the air surrounding it moved up after being heated and the replacement air was cooler and therefore absorbed more heat from the pie. There are two types of convective transfer: free, or natural, convection and forced convection. Free convection is due to the fluid's change in density as its temperature changes. Air warmed over a hot surface, such as a radiator, becomes less dense and rises. Forced convection takes place when a fluid is forced, by a fan or pump, past a surface at a different temperature than the fluid. Forced convection is the mechanism of heat transfer in an automobile heater. A fan blows air over tubes heated by water from the car's engine. Forced convection can move more heat than free convection for a given temperature difference, which is why small electric space heaters have a fan.

The equation describing convective heat transfer is similar to that describing conductive heat transfer, but we use another coefficient instead of "k," and there is no thickness to consider.

The equation is as follows.

$$Q_c = h_c \cdot A(T_s - T_f)$$

Where

Q_c = the rate of heat transfer

h_c = the average convection heat transfer coefficient

A = the surface area in contact with the fluid

T_s = the surface temperature

T_f = the fluid temperature

For air in free convection the convective heat transfer coefficient is between 5 and 30 watts per meter^2 per degree Celsius and about 3 BTUs per second per feet^2 per degree Fahrenheit. Table 7 to this article list some convective heat transfer coefficients.

TABLE 7
Convective heat transfer

Wind Speed (MPH)	Convective Heat Transfer Coefficient (Btu/(hr × sq ft × °F))
0	1
2	1.6
4	2.2
6	2.8
8	3.4
10	4
12	4.6
14	5.2
16	5.8

So why do we care? One example is heat loss from a swimming pool. A comfortable temperature for the water might be 25°C. The pool might be exposed to moving air at 10°C, corresponding to a cool breeze blowing over it. We will take the heat transfer coefficient to be 25 W/m^2°C. The following equation gives us the heat lost per second from each square meter of water surface.

$$q/A = 25\ (25 - 10) = 375\ \text{Watts/square meter}$$

The water surface area of a small swimming pool might be 10 meters by 5 meters. In this case, the pool loses 18.75 kilowatts to the wind—equivalent to about 187 lightbulbs—which is why swimming pool heaters use much energy. The situation is actually worse because only about half the energy from a pool is lost by convection. Convective heat transfer is also important when estimating heat loss from a window or when calculating the efficiency of a solar collector.

TOTAL HEAT LOSS FROM A BUILDING

Now we will start on a major calculation—the amount of heat lost in a building. We do this for several reasons. One is to estimate how much heat we must supply to keep the house comfortable, and another is to understand where the heat is going so we can judge where to fix up the house. Heat flows out of the walls, the roof, the windows, the door, the foundation, and through infiltration. In most cases only a little heat flows out directly into the ground because dry earth is a good insulator, so we can disregard that heat loss. We need to do one more thing before we are ready to attack the house—we must look at the walls more carefully.

HEAT FLOW THROUGH A TYPICAL WALL

Most walls, instead of being made of a single material, usually have several different materials forming a sandwich. A typical insulated wall for a house is illustrated in Figure 58. The exterior is sheathed with wood 18 mm (.75 in.) thick. The interior wall is plaster and the middle is filled with fiberglass. Each of the materials that make up the wall have different conductivities; therefore, we need a way to combine the conductivities to find the total heat loss.

The effectiveness of a piece of insulation, whether a single material or a sandwich, is specified by an R-value—which includes both the conductivity and the thickness. To understand the R-value, we need to digress a little. The R-value is a material's resistance to conductive heat transfer and is defined as

$$R = \frac{\text{thickness}}{\text{thermal conductivity}} = \frac{t}{k}$$

The nice thing about using thermal resistance R is that it makes multimaterial calculations easier because R-values add for the different material making up a wall. It makes sense for R-values to add, as the R for each material is directly proportional to the thickness of the material. To get the total thickness of a wall we would add the thicknesses of each layer of the wall. Heat flow, however, depends not only, on thickness but also on thermal conductivity—a thin material with a low thermal conductivity conducts heat poorly, acting like a thicker material with a higher heat conductivity. So to find the total effective thickness, analogous to resistance, we add the effective thicknesses of the layers, these effective thicknesses being the actual thickness weighted by the conductivity.

To express heat transfer in terms of the resistance R, we combine our original equation for heat transfer with our equation for R, which is rewritten here for convenience.

$$Q = \frac{A \cdot k}{t} \cdot (T_{\text{in}} - T_{\text{out}})$$

Our equation for heat transfer becomes as follows.

$$Q = \frac{A}{R} \cdot (T_{\text{in}} - T_{\text{out}})$$

The total resistance R in the above equation is the sum of the Rs for the material forming the walls.

Now we can go back to our house example. First we will calculate the Rs for the various materials using metric units shown in Tables 8a and 8b. Values for k and R can be found in handbooks (see the References). These have been copied here.

Using these values, the value of R for the wall is

$$R \text{ (Wood)} + R \text{ (Fiberglass)} + R \text{ (Plaster)} = .164 + 2.37 + .026 = 2.56$$

Note that in this wall the fiberglass blocks the heat most effectively. If the temperature difference is 18°C, then the heat loss is 7.03 watts per square meter.

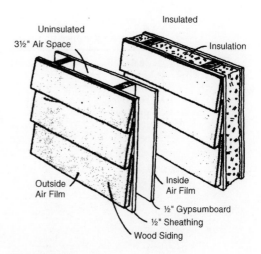

FIGURE 58
Typical residential wall construction.

TABLE 8a
Thermal resistance in metric units

Material	Thickness (m)	k (W/m°C)	R (m^2°C/W)
Wood	.018	.147	.164
Fiberglass	.090	.038	2.37
Plaster	.012	.47	.026

TABLE 8b
Thermal resistance in english units

Material	Thickness (in)	k (BTU in. /hr ft^2°F)	R (hr ft^2°F/BTU)
Wood	.75	.8	.937
Fiberglass	3.5	.27	12.69
Plaster	.5	8.0	.062

If we do the same problem using the English system of units the thickness, values of k and R become those in Table 8b.

Therefore,

$$R = .937 + 12.69 + .062 = 13.689$$

With a temperature difference of 32°F we have a heat loss of 2.34 Btu/hr-square foot.

CALCULATION OF HEAT LOSS FOR A SIMPLE HOUSE

Let us consider a very simple house, shown in Figure 59. The back wall has no windows, and the other side wall is identical with the one shown.

We will estimate the heat flow out of the house in Btu/hr when the outside temperature is 32°F—freezing—and the inside temperature is 65°F—about as cool as we would like it.

Heat, as described here, flows out the roof, the windows, the door, and the foundation and through infiltration—heated air escaping through cracks. Infiltration is a special problem to be looked at after we do the conductive heat losses. Heat loss through the foundation is also somewhat different because most of the heat flows through the top edge of the foundation, so the length of the foundation, not its area, is what determines the heat loss. The length of the foundation is the perimeter of the house. The R-value for the foundation is in terms of feet, not square feet. We might start by looking up R-values for the materials used in the house construction. To simplify matters, this research has been done, and the results are shown in Table 9. From these R-values we will estimate the heat losses.

Walls: The R-number given in Table 9 assumes the walls are made as shown in the insulated wall of Figure 58. Previous equation can be used, but we must find the area of the walls first.

$$\begin{array}{ll} \text{Front } 12 \times 20 - (8 \times 3) - (2 \times 3) = 210 \text{ ft}^2 \\ \text{Sides } 2 \times (12 \times 12 - (2 \times 3)) \quad = 276 \text{ ft}^2 \\ \text{Back } 12 \times 20 \qquad\qquad\qquad = 240 \text{ ft}^2 \\ \qquad \text{Total} \qquad\qquad\qquad\quad\; 726 \text{ ft}^2 \end{array}$$

TABLE 9
Typical R-values

Material	R-value
Walls	13.7
Roof (with 6" fiberglass batting)	25
Window (single glazing)	.88
Window (double glazing)	1.54
Door	2.3
Foundation (with foam insulation)	20

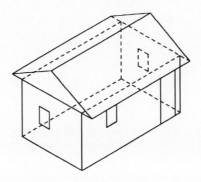

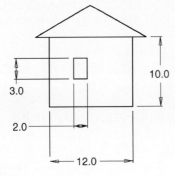

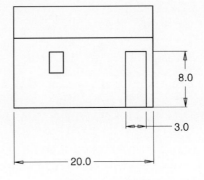

FIGURE 59
Simple house.

Using our equation for heat transfer, we have

$$Q_{\text{wall}} = \frac{A}{R} \cdot (T_{\text{in}} - T_{\text{out}}) = \frac{276.33}{13.7} = 665 \text{ BTU/hr}$$

Roof: The calculation is the same. The area is 240 ft^2.

$$Q_{\text{Roof}} = 316 \text{ BTU/hr}$$

Windows: We assume single glazing has been chosen for all three windows. It should be noted that convection plays an important role in heat transfer through windows in the winter, as air flows up along the inside of the window as it is warmed. This convection was considered when the R-values shown in Table 9 were calculated.

$$Q_{\text{Windows}} = 675 \text{ BTU/hr}$$

Door:

$$Q_{\text{Door}} = 343 \text{ BTU/hr}$$

Of course, this value of Q does not account for the large heat loss when the door is open.

Foundation: Heat loss through the foundation is done the same way except we use the length of the foundation, the perimeter of the house. The perimeter of the house is 64 feet.

$$Q_{\text{Foundation}} = \frac{64.33}{20} = 105 \text{ BTU/hr}$$

Infiltration: Cold outside air blows into the house through cracks, and this air must be heated to make the house comfortable. If air did not blow in, the house would be uncomfortable for another reason—buildup of odors and CO_2. Less than one air change every 2 hours, or 0.5 air changes an hour, will make the house uncomfortable. We assume, then, that in every hour, half the air in the house must be heated from 32° to 65°. It takes 0.018 BTUs to raise 1 cubic foot of air 1°F. (This number is called the specific heat.).

$$Q_{\text{Infiltration}} = \text{Volume} \cdot \text{Specific heat} \cdot \text{Temperature rise} \cdot \text{Number of air changes}$$

$$Q_{\text{Infiltration}} = (12 \cdot 12 \cdot 20) \cdot (0.018) \cdot 33 \cdot 0.5 = 855.36 \text{ BTU/hr}$$

The total heat loss is found by adding all these losses.

$$Q_{\text{Total}} = Q_{\text{Walls}} + Q_{\text{Roof}} + Q_{\text{Windows}} + Q_{\text{Door}} + Q_{\text{Foundation}} + Q_{\text{Infiltration}} = 4044 \text{ BTU/hr}$$

Why do we care about this calculation of heat loss? For one thing, it shows us how we can most effectively reduce that loss. In our example, it makes much more sense to reduce losses through the windows than through the roof.

Second, the calculation shows us how much heat we must bring into our house every hour. Actually, if we are using solar heat, we can bring heat in only when the sun is shining, so we must be able to store heat. To determine how much heat we need to bring in and store, we need to calculate our daily heat losses. To find our daily heat loss, we multiply our average hourly loss by 24. To find the average hourly loss, we use an average daily outside temperature. In our case, if we assume the average outside temperature cycles between 28° and 36°, and we want a constant 65° inside. Then we can use, as we did above, 32° as our outside temperature and 33° as the temperature difference. In this case the total daily heat loss is

$$Q_{\text{Daily}} = Q_{\text{Total}} \cdot 24 = 70{,}784 \text{ BTU/day}$$

Actually, we do not need to supply all this heat because the occupants and electrical appliances will supply some. A person produces about 250 BTU/hr and a large dog 200 BTU/hr. Electrical appliances, such as lights or toasters, also supply heat: 1 watt-hour is 3.4 BTU. So a 100-watt bulb on for 10 hours supplies 3,400 BTU. If our house held 1 person and 1 dog and used 200 watts for 10 hours, the energy supplied would be

$$Q_{\text{Supplied}} = (250 \cdot 24) + (200 \cdot 24) + (200 \cdot 10 \cdot 3.4) = 17{,}600 \text{ BTU/hr}$$

So the additional heat needed is

$$Q_{\text{Needed}} = Q_{\text{Daily}} - Q_{\text{Supplied}} = 53{,}184 \text{ BTU/day}$$

In the next section we consider how to get this much heat from the sun. We start by looking at a third way heat travels—radiation.

HEAT RADIATION

Most of us have seen a greenhouse or cold frame and know that in bright sun it is warmer inside than outside. Why is this? Why does putting a glass cover over something allow it to be warmer? To answer this we need to understand how heat radiates. Suppose we heat a steel bar in a furnace. As it gets hotter it begins to glow, first dark red, then orange, then yellow, and finally almost white—too bright to look at. The glow we see is energy being radiated. The sun is a perfect example of something glowing "white hot" and radiating energy. In an attempt to analyze the behavior of radiated heat (and light) scientists developed the theory of electromagnetic radiation. A central idea of that theory is that radiation travels in waves. If it were possible to isolate and make visible a single wave from the many waves around us, the single wave would look like Figure 60. As you can see, there are two quantities that describe the wave: amplitude and wavelength. Amplitude gives a number to the intensity of the heat or light. Wavelength gives a number to the "color" of the wave. The color violet (the shortest wavelength people can see) has a wavelength of about .4 microns (1 micron $= 10^{-6}$ meters). Red (the longest visible wavelength) has a wavelength of about .6 microns. Thermal radiation ranges from about 1,000 microns to .25 microns.

Electromagnetic radiation exists in a broad spectrum of wavelengths. Figure 61 shows the names given to various parts of the spectrum.

Energy from the sun is concentrated in the visible wavelengths. This means that most of the sun's heat and light comes in these wavelengths. Figure 62 shows the spectrum of sunlight—how much of the sun's energy is at each frequency.

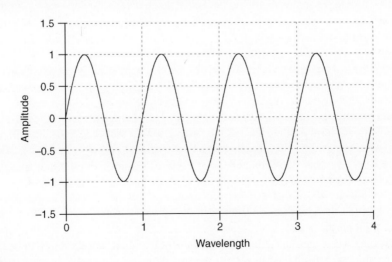

FIGURE 60
Typical wave.

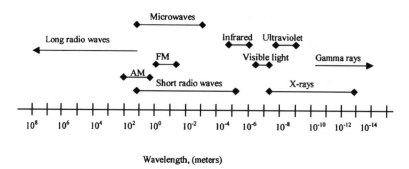

FIGURE 61
Electromagnetic spectrum.

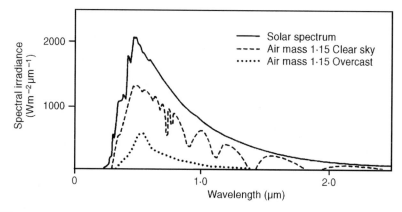

FIGURE 62
Solar spectrum.

Fortunately for our understanding, heat radiation behaves very much like light; it can be reflected, transmitted, absorbed, and emitted. Our experience with light gives us an intuitive understanding of heat radiation. Recall our steel bar in the furnace: Before it glowed, it emitted radiation at wavelengths longer than we can see. The wavelength of the radiation that a body emits depends on its temperature. The higher the temperature the shorter the wavelength. The wavelength also depends on the emitter material and its surface (polished, dull, rough, etc.). Materials respond differently to different wavelengths.

The reason the interior of a greenhouse becomes warm when sunlight hits it is the differences in the response of the glass to different wavelengths. The glass of our greenhouse is transparent to the wavelengths in sunlight. The glass, however, absorbs—retains rather than trans-mits—radiation of longer (and shorter) wavelengths. The plants and soil inside the greenhouse, because they are cooler than the sun, radiate heat at longer wavelengths. Thus, the heat from the sun passes through the glass, is absorbed by the plants and soil, and is reradiated. This reradiated energy, because of its longer wavelength, does not pass through the glass but is absorbed and reradiated again. Some of the reradiated energy goes to the inside of the house and some to the outside. For a typical greenhouse about 83 percent of the sun's heat is transmitted through the glass but only about 0.003 percent of the radiation from the interior passes back through the glass.

What happens when thermal radiation hits an object? There are three possibilities. It may either be transmitted, reflected, or absorbed. We saw with glass that what happens is dependent on the wavelength of the radiation—visible light is transmitted and thermal radiation is absorbed.

What happens also depends on the surface—a rough surface will absorb more energy than a shiny one—and on the internal absorption of the material. High internal absorption means the material is opaque to the incident radiation. Low absorption means that the material is transparent. Again glass is a good example. Impurities can raise the internal absorption and make the glass opaque. Some glass is made with a rough surface so it does not transmit light.

SOLAR RADIATION

To estimate how much heat we can actually get from the sun, we need to know how much energy the sun radiates or, more precisely, how much energy is received from the sun on the earth's surface. This has been measured. The radiant power reaching earth's outer atmosphere varies between 1.32 kW/meter^2 in early July to 1.42 kW/meter^2 in early January. The sun is actually closest to the earth in January. The solar constant is defined as the average power reaching a surface, outside the earth's atmosphere, one square meter in area facing the sun directly. (Facing the sun directly means the surface is perpendicular to the solar radiation.) The solar constant is 1.36 kW/meter^2. The equivalent number in "English" units is about 430 BTU/hr-ft^2.

All this energy will not be available for heating. For one thing, some of the energy will be lost, reflected or absorbed, as it goes through the earth's atmosphere. For another thing, to collect the maximum amount of radiation the collecting surface must face the sun directly. It is convenient in making solar designs to look at four separate factors that determine the amount of solar energy received: geographic location, collector orientation, time of day and year, and atmospheric conditions.

GEOGRAPHIC LOCATION

The geographic location, essentially the distance from the equator, determines the amount of atmosphere the radiation must pass through before hitting the collector. The greater the latitude, the more atmosphere the radiation must go through, as shown in Figure 63. More atmosphere traversed results in more energy lost. The distance from the sun is also slightly longer further from the equator, but the sun is so far away that this increase in distance is insignificant. Location also determines the weather patterns. Obviously a cloudy climate will be less productive than a sunny one.

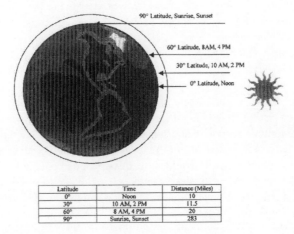

Latitude	Time	Distance (Miles)
0°	Noon	10
30°	10 AM, 2 PM	11.5
60°	8 AM, 4 PM	20
90°	Sunrise, Sunset	283

FIGURE 63
Latitude and path through atmosphere.

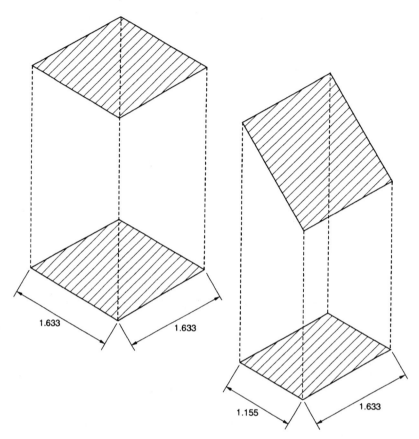

FIGURE 64
Effective area and incident angle.

COLLECTOR ORIENTATION

The more nearly perpendicular the face of the collector is to the sun's rays, the more radiation the collector intercepts (see Figure 64). Another way of visualizing the affect of collector orientation is to realize that the larger the shadow cast for a given area of collector, the greater the radiation incident on the collector.

TIME OF DAY

During the day the insolation (incident solar radiation) at a particular location reaches its peak at solar noon with minimums at sunrise and sunset. This is because at noon the location is facing the sun directly, while at sunrise and sunset the location is facing 90 degrees from the direction of the sun. (A drawing like Figure 63. would show this change in collected radiation as the earth rotates and the reader might want to make the sketch.) Not only does is the direction of the radiation best at noon but also the radiation has to go through less atmosphere then. Again the idea is illustrated in Figure 63.

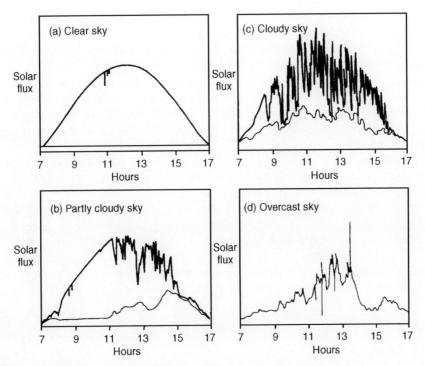

FIGURE 65
Insolation with respect to time of day and time of year.

TIME OF YEAR

There is more solar energy available to collectors during the summer because the days are longer. The collector faces the sun more directly then also. A second factor is that seasonal weather patterns affect the radiation reaching the collector, and summer weather tends to have fewer clouds. Figure 65 shows the change in insolation with respect to time of day and time of year. The dips in the graphs are clouds passing. A "Langley" is a unit of solar radiation, equal to one gram calorie per square centimeter.

ATMOSPHERIC CONDITIONS

The cloud cover is the most important condition affecting the amount of radiation reaching the earth's surface or house or solar collector. Clouds reduce incoming radiation by as much as 90 percent because they reflect that amount of radiation back into space. Ozone absorbs much energy in the infrared region. Carbon dioxide, oxygen, other gases, and dust all absorb radiation to some degree. The atmosphere scatters radiation as well as absorbing and reflecting it. Scattering is wavelength dependent and affects short wavelengths most. The reason the sky is blue is that short wavelengths are seen as blue, and these wavelengths are the ones scattered.

Global warming, which is discussed often in newspapers and magazines, is a result of the reflection of thermal radiation by cloud cover and other gases in the atmosphere. In this case the concern is with thermal radiation produced on the earth and reflected back to the earth's surface, warming that surface. A major problem is suspected to be carbon dioxide, a gas produced when burning wood, coal, oil, and other organic fuels. Increased amounts of carbon dioxide in the atmosphere will increase the amount of heat reflected back and increased reflected heat, warming the earth. This warming will produce significant changes in climate. One forecasted change is the melting of the ice in the Arctic and Antarctic that will raise the level of the oceans and flood coastal areas.

The net effect of the atmosphere is to reduce the energy actually available on a flat surface on the ground, facing the sun directly, from 1.36 kW/meter^2 to about 1.0 kW/meter^2 or from 430 BTU/hr foot^2 to about 300 BTU/hr foot^2.

How can we parlay this information into something useful—like designing a house to be heated by solar energy? Let's think of the things necessary to heat our home successfully from the sun. Before we can use the heat we have to capture it. We'll need a surface that is a good absorber of solar wavelengths. A good absorber has low reflectivity and emissivity—emissivity is a measure of how readily the surface radiates energy. What do you do when the sun goes down or on a cloudy day? Obviously we need some way of storing the energy.

PASSIVE SOLAR HEATING

In designing a solar heated house we must worry about how the solar energy is collected, absorbed, stored, and distributed. We must also worry about how we will regulate the temperature of the house, both in summer and winter. In a typical solar system, radiation enters the house through large south-facing windows to be absorbed and stored by masonry walls and floors, water-filled containers, or rock beds under the house. Natural convection then distributes the heat to the rest of the living areas. A passive design is one in which the heat flows naturally through the space in contrast to an active system where a fluid is heated and is then pumped through the house.

In a passive solar system both the design and the materials are chosen to make best use of the solar radiation. Because passive solar is integrated into the building rather than being a separate system (like conventional heating and cooling) its incorporation requires an integrated approach to the design. We have been thinking about heating the house in winter, but we should also think about cooling. In summer, passive solar design minimizes the amount of sunlight entering the house and provides ventilation. In general, siting, landscaping, floor plan, window placement, and building materials all affect the performance of a passive solar heated house.

THE FIVE COMPONENTS OF A PASSIVE SOLAR SYSTEM

A passive solar system will combine five interrelated components: collector, absorber, storage, distribution, and regulation. We will go through these one by one.

The collector is the portion of the building that actually lets in the sunlight. A collector can be south-facing windows, skylights, or panels especially made to collect and absorb. To be beneficial, collectors should be oriented to the south within 30 degrees. They should be tilted relative to the earth's surface to face the sun directly. The angle of tilt should be approximately the latitude of their location, as Figures 63 and 64 show. (Latitude is 0 degrees at the equator and 90 degrees at the poles.)

The absorber is the recipient of the solar radiation. It is warmed by the radiated heat from the sun and makes that heat available for direct heating or storage. Absorbers are usually painted black because black surfaces have the lowest emissivity of any color and so reradiate the least heat. Special paints are available with very low emissivity, but these cost more than ordinary paints.

Storage is usually done in dense materials such as brick, rock, water, adobe, and concrete. Storage may be incorporated in a wall, floor, or room divider. A masonry or slate floor in the room with south-facing windows would be an effective storage medium. So would the back wall of the room. In these cases the surface of the floor or wall is the absorber. The storage needs to be of the proper size to hold sufficient heat for the particular use—for example, a slate floor thick enough to keep the room warm until midnight might be chosen. A rule of thumb for residential spaces is 1 cubic foot of masonry for each square foot of window area. Because it is difficult to move heat a long distance by natural convection, storage is usually placed adjacent to the rooms it is intended to heat.

The distribution system moves the heat from the storage area to the rest of the building. Distribution is, in most houses, by natural convection. The designer can put in passages for naturally flowing air currents so the warm air goes into the living areas. Occasionally, small fans or pumps are used to augment natural convection.

Regulation in a passive solar system is usually done by adjustable vents that can control the movement of heated air through the building. Another type of control device is insulation to cover windows at night to prevent stored heat from escaping. Regulation is more difficult in a passive system than in a conventional heating system with automatically controlled blowers to force the heat around—of course, the automatic system is more expensive.

DIRECT GAIN

Three basic techniques are used in passive solar systems—direct gain, indirect gain, and isolated gain. We consider direct gain first.

In a direct gain system sunlight goes directly into the space to be warmed—for example, solar radiation entering a room through south-facing glazing. The radiation is absorbed and stored in masonry walls and floors and any other masses that are directly warmed by the sun. Direct gain warms the living space immediately when the sun shines, provides good natural lighting, and allows an outdoor view. The major drawback of direct gain is that it sometimes works too well and overheats the living space. Because storage is in the same place as the heat is used, one has less control over the temperature than if the heat were stored separately from where it is used. Figure 66 illustrates direct gain.

Let us estimate how big the window would have to be for the house in Figure 59. As mentioned in the first part of this article, each hour about 300 BTU per square foot come from the sun through the atmosphere. We need 53,184 BTUs in a day. The room collects at least some sun for about 8 hours a day—from 8 A.M. to 4 P.M. or so. Not all of the sun's energy gets through the glass; some is reflected back out the window, and some is absorbed by the carpet and draperies, which do not radiate heat efficiently. Actually only about 20 percent of the peak

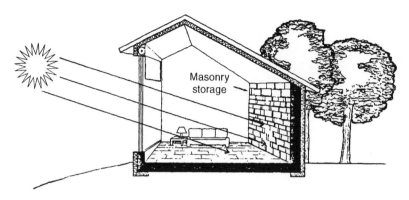

FIGURE 66
Direct gain.

energy—the energy available at solar noon—outside the window actually heats the house (Shapiro 1985; Johnson 1981).

Now let us work out the area of the window.

$Q_{\text{In}} = Q_{\text{Daily}}$
$Q_{\text{In}} = 300 \cdot \text{Area} \cdot 8 \text{ hours} \cdot 0.2$
$Q_{\text{Daily}} = 79{,}465 \text{ BTU (Eq. 6)}$
$\text{Area} = 165 \text{ ft}^2$

Actually this is the area perpendicular to the sun's rays, so a somewhat larger window would be useful, perhaps 10 feet by 14 feet.

Such a window would increase heat loss, so double glazing would certainly be called for. In a serious design we would go back and recalculate our heat loss through these larger double glazed windows and see if we would have to make them even larger to make up for the additional loss. We might as well provide for insulation at night over the window, which probably means the owner must remember to put up that insulation each night. The heat input would be less, perhaps half, on cloudy days, so a backup heater, perhaps a wood stove, is also needed.

Sufficient heat storage is also necessary. Using our rule of thumb of 1 cubic foot of masonry per square foot of window and deciding to use masonry—floor and back wall—6 inches thick, we calculate that a total area of 280 feet^2 is needed. This area might come from a floor and back wall 14 feet by 20 feet—a reasonable size.

INDIRECT SOLAR GAIN

Indirect gain collects radiation through the same south-facing glazing as direct gain, but the absorber and the storage are separate from the living area. A commonly employed indirect gain system is an 8- to 16-inch-thick masonry wall built as the south wall of the house. The exterior surface of the wall is painted black to absorb the radiation, and the entire wall is glazed—that is, a large glass window is between the wall and the outdoors. This wall is known as a Trombe wall, shown in Figure 67.

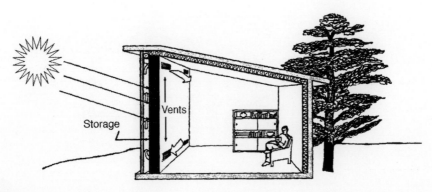

FIGURE 67
Indirect gain, Trombe wall.

A Trombe wall provides heat in two ways. The first way is through convection. Two sets of vents are installed in the wall, one at the floor level and another at the ceiling. When the sun shines on the wall, the air in the cavity between the wall and the glazing is heated and rises, flowing through the ceiling vents. Cool air from the adjoining room is drawn in through the floor vents. The path taken by the air is called a natural convection loop.

The Trombe wall also provides heat by absorbing solar radiation and storing it. The wall is warmed by the sun and reradiates the heat to the rest of the house as long as the temperature of the wall is greater than that of the living space. The stored heat allows the wall to continue to provide heat after the sun has set. This time lag gives indirect gain systems the advantage of supplying heat when it is needed without making the living space uncomfortably warm.

ISOLATED GAIN

In an isolated gain system solar radiation is captured in a separate, glazed, otherwise unheated space and then transferred to the living space. Atriums and attached greenhouses are examples of isolated gain. Solar greenhouses are the most practical method for retrofitting existing housing with passive solar heating and are the usual way of employing isolated gain. A greenhouse is advantageous for three reasons. First, the solar heat can be stored without having unsightly or bulky masonry or water tanks in the living areas, and the heat can be distributed effectively to where it is needed. This distribution is done by vents and ducts. Second, the greenhouse provides additional living space. Third, the greenhouse can serve as a buffer between the living space and the outdoor, thus reducing heat losses. Shapiro's (1985) book describes how to design attached greenhouses. An example of isolated gain is shown in Figure 68.

COOLING

Although our main objective in passive solar design is to heat a house, for the house to be livable, it will also require some type of cooling. The effectiveness of passive cooling depends on how well heat gain is controlled and on adequate ventilation. Ventilation removes heat generated inside the house—for example by the people living there. Moving air also feels more comfortable than still air. Heat can be kept out by shading the collection area with overhangs

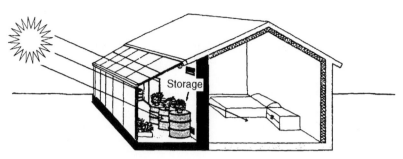

FIGURE 68
Isolated gain.

or vegetation. Moveable insulation can be used to reduce heating through windows. House design can promote natural ventilation. Such design requires a clear path for summer breezes and use of the chimney effect to pull cooler air into the house. The chimney effect is produced by warm air rising and flowing out of the house, as smoke rises and flows out of a chimney. Traditional housing in the Middle East and in West Africa have made remarkably effective use of natural ventilation for centuries.

SITING A HOUSE

Careful placement of a house will make passive solar heating more effective. The ideal site for a passive solar home is on a south-facing slope with evergreen trees to the north of the house and deciduous trees in front. By building the house into the slope, we provide protection from winter winds attacking the back while giving the front of the house access to the sun. If the site is flat, the same effect can be accomplished by building up the earth on the north and sometimes east and west sides. This partial burying of the house is called earth berming and protects the house from air infiltration and smoothes out the temperature extremes. A further extension of berming is to bury the house into the hill.

The amount of water in the soil also influences the effectiveness of solar heating. A low water table and good drainage will reduce the loss of heat to the soil. Wet soil will cause damp basements and floor slabs and will conduct heat away from the house much faster than dry soil. Soil is a heat sponge when it is wet; it is a fairly effective insulator when dry. Not much can be done about a damp site except to avoid it.

Trees do more than simply add beauty. The size, location, and number of existing trees is important when one is selecting a site—waiting for new trees to grow can take a long time. A healthy stand of evergreens to the north, northeast, and northwest will provide a much-needed windbreak. Deciduous trees to the south can shade the collector area during warmer weather. It may be possible to plant deciduous trees in such a way as to channel summer winds into the house. In general, topography, soil, drainage, and vegetation must be considered to make the best design possible.

ENVIRONMENTAL FACTORS

In our design calculations so far we have simply assumed a particular outside temperature and availability of sunlight. These temperature and sunlight data would probably be the averages expected at the site. In a more thorough design we would worry about how much and how long conditions varied from the average. The available sunlight; the number and pattern sequence of cloudy days; the direction and speed of the prevailing wind; the average, maximum, and minimum daily temperatures; and the relative humidity are all factors that must be taken into account when proceeding with a passive solar design. (Dampness can decrease the comfort level on both hot and cold days.) Climatic data about a particular site can be found in a number of sources (Johnson 1981; Leckie et al. 1981).

A particular number often given when describing climatic conditions at a particular site is the number of degree-days. Degree-days are accumulated daily throughout a heating season by subtracting the average outdoor temperature from a base of 65°F. For example, one day with an average outdoor temperature of 20°F produces 45-degree days; ten such days produce 450 degree days. A house in Caribou, Maine, will not have the same requirements as a house in Oak Ridge, Tennessee. In Caribou, the heating demand is 9,767-degree days. In Oak Ridge the heating demand is 3,800-degree days. The outside temperature in Oak Ridge is closer to 65°F and closer for more days.

The primary function of the design for a house in Caribou, based on the number of degree-days, will be to provide adequate heat for a long cold winter. In Oak Ridge the goal will be to provide a balance between cooling for the summer and heating for a relatively mild winter. A major factor in determining how passive solar design is used is the climate.

SOLAR ACCESS

Passive solar design requires adequate access to sunlight. Usually this is not difficult for single-family houses, but multifamily units and closely spaced developments will need careful planning so each unit receives sun. North of the equator, the winter sun is never overhead. It is always in the southern sky. A solar house ideally should face south, and if it can't face due south, it should always face within 30 degrees of south. During the peak radiation period from 9 A.M. to 3 P.M. the collection area should not be shaded by trees, hills, or other buildings. Sunlight collected before 9 A.M. and after 3 P.M. is also of value but will come from the east or west, so if one wants to collect the early morning or late afternoon sun one must have east- or west-facing windows to collect it. Early morning sunlight, for example, can provide quick heat for an east-facing breakfast area. The house designer needs to work out the solar window—the portion of the sky from which the house receives sunlight—and see if a significant portion of that solar window is blocked to determine if the house will receive sufficient sun.

ENERGY CONSERVATION

An important part of a passive solar design is reducing the heat loss from the house. Insulation and double glazing, as described when we were calculating the heat flow out of a house, will certainly reduce the heat loss, and a reduction in the heat loss will in turn reduce the size of the collection system. The design of the house also influences the heat loss. The major issues for a passive solar house are the shape and orientation of the building and the layout of the interior rooms. The shape of the building determines its solar gain and heat loss. The ideal shape would have a large area facing south, filled with a window, and everywhere else a small, well-insulated area.

The orientation of the house should be chosen so that the long axis of the house runs east to west, allowing a south-facing solar collection area that is large enough to be shared with most of the living space. The orientation of this south-facing wall must be chosen to collect solar radiation effectively—within 30 degrees of true south.

Rooms should be laid out so as to reduce heat loss and take advantage of solar gain and natural convection. A good design puts the low-use and unheated areas, such as the garage and utility room, on the north side of the building, thus making a buffer between the heated living space and the north wall. Bathrooms, with small windows, also fit on the north side. The kitchen might be placed on the east end of the house, where it will catch the morning sun.

To reduce heat loss it is necessary to insulate and seal the building envelope, as previously discussed. Maintaining a low exterior-surface-to-interior-volume ratio and building a compact structure improves heat conservation. A major heat sink is the leakage of air through cracks around doors and windows and through joints in the walls and roof—the infiltration we discussed. Weather stripping and caulking will minimize these leaks. A vapor barrier will also help reduce infiltration. There is one caution: If a house is made too tight, condensation and stagnant air will result. A rate of 0.5 air changes per hour is recommended. Eliminating or minimizing the windows on the north side and using double and triple glazing elsewhere will decrease the losses through windows.

Walls, ceilings, and floors all lose heat by conduction. The proper installation of insulation in walls, floors, crawl spaces, roofs, and foundations will reduce this loss to acceptable levels.

Another area to be addressed in the design phase is the entrance. Every time an outside door is opened and closed, cold air from outside is exchanged for warm air from inside. An "air-lock" or unheated vestibule that creates a two-door entry will allow only the cool air in the entry to be exchanged.

SOLAR HOTWATER SYSTEMS

The next few sections discuss the design and operation of a flat plate hot water heating system. The discussion has three parts: flat plates, systems, and economics. A flat plate collector could also be used to heat a house, but it is usually cheaper to use direct, indirect, or isolated gain for space heating, rather than to build a collector.

The essential parts of a flat bed collector are an absorber to collect heat and fluid to transfer heat from the absorber to where it is useful. The sophistication of the design of flat plate collectors runs from trickling water down a sheet of corrugated steel into a trough to a triple selectively glazed all copper absorber with optimized coating. The design chosen depends on the application—economics, climate, and required performance.

TRICKLE TYPE COLLECTOR

The trickle type collector uses a sheet of corrugated metal, painted black and tilted toward the sun. Once the sun warms the plate, water is trickled from a tube at the top of the collector. The water is heated by the plate as it flows down into a trough at the bottom. The warm water in the trough is piped to storage. Figure 69 illustrates a typical trickle collector. Glazing (the glass or plastic that covers a collector) is often omitted from this type of collector because the heat loss by radiation from the heated water is not significant. The beauty of a trickle collector is its ease of construction and economy. An unglazed trickle collector could be built for less than $20.

TUBE IN PLATE COLLECTORS

As the performance requirements of the application increase, so does the complexity of the collector. More prevalent than trickle collectors are "tube in plate" absorbers. A generic tube in plate is shown in Figure 70. In this design, the fluid-carrying tube is incorporated in the absorber plate.

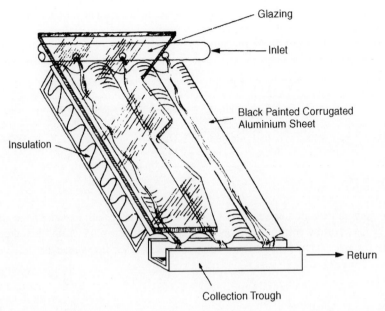

FIGURE 69
Trickle-type hot water heater.

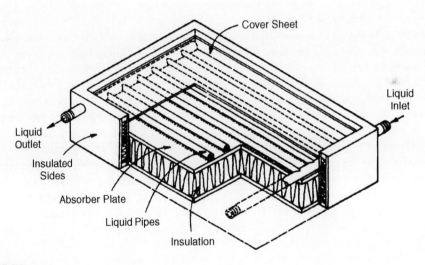

FIGURE 70
Tube-in plate collector.

This offers several immediate advantages over the trickle collector. The water is not exposed to airborne contamination, there is no evaporation, and the fluid is warmed by a larger heated surface. Several companies offer absorber plates that have tubes formed in them during the manufacturing process. User built plates may have tubes bonded to plates by soldering or some mechanical fastener.

The efficiency of a collector is determined by the thermal conductivity of the tube and plate material, the losses to the surroundings, the ability of the plate to absorb solar energy, and the path of the fluid through the plate. An absorber would be most efficient if its temperature were constant over the whole surface. Efficiency is greater when the temperature of the material losing heat and

TABLE 10
Material properties

Material	k (W/m°C)	Cost	Corrosion Resistance	Coefficient of Thermal Expansion	Availability
Silver	407	Highest	Good	1.84×10^{-5}	Poor
Copper	386	High	Good	1.62×10^{-5}	Excellent
Aluminum	204	Medium	Fair	2.22×10^{-5}	Good
Steel	54	Medium	Poor	1.14×10^{-5}	Good
Plastic	.195	Low	Excellent	6.75×10^{-5}	Good

the material gaining heat are nearly equal. Obviously the temperature cannot be the same all over the plate if we are putting cold water in one side and taking warm water out the other. Efficiency is improved, however, if the input water warms up quickly as it flows through the collector.

MATERIALS

The choice of tube and plate material is a tradeoff between thermal characteristics, cost, and joining. An added constraint is that the tube material not corrode nor contaminate the water. This constraint limits the tube selection to copper or plastic, materials used in plumbing. Table 10 gives some properties for a variety of materials.

Let us consider the implications of the several parameters in Table 10. Thermal conductivity, you'll recall, is how fast a material transfers heat from a hot region to a cooler one. The higher the thermal conductivity, the faster heat is transferred to the incoming fluid, the faster that fluid warms, the more uniform the temperature, and therefore the more efficient the collector.

The second parameter in Table 10 is the cost of the material, which determines the initial cost of the project. If the initial cost is too high, the project may not be possible for the homeowner. But the homeowner does need to consider long-term benefits and problems of the project in addition to the initial cost. In many cases a solar hot water heater will pay for itself within six years. Copper tubing is expensive, but it is rugged and easy to solder.

The last two parameters in Table 10 measure corrosion resistance and thermal expansion. Corrosion resistance determines life of the panel and whether the tube will contaminate the fluid. The coefficient of thermal expansion is included in the table because absorbers require some sort of bond between the tube and the plate. For good thermal performance we would like to maximize this area. If the tube has a coefficient of thermal expansion very different from that of the plate, the two will expand at different rates and may break the bond.

The implication of the information in Table 10—the bottom line—is that copper plates and tubes are most often used. Although their cost is higher than the other materials shown (except for silver, which is prohibitively costly), the improved efficiency and the ease of construction usually outweigh the cost. Second most common in usage is plastic, used primarily in applications where economy is the most important factor.

CONSTRUCTION OF THE ABSORBER

Two design decisions are made in constructing the absorber part of the collector. The first is the path the fluid takes in going from input to output. The second is what kind of a coating to put on the absorber surface.

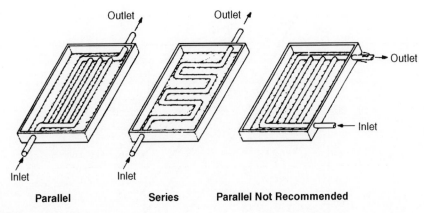

FIGURE 71
Paths of fluid through collectors.

The path of the fluid through the plate will affect the uniformity of temperature in the plate, which influences the efficiency of the collector. If we use one tube, snaked back and forth across the surface, the plate temperature will be lowest at the inlet and highest at the outlet, increasing as the fluid moves from inlet to outlet. This will not be a uniform distribution—the price of being easy to construct. A better arrangement is to have a series of tubes in parallel, connected to inlet and outlet headers. This is more difficult to build but provides a more even temperature. Figure 71 shows these paths. The pressure drop through the panel determines the size of the pump needed to move the fluid. A parallel path has less of a pressure drop than a series path and thus will require a smaller pump.

The spacing of the tubes across the panel also affects the efficiency. Wide spacing will give hot spots far from the tubes and cold sections near the tubes because the fluid in the tubes will remove heat from the section of the absorber in contact with the tubes. The thermal conductivity comes into play here. A copper plate will transfer heat much more readily across the absorber than steel, therefore decreasing the temperature difference between different parts of the absorber. Figure 72 shows efficiency as a function of tube spacing for various materials. Of course, the closer the spacing, the more tubes must be used and thus the more expensive the collector will be.

A shiny copper absorber surface will reflect much of the incoming solar energy back into space. This won't help raise the temperature of our water. An obvious solution is to paint the surface with a flat black paint that will withstand high temperatures. The paint will typically absorb over 90 percent of the incoming energy. Absorption, though, is only part of the story. The reader will recall that if a body is warmer than its surroundings, it radiates heat to the surroundings. Flat black paint has an emittance almost equal to the absorptance at the wavelengths of concern, so an absorber painted with a flat black paint will emit nearly as much energy as it absorbed. If we can afford the cost, we should choose a coating that offers both high absorbtance and low emittance. Figure 73 shows the spectral properties of two coatings and their relative costs.

CONTROLLING LOSSES

Once we've done what we can to get the solar energy into the absorber panel, we need to keep it there to heat our water. Heat can leak out the top of the collector, through the glazing, or through the sides and bottom.

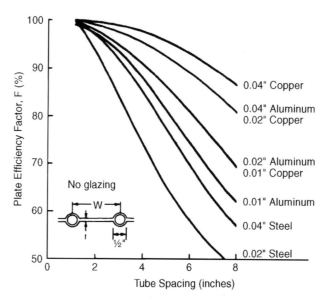

FIGURE 72
Collector efficiency versus tube spacing.

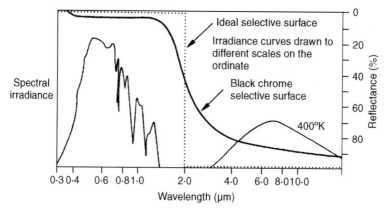

FIGURE 73
Spectral properties of several coatings.

GLAZING

Collectors that operate at a plate temperature less than 20°F above ambient require no glazing because radiation is low. When the temperature difference is greater than 20°F, glazing is required to keep most of the heat in. The factors to be considered in choosing a glazing material, in addition to being transparent to incoming radiation, include cost, appearance, resistance to impact and weather, low transmittance of reradiated wavelengths, and ease of working. Figure 74 shows plate efficiency factors with several glazing options.

Glass is the common product for glazing. Table 11 gives some of the properties of commercially avaliable glass. Although glass is most common, it has some drawbacks. It is difficult for an amateur to cut and handle. Even tempered glass has limited impact resistance. The other option is plastic. There are a variety of plastics on the market that give almost as much transmittance

TABLE 11

Properties of glazing materials

Material	Solar Transmittance (%)	Reflectance (%)	Vertical Shading Coefficient
Acrylic	85		0.98
Polycarbonate	90		
Clear Glass	83	8	0.99 Single
Starphire Glass	90	8	1.05 Single
Solex Glass	61	6	0.81 Single
Clear Glass	69	13	0.88 Double
Solex Glass	51	9	0.69 Double

The shading coefficient is the ratio of the total amount of solar energy that passes through a glass relative to 1/8-inch-thick clear glass.

as glass while avoiding some of the drawbacks. High operating temperatures require that the type of plastic be carefully chosen to avoid softening and warping. Figure 75 shows the transmission through some glazing options, both glass and plastic.

INSULATION

Because the greatest heat loss is through the glazing, it is usually cost effective to limit insulation to the equivalent of 3 inches of fiberglass on the back and 1 inch on the sides.

On a warm summer day, a noncirculating collector may reach temperatures greater than 200°F. The high temperature implies that the insulation material must withstand this high heat without degradation. One option is fiberglass with a low binder content. Binder is added to fiberglass to give it some shape; however, the binder tends to outgas at high temperatures. The result of outgassing is a cloudy coating on the glazing that reduces the efficiency of the system. The only foam insulation suitable for this application is isocyanurate, which is rated for 400°F and will work with painted absorbers.

HOUSING

The collector housing has several functions. It provides mechanical support for the absorber plate and glazing. It protects the plate from the weather. Housings are normally constructed of steel, aluminum, or weather-resistant wood. The housing should be designed to give a watertight seal between

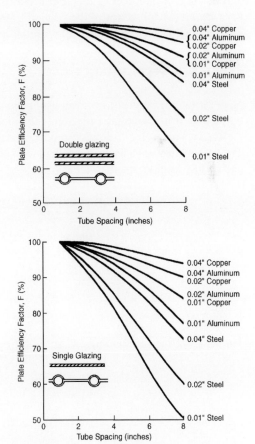

FIGURE 74

Plate efficiencies for glazing/spacing material options.

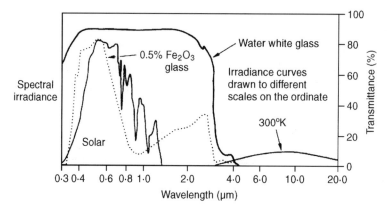

FIGURE 75
Spectral properties of glazings.

the glazing and the housing—water vapor between the glazing and the absorber will itself absorb radiation. The housing should be constructed in such a way that the glazing, the most fragile part of the collector, can be replaced if it is damaged. The weight of the housing is important because the overall weight of the system determines how strong the supports must be and what kind of equipment will be needed to lift and install the system.

COLLECTOR PERFORMANCE

Now for the moment of truth. We need some estimate of the performance of the system. This will tell us if it is worth the time and effort to install a flat plate water heater.

What we are trying to do is absorb incoming solar radiation and transfer it to a working fluid. The rate of heat transfer is the difference between the energy absorbed by the plate and the energy lost from the plate to the surroundings.

Rate of transfer (Q_u) = energy absorbed (Q_{abs}) − energy lost (Q_{loss})

The energy absorbed is the product of the energy striking the surface times the transmission of the glazing times the absorbtance of the plate.

Q_{abs} = Insolation (I) · Area (A) · transmittance (t) · absorbance (a)

$$Q_{abs} = I \cdot A \cdot t \cdot a$$

The losses to the surroundings are given by

Q_{loss} = collector heat loss coefficient $(U1)$ · Area (A) · difference in temperature between plate and ambient $(T_p - T_a)$.

$$Q_{loss} = U1 \cdot A \cdot (T_p - T_a)$$

If we combine the equations for energy absorbed and energy lost, we arrive at

$$Q_u = A \cdot (I \cdot t \cdot a - U1 \cdot (T_p - T_a))$$

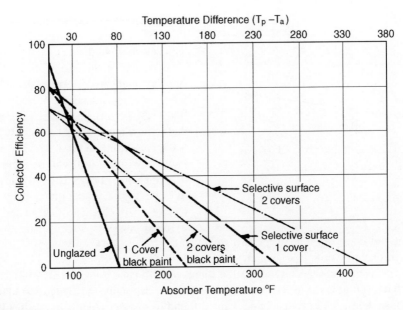

FIGURE 76
Collector efficiencies.

The efficiency of the system is

$$\eta = \frac{\text{energy collected}}{\text{energy incident}} = \frac{Q_u}{I \cdot A}$$

or

$$\eta = t \cdot a - \frac{U1 \cdot (T_p - T_a)}{I}$$

Figure 76 shows some typical efficiencies of various collector types. This graph shows several things.

- The efficiency of all collectors decreases as the temperature difference between the outside air and the plate decreases.
- There is no universally efficient collector.
- Glazing reduces efficiency at small temperature differences because it reduces incoming insolation.

For example, if we are heating water for a swimming pool and want a temperature of 80°F when the ambient temperature is 65°F, we are probably better off with an unglazed collector delivering water at 85°F rather than a sophisticated collector delivering water at 140°F. If our task is to supply domestic hot water at 140°F (a temperature difference of 80°F), then a glazed collector with a selective coating is required.

To design an effective collector there are two questions we need to answer.

1. What is the temperature difference under normal operating conditions?
2. What size collector must we install to meet our needs?

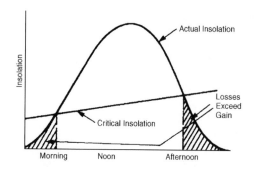

FIGURE 77
Insolation versus time.

The temperature difference will determine the type of collector. Figure 76 can be used as a selection guide.

Determining the size of the collector is a little more involved. So far our efficiencies have been instantaneous. What we need to find is the performance over a day or several days—how much water can be heated during a day or during several days? We need to know how much heat will be collected. The first thing to consider is how many hours a day we will operate the collector. When energy absorbed is less than the energy lost, it makes no sense to operate the collector—that is, we don't run a collector at night. Figure 77 shows insolation versus time and the minimum insolation ("critical insolation" in the figure) needed for operation. Actual values for insolation versus time of day can be found in Johnson (1981) and Leckie et al. (1981). Once the turn-on and turn-off times have been estimated, we need to determine the total expected insolation between those times. Again actual values can be found in weather records in handbooks.

Time	Insolation (BTU hr/ft^2)
7 A.M.	0
8	123
9	212
10	274
11	312
12 noon	324
1 P.M.	312
2	274
3	212
4	123
5	0
Total	2166

Massaging the equations given previously, we can arrive at

$$\text{Area} = \text{efficiency} \cdot \frac{\text{energy required}}{\text{insolation}}$$

This equation should make intuitive sense—area equals the total energy required divided by the energy gained per unit area.

Let's try an example to clarify these ideas. Suppose we have a family of four requiring 20 gallons of hot water (140°F) per person per day. The ambient temperature is 50°F. Recall that a BTU is the amount of heat required to raise the temperature of 1 pound of water 1 degree. A gallon of water weighs 8.34 pounds. The energy required to heat the water is

$$Q = 20 \, \frac{\text{gallons}}{\text{person} \cdot \text{day}} \cdot 4 \text{ people} \cdot 8.34 \, \frac{\text{pounds}}{\text{gallon}} \cdot (140 - 50)°F$$

$$= 60{,}048 \, \frac{\text{BTU}}{\text{day}}$$

From a handbook we find the insolation versus time for January 21 at 32 degrees north latitude—this day is chosen because it will be a difficult one.

With a temperature difference of 90°F we would probably select a collector with a selective surface and one layer of glazing—see Figure 76. From Figure 76 we determine an efficiency of about 55 percent. Therefore, the area is

$$A = 60,048 \text{ Btu/day}/(2166 \text{ Btu/sq ft day} \times .55 \text{ (efficiency)}) = 50.4 \text{ square feet}$$

A collector with 50.4 square feet of absorber area would provide the energy required to heat 80 gallons of water from 50°F to 140°F.

COLLECTOR PLUMBING

Now that we've devised a way to capture heat from the sun, we need to move the heat somewhere useful. Because the demand for hot water does not usually coincide with peak insolation, some storage is also required. A method of preventing freezing is also necessary.

The simplest approach to these problems is a thermo-siphoning system. Thermo-siphoning applies the principle that warm fluid is less dense, and therefore lighter than cold fluid. Warm fluid will rise to the top of the system and cold fluid will sink to the bottom. We can take advantage of this rise of warm fluid by placing our storage tank above the collector. Figure 78 shows a thermo-siphoning arrangement. Cold water is let into the bottom of the storage tank. If the collector is warm, the coldest water, being densest and therefore heaviest, will drop to the lowest point in the system—the bottom of the collector. As the cold water goes into the collector, it forces the warmed water that had been in the collector, into the upper part of the storage tank. The flow is continuous even if no cold water is added or hot water taken out. As the water in the storage tanks cools, it becomes heavier and flows down into the collector, displacing the lighter warm water that goes to the storage tank.

When a hot water tap is opened, hot water flows out of the auxiliary tank. Hot water from the top of the storage tank, flows into the auxiliary tank, where it will be brought up to the required temperature if necessary. Cold water then can flow into the storage tank. Thermo-siphoning will

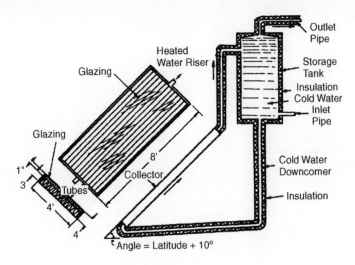

FIGURE 78
Thermosiphoning hot water system.

continue in the collector and storage tank, whether or not there is any demand, until the water in the storage tank is as warm as in the collector.

This system is simple, requires no pump, and works well if properly designed and maintained. However, there are a couple of problems related to tank placement, to a possible spell of cloudy weather, and to freezing. We consider tank placement first.

If the collector is roof mounted, the storage tank must be placed in the attic or rafters. The tank, when full, will weigh about 700 pounds and will require special structural support. If it leaks, the damage may be considerable.

If one adds a small pump to raise the cold water to the collector, one is freed from the storage tank placement constraints of the thermo-siphoning system. One does not have to depend on natural siphoning, which requires the tank above the collector, to move the water. Pumping the water allows a more conventional arrangement with the collector on the roof and the storage tank in the basement or utility room.

Now we will consider how to avoid the problem of a spell of cloudy weather. The hitherto unexplained auxiliary heater is our concession to the reality that our collector cannot always provide hot water. Cloudy days require that we have some sort of backup heating. The alternative to an auxiliary heater would be an impracticably large collector and storage tank. The auxiliary heater is more economical and also makes the system more reliable by providing a backup heater if something goes wrong in the collector.

The system shown in Figure 78 will be damaged if the water in the collector freezes. One solution is to drain the collector when there is any danger of freezing—a significant nuisance. An alternative to draining the system each time there is a threat of freezing is to circulate antifreeze rather than water. Obviously, one does not want to shower in antifreeze. To avoid that discomfort, the antifreeze is circulated through a heat exchanger in the storage tank. A heat exchanger is a length of tubing (usually copper), sometimes with fins, that transfers heat from one fluid to another—that is, it provides a path for efficient heat transfer but prevents mass transfer. The hot antifreeze from the collector flows through the heat exchanger, which warms the water in the storage tank.

The variety of systems and components available for hot water heating allows users a great deal of flexibility in choosing the system that best meets their lifestyle, the amount of money they want to spend, and the amount of hot water they require.

HOUSE HEATING

It was mentioned that collectors are not usually used for heating a house. If for some reason one did want to use a collector for space heating, then one could calculate the required area as we did. However, the efficiency factor should be changed from 0.20 because collectors are more efficient than direct gain absorbers. Efficiencies could be estimated from Figure 76.

REFERENCES

Johnson, Timothy E. (1981). *Solar Architecture—the Direct Gain Approach*. McGraw-Hill Book Co.

Leckie, James, Gil Master, Harry Whitehouse, and Lily Young. (1981). *More Other Homes and Garbage*. San Francisco: Sierra Club Books.

Shapiro, Andrew M. (1985). *The Homeowners Complete Handbook for Add-on Solar Greenhouses and Sunspaces*. Emmaus, PA: Rodale Press.

Department of Housing and Urban Development. (1982). *Passive Solar Homes*. Facts on File, NY.

AN EVAPORATIVE COOLER

Myra Wong

The purpose of the evaporative cooler described is to decrease the indoor temperature without electricity. It is aimed at poor households in hot, but preferably dry, regions. In these places there are occasionally life-threatening heatwaves that cause some of the elderly and sick to die. In 1995 a heatwave in northern India hundreds of people either died or were hospitalized from heatstroke. At other times it may still be very hot indoors and cooling simply be a matter of comfort. The need is for an inexpensive, easily crafted method of cooling a room without the use of valuable electricity in a fan or air conditioner (or no electricity may be available at all).

Presently, there are evaporative coolers that use an electric fan to move air across the surface of water and evaporative humidifiers that use an electric fan to move air through a wick. All of these rely on the evaporation of water from a reservoir into the indoor air. The evaporation is an endothermic process and thus "saps" energy (in the form of heat) from the room in order to convert the water-to-water vapor. In our design the outdoor air is moved through the cotton wick by the breeze though the open window, removing the need for a fan while still cooling the air before entry.

Our design (see Figure 79) is simple and intuitive, so that if anything breaks or needs replacement the repair can be done by nearly any layperson and definitely by the local craftsperson. The cooler costs less than $10, not including the lumber. The construction is very simple, and the cheap materials are easily substituted with locally available materials, making it even more likely that capital costs will not be a large problem. For example, a cotton cloth was used in our prototype because it takes up water well and is available in most areas of the world. Further, if the water for the reservoir is heavy in minerals, the deposits onto the wick will build up and impair the uptake of water. Thus, a very cheap and disposable wick material is advantageous. But any type of natural fiber (from old, torn clothes, scrap cotton and wool, etc.) can be used in its place. I like the idea of a more attractive and durable macramé or woven wick because pride in local art practices will encourage use and a feeling of ownership. Macramé or woven material can also be washed of mineral deposits more frequently.

Wood, scrap metal, pottery, bamboo, strong sticks, and plastic are all suitable frame materials. The frame need not be very strong. In place of the plastic lining, a ceramic bowl, large, waxy leaves, waterproofing wood varnish, or bamboo would probably work equally well.

Alternatively, less moist air can be circulated by a mobile constructed of a bamboo frame and colorful macramé ropes or woven strips hanging down. Another design on the same principle is a sturdy woven bag for a brick, which is then swung indoors (where it won't hit a child!), with both the brick and macramé being dunked in the water to be evaporated.

Although the design is easy to make, it is still relatively labor intensive—for example cutting the strips of cotton into one-inch segments (or weaving/knotting string). Environmental impacts should be small, depending on the frame materials, but processing methods on the raw materials would influence the amount of pollution created (unnecessary bleaching of fabric and string, water used for wood).

A small amount of water added to initially very dry air (humidity less than 10 percent) should not overly impair the evaporation of perspiration. I would like to stress that the evaporative cooler is only one part of the preventative effort and not a priority when someone is in medical distress.

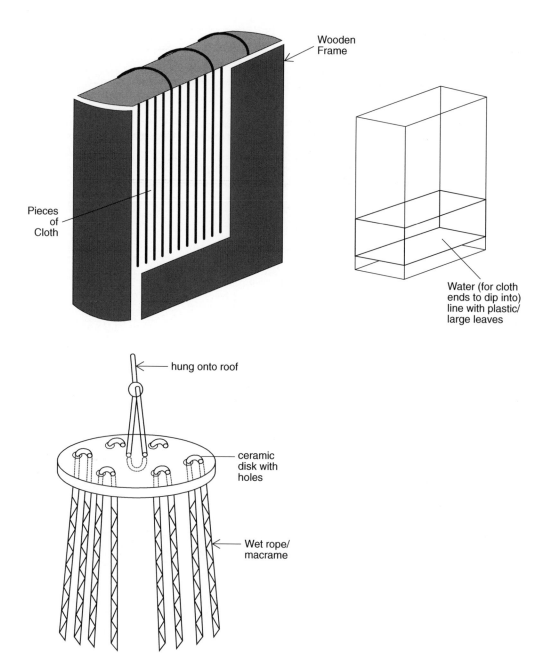

Wooden Frame

Pieces of Cloth

Water (for cloth ends to dip into) line with plastic/ large leaves

hung onto roof

ceramic disk with holes

Wet rope/ macrame

FIGURE 79
Evaporative cooler.

REFERENCE

Rusten, Eric. (1985). *Understanding Evaporative Cooling*. Technical paper #35. Volunteers in Technical Assistance, 1600 Wilson Boulevard, Suite 710, Arlington, Virginia 22209, USA. Telephone: 703-276-1800. Fax: 703- 243-1865. E-mail: vita@vita.org.

METHANE DIGESTERS

A methane digester is basically a large tank in which manure and other plant wastes ferment, producing biogas—a fuel that burns—and an odorless effluent that can be used as fertilizer. Biogas can replace natural gas or gasoline, and the effluent is easier to apply as fertilizer than the raw manure. The fermentation process is not difficult to operate. An early proponent was a pig farmer, John Fry, who realized that he was spending most of his time shoveling manure from the barn to a compost heap and from the compost heap to the field. He built a digester primarily to reduce shoveling but found that the amount of gas generated was sufficient to power all the machinery used on the farm.

BIOGSAS

Biogas is primarily methane (hence, biogas generators are also called methane digesters) with some carbon dioxide, carbon monoxide and traces of other gases. Methane consists of carbon and hydrogen. The chemical symbol is CH_4. The generation process employs bacteria that work under anaerobic conditions—that is, without oxygen (oxygen kills the bacteria). The raw material commonly used is livestock manure, straw, weeds, sewage wastes, and so forth. The methane-producing bacteria are found naturally in all these raw materials, especially in livestock manure.

There are several reasons why biogas technology is useful as an alternate form of technology: Primarily, the raw material used is very cheap, and to farmers it is practically free; the biogas can be utilized for many household and farming applications; the effluent obtained from the digester is a good fertilizer for crops, not having lost any of the nutritional value of the original raw material, but is odorless and usually germ free; the burning of biogas does not produce harmful gases, so it is environmentally clean; and the technology is relatively simple and can be reproduced in large or small scale by many people without the need for a large initial capital investment (Palmer 1981).

RAW MATERIAL

There are several conditions in the digesting tank that must be maintained for efficient gas production. The bacteria need nutrients. These are provided by the raw material, which can be a combination of things, such as cow manure, straw, and night-soil (human wastes) combined with water. Essentially any organic material can be used as input, except for mineral oil and lignin (the "woody" part of plants and trees), neither of which will ferment. Animal manure, sludge, and effluents from fermentation factories, such as leather processors, are among the organic materials that easily produce biogas. Some materials, like leaves, stalks, and straw, which are high in cellulose content, are harder to "digest" by the methane-producing bacteria. Such material can be pretreated, however, and then fed in as raw material.

There are two ways to pretreat raw material like straw and leaves: (1) feeding it to animals and using the resulting manure for the biogas generator or (2) composting the material by mixing it with manure and a little limewater (to make it less acidic) and leaving the mixture in a heap for several days. Either process will break up the cellulose in the stalks. Without this pretreatment, the digestion in the generator will be slow and the production of biogas will decrease considerably. It is evident that the first method of pretreatment is easy and has other benefits, such as better-fed animals.

TABLE 12
C/N ratio of common raw materials

| Raw Materials | C | N | C/N ratio |
	(dry wt %)		(dry wt %)
Stalks			
Straw (wheat)	46	0.53	87:1
Straw (rice)	42	0.63	67:1
Stalks (corn)	40	0.75	53:1
Fallen Leaves	41	1.00	47:1
Stalks (soybean)	41	1.30	32:1
Weeds	14	0.54	27:1
Stalks and Leaves (peanut)	11	0.59	19:1
Manure			
Sheep	16	0.55	29:1
Cattle	7.3	0.29	25:1
Horse	10	0.42	25:1
Swine	7.8	0.65	13:1
Human Feces	2.5	0.85	2.9:1

The raw material is the nutrient for the bacteria, and it is important that the bacteria's diet be balanced. The raw material must contain carbon and nitrogen. The methane-forming bacteria use the carbon for fuel and the nitrogen for building their cell structures, so the carbon to nitrogen ratio (C/N) is important. The bacteria need about 25 to 30 times more carbon than nitrogen. In other words, the C/N ratio of the raw material should be approximately 25–30:1. Table 12 is a list of raw materials and their C/N ratios. The combination mentioned earlier has a C/N ratio approximately 27:1 (cornstalks: 53:1; cattle manure: 25:1; human feces: 2.9:1—a mixture of equal parts by weight will give an average $(53 + 25 + 2.9):3 = 27:1$). To summarize, optimum biogas production requires proper raw materials in the proper ratio. The raw materials should also be diluted with water approximately 50-50 so the mixture is fluid.

MANAGING THE PROCESS

There is a close relationship between biogas production and temperature because the efficiency of methane-producing bacteria depends on the temperature. There are two types of methane-producing bacteria: thermophyllic and mesophyllic. The first type of bacteria thrive at temperatures between 47°–55 °C, whereas the latter type of bacteria operate best between 27°–38 °C. Thermophyllic bacteria produce gas at a daily rate of 2.5 cubic meters of gas/cubic meter of digester volume at their optimum temperature, whereas mesophyllic bacteria produce 1.0–1.5 m^3/m^3 per day at their optimum temperature. Although thermophyllic bacteria produce biogas at a higher rate, it may be costly to keep the generator heated. Some heat is produced by the digestion but not enough. Thus, mesophyllic bacteria are used in most digesters.

Even mesophyllic bacteria produce faster when the temperature is above normal outdoor temperature. At 25° a typical generator will produce about 1 cubic meter of gas per day for each cubic meter of tank volume. At 15 °C, the production rate goes down to about 0.3 cubic meter of gas per day. Cattle manure is completely digested in a generator at 24 °C in 50 days. At 35 °C, the fermentation period is only 28 days. Thus, by raising the temperature by 11 °C, one can almost double the rate of gas production, but the total amount of gas is about the same.

A faster rate is useful even if more gas is not produced because a faster rate means a smaller volume tank is needed for the same output. A digester can be heated by burning some of the gas produced, but this partly defeats the purpose if the gas could be used elsewhere. Solar heating is possible. To take advantage of the sun's heat to the fullest extent, the digester should be kept in a sunny place with a wind barrier around it. Insulation can be used, which keeps heat produced inside the digester but also keeps out the sun's energy.

Another factor influencing the operation of the digester is pH, a measure of acidity—lower pH corresponds to higher acidity. Biogas fermentation takes place in a slightly alkaline environment (pH value between 7.0 and 8.0). It has been observed that during the initial stage of biogas fermentation, the pH level drops slightly before it stabilizes—that is, the mixture becomes acidic. During this initial stage, it may be helpful to add alkaline materials to raise the pH level back between 7.0 and 8.0. Adding burnt ashes to manure is one way of raising the alkalinity of the digestive material. Limewater may be used, but care has to be taken to use a very dilute solution of limewater, or the pH level may become higher than 8.0, and methane-producing bacteria are killed. Generally the pH level adjusts itself to within the functioning range after the initial stage. If a digester is not producing as much gas as expected, then the pH may be too low and ashes should be added.

Each time new material is fed to the digester the rate of biogas production drops slightly. The reason for the decrease in gas production is that the raw material entering the digester contains relatively few methane-producing bacteria, whereas the sludge that is removed from the digester is rich in such bacteria. To compensate for the reduction in bacteria after feeding, one may prerot the raw material by piling it in a heap mixed with a small amount of sludge from a digester. This process of prerotting the raw material with "starter" bacteria is called enrichment.

Adding bacteria with new raw material makes a big difference. A large-scale biogas generator will not produce any methane until four weeks after it is filled with fresh raw material. On the other hand, when prerotted raw material is used along with some digester sludge, the methane content of the gas generated will be 50 percent of full production on the sixth day and 75 percent by the end of four weeks. Besides sludge from other digesters, sewage from slaughterhouses (or any house for that matter) is a common additive to introduce the bacteria. Using additives greatly reduces the retention period and increases the early gas yield and its methane content.

It is not hard to visualize the inside of a biogas generator. The contents divide themselves into three layers. The raw material settles in the bottom of the digester. A murky watery layer in the middle contains the bacteria. The top layer is a thick crust of scum. This layer prevents the formation of gas at an optimum rate because when the gas is formed, it has a hard time escaping to the gas collector through the crust. The remedy is stirring the digester frequently without, of course, letting oxygen in. Stirring ensures an even distribution of raw materials, extends the contact surface area of the raw materials with the bacteria, and speeds up fermentation, thereby increasing the gas yield. Experiments show that stirring can increase the gas yield by 10 to 15 percent.

The pressure of the gas inside the digester is important. Anyone who has opened a bottle of carbonated drink knows that once the cap is removed, the drink becomes visibly "fizzy" as pressure is released inside the bottle. Before the cap was opened, the pressure inside the bottle prevents the dissolved carbon dioxide from forming bubbles. Pressure can play a similar role inside a biogas generator—high internal pressure will not let gas bubbles form. Internal pressure less than 5 percent above atmospheric pressure will not affect biogas production significantly. However, any increase of pressure beyond that will reduce gas production. Thus, it may be useful to have a manometer (a pressure measuring device) attached to a biogas generator. A way to control the internal pressure should also be included in the design of a generator.

PROBLEMS IN DIGESTER OPERATION

One may design and build a biogas generator and feed it raw material properly but still obtain no methane in the gas. If, after checking the generator for leaks and checking the raw material for correct feeding, one still finds no production of gas, then the digester may have been contaminated with fermentation inhibitors—perhaps from something sprayed on the field from which the straw was taken. Biogas formation is a microbiological process and certain materials may inhibit the biological activity of the microbes. If these "poisons" contaminate the digester badly enough, the tank must be emptied and washed. Care should thus be taken as to where and what sort of raw material is fed into the digester. None of these inhibitors are likely to be in manure or household wastes.

Another type of "poisoning" may occur if the concentration of volatile acids released by the fermentation is too high. The concentration of volatile acids may rise for several reasons: (1) too high a concentration of the raw materials—not enough diluting water; (2) lack of methane bacteria; (3) too much acidic raw material—for example, manure with a high proportion of cow urine; or (4) too little fresh raw material being fed into the digester, making the mixture inside stagnate.

A major possible problem is an explosion. Care must be taken to keep air away from the biogas until one wants to burn it. Leaks must be avoided. A big explosion could be fatal, and a small one certainly would leave a disagreeable mess.

A DIGESTER USED IN CHINA

Figure 80 is the cross-sectional diagram of a digester used in the Sichuan Province of China. The internal volume of the fermentation chamber can be between 6 and 123 cubic meters. These sizes are practical for household use. Small plantations and pigsties in communes have digesters with internal volumes between 500 and 1,000 cubic meters.

The digester is built underground, using stones, bricks, lime, and some cement. It is circular, small, and shallow. There are advantages of placing the digester underground: (1) saving farmland as food can be grown over the tank; (2) using the soil as structural support; (3) improving thermal insulation, thus preventing thermal cracking due to quick changes in temperature and moisture; and (4) simplifying operation, since feeding of the digester and discharging can occur at ground level. The reason the design uses a shallow container is to increase surface area for

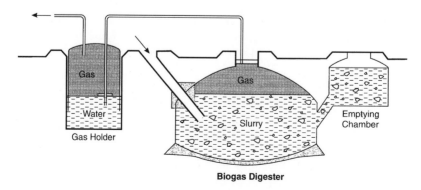

Biogas Digester

FIGURE 80
Digester used in China.

the gas to escape. Of course, the effluent must be pumped out or removed with buckets, although the pressure of the gas raises some effluent into the emptying chamber.

With this design about 1,530 liters of biogas are produced each day for every cubic meter of digested material with temperatures between 15° and 30°C. These are the main components of the digester in Figure 80.

- A straight inlet, convenient for adding the material without clogging. The rush of material from the inlet also mixes the material inside the digester.
- An emptying chamber placed so the effluent flows out from the middle of the tank, where 90 percent of the parasite eggs in the mixture have been killed. The emptying chamber has a volume of about 10 percent of the digester volume.
- A removable cover that allows repairs to be made inside a digester. The cover is tested underwater when sealing the digester, thus ensuring that the digester is airtight.

In China many villages use methane digesters to decompose night soil. If the digester is well insulated, its temperature will exceed 100°F, which is sufficient to kill nearly all the harmful pathogens, so the effluent is generally safe as a fertilizer. One should not use it, however, on leafy vegetables that will be eaten raw.

METHANE DIGESTER FOR A FARM IN THE UNITED STATES

Assume a farmer in the United States has 10 cows and would like to build a digester. How big should it be? How much methane will be produced? This example is based on one in Palmer (1981).

Table 13 will be used in our design. To make sure the table is clear, let us consider the top row. It says that one cow will produce 1.5 cubic feet of manure each day; that in loading the digester one should add 25 percent additional water; and that the manure will give 50 cubic feet of biogas. The additional water in this case is just to flush the manure into the tank, but in the case of the pigs or chickens, the manure needs dilution to spread the bacteria around or to reduce the concentration of ammonia and other harmful material produced in the digestion. The hydrogen in the methane gas comes from the water, so some water is essential.

First, we calculate the size of the tank. The amount of manure from 10 cows is 15 ft^3. We need 25 percent more water, so the total amount of manure and water is

$$15 \times 1.25 = 18.75 \text{ ft}^3$$

We saw above that if the tank is not heated, it can take 50 days for manure to digest. We add 18.75 ft^3 of manure and water each day. The size of the tank then is

$$18.75 \times 50 = 937.5 \text{ ft}^3$$

TABLE 13
Manure and biogas produced by 1 animal

	Manure Produced by 1 Animal ft^3/day	Additional Water Needed, % of Manure	Biogas Resulting ft^3/day
Cow	1.5	25	50
Pig	.2	200	7.8
Chicken	.004	800	0.4

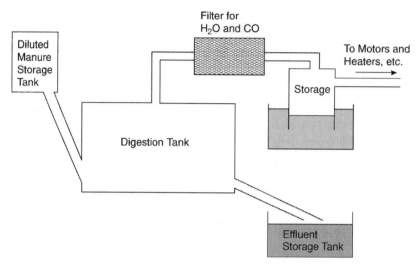

FIGURE 81
Methane digester—overall system.

We should add 10 percent for gas collection—a space above the liquid where the methane accumulates.

$$937.5 \times 1.1 = 1031.25 \text{ ft}^3 \qquad \text{(Total Tank Volume)}$$

Calculating the amount of methane gas from the 10 cows is straightforward, using the last column of Table 13.

$$10 \times 50 = 500 \text{ ft}^3 \text{ of Biogas/day}$$

A block diagram of the overall system is shown in Figure 81. Holding tanks for both the manure and the effluent are shown in case one does not want to load the tank or fertilize the field each day. The filter removes hydrogen sulfide, generated from sulfur compounds in the manure. Hydrogen sulfide is corrosive and can damage engines. The filter consists of iron filings mixed with wood shavings. Limestone is also used in the filter to remove carbon dioxide produced in the gas.

If a faster reaction—which means a smaller tank and less retention time—is desired, then the tank should be heated. One way is to use the sun's energy, as in passive solar heating. Another way is to burn some of the biogas to heat water. The heated water is pumped through pipes inside the digester. At most 20 percent of the biogas might be required for heating the mixture. The loss of gas in heating needs to be balanced with the savings in building a smaller tank.

WHAT TO DO WITH THE BIOGAS

Table 14 shows estimates of the amount of biogas required to do certain things. A mantle, as in a "Coleman" lantern, is used when gas is burned for illumination.

The amount of biogas required by a family is estimated in Table 15. We assume two burners are on for two hours each and four mantles are on for four hours each. The refrigerator and the engine are both on for four hours. We see that the 10 cows will just suffice.

TABLE 14
Quantities of biogas used by
various appliances

	ft^3/hr
Cooking (one large burner)	20
Cooking (one small burner)	10
Lighting (one mantle)	3
Refrigerator	20
Engine (5 hp)	80

Note: 1 ft^3 = 28.3 liters.

TABLE 15
Total biogas use by a family

Cooking	80 ft^3/day
Lighting	16 ft^3/day
Refrigeration	80 ft^3/day
Engine	320 ft^3/day
TOTAL	496 ft^3/day

The amount of biogas generated by each cow is variable, so the numbers shown are not precise but are reasonable estimates for the United States.

In principle, any gasoline or diesel engine can be easily adopted to run on biogas. Actually biogas is only 60 percent methane, so one needs to use more biogas than natural gas, which is nearly pure methane. People have driven tractors, even automobiles, using biogas, but a large storage tank is needed—1 gallon of gasoline is replaced by 213 cubic feet of biogas. Biogas, of course, can be compressed into a smaller volume, but the compressor takes energy. Stationary engines make the most sense. Electric generators could be driven by these engines, and a large farm could sell electricity back to the utility.

APPROPRIATENESS OF METHANE DIGESTERS

Methane digesters have been built and used successfully both in the United States and in the Third World. Like many energy-producing projects, their value depends on the cost of fuel. If oil prices go up, then the biogas is more valuable. Along another line, if a farmer has a great deal of manure to dispose of, then digesters look more attractive—to reduce odor if nothing else. In actual fact, though, methane digesters are not in common use, either in the United States or in the Third World. It is not clear why. Perhaps people do not know about them. Perhaps the collection of manure from fields is too difficult.

An issue not addressed here is the alternative use of the manure if a digester is not built. In the United States it would probably be put on a compost heap and then spread on the fields. The digester makes the handling easier, and it does kill most of the germs, so it is beneficial. In parts of the Third World, the manure may be collected by the very poor and used as fuel, so use of a digester may hurt the very poor and help the better-off farmers. On the other hand, dried manure is not a good fuel—its smoke irritates the eyes and can cause blindness.

REFERENCES

Chawla, O.P. (1986). *Advances in Biogas Technology.* Indian Council of Agricultural Research, New Delhi.

Mother Earth News. (1983) *Mother's Energy Efficiency Book.* Henderson, NC.

National Academy of Science. (1977). *Methane Generation from Human, Animal, and Agricultural Wastes.* Report of the Ad Hoc Advisory Committee on Technology Innovation, National Academy of Sciences, Washington, DC.

Palmer, David G. (1981). *Biogas, Energy from Animal Waste.* The Superintendent of Government, US Government Printing Office, Washington, DC 20402.

CHARCOAL PRODUCTION

This article is based on Application of Biomass-Energy Technologies, United Nations Centre for Human Settlements (Habitat) Nairobi, 1993 United Nations, HS/287/93 E ISBN 91-1-131210-8 and Energy Options, Drummond Hislop, editor, Intermediate Technology Publications, 103–105 Southampton Row, London WC1B 4HH, 1992

Charcoal is usually produced by heating wood in a kiln. Charcoal an important household fuel in many less industrialized countries, especially in urban areas. It is superior to wood because it weighs less than wood for the same heat produced (30 MJ/kg as compared with 15 MJ/kg for wood fuel). It also is easier to store, has lower levels of smoke emissions, and resists insects attacks better than wood fuel.

The production process consists of heating the wood in an enclosure called a kiln. Other raw materials such as coconut husks can be used, but wood is by far the most common. A purpose of the kiln, which excludes air, is to prevent the wood from burning. Heating the wood drives off water and other volatile constituents. Heat for the process is obtained by burning the wood or the gases given off in the reaction.

The production and distribution of charcoal consist of seven major stages.

1. Preparation of wood
2. Drying—reduction of moisture content
3. Precarbonization—reduction of volatiles content
4. Carbonization—further reduction of volatiles content
5. End of carbonization—increasing the carbon content
6. Cooling and stabilization of charcoal
7. Storing, packing, transport

The first stage consists of collection and preparation of wood. For small-scale and informal charcoal makers, preparation consists of simply stacking odd branches and sticks either cleared from farms or collected from nearby woodlands. Stacking may assist in drying the wood, facilitating the carbonization process. More sophisticated charcoal production systems entail additional wood preparation, such as debarking the wood to reduce the ash content of the charcoal produced. It is estimated that wood that is not debarked produces charcoal with an ash content of almost 30 percent. Debarking reduces the ash content to between 1 and 5 percent, which improves the combustion characteristics of the charcoal.

The next five stages all take place within the kiln, as the temperature increases and then decreases. The second stage, drying, is carried out at temperatures ranging from 110 to 220°C. This stage consists mainly of reducing the water content by first removing the water stored in the wood pores, then the water found in the cell walls of wood, and finally chemically-bound water.

The third stage, precarbonization, takes place at higher temperatures of about 170° to 300°. In this stage pyroligneous liquids in the form of methanol and acetic acids are expelled, and a small amount of carbon monoxide and carbon dioxide is emitted. The fourth stage, carbonization, occurs at 200° to 300°, where a substantial proportion of the light tars and pyroligneous acids are produced. The end of this stage produces charcoal, which is in essence the carbonized residue of wood. The fifth stage takes place at temperatures between 300° and a maximum of about 500°. This stage drives off the remaining volatiles and increases the carbon content of the charcoal. The sixth stage involves cooling the charcoal for at least 24 hours to enhance its stability and reduce the possibility of spontaneous combustion. A skilled operator can control the rate of combustion so each of the stages listed above are completed, producing a high yield.

The final, seventh stage consists of removing the charcoal from the kiln, packing it and transporting it for bulk and retail sale to customers. The final stage is a vital component that affects the quality of the finally delivered charcoal. Because of the fragility of charcoal, excessive handling and transporting over long distances can increase the amount of fines to about 40 percent, thus greatly reducing the value of the charcoal. Distribution in bags helps to limit the amount of fines produced in addition to providing a convenient measurable quantity for both retail and bulk sales.

Charcoal kilns can be broadly classified into five categories.

1. Earth kilns
2. Metal kilns
3. Brick kilns
4. Cement or masonry kilns
5. Retort kilns

Earth kilns are either holes in the ground into which the wood is placed or piles of wood stacked on the ground and covered with soil. Metal kilns can be made in several detachable pieces so as to be portable—possibly rolled along the forest floor. Brick kilns are half-spherical domes with openings for loading the wood and for retrieving the charcoal. Use of a chimney that ensures optimum draught conditions can increase yields significantly, especially for an earth kiln.

The critical factors in improving yield appear to be the skill of the operator and moisture content of the utilized wood. Some specific causes of low yields are the following.

1. Use of raw materials that have not been dried sufficiently
2. Poor stacking methods that allow excess air inside the kiln
3. Kiln leaks and poor control of combustion, allowing in too much air, resulting in complete combustion of some of the wood
4. Adulteration of the charcoal with stones, earth, and dirt—especially serious with earth kilns
5. Use of soft wood, which gives a crumbly, low-grade charcoal and results in a high proportion of unusable fines

FUEL CELLS

In principle, a fuel cell operates like a battery. Unlike a battery, a fuel cell does not run down or require recharging. It will produce energy in the form of electricity and heat as long as fuel is supplied. A fuel cell consists of two electrodes sandwiched around an electrolyte. Oxygen passes over one electrode and hydrogen over the other, generating electricity, water, and heat.

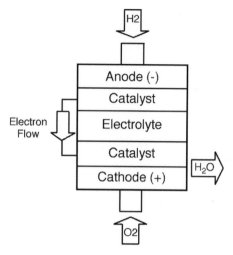

FIGURE 82
Schematic of a fuel cell. (Courtesy of Fuel Cells 2000)

Hydrogen fuel is fed into the "anode" of the fuel cell. Oxygen (or air) enters the fuel cell through the cathode. Encouraged by a catalyst, the hydrogen atom splits into a proton and an electron, which take different paths to the cathode. The proton passes through the electrolyte. The electrons create a separate current that can be utilized before they return to the cathode, to be reunited with the hydrogen and oxygen in a molecule of water (see Figure 82).

A fuel cell system, which includes a "fuel reformer," can utilize the hydrogen from any hydrocarbon fuel—from natural gas to methanol and even gasoline. Since the fuel cell relies on chemistry and not combustion, emissions from this type of a system would still be much smaller than emissions from the cleanest fuel combustion processes.

Phosphoric Acid

This is the most commercially developed type of fuel cell. It is already being used in such diverse applications as hospitals, nursing homes, hotels, office buildings, schools, utility power plants, and an airport terminal. Phosphoric acid fuel cells generate electricity at more than 40 percent efficiency—and nearly 85 percent if steam this fuel cell produces is used for cogeneration—compared to 30 percent for the most efficient internal combustion engine. Operating temperatures are in the range of 400°F. These fuel cells also can be used in larger vehicles, such as buses and locomotives.

Proton Exchange Membrane

These cells operate at relatively low temperatures (about 200°F), have high power density, can vary their output quickly to meet shifts in power demand, and are suited for applications—such as in automobiles—where quick startup is required. According to the U.S. Department of Energy, "They are the primary candidates for light-duty vehicles, for buildings, and potentially for much smaller applications such as replacements for rechargeable batteries in video cameras."

Molten Carbonate

Molten carbonate fuel cells promise high fuel-to-electricity efficiencies and the ability to consume coal-based fuels. This cell operates at about 1,200°F. The first full-scale molten carbonate stacks have been tested, and demonstration units are being readied for testing in California in 1996.

Solid Oxide

Another highly promising fuel cell—the solid oxide fuel cell—could be used in big, high-power applications, including industrial and large-scale central electricity-generating stations. Some developers also see solid oxide use in motor vehicles. A 100-kilowatt test is being readied in Europe.

Two small, 25-kilowatt units are already on line in Japan. A solid oxide system usually uses a hard ceramic material instead of a liquid electrolyte, allowing operating temperatures to reach 1,800°F. Power generating efficiencies could reach 60 percent. One type of solid oxide fuel cell uses an array of meter-long tubes. Other variations include a compressed disk that resembles the top of a soup can.

Alkaline

Long used by NASA on space missions, these cells can achieve power-generating efficiencies of up to 70 percent. They use alkaline potassium hydroxide as the electrolyte. Until recently they were too costly for commercial applications, but several companies are examining ways to reduce costs and improve operating flexibility.

Clean and Efficient

Fuel cells could dramatically reduce urban air pollution and decrease oil imports. Fuel cells running on hydrogen derived from a renewable source will produce nothing but water vapor.

What Sort of Fuels Can Be Used in a Fuel Cell?

Hydrogen—the most abundant element on Earth—can be used directly. Fuel cells can also utilize fuel containing hydrogen, including methanol, ethanol, natural gas, and even gasoline or diesel fuel. Fuels containing hydrogen generally require a "fuel reformer" that extracts the hydrogen. Energy also could be supplied by biomass, wind, solar power, or other renewable sources. Fuel cells today are running on many different fuels, even gas from landfills and wastewater treatment plants.

How Much Do Fuel Cells Cost?

One company commercially offers fuel cell power plants for about $3,000 per kilowatt. At that price, the units are competitive in high value, "niche" markets, and in areas where electricity prices are high and natural gas prices low. A study by Arthur D. Little, Inc., predicted that when fuel cell costs drop below $1,500 per kilowatt, they will achieve market penetration nationwide. Several Companies are selling small units for research purposes. Prices vary.

Where Can I Buy a Fuel Cell?

The following companies offer a variety of fuel cell products, including prototype demonstration systems, low-wattage systems, beta-testing systems, and fuel cell-powered products. You will need to check with the individual companies to see if their systems/products are suited to your needs.

Avista Laboratories, http://www.avistalabs.com/ (PEM fuel cells for residential applications)

Ball Aerospace & Technologies Corp., http://www.ball.com/ (portable PEM fuel cell power systems)

BCS Technology, Inc., http:// www2.cy-net.net/~bcstech (small PEM fuel cell systems)

DAIS-Analytic Corporation, http://www.daisanalytic.com (small PEM fuel cell systems)

DCH Technology, Inc., www.dch-technology.com (small PEM fuel cell systems)

EcoSoul, Inc., http://www.ecosoul.org/ (small, educational regenerative fuel cell kits)

ElectroChem, Inc., http://www.fuelcell.com (small PEM fuel cell systems)

Electro-Chem-Technic, http://www.i-way.co.uk/~ectechnic/HOME.HTML (educational fuel cell kits)

Energy Partners, Inc., http://www.energypartners.org/ (PEM fuel cell systems in a variety of sizes)

H Power Corporation, http://www.hpower.com/ (a variety of PEM fuel cell powered products, including backup power system)

H-TEK, Inc., http://www.h-tek.com/ (educational fuel cell kits)

Heliocentris Energiesysteme, http://www.heliocentris.com (educational fuel cell kits)

ONSI Corporation, http://www.onsicorp.com/ (200kW PAFC power plants)

Plug Power, LLC, http://www.plugpower.com/ (PEM fuel cells for residential applications)

Warsitz Enterprises, http://www.warsitz.com (small PEM fuel cell systems for portable power, experiment kits)

What's Holding Back Use of Fuel Cells?

Fuel cells are still a young technology. Many technical and engineering challenges remain; scientists and developers are hard at work on them. The biggest problem is that fuel cells are still too expensive. One key reason is that not enough are being made to allow economies of scale. When the Model T Ford was introduced, it, too, was very expensive. Eventually, mass-production made the Model T affordable.

How Can I Build My Own Fuel Cell?

There is a PDF file available on the Internet that provides step-by-step instructions on how to build a fuel cell from scratch.

http://www.humboldt1.com/~michael.welch/extras/fuelcell.pdf

REFERENCES

http://www.dodfuelcell.com/

BATTERIES

The easiest way to store electricity is with batteries, but only dc can be stored in batteries. A battery converts chemical energy into electrical energy. A rechargeable battery, when charging, also converts electrical energy into chemical energy. Batteries generally consist of two metallic plates—electrodes—in a chemical solution. Small batteries, such as used for flashlights, use a carbon rod instead of one of the plates and use only enough chemical solution to dampen an absorbent layer. Because the solution is not apparent, these batteries are called dry cells.

VOLTAGE AND AMPERE-HOURS

Two electrical parameters are important in choosing a battery: the voltage produced and the ampere-hour capacity. The voltage of a battery depends on the chemical reaction inside and is determined by the material used. Typically it is about 1.5 to 2.0 volts. The voltage of a new dry cell can be as high as 1.6 volts. It decreases as the battery is used. To get more than 1.5 volts, several batteries can be put in series. In an ordinary flashlight two batteries are placed in series, and the voltage applied to the bulb is about 3 volts. Automobile batteries are made of 6 cells, each producing 2 volts. The ampere-hour capacity determines how much current can flow over time: 100 ampere-hours means, in principle, that 100 amperes could be drawn from the battery in 1 hour or 2 amperes for 50 hours or 1 ampere for 100 hours. Actually the batteries work best when the current drawn is small, so such a battery probably would not produce 100 amperes for a full hour.

DRY CELLS

Dry cells come in many varieties. Some can be recharged. The chemical reaction discharging the cell will even go on slowly if the battery is not connected to a load, so the so-called shelf life is limited. Modern batteries have a longer shelf life than older ones, but one should be suspicious of batteries older than a year. The self-discharge is accelerated at higher temperatures, so batteries should be stored in a cool place. In a dry cell one of the two electrodes is the case, so when the battery discharges, the case becomes thin and can eventually leak. The leakage material can damage equipment, so ordinary dry cells should not be left in unattended equipment. The small dry cells used in watches and other electronic equipment have a strong metal case, so leakage is not a problem. Aside from keeping dry cells cool and watching for leaks, no maintenance is necessary. Incidentally, when dry cells are being discharged heavily, gas forms around the carbon electrode, lowering the output voltage. If the dry cell is "rested," this gas dissipates, and the dry cell recovers some of its voltage. Dry cells have limited capacity—not much current can flow through the small amount of fluid available so larger batteries with a liquid are used in automobiles and other high current applications. These are called storage batteries.

STORAGE BATTERIES

Storage batteries are a relatively mature technology—in other words, most of the complications have already been worked out. Automobiles use lead-acid batteries—the plates are lead and the fluid is sulfuric acid. When charging, a chemical reaction takes place at the plates. At discharge the reverse reaction takes place, letting current flow out. Nickel-cadmium batteries are similar, somewhat smaller and more reliable, but more expensive. Lead-acid batteries are probably the most practical for an electrical system in the Third World because they are readily available. Automobile batteries actually are designed to give a large amount of current for a few seconds to start the engine, which is a different application than giving sustained current for several hours or even days, so in many power applications a "Deep Cycle" battery, designed for golf carts, recreational vehicles, or small boats, is used. A typical deep-cycle battery might supply 9 amperes for 10 hours or 15 amperes for 5 hours before needing recharging.

Storage batteries do need care—mostly checking the fluid level. Water may evaporate from the fluid, and some water may be decomposed into oxygen and hydrogen, which pass into the atmosphere. Sealed batteries are available; they do not lose fluid, but they are more expensive. Storage batteries work best when warm and need protection from freezing. The chemical reaction when the battery is discharged produces water, which dilutes the sulfuric acid. The level of charge of the battery can be estimated from the specific gravity of the battery fluid; sulfuric acid is heavier than water. Battery testers measure that specific gravity. Batteries fail when the lead forming the plates is damaged either by a full discharge, by high temperature caused by drawing too much current, or by impurities in the replenishment water. Only distilled water should be used to top off storage batteries. Unusable storage batteries can be recycled; battery recycling is a viable industry in parts of the Third World.

REFERENCE

Merriman, Lee. (1985). *Understanding Batteries*, VITA. 1600 Wilson Blvd., Arlington, VA 22209. Most manufacturers have helpful Web pages.

ALCOHOL AS A FUEL

(based on Application of Biomass-energy Technologies, *United Nations Centre for Human Settlements (Habitat) Nairobi, 1993 HS/287/93 E ISBN 91-1-131210-8)*

Alcohols can be used as a liquid fuel in internal combustion engines either on their own or blended with petroleum. Therefore, alcohol has the potential to change and/or enhance the supply and use of fuel (especially for transport) in many parts of the world. There are many widely available raw materials from which alcohol can be made, using already improved and demonstrated existing technologies. Alcohols have favorable combustion characteristics, namely clean burning and high octane-rated performance. Internal combustion engines optimized for operation on alcohol fuels are 20 percent more energy-efficient than when operated on gasoline, and an engine designed specifically to run on ethanol can be 30 percent more efficient. Furthermore, there are numerous environmental advantages, particularly with regard to lead, CO_2, SO_2, particulates, hydrocarbons, and CO emissions.

Global interest in ethanol fuels has increased considerably over the last decade despite the fall in oil prices after 1981. In developing countries interest in alcohol fuels has been mainly due to low sugar prices in the international market and also for strategic reasons. In the industrialized countries, a major reason is increasing environmental concern and also the possibility of solving some wider socioeconomic problems, such as agricultural land use and food surpluses. As the value of bioethanol is increasingly being recognized, more and more policies to support development and implementation of ethanol as a fuel are being introduced. A number of countries have pioneered both large-scale and small-scale ethanol fuel programs. In the United States, the current fuel ethanol production capacity is over 4.6 billion litres, and there are plans to increase this capacity by more than 2.3 billion litres. Worldwide, fermentation capacity for fuel ethanol has increased eightfold since 1977 to about 20 billion litres per year in 1989.

Latin America, dominated by Brazil, is the world's largest production region of bioethanol. Countries such as Brazil and Argentina already produce large amounts, and there are many other countries such as Bolivia, Costa Rica, Honduras, and Paraguay, among others, that are seriously considering the bioethanol option. Alcohol fuels have also been aggressively pursued in a number of African countries currently producing sugar—Kenya, Malawi, South Africa, and Zimbabwe. An example showing the use in Malawi is at the end of this article. Others with great potential include Mauritius, Swaziland, and Zambia. Some countries have modernized sugar industry and have low production costs. Many of these countries are landlocked, which means that it is not feasible to sell molasses as a by-product on the world market, while oil imports are also very expensive and subject to disruption. The major objectives of these programs are diversification of the sugarcane industry, displacement of energy imports and better resource use, and, indirectly, better environmental management. These conditions, combined with relatively low total demand for liquid transport fuels, make ethanol fuel attractive.

PRODUCTION OF ETHANOL

The production of ethanol by fermentation involves four major steps: (1) the growth, harvest, and delivery of raw material to an alcohol plant; (2) the pretreatment or conversion of the raw material to a substrate suitable for fermentation to ethanol; (3) fermentation of the substrate to alcohol and purification by distillation; and (4) treatment of the fermentation residue to reduce pollution and to recover by-products. Fermentation technology and efficiency has improved rapidly in the past decade and is undergoing a series of technical innovations aimed at using new alternative materials and reducing costs. Technological advances will have, however, less of an impact overall on market growth than the availability and costs of feedstock and the cost of competing liquid fuel options.

The many and varied raw materials for bioethanol production can be conveniently classified into three types: (1) sugar from sugarcane, sugar beet, and fruit, which may be converted to ethanol directly; (2) starches from grain and root crops, which must first be hydrolyzed to fermentable sugars by the action of enzymes; and (3) cellulose from wood, agricultural wastes and so forth, which must be converted to sugars using either acid or enzymatic hydrolysis. These two latter systems are, however, at the demonstration stage and are still considered uneconomic. Of major interest are sugarcane, maize, wood, cassava, and sorghum and to a lesser extent grains and Jerusalem artichoke. Ethanol is also produced from lactose from waste whey—for example, in Ireland to produce potable alcohol and also in New Zealand to produce fuel ethanol. A problem still to be overcome is seasonality of crops, which means that quite often an alternative energy source must be found to keep a plant operating all year round.

RAW MATERIAL

Sugarcane is the world's largest source of fermentation ethanol. It is one of the most photosynthetic efficient plants—about 2.5 percent photosynthetic efficiency on an annual basis under optimum agricultural conditions. A further advantage is that bagasse—straw, a by-product of sugarcane production, can be used as a convenient on-site electricity source. The tops and leaves of the cane plant can also be used for electricity production. An efficient ethanol distillery using sugarcane by-products can therefore be energy self-sufficient and also generate a surplus of electricity in addition to CO_2 for industrial use, animal feeds and a range of chemical-based products. The production of ethanol by enzymatic or acid hydrolysis of bagasse could allow off-season production of ethanol with very little new equipment

Methanol, that can be obtained from biomass, and coal, that is currently produced from natural gas, has only been used as fuel for fleet demonstration and racing purposes and, thus, will not be considered here. In addition, there is a growing consensus that methanol does not have all the environmental benefits that are commonly sought for oxygenates and that can be fulfilled by ethanol.

EXAMPLE OF ETHANOL PRODUCTION

Malawi is entirely dependent on an agricultural economy for its export earnings. A major reason for embarking on the production of fuel ethanol has been the continuous deterioration of the regional transport system and the uneasy security situation with regard to Mozambique, both of which have caused frequent petrol shortages. Malawi commenced its bioethanol program in 1992 utilizing ethanol from a distillery located at Dwangwa sugar mill with a capacity to produce 10 million litres/year. The Ethco (Ethanol Company Ltd) produces ethanol from molasses and raw sugar efficiently and profitably. Ethco has also provided the driving force for the exploration of wider applications of ethanol as neat fuel, diesel fuel substitute, and illumination fuel for paraffin lamps. It has sought to expand the options for feedstock with work on cassava and wood chips.

Ethco currently produces ethanol to supply a national blend of 15 percent (v/v) ethanol, which could be increased to 20 percent. A production of 20 million litres/year could be achieved with minimal capital investment by operating the present fermentation/distillation plant all year round. A further option under consideration is the construction of a second plant near the Sucoma estate, whose by-product, molasses, is of little or no opportunity value. The potential exists to double ethanol production immediately and, in the longer term, to produce sufficient to displace the country's entire gasoline imports. The annual demand is approximately 60 million litres of gasoline and 80 million litres of diesel oil (Moncrieff & Walker 1988). If extended applications of neat ethanol being tested in a small fleet of government Land Rovers (approximately 1500) indicate that only a modest substitution of diesel fuels in transport and agriculture can be achieved, even then, Malawi could displace as much as 10 to 20 million litres of imported petroleum with ethanol in the medium term.

In terms of feedstock for new ethanol production in Malawi, a report (Steinglass et al. 1988) estimates that surplus molasses and the sugar sold at world market prices would be the cheapest feed stocks and would yield approximately twice the current amount of ethanol produced. Beyond this level alternative feedstocks would have to be considered if the ethanol market expands sufficiently.

REFERENCES

Borges, J. M. M., and R. M. Campos. (1990). "Economic Feasibility of the Brazilian Fuel Alcohol Programme." In G. Gosse Grassi, and G. dos Santos (Eds.), *Biomass for Energy and Industry, 5th EC Conference on Biomass for Energy*. London: Elsevier Applied Science.

Gotelli, C. A. (1988). "Ethanol as Substitute of Tetraethyl Lead in Gasoline." *Proceedings of the VIII International Symposium on Alcohol Fuels*. Tokyo, NEITDO.

Hall, D. O., F. Rosillo-Calle, and P. De Groot. (1992a). "Biomass Energy–Lessons from Case Studies in Developing Countries." *Energy Policy*. Vol. 20, pp. 62–73.

Moncrieff, I. D., and F. G. B. Walker. (1988). "Retrofit Adaptation of Vehicles to Ethanol Fuelling: A Practical Implementation Programme." *Proceedings of the VIII International Symposium on Alcohol Fuels*. Tokyo, NEITDO.

NAS (National Academy of Sciences). (1983). *Alcohol Fuels—Options for Developing Countries*. Washington, DC: National Academy Press.

Rosillo-Calle, F. (1990). "Liquid Fuels." In J. Pasztor and L. A. Kristoferson (Eds.), *Bioenergy and the Environment*. Boulder: Westview Press.

Steingass, H., et al. (1988). *Electricity and Ethanol Options in Southern Africa*, Report No. 88-21 USAID, Office of Energy, Bureau for Science and Technology.

Zabel, M. (1990). "Utilization of Agricultural Raw Material as an Energy Source—A Case Study of the Alcohol Industry in Sao Paulo State, Brazil." In A. A. M. Sayigh (Ed.), *Energy and the Environment into the 1990s. Proceedings of the 1st World Renewable Energy Congress*. Oxford: Pergamon Press.

TIDAL POWER

Obtaining energy from tides appears to be an attractive idea, but successful systems are rare. One approach is to locate a place where at high tide water fills a basin. A dam is built across the channel leading to the basin. When the tide recedes, water flows through the dam from the basin into the ocean. The force of the moving water turns turbines in the dam. These turbines drive electric generators. The system can be built so water drives the turbines both when the tide is rising and when it is falling. The estimated cost of building such a system is very large, so few systems have been attempted. Another restraint is environmental—the installation is sometimes deemed less attractive than the natural channel. The restriction of the waterways may also be harmful to wildlife. A few places in the world have especially high tides, and in such places a tidal system may make sense, but the idea has not been proven effective. It appears also that much power exists in waves and that such power could be commercially extracted, but a working system has not been demonstrated. One of the problems is building a system sensitive enough to extract energy on relatively calm days and rugged enough to survive storms.

CHAPTER 3

Food

Edited by TUARIRA ABDUEL MTAITA
Africa University

INTRODUCTION TO THE FOOD CHAPTER

The articles in this chapter fit into three categories: technology, plants, and animals. The technology articles deal with tools used by farmers and with approaches to producing food—urban gardens, hydroponics and related ways of growing plants, and sustainable agriculture. Also included are articles on common food processing equipment: a hammer mill and an oil press. The series of articles about plants starts with notes on growing as much produce as possible in a small space and then presents articles on farm-scale production of cereals, vegetables, fruits, and trees. Rice is grown somewhat differently, so its production, both in inundated fields and dry fields, is discussed in a separate article. Tools and equipment used for rice cultivation and harvesting are discussed in a separate article in the technology chapter. Included with the plants articles are discussions of several crops of special interest in the Third World: amaranth, cassava, coconuts, and mushrooms. The animal articles section begins with aquaculture, followed by articles on familiar animals, including ostriches, and end with an article on beekeeping. A final article discusses nutrition.

AGRICULTURAL MACHINERY

Small-scale farming equipment—such as planters, hoes, and grinding mills—can speed up production significantly. Some artisans have made profitable businesses, for example, making hoes and axes. Several nonprofit organizations design and manufacture machinery for small farms. Four prominent such organizations are ApproTEC in Kenya, the International Rice Research Institute outside of Manila in the Philippines, Intermediate Technology Development Group in Zimbabwe, and the Rural Industries Innovation Centre in Kanye, Botswana. Addresses are given in the Reference section of this article.

Design of such equipment is in general straightforward, so it may make sense for village artisans to do their own design and fabrication. If the intention is to sell equipment, careful attention should be paid to the user's desires. It is tempting for a designer to add extra features, each of which costs only a little and increases weight only slightly. The sum of all these extras may make the cost more than the buyer can afford and make the device too heavy for a person or animal to propel. Users tend to want simple machines that are easily repaired and maintained rather than ones with many features not often needed. The most serious problem faced by village artisans, however, is probably gaining access to a market of significant size. Successful artisans appear to be those who are willing to change from one product to another as demand changes.

This article discusses various machinery in the order they are used in growing a crop: preparing the soil and planting first, then harvesting, then processing the harvested seed or plant, and finally drying. Another article, "Rice Paddy: Harvesting, Threshing, Cleaning, and Handling," explains the entire process of harvesting and threshing rice. Two more articles supply the design of a hammer mill made in a small workshop and the process of extracting vegetable oil.

SOIL PREPARATION AND PLANTING

Improved plows, used to open dry soil for planting, have been designed at many national agricultural research stations. These are generally made of steel and pulled by oxen. These same plows can be used in wet soil—for rice cultivation—but other alternatives, using discs or caged wheels, can be more effective in mud. Water buffaloes are often used to pull plows in rice cultivation. Hand tractors, self-propelled by a small (10 horsepower) gasoline motor and guided by a person walking behind, are an alternative.

A manual seed planter is shown in Figure 1. The purpose of the planter is to space the seeds along a row. A wheel with teeth rotates in the bottom of the seed hopper. Only a single seed fits between two teeth. As the wheel rotates, these single seeds drop through the seed tube to the ground. The press wheel covers the seed. This design has a small plow in front of the seed tube to loosen the soil. An alternative planter is attached to a plow that follows it and drops seeds into the furrow. Similar machines can be used to spread fertilizer or a single machine can be designed to deposit both seeds and fertilizer.

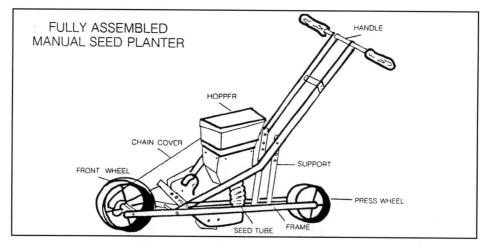

FIGURE 1
Seed planter. (from Appropriate Technology, Vol. 14, no. 1)

Wetland rice plants are usually started in a controlled area and transplanted to fields. (Rice production is described in another article in this section.) The transplantation is done by hand, but machinery, basically a moveable rack, can lessen the burden by holding the seedlings conveniently near where they will go. Machines similar to plows can be used to "harrow"— disturb—the soil between rows of growing plants to remove weeds.

HOES

One widely used tool that can be made locally is a hoe. In Africa the typical local hoe is a blade with a sharpened post—the "tang"— welded to it. The tang fits into a wooden handle, sold separately in village markets. A small hoe, used for close-in weeding, is also shown in Figure 2.

Hoes of this kind are made by cutting the blade from 3 mm steel sheet. Scrap steel can be used if it is of proper thickness. Cutting the blade is the most time-consuming task. The traditional way of cutting is by means of a hammer and a chisel, but alternative cutting machinery would make the process quicker and easier. The tang can be forged from a flat bar or by collapsing a steel pipe. Alternatively, a tang can be retrieved from

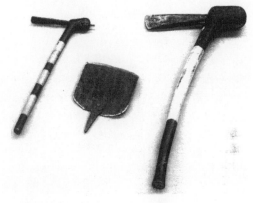

FIGURE 2
Southern african tools: small hoe, large hoe blade, axe.

an old hoe, since the blade usually breaks before the tang. The blade should be welded onto the tang. The alternative, which makes a significantly weaker hoe, is to use rivets. The handle is carved from the forked branch of a tree—the twisted grain at the fork makes the head, where the tang fastens, stronger. In many markets the blade/tang and the handles are sold separately.

Other implements, such as axes, adzes, and various carpentry tools, are made similarly. Scrap auto leaf springs are a good source of metal—too narrow for a hoe but suitable for an axe. These can be tempered by heating and then plunging into a bucket of water. A chisel for cutting the hoe blade, as just described, can be made in this way. The Blacksmithing article in the Tools Chapter gives more information about making metal objects.

REAPING, THRESHING, DEHULLING, GRINDING

Reaping (harvesting) can be done by hand using sickles or scythes. Self-propelled machines with long knives in front can also be used. An alternative to cutting the stalk is to shake the plant so the grain or fruit falls off into a container on the reaper.

Grain crops—corn, rice, sorghum, wheat, and others—need processing before use or sale. The seed must be separated from the stalk or cob which is called threshing; the hard outer coating of the seed must be removed, dehulling; and the grain must be ground. Some purposes of grinding are to make a marketable flour, to decrease drying time, or to prepare the product for mixing as, for example, an animal feed.

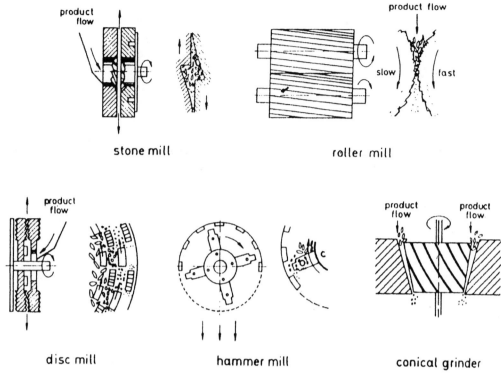

FIGURE 3
Grinding mills.

Dehulling is done manually by pounding and then winnowing the pounded mixture to remove the separated coating. Manual pounding is done in a large mortar, using a pole of about 3 inches in diameter as a pestle. Two names for the separated seed coating are "bran" and "chaff." Bran usually refers to material that will be eaten either by people (fiber) or animals. Winnowing is done by tossing the mixture into the air and letting the wind blow the chaff while the grain falls directly down. The Rural Industries Innovation Centre makes a dehuller using abrasive stones. This dehuller requires a 10-horsepower motor.

The dehulled grain is then ground to a fine powder. Various grinding mills are shown in Figure 3. A hammer mill, perhaps the simplest and most versatile, is described in a separate article. Mills can be designed and built by experienced village artisans, but they have received significant attention from several appropriate technology groups and private companies, so it may make sense to consider purchasing one.

The parameters used to decide which mill is appropriate in a given situation include the price, of course, as well as the throughput—how much grain can be ground in an hour. Three other factors to consider are the fineness of the flour produced, the hardness of the grain to be ground, and ease of cleaning and maintenance. Input power needs to be considered also; some alternatives are manual cranking, electric power, if available, or a small gasoline engine. A final consideration is how the mill matches the other machinery in the process. A mill that can grind 20 kilograms an hour is not needed if the dehuller or thresher can only do 8 kilograms an hour. It is most efficient, of course, if the outlet of the mill can go directly to the bags or other containers that will hold the grain. Similarly, it makes sense to think about how best to get the grain into the mill from the previous operation.

Plant stalks for animal feed are cut rather than ground, but the machinery is similar. In general, coarse chopping or grinding is preferable for animal feed, partly because finely chopped food passes too quickly through the animal and oxidizes quickly and partly because spillage of finely ground feed is greater. Collecting crop residues from fields and chopping it for animal food allow the farmer to reduce the amount of food that must be purchased during the dry season. The process of feeding chopped forage is more efficient than allowing animals to graze freely because it allows grass to grow quicker and gives the farmer more control over the animals' diet.

STORAGE

The concern when storing foodstuff is to prevent loss through spoilage or predation by insects, birds, and rodents. Before storage most foodstuff should be processed. Spoilage of grain is caused by fungal action, germination, and respiration, which depletes sugars. Fungi need moisture, as do insects, so grain should be dried before storage. Germination can be prevented by heating the grain to 46°C.

Drying is done by blowing hot air through a pile of grain. The simplest way, of course, is just to use smoke from a fire, but smoke damage may be significant, and the fire may burn the grain. More elaborate systems use fans to propel the air. The air is heated in a chamber and does not contact the fire. In one design, rice husk—chaff—is used for fuel. After drying or heating, the grain should be allowed to cool and stored at ambient or cooler temperature. Vegetables and fruit will rot if moist. Bruises allow harmful organisms to enter the vegetable, so they should be avoided. Most vegetables can be dried and then successfully stored. Potatoes and yams should be cured by heating to a high temperature—but lower than boiling—in a high-humidity environment and then stored in cool, dry place.

Storage of grain should be done in a dry, dark place, with ventilation—to carry off water vapor. One problem with ventilation of a large pile of grain when the grain is warmer than the outside air is that convective air currents can leave moisture near the top of the pile as warm air rises, and some of the water vapor carried in warm air condenses when the air cools at the top of the pile. The pile should be stirred regularly to dissipate this moisture. Darkness retards germination of most plants and respiration. Protection from insects, birds, and rodents is essential. In some countries a significant percentage of crops is lost to various creatures.

PROTECTION AGAINST PESTS

Insects are usually the major pests. Three approaches are possible to control insect infestation: preventive measures in the field and granary, using insecticides with the stored product, and storage in a sealed container.

Preventive measures include crop rotation and intercropping, choice of seed varieties less susceptible to insects, thorough cleaning of the storage space before the grain is deposited, and removal of infected grain before storage.

Synthetic insecticides can be effective but their use on small farms is risky. Some of the dangers are toxicity, improper use, improper storage, and improper choice of insecticide. Chemical insecticides are powerful poisons, and people who will be using them need training, which farmers and farm supply storekeepers often do not have. Before promoting such chemicals one should be sure they will be used safely and effectively. In most cases insects will not seriously damage foodstuffs that will be stored for three months or less, and insecticides are not really necessary. Wood ashes or plant product such as Annona, Hyptis spicigera, Lantana, or Neem can be as effective as synthetic chemicals and are safer as well as cheaper. If chemicals are to be used, dustable powders are the most widespread formulations for protection. On the farmer's level the application of dusts is comparatively easy and safe. The manufacturers guidelines must be followed. If such cannot be ensured, the product should not be recommended. The dusts should be used on the empty storage container before it is filled and mixed with the grain or placed in layers.

Sealed storage eliminates the oxygen that insects and molds require for their growth. In tropical countries with a high relative humidity, promoting the growth of airtight storage is difficult. Thorough drying before storage and prevention of condensation is necessary. Before storage—sealed or not—insects and larvae should be removed as much as possible. This can be accomplished by winnowing, as well as visual inspection.

Rodents can be as harmful as insects, with rats and mice the most common culprits. They can eat a substantial amount of grain and contaminate more with their wastes. Rodents carry diseases that are harmful to humans, and they can damage buildings by gnawing or by causing electrical fires. Prevention of rodents is done by keeping the storage area clear of split grain, by keeping an open space around the grain that rodents do not like to cross, and by keeping the area free of any water that the animals can drink. A tight building will keep rodents out. Particular attention should be paid to doors, ventilation openings, missing bricks, and the junctions between the roof and the walls. Traps, cats, and poison can control rodents. The first two are effective if not many rodents are present. Poison is effective but must be used cautiously.

REFERENCES

Catalogs are available from the following four organizations.

ApproTEC, PO Box 64142, Nairobi, Kenya. Tel/Fax: 783046, 787380/1, 796278. approtec@nbnet.co.ke
International Rice Research Institute, PO Box 933, Manila, Philippines.
Intermediate Technology—Zimbabwe, PO Box 1744, Harare. Tel.: 263-4-496745/6. Fax: 263-4-496041. itdg@harare.
    iafrica.com
Rural Industries Innovation Centre, Private Bag 11, Kanye, Botswana, Tel.: 340392. Fax: 340642.
Asiedu, J. J. (1989). *Processing Tropical Crops.* London, NY: MacMillan.
Crutchley, Victor. (1996). *Inventors of Zambia.* Bridport, UK: Eggardon Publications.
Gwinner, Joost, Rüdiger Harnisch, and Otto Mück. (1996). *Manual on the Prevention of Post-Harvest Grain Losses.*
    Post-Harvest Project, Deutsche Gesellschaft für Technische Zusammenarbeit (GTZ), GmbH, Postfach 5280 D-65726
    Eschborn, FRG.
Hall, Carl W. (1988). *Drying and Storage of Agricultural Crops.* Westport, CT: AVI Publishing.
Whitby, Garry. (1987). "Making Hoes in Malawi." *Appropriate Technology*, Vol. 14, No. 1, 17–19. IT Publications, 9
    King Street, London WC2E 8HW, UK.

HYDROPONICS, ROOFTOP, AND RELATED GARDENS

ROOFTOP GARDENS

People have succeeded in growing vegetables on roofs in cities. Major problems are drying winds and the baking sun. Windscreens and frequent—at least daily—irrigation are practical solutions to these problems. Of course, one must be careful to verify that the building itself can withstand the weight of the soil and water. A compost pile may not be feasible, so other ways of conditioning the soil must be provided. Rooftop gardening has been successful on apartment houses in St. Petersburg, in Russia, where families grow food for their own use. The same techniques can be used at street level—growing plants above ground. Discarded automobile tires have been used successfully as containers in San Salvador.

Soil can be used, but lightweight materials such as pine needles, coconut husks, or coarsely granulated plastic packing may be better. If an inert planting medium is used, then nutrients are supplied through the water. Alternatively, a small amount of compost can be added to the planting medium. As organic material decays, it takes nutrients from the water, so heavier fertilizer is needed—commercial farm fertilizer is effective. The depth of the container need be only 3 inches. Lettuce, onions, and herbs seem to be the most successful crops. Experimentation with different growing media and plants is necessary.

HYDROPONICS

Hydroponics is basically growing plants in only water. Garden supply stores sell racks in which trays of plants can be held, connected by tubes carrying the nutrient solution. On the other hand, you can build your own system. A hydroponic system is shown in Figure 4. In the United States hydroponic projects have earned much money because they can produce fresh vegetables all year round that are sold in upscale markets. These systems usually require much electricity to circulate the nutrient fluid and for lighting to compensate for the short daylight. It is not clear that hydroponic systems make sense in the Third World, where the demand for high-priced vegetables is low.

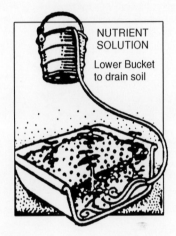

FIGURE 4
Hydroponic system.

In the system shown in Figure 4, the container containing the nutrient solution is raised several times a day. When the container is lowered, the solution drains back. The growing medium can be sand, gravel, or plastic particles. The nutrient solution can be chemicals or natural fertilizers, such as water from a fish tank. In a high-production system the circulation would be done by computer-controlled pumps.

More involved hydroponic systems use rafts of plants floating in fertile water, such as in a fish pool. It is necessary to raise the plants 10 cm or so above the water so air—oxygen—can reach some of the roots. An advantage of hydroponics is that one has control over the growing medium and the nutrient solution, although use of fish pond water reduces that control. The book by Shane Smith (1982) describes hydroponics more fully, and the Meitzner and Price (1996) book points out some of the problems in the Third World.

GREENHOUSES

A small solar-heated greenhouse can produce vegetables year round in northern climates, although winter crops are probably restricted to leafy vegetables such as lettuce, spinach, and (for the experienced) celery. Tomatoes will grow in fall and spring. The limiting factor seems to be the amount of sunlight available. A greenhouse does not have to shut down in the summer, but one has to be careful about overheating. Soils need more attention in a greenhouse than outside because propagation is more intensive and the ecosystem is probably incomplete. At best, yields per square foot in a greenhouse can be the same as in an intensive garden. Insect pests can be more of a problem because natural controls are absent and the compactness facilitates movement from plant to plant.

REFERENCES

Meitzner, Laura S., and Martin L. Price. (1996). *Amaranth to Zai Holes: Ideas for Growing Food Under Difficult Conditions.* ECHO, North Fort Myers, Florida, USA. ECHO@xc.org

Smith, Shane. (1982). *The Bountiful Greenhouse.* Santa Fe, NM: John Muir Publications.

The New Alchemy Institute built and studied greenhouses in the 1970s. The successor organization has some of those reports online and sells hard cover versions of others. The Green Center, 237 Hatchville Road, East Falmouth, MA 02536. http://www.fuzzylu.com/greencenter/

VITA. (1989). "Understanding Hydroponics." Volunteers in Technical Assistance, 1600 Wilson Boulevard, Suite 710, Arlington, Virginia 22209 USA. Tel.: 703-276-1800. Fax: 703-243-1865. vita@vita.org. http://idh.vita.org/pubs/docs/uhn.html

---

URBAN GARDENS

Aubrey Ludwig

INTRODUCTION

In urban areas around the world people have begun to imagine more ways to incorporate green space into the city. Whether for recreation, beautification, or economic sustainability, city residents have taken the initiative to create and reinvent *public* and community gardens, as well as urban farms. These sites are filling vacant lots, creeping in between buildings, in schoolyards, on rooftops, and behind houses, reminding urbanites that the natural world has much to offer. A city garden can provide physical nourishment, an outdoor classroom for environmental education, a community meeting place, and a laboratory personal development. Urban spaces are alive; they add character and color to the city—they add soul. For all of these reasons, the construction of urban gardens should be encouraged.

You have committed to start an urban garden and do not know where to begin. While many sites are first initiated by city governments, community gardening organizations, or neighborhood associations, individuals also play a key role in the greening process. This is particularly true in the developing world, where formal networks are often not in place. In the following pages I outline the general considerations one must review before launching an urban gardening project. I highlight the social and environmental factors that must be accounted for before taking action and suggest a few techniques that have been particularly practical in the urban site. Finally, a list of resources is provided for more specific information on the history, philosophy, and mechanics of urban gardening.

A PURPOSEFUL DESIGN

Selecting a clear and specific purpose is the first step of your project. As urban gardens may serve varied functions, acknowledging your intentions and outlining your basic goals is the only way you will be able to carry the project through successfully. Defining the purpose of the garden will shape what kinds of considerations should follow. To illustrate this point, imagine you are building an ornamental flower garden adjacent to a community center to be used for educational purposes. In this case, the most crucial considerations would be design, plant diversity, and ensuring garden maintenance. If you are instead initiating a community vegetable garden for a low-income neighborhood, your primary concerns would be finding an enthusiastic community, ensuring that the site is environmentally viable, and garnering ongoing support. Every factor, from necessary supplies to upkeep to design and participation, will revolve around the fundamental question of intent.

A FEW BASIC CONSIDERATIONS

Do not begin any greening project without first considering the control of human resources. Green areas are alive, and as such they require a great deal of maintenance and attention. A successful garden necessitates a sustained commitment to the process from weeding and watering, to pruning, harvesting, pest management, and cleanup; there is always something to be done in the garden. Therefore, you must be certain that there are people who are willing to continue the process. if you are initiating a community garden, be sure that the neighborhood sees a need for a green space, assess what that need is, and include the participants in all stages of planning. The resulting sense of ownership will guarantee the community's active involvement in preserving the garden.

Where will your basic resources come from? The creative nature of gardening provides a lot of leeway for the acquisition of needed resources. Garden centers, nurseries, and local farms are often willing to donate plant material for urban gardens. Tools can be shared within a neighborhood, and other needed supplies can be recycled from the trash or abandoned buildings. Do not let a lack of resources deter you from continuing the project; it can undoubtedly be accomplished with what you have. After outlining the goals of your urban garden you will be ready to move on to the physical plan.

NEGOTIATING WITH NATURE

To some degree the overall success of the garden is beyond the control of human hands. Each site and each season brings its own set of problems, and it is important to recognize from the beginning that you cannot account for every natural variable. With that in mind, you can start by focusing on what is known about growing plants successfully. Since plants are most dependent on water, light, and the nutrients found in soil, ensuring that these three elements are readily available at your site is the first step. Your environmental assessment should consist of a thorough site evaluation, which will include careful observations and tests to determine the availability and quality of the essential natural elements. From this point you can determine what kind of human intervention will be needed to create the ideal growing conditions. I will briefly discuss the specific things you should look for to ensure a flourishing urban garden.

Water. Most plants depend on water; therefore, locating a reliable water source is essential. First, observe or research the average level of rainfall, the seasonal temperature changes, and the lengths of seasons. To what degree will the site need irrigation? There are many easy and cost-effective ways to ensure that your garden will be well watered. Can the water be collected from rain, or does it need to be pumped in from an outside water system? Look into the most efficient uses of water, such as drip hoses and water collection systems. On the other hand, too much water can be a detriment to the garden as well. If in your initial evaluation you note flooding or water collecting on the surface of the soil, you may need to consider drainage systems or abandoning the site in favor of a dryer area.

Soil. As the primary medium through which nutrients and water are transferred to the plants, healthy soil is imperative. Your site evaluation should include thorough soil tests to determine the availability of necessary micronutrients. Various organic materials can be purchased to amend soil lacking requisite components. Additionally, in urban areas tests should be conducted not only to determine what is missing but also to ascertain what the city environment has added to the soil. It is quite common that soil in vacant lots is contaminated with lead, arsenic, or other toxic chemicals that could be dangerous if food plants are being grown on the site. A polluted site should not prevent the project. There are many ways to address soil contamination, including the removal of chemicals (often very costly and difficult to accomplish), the construction of above ground beds, container gardening, hydroponic gardening, and so forth.

Light. Depending on the type of garden you are trying to establish, the amount of available light is another critical ingredient. City environments often make light a limiting factor, since tall buildings and trees in small spaces may create enough shade to overwhelm a garden. Considering the amount of light on a particular site will help determine the design of the garden and what plants you will choose to grow.

Other questions you should ask yourself to aid in the site evaluation are How much foot traffic does this area get? Is the site situated near a road? Are animals a threat to the plant growth? Is the site manageable for the given resources? Having completed the site evaluation, you are now ready to move on to building the garden.

USEFUL TECHNIQUES FOR THE URBAN GARDENER

There are an endless number of tools you can use to implement your garden, many of which are very simple to imagine and use. While various technologies have been designed to keep a garden adequately fertilized, watered, and free of pests and weeds, the most useful tools are by and large those that have been used for years. The most viable tool is the imagination: Old tires can be used for planters, barrels can be used for water collection, old sinks can be rigged as birdbaths and placed in the midst of a flower garden designed to attract butterflies. Whatever you are able to imagine can be created. If you are having trouble envisioning the site, it is always useful to look around and observe how other people in the vicinity are gardening. Often what works well for one person will be successful for someone else in the same area. There many ways to accomplish one gardening task. See which one works best for you.

Having made my plea for creative thinking, I would make two general suggestions to anyone initiating an urban gardening project. The first is build raised beds. If the site is sloped or very large, this might not be feasible. Raised beds are a good way to use space effectively and differentiate between plots (you may want to create plots for individual gardeners or families). Most important, raised beds create adequate space for *roots* to grow deep down, while promoting water drainage and greater accessibility to gardeners. Additionally raised beds allow soil to warm up faster and thus create the conditions for a longer planting season. They can be built out of any material and can be any length, width, or height, although the greater the width, the harder it may be to reach things grown in the middle of the bed. Raised beds also alleviate the problem with soil pollution because they create space without requiring that plants be planted directly in the ground.

The following formula calculates how much soil is needed to fill a raised bed.

Length (ft) × width (ft) × depth (ft) = cubic feet

Number of cubic feet divided by 27 = Number of cubic yards

(soil is most often sold in cubic yards)

(One standard trashcan is equal to 135–162 cubic feet or 5–6 cubic yards.)

The second useful technique I would highlight is compost. Compost is an easy solution to two problems that all gardens face: the maintenance of soil fertility and structure and the disposal of organic waste materials. Compost is a system that recycles plant nutrients by converting waste into a fertilizer in which all nutrients are readily available to growing plants. Establishing a compost pile is quite simple and can be done in a variety of ways. Almost any organic material can be added: weeds, leaves, lawn clippings, fruit and vegetable garbage, sawdust, and shredded paper. For the process to work most effectively, it is advisable to keep the pile moist, to turn it frequently, and to keep it in partial shade. The type of material composted, the frequency of turning, and the season of the year all determine how long it will take for the pile to be converted into useable soil. Composting is both environmentally and economically efficient and should be done wherever possible.

Other things to think about: Do you want fencing around your garden (this can be useful in deterring theft or vandalism but may also deter people from coming into the garden at all)? Do you have the most basic tools, including shovels, hoes, and rakes? How can you maximize the space while keeping in mind light and shade, the water source, and the accessibility of the plots? What will you put in the nonplanted areas of the gardens (such as pathways or any other areas that are not being used)? The most commonly used materials are grass, wood chips, ground cover, or mulch.

As you can see, there are many variables to consider in starting a city garden. Fortunately, for as many factors as there are to consider, there are ways of accomplishing the tasks at hand. Don't expect instant victory; gardens take a few seasons to become well established. And if you do not succeed at first, keep trying—imagination and the ability to problem solve are the most important skills a gardener can have.

REFERENCES

Hynes, Patricia H. (1996). *A Patch of Eden: America's Inner-City Gardener.* VT: Chelsea Green Publishing Company.

American Community Gardening Association, 100 N. 20th Street, 5th Floor, Philadelphia, PA 19103-1495. Tel.: 215-988-4785. www.communitygarden.org.

www.cityfarmer.org

www. gardenweb.com/forums/commgard

Minnesota State Horticultural Society, Minnesota Green, 1970 Folwell Avenue #161, St. Paul, Minnesota 55108.

SUSTAINABLE AGRICULTURE

Zephirin Ndikumana

INTRODUCTION

From the hunter-gatherer communities to the modern societies, the environment has undergone various changes as people relied on it to provide for their own needs. As hunter-gatherers settled and started domesticating plants and animals, their human nature changed to greed and excessive hunger for wealth. Unlike primitive societies, whose relationship with nature was sustainable, current generations have exploited nature to unsustainable levels. The increase in population, coupled with the diminishing land fertility, have resulted in a decline in per capita food production. This state of affairs has led to continued desertification, land degradation, soil erosion, and environmental pollution. In Zimbabwe, land degradation and soil fertility decline have resulted in an increased utilization of less productive lands leading to a wide-scale scramble for better and fertile soils. This struggle has raised a lot of concern from the government and certain nongovernmental organizations. While the government has advocated the land acquisition and redistribution exercise to solve the problem, some nongovernmental organizations have approached it from a different angle. They have initiated a process of developing approaches and technologies that promote the earth's self-renewal. Emphasis is on the development and management of natural resources to regain or maintain their productivity. The concept is termed *sustainable agriculture*.

ZIMBABWE'S SOCIOECONOMIC PROFILE

Zimbabwe extends on 390,624 km^2 of land area with an 11,002,000 population (1992 census). The population is 51.2 percent female and 48.8 percent male with a growth rate of 3.2 percent per annum. The life expectancy that was estimated at 54 years in 1995 is now pegged at 38 years due to the HIV/Aids epidemic. Zimbabwe is a low-income country with a gross national product (GNP) of U.S.\$490. The major economic sectors are agriculture (13 percent of GNP), manufacturing (25 percent of GNP), forestry, mining, services, and tourism. There is an imbalance in land resource ownership in Zimbabwe. The majority of black Zimbabweans, who practice communal farming, depend on depleted soils, while their white counterparts practice commercial farming on very fertile lands. The economy is relatively diversified with a good infrastructure and strong financial and manufacturing sectors. The main exports include tobacco, gold, nickel, cotton, clothing, and textiles. Imported products include machinery, transport equipment, chemicals, and fuel. Zimbabwe is highly indebted with an external debt exceeding U.S. \$4.368m in 1994.

THE PRINCIPLES OF SUSTAINABLE AGRICULTURE

The concept of sustainable agriculture involves the utilization of natural resources to meet the needs of the present generation while maintaining their productivity to sustain future generations. The indicators considered for intervention may be classified as biophysical, social, and political factors. The choice of intervention technologies is influenced by the natural resource base of a community, the soil fertility status, and crop yields. Different areas show different severity levels in terms of land degradation, and therefore diversified land use management practices are needed in order to regain lost fertility and maintain fertility.

Biophysical Factors. These are natural resources available that are used as a means of production. They include soil, water, trees, and renewable resources such as manure, organic matter, and soil nutrients. Factors affecting crop yields are important because they are an indication of whether a community can produce enough food and services to satisfy its own needs without importing food-related resources from other communities.

Social and Political Factors. A community with a well-defined administrative structure is usually organized and can coordinate activities that are done collectively to enhance development and self-sufficiency. In this respect, the role played by traditional leadership is vital. Chiefs and village heads are watchdogs of traditional norms and values that in most cases protect the environment. Sustainability is promoted in instances when laws are put in place to prevent people from clearing sacred bushes, to foster conservation of community woodlots, and to reduce misuse of water sources. Social factors that are of great concern deal with community income, health, and equity issues. The ownership of resources plays an important role in development. This aspect has impacted negatively on development in most African countries, where people who live on the land have no ownership rights. People resist implementing soil and water conservation projects on land that does not belong to them.

SUGGESTED INTERVENTION TECHNOLOGIES

Organic Farming

Organic farming is an agricultural system that uses natural methods to keep the soil fertile, a process that produces healthy crops and livestock. Nutrients present in organic matter are recycled by use of compost, animal manure, and green manure. The use of artificial fertilizers and pesticides is not permitted. This practice is usually referred to as permaculture. In a permaculture garden, plants and domestic animals are placed, arranged, or linked to benefit from each other in a sustainable manner. Plants are intercropped to produce yields all year round. Unlike conventional farming that continually exploits and destroys the land, permaculture (organic farming) aims at recharging the earth with all useful elements including nutrients, organic matter, and organisms.

The following activities are carried out.

- Land preparation: Done by minimum tillage using hoes and other hand garden tools. The aim is to make the minimum possible soil and plant roots disturbances.
- Soil protection: Done by protecting the soil from direct sunlight that may disturb micro-organisms' activity. These microorganisms decompose organic matter, thereby playing a critical role in nutrient recycling, soil structure improvement, and environmental cleanup. The soil should also be stable and adequately protected from direct raindrops, which result in runoffs and soil capping. This protection is achieved by planting ground covers, mulching, and practicing intercropping.
- Soil fertility improvement: Composting and manuring are essential in an organic farming system. Stovers and organic materials of all harvested plants are heaped together to make

composts that will be applied into the fields. Animal manure should be applied, while specific green annuals are continually cultivated and ploughed in the soil to improve its structure and fertility. Nitrogen-fixing plants and multipurpose trees are planted in the garden to provide a wide range of benefits. Their symbiotic relationships with the rhyzobia give them the ability to fix the atmospheric nitrogen into the soil. They produce fodder for domestic animals together with other by-products. Using these plants and trees enables the permaculturist to take advantage of the natural interactions between garden plants and small animals.

- Liquid manure making and worm farming: Liquid manure making and worm farming are indispensable components of organic farming. The liquid manure is prepared by mixing special green garden herbs, leafy vegetables with fresh or dried animal manure, and water. These are allowed to soak and decompose to produce an organic solution. This solution, rich in nutrients, is applied to growing vegetables and fruit trees. Worms are used for quick organic matter decomposition.

- Pest control: Insects are hand picked or sprayed with an organically made solution of herbs and indigenous plants. This is a highly safe and effective mode of pest control with no harmful residual effects. It results in the production of the most nutritious vegetables.

Soil and Water Conservation

- Soil conservation: Indiscriminate cutting of trees and random grazing are number one agents of land degradation. They are the root cause of soil erosion, resulting in galley formation and poor soils. Soils should be protected from natural forces such as wind, rainfall, and bad tillage practices that render them unproductive. Soil conservation techniques encourage water infiltration in the soil or harvest water in infiltration pits and dams rather than having the water run on the land surface and carry with it the topsoil and organic matter. Soil conservation consists of protecting the soil from water runoff through mulching, planting of cover crops, and contour ridges construction. Tied ridges are also used in various farming systems. Among all these soil conservation techniques, vetiver grass is a prominent feature, since it is the most resilient and suitable grass for stabilizing the soil.

- Water harvesting techniques: These consist of harnessing water runoffs and putting it to good use. Water management is based on four principles: spread, sink, store, and protect.
  —Spreading: Water is trapped by use of swales and gabions, contour ridges, and tied ridges and is allowed to spread at the edges into infiltration pits. Trees and vetiver should be planted around the pits to prevent further runoffs that may result in formation of gullies.
  —Sinking: Water is encouraged to infiltrate the soil by preventing runoffs through application of organic matter, planting of ground cover, trees, and mulching.
  —Storing: Dams and ponds are constructed to store water runoffs for use during the dry season. The water can be used for irrigation, household needs, and consumption by livestock.
  —Protecting: The concept involves protecting water sources to avoid siltation.
    a. Boreholes: Water may be protected by building a casing around the water source.
    b. Springs and dams: Plant well-suited trees, such as water berries, reeds, and so on, around springs and dams. Fence the area or divert the water to tanks where a tap can be attached.

Community Organization and Mobilization

This is an empowering process of human development designed to stimulate active participation of local institutions, communities, and local groups. The concept aims at implementing developmental strategies by analyzing the root causes of poverty and planning a course of action to redress the situation. In sustainable agriculture, the communities involved are trained to take up challenges. They are originators of the development project.

The following activities are performed.

- Conducting training workshops for rural communities to build self-awareness and self-confidence.
- Carrying out community resource analysis in order to come up with appropriate intervention strategies that respond to the real needs of communities. Projects should be initiated by the people who will be the beneficiaries. This approach is referred to as the bottom-up approach. It is participatory, and it enhances resource ownership by community people.
- Learning appropriate ways of approaching communities in order to be an effective community worker, capable of influencing change. The top-down approach should be avoided. Decisions should not be made by a few privileged individuals and imposed on the rest of the community. It is important to involve all community members in order for them to understand that development projects belong to the community and donors are only empowering agents.
- Identifying and analyzing various community governance structures and their relevance to ecological land use management.

The development worker or extension officer should be appointed from the same community in which he or she serves. He or she will then have an advantage of communicating with communities in their own language, with an added advantage of knowing the beliefs and values that community people treasure. Values and attitudes make people behave the way they do.

CONCLUSION

The issues pertaining to sustainable agriculture are often talked about and highly debated in international workshops, donor conferences, regional summits, national strategic planning meetings, and on down to district and household levels. Although the issues may be debated by people with various perspectives, concerns of common interest relate to ecology, economy, and the lives of people who will be beneficiaries and end users. An environment that is sustainable encourages biodiversity and is free from pollution. The economic aspects ensure that system outputs, goods, and services are valuable compared to the inputs utilized to produce them. Sustainable development makes it possible for people and their descendants to have a fruitful and full life in their home place.

HAMMER MILL OF THE ELC CAR

Philip Nelson

The Hammer Mill of the Village Development Program of the Evangelical Lutheran Church in Central African Republic is a good example of an intermediate technology that is appropriate to the situation in Central African Republic. It has been produced by the Village Development Program of the church since 1989. It is of rugged construction, and the materials can be found easily nearly anywhere in the world.

The impetus for building the mill came about for two reasons. First the women were already heavily burdened with the other chores around the house especially the preparation of food, general housecleaning, and the transportation of drinking and cooking water for the family. Since these tasks were often performed after working in their fields, the 20 to 30 minutes that were necessary for the pounding of manioc into flour by the traditional mortar and pestle method were some of the more tiring minutes of the day. This required the raising and precise lowering of a 5- to 10-pound wooden pestle. Although this made for very strong arms among the women, it also became a symbol of the drudgery of the woman's work in the house. Many husbands would jokingly complain that since we had started producing these mills the women had become lazy and would no longer put out the effort to properly pound the manioc as before.

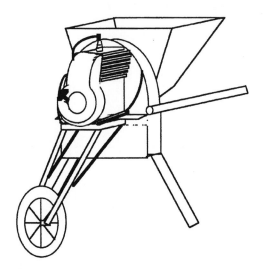

FIGURE 5
Hammer mill.

Also, even though grinding mills were seen as a business, it was a very desirable one for all involved because of the reduction of time and drudgery that resulted. Because of this, when the government-sponsored women's cooperatives were able to accumulate enough in savings, the first thing they bought were these hammer mills, which they ran as part of their money-making efforts (see Figure 5).

The second reason was that a market developed for imported mills at a price of two to three thousand dollars. This money, of course, went to the companies that produced them overseas. This resulted in the loss of scarce money that could have been employed inside the country for other things if it had stayed in circulation there. We were able to produce these mills for less than half of the imported price. This allowed people with less capital to be able to buy one and make a business out of it, passing on the savings to their clients, since it was now possible to grind in about 2 minutes the same amount of manioc that once took 20 to 30 minutes for only about 10 cents. Some good managers came back with enough money to buy a second mill in four to six months.

Most of the imported mills were the stone grinding against stone and were often so large that they needed to have a shack or some other building built with a good foundation that could withstand the vibration of the one-cylinder diesel motor that ran them. One of the most appreciated features of these hammer mills was the wheel barrow style that lent itself to easy transportation to and from the workplace, which was most often the marketplace (wherever that happened to be that day).

The materials used in the hammer mill include $4 \times 40 \times 40$ mm and $4 \times 30 \times 30$ mm angle iron, 10 mm, 20 mm, 30 mm sheet steel, 10 mm and 6 mm reinforcing rod, 25×19 black pipe, 4×40 mm flat bar, 10 mm, 8 mm, 6 mm threaded rod with the necessary burrs that go with them, a few miscellaneous bolts, paint, and a wheel barrow wheel. The tools necessary for construction are a welder (this would have to be with a generator if there is no electricity in the locale of construction), a side grinder with cutting wheels, a drill or drill press with bits, hacksaw, files, wrenches, a tape measure, and a system for balancing the hammer (a knife edge will work as the simplest).

The only part that is imported is the motor. Its cost is the deciding factor in the decision to buy or construct one. In the Central African Republic (CAR) it costs about $700 to $900 U.S., which is 2.5 to 3 times the price of the mill itself. The motors available in CAR are Honda, Yamaha, Kubota, or Robin 3- to 5-horsepower motors that were made to power water pumps. They are very well adapted to the hammer mill because the hammer can be bolted directly on to the crankshaft of the motor. This is the secret of the simplicity and safety, since there are no moving parts outside the milling chamber.

The hammer is the only part that must be very carefully made, since it must be balanced so as not to cause vibration when the motor gets up to speed. We made it from the 4×40 mm flat bar and welded a 50 mm piece on each end so that from the edge, it looked like an "I" that was attached to the shaft of the motor in the middle. The hammer ends swing within 15 mm of the sieve to make sure it stays clear even if the manioc or grain is not totally dry.

The sieve can be made by drilling 2 mm holes every 5 or 6 mm on a strip of 1 mm sheet steel 7 cm wide by 55 cm long. Approximately 3 cm on the ends of these strip can be left without holes so it can fastened below the spinning hammer.

Figure 6 shows the two most common methods of attaching the hammer to the shaft of the motor. The one on the left is used when the shaft is keyed with a threaded hole in the end. A short piece of pipe that fits over the shaft is fitted with a key and lightly welded to the hammer while it is bolted to the shaft. The hammer is then removed from the shaft, and the short piece of pipe is welded solidly in place.

The second method (on the right) is the one used on the motors that have been adapted for use with pumps. The end of the shaft is usually threaded already and need

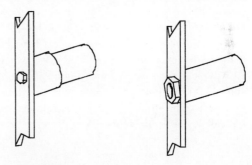

FIGURE 6
Fastening hammer to shaft.

only a slot ground into one side to serve as a keyway. A small amount of weld is added to the hammer and ground down to fit the keyway that was ground into the shaft.

There is a 12×12 hole at the center of the back that leads from the hopper to the milling chamber, and this is covered with a sliding door that can be opened for access to the end of the motor shaft so that the hammer can be fastened or unfastened from there. The flow of grain or manioc from the hopper is controlled by this sliding door that is held onto the housing of the mill by two strips of sheet steel bent to accommodate the door and welded in place. The dimensions of the mill are not so very critical—just so everything fits together. This is perhaps the area where there is the most room for improvement, but if the mill is built in this manner, it allows for simplicity of design and imprecision of the tools in the shop. Another drawback is that spare parts cannot be made to a standard and must be made to fit the particular machine in question.

The piece that most often needed replacement was the hammer, and this could usually be replaced if the old one was sent in to show the particularities of that machine. The motors not being one standard brand and model also produced some differences between different machines, especially in their mounting and their placement in relation to the center hole of the mill. This produced the problem that all mills had to be made for the motor that was to be placed on it.

In the eight years that it has been in production, the Village Development Program produced some 800 of these mills, even though there was a very large dropoff in orders when the CFA was cut in half against the French franc in 1994. This control of the economy from the outside was very hard on manufacturing in the CFA countries in Africa.

Besides the more obvious benefits of the hammer mills, there was also a change in the status of the Lutheran Church in CAR. Before their production the Lutheran Church was a little-known church in the western quarter of the country. Afterwards it was famous as "the church where they made those hammer mills."

OIL PRESS

Based on "Zopper + 30 Users' Manual
Using and Maintaining Your Ram Press"
By Jonathan Herz, revised by Munyaradzi Mundava, 1998
Zimbabwe Oil Press Project, Enterprise Works, Harare

Oil presses are used to extract vegetable oil. Usually sunflower seeds are used but groundnuts, sesame, and shredded coconut (copra) have been pressed successfully. The end product from the press is vegetable oil plus seed cake, which is an animal feed.

A typical press is shown in Figure 7. This press, called a Ram Press, generates high pressure—2,000 pounds per square inch—so it is difficult to make by a village artisan without precision tools. Simpler presses based on automobile tire jacks have been successfully fabricated without sophisticated tools.

PREPARING THE SEED

The ideal seed has high oil content and a soft shell. Dry seeds work best because moist seeds may get moldy and clog the cage. The seed should be clean, since dirt and dust will clog the cage. Cleaning can be done by winnowing. Warm seed will yield the most oil for the least effort. Spreading the sunflower seed and having it lie in the sun for an hour will warm it sufficiently. Sesame seed needs only 30 minutes. Groundnuts should be heated in a double boiler for 15 minutes. Hard sunflower seeds can be softened by mixing a liter of oil with each 10 kg of seed.

USING A RAM PRESS

Some experimentation is necessary to get the optimum setting of the cone. If the cone is set too tight, the press may be damaged. If it is set too loose, more oil will remain in the cake than necessary. When starting each day the first kilogram of seed will probably not be well pressed, so it should be returned to the hopper for repressing.

A good procedure is to have two operators working with the press at the same time. They can change jobs every half hour or so—one doing the pressing and the other heating the seed, loading the hopper, removing the oil and cake, and cleaning up. For sunflower seeds use 15 strokes a minute, for sesame seeds use 4 strokes a minute, and for groundnuts 3 strokes a minute. Expect about 12 liters of oil from one bag, 50 kg, of sunflower seed. Pressing the seed in one

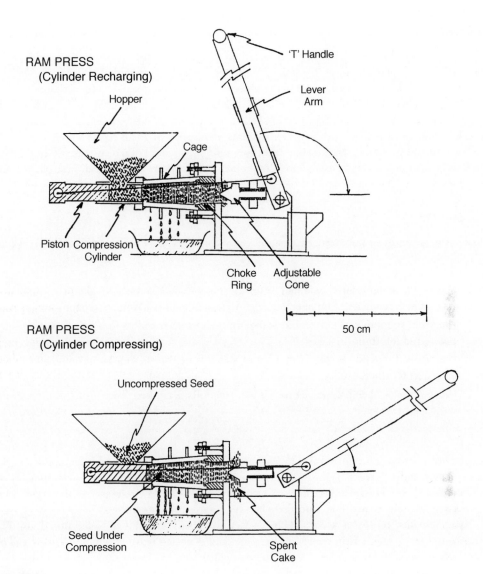

FIGURE 7
Ram press.

bag should take about 7 hours. A bag of sesame or groundnuts takes about twice as long to press and yields about 50 percent more oil.

Filtering

The oil should be filtered through a cloth filter. The filter works best when it is covered with the black sediment in the oil, so when starting with a new filter cloth, use oil directly from the press. Once the filter is covered with black particles, it is best to let the oil settle for a day before filtering. Of course, do not filter the sediment at the bottom of the settling container. Clean the filter with hot water. Do not use soap because it can leave a residue that can be tasted in the oil.

Storage

Oil should be stored in a glass or plastic container—metal can cause the oil to go rancid. Never use pesticide or fuel containers—the risk of improper cleaning is too high. Make sure the storage containers are thoroughly dry. Oil stored in tightly closed containers will last for six months or so.

Seed cake should be kept dry before being used. Moist cake will mold, and wet cake will rot. Moldy cake is unpalatable to animals and can make them sick. Mold on groundnuts can produce aflatoxin, which is carcinogenic. Rodents and insects will eat seed cake.

Cows should be given 2 kg of cake daily, no more. Pigs can be fed a mixture of equal parts cake and cereal grain. Chickens should get one part cake to three parts cereal. Free-range adult chickens should get a small handful of cake each day.

Coconut Oil

Coconut palm oil is an important product in many countries. It is produced from the coconuts as follows.

1. The shells or hulls are separated from the nut or seed kernels to obtain a mass with maximum oil content. Palm nuts from the African oil palm or American palms are cracked.
2. The oil-containing kernel material is then milled between rollers to obtain a well-crushed material in the form of flakes.
3. The crushed mass is "cooked" in a set of steam-heated pans in a humid atmosphere and subsequently dried.
4. The dry mass is then pressed, a process that generally is applied twice—that is, prepressing and deep pressing.
5. Finally, the oil is filtered.

A common press for coconut oil is a hand-operated screw press. The mass to be pressed is put in a cylinder, and a piston is screwed down on it. Hand presses may have long arms attached to the threaded shaft so two people can exert torque by walking in a circle around the press. Such a press can be built locally—the threaded shaft may have to be purchased.

Two alternative processes are available. In wet processing the fruit is boiled in a drum. The oil rises to the top and is skimmed off. Often the fruit is grated before cooking to make the oil more accessible. The yield is low, but the equipment is simple.

The second alternative uses motorized expellers. Expellers, or continuous screw presses, are used for the expression of oil from copra but also for palm kernels, peanuts, cottonseed, flaxseed, and almost every other variety of seed—wherever there is a large enough seed supply to justify a continuous operation. Expellers achieve the pressure needed to express the oil by means of an auger that turns inside a barrel. The barrel is closed, except for an opening through which the oil drains. An expeller can exert much greater pressure on the seed cake than a batch press can. This increased pressure permits the recovery of a larger proportion of the oil. Generally, about 3 to 4 percent of the oil is left in the cake with an expeller, compared to 6 to 7 percent with a manual press.

REFERENCE

Wiemer, Hans-Jürgen, and Frans Willem Korthals Altes. (1989). *Small Scale Processing of Oil Fruits and Oilseeds.* Deutsches Zentrum fur Entwicklungstechnologien—GATE in Deutsche Gesellschaft fur Technische Zusammenarbeit (GTZ) GmbH.

VEGETABLE GARDENING

Tuarira Abduel Mtaita

A common prescription for improving nutrition, especially among children, is to promote the addition to people's diet of vegetables grown in home gardens. Vegetables form an indispensable part of the human diet. Above all, they are an important source of vitamins and minerals. About 100 g of fresh vegetables are necessary for the supply of the daily minimum requirements of vitamins and minerals. Many vegetables contain protein as well. This is of great importance, especially in the tropics where the diet is often poor in proteins. If consumed in fairly large quantities (up to 500 g a day), vegetables can furnish a large part of our needed proteins.

Producing one's own vegetables is an interesting and rewarding hobby. Not only are there economic rewards, but also one can enjoy the test and quality of fresh vegetables, rarely available otherwise, and grow various kinds of vegetables seldom marketed commercially.

Setting up vegetable gardens and marketing the produce is not as easy as it sounds. Indeed, if it were simple, vegetable gardening would be more common, since people generally like to vary their diet, and vegetables can add variety.

Planning is necessary to get the maximum production and enjoyment from vegetable gardens. Decisions must be made on kinds and varieties of vegetables to grow, how much of each to plant, when they should be planted, and how they should be placed in the available space.

Climate influences the kinds of vegetables grown and when they may be planted. The number of days from planting to maturity, as usually suggested in seed catalogs, can be used as a general guide for planning, although the actual growing times will vary.

GROWING CONDITIONS

There is no simple formula for selecting a site for vegetable production. Physical factors such as climate and soil type play a major role in determining which species can be grown and the potential productivity of a particular site. Social and economic factors can modify decisions on site choice. The desire to be self-sufficient in a particular crop, for example, can have a bearing on site selection. This, in part, may override otherwise unsuitable sites.

The cropping possibilities of a number of vegetables under various climatic conditions are indicated in Table 1.

RAISING VEGETABLE SEEDLINGS

Whatever method is used, vegetable seedlings should be physically sound—that is to say free of diseases and undamaged by insects. They should also have abundance of flower buds, which for fruit vegetables will assure a good and sufficient harvest. Because most vegetables are transplanted from the seedbed nursery to the crop-growing field, they should suffer as little damage as possible during the actual transplantation process. Good seedlings to be used for transplanting must be large and strong enough to set well in the field. They must have adequate height, thick stems, good leaves, and a good root system. Good seedlings should also have low top weight in relation to their root-weight ratio.

TABLE 1

Cropping of vegetables under various climatic conditions

Crops	Humid Tropics 28–35	Hot Season 30–40	Cool Season 20–30	Mountainous (>700 m) 15–28	Summer–Winter 30–40	Summer–Winter 15–25	Summer 15–25
Greens							
Chinese cabbage (*Brassica chinensis*)	+	+ +	+ +	+ +	+ +	+	+
White cabbage (*Brassica oleracea*)	–	–	+	+ +	–	+ +	+ +
Lettuce (*Lactuca sativa*)	–	–	+	+ +	–	+ +	++
Spinach (*Spinacia oleracea*)	– –	– –	– –	–	–	+	+ +
Fruits, tubers, roots, etc.							
Onion (*Allium cepa*)	–	– –	+	+	+	+ +	+ +
Sweet pepper (*Capsicum sp.*)	+ +	+ +	+ +	+	+ +	+	–
Cucumber (*Cucumis sativus*)	+	+	+ +	+ +	+ +	+	+
Carrot (*Daucus carota*)	– –	–	+	++	–	+ +	+ +
Fresh beans (*Phaseolus vulgaris*)	–	–	+	+ +	–	+ +	++
Tomato (*Lycopersicon esculantum*)	+	+	+ +	+ +	+	+	+
Eggplant (*Solanum melongena*)	+ +	+ +	+ +	+ +	+ +	–	–
Potato (*Solanum tuberosum*)	– –	– –	+	+	– –	+ +	+ +
Legumes							
Beans (dry) (*Phaseolus vulgaris*)		–	+ +	+ +	– –	+ +	+ +
Peas (*Pisum sativum*)	– –	– –	–	+	– –	+	+ +

Key: + + = suitable; + = fairly suitable; – = hardly suitable; – – = unsuitable.
Source: Grubben (1975).

Several methods for raising seedlings have been developed through a "devoted farmers technology." These include the use of seed boxes, nursery beds, pots, soil blocks, hydroponics, carbonized husk, and grafting to raise seedlings.

NURSERY SOIL

The bed soil must be good in order to raise strong seedlings that can withstand undesirable conditions such as drought, low temperature, high temperature, and little sunlight. Good bed soil should satisfy the following requirements.

1. The soil should have sufficient fertilizer/nutrients. Mix soil with organic fertilizers such as compost or stable manure from the previous season.

2. The soil should be free from pests. Use soils from paddy fields or noncultivated areas or subsoil that is relatively uncontaminated. If soils seem contaminated by damping off, fusarium, or nematodes, use resistant cultivators, sterilize the soil by chloropicrin, methylbromide, or steam (80°C).

3. The soil should have adequate soil pH. Apply slaked lime to adjust soil pH to 5.5–6.0 if soil has low acidity.

4. The soil should have good aeration, moisture, and drainage. Use adequate compost and mulching.

PLANTING

Correct depth of planting and crop spacing can be achieved with hand planting in well-prepared, previously worked land, but rapid work rates are difficult to maintain. Plant populations and spacing have a great influence on yield. Growers should reduce plant populations from the maximum so that crops may be grown successfully in parts of the world with water shortages, poor soil fertility, or limited fertilizer use. Market requirements also influence the density at which vegetables are grown, since population affects individual plant size and the time taken to reach maturity. Amount of seed required can be estimated using the following formula.

$$\text{Weight of seed required (kg/ha)} = \frac{\text{Number of plants required per m}^2 \times 1{,}000}{\text{Number of seed per g} \times \% \text{ laboratory germination} \times \text{field factor}}$$

ROTATION

The aim of crop rotation is to retain soil fertility and prevent the buildup of disease, insect infestation, and weeds, which will lead to uneconomic production. Crop rotation is not really complex on large farms and for long-term crops. The smaller the available soil area and the more short-term crops (vegetables that are cultivated in three months or less) that are cultivated, the more complicated crop rotation becomes.

The procedure of a crop rotation in cases where several vegetable crops are to be grown in a small area of approximately 1 hectare (1 ha = 2.47 acres) may be as follows.

1. Divide the soil in blocks to accommodate all the vegetables.
2. See that related crops do not follow each other in a specific block.
3. Work in four-year cycles because most of the disease-causing organisms disintegrate in the soil after a few years, four years at a maximum.
4. Keep in mind which crops should be grown in summer or winter and what is the total growth period of each crop—that is, how long a specific crop is going to occupy the land.
5. Make use of green manure plants.
6. Keep records of rotation details.

A typical example of a vegetable rotation system is shown in Table 2.

TABLE 2
Crop rotation system over four years for 1 ha of land

Year	Season	Block Number			
		1	2	3	4
2000	Summer	Green beans	Tomatoes	Green maize	Sweet potatoes
	Winter	Carrots	Green manure	Beetroot	Cabbage
2001	Summer	Sweet peppers	Green maize	Sweet potatoes	Green beans
	Winter	Green manure	Beetroot	Cauliflower	Carrots
2002	Summer	Green manure	Sweet potatoes	Green beans	Tomatoes
	Winter	Beetroot	Cabbage	Carrots	Green manure
2003	Summer	Sweet potatoes	Green beans	Sweet peppers	Green maize
	Winter	Cauliflower	Carrots	Green manure	Beetroot

FERTILIZATION

Vegetable growers should maintain good soil fertility to attain good yield. Good soil is not only rich in the basic nutrients; it is also physically well structured and biologically active. When plants get minerals, they become robust and grow well. Ideally, organic matter should be added and returned to the soil. The amount of humus in the soil decreases through mineralization; thus, resupplying lost humus every year is a must for maintaining soil fertility and good quality. Approximately 37 tonnes (metric tons) of humus per hectare are necessary for this purpose. Organic matter can be added to the soil using liquid manure, mulching, farmyard manure, green manure, improved fallow in the rotation compost, permanent trees, and grasses.

IRRIGATION

The supply of water forms a basic condition for plant production. In order to ensure optimum production, an adequate supply should be provided. The type of irrigation needed is determined by the type of vegetable grown, the length of the growing season, the stage of growth of the vegetable, and the climatic condition during that season.

WEED CONTROL

Successful weed control is a challenge extended to one's personality as well as talents. Prevention of weed infestation is more important than the control of weeds that have established themselves in the field. The following are some means of preventing infestations.

1. Establish and comply with strict quarantine rules.
2. Use seed-lots that are free of weed seeds.
3. Cut weeds before they flower and before their seeds ripen.
4. Burn threshing residues and chaff.
5. Clean farm tools that may have been infested previously.

To control weeds, a gardener may use any of the following methods.

1. Mechanical and cultural control: Hoeing, tilling, flooding, burning, or smoothening.
2. Chemical control: Chemicals are generally used to control weeds in combination with tillage operations and hand weeding.
3. Biological control: Parasites, predators, or pathogens may be used to maintain a weed population density at a lower average than it would occur in the absence of these agents. This employs the natural enemies of weed plants to control weeds.

MATURATION AND MATURITY INDIXES

The degree of maturity at harvest has an important bearing on the way vegetables are harvested, handled, transported, and marketed, and on their storage life and ultimate quality. For many vegetables, the optimum eating quality is reached before full physiological maturity (before they are fully developed). These vegetables may include leafy vegetables and imma-ture fruits (cucumbers, sweet corn, green beans, peas, etc.). For these vegetables, the problem is delayed harvest, which results in overmaturity and consequently lower quality. Fruits and vegetables consumed when ripe attain their best quality when ripened on the plant. Maturity indices for selected vegetables that have been proposed or are presently in use are shown in Table 3.

TABLE 3
Maturity indices for selected vegetables

Index	Example
Harvest units during fruit development	Peas, sweet corn
Development of abscission layer	Muskmelons
Drying of foliage	Potatoes
Drying of tops	Garlic, onions
Surface morphology and structure	Muskmelon skin netting Tomatoes-cuticle development
Size	Most vegetables
Color (external)	Most vegetables
Specific gravity	Watermelons, potatoes
Shape, compactness	Cauliflower, broccoli
Internal color structure	Tomatoes
Tenderness	Peas
Toughness	Asparagus
Solidity	Cabbage, lettuce, Brussels sprouts

Source: Swiader et al., 1992.

CHALLENGES

The most important challenges to a smallholder vegetable producer include the following.

1. Vagaries of weather: Because many vegetable crops have rather specific climatic requirements, garden vegetables should be selected to suit different ecological zones. In vegetable gardening it often pays to modify the climate: wind screens to reduce wind speed and evaporation, shade to reduce light intensity and temperature, plastic covers to increase humidity for propagation, mulches to maintain more equitable soil moisture and soil temperature regimes.

2. Shortage of good quality seed: Production of seed should be done locally rather than relying on imports. Seed production first requires development and selection of appropriate cultivars under local conditions. Progressive farmers should be identified and trained in vegetable seed production so that they might serve as effective and reliable sources of high-quality seed/seedlings for other growers.

3. Lack of knowledge of the potential value of production technologies. There is a need for gathering information on what has already been done on vegetable production techniques and disseminate information to vegetable growers. Necessary areas of research may be identified and carried out. Much effort should be directed toward breeding improved varieties, optimizing fertilizer and crop water requirements, better pest control, and aspects of good farm management. There is also a need to educate the public on the value of vegetables and to promote their consumption

4. Inadequate transport and poor market intelligence. Rural road networks need to be improved. It is important to have an exchange of information between the farmers and the market personnel that will enable supply and demand to be understood by all. Mechanisms should be put in place to ensure that farmers have knowledge on the potential outlets for their vegetables. Potential buyers should be informed as to what is available, where, and when. The three links in the chain—production, delivery, and sale—must all be given due attention and developed simultaneously.

5. Limited access to reliable sources of water for irrigation. Water harvesting techniques should be promoted. Dams need to be constructed, and water catchment areas should be conserved. Simple water harvesting methods from rocks and roofs have to be studied and promoted.

6. Lack of simple technologies for postharvest handling methods. In developing countries, vegetable production is seasonal and characterized by periods of too much and too little. Wastage is high during the peak production season, and consumer prices are exorbitant during the off-season despite the poor quality of the produce. If proper postharvest handling procedures and processing were introduced and practiced, year-round availability of better quality vegetables will exist at affordable prices. Serious attention must be given to constraints relating to the transportation, storage, and processing and marketing needs. Good postharvest handling is crucial and must be addressed together with the other bottlenecks hampering increased vegetable production.

REFERENCES

Fordham, R., and A. G. Biggs. (1985). *Principles of Vegetable Crop Production*. London, Collins.

FAO. (1988). *Vegetable Production Under Arid and Semi-Arid Conditions in Tropical Africa*. FAO, Rome.

Swader, J. M., W. W. George, and J. P. Mccollum. (1992). *Producing Vegetable Crops*. Danville, IL: Interstate Publishers, Inc.

INTENSIVE GARDENING

"Intensive gardening" usually refers to gardening intended to grow a lot in a small area. It is most common in urban or suburban noncommercial vegetable gardens, but the techniques can be applied in other situations. The system described here is based on "French intensive" gardening, used in Paris suburbs toward the end of the nineteenth century, and "biodynamic" gardening, used later throughout Europe. The central idea is to use raised beds four or five feet wide and plant intensively so all the ground is covered. The width chosen makes tending the plot easier. The raised bed, only six inches or so above the ground, catches the sunlight effectively, and the height clearly demarcates the boundary. If all the ground is covered, the sunlight is captured effectively, and weeds have difficulty growing.

Beds are prepared by digging deeply, twelve inches or so, but not below the topsoil. Compost, described below, is added to the soil and mixed in. A thin layer of compost can also be put around the sides of the beds to keep the soil in place.

Compost is decayed vegetable matter. It can be made in many ways. A pile of leaves will turn into compost within a year. The process is hastened by turning the pile over every month or so. Table scraps from the kitchen and other vegetable wastes can be added. Manure is a good source of nutrients and speeds the decay of the other material. Manure can be spread directly on the garden, but raw manure is not good for a garden, since it takes nutrients out as it decays. One should let the manure rot, either in the compost pile or by itself, before putting it on the garden. The effluent from a methane digester described in the Energy Chapter is another good fertilizer.

The best seeds can be found by experimenting, asking people with similar soil and farming objectives, or by contacting (in the United States) the local Cooperative Extension Service. Some varieties are more pest and disease resistant than others. Planting several varieties of the same vegetable also lowers the risk of disease spread. One can save seeds from successful plants from the previous year, but offspring of hybrid seeds may not grow back to be the same as the parents.

"Companion planting," also call "intercropping," is a sensible option. Several different vegetables that will not compete are planted in the same plot. One standard combination is corn, beans, and squash. The squash hugs the ground, the corn grows vertically, and the beans fill up the intermediate space. Some combinations reduce pests. Marigolds in particular are protectors of many plants. Garden books, such as *Amaranth to Zai Holes*, will give advice on which plant combinations are optimum. In Africa, where too much sun can be a problem, corn is planted among trees. The UNICEF handbook also suggests planting trees with vegetables. Legumes, such as beans and peas, add nitrogen compounds to the soil, and these compounds help nourish other plants.

REFERENCES

Maingay, Hilde. (1977). "Intensive Vegetable Production." *The Journal of the New Alchemists*, No. 4. The New Alchemy Institute. Available from The Green Center, 237 Hatchville Road, East Falmouth, MA 02536. http://www.fuzzylu.com/greencenter/.

Meitzner, Laura S., and Martin L. Price. (1996). *Amaranth to Zai Holes: Ideas for Growing Food Under Difficult Conditions*. ECHO, North Fort Myers, Florida, USA. ECHO@xc.org.

UNICEF. (1982). *The UNICEF Home Gardens Handbook*. United Nations, NY.

SOIL FERTILITY AND FERTILIZER USE

Fanuel Tagwira

Soil fertility management is very important for sustainable food production and maintenance of the environment. Fertilizer use and improved varieties were some of the major ingredients of the green revolution in Europe, Latin America, and Asia. Africa is the remaining region in the world with decreasing food production per capita. The worst levels of poverty and malnutrition in the world exist in this region (Sanchez et al. 1997). A total of 660 kg N/ha, 75 kg P/ha and 450 kg K/ha is said to have been lost in the last 30 years in sub-Saharan Africa. These figures represent the difference between nutrient input and nutrient output (Stoorvogel and Smaling 1990, Sanchez et al. 1997 and Tagwira 1998). Over the last 30 years, Europe had a net positive nutrient balance of 2000 kg N/ha, 700 kg P/ha and 1000 kg K/ha (Frissel, 1978, Sanchez, 1994), while Africa lost 4.4 million metric tons of N, 0.5 million tons of P, and 3 million metric tons of K every year from cultivated lands, with a fertilizer input of 0.8 million metric tons of N, 0.26 million metric tons of P and 0.2 million metric tons of K (FAO, 1995, Sanchez et al. 1997). Africa is therefore experiencing a nutrient depletion. The results of a recent study (Chibudu, 2001) showed a serious soil fertility decline in Zimbabwe. The difference between nutrient input and nutrient output was greater in fields further away from the farmer's homestead (Table 4). The number of abandoned smallholder farming areas in Zimbabwe illustrates the extent of the problem. After working a piece of land for a number of years farmers discover that the productivity level of the land is sub-optimal and decide to abandon the land.

Soil fertility surveys in Zimbabwe have shown that only 20% of the soils in smallholder farming areas have adequate P and only 15% are in the suitable pH, while 70% of the soils are strongly acidic.

It is generally believed that the breakdown of traditional soil fertility maintenance strategies, (fallow land, intercropping cereals and legumes, mixed crop livestock farming, opening of new land) have been part of the cause. Population pressure has reduced farm sizes. Poor roads, market infrastructure, low producer prizes, and ineffective extension have affected yields. Fields that used to yield 2–4 t/ha now yield 1 t/ha. In Kenya they found out that one red soil had lost 1 t/ha of soil organic matter N, 100 kg P/ha of organic P over 18 years when it was put to maize bean rotation with no inputs. Maize yields decreased from 3 t/ha to 1 t/ha (Qureshi 1991, Swift et al. 1994, Kapkiyai 1996, Sanchez 1997, and Bekunda et al. 1997). Nutrient losses are even greater in sandy soils.

FERTILIZER USE

Fertilizer use is able to bring large-yield returns for the farmer compared to other management practices and therefore is the obvious way to improve soil fertility and productivity of land as has happened in Europe, Asia, and Latin America. Despite this important fact, fertilizer use has

TABLE 4
Nutrient balance (kg ha^{-1}yr^{-1}) of N and P in fields near the homestead (homefields) and fields away from the homestead (outfields)

Nutrient Element	Niche	Input	Output	Nutrient Balance
N	Homefields	45.5	50.7	5.2
N	Outfields	16.4	51.7	35.3
P	Homefields	7.8	5.5	2.3
P	Outfields	4.6	7.9	3.3

(Chibudu, 2001).

not increased in many developing countries, particularly in Africa, for several reasons, which include cost of fertilizer, lack of credit, delays in delivery of inputs, and low and variable returns, particularly on dry land. Those farmers who use fertilizers in some cases are not able to apply it at recommended rates because of lack of the necessary financial resources. The price of fertilizer in Africa is twice international price (Bumb & Baanante 1996). Transport costs seven times what it costs in Europe and America. In the last 25 years the international cost of fertilizers decreased by 38 percent for N and 50 percent for P (Donovan 1996), but in Africa fertilizer increases continuously. Maintenance of soil fertility requires us to return nutrients we have removed from it due to harvests, runoff, erosion, leaching, and other losses. Currently we are simply mining the soil like we do minerals. Fertilizer application by farmers for food crops is not profitable due to higher fertilizer costs, low crop prices, and high risk of rain-fed agriculture and lack of credit.

The most limiting nutrients are N and P. While K depletion is six times that of P, you rarely see deficiency due to the ability of soil to replenish the K.

Since water is usually a limiting factor to increased productivity, there is need to make sure that increased fertilizer application is accompanied by adequate moisture. Fertilizers are also known to improve the plant rooting depth and therefore enable plants to use moisture at greater depth. Where moisture is critical the response to fertilizer application is likely to be determined by the amount of moisture.

Fertilizer application must be combined with other improved management practices if full yield is to be obtained, particularly in rain-fed crops. Unless other conditions of growth are optimum, limiting factors like pests, diseases, and weeds tend to have major depressing effects on growth and yield. Soil and water conservation measures, optimum plant population, and effective management of pests, weeds, and diseases contribute substantially to improved fertilizer use efficiency. Tandon (1980) showed that improved management practices could give tenfold increase in yield of sorghum and increase fertilizer use efficiency 3.5 times.

NITROGEN (N)

Nitrogen applied to a crop is either taken up by plants, leached to lower levels, or lost through denitrification or erosion. The recovery of nitrogen by crops is rarely more than 50 percent. In some experiments recovery rates of 20 to 50 percent have been observed (Harmsen et al. 1983). Recovery of nitrogen added could be improved by using suitable crop cultivars and sowing dates, appropriate methods of fertilizer application, applying organic manures, and adopting adequate soil and water conservation measures.

Improved high-yielding cultivars utilize N for production of grain more than unimproved cultivars. Optimum response to fertilizers as well as overall yield levels depend on time of sowing. It has been observed that in wheat and rapeseed, for example, any deviation in time of sowing will result in reduced fertilizer use efficiency (De 1981).

Nitrogen Fertilizer

Ammonia is the source of most nitrogen fertilizer. It is the cheapest and most concentrated form of N but needs special equipment and handling techniques for its use. Ammonia is 82%N. Ammonium sulphate contains 21%N. Ammonium nitrate is 35%N, but with impurities its concentration is 34%. The most concentrated solid N is urea.

Grass, pastures, and other no-till cropping systems are fertilized without disturbing the soil and therefore need fertilizers that can be applied on the surface without volatilization. Ammonium nitrate, ammonium phosphate, and ammonium sulphate are soluble forms that resist volatilization. Anhydrous ammonia can only be used where it's injected into the soil. Urea is subject to volatilization when applied on the surface. Urea hydrolyzes and releases ammonia if left on the surface and in contact with moisture. Losses of up to 50 percent have been recorded.

$$(NH_2)_2CO + H_2O ===== 2NH_3 + CO_2$$

Rates of application of N depend on the crop being grown and the soil. Soil testing for available N is considered unsatisfactory, since N tends to leach and levels in the soil change over short periods of time. Some laboratories measure the supply capacity rather than available N. The sample is incubated for 14 days to allow mineralization of N to occur. Most laboratories, however, base N requirements on balance sheet approach instead of soil tests. They take into consideration crop history, yield level, previous fertilizer applied and type of soil. The amount of carryover N is estimated from these factors and fertilizer recommendations are made.

Timing of N Application

The time of N fertilizer application has an effect on yield of rain-fed crops. It is generally recommended to split the fertilizer application so that half is applied at sowing and the other half 20 to 45 days after sowing, depending on soil moisture availability and type of crop. The type of soil is also very important in determining the time of fertilizer application. In Zimbabwe nitrogen application to maize on a clay soil is split into application at planting and at six weeks after emergence or when the maize is knee high. In sandy soil it is recommended that fertilizer be split into three parts, with applications at planting, at six weeks after emergence, and at tasselling. Split application is very important, since nitrogen is mobile in the soils, particularly soils of low clay content, and therefore losses to leaching tend to be high. Split applications also reduce the farmer's losses in the event of a low rainfall year because he can skip the second application. In rain-fed crops, the uncontrollable water supply results in alternate wetting and drying patterns and therefore in volatilization, leaching, and denitrification losses of nitrogen. In rain-fed lowland rice, deep placement of N fertilizer is recommended as opposed to split application. De Datta (1986) observed that sulphur-coated urea broadcast and incorporated and urea supergranules deep placed at seeding time were more effective than split applied prilled urea. To obtain a 1t/ha increase in yield, 57 percent less N was needed in the form of sulphur coated urea and 62 percent less in the form of urea supergranules relative to split prilled urea.

Methods of N Fertilizer Application

Band applications of urea and ammonium nitrate have been observed to be more effective than broadcast application. The fertilizer can be left on the surface or incorporated. Experiments in vertisols have shown that the loss of N from broadcast urea was 27.7 percent when left on surface and 25.9 percent when incorporated, whereas the loss from split band application was only 5.8 percent. The comparative figures on an alfisol were 12.6, 16.8 and 10.4 percent. The recovery of nitrogen was 55.6 percent from a split band application method and 30.5 percent from surface application (Kanwar and Rego 1983). Application of nitrogen by drilling below the seed was found to result in the same yield as banding the fertilizer in Zimbabwean oxisols. Umrani and Patil (1983) observed that drilling fertilizer below the seed increased N fertilizer use efficiency in sorghum.

In some countries application of N fertilizer through foliar spray has been found to be as effective as soil application. However, normally foliar application is of relevance where the farmer has not been able to apply any fertilizer at sowing or subsequent soil N application is operationally difficult.

Organic Matter Application

Utilization of inorganic fertilizers is very low in Africa. This is due to high fertilizer costs and delayed deliveries to the farmer. Because of this organic N accounts for one-third of the N inputs. The use of organic matter is considered a cheaper alternative to inorganic fertilizer. Organic matter, however, tends to have low nutrient levels, and the quality is as good as the soil on which the organic material was growing. Manure is one form of organic matter used as an N source by farmers in Africa. The amount of manure is, however, limited and the quality is very variable, depending on the source. High-quality manure improves yield, and low-quality manure decreases yield through N immobilization. Critical threshold value is 1.25%N. In Zimbabwe the quality of manure has been observed to be as good as the pasture on which the cattle are fed. Manure from overgrazed areas tends to be very poor (Mugwira & Mukurumbira 1986). It takes soil fertility to grow the organic inputs. The use of manure and other forms of organic matter helps improve organic matter content of the soil apart from providing nutrients to the plants. The organic matter helps improve the nutrient and moisture-holding capacity of the soil, and also other physical properties of the soil, like the structure. Most soils of the tropics are, however, low in organic matter content due to high temperatures, which increase organic matter decomposition, and also because the stover left after harvest is either burned or removed from the field and fed to cattle without returning the resultant manure to that field. Where crop residues are given to cattle, the manure generated should be returned to the field to restore the fertility.

Hedge (1980) showed that incorporation of crop residues after harvest increased organic matter content from 0.55 to 0.90 percent, while yields of cowpeas and pearl millet were increased by 14 and 19 percent, respectively. When the residues were returned, fertilizer use efficiency increased by 8 to 16 percent, respectively, relative to fertilizer application without crop residues. The cycling of nutrients in the fields is one way to improve the nitrogen content of the soil. Application of crop residues with high carbon-nitrogen ratio should, however, be discouraged as this may result in nitrogen immobilization. The use of low-quality manure can also result in nitrogen immobilization. Manure as a general rule has been observed to be a good source of K and trace elements but has low levels of N and P. The integration of N and P fertilizers into manure to improve its fertilizer value has been studied extensively, and the results are mixed. In most experiments no interaction has been observed between the fertilizer and the manure.

Another means of incorporation of organic matter is the growing of leguminous green manure crops in situ with or without a legume crop and incorporating the residue into the soil.

Timing of organic matter application for rain-fed crops is very important and can influence the benefits to soil fertility and crop yields. There is need to synchronize application of crop residues and the release of nutrients to the soils in such a way that the nutrients are released at a time they can be used by the growing crop.

Raising Soil N Status Through Legumes

Herbaceous legume cover crops have been observed to produce enough N inputs through biological nitrogen fixation (BNF) to meet the needs of subsequent maize crop. The main species used include genus mucuna, Crotolaria, Dolichos, and desmodium. Improved herbaceous and woody leguminous fallows can provide excellent options for managing N biologically, provided farmers are willing to make land available for crop fallow rotations. Where high yields are expected, there may be need to supplement the N from the BNF with fertilizer N. In Chipata, Zambia farmers rely on two-year sesbania or tephrosia fallows to provide basal N, and then they use N fertilizer at topdressing time. Increasing N supply in the short and medium term should come from regular application of N inputs in form of organic inputs and or fertilizer (Giller et al. 1997).

PHOSPHORUS (P)

Supplying adequate P is very important in summer rain-fed crops because of the short growing season and because most tropical soils have low inherent P fertility. Crops have been observed to differ in the amount of available P required for their optimum growth. When P is added to P-deficient soils, growth is stimulated and anthesis and seed maturity brought forward. This is important because it helps rain-fed crops to complete their cycle before the end of the rainy season.

Many tropical and subtropical soils have low available phosphorus. P availability varies with solubility and the amount of water present and the distance the ion should move to reach the plant root that will absorb it. The solubility of P depends on pH. Liming of acid soils is very important for P availability. Optimum pH depends on type of clay, but generally optimum pH is 6.5 to 7.0. A soil pH of 6.0 and above precipitates all the iron and Al present as hydroxide. This means less P is precipitated and made unavailable. Crops generally differ in their P requirements. The overall yield potential of the crop being grown determines whether P supply will limit growth or not. P supply to deficient soils stimulates initial vegetative growth and brings forward athesis and seed maturity. This enables rain-fed crops to complete their growth cycle within the period of moisture availability. The P level in the soil can be measured through soil analysis.

Growing crops without addition of P, as happens in many African countries, results in serious P depletion. Kang and Osiname (1979) reported that after growing five crops of maize on an alfisol in Nigeria, yields without P declined from 4066 to 2050 kg/ha. Experiments carried out in many African countries show response to P application to maize, wheat, sorghum, and other crops. Typical removal of P per tonne of cereals, soybean, and groundnuts is 3.2–5.3, 7.3, and 10.3 kg/ha, respectively.

Plant dry matter at 4t/ha contains 8 to 12 kg P/ha. This is half of the P required to grow 4t/ha of maize, which accumulates 18 kg P/ha in its tissues (Sanchez 1976).

P is lost in the soil through crop removal and soil erosion (Smaling 1993). Most P in cereal crops and in grain legumes is concentrated in grain and therefore gets removed from the field at harvest. The proportion of P cycled back to soil in grain crops, assuming complete crop residue return, is in the order of 40 percent in contrast with 50 to 70 percent for N and 90 percent for K (Sanchez 1976; Sanchez & Benites 1987). In African smallholder most residues are not returned where they came from resulting in 100 percent removal. In Zimbabwe crop residue in many cases are left in the field to be grazed by animals communally. These animals are put in pens at night, thereby providing some manure to the cattle owner. Those who do not own cattle tend to lose because all their stover is eaten by their neighbor's cattle, resulting in nutrients in the manure being carried to their neighbor's fields.

Soil Analysis

P availability for plant growth can be measured through soil analysis. No soil test method is universally applied. Different methods are used on different soils. Which chemical extractant extracts the P that is close to what the plant takes is a question that has not received consensus among scientists. The resin extraction method is generally considered to extract approximately the P that will be available for plant growth, but the method is not favored because of the cost and time required to do the test. Bray No 1. extracting solution is the most widely used. Interpretation of results requires knowledge of the soil being used.

Phosphate Fertilizer

Phosphorus lost from the soil can be replaced by addition of P fertilizer. The P can be added as one large input or as many small applications. In Africa, Angola, Burkina Faso, Mali, Niger,

Senegal, and Togo have medium to highly reactive sedimentary phosphate deposits, while Tanzania and Madagascar have highly reactive biogenic phosphate rock deposits. Burundi, Congo, Kenya, South Africa, Uganda, Zambia, and Zimbabwe have igneous deposits that are not very suitable for direct application.

In soils that are phosphate fixing, one-time application is more favorable. One-time application is also favorable in clay soil. Small P applications in high P-fixing soils result in most of the P being fixed, leaving very little for the subsequent crop. The phosphate application can be made as processed fertilizer or as phosphate rock. Acid soils are more favorable for phosphate rock applications because this results in more dissolution of the rock than in alkaline soils. In some cases phosphate rock can be composted with organic matter, resulting in better P dissolution. Phosphate as processed fertilizer can be applied as compound fertilizer or as straight phosphate, as in single superphosphate, double superphosphate, and triple superphosphate.

P application usually has a measurable residual effect, and the following crop tends to benefit from the previous season's application. When P fertilizer is applied to the soil, it tends to be immobilized. The rate of immobilization depends on various soil characteristics. Generally clayey soils tend to fix more P than sandy soils.

Methods of P Application

When fertilizers are applied to the soil, some soluble phosphate is converted to less soluble forms through reactions with the soil, especially in highly phosphate-fixing soils. Minimizing the volume of soil with which the fertilizer comes into contact can reduce the rate of immobilization of the phosphate.

This volume minimization can be done through a number of ways. Placement and granulation of fertilizer are some of the ways. It is important that the phosphate should be placed as close to the growing crop as possible, since it moves very slowly in the soil. Placing the phosphate near the growing crop enables the roots to gain access to it. This is particularly important in dry soils where root growth is restricted. In highly phosphate-fixing soils, phosphate in powder form tends to be fixed faster than phosphate in granular form. The powdered fertilizer exposes greater surface areas to the soil, and therefore more P fixation takes place. Phosphate fertilizer can be applied as broadcast, band, or spot placement. Broadcasting has the advantage that the whole field receives some P but has the disadvantage that in phosphate-fixing soils more P is likely to be fixed. Banding and applying P in the planting hole (spot placement) have the advantage that more P is applied in one place, and there is a reduction in the mixing of soil and fertilizer, and hence, less P is likely to be fixed. In all cases it is important that P should be incorporated into the soil to increase its availability to the roots of the growing crop. If P is applied on the surface, it is unlikely to move to the growing roots within the growing season.

There is generally little or no response to N fertilizer if crop cannot obtain sufficient P. The reverse is also true. The lack of response is observed more in soils that are deficient in both nutrients. Therefore, it is important that adequate N and P should be made available for good plant growth.

POTASSIUM (K)

In most regions of the tropics and subtropics crop response to K application is not widespread. In Zimbabwe there has rarely been a response to K application in the soil. The K-bearing rocks in most soils are able to quickly replenish the K that has been removed from the soil. Also, where stover is returned to the soil, less K is removed through crop growth. About 90 percent of the K in maize, for example, is found in the stover and only 10 percent is in the harvested grain. The low yields of most African farmers means less K is removed. A lot of debate has surrounded the question of K application to a soil where K response has not been identified. The fact is that growing crops without addition of nutrients would result in nutrient mining and at some stage K deficiency would occur. Therefore, farmers should be encouraged to apply what they take out.

Potassium deficiency is more likely to occur in acid soils than in neutral soils. Liming can cause K availability to increase because K is replaced from exchange surface by calcium ions and goes into solution.

Many researchers have shown high correlation between levels of exchangeable K and crop yield. The exchangeable K is extracted with neutral and acidified ammonium acetate. Good correlation between plant uptake and this test has been observed. Plants require abundant supply of K during the time when the photosynthetic rate and uptake of nitrates are high. Sometimes potassium tends to be taken up in luxury amounts.

Potassium Fertilizer

Most of the potassium in use today is KCl, commonly known as muriate of potash. Potassium sulphate, commonly known as sulphate of potash, is used where sulphur is needed in the crop.

Potassium does not move far in the soil and should be placed where plant roots will reach it. Topdressed applications are satisfactory for established permanent vegetation that has roots close to the surface. Potassium fertilizer for tilled crops should be placed within the soil. It may be ploughed under or disced in, but there is advantage in band application to reduce soil contact in soils that fix potassium. Potassium mobility within plants is high enough that a small part of the root can absorb as much K as the plant needs.

SECONDARY NUTRIENTS

Calcium, magnesium, and sulphur are normally considered secondary nutrients. Many times the elements are not considered fertilizers, and yet they are present in most fertilizers, and large amounts are added to the soil with fertilizers and other soil amendments like calcitic and dolomitic lime and gypsum. Deficiency of calcium, magnesium, and sulphur is likely to increase in the future as the percentages of Ca, Mg, and S in fertilizers have declined. The trend has been to increase the content of N, P, and K at the expense of other fertilizer constituency. Sulphur deficiency also occurs in many tropical regions. Soils of low organic matter generally tend to have low sulphur. Other sources of inputs are atmospheric deposition and fertilizers.

Calcium supplies are smaller in acid soils than in alkaline soils. The amount of Ca required by plants differs from crop to crop. Legumes generally require more Ca than grain cereals. Lime ($CaCO_3$) and gypsum are the two materials likely to be considered if and when specific needs of calcium fertilizer are identified. Lime is considered when pH needs to be raised and gypsum when neutral salt is desired.

Most magnesium applied is contained in dolomitic limestone or pure dolomite. Ratios of calcium to magnesium normally found in soils range from 2:1 to 6:1. If the calcium to magnesium ratio is very high, then magnesium will need to be added. The principle magnesium fertilizer is magnesium sulphate ($MgSO_4$), sometimes known as Epsom salts.

Sulphur is normally added as ammonium sulphate (24%S). Superphosphate contains about 11%S in the form of gypsum. Potassium sulphate contains 18%S. The modern trend is the use of higher analysis fertilizers, which contain N, P, and K but less sulphur. High yields from these higher analysis fertilizers, however, increase demand for sulphur. Sulphur is also supplied through the application of gypsum and elemental sulphur. Manure and other forms of organic matter are able to provide adequate levels of Ca, Mg, and S.

MICRONUTRIENTS

The availability of micronutrients is variable and depend to a large extent on the soil properties. In alkaline soils zinc, copper, iron, and manganese tend to be deficient. Crops are generally different in their susceptibility to micronutrient deficiency. Cropping intensity also affects long-term availability of micronutrients. Where high crop yields or frequent cropping occurs, there is likely to be an increase in micronutrient deficiency. Zinc deficiency has been found to be widespread in Zimbabwe. In research work carried out, Tagwira (1992) observed low levels of zinc and copper in Zimbabwean soils. Zinc deficiency when soil was limed was also found to be prevalent.

Micronutrient fertilizers can be divided into three groups: inorganic salts, frits, and chelates. Inorganic salts are the cheapest. Soluble salts are normally chosen for quick response, obtained by spraying a foliar application directly on plants. Foliar applications require the use of low application rates to avoid toxicity problems. Care is needed in application of micronutrients, especially those containing boron to make sure toxicity is not caused by overapplication. Distribution of the micronutrient is also important. The usual solution to this problem is to apply micronutrients along with macronutrients in mixed fertilizer.

Chelates are an important group of natural and synthetic organic compounds, which are used as sources of micronutrients. The best-known synthetic chelating compound is ethylene diamine tetraacetic acid (EDTA). Zn-EDTA and Fe-EDTA are used as fertilizer quite a lot. Copper and manganese can also be chelated. The most widely used chelate is Fe-EDTA.

Zinc

Zinc availability is highly pH dependent. Severe zinc deficiency is noticed in soils of high pH. Organic matter decomposition gives rise to certain chelates, which contribute to zinc availability to plants. Soils low in organic matter are prone to deficiency. Coarse-textured soils contain and retain less zinc. Very high P availability negatively affects zinc availability. Some crops are more sensitive to zinc deficiency than others.

Zinc can be applied as a fertilizer from various sources, which include zinc sulphate, zinc oxide, zinc chelates, zinc frits, and zinc blends with other fertilizers, micronutrient mixtures, and organic manures. Among these various sources zinc sulphate is by far the most widely used.

Several methods of zinc application are used. These include soil application, foliar application, dusting seed with zinc powder or soaking in zinc solution, swabbing foliage or rubbing wounds with zinc paste or solution, dipping roots of transplanted crops in solution suspensions of zinc salts, and pushing galvanized nails or pieces of metallic zinc into tree trunks. Among these, soil treatment and foliar spray are the most common.

Boron

Boron availability is influenced by soil pH. B availability decreases with pH increase. Boron availability is enhanced by organic matter, which prevents its leaching. Coarse-textured soils are inherently deficient in boron. Calcium is known to antagonize the availability of boron to crops. The main sources of boron are borax, sodium tetraborate, boric acid, and boron frits. The first two are the most commonly used. The methods of application are mainly soil and foliar application (Troeh & Thompson 1993).

Molybdenum

Molybdenum availability is affected by pH. Molybdenum is highly available in alkaline soils. Organic matter is capable of complexing molybdenum. The complexes make Mo more available and protect it from being fixed in the soil. Coarse-textured soils are more liable to lose their molybdenum very quickly. Molybdenum increases with moisture content. Sensitivity of crops to molybdenum differs greatly. Sources of molybdenum include sodium molybdate and ammonium molybdate, molybdenum trioxide, molybdenite, and molybdenum frits. Methods of application include soil and foliar application, seed treatment, and seedbed application. In many countries with molybdenum deficiency, seed is treated with molybdenum before it is sold to farmers.

SOIL pH

Most soils found in tropical countries have low pH and need to be limed. The acidity in these soils is caused by leaching of bases by percolating water. Liming a soil helps to maintain soil fertility. Fertilization often increases the need for liming, since many fertilizers decrease soil pH.

The liming material generally used is ground limestone rock because it best meets requirements. Limestone is mainly calcium carbonate and dolomitic varieties containing magnesium carbonate. The use of dolomitic lime is important where magnesium deficiency occurs. The amount of lime needed depends on the crop, the soil, and the effectiveness of the liming material. The pH requirements of different crops differ. Legumes are generally known to utilize more calcium and respond well to liming.

The amount of lime required for a soil can be determined using the pH base saturation method and the buffer solution method. In both methods the desired pH should be known.

Applying lime can be done yearly or during an optimum year in a rotation. The lime can be applied early in the season before crop establishment or during the time of turning in the stover before the next crop. This gives the soil a chance to react with the lime before the next crop. Applied lime will not move significantly in the soil and therefore should be mixed. Crop removal, leaching, and the effect of N fertilizers make soil acid yearly, and therefore it is necessary to apply lime periodically.

Too much lime can be as bad as not enough. Excess lime reduces availability of iron, phosphorus, manganese, boron, zinc, and sulphur. Too much lime suppresses K availability as well.

The purpose of lime in tropical countries is to neutralize exchangeable aluminium when aluminium-sensitive crops are grown. This is achieved by bringing the pH to 5.5 where manganese toxicity is an additional problem. In this case the pH should be raised to 6.0 (Ahn 1992).

The amount of lime that must be applied to neutralize exchangeable aluminium can be calculated if exchangeable aluminium is determined in the laboratory. Overliming can be harmful to the plants. One harmful effect of overliming is reduction of availability of micronutrients, except for molybdenum, which is more available in acid soils than in neutral or alkaline soil. In Zimbabwe it was observed that in soils of low zinc availability liming tends to increase the incidence of Zinc deficiency (Tagwira et al. 1993).

Increased crop productivity requires not only improvement of soil fertility but also other accompanying technologies and policies to be effective in raising and sustaining food production.

The following are some of them.

1. Soil conservation
   - Erosion control to reduce loses of nutrient capital
     - —Minimum tillage
     - —Tied ridging
2. Sound agronomic practices
   - Improved seed
   - Integrated pest management
   - Crop residue returns
   - Crop rotations
   - Supplemental irrigation
   - Maintenance fertilization

REFERENCES

Ahn, P. M. (1993). *Tropical Soils and Fertilizer Use*. Intermediate Tropical Agriculture Series. Longman Scientific and Technical.

Bekunda, M. A., A. Bationo, and H. Ssali. (1997). "Soil Fertility Management in Africa. A Review of Selected Research Trials." In R. J. Buresh, P. A. Sanchez, and F. Calhoun (Ed.), *Replenishing Soil Fertility in Africa*. SSSA Special Publication Number 51.

Bumb, B. L., and C. A. Baanante. (1996). *The Role of Fertilizers in Sustaining Food Security and Protecting the Environment to 2020*. Food, Agric. and the Environ. Dis. Pap. 17. Int. Food Policy and Res. Inst., Washington, DC.

Chibudu, C. (2001). "Partial Nutrient balances in Zimbabwean Smallholder Farming systems." *Target* 27: 8–9.

De, R. (1981). *Agronomic Management of Crops and Cropping Systems for Arid and Semi-Arid Lands*. New York: Academic Press.

De Datta, S. K. (1986). "Fertilizer Management and Other Cultural Practices for Rainfed Lowland Rice in South and Southeast Asia." World Phosphate Institute 2nd Regional Seminar on Crop Production Techniques and Fertilizer Management in Rain-fed Agriculture in Southern Asia. New Delhi. 22–25 January 1986.

Donovan, W. G. (1996). "The Role of Inputs and Marketing Systems in the Modernization of Agriculture." In S. A. Breth (Ed.), *Achieving Greater Impact from Research Investments in Africa*. Sasaka Africa Assoc. Mexico City.

FAO. (1995). *FAO Fertilizer Yearbook, 1994*. Vol. 44. Food and Agriculture Organization of the United Nations, Rome.

Frissel, M. J. (Ed.). (1978). *Cycling of Mineral Nutrients in Agricultural Ecosystems*, Elsevier, Amsterdam, The Netherlands.

Giller, K. E., G. Cadisch, C. Ehaliotis, E. Adams, W. D. Sakala, and P. L. Mafongoya. (1997). "Building Soil Nitrogen Capital in Africa." In R. J. Buresh, P. A. Sanchez, and F. Calhoun (Ed.), *Replenishing Soil Fertility in Africa*. SSSA Special Publication, Number 51.

Harmsen, K., K. D. Shepherd, and A. Y. Allan. (1983). "Crop Response to Nitrogen and Phosphorus in Rain-Fed Agriculture." In *Nutrient Balances and the Need for Fertilizers in Semi-Arid and Arid Regions*. International Potash Inst., Berne.

Hedge, B. R. (1980). "Agronomic Practices Suited to the Dryland Conditions of Karnataka and Andhra Pradesh." In *Third FAI Specialized Training Programme on Management of Rain-Fed Areas*. FAI, New Delhi.

Kang, B. T., and O. A. Osiname, "Phosphorus Responses of Maize Grown on Alfisols of Southern Nigeria." *Agronomy. J.*, 71: 873–877.

Kanwar, J. S., and T. J. Rego. (1983). "Fertilizer Use and Watershed Management in Rainfed Areas for Increasing Crop Production." *Fert. News*, 28(9): 33–42.

Kapkiyai, J. (1996). "Dynamics of Soil Organic Carbon, Nitrogen and Microbial Biomass in a Long-Term Experiment as Affected by Inorganic and Organic Fertilizers." M.S. thesis. Dept. of Soil Sci., Univ. of Nairobi, Kenya.

Mugwira, L. M., and L. M. Mukurumbira. (1986). "Nutrient Supplying Power of Different Groups of Manure from Communal Areas and Commercial Feedlots." *Zimbabwe Agric. J.* 83: 25–29.

Qureshi, J. N. (1991). "The Cumulative Effect of N-P Fertilizers, Manure, and Crop Residues on Maize and Bean Yields and Some Soil Chemical Properties at Kabete." In *Recent Advances in KARI's Research Programs*. Proc. of the KARI Annual Scientific Conference. 2nd Nairobi 5–7 Sept. 1990. Kenya Agric. Res. Inst. Nairobi, Kenya.

Sanchez, P. A. (1976). *Properties and Management of Soils in the Tropics*. New York: Wiley Interscience.

Sanchez, P. A., and J. R. Benites. (1987). "Low Input Cropping for Acid Soils of the Humid Tropics." *Science* (Washington, DC) 238: 1521–1527.

Sanchez, P. A. (1994). "Tropical Soil Fertility Research, Towards the Second Paradigm." In *Trans. 15th World Congress of Soil Science*. Acapulco, Mexico. 10–16 July 1994. ISSS, Wagenigen, The Netherlands.

Sanchez, P. A., K. D. Shepherd, M. J. Soule, F. M. Place, R. J. Buresh, A. M. N. Izac, A. U. Mokwunye, F. R. Kwesiga, C. G. Ndiritu, and P. L. Woomer. (1997). "Soil Fertility Replenishment in Africa: An investment in Natural Resource Capital." In R. J. Buresh, P. A. Sanchez, and F. Calhoun (Ed.), *Replenishing Soil Fertility in Africa*. SSSA Special Publication Number 51.

Smaling. (1983). "Caulculating Soil Nutrient Balances in Africa at Different Scales." *Fertilizer Research* 35: 237–250.

Stoorvogel, J. J., and E. M. A. Smaling. (1990). "Assessment of Soil Nutrient Depletion in Sub-Saharan Africa." *1983–2000* Rep., No. 28, Vol. 1–4. Winand Staring Ctr., Wageningen, the Netherlands.

Swift, M. J., P. D. Seward, P. G. H. Frost, J. N. Quireshi, and F. N. Muchena. (1994). "Long-Term Experiments in Africa: Developing a Database for Sustainable Land Use Under Global Change." In R. A. Leigh and A. E. Johnson (Ed.), *Long-Term Experiments in Agricultural and Ecological Sciences*. Wallingford, England: CAB Int.

Tagwira, F., T. Oloya, and G. G. Nleya. (1992). "Copper Status and Distribution in the Major Zimbabwean Soils." *Commun. Soil Sci. Plant Anal.*, 23(7&8): 659–671.

Tagwira, F., M. Piha, and L. Mugwira. (1993). "Zinc Studies in Zimbabwean Soils: Effect of Lime and Phosphorus on Growth, Yield and Zinc Status of Maize." *Comm. Soil Sci. Plant Anal.*, 24(7&8): 717–736.

Tagwira, F. (1998). "Soil Erosion and Conservation Techniques for Sustainable Crop Production in Zimbabwe." In A. G. M. Ahmed, and W. Mlay (Eds.), *Environment and Sustainable Development in Eastern and Southern Africa, Some Critical Issues*. Book edited by A. G. M. Ahmed and W. Mlay was published in great Britain by Macmillan Press Ltd, London (ISBN 0-333-69942-4) and also by St Martin's Press Inc. New York. (ISBN 0-312-21047-7).

Tandon, H. L. S. (1980). "Fertilizer Use in Dryland Agriculture." *Fert. Inform. Bull. No.* 13. European Nitrogen Service Programme, New Delhi 110024.

Troeh, F. R., and L. M. Thompson. (1993). *Soils and Soil Fertility*, 5th Edition. Oxford: Oxford University Press.

Umrani, N. K., and N. D. Patel. (1983). "Advances in Fertilizer Management for rainfed Sorghum." *Fertilizer News*, 28(9): 57–61.

AMARANTH: GRAIN AND VEGETABLE TYPES

Based on G. Kelly O'Brien and Dr. Martin L. Price
ECHO, 17430 Durrance Rd., North Ft. Myers FL 33917
Published 1983

INTRODUCTION

Amaranth (*Amaranthus hypochondriacus: A. cruentus*, a grain, and *A. tricolor*, a vegetable) is a plant with an upright growth habit, cultivated for both its seeds, which are used as a grain, and its leaves, which are used as a vegetable. Both the leaves and seeds contain protein of an unusually high quality. The grain is milled for flour or popped like popcorn. The leaves of both the grain and vegetable types may be eaten raw or cooked. The amaranths that are grown principally for vegetable use have better tasting leaves than the grain types.

Amaranths are moderately branched from a main stem. Grain types form large, loose panicles at the tips of the stems. Vegetable types form flowers and seeds along the stems. They are indeterminate in growth habit but may set seed at a smaller size during short days. Grain amaranth grown in winter at ECHO (southern Florida) began flowering at less than half of the height of amaranth growing in May. Grain types may grow 1 to 2 meters tall and produce yields comparable to rice or maize. Amaranth has the "*CA*" photosynthetic pathway (along with such plants as corn and sorghum), which enables it to be uniquely efficient in utilizing sunlight and nutrients at high temperatures. It is more drought-resistant than corn.

Amaranth is quite nutritious. Amounts of vitamin C, iron, carotene, calcium, folic acid, and protein are especially high. There are reports that the incidence of blindness in children due to poor nutrition has been reduced with the use of 50 to 100 g of amaranth leaves per day. On a dry weight basis, the content of protein in leaves is approximately 30 percent.

The presence of rather high amounts of oxalic acid and nitrates places some limitation on the quantity of amaranth leaves that can be consumed daily. The amount of oxalic acid is roughly the same as that found in spinach and chard. Excessive amounts (over 100 g per day?) may result in a level of oxalic acid that begins to reduce the availability of calcium in humans. This is especially a concern if calcium intake levels are low to begin with. Nitrate in vegetable portions of amaranth is a concern because it is hypothesized that nitrates may be chemically changed in our digestive tracts into poisonous nitrosamines. Evidence for this is lacking at the present time. Nevertheless, over 100 g per day may be an unsafe amount to eat, according to scientists. Boiling the leaves like spinach and discarding the water reduces the levels of both oxalic acid and nitrates.

Amaranth grain has more protein than corn, for example, and the protein is of an unusually high quality. The protein is high in the amino acid lysine, which is the limiting amino acid in cereals like maize, wheat, and rice. The protein is also relatively rich in the sulfur-containing amino acids, which are normally limiting in the pulse crops (such as beans). The "protein complement" of amaranth grain is very near to the levels recommended by FAO/WHO. It has a protein score of 67 to 87. Protein scores are determined by taking the ratio of the essential amino acids to the level for those amino acids recommended by FAO/WHO and multiplying by 100. By comparison, wheat (14% protein) scores 47, soybeans (37%) score 68–89, rice (7%) scores 69, and maize (90%) scores 35. Although amaranth is theoretically close to the ideal, combining it with another grain increases the quality to very close to the FAO/WHO standards.

Weight gain studies with rats point out, however, that the actual nutritional value is considerably less than would be expected from the above considerations. This is apparently due to certain antinutritional factors in raw amaranth. Performance is improved somewhat by cooking. For example, Dr. Peter Cheeke at the University of Oregon compared the rate of weight gain by 120-gram rats fed a corn-soybean diet to rats fed a diet of corn and seed from A. *hypochondriacus*, either raw or cooked. The average daily gain for rats on the corn-soybean diet during the first 20 days was 3.9 grams. Rats fed the corn-amaranth diet gained only 0.3 grams per day. The average daily gain for rats fed corn and cooked amaranth was 1.6 grams. Raw amaranth seed is extremely unpalatable to rats. Cooked seed also does not seem to be very palatable, though it smelled good to Dr. Cheeke. In another study, Dr. Cheeke found that after 11 days on a corn-amaranth diet, rats (which weighed 120 g initially) "had an unthrifty hunched-up appearance, and exhibited symptoms typical of semi-starvation."

We phoned Dr. Cheeke to get his perspective on the seriousness of these negative results. He told us that there are definitely toxins and/or antinutritional factors in the raw grain and that it is less of a problem with cooked grain. He said that a scientist in Australia had been feeding raw amaranth seed to poultry as the major component of the diet. He found that chickens went into spasms and convulsions, and finally died. This unidentified factor causes liver damage. Other problems are caused by saponins, including unpalatability. But to keep this in perspective, Dr. Cheeke pointed out that there are few raw foodstuffs that do not have problems. Raw soybeans contain ten kinds of toxins. Raw kidney beans will kill rats, but cooking eliminates the problem. The key seems to be to use the grain in moderate amounts and to cook it. We asked whether we could say that there would be no problem unless people had little other than amaranth to eat. He thought that this was probably a fair statement. It is our opinion that more research needs to be done before we can recommend amaranth grain as a major ingredient in animal feed. To our knowledge it has not been shown whether these factors decrease the value of amaranth in human nutrition. Until more work is done, however, the feeding trial results must moderate our otherwise enthusiastic promotion of grain amaranth.

CULTIVATION

Vegetable Types

There appears to be considerable latitude in choice of plant densities. One approach is to plant dense stands (5–10 cm spacing) and harvest by uprooting when the plants are five to seven weeks old. Another common approach is to sow less densely (15–30 cm spacing) and harvest by cutting the stem tips and plucking tender leaves periodically, beginning when the plants are about 15 cm tall (four to six weeks old).

Seeds may be planted in a nursery for subsequent transplanting or sown directly where plants are to be grown. Transplanting is a very efficient use of seeds and allows the growing area to be weeded just before the seedlings are transplanted. The very small size of the seeds, however, means that a few seeds go a long way. The number of seeds saved is probably not a sufficient justification for the extra work involved in transplanting. On the other hand, gaining a two-week jump on the weeds can be significant because amaranth seedlings are not vigorous growers when very young. Planting in a nursery also reduces risk of loss due to disease such as damping off.

Direct seeding involves much less labor but incurs a greater risk of poor stand due to diseases and predators of young seedlings and to poor competition with weeds in the crucial initial couple of weeks. If direct seeding is used, sowing should probably be in rows to facilitate cultivation.

Whether sown in the nursery or field, seeds need to be planted about 4 mm deep (or covered with 4 mm of soil) for good germination. Because of the shallow depth, special care must be taken to prevent drying out of the soil until plants are established. Transplanting or thinning may be done in about two weeks when plants should be 5–10 cm tall. Delay in transplanting for even one week can reduce total yield.

When harvesting by repeated clippings, a two- or three-week interval is common through the end of the season (usually the shortened days of fall). Both the yield and quality of leaves are higher with more frequent clippings.

When the vegetative stage ends and flowering begins, subsequent harvests are lower in both quality and quantity. Short days, water stress, or other environmental stresses may promote flowering. The stress that comes with delayed transplanting also can cause the plants to flower more quickly. It is reported that plucking flower heads from the plant may prolong the vegetative phase of growth.

Amaranth is generally considered tolerant of nematodes and is even recommended as a rotation crop to reduce nematode populations for subsequent crops. However, one article reports the presence of root knot nematodes on amaranth roots. Control of nematodes is such a serious problem that it is important to know whether amaranth can be used to control them and/or whether it can be planted where nematodes are a problem. We will include this question in our list of research projects that could be done. It is possible that the discrepancy in reported results is because varieties differ in their susceptibility to nematodes.

Amaranth is susceptible to damping-off disease, root rot, and caterpillars and stem borers. It thrives in 30–35 °C temperatures. It tolerates poor fertility and drought. However, plant quality is poor under stressful conditions. There is good response to fertilizer.

Grain Types

Recommendations for plant spacing vary widely for grain amaranth. One recommendation is to space 23 cm between plants and 75 cm between rows. This corresponds to a planting density of 38,000 plants per hectare (15,400 per acre). Seeding rates up to *nine* times this density have been used successfully! It would seem that if harvesting is to be done by hand, the less dense spacings are advisable. This results in fewer but larger heads that can be harvested more quickly. Closer plant spacing may be advisable for mechanical harvesting.

The decision whether to transplant or direct seed is subject to the same considerations that were discussed for vegetable amaranth. Cultivation is essential until plants have reached a size where the leaf canopy can shade out weeds. After the plants are about 30 cm tall, it is helpful to mound soil from the centers of the rows up around the plants. This helps to reduce lodging (plants blowing over in the wind), suffocates weeds around the plant, and uproots weeds between rows.

Grain amaranth is grown from tropical lowlands to 3,500 m in the Himalayas. In the tropics, altitudes above 1,000 m are considered best. Although it tolerates droughts and low fertility, it does much better under conditions that are considered ideal for maize (corn). It may be intercropped with maize, beans, peppers, or squash. In some pure stands it has yielded as well as the world average yields for maize or rice (2,000 kg/ha). Loss of the tiny seeds by shattering before or during harvest can be a problem, especially with mechanical harvesting. (There are approximately 1,100 seeds per gram of amaranth.)

The seeds are mature when they can be easily separated from the heads by rubbing between the hands. Seeds can be chewed to test whether they have passed beyond the "dough stage." Heads should be cut from the stalk and side branches as soon as possible after they have reached maturity. Heads should be dried if necessary, keeping green plant parts to a minimum. Once dry, the seeds are knocked from the heads, sifted through an ordinary window screen, and winnowed to remove chaff. There appear to remain serious problems with mechanical harvesting. Primary among these problems are the tendency for plants to lodge and the loss of grain during harvesting.

Grain should be dried to about 9 percent moisture for safe storage. It is reported that grain remains viable for up to seven years. We left heads stacked in a building for five summer months (high humidity and temperatures in the 90s). Viability still appears to be high.

HARVEST

Basically, you must thrash it as mankind has always done until the invention of the thrashing machine. The three stages include (1) letting the heads dry out, (2) knocking the grain from the heads, and (3) winnowing the grain. Here is what we do with small quantities of seed (which must be kept separate from other varieties).

- Cut the heads when the grain appears to be mature, and put them somewhere to dry. If left too long, much of the grain may shatter (fall to the ground).
- Grain easily shatters from the dried heads. Put a few heads in a burlap bag and beat it against the cement floor a few times to knock it loose, or strike the bags with a stick. Then place the grain in a five-gallon bucket (many other containers would be suitable). You will notice that a lot of chaff comes along with the grain. This is why winnowing is necessary.
- Place an empty five-gallon bucket in front of a fan and, cautiously at first, pour some grain and chaff into the empty bucket. A steady wind will accomplish the same thing as the fan, but a gusty wind will cause problems. The grain is denser and will fall closer to the fan than the chaff. One should quickly begin to get a feel for how far the buckets should be from the fan and at what height to hold the one bucket in order for the grain to land in the empty bucket and the chaff to blow far enough to miss it. Pour the grain back and forth until it appears to be clean.
- Final cleanup can be done by swirling and shaking the grain around. Remaining chaff will "float" to the top like ice in water, and it can be removed by hand.

PREPARATION

Vegetable amaranth leaves and stems or entire plants may be eaten raw or cooked as spinach. As discussed earlier, however, cooking and discarding the water will remove potentially harmful oxalates and nitrates.

The seeds from grain amaranth can be ground for use as a good quality flour for breads or pastries. It must be combined with wheat flour for a yeast dough. The Organic Farming and Research Center (Rodale) has used a 50:50 ratio successfully, but they suggest that the percent of amaranth could be even greater if desired. They state, "Amaranth flour contributes to the sweetness and moistness of a baked good."

Alternatively, seeds can be popped like popcorn. The people at Rodale say that popped amaranth can be used in confections bound with sorghum, molasses, or honey; in high-energy granola and granola bars; in cheese spreads; to flavor salad dressings; in breading for chicken and fish; in crackers, pie crusts, and breads; and as toppings for casseroles and desserts.

Several recipes can be found in the book *Amaranth: from the Past, for the Future* by Rodale Press. Rodale Press, Emmaus, PA 18049.

CASSAVA PRODUCTION

Brian Mangwiro, Stanslous Thamangani, and Tuarira Abduel Mtaita

INTRODUCTION

Cassava (Manihot esculenta Crantz) is a high-energy starchy root crop. It is widely grown in most tropical regions, 25° north and south of the equator, where it forms the basic diet for the indigenous communities. Cassava is ranked fourth after rice, sugarcane, and maize as a source of calories for human consumption. In tropical Africa, the crop is the single most important source of calories in the diet. The world over, it is the major carbohydrate for more than 500 million people. The leaves are eaten as a vegetable in parts of Asia and Africa, and they provide high nutritional levels of vitamins and protein. It is also important as animal feed and a source of starch and starch derivatives.

The nutritional capacity is comparable to that of potatoes, except that it contains approximately twice the fiber content and a higher level of potassium. Cassava is an attractive crop to the small-scale farmers with limited resources for the following reasons.

- It is highly efficient in producing carbohydrates.
- It can tolerate low soil fertility and drought, and it is able to recover well when attacked by pests and diseases.
- It fits well to multispecies agricultural systems.
- The edible roots can be left in the ground for long periods, thus providing a source of insurance against famine.

CLIMATIC AND EDAPHIC REQUIREMENTS

Cassava is adapted to a wide range of environmental conditions. The relative size of the tubers is dependent on the combination of climatic and edaphic factors. These factors and the stage of maturity may also influence the amount of HCN in the tubers.

The ideal soil pH range is 5.5–8.0, though superior yields are obtained with fertile free draining loamy soils with a pH of 6–7. High nutritional regimes result in excessive vegetative growth at the expense of tuber growth. The resultant lush growth may also increase susceptibility of the crop to disease.

Cassava performs well in ambient temperatures between 18°C and 30°C, and annual rainfall between 500 and 2500 mm. A high relative humidity between 65 and 80 percent encourages plant growth, even when the soil water regimes are relatively low.

VARIETIES

Cassava varieties can be divided into sweet and bitter types, depending on the level of cyanogenic glycoside contained in the tubers. Sweet types have very low levels of the glycoside in their flesh and higher concentrations in the peel, while the bitter types have a high content of glycoside in both peel and flesh. Cassava may yield as much as 25 tonne/ha in approximately nine months.

PROPAGATION TECHNIQUES

The performance of the resultant plant population is dependent on the initial quality of the planting material. The best establishment and performance is obtained from pest-free stem cuttings with four to eight nodes. A fungicide treatment prior to planting is important to minimize the risk of disease and pest damage.

Multiplication of Propagules

In large-scale plantings, rapid propagation is achieved by taking cuttings 30 to 60 cm long and arranging them longitudinally in a propagation container containing a moist growing medium. Initial fresh shooting occurs within 7 to 14 days. These sprouting shoots are removed after they reach 10–15 cm in length. The shoots are then placed with their cut end in moist medium until rooting occurs. Rooted shoots are hardened and then directly transplanted to the field.

Land Preparation

The seedbed should be pulverized to a fine and level tilth for optimal tuber growth and development. The variety and density of planting would determine the spacing between ridges.

Adequate amounts of limestone, phosphorous fertilizer, and/or organic matter are applied to the soil 2 to 4 weeks before land preparation. The exact requirements needed for the crop growth are determined from the necessary soil analysis.

Crop Establishment

Fungicidal and insecticidal pretreatments are important to minimize respective field losses. The common chemicals in use include diazinon and Nutrex at the manufacturers' recommended rates. Captafol at 4,000 ppm is often recommended as a dip for the eradication of super-elongation disease.

Planting and Spacing

High plant density results in high overall yields but low yield per plant. The spacing to be used depends on the variety used, the soil type, soil fertility and water availability, and the size of the tubers desired. Spacing range from 0.6 m × 0.6 m to 1 m × 1 m.

The cuttings can be planted vertically, horizontally, or inclined. Horizontal planting of 5–10 cm in depth are recommended for relatively dry climates, while vertical or inclined planting may be preferable for wet conditions. Planting depth impacts tuber size and shape and the efficiency of harvesting.

Plant Nutrition Requirements

Cassava extracts large quantities of macronutrients from the soil; for example, a 25-tonne/ha crop removes 122 kg N, 27 kg P, 145 kg K, 45 kg Ca, and 20 kg Mg. Therefore, adequate fertilization may be important to replenish the depleted soil nutrients. Whenever possible, fertilizer recommendations should be based on soil analysis. On soils that are moderately deficient in P and K, a $N:P_2O_5:K_2O$ ratio of 1:1:2—that is, 40–80 kg N, 40–60 kg P_2O_5, and 80–150 kg K_2O—per ha is recommended.

Care must be taken not to overfertilize the crop. Excessive N application may lead to high HCN content and bitterness of the tubers. Inadequate K results to excessive vegetative production at the expense of tuber growth. Added K also reduces the HCN content of the tubers. Cassava appears to be particularly susceptible to zinc deficiency.

Weed Control

This is particularly important during the first few months before the leaf canopy has closed over. In addition to hand control, preemergence herbicide treatments may be used to avoid weed damage.

PEST AND DISEASE CONTROL

Common pests and diseases include the following.

Cassava Bacterial Blight. This is a bacterial disease. It is spread mainly through the use of infected planting material and contaminated soil water. The major symptom of the disease is rapid wilting and death of the plant. The disease is controlled by use of clean planting material.

Shoot Fly. The cassava shoot fly attacks the young growing areas of the cassava shoot. Control measures are normally required when a severe attack occurs in the early stages of crop growth. These flies can be controlled by the application of an appropriate insecticide (such as Perfekthion or Malathion) every three to four weeks.

Red Spider Mite. Spider mites are more prevalent during dry seasons. During the wet season, raindrops wash the mites off the leaves, giving a measure of natural control. Use of a mitecide such as Actellic may control the mites.

Hornworm (Errnis ello L.). This is more prevalent in poor managed fields where it causes severe defoliation. *Trichomma sp.* parasites the eggs, while ground beetles and Jack Spaniard wasps feed on larvae and pupae.

Cassava Mites. These cause mottling, leaf drop, and severe yield reduction. Biologically they can be controlled by lace bugs and ladybird beetles.

Fusarium Solani (Mart) Sacc. This often accompanies nematode damage. Brown dry rot begins at the center. To control the disease, one should take care of the nematodes.

Stem Rot (Glomerella cingulata). This is the most common stem rot on stored cassava cuttings. Treating the cuttings with fungicides controls the rot.

MATURITY, HARVESTING, AND POSTHARVEST TREATMENTS

It is essential that harvesting be carried out at the appropriate stage of maturity. Immature roots are very susceptible to physical damage, and overmature roots have reduced cooking qualities. If the crop is to be used for human consumption, harvesting should begin about 6 months after planting. Some varieties mature approximately 8 to 12 months after planting. However, periods of at least one year are recommended for good yields for use in animal feed. Harvest indicators are date to maturity (which is specific to variety) and the yellowing and falling of leaves from the plant.

It is critical to maintain a high quality during field handling operations to minimize the risk of physical damage. Hand harvesting is the commonly traditional method used. This is a largely seen in smallholding up to five acres. Under larger plantations, mechanical methods are recommended. Sometimes plants may be pruned three weeks prior to the harvest date. Once harvested, roots should not be left exposed for long periods in direct sunlight, since water loss causes vascular streaking and spoilage.

Only roots with minimum physical damage should be selected for storage. With respect to root size, producers must know their market, since different consumers have different size preferences—for example, homemakers vs. processors.

Curing

At 80 to 85 percent relative humidity and temperatures between 25° and 40°C tuberization occurs in one to four days and a new cork layer forms around three to five days later. At 40°C and above, primary deterioration usually takes place before wound healing. (Hahn, 1998)

Storage

Field storage is one of the most economical ways of keeping the crop. However, for short-term storage (14–35 days), cassava may be stored in polyethylene bags, boxes, and freezers or in well-ventilated rooms after waxing. Care must be taken to avoid physical, physiological, insect and microbial deterioration.

Roots that are intended for use as a boiled vegetable should not generally be stored for more than two weeks. With prolonged storage, changes in the texture of the roots occur such that they will not cook to a soft, mealy texture but rather turn hard when boiled.

REFERENCES

Rosling, Hans. (1993). *Cassava Toxicity and Food Security.* International Child Health Unit, Dept. of Pediatrics, S-751, 85 Uppsala, Sweden.

Food and Agriculture Organization of the United Nations. (1984). *Roots and Tubers.* Published by arrangement with the Institut africain pour la développement économique et social, B.P. 8008, Abidjan, Côte d'Ivoire, Rome. FAO Economic and Social Development Series No. 3/16.

Hahn, S. K. (1998). Trandiqional Processing and Utilization of Cassava in Africa, IRG. 41, IITA, PMB 5329 Ihadan, Nigeria.

CEREAL CROP PRODUCTION

Based on Traditional Field Crops, David Leonard, edited by: Marilyn Chakroff and Nancy Dybus, Illustrated by: Marilyn Kaufman Produced for Peace Corps by the Transcentury Corporation, Washington, D.C. December 1981

There are several reasons why the six reference crops—maize, grain sorghum, millet, peanuts, field beans, and cowpeas—are grouped together in one article. All of the reference crops are row crops (grown in rows), and because of this, they share a number of similar production practices. Also, in developing nations, two or more of the crops are likely to be common to any farming region and are frequently interrelated in terms of crop rotation and intercropping. In addition, all of them are staple food crops. The developing countries are the major producers of these crops, with the exception of maize, which is grown widely.

THE "PACKAGE" APPROACH TO IMPROVING CROP YIELDS

In most cases, low crop yields are caused by the simultaneous presence of several limiting factors, rather than one single obstacle. When a specially developed and adapted "package" of improved practices is applied to overcome these multiple barriers, the results are often much more impressive than those obtained from a single factor approach. A crop "package" consists of a combination of several locally proven new practices. (Few packages are readily transferable without local testing and modification.) Most include several of the following: an improved variety; fertilizer; improved control of weeds, pests, and diseases; improvements in land preparation; water management; harvesting; and storage.

It should be stressed that a package does not always have to involve considerable use of commercial inputs. In fact, an extension program can focus initially on improvement of basic management practices that require little or no investment such as weeding, land preparation, changes in plant population and spacing, seed selection, and timeliness of crop operations. This helps assure that small farmers benefit as least as much as larger ones, especially in those regions where agricultural credit is poorly developed.

CEREAL CROPS VERSUS PULSE CROPS

Maize, grain sorghum, and millet are known as cereal crops, along with rice, wheat, barley, oats, and rye. Their mature, dry kernels (seeds) are often called cereal grains. All cereal crops belong to the grass family (Gramineae) that accounts for the major portion of the monocot

(Monocotyledonae) division of flowering (seed-producing) plants. All monocot plants first emerge from the soil with one initial leaf called a seed leaf or cotyledon.

A germinating maize seedling is shown in Figure 8. Note that it has only one seed leaf, which makes it a monocot. Monocots emerge through the soil with a spike-like tip. They generally have fewer problems with clods and soil dusting than dicots.

Peanuts, beans, and cowpeas are known as pulse crops, grain legumes, or pulses, along with others such as lima beans, soybeans, chickpeas, pigeon peas, mung beans, and peas. The pulses belong to the legume family (Leguminosae), whose plants produce their seeds in pods. Some legumes like peanuts and soybeans are also called oilseeds because of their high vegetable oil content.

A germinating bean plant is shown in Figure 9. Note the two thick cotyledons (seed leaves) that originally formed the two halves of the seed. The pulses belong to the other major division of flowering plants called dicots (Dicotyledonae). Unlike the monocots, dicot plants first emerge from the soil with two seed leaves.

In addition, the pulses have two outstanding characteristics for farmers and for those who consume them.

1. They contain two to three times more protein than cereal grains.
2. Legumes obtain nitrogen for their own needs through a symbiotic (mutually beneficial) relationship with various species of Rhizobia bacteria that form nodules on the plants' roots. Nitrogen is the plant nutrient needed in the greatest quantity and is also the most costly when purchased as chemical fertilizer. The Rhizobia live on small amounts of sugars produced by the legume and, in return, convert atmospheric nitrogen (ordinarily unavailable to plants) into a usable form. This very beneficial process is called nitrogen fixation. In contrast, cereal grains and other nonlegumes are totally dependent on nitrogen supplied by the soil or from fertilizer.

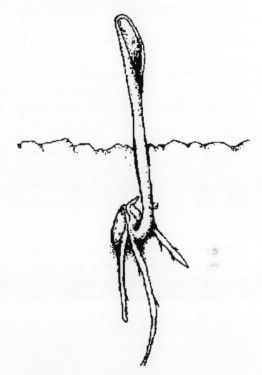

FIGURE 8
Germinating maize seedling.

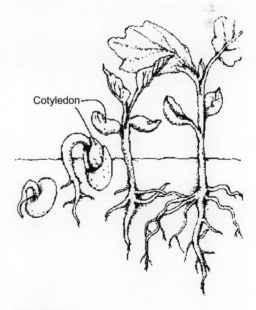

FIGURE 9
Geminating bean seed.

THE NUTRITIONAL VALUE OF THE REFERENCE CROPS

The cereal grains, with their high starch content and lower prices, make up a major source of energy (calories) in developing countries. There, cereal consumption is high enough to contribute a substantial amount of protein to the diets of older children and adults (although still well below quantity and quality requirements). Another plus is that cereal grains contain a number of vitamins and minerals, including vitamin A, which can be found in the yellow varieties of maize and sorghum. Infants and children, who have much higher protein needs per unit of body weight and smaller stomachs, do not get as much protein from cereals as adults. Studies have also shown that some reference crops lose vitamins and protein in substantial amounts with traditional preparation methods (milling, soaking, and drying).

The pulses have considerably higher protein contents than the cereal grains (17 to 30 percent in the reference pulses) and generally higher amounts of B vitamins and minerals. Unfortunately, they also may have some deficiencies in amino acids. All animal proteins (meat, poultry, fish, eggs, milk, and cheese) are complete proteins (contain all essential amino acids), but their high cost puts them out of reach of much of the population in developing nations.

Fortunately, it is possible to satisfy human protein requirements without relying solely on animal protein sources. The cereals and pulses, though not complete proteins in themselves, can balance out each other's amino acid deficiencies. Cereals are generally low in the essential amino acid lysine but relatively high in another: methionine. The opposite is true for most of the pulses. If eaten together or within a short time of each other and in the right proportions (usually about a 1:2 ratio of pulse to cereal), combinations like maize and beans or sorghum and chickpeas are complete proteins. In most developing countries, however, pulses are more expensive than the cereal grains, which creates difficulties in achieving a balanced diet.

AN INTRODUCTION TO THE INDIVIDUAL CROPS

Maize (Zea mays)

In terms of total world production, maize and rice vie for the number two position after wheat. Several factors account for the importance of maize: Maize can adapt to a wide range of temperatures, soils, and moisture levels, and resists disease and insects; it has a high yield potential; and it is used for both human and animal consumption. Many people believe that yellow maize has more protein than white maize, but the only nutritional difference between the two is the presence of vitamin A in the yellow variety (also called carotene).

An ear of maize is shown in Figure 10. Each silk leads to an ovula (potential kernel) on the cob. Varieties vary in length and tightness of husk covering, which determines resistance to insects and moisture-induced molds that may attack the ear in the field.

Maize Yields. Top farmers in the U.S. Corn Belt will get 10,000 or more pounds per acre. The U.S. average is about half this. The average in less industrialized countries is 1,000 pounds per acre. With improved practices the feasible yield could be as much as 5,000 pounds per acre.

Climatic Requirements of Maize. *Rainfall:* Nonirrigated (rain-fed) maize requires a minimum of around 500 mm of rainfall for satisfactory yields. Ideally, the bulk of this should fall during the actual growing season, although deep, loamy or clayey soils can store up to 250 mm of preseason rainfall in the future crop's root zone. Maize has some ability to resist dry spells but is not nearly as drought tolerant as sorghum and millet.

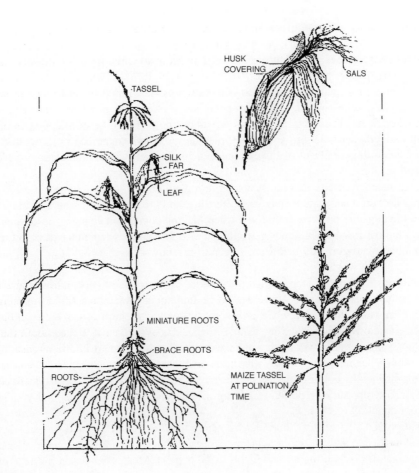

FIGURE 10
Maize plant and an ear.

Temperature. The optimum growth rate of maize increases with temperatures up to about 32 to 35°C if soil moisture is abundant but decreases slightly with temperatures around 27 to 30°C when soil moisture is adequate. If soil moisture is low, the optimum growth rate temperature drops to 27°C or below. At temperatures of 10°C or below, maize grows slowly or not at all and is susceptible to frost. However, daytime temperatures in excess of 32°C will reduce yields if they occur during pollination. Yields are also reduced by excessively high nighttime temperatures, since they speed up the plant's respiration rate and the "burning up" of the growth reserves.

Soil Requirements. Maize grows well on a wide variety of soils if drainage is good (no water logging). It has a deep root system (up to 185 cm) and benefits from deep soils that allow for improved moisture storage in dry spells. The optimum pH for maize is in the 5.5–7.5 range, although some tropical soils produce good yields down to a pH of 5.0 (very acid).

Planting. Overly high populations cause increased lodging, barren stalks, unfilled ears, and small ears. Dry husked ears weighing less than 270–310 g indicate that plant population was probably too low for the conditions and that yields might have been 10 to 20 percent higher. Ear size of prolific (multiple-ear) varieties will not vary as much with changes in plant density as will single-ear varieties; rather, the number of ears per plant will decrease as density increases.

Hill versus drill planting: Numerous trials with maize have shown yield increases of 10 to 13 percent when drill planting (one seed per hole) was substituted for hill planting at two to three seeds per hole. However, lodging appears to be more of a problem with drill planting. Farmers who are hand planting four to six seeds per hole should be encouraged to switch to two to three seeds per hole and space the holes close enough together to achieve the desired plant population. It is doubtful that switching to drill planting is worth the extra labor involved under hand planting.

Under adequate moisture and fertility, optimum plant populations vary from about 40,000 to 60,000 per hectare. Plant size, management, fertility, and the variety's tolerance to plant density and available moisture must be considered when making population changes. Studies also show that overly high populations have a negative effect on maize yields when moisture is low. Planting depth should be between 3.75 and 8 cm—deeper in sandy soils.

Maize Stages of Growth. Depending on the variety and growing temperatures, maize reaches physiologic maturity (the kernels have ceased accumulating protein and starch) in about 90 to 130 days after plant emergence when grown in the tropics at elevations of 0 to 1,000 meters. At higher elevations, it may take up from 200 to 300 days. Even at the same elevation and temperature, some varieties will mature much earlier than others and are known as early varieties. The main difference between an early (90-day) and a late (130-day) variety is the length of time from plant emergence to tasseling (the vegetative period). This stage will vary from about 40 to 70 days. The reproductive period (tasseling to maturity) for both types is fairly similar and varies from about 50 to 58 days.

Pollination is a very critical time, during which there is a high demand for both water and nutrients. One to two days of wilting during this period can cut yields by as much as 22 percent, and six to eight days of wilting can cut yields by 50 percent. Maize is also very prone to moisture stress (water deficiency) when the kernels are first developing. The water requirement is up to 10 mm per day under very hot and dry conditions.

Ears are left on the stalks to dry for two or three weeks after reaching maturity. If ears were harvested immediately after reaching maturity, then they would be damaged in the harvesting process. Sweet corn, for immediate eating, is harvested at maturity or earlier.

Grain Sorghum (Sorghum bicolor)

Although grain sorghum accounted for only 3.6 percent of total world cereal production in 1977, several factors make it an especially important crop in the developing world: It is drought resistant and heat tolerant and particularly suited to the marginal rainfall areas of the semiarid tropics (such as the savanna and Sahel zones of Africa where food shortages have been critical).

Sorghum Yields. Top farmers in the United State using irrigation will get 10,000 or more pounds per acre. The U.S. average is about one-third of this. The average in less industrialized countries is 600 pounds per acre. With improved practices the feasible yield could be as much as 5,000 pounds per acre.

Grain sorghum exhibits greater yield stability over a wider range of cropping conditions than maize. Although it will outyield maize during below-normal rainfall periods, the crop might suffer some damage under very high rainfall.

Protein content vs. yield: The protein content of sorghum kernels can vary considerably (7 to 13 percent on soils low in nitrogen) due to rainfall differences. Since nitrogen (N) is an important constituent of protein, kernel protein content is likely to be highest under very low rainfall that cuts back yields and concentrates the limited amount of N in a smaller amount of grain. Protein fluctuation is much less on soils with adequate nitrogen.

Climatic Requirements of Sorghum.
Grain sorghum tolerates a wide range of climatic and soil conditions.

Rainfall: The sorghum plant, aside from being more heat and drought resistant than maize, also withstands periodic waterlogging without too much damage. The most extensive areas of grain sorghum cultivation are found where annual rainfall is about 450 to 1,000 mm, although these higher rainfall areas favor the development of fungal seedhead molds that attack the exposed sorghum kernels. The more open-headed grain sorghum varieties are less susceptible to head mold.

Several factors account for the relatively good drought tolerance of grain sorghum.

- Under drought conditions the plants become dormant and will curl up their leaves to reduce water losses due to transpiration (the loss of water through the leaf pores into the air).
- The leaves have a waxy coating that further helps to reduce transpiration.
- The plants have a low water requirement per unit of dry weight produced and have a very extensive root system.

Temperature and soil requirements: Although sorghum withstands high temperatures well, there are varieties grown at high elevations that have a good tolerance to cool weather as well. Light frosts may kill the above-ground portion of any sorghum variety, but the plants have the ability to sprout (ratoon) from the crown. Sorghum tends to tolerate very acid soils (down to pH 5.0 or slightly below) better than maize, yet it is also more resistant to salinity (usually confined to soils with pHs over 8.0).

Response to daylength (photosensitivity): Most traditional sorghum varieties in the developing countries are very photosensitive. In these photosensitive types, flowering is stimulated by a certain critical minimal daylength and will not occur until this has been reached, usually at or near the end of the rainy season. This delayed flowering enables the kernels to develop and mature during drier weather while relying on stored soil moisture. (This is actually a survival feature that allows seedheads to escape fungal growth in humid rainy conditions.) These local photosensitive varieties usually will not yield as well outside their home areas (especially further north or south), since their heading dates still remain correlated to the rainy season and day length patterns of their original environment.

Other sorghum characteristics—ratooning and tillering ability: The sorghum plant is a perennial (capable of living more than two years). Most forage sorghums and many grain varieties can produce several cuttings of forage or grain from one planting if not killed by heavy frost or extended dry weather. New stalks sprout from the crown (this is called ratooning) after a harvest. However, ratooning ability has little value in most areas where nonirrigated sorghum is grown. In these areas, either the rainy season or frost-free period is likely to be too short for more than one grain crop or too wet for a mid–rainy season first crop harvest without head mold problems. Forage sorghums take good advantage of ratooning, since they are harvested well before maturity, usually at the early heading stage. Cattle farmers in El Salvador take three cuttings of forage sorghum for silage making during the six-month wet season. In irrigated tropical zones with a year-round growing season, such as Hawaii, it is possible to harvest three grain crops a year from one sorghum planting by using varieties with good ratooning ability.

Some grain sorghum varieties have the ability to produce side shoots that grow grain heads at about the same time as the main stalk (this is called tillering). This enables such varieties to at least partially make up for too thin a stand of plants by producing extra grain heads.

Young sorghum plants or drought-stunted ones under 60 cm tall contain toxic amounts of hydrocyanic acid (HCN or prussic acid). If cattle, sheep, or goats are fed on such plants, fatal poisoning may result. Fresh, green forage, silage, and fodder (dried stalks and leaves) are usually safe if over 90–120 cm tall and if growth has not been interrupted. The HCN content of sorghum plants decreases as they grow older and is never a problem with the mature seed. An intravenous injection of 2–3 grams of sodium nitrite in water, followed by 4–6 grams of sodium thiosulfate is the antidote for HCN poisoning in cattle. These dosages are reduced by half for sheep.

Planting. Optimum plant population varies markedly with available water, plant height, tillering ability, and fertility. In varieties that tiller well, plant population is less important than with maize, since the plants can compensate for overly low or high populations by varying the production of side shoots.

In West Africa, the improved long-season photosensitive and the improved short-season non-sensitive varieties are sown at the rate of 40,000–80,000/ha under good management. The dwarf photosensitive, long-season varieties are sown at rates of 100,000/ha or more. All these populations are based on monoculture. The planting depth should be 3.75–6 cm—deeper in sandy soils.

Nutritional Value and Uses of Sorghum. Nearly all grain sorghum used in the developed world is fed to livestock (mainly poultry and swine). However, in developing countries it is an important staple food grain and is served boiled or steamed in the form of gruel, porridge, or bread. In many areas it is also used to make a home-brewed beer. In addition, the stalks and leaves are often fed to livestock and used as fuel and fencing or building material.

Like the other cereals, grain sorghum is relatively low in protein (8 to 13 percent) and is more important as an energy source. If eaten along with pulses in the proper amount (usually a 1:2 grain:pulse ratio), it will provide adequate protein quantity and quality. Only those varieties with a yellow endosperm (the starchy main portion of the kernel surrounding the germ) contain vitamin A.

Because sorghum is very susceptible to bird damage during kernel development and maturity, bird-resistant varieties have been developed. Because these varieties have a high tannin content in the seeds, stalks, and leaves, they are partly effective in repelling birds from the maturing seedheads. However, these high-tannin varieties are more efficient in the essential amino acid lysine than ordinary varieties, which has consequences for humans and other monogastrics like pigs and chickens. In the United State, this is overcome by adding synthetic lysine to poultry and swine rations that are made from bird-resistant sorghum grains. In developing countries a slight increase in pulse intake can overcome this problem in humans.

Sorghum Growth Stages. Depending on variety and growing temperatures, nonphotosensitive grain sorghum reaches physiologic maturity in 90 to 130 days within the 0–1,000 m zone in the tropics (see Figure 11). However, the local, daylength-sensitive varieties may take up to 200 days or more because of delayed flowering. At very high elevations, all varieties may take 200 days or more.

As with maize, the main difference between a 90-day and 130-day sorghum variety is in the length of the vegetative period (the period from seedling emergence to flowering). The grain-filling

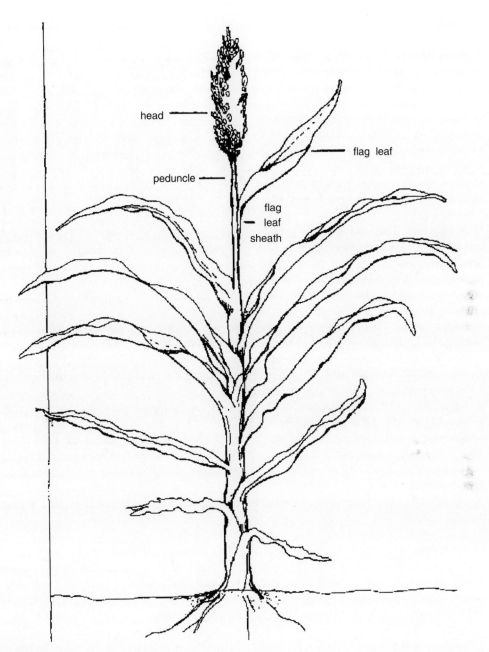

FIGURE 11
Grain sorghum plant nearing maturity.

period (pollination to maturity) is about the same for both (30–50 days). These principles remain the same no matter what variety is grown.

For the first 30 days or so, the growing point, which produces the leaves and seedhead, is below the soil surface. Hail or light frost is unlikely to kill the plant, since new growth can be regenerated by the growing point. However, regrowth at this stage is not as rapid as with maize.

The Millets

Types of Millet. The millets comprise a group of small-seeded annual grasses grown for grain and forage. Although of little importance in the United State and Europe, they are the main staple food grain crop in some regions of Africa and Asia and are associated with semi-arid conditions, high temperatures, and sandy soils. Of the six major millet types listed below, pearl millet is the most widely grown (see Figure 12).

Pearl millet: The main areas of production are the semi-arid plains of southern Asia (especially India) and the Sahel (sub-Saharan) region of Africa. This is the most drought and heat tolerant of the millets, more prone to bird damage than finger millet.

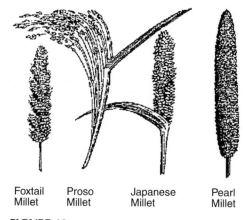

Foxtail Millet   Proso Millet   Japanese Millet   Pearl Millet

FIGURE 12
Four types of millet.

Finger millet: The main areas of production are the southern Sudan, northern Uganda, southern India, and the foothills of Malaysia and Sri Lanka. Unlike other millets, it needs cool weather and higher rainfall; it is higher in protein than the others.

Proso millet: The main areas of production are Central Asia and Russia. It is used mainly as a short-duration emergency crop or irrigated crop.

Teff millet: The main area of production is mainly the Ethiopian and East African highlands up to 2,700 m, where it is an important staple food.

Japanese or barnyard millet: The main areas of production are India, East Asia, parts of Africa, and as a forage in the eastern United State. An important characteristic is its wide adaptation in terms of soils and moisture; it takes longer to mature (three to four months total) than the others.

Foxtail millet: The main areas of production are the Near East and mainland China. It is very drought resistant.

Millet Yields. Average millet yields in West Africa range from about 300 to 700 kg/ha, the number of pounds per acre is about 10 percent less. They tend to be low due to marginal growing conditions and the relative lack of information concerning improved practice. Compared to maize, sorghum, and peanuts, research efforts with millet have only yielded 1,000–1,500 kg/ha and improved varieties have produced up to 2,000–3,500 kg/ha.

Climatic Requirements of Millet Rainfall. Pearl millet is the most important cereal grain of the northern savanna and Sahel regions of Africa. It is more drought resistant than sorghum and can be grown as far north as the 200–250 mm rainfall belt in the Sahel where varieties of 55–65 days' maturity are grown to take advantage of the short rainy season. Although pearl millet uses water more efficiently and yields more than other cereals (including sorghum) under high temperatures, marginal rainfall, suboptimum soil fertility, and a short rainy season, it does lack sorghum's tolerance to flooding.

Soil: Pearl millet withstands soil salinity and alkaline conditions fairly well. It is also less susceptible than sorghum to boring insects and weeds, but it shares sorghum's susceptibility to losses from bird feeding, which damages the maturing crop.

Nutritional Value and Uses of Millet. Pearl, foxtail, and proso millets all contain about 12 to 14 percent protein, which is somewhat higher than most other cereals. The most common method of preparing pearl millet in West Africa is as "kus-kus" or "to," a thick paste made by mixing millet flour with boiling water. Millet is used also to make beer. The stalks and leaves are an important livestock forage and also serve as fuel, fencing, and building material.

Growing Millet. The traditional West African pearl millet varieties are generally 2.5–4.0 m tall with thick stems and a poor harvest index. They are usually planted in clumps about a meter or so apart, very often in combination with one to three of the other reference crops, usually sorghum, cowpeas, and groundnuts. Many seeds are sown per clump, followed by a laborious thinning of the seedlings about two to three weeks later. The tiny millet seeds are low in food reserves, which become exhausted before the seedlings can produce enough leaf area for efficient photosynthesis and enough roots for good nutrient intake. Therefore, as with sorghum, the growth rate is very slow for the first few weeks.

Many of the traditional millets produce abundant tillers (side shoots produced from the plant's crown). However, this tillering is nonsynchronous—that is, tillering development lags behind that of the main stem. As a result, these secondary shoots mature later than the main stem. If soil moisture remains adequate, two or more smaller harvests can be taken.

Aside from the normal rainfed millet production, the crop is also planted on flood plains or along river borders as the waters begin to recede. This system is referred to as recessional agriculture and also may involve sorghum. Planting depth should be 2–4 cm—deeper in sandy soils.

Peanuts (Arachis hypogea)

Peanuts are an important cash and staple food crop in much of the developing world, particularly in West Africa and the drier regions of India and Latin America. The developing nations account for some 80 percent of total world production, with two-thirds of this concentrated in the semi-arid tropics.

Types of Peanuts. There are two broad groups of peanuts (see Figure 13).

Virginia group: Plants are either of the spreading type with runners or of the bunch (bush) type. Their branches emerge alternately along the stem rather than in opposed pairs. The Virginia varieties take longer to mature (120–140 days in the tropics) than the Spanish-Valencia types and are moderately resistant to Cercospora leaf-spot, a fungal disease that can cause high losses in wet weather unless controlled by fungicides. The seeds remain dormant (do not sprout) for as long as 200 days after development, which helps prevent premature sprouting if they are kept too long in the ground before harvest.

Spanish-Valencia group: Plants are of the erect bunch type and nonspreading (no runners). Their branches emerge sequentially (in opposed pairs), and their leaves are lighter green. They have a shorter growing period (90–110 days in warm weather), are highly susceptible to Cercospora leaf spot, and have little or no seed dormancy. Preharvest sprouting can sometimes be a problem under very wet conditions or delayed harvest. They are generally higher yielding than the Virginia variety if leaf spot is controlled.

Peanut Yields. Average peanut yields in the developing countries range from about 500 to 900 kg/ha (kilograms/hectare) of unshelled nuts, compared with the U.S. average of 2,700 kg/ha, based on 1977 FAO data. Farmers participating in yield contests have produced over 6,000 kg/ha under irrigation, and yields of 4,000 to 5,000 kg/ha are common on experiment station plots

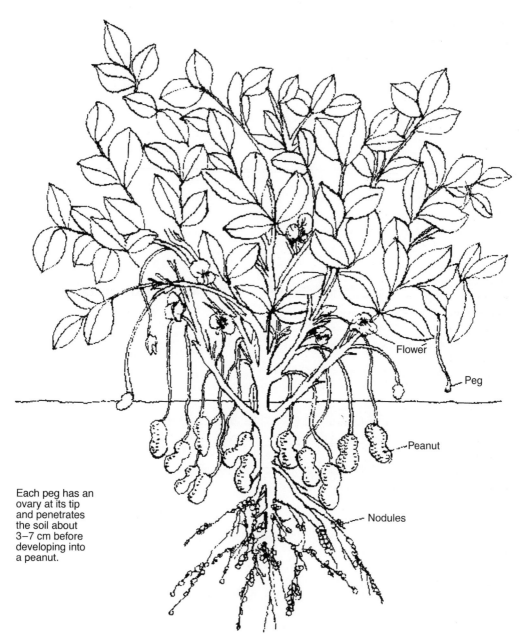

Flower

Peg

Peanut

Nodules

Each peg has an ovary at its tip and penetrates the soil about 3–7 cm before developing into a peanut.

FIGURE 13
Peanut plant.

throughout the world. Feasible yields for small farmers who use a suitable combination of improved practices are in the range of 1,700 to 3,000 kg/ha, depending on rainfall.

Climatic and Soil Requirements of Peanuts. *Rainfall*: Peanuts have good drought resistance and heat tolerance. They mature in 90–120 days in warm weather, which makes them especially well suited to the short, wet season of the northern savanna zone of West Africa. They can be grown in moister climates if diseases (especially leaf spot) can be controlled and if planted so that harvest does not coincide with wet weather.

Temperature: During the vegetative (leaf development) phase, temperature has little effect on yields. However, the rate of flowering and pollen viability is greatly influenced by temperatures during flowering (about 35–50 days after emergence). Pod production is adversely affected by temperatures below 24°C or above 33°C. At 38°C, for example, flowering is profuse, but few pods are produced.

Soils: Peanuts do not tolerate water logging, so good soil drainage is important. Soils that crust or cake are unsuitable, since penetration of the pegs is unhindered. Clayey soils can produce good results if well drained, but harvest (digging) losses may be high due to nut detachment if the plants are "lifted" when such soils are dry and hard. On the other hand, harvesting the crop on wet, clayey soils may stain the pods and make them unsuitable for the roasting trade.

Peanuts grow well in acid soils down to about pH 4.8, but they do have an unusually high calcium requirement, which is usually met by applying gypsum (calcium sulfate).

Planting. In parts of West Africa, peanuts are frequently interplanted in combination with sorghum, millet, and maize. Because peanuts are the most valuable, the tendency is to keep the cereal population down to about 3,000 to 6,000 hills per hectare and the peanut density high at about 30,000 hills per hectare, or about the same as undersole cropping.

In West Africa, the recommended plant population for improved varieties grown alone ranges from about 45,000 to 100,000/ha. Rows vary from 24 to 36 inches (40 to 60 cm) and seed spacing in the row from 15 to 25 cm. For the Virginia types populations of 45,000 to 60,000/ha have been found to be optimum, with higher populations recommended for the Spanish-Valencia types.

In the northern savanna areas of West Africa, peanuts are generally planted in June and harvested in September or October. In the southern, higher rainfall sections of the savanna, it is often possible to grow two crops (April or May until August for the first, and August or September to November or December for the second). Most of the local varieties, especially in the more humid areas, are of the Virginia type that has much better leaf-spot resistance. Planting depth should be 3–8 cm—deeper in sandy soils.

Nutritional Value and Uses of Peanuts. The mature, shelled nuts contain about 28 to 32 percent protein and vary from 38 to 47 percent oil in Virginia types to 47 to 50 percent oil in Spanish types. They are also a good source of B vitamins and vitamin E. Although lower in the essential amino acid lysine (a determinant of protein quality) than the other pulses, peanuts are a valuable source of protein.

In the developing nations peanuts are consumed raw, roasted, or boiled, or used in stews and sauces. The oil is used for cooking and the hulls for fuel, mulching, and improving clayey garden soils.

Commercially, the whole nuts are used for roasting or for peanut butter. Alternatively, the oil is extracted using an expeller (pressing) or solvent method, and the remaining peanut meal or cake (about 45 percent protein) is used in poultry and swine rations. Peanut oil is the world's second most popular vegetable oil (after soybean oil) and can also be used to make margarine, soap, and lubricants. The hulls have value as hardboard and building block components.

Plant Characteristics of Peanuts. Peanuts are legumes and can satisfy all or nearly all of their nitrogen needs through their symbiotic relationship with a species of Rhizobia bacteria. A characteristic of the peanut plant is that the peanuts themselves develop and mature underground.

Peanut Stages of Growth. Depending on variety, peanuts take anywhere from 90 to 110 days to 120 to 140 days to mature. The peanut plant will flower about 30 to 45 days after emergence and will continue flowering for another 30 to 40 days. The peanuts will then mature about 60 days after flowering.

It is important to note that the fruits do not all mature at the same time, since flowering occurs over a long period. An individual fruit is mature when the seed coats of the kernels are not wrinkled and the veins on the inside of the shell have turned dark brown. Harvesting cannot be delayed until all the fruits have matured or heavy losses will result from pod detachment from the pegs and from premature sprouting (Spanish-Valencia types only). Choice of harvesting date is an important factor in obtaining good yields.

Common Beans and Cowpeas

Along with peanuts, this group makes up the bulk of the edible pulses grown in tropical and subtropical developing nations. Aside from their importance as a protein source, the crops play an important role in the farming systems of these areas.

- They are especially well suited to climates with alternating wet and dry seasons.
- Being legumes, they are partly to wholly self-sufficient in meeting their nitrogen requirements.
- They are the natural partners of the cereals in intercropping and crop rotations.

This section deals with common beans and cowpeas (dry beans). Other useful pulses are pidgeonpeas, chickpeas, lima beans, mung beans, soybeans, and winged beans.

Common (Kidney) Beans (Phaseolus vulgaris)

Other names are field beans, frijoles, haricot beans, string beans (immature stage), and snap beans (immature stage).

Types. Bean varieties can be classified according to three basic characteristics: seed color, growth habit, and length of growing period. Varieties can be erect bush, semi-vining or vining types. The latter have a vigorous climbing ability and require staking or a companion support crop like maize. Bush varieties flower over a short period with no further stem and leaf production afterwards. These are called determinate; indeterminate plants flower over several weeks. In warm weather, early varieties can produce mature pods in about 70 days from plant emergence, while medium and late varieties take 90 days or more. Time to first flowering ranges between 30 and 55 days. With some exceptions, the erect bush types reach maturity earlier than the vining, indeterminate types. Plant breeders are developing indeterminate varieties with shorter growing periods and more compact maturity (see Figure 14).

Climatic Requirements of Beans. *Rainfall*: Common beans are not well suited to very high rainfall areas (such as the humid rainforest zones of tropical Africa) because of increased disease and insect problems. Ideally, planting should be timed so that the latter stages of growth and harvest occur during reasonably dry weather.

Temperature: Compared to sorghum and millet, beans do not tolerate extreme heat or moisture stress well. Few varieties are adapted to daily mean temperatures (average of daily high and low) over 28°C or below 14°C. Optimum temperatures for flowering and pod set are a daytime

The flowers develop into pods after pollination.

Part of a bean plant with flowers.

A bean pod.

FIGURE 14
Bean plant and pod.

high of 29.5°C and a nighttime low of 21°C. Blossom drop becomes serious over 36°C and is aggravated also by heavy downpours.

Soil: The plants are very susceptible to fungal root rot diseases, and good drainage is very important. They usually grow poorly in acid soils much below pH 5.6, since they are especially sensitive to the high levels of soluble manganese and aluminum that often occur at the lower pH levels.

Planting. Bush beans grown alone give highest yields in spacings of either 30 cm between rows and 9 cm between plants or 45 cm between rows and 60 cm between plants (equivalent to about 400,000 seeds/hectare). A yield plateau is usually reached around 200,000–250,000 actual plants per hectare, but stand losses from planting to harvest are often in the 25 to 40 percent range, meaning that considerable overplanting is necessary. High-density plantings also appear to increase the height of the pods from the ground, which lessens rotting problems. However, very narrow rows aggravate Sclerotium stem rot where it occurs.

Trials with climbing beans show that final plant populations of 100,000 to 160,000/ha are optimum, whether grown alone with trellising or with maize. Planting depth should be 3–8 cm—deeper in sandy soils.

Nutritional Value and Uses of Beans. Common beans contain about 22 percent protein on a dry seed basis. They provide adequate protein quality and quantity for older children and adults if eaten in the proper proportion with cereals (about a 2:1 grain:pulse ratio). In the green bean form, they provide little protein but are a good source of vitamin A. The leaves can be eaten like spinach and also are used as livestock forage.

Cowpeas (*Vigna sinensis, V. unguiculata, V. sesquipedalia*)

Types. Cowpeas have much the same variations in seed color, growth habit, and length of growing period as common beans except that cowpea seeds are usually brown or white. Most traditional varieties tend to be late maturing (up to five months' and vining). Improved bush (little or no vining) types are available and capable of producing good yields in 80 to 90 days.

Growing Practices and Yields of Cowpeas. Traditional practices and yield constraints of cowpeas are similar to those of common beans. Average yields in the developing countries run from 400 to 700 kg/ha of dry seed, compared to a California (U.S.) average of about 2,200 kg/ha under irrigation. Field trial yields in Africa and Latin America are largely in the 1,500 to 2,000 kg/ha range with some over 3,000 kg/ha. In West Africa, improved cowpea varieties of the vining type are grown at population densities ranging from 30,000 to 100,000/ha in rows 75–100 cm apart.

Climatic Requirements of Cowpeas. *Rainfall*: Cowpeas are the major grain legume (peanuts excluded) of the West African savanna (zone). However, they also are grown in many other regions. They have better heat and drought tolerance than common beans, but the dry seed does not store as well and is very susceptible to attacks by weevils.

Temperature: High daytime temperatures have little effect on vegetative growth but will reduce yields if they occur after flowering. High temperatures at this time can cause the leaves to senesce (die off) more quickly, shortening the length of the pod-filling period. High temperatures will also increase the amount of blossom drop. As with common beans and most crops, humid, rainy weather increases disease and insect problems. Dry weather is needed during the final stages of growth and harvest to minimize pod rots and other diseases.

Soil: Cowpeas grow well on a wide variety of soils (if they are well drained) and are more tolerant of soil acidity than common beans.

Nutritional Value and Uses of Cowpeas. The dry seeds contain about 22 to 24 percent protein. The immature seeds and green pods also are eaten. They are considerably lower in protein than the mature seeds, but are an excellent source of vitamin A while green, as are the young shoots and leaves. The plants are a good livestock forage and are sometimes grown as a green manure and cover crop.

FERTILIZERS

Fertilizer use is often the management factor producing the largest increases in reference crop yields. Aside from water, sunlight, and air, plants need 14 mineral nutrients that are usually grouped as follows.

Macronutrients

Primary   *Secondary*
Nitrogen (N)   Calcium (Ca)
Phosphorus (P)   Magnesium (Mg)
Potassium (K)   Sulfur (S)
Micronutrients (not primary or secondary)
Iron (Fe)
Zinc (Zn)
Manganese (Mn)
Boron (B)
Copper (Cu)
Molybdenum (Mo)

The macronutrients make up about 99 percent of a plant's diet. Nitrogen, phosphorus, and potassium account for about 60 percent and are definitely the "Big Three" of soil fertility, both in terms of the quantity needed and the likelihood of deficiency.

Fertilizer Types and How to Use Them

Chemical (inorganic) fertilizers are frequently accused of everything from "poisoning" the soil to producing less tasty and nutritious food. Should the extension worker encourage client farmers to forget about chemical fertilizers and use only organic ones (compost, manure)? The "organic way" is basically very sound because organic matter (in the form of humus) can add nutrients to the soil and markedly improve soil physical condition (tilth, water-holding capacity) and nutrient-holding ability. Unfortunately, some misleading and illusory claims on both sides of the issue cause a lot of confusion.

Chemical fertilizers supply only nutrients and exert no beneficial effects on soil physical condition. Organic fertilizers do both. However, compost and manure are very low-strength fertilizers; 100kg of 10-5-10 chemical fertilizer contains about the same amount of N-P-K as 2,000 kg of average farm manure. The organic fertilizers need to be applied at very high rates (about 20,000 to 40,000 kg/ha per year) to make up for their low nutrient content and to supply enough humus to measurably improve soil physical condition.

Recommended Fertilizer Rates

Since the efficiency of fertilizer response declines as rates go up, the small farmer with limited capital is usually better off applying low to medium rates of fertilizer. He or she will end up with a higher return per dollar invested, be able to fertilize more land, and have money left over to invest in other complementary yield-improving practices.

Keeping in mind the many factors that determine optimum fertilizer rates, Table 5 provides a very general guide to LOW, MEDIUM, and HIGH rates of the "Big Three" based on small

TABLE 5
Recommended fertilizer rates

	Low (Lbs./acre or kg/hectare)	Medium (Lbs./acre or kg/hectare)	High (Lbs./acre or kg/hectare)
N	35–55	60–90	100+
P_2O_5	25–35	40–60	70+
K_2O	30–40	50–70	80+

farmer conditions and using localized placement of P. The "high" rates given here would be considered only low to medium by most farmers in Europe and the United States, where applications of 200 kg/ha of N are not uncommon on maize and irrigated sorghum.

WEED CONTROL

Numerous trials in the United States have shown maize yield losses ranging from 41 to 86 percent when weeds were not controlled. One trial in Kenya yielded only 370 kg/ha of maize with no weed control compared to 3,000 kg/ha for clean, weeded plots. A CIAT trial with beans in Colombia showed a yield drop of 83 percent with no weeding. Of course, all farmers weed their fields to some extent, but most of them could significantly increase their crop yields if they did a more thorough and timely job. A University of Illinois (U.S.) trial showed that just one pigweed every meter along the row reduced maize yields by 440 kg/ha. By the time weeds are only a few inches tall, they have already affected crop yields.

Weeds lower crop yields in several ways.

- They compete with the crop for water, sunlight, and nutrients.
- They harbor insects, and some weeds are hosts for crop disease (especially viruses).
- Heavy infestations can seriously interfere with machine harvesting.
- A few weeds like Striga (witchweed) are parasitic and cause yellowing, wilting, and loss of crop vigor.

Relative competitive ability of cereal crops: Slow starters like peanuts, millet, and sorghum compete poorly with weeds during the first few weeks of growth. Low-growing crops like peanuts, bush beans, and bush cowpeas, however, are fairly effective at suppressing further weed growth once they are big enough to fully shade the inter-row spaces. However, tall-growing weeds that were not adequately controlled earlier can easily overtake these "short" crops if allowed to continue growing.

Weed control methods include burning, mulching, shading (arranging crops in rows gives better shade competition against weeds), hoe and machete cultivation, animal and tractor-drawn cultivation, and herbicides.

METHODS OF INSECT CONTROL

Biological control is the purposeful introduction of predators, parasites, or diseases to combat a harmful insect species. About 120 different insects have been partially or completely controlled by this method in various parts of the world. Microbial insecticides such as Bacillus thuringiensis (effective against a few types of caterpillars) are now commonly used by farmers and gardeners in many areas. Unfortunately, biological control measures are presently effective against a very small portion of harmful insect species.

Cultural controls: Cultural controls such as crop rotation, intercropping, burying crop residues, timing the crop calendar to avoid certain insects, and controlling weeds and natural vegetation that harbor insects are all effective control methods for some insects. In most cases, however, cultural controls need to be supplemented by other methods.

Varietal Resistance: Crop varieties differ considerably in their resistance to certain insects. For example, maize varieties with long, tight husks show good resistance to earworms and weevils. Screening for insect resistance is an important part of crop breeding programs.

"Organic" controls: "Organic" control refers to nonchemical methods in general. These include the application of homemade "natural" sprays made from garlic, pepper, onions, soap, salt, and so on and the use of materials like beer to kill slugs and wood ashes to deter cutworms and other insects. Some of these "alternative" insecticides are slightly to fairly effective on small areas like home gardens and where insect populations are relatively low. They are seldom feasible or effective on larger plots, especially under tropical conditions that favor insect buildup.

Chemical control: Chemical control refers to the use of commercial insecticides in the form of sprays, dusts, granules, baits, fumigants, and seed treatments. While some of these insecticides like rotenone and pyrethrin are naturally derived, most are synthetic organic compounds that have been developed through research. Insecticides are relatively inexpensive, and their proper usage can often return $4 to $5 for every $1 spent. Insecticides do have drawbacks. Insect resistance to pesticides is a growing problem. Few insecticides kill all types of insects, and some actually promote the increase of certain pests. For example, continual use of Sevin (carbaryl) in the same field may increase problems with some types of aphids, which it does not control well. Insecticides damage nontarget species. These include beneficial predators such as bees and wildlife. Some insecticides present a residue hazard—DDT, Aldrin, Endrin, Dieldrin, and Heptachlor are highly persistent in the environment and may accumulate in the fatty tissues of wildlife, livestock, and humans. Many other insecticides, however, are broken down into harmless compounds fairly rapidly. Also, some insecticides are extremely toxic in small amounts to humans and animals—insecticides vary greatly in their toxicity. When applying insecticides, the manufacturer's guidelines should be followed carefully.

INTEGRATED PEST CONTROL

The disadvantages of total reliance on insecticides have given rise to integrated pest control or pest management that involves the judicious use of these chemicals based on the following guidelines and principles.

- The development and use of cultural and other nonchemical control methods to avoid or reduce insect problems.
- Determining crop tolerance to pest damage based on the principle that complete freedom from pests is seldom necessary for high yields. Nearly all plants can tolerate a surprising amount of leaf loss before yields are seriously affected.
- The appropriate timing and frequency of treatments to replace routine, preventative spraying. Treatments are not initiated before the particular insect has reached the economic damage threshold, which will vary considerably with the species.
- Insect scouting—looking for related kinds and number of insects and their density and population counts—is an essential part of this system.

Integrated pest control of cereal crops is still in the very early stage, especially in the Third World.

BIRD AND RODENT CONTROL

Birds: Perhaps the most effective control method is continuous string flagging, which uses cloth or plastic streamers 5–6 cm wide and 50–60 cm long. The streamers are attached at 1.5 m intervals to strong twine, which is strung along heavy stakes at least 1.2 m tall spaced about 15 m apart.

Rodents: Rodents can be discouraged from entering fields by maintaining a 2.0 to 3.0-m-wide cleared swath around the field borders from planting until harvest. Fences made from oil palm fronds or split bamboo are also effective, especially if traps or wire snares are set in the gaps. Good weed control in the field is helpful. Leaning or fallen plants should be propped up and the dry lower leaves stripped off to help deter climbing. Many villages carry out organized killing campaigns. The best time for such campaigns is during the dry season when the rats congregate in the few remaining pockets of green vegetation. Repellants like Nocotox 20 may be partly effective. Rats should be prevented from gaining access to stored grains and other food that can cause a buildup in populations during the dry season. Poison baits can be used, but killing rats in the field with poisons, traps, and other methods is usually not a very effective long-term solution. The best approach is to prevent a rat population buildup; this requires area-wide coordination.

REFERENCES

Price, Martin L. (1990). "Pigeon Pea." Echo Technical Note, ECHO, 17391 Durrance Rd., North Ft. Myers, Fl 33917. Tel.: 914-543-3246. echo@echonet.org, http://www.echonet.org

Stephen, Roy M., and Betsey Eisendrath. (1986). "Understanding Cereal Crops Maize, Sorghum, Rice, And Millet." Volunteers in Technical Assistance, 1600 Wilson Boulevard, Suite 710, Arlington, Virginia 22209, USA. Tel.: 703-276-1800. Fax: 703-243-1865. vita@vita.org, http://idh.vita.org/pubs/docs/uhn.html

---

COCONUT PRODUCTION

Tuarira Abduel Mtaita

INTRODUCTION

The coconut palm (Cocos nucifera L.) is also known as a tree of abundance, tree of life, and tree of heaven. It is a beautiful palm that can reach a height of 30 m. Coconut palms bear fruit at from 6 to 10 years, though full production is not reached until the trees are 15 to 20 years old. Fruits require about one year to develop.

Coconut is used mainly for oil production, and copra is the richest source of vegetable oil, with an oil yield of about 64 percent. Coconut oil is used for making soaps, creams, confectioneries, and cooking oil. Coir or coconut fiber is used to make mats, floor coverings, brushes, furniture, upholstery filling, and ropes that are resistant to water. The shells are used as fuel for drying kilns. Leaves are used for roofing and mat making. Roots are extensively used in traditional medicines and in the production of dye. The sweet sap obtained from the inflorescences before the flower matures is used to make wine. The trunk may be used for house and boat building. Despite of the shining virtues of the coconut palm, its production standards remains, for the most part, rather low.

ENVIRONMENTAL REQUIREMENTS

Latitude: Coconut is essentially a crop of humid tropics. Important centers of production lie in a zone no wider than 15° of latitude. However, within this zone, it shows a fairly wide adaptation to various climatic changes.

Altitude: Coconut palm grows well from sea level up to 1,100 m. However, the highest yields are generally obtained from trees grown at 600 m to 900 m above sea level. Within these limits altitude is less important than constant water supply.

Rainfall: Coconut requires a well-distributed annual rainfall of about 1,250 mm. Droughts cause abortion of the inflorescence and a reduction in the number of nuts and the amount of copra per inflorescence. In climates with heavy rainfall (4,000 mm per annum), insulation may become a limiting factor.

TEMPERATURE

The ideal temperature for coconut production is around 27°C, with average diurnal variation between 5°C and 7°C. Low temperatures are more limiting than high temperatures. Temperatures below 20°C may cause abnormalities to the fruit. In fact, frost is lethal to coconut.

SOILS

Coconut palms need free draining soils, which are deep enough to allow root development. It does well in sandy and saline soils that would support few other crops. It is a heavy feeder of potassium, and this element should be amply available.

VARIETIES

The varieties are grouped into tall and dwarf cultivars. Dwarf types are planted for ornamental purposes because they produce little copra and limited wine though excellent immature nuts.

MANAGEMENT PRACTICES

Nursery: Seed nuts may be placed in a germinating bed before planting out in the nursery. In the germination bed, the nuts are planted much closer than in the nursery. Germination begins about 12 weeks from planting and reaches maximum at 17 to 18 weeks. Seedling selection is restricted to healthy and vigorous seedlings.

The selected seedlings are then transplanted to the nursery at a spacing of 25 cm × 25 cm apart, using a triangular layout. Manuring is not necessary, since the seedlings have enough reserve to last for 12 months. Termites may cause considerable damage to the seedlings, and to guard against this chemical control may be necessary. Diseased seedlings are normally removed and destroyed to avoid spreading disease to healthy seedlings.

Land Clearing: The clearing of new land is not different from that of any other crop. Total burning of the organic mass, after felling and drying, is a common practice in tropical agriculture. However, burning plant residues is advisable only when absolutely necessary, since this lead to a loss of potash, nitrogen, and other plant nutrients.

When clearing an old coconut plantation, stumps, stems, and leaves must be either buried deeply or burned to discourage rhinoceros beetles from breeding in the decomposing mass. These beetles may cause serious damages to the young seedlings.

PLANTING

Planting holes are prepared a few months before planting. Holes should be larger than the polybag in which the seedling is planted. On lighter soils, for example, holes may be 1 m × 1 m × 1 m deep. After they have been excavated, the holes should be filled with a rich loamy soil or well-rotted compost to add initial nutrients to the soil.

When transplanting, the nut should be about 30 cm below the soil surface, and it should be earthed up only to the base of the shoot. The rest of the hole can be filled in gradually as the stem grows. This practice sets the bole well into the ground.

Spacing used depends on water availability, type of the soil, variety grown, and whether pure culture or intercropped. Spacing of 8 m × 8 m is normally recommended. When interplanting with cashew, mangoes, bananas, or cassava, adequate spacing of both coconut palms and fruit trees should be maintained. Interplanting with annual crops is a common practice to cover the cost of establishment.

FERTILIZATION

Generally, coconut palms are planted on rather poor soils. Even in better soils, the continuous harvesting of nuts and sometimes leaves will deplete the soil of one or more elements. It is therefore important to apply 25 to 50 kg of cattle manure or 25 kg of compost per tree per year. In the absence of organic fertilizers, 0.9 to 1.2 kg of NPK mixture per tree per year may be applied from two to four years after planting. Each adult palm will require 0.9 kg of muriate of potash plus 0.9 kg of ammonium sulfate plus 0.5 kg of double super-phosphate per year. Fertilizers should only applied in conjunction with appropriate weed control measures.

WEED CONTROL

Weeding is necessary to avoid yield reduction. Weed control practices in coconut plantations depend on local conditions such as soil, climate, weed species, and the age of the plantation. However, for the first year of planting, it is only necessary to maintain a weed-free circle around each tree. For mature trees, weed growth and bush regeneration should be checked throughout. Slashing and grazing may also be used as an alternative to clean weeding. Leguminous cover crops are often planted to check weed growth, prevent nutrient loss, and add nitrogen to the soil.

IRRIGATION

In the absence of rain or in nonirrigated plantations, watering is required until the seedlings form roots and become established. In general watering for two to three years after transplanting is recommended. The economics of irrigating coconut plantations depends on the availability of water, the irrigation method, and the overall cost. Palms with a high yielding potential have been found to respond to irrigation better then the poor yielding palms. As with fertilization, the first response of the palm to irrigation is the reduction of premature nut-fall and an almost direct yield increase within a year.

HARVESTING

Harvesting is done by skilled climbers and sometimes trained monkeys. Immature nuts are harvested nine or ten months after flowering when they have high water content. Fully mature nuts are harvested for the best yield of copra and oil content. They are picked at two-month intervals. At this rate, two bunches per tree should be ready at each picking. In the absence of theft, they may be allowed to fall to the ground. Cutting steps in the trunk should be discouraged because it harms the tree.

COPRA DRYING

Copra is either sun or kiln dried. In sun drying, the husks are removed by impaling the nuts on a sharpened stake whose other end is fixed firmly in the ground. The halves are laid in the sun with the flesh facing upwards. Copra should be dried in a clean surface. After two to three days, the flesh should have contracted enough to be removed from their shells. Moisture is reduced from 45 percent when they are split to 6 percent when they are fully dry. In fine, dry weather this takes about five days after removing them from their shells. It is important to cover them at night and from the rain.

Kiln drying: Most kiln designs consist of a pit for making fire, a wire mesh platform over the fire to support the copra, and a roof to protect the copra from the rain. Coconut shells are the common source of fuel because they are readily available and they produce a considerable amount of heat with a limited smoke. Kiln drying takes about four days.

Proper dried copra from mature nuts is brittle, clean, and white, and it smells fresh. It has moisture content of about 6 percent, oil content of 65 to 70 percent, and free fatty acid content of below 2.5 percent. Immature nuts give flabby and elastic copra that results in poor oil extraction during milling.

PESTS AND DISEASES

Rhinoceros beetle (Oryctes monoceros): The beetle attacks the terminal buds in the crown, eating unopened leaves and occasionally destroying the growing point. Chemical control is not practical because the crowns are so inaccessible. The best approach is to destroy all decaying coconut palms. This involves splitting and drying the trunks to make them burn, since they are very difficulty to burn whole. Poking a piece of wire into their entry points can kill adult beetles.

Coreid bug (Pseudotheraptus wayii): This small bug punctures the very young nuts, causing the nuts to drop prematurely. Predators like Oecophylla longinoda can check them. Planting cashew or citron trees between the coconut palm may encourage this particular predator.

Red legged ham beetle (Necrobia rutipes): This beetle attacks moist copra, causing loss in weight and large amounts of useless copra dust.

DISEASES

Bole rot caused by the fungus Marasmiellus cocophilus is soil borne and enters the plant via the roots, causing wilting and yellowing or bronzing of leaves. As a precaution, transplanting should be done at early stages without damaging the roots. Seedlings from infested nurseries and implements that injure the roots during cultivation should be avoided.

YIELD

The yield potential is about 100 to 140 nuts per palm per year. However yields as low as 15 to 20 nuts per palm per year is common in poorly managed plantations. Two mature nuts should yield 0.45 kg of good quality copra.

REFERENCES

Ackland, J. D. (1971). *East African Crops.* FAO/Longman.
Green, A. H. (1991). *Coconut Production.* Washington, DC: The World Bank.

GROWING FRUIT UNDER MARGINAL CONDITIONS

Tuarira Abduel Mtaita

INTRODUCTION

The importance of fruits in providing valuable nutrients, particularly vitamins and minerals, in the human diet is well known. Nutritionists advise a daily intake of at least 100 g and as much variety as the season permits. Because our bodies cannot store vitamin C, we must eat fresh fruit every day. In many tropical countries there is a serious shortage of fresh fruit, at least during part of the year. To correct this shortage, the production will have to be raised through increasing orchard efficiency and by bringing more farming areas under fruit crops.

Traditionally, fruit growing has been associated with fertile wetlands and garden lands with assured irrigation. In marginal lands of arid and semi-arid regions with little or no supplementary irrigation, the grower may create microclimates for planting and use special practices necessary for optimal fruit production.

MANAGEMENT PRACTICES

Using modern planting methods, one of the most important factors enabling high productivity is efficient planting and early care of the young trees. Some of the most important management practices of value to fruit tree growing under marginal areas follow.

Sowing

Direct sowing should immediately follow seed treatment or the germination rate is reduced. The methods of sowing are in situ and field nurseries.

Sowing in Situ. In this method, seeds remain at their planting station for their lifetimes. Planting holes for fruit trees are dug 60 cm × 60 cm × 60 cm, using spacing suitable for the given species. The soil is pulverized and filled to three-quarters of the depth. The remainder is filled with soil mixed with organic matter in a 1:1 volume ratio. In addition, 5–10 g of a slow release fertilizer is added. The seed is later planted at 1 to 20 seeds per station, depending on the crop being grown. For example, coconut is sown at one seed per station, while with papaya one can use up to 20 seeds per station. The plot should be kept moist at all times to encourage germination and growth of the seedlings. Rouging should be based on the trueness to type, vigor, and health.

Field Nurseries. Two types of nurseries are used.

- *Seedbed culture*: Fertile and well-drained soils are necessary for good and healthy seedlings. In addition, shrinking and cracking soils must be avoided, since they result in extensive root damage, which might reduce the rate and efficiency of the seedling establishment. The soil must be disked to 25–30 cm and, in addition, pulverized to a fine-level tilth. Nursery beds are then made, the size depending on the intended production scale. It is important that these beds are raised about 2 to 3 cm above ground level to reduce the incidence of soil-borne pathogens. Organic matter may be incorporated at 5–10 kg/m^2, or alternatively, 10 kg/ha N, 10 kg/ha P_2O_5, and 5–10 kg/ha K_2O are applied. On planting, the seed is either drilled or broadcast close enough to give a high plant population, avoiding overcrowding that might result in competition among plants and consequently economic losses. The seedbed should be kept moist all times to improve seedling performance. Further nutrient dressing should be performed based on appearance of deficiency symptoms.
- *Nursery-row culture*: Land preparation is similar to seedbed culture. The rootstock is established in the nursery at 1.2 × 1.2 m spacing. The desired variety is then budded in place, and the resultant plants are planted out in the field at two- to three-year intervals. Citrus and mangoes are commonly raised in this manner.

Grafting: This is a process in which a detached plant part is united with another to form a composite plant. It is mainly employed in changing the flowering or fruiting variety in fruits like citrus, mangoes, avocados, grapes, apples, and peaches. It is also used to propagate plant types that cannot be propagated directly from seed—for example, Washington navel oranges. Furthermore, rootstock characteristics like tolerance to salinity, sodicity, heat, frost, and specific diseases are utilized, a feature that is lacking in direct seeding and in cuttage. These characteristics permit fruit production in areas of adverse conditions. Graftage is also used in reducing the period of juvenile growth and hence assists early yields in fruit crops.

Layout: For the optimum development of bearing wood, fruit trees require maximum exposure to light. The shading of any part of the tree, especially the lower portion, should therefore be avoided. Several planting systems are used, of which the contour and rectangular systems are recommended.

- Contour: In this planting system the trees are planted according to the slope of the land and its contour. It is the only method for hilly areas where no rectangular system can be used. This is the common method for smallholder farmers.

- Rectangular system: This method is highly recommended in flat lands. To calculate the number of tree needed for a certain plot, divide the area of the plot by the tree spacing.

Transplanting: This is done two- to three-years after nursery establishment and is after two- to three-week hardening, depending on the species. Transplanting also offers an opportunity for removing inferior plants. In addition, root and shoot pruning increase the rate of establishment of the transplants.

Planting depth should be approximately to the same level in the nursery. Firm planting should be the rule, and plants should be well watered. If a grafted or budded seedling is used, make certain that the tree is not placed deeper than it was in the nursery. When establishing unbranched "whips" like those of deciduous fruit trees that have not been pruned, one-half of the tops of such plants should be cut off to balance the top with the remaining root system.

Deblossoming: Some young trees flower soon after transplanting at the expense of vegetative growth. The flowers should be continuously nipped off until a strong basal frame is formed.

Provision of micro-catchment around the tree: In areas where the average annual rainfall is very low, catchment areas for each tree may be developed, the size determined by the slope of the land, water requirement, runoff coefficient, and the feeding zone.

Mulching: A mulch material of straw, fern fronds, leafy branches, cut grass, dry leaves, or plastic should be laid round each plant to minimize the evaporation of water from the soil. Mulch also suppresses weeds, prevents soil erosion, and adds organic matter to the soil. A few centimeters around the tree should be left uncovered to prevent the base from becoming excessively wet and prone to attack by disease organisms and to discourage small rodents from chewing the bark.

Manuring: The manuring of trees is simply accomplished by scattering fertilizers around each tree over an area corresponding to the spread of the root system.

Crescent bunding and opening of catch pits: Curved bunds with a diameter of 5 to 10 meters and catchpits are prepared on the upper side of the slope. Crescent bunds help collect rainwater, while the catchpits conserve the same.

Trench planting: Under dry conditions, deep trenches (0.5–2 m) are dug to collect rainwater along with silt and organic matter. Planting in these trenches promotes tree growth.

Windbreaks: Living windbreaks should be composed of well-anchored tall trees, preferably leguminous species. In addition, an inward row of any fruit crop may also be planted to reduce wind velocity and maintain humidity, thus ensuring water conservation and a microclimate for good growth.

Irrigation: Specific details are subject to variation depending on the species. The grower should ensure an adequate supply during the critical periods such as flowering and fruit set.

Disease and pest control: Methods used in the control of diseases and pests include the following.

1. Sanitation: eradication, disinfection, rotation
2. Legislation: quarantine laws
3. Resistance: use of resistant or tolerant cultivars
4. Mechanical means: handpicking, flouring, banding
5. Biological means: predators, parasites
6. Chemical means: spraying, dusting
7. Integration: a combination of methods

Pruning. Trees are pruned to remove only undesirable wood without upsetting the balance of the tree or going against the natural habit. The primary objective is to direct the growth of the plant to better serve the purpose of the pruner. To maintain health and safety, dead, broken, and disease- or insect-infested wood should be removed. Careful pruning helps to control the form and size of the fruit tree, stimulates vigorous growth, and brings the shoot system into balance with the root system. To accomplish these objectives, or others that might be applicable, one must have some knowledge of the natural form and growth habit of the tree, the time of bud initiation, flowering, and fruiting, as well as the susceptibility of the species to winter or other injury.

Establishment of maturity indices: These are measurements used to determine whether a fruit is mature. They are important in trade regulations, especially in strategizing the market and for efficient use of labor and resources.

Harvesting. The aim should be to gather fruits from the field at the proper level of maturity with a minimum of damage and loss, as rapidly as possible, and at a minimum cost. Today as in the past these goals are achieved through hand harvesting.

CHALLENGES

Environment. The protection of the environment and improvement of the human environment should be one of the major considerations in any agricultural activity. In growing fruit trees, choices must be made: the needs to be satisfied, the methods to satisfy those needs, and the environmental and other side effects that those methods imply. Riverbank cultivation should be avoided at all costs. An area that is barren and without value may be reclaimed through hard work and patience. Earthen barriers bunds may be constructed to retain rainwater and replenish wells. Multipurpose trees, especially leguminous plants, may be intercropped with valuable fruit trees like mango, jack fruit, cashew, papaya, and bananas to improve production.

Delayed foliation: Most deciduous fruit trees grown under subtropical climates suffer in varying degrees from prolonged dormancy. A reason for prolonged dormancy is absence of a cold season and then a subsequent warm spring. Insufficient chilling may be compensated by the following.

1. Defoliation of trees immediately or up to four weeks after the harvest.
2. Spraying trees when dormant with Dinitrocresol spray in oil. A locally made oil may be used, the stock emulsion of which contains 1 liter linseed (raw), 8 liters water, and 200 grams soft soap. One liter of stock should be sprayed mixed with 4 liters of water.
3. Choosing appropriate sites in relation to varieties. Pome fruits, which bloom after the frost, may be planted in the so-called frost pockets (in valleys) to get greater winter chilling. Those that bloom early should be planted in a frost-free area. Stone fruits can be grown successfully at lower altitudes, since they need far less chilling in order to break their dormancy.

Fruit culture. There is a great need for standardization of techniques for faster multiplication of planting materials and establishment of nurseries to raise and supply much-needed planting materials. Extension activities to educate farmers on fruit growing under suboptimal conditions should be strengthened. Suitable technologies to reduce postharvest losses need to be introduced.

Maintaining soil fertility. Due to the escalating costs of inorganic fertilizers, fruit growers should be trained on how to maintain soil fertility using compost, green manure, worm compost, multipurpose trees, farmyard manure, liquid manure, and comfrey.

REFERENCES

Sampson, J. A. (1986). *Tropical Fruits*. London: Longman Scientific and Technical.

Van Ee, Simone. (1999). *Fruit Growing in the Tropics*. AGROMISA, PO Box 41, 6700 AA Wageningen, The Netherlands. Fax: +31 317 419178. agromisa@worldaccess.nl

---

GROWING RICE

Based on Wet Paddy or Swamp Rice, Food and Agriculture Organization of the United Nations, Rome, 1984, ISBN 92-5-100622-9 and Upland Rice, Food and Agriculture Organization of the United Nations 1977 Rome, 1970, ISBN 92-5-100621-0

Rice offers much to the Third World. It is an excellent energy food, and it keeps well. Cassava, yams and bananas rot quickly, but rice does not rot if it is protected from dampness. Rice can be kept a very long time if it is protected from rats and insects. There are many regions of Africa where rice can be grown: along the banks of streams and rivers, around dams, and at the bottom of valleys. Rice can be grown through wet paddy cultivation or upland rice cultivation. The two methods of cultivation are quite different. Another article in this section, "Rice Paddy: Harvesting, Threshing, Cleaning, and Handling," focuses on the technology of rice cultivation.

DIFFICULTIES WITH TRADITIONAL WET PADDY RICE CULTIVATION

Wet paddy, or swamp rice, is grown on land that is covered with water when the water rises at the beginning of the rainy season and falls at the end of the rainy season. It is also grown on lowland plains that are covered with water most of the year. When these plains are covered with very deep water, people sow floating rice. The stems of floating rice grow longer when the water rises. Yield varies a great deal depending on how much it rains. With traditional methods it is impossible to give rice the amount of water it needs at the appropriate time. It is impossible to make the water go away or to drain the land. Rice does not need the same depth of water all the times while it is growing. The farmer has no control over the depth of the water cover.

To get good yields from wet paddy cultivation, the farmer must do the following.

- Be in control of the water—have enough water when the rice is growing, add more water at the right time, take water away (drain) at the right time, and have the right depth of water.
- Make good nurseries so as to have fine seedlings for transplanting.
- Transplant into rows at the right date.
- Prepare the soil of the rice field well by tillage and levelling.
- Tend the rice field carefully. Cultivate whenever weeds have grown. Apply fertilizers, flood the field, protect the rice from pests.
- Harvest with care, and dry the paddy grains well.

CONTROLLING WATER

Extensive water works must be built to control the water irrigating the rice field. Start with a tract of low-lying land with a stream or small river flowing through it. Dam the stream—concrete dams last longer. Build irrigation channels around the dam so water flows into the field evenly. At the downstream edge of the field build another dam across the stream. Build a drain channel through the field, leading to the stream flowing out of the field. Build smaller channels to take water from the main irrigation channels and other smaller channels to take water from the field to the drain channel. This network of channels will divide the field into plots. These plots should be less than 20 m in each direction. Make sure every channel has sufficient slope so water will flow—a slope of 1 in 200 should be adequate. The channels should be straight so the water does not erode the banks. Earth that is dug up when forming the channel is put next to the channel walls, forming footpaths through the field. If the field is not level, divide it into small plots and build walls around the plots so the water level will be same throughout the plot.

FLOODING AND DRAINING

When you want water to flood any particular plot, block up the main irrigation channel with earth just below the distribution channel leading to that plot. When there is a good depth of water in the plot, open the irrigation channel and flood the next plot.

In the same way, when you want the water to flow out of a plot, you can make a hole in the earth wall that separates the plot from the feeder drain. You can also push a pipe through the wall—for example, a piece of bamboo—to connect the plot with the drain.

In order to produce a high yield, the rice must constantly stand in water, but it is wrong to flood the field to always the same depth of water. For the first six to eight days after transplanting, leave the soil a liquid mud. If the soil becomes dry, let in only a little water. About a week after transplanting, when the rice has begun to grow, flood the field with 2 to 3 centimeters of water and leave it like that for 45 days, but twice during these 45 days, drain the field in order to apply fertilizers. Each time, drain for two days. After 45 days (two months after transplanting), increase the depth of water to 10 centimeters. When the panicles (see below) have formed and are turning yellow, the rice field must always be well flooded. It should have about 20 centimeters of water. Afterwards, gradually make the water less deep. Ten days before harvesting, drain away all the water.

MAKE A NURSERY FOR YOUNG PLANTS

You should make a nursery on one of the plots of the rice field or close to your house. If the nursery is near the house, it is easier to look after. If the nursery is on a plot, it needs less work to transport the seedlings and to water. The right size for the nursery is one-tenth the size of the rice field to be planted.

The soil of the nursery must be tilled so the earth is loose, without weeds, and moist and fertile. This work is done with a hoe one month before sowing in the nursery (two months before the right time for transplanting). Put in 200 kilograms of manure per 100 square meters and turn over the soil. Manure improves the soil structure and adds mineral salts to the soil. Then let the soil rest for two weeks. Break up clods of earth and remove weeds. When the soil is well tilled and quite loose, divide the nursery into strips 1.5 to 2 meters wide, 10 to 20 meters long, and 20 centimeters high.

Before sowing, break up all small lumps of earth, fill up hollows, and remove small stones. Apply chemical fertilizers. For each 100 square meters, apply 1.5 kg ammonium sulfate, 1 kg declaim phosphate, and 1 kg potassium chloride.

Spread these fertilizers all over the nursery. For example, on a bed 20 meters long and 2 meters wide, apply 600 grams ammonium sulfate, 400 grams declaim phosphate, and 400 grams potassium chloride. This nursery of 40 square meters will give you enough seedlings for a rice field of 400 square meters.

Always sow unhusked rice grains—that is, paddy. Agricultural extension workers or successful farmers in the region can suggest appropriate varieties to sow.

Before sowing, it is best to pregerminate the seeds and disinfect them. Pregermination will result in more seeds growing and faster growth. Rats and birds will not eat the seeds so easily. To pregerminate seeds, put the paddy grains into a bucket. Fill the bucket with water. Leave the grains in the water for 24 hours. Remove all the grains that rise to the surface, since they will not grow. Afterwards, put the grains into baskets or sacks, and leave them there for one to two days. The grains then begin to germinate. You will see a tiny white spot on the grains. Never pregerminate the grains more than three days before sowing. To disinfect the seeds, put a little disinfectant, such as Panogem, into the water that you use for pregerminating.

Sow the rice grains in the nursery one month before the date of transplanting the seedlings. Broadcast your seed. Use 6 kilograms of paddy for each 100 square meters. For example, if the size of the nursery is 40 square meters, use 2.5 kilograms of paddy. Then cover the grains with very fine earth.

The nursery must be protected against animals, rats, and birds. If you have put your nursery close to the nurseries of other farmers, protection is easier. There needs to be a watchman, especially during the first week. Put straw on the beds. Straw protects the baby seedlings against birds and against the sun. Remove all weeds. If the earth gets too hard, water the nursery in the morning and in the evening.

PREPARING THE PLOTS

Till the soil of the plot two months before transplanting. Remove any weeds that have sprouted. When tilling, mix manure into the soil. Well-tilled soil retains water better. Till to an average depth of 15 centimeters.

Level the soil in the plots. This is done by flooding the soil with only a little water. This way you can see better where there are humps (places where there is little water), and where the water is too deep. Remove the humps, and put the earth into the hollows. When you have done this, go over the ground with a leveling board to make sure the soil is quite flat everywhere in the rice field. It is important to be able to drain the soil well. If the soil is not level, water will remain in the hollows rather than drain. If you have leveled the soil well, the water will cover all the soil, and weeds cannot grow. The ideal plot is shown in Figure 15.

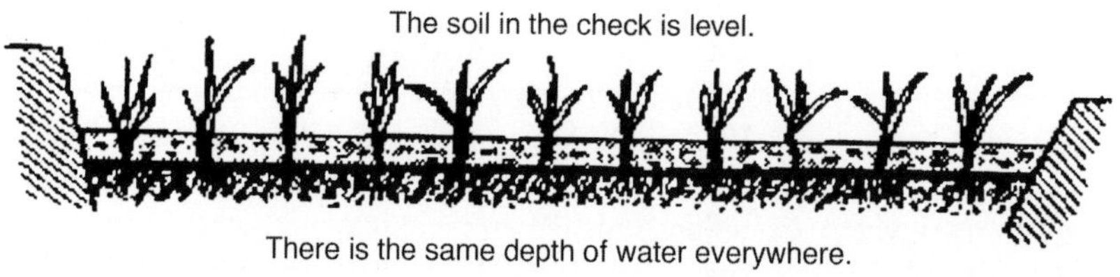

The soil in the check is level.

There is the same depth of water everywhere.

FIGURE 15
A level plot.

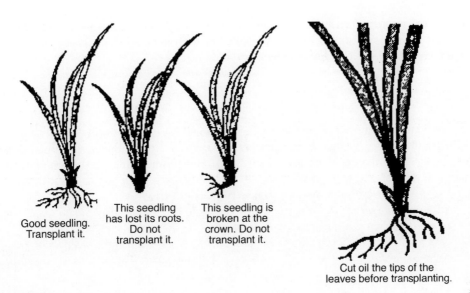

Good seedling.
Transplant it.

This seedling
has lost its roots.
Do not
transplant it.

This seedling is
broken at the
crown. Do not
transplant it.

Cut oil the tips of the
leaves before transplanting.

FIGURE 16
Good and poor seedlings.

REMOVING SEEDLINGS

Take the seedlings out of the nursery 30 days after sowing. Do not leave the seedlings in the nursery more than 30 days. If you wait too long, the seedlings will grow less well later, but if you do not wait long enough and the seedlings are too young, you will spoil them when you take them out. Take out the seedlings when they have four or five leaves. To get your seedlings out without damaging them, water the nursery beforehand. When the soil is very damp, seedlings come out more easily.

It is very important to sort out seedlings before transplanting them. Throw away any seedlings broken at the crown, seedlings that have no roots, and seedlings that are too small. If you sort your seedlings well, all those you transplant will grow and grow better. When you have sorted out your seedlings, tie them together in small bundles so you can transport them more easily.

Next, prepare the seedlings by cutting off the tips of the leaves. This way the leaves will stay straight and will not touch the ground. The transplanted seedlings will grow better. Do not wait longer than two or three days between taking out the seedlings and transplanting them. Seedlings are shown in Figure 16.

TRANSPLANTING

To make the seedlings grow well, they must be transplanted into very wet soil. The soil is right for transplanting when the mud comes up to your calf when you walk in the field. Plant the seedlings in straight rows. This makes it easier to weed and to apply fertilizers. Mark out the rows with a string tied to two pegs. Leave 20 centimeters between rows and, in each row, 20 centimeters between each seed hole. Put two seedlings in each seed hole. If you have left the seedlings longer than 30 days in the nursery, however, put four or five seedlings in each seed hole. Take the roots between your fingers and push your fingers two or three centimeters into the soil so that the roots will be at the right depth in the earth.

If you see seedlings that have not grown six to ten days after transplanting, replace them. Take some seedlings out of the nursery and plant them in the field in the places of those that have not grown. Do not wait too long before you do this.

WEEDING

Weeds take mineral salts out of the soil, so the rice cannot use them and will be undernourished. Weeds get in the way of harvesting and prevent the rice from growing and tillering well. "Tillering" is when the rice develops several stems on the same plant. The buds at the bottom of the stem develop and make new stems. After tillering, each grain you have sown will have several stems, and every stem makes a panicle (or head) of rice.

When should you weed? If you have flooded your field before transplanting, no weeds should have grown. But when you transplant, you drain away the water, so then weeds can grow. Weed two weeks after transplanting, pulling out the weeds by hand. Later more weeds will grow, so you must weed again. Every time you see weeds, pull them out. This is a lot of work, but at least it doesn't cost anything.

Two weeks after transplanting, pull out the weeds. Drain all the water away and apply 100 kilograms of ammonium sulfate for every hectare of rice field. This fertilizer contains nitrogen. When the panicles are forming, drain and apply 50 kilograms of ammonium sulfate for every hectare.

CUTTING THE RICE

You can get a better price for paddy if you cut your rice only when it is quite ripe (a ripe grain makes a crunching noise when you bite it); when the paddy is very clean, without any little stones or weed seeds; and when it is quite dry. If the heads are left on the ground too long, the grains may rot or germinate. Cut the rice with a sickle when the heads are yellow. Either cut the stems close to the ground or cut only the panicle. See also the article "Rice Paddy: Harvesting, Threshing, Cleaning, and Handling" in this chapter.

When you have cut the rice, make sheaves by binding a lot of stems together. Either stack the sheaves so they lean against each other, standing upright with heads upward, placing one sheaf over the top of the heads to protect the grains from the rain, or lean the sheaves against a stick supported by two poles. Either way the rice can dry well. Leave the rice to dry for three or four days before threshing.

THRESHING AND WINNOWING

There are three methods of threshing.

1. Put the rice on a clean, hard, dust-free piece of ground or cover the ground with mats, and beat the heads with a stick.
2. Beat the rice against a large stone or a tree trunk.
3. Use a small thresher. You can join with a few other farmers and buy a small thresher together.

It is important that the rice grains be very clean and free of dirt and stones. When you have threshed your rice, winnow it to make it quite clean. For winnowing, use a sieve or pour the rice from one flat bowl into another. The wind blows the dirt away.

GROWING UPLAND RICE

To improve upland rice cultivation, do the following.

- Use new implements. With animals and a plow the soil can be prepared better and more quickly. It is also possible to cultivate larger fields.
- Plant on a well-prepared soil so the seed can be sown in rows. This makes it easier to remove weeds.
- Apply manure and fertilizers.
- Protect the rice against pests.

PREPARING THE SOIL

Most often, rice is sown on a field that has already been cultivated, perhaps after a crop of yams, groundnuts, or cotton. Before sowing, you must prepare the soil by tilling it. (Tilling is turning over the soil.) If you till 15 to 20 centimeters deep, this will stir the earth very well. Do this with a plow or hoe. Tilling loosens the soil so air and water can get into it. Tilling enables you to mix the herbage with the soil. When the herbage rots, it makes humus. On flat land, if a soil has been well loosened by tilling, the water penetrates well and stays for a long time. Therefore, till at the beginning of the rainy season so the soil holds water. This first tilling is very important. Do it as soon as you can move the soil. Slopes should not be tilled where there is a danger of rain carrying away the soil.

Tilling can be done with a hoe, a spade, or a digging fork, but this is slow and tiring work. Today, people use a plow drawn by donkeys or oxen. Often, people use a simple plow consisting of a plowshare, a moldboard, and two handles.

Plowing often does not leave the soil flat. There remain clods of earth that must be broken up with a harrow. If you do not have a harrow, you can let an animal draw big branches from trees over the field. The branches will crush the clods.

CHOOSING THE SEEDS

If you have already grown a rice crop, choose the best seeds from your own harvest. Remove broken and misshapen grains and grains attacked by insects. The extension services and research centers have selected rice varieties best suited to the climate of each region. They are disease-resisting varieties that provide high yields. Use the finest seeds of your own harvest for sowing in the following years.

DISINFECTING THE SEEDS

Disinfectant is usually available from the extension services. Mix the seeds and the disinfectant very well so the disinfectant covers all the seeds. For example, you might mix 200 grams of a disinfectant such as Cérégan with 100 kilograms of rice seed. Disinfected seeds are not eaten by insects, and they do not rot easily.

Disinfectant is poisonous, and you should be very careful when handling it. Wash your hands well after touching the disinfectant. Don't leave disinfectant where children can get at it, and never feed disinfected seeds to animals.

SOWING IN ROWS

Farmers often broadcasting their seed, but this can make it very difficult to remove weeds later. If the seed is sown in rows, it is easier to remove weeds. The animal that draws the hoe can walk between the rows. On flat soil, you can trace your rows with a marker. Leave 40 centimeters between rows. The spikes of the marker make little furrows where you can put the rice seeds. Leave 1 to 2 centimeters between seeds. Cover the seeds with a little earth. You will need between 30 and 50 kilograms of seed for 1 hectare. You can use a seed-planting machine (see the "Agricultural Machinery" article in this chapter). Some of these planters are drawn by a donkey or by oxen. The machine makes a furrow and inserts the seeds in the soil at the same distance from each other and all at the same depth. With some machines the fertilizer can be applied at the same time.

If the field is on a slope, make the seed rows along the contour lines (a contour line is a line across the slope running always at the same height) and leave barrier strips between the different levels of soil. This helps to reduce erosion. Fast-flowing water carries away some soil. When you slow up the water, the soil in the water drops to the ground. The water becomes cleaner and the soil is not lost.

A barrier strip is an uncultivated strip of land. Grass grows on this strip and holds back the water so that the water sinks into the ground. The barrier strips must also follow the contour lines. A barrier strip should be about 2 meters wide. To hold back the water better, you can plant tall grasses. If the slope is very gentle, you can leave 30 to 40 meters between barrier strips. If the slope is steeper, leave only 10 to 20 meters between barrier strips. Do not grow rice if the slope is very steep.

CULTIVATING

Cultivating means removing weeds by hoeing. Cultivate 15 to 20 days after sowing, and again whenever fresh weeds have grown. When you cultivate well, the buds at the bottom of the main stem can develop and make new stems—tillering. You can cultivate either with a hand hoe or with an animal-drawn cultivator, which is faster and makes it possible to cultivate more often. Pull out the weeds that have grown between the rows. If any weeds grow in the rows, pull them out by hand.

ORGANIC MANURE AND CHEMICAL FERTILIZERS

Organic manures are animal manure and green manure. Organic manures improve the structure of the soil. Plants grow better in a soil of good structure, and the chemical fertilizers are used better. Organic manuring should therefore be done at the beginning of the rotation—that is, before growing the first crop on a field. For example, if in the first year after clearing the field you grow a crop of yams and the second year you grow rice, you must apply organic manure in the first year before you plant your yams.

Above all rice needs nitrogen (N). The nitrogen fertilizer most often suitable for the soils of Africa is ammonium sulfate. But rice also needs phosphorus (P) and potassium (K). If the rice cannot take enough phosphorus and potassium out of the soil, the stems will be weak so they bend down to the earth and the grains cannot form and ripen well. Ask advice on the amount of fertilizer needed from the extension services in your area.

It is best to apply nitrogen—ammonium sulfate—in three separate applications. For example, if you have to give your field 100 kilograms of ammonium sulfate, apply (1) 40 kg before sowing, (2) 30 kg after the first cultivation, and (3) 30 kg when you see the panicles are forming.

Be careful not to let ammonium sulfate get onto the leaves. The fertilizer can burn them. Phosphoric acid and potassium are applied before sowing.

It is difficult to keep away rats and birds. You can have a watchman near the field. Noise will frighten away birds. Rice fields must be watched, especially when the grain begins to ripen.

There are also certain insects that damage rice—for example, rice borers that lay their eggs on the leaves. When the borers grow, they eat through the stem. When the stems go white, apply BHC (benzene hexachloride) or safe products that can kill these insects. Ask your extension service for information and products.

HARVESTING

Cut the rice when it is ripe. Wait until the heads are almost entirely yellow. You can cut the rice more quickly with a sickle. Dry and store upland rice exactly as you would paddy rice.

RICE PADDY: HARVESTING, THRESHING, CLEANING, AND HANDLING

Philip B. Cedillo

HARVESTING

Harvesting is the process of collecting the crops from the field. In the case of rice, it refers to the gathering in of mature rice panicles. Harvesting can be done either manually, with the use of hand harvesting tools, or mechanically, with the use of various harvesting machines. Generally harvesting includes all on-farm operations—that is, cutting, stacking, handling, threshing, cleaning, and hauling.

Harvesting Tips

The best time to harvest is when crop conditions are ideal. Standing or upright crops should have uniform height (no drooping panicles). The rice must not be underripe or overripe, with at least 85 percent of the grain straw-colored.

Harvesting an underripe crop results in a higher percentage of immature and unfilled kernels. Immature grains result in lower head rice recovery when milled. An overripe crop results in higher shattering losses. Stems are weaker and more susceptible to lodging. Overripe rice is usually too dry, and the grain may be more susceptible to damage. The moisture content should be 20 to 24 percent. There should be few pests or weeds present.

LOSS TERMINOLOGY

Preharvest loss: Loss that occurs before harvesting, due to pests or weather conditions.

Unharvested loss: Panicles missed by harvester.

Shatter loss: Loss that occurs during the actual cutting of panicles.

Stacking loss: Loss that occurs when crop is laid on the ground after cutting.

Handling loss: Loss that occurs each time the cut crop is moved from one place to another.

Piling loss: Loss that occurs when crop is piled prior to threshing.

Windrow loss: Loss that occurs when crop is laid on the ground after cutting.

Conveying loss: Loss that occurs when the crop is conveyed from the cutter bar to the threshing cylinder.

Separation loss: Threshed and unthreshed grain that exits through the straw outlets.

Cleaner loss: Threshed and cleaned grain that exits along with the trailings.

METHODS OF HARVESTING RICE

Rice can be harvested by both hand and mechanical methods (see Table 6).

Hand Harvesting

In manual reaping, one grasps the nice plants and cuts them using a sickle or scythe. In panicle harvesting, only the panicles are cut using a small hand knife (see Figure 17).

Machine or Mechanical Harvesting

The following machines can be used for harvesting.

- *Mechanical reaper*: Cuts the crop and lays the cut plants in long rows called "windrows." Large amount of labor are needed to collect and stack the crop before threshing
- *Stripper gatherer*: The stripper rotor strips or combs the grain from the straw while the plants are still standing. Stripped material is put into a removable collection container and emptied into a canvas for rethreshing.

THRESHING

Threshing is the process of separating the grain or seed from straw or chaff. This can be done either manually or mechanically. Three different types of manual threshing are used (see Figure 18)

1. *Foot threshing (trampling)*: Using bare feet to thresh the crop. The rice is usually spread over a mat or canvas and workers trample the crop with their feet.
2. *"Hampasan" or threshing rack*: The farmer holds the crop by the sheaves and thrashes it against a slatted bamboo or wooden platform.
3. *Flail*: The use of a flail or stick for thrashing the crop.

TABLE 6
Rice harvesting methods

Technology	Advantages	Disadvantages
Manual Harvesting	• Efficient • Most effective in lodged crop	• High labor cost • Skill dependent
Mechanical Harvesting		
Mechanical Reaper	• 2 ha/day throughput (cutting only) • Windrows can be left to dry before threshing • Relatively low cost	• Less effective in partially lodged crop • Cannot be used in very muddy fields • Relatively high labor required for gathering cut crop
Stripper Gatherer	• 1 ha/day throughput • Can harvest partially lodged crop • Can be used in moist crop conditions • Less power required for subsequent threshing and cleaning operations • Relatively low cost	• Cannot be used in very muddy fields • Cannot be used for crops taller than 1 m • Operator dependent
Western Reaper-Combine	• High throughput	• High cost • Less effective in partially lodged crop • Not suited for small, muddy Asian fields • Threshing system not suited for moist crop
Western Stripper-Combine	• Highest throughput • Less power required for subsequent threshing and cleaning operations	• High cost • Not suited for small, muddy Asian fields • Threshing system not suited for moist crop
Japanese Head-Feed Combine	• High mobility • Can operate in small fields • Less power required for threshing and cleaning system	• High cost • Low throughput • Pickup reel not suited for indicated high-shattering varieties
Thai-Axial Flow Combine	• High throughput • Axial-flow threshing system ideal for tropical conditions	• High cost but lower than Western and Japanese combines

Feeding-Type Mechanical Threshers

The following mechanical threshers have the rice inserted into them (see Figure 19).

Hold-on–pedal/treadle: A manually operated or motorized machine. The farmer holds the crop and places the panicles over a rotating wire-looped drum, which strips the grain from the straw.

Throw-in-tangential flow: A motor-driven thresher that consists of a rotating drum with either a wire loop or peg teeth. The whole crop is thrown into the machine in a single pass.

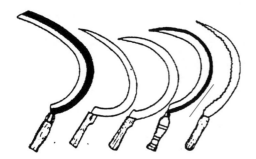

FIGURE 17
Sickles and panicle harvester.

"Hampasan" or threshing rack.   Threshing with the flail.

FIGURE 18
"Hampasan" and threshing with a flail.

> *Axial flow*: A throw-in type of machine that consists of a rotating peg tooth threshing drum. Inclined louvers inside the cover allow the crop to move tangentially or along the axis of the rotor shaft, and straw is ejected on the other end.

Types of Threshing Teeth

Figure 20 shows the types of threshing teeth.

> *Peg tooth*: Commonly used by most axial flow threshers.
> *Wireloop*: Usually used on hold-on type threshers.
> *Raspbar*: Commonly used for threshing wheat.

Thresher Capacity and Efficiency

Several factors affect the success of the threshing process. The moisture content of the rice determines the efficiency of the thresher operation. Higher moisture content requires more power. Moist chaff may clog the slots in the concave or in the straw outlet. The higher the feed rate, the faster the output, but feed rate is limited by the engine capacity. The larger the amount of straw, the higher the load. Long straw may wrap around the threshing cylinder, causing clogging. Some varieties of rice are easier to thresh (shattering variety) than others. Operator skill is an important factor.

Hold-on thresher.

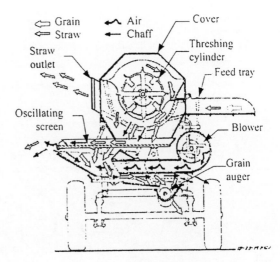

Throw-in thresher.

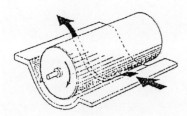

Tangential flow thresher.

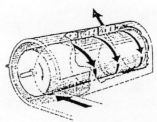

Axial flow thresher.

FIGURE 19
Mechanical threshers.

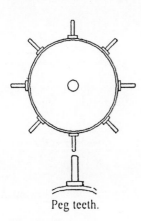

Peg teeth.

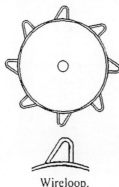

Wireloop.

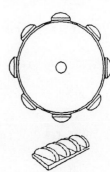

Raspbar.

FIGURE 20
Types of threshing teeth.

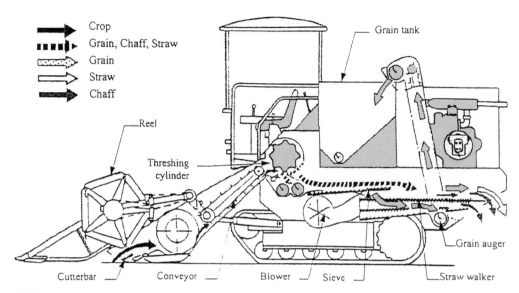

Crop
Grain, Chaff, Straw
Grain
Straw
Chaff

Grain tank
Reel
Threshing cylinder
Grain auger
Cutterbar
Conveyor
Blower
Sieve
Straw walker

FIGURE 21
Schematic of a combine harvester.

CLEANING

Cleaning is the removal of materials other than grain (MOG), such as straw, chaff, immature grains, unfilled grains, and so forth. Foreign material may affect grain quality during storage and milling.

COMBINE HARVESTING

A combine is a single machine that harvests, threshes, and cleans grain all in one operation. One is shown in Figure 21.

TWO METHODS OF HANDLING PADDY

Bagging: Clean grain is placed in stacks for transport to storage areas or for further processing.
Bulk handling: Usually used in highly mechanized systems, where the harvested grain is conveyed directly into large collection bins instead of bagging.

FACTORS TO BE CONSIDERED

When determining which harvesting system is preferable, one should consider the following.

1. Field conditions: topography, soil type, farm size and layout, accessibility, cultural practice (upland or wetland), water source, drainage
2. Crop conditions: rice variety, planting method

3. Available labor
4. Social practice
5. Available capital

SHIITAKE MUSHROOMS

Jeffrey Doff

I wanted to grow my own shiitake mushrooms. It felt "right" somehow, to begin mushroom growing as part of my northern gardening, adding another element as a source of food. I learned that shiitake mushrooms have an impressive store of nutrition as well as an incredible list of medicinal effects. Although shiitakes are expensive in the supermarket ($10 a pound where I live), the quality I would get from growing my own would be even better. I could be more self-sufficient and more connected with where my food came from. I hope that, although I am not a shiitake-growing expert, I can help other beginners. The questions I had in the beginning are quite possibly similar to the questions others might have. My spawn supplier was extremely informative and helpful in answering my questions along the way. I suggest that if you do decide to grow shiitakes, you cultivate a relationship with your spawn supplier.

WHY GROW SHIITAKE?

- Shiitakes are a delicious source of protein, B vitamins (including B_{12}), iron, and calcium.
- You will have the advantage of local and organic, self-reliant food production. Most shiitakes available in grocery stores are grown in sawdust blocks, sometimes called artificial logs. Often, these blocks are soaked in water containing pesticides, fungicides, and chemical and nutritional additives. By growing your own shiitakes, you will have fresher, more natural, and healthier mushrooms than you find in a store.
- It's fun!
- Shiitakes have amazing healing properties. Five powerful immune-boosting actions reverse the T-cell suppression caused by tumors, making it a valuable ally against cancer, leukemia, lymphosarcoma, and Hodgkins disease. Antiviral actions, due to substances present in spores and mycelia, inhibit division of viruses, impeding the spread of flus and other infections. One shiitake mushroom eaten with a tablespoon of butter actually reduces serum cholesterol. In Japan a shiitake mushroom is taken to prevent heart disease because it regulates both high and low blood pressure. As an anti-inflammatory, it improves stomach and duodenal ulcers, neuralgia, gout, constipation, and hemorrhoids. Shiitakes also counteract fatigue, generate stamina, and improve the complexion.
- It can be an extra source of income. This wasn't a reason for me, but it is definitely possible. On the small scale, you could barter or sell extra mushrooms from your harvest. On the larger scale, you could set up a real commercial enterprise. There are books and other resources for growing shiitakes commercially. If you are interested, check them out.

LOG CULTIVATION

I grow shiitakes with the log cultivation method. Here is the basic process: First you need to acquire logs. Then you inoculate your logs. Then you stack your logs in a shady place outdoors and wait 6 to 12 months. During this time, you will water your logs periodically. And then the logs will produce mushrooms!

Log-grown shiitakes reportedly do best in oak. (You can use some other hard hardwoods; ask your spawn supplier or folks who have done it.) The wood should be freshly cut (ideally within the past four weeks) but should also be cut when the sap is not running. If you do not need to cut any trees or if you do not know how to cut trees, you can go to a local farmer or tree cutter. Whoever does the cutting, it is important to treat the logs gently—the integrity of the bark will be important for your growing. It is the bark that keeps the moisture in the logs and keeps out contaminants. The standard log length is between three and four feet; some growers specifically recommend 40 inches. My logs are all 40 inches long. As far as diameter goes, mushroom people (including my spawn supplier) recommend between two and six inches. I chose to go from three to six inches. As you'll see below, the bigger the log, the longer it is expected to produce mushrooms. On the other hand, bigger logs may take longer before their first fruiting (production of mushrooms) and are more difficult to move around.

EXPECTED YIELDS

A four-inch-diameter log is projected to produce about one-quarter pound of mushrooms each spring and fall for approximately four years. You can also receive summer harvests (and get the logs to fruit on your schedule) by "forced fruiting," which involves managing your logs in a cycle of soaking, harvesting, and resting them. Logs are reported to last as many years as their diameter, and logs larger than four inches are expected to have harvests greater than one-quarter pound.

INOCULATION

Inoculation is basically a three-step process: Drill holes, fill them with shiitake spawn, and cover holes with hot wax. We drilled our logs on two sawhorses. We found that the logs were rolling around on the horses, so we attached some small pieces of wood onto the horses that the logs could rest between. We set up one resting place on each side of the horses. One person drilled on one side, another person inoculated on the other side, and the extra logs sat in the middle. If you have another set of horses, or a long table, you can get a third person to do the waxing at the same time. Whatever you do, make sure you have a sturdy setup. The holes should be six to eight inches apart on the horizontal and two to three inches apart in the circular dimension. (This is because the spawn travels more quickly along the grain of the log and thus can start out spaced farther apart.) We would mark the end of a log with a pen (four to eight marks, depending on the size of the log) two inches apart from each other. These marks denote the rows. Then we would use a measuring stick to decide where to drill the holes. The stick was marked with two different colors. The green marks were six inches apart from each other. The red marks were staggered exactly in the middle between the green marks, also six inches apart from each other.

Using the stick and the green marks for the first row, we'd rotate the log to the next two-inch mark and for the next row use the red marks. The effect of this is to stagger the holes in a diamond pattern on the log. If you make circles of holes around the log in the same place, the bark is likely to weaken and fall off. The diamond pattern keeps the integrity of the bark, and the stick is a fast way to make the diamond pattern.

The diameter and depth of your holes will depend on whether you use dowels or sawdust spawn. Your spawn supplier will be able to tell you how big and how deep to drill. You can purchase spawn in the form of dowel "plugs" inoculated with shiitake culture, or you can buy a bag of sawdust spawn. With plugs, you drill your holes, tap in the plugs with a hammer, and seal the holes with hot wax. I went with the sawdust spawn for two reasons. First, it is cheaper. Second, you can grow your own shiitake culture and thus make your own sawdust spawn.

When inoculating, you'll want to keep your spawn bag closed and out of the sun. To hold a workable amount of spawn, you can use yogurt containers, the bottom of soymilk containers (these worked well and did not crack but would wear out by the end of the day), or some other makeshift container. If you use sawdust spawn, I recommend that you purchase an inoculating tool. You fill the spring-loaded tool by jabbing it into your pouch of spawn and use the thumb button to push the spawn into the drilled holes. The tool picks up the right amount of spawn to fill one hole per jab. It makes the process go quickly and more comfortably than trying to pack the holes with your fingers.

Once all the holes on a log are drilled and inoculated, you seal the holes with hot wax. This is to keep moisture in and competing fungi out. Cheese wax is excellent in that it does not become brittle when it dries. If you have paraffin, you could try adding some mineral oil to it to make it more pliable, but that did not work very well for me. I was advised to apply the wax at a very hot temperature, between 350 and 400°F. This "provides the best, thinnest seal" and "applies quickly and cools quickly" (Kozak & Krawczyk 1993). To get the wax hot, we used an electric hotplate and an old saucepan. At first we tried measuring the temperature with a candy thermometer, but that did not seem to be accurate. We ended up using the sizzle test. If the wax sizzled when we applied it to the log, it was hot enough. After a while, we could tell from the slight wisp of smoke when the wax in the pot would be hot enough to sizzle. We applied the wax with natural bristle paintbrushes. They worked well, except when we accidentally left a brush in the wax pot and the bristles melted. After covering each hole with wax, we would carefully check each and every hole to make sure it was well sealed. If there were any holes in the wax seal, we would give a gentle second daub of wax.

Be careful with your wax! When it gets hot enough (450°F), the wax bursts into flame. You can avoid this by staying aware of how hot and smoky the wax is getting. You should definitely have a grease fire extinguisher and a metal lid on hand. My wax ignited once: I took the pot off the burner and covered it with a metal lid. Without oxygen, the fire went out. Knowing this as a possible danger, do not heat up your wax with combustible material above your workstation. Also, if the wax does ignite, never put water on it. Like other grease fires, water will spread out the fire and make it much worse.

We found the inoculation to take about an hour per log, depending on the size of the log. (With two people, we could inoculate two to three logs per hour.) The great thing about inoculating a log is that you only have to do it once!

EFFICIENCY AND SAFETY

Kozak and Krawczyk point out, "As you start to inoculate, the inefficiencies of your system will become clear. Take care of minor aggravations as soon as you notice them—anything that makes the job faster and easier is worth it, and the sooner you make corrections, the better off you will be." I would add to this that if anything seems unsafe or even potentially unsafe, take care of it right away. You are working with power tools, cords, and hot wax. Better to make sure your wax station is not wobbly before the wax is superhot and you are busy moving logs around or trying to fill drilled holes.

ALTERNATIVE INOCULATION METHODS

At Aprovecho Research Center in Oregon, I hear that they have experimented with inoculation by osmosis, so to speak. Rather than drilling holes, they pack sawdust spawn around the logs and let it colonize the logs on its own. If that works, it would save a great deal of time on drilling, filling, and waxing. If you are interested in trying this kind of inoculation, you can get in touch with them. (See the References).

TAG AND STACK

Once your logs are inoculated, you tag and label them. For tags, I used leftover pieces of aluminum flashing, cut into 1" × 2" pieces. (You can also buy tags from a shiitake supplier.) I labeled my logs with a pen—the day/month/year of inoculation and the strain of the spawn. (There are cold weather, warm weather, wide range, and other strains of spawn. Your friendly supplier will explain this to you.) Then I attached my tags using a hammer and a one-inch roofing nail.

For the first couple of months, you can deadstack your logs (like firewood). They have enough moisture for now, and they should not start rotting. Be sure to put them in a shady place and cover with a tarp. Leave the ends of the tarp open so they do not get too much moisture buildup.

As it starts to get warmer outside, I stacked my logs in the lean-to pattern. This pattern is diagrammed and described in the *Mushroompeople* catalog and in Kozak and Krawczyk's book. The main purposes of this stack pattern are to allow some air flow around the logs, keep a low angle so water does not run off the logs too quickly, and keep the smallest possible water shadow (or to allow the water to fall on the greatest possible area of logs).

WATERING

Watering shiitake logs is very important. It ensures that the spawn lives and lives well. The key is to soak logs thoroughly but let the bark dry between soakings. If the bark does not get the chance to dry between waterings, other fungi can appear on the logs. For the first 6 to 12 months after inoculation, logs need watering twice a week, if it hasn't rained—twice a day in particularly hot and dry spells. You can do this by hosing the logs by hand, or you could set up a system that works well for you. My friend Wiley watered his logs by setting up a hose with a small sprinkler attached to the end of it. Some people use lawn sprinklers or soaker hoses. I did hand watering for several months, and then I set up a hose with small holes in it that created a misty sprinkling of water over my logs. Another possibility, if you do not have many logs or do not mind the lifting, would be to soak your logs under water for five or ten minutes (submerged). Another experimental possibility would be to stand logs with their ends in water for a period of time, relying on the capillary action of the tree to suck up the water.

EATING SHIITAKE MUSHROOMS

I have read that in recipes you can substitute shiitakes for white button mushrooms. They can be used in stir-fry, soups, omelets, pizza, nut loaf, and salads. You can eat shiitake baked, sautéed, barbecued, boiled, or raw (see Kilby and the www.mushroomcouncil.com Web sites). If you want some guidance from those experienced in the culinary delights of the shiitake, there are shiitake mushroom cookbooks available.

REFERENCES

Aprovecho Research Center, 80574 Hazelton Road, Cottage Grove, Oregon 97424. Tel.: 541-942-8198. apro@efn.org, www.efn.org/apro

Kilby, Suzanne. (2000). "Shiitake—The Healing Mushroom." www.wvi.com/ki1by/suzi.html.

Kozak, M. E., and J. Krawcyzk. (1993). *Growing Shiitake Mushrooms in a Continental Climate*. 2d Ed. Peshtigo, WI: Field & Forest Products.

Mushroom Council, www.mushroomcouncil.com, 11875 Dublin Blvd, Suite DZ6Z, Dublin CA 94568 info@mushroomcouncil.com

"Mushroom Cultivation and Marketing: Horticultural Production Guide." Appropriate Technology Transfer for Rural Areas (ATTRA), University of Arkansas in Fayetteville, PO Box 3657, Fayetteville, AR 72702. Tel.: 800-346-9140. (Several of the books and most of the spawn suppliers I found in ATTRA's guide.)

Przybylowicz, Paul, and John Donoghue. (1990). *Shiitake Growers Handbook*. Dubuque, IA: Kendall/Hunt Publishing Co.

Mushroompeople, 560 Farm Road, PO Box 220, Summertown, TN 38483. Tel.: 31-964-2200. E-mail: mushroom@the-farm.org. www.thefarm.org/mushroom

---

AGROFORESTRY PRINCIPLES

Franklin W. Martin and Scott Sherman
Echo Technical Note, ECHO: North Ft. Myers, FL. 1992

INTRODUCTION

What is agroforestry? In simplest language, agroforestry is the production of trees and nontree crops or animals on the same piece of land. The crops can be grown together at the same time, can be grown in rotation, or can even be grown in separate plots when materials from one are used to benefit another. However, this simple definition fails to take into account the integrated concepts associated with agroforestry that make this system of land management possibly the most self-sustaining and ecologically sound of any agricultural system. Thus, a second definition of agroforestry would be the integration of trees, plants, and animals in conservative, long-term, productive systems. Agroforestry can be considered more as an approach than as a single finished technology. Although several finished systems have been devised and tested, such technology may require adjustment for particular situations. The flexibility of the agroforestry approach is one of its advantages.

Why agroforestry? Agroforestry systems make maximum use of the land. Every part of the land is considered suitable for plants that are useful. Emphasis is placed on perennial, multiple-purpose crops that are planted once but yield benefits over a long period of time. Furthermore, systems of agroforestry are designed for beneficial interactions of the crop plants and to reduce unfavorable interactions. They are designed to reduce the risks associated with agriculture, small-scale or large, and to increase the sustainability of agriculture. Agroforestry practices normally help conserve, and even improve, the soil.

Agroforestry includes a recognition of the interactions of crops, both favorable and unfavorable. The most common interaction is competition, which may be for light, water, or soil nutrients.

Competition invariably reduces the growth and yield of any crop. Yet competition occurs in monoculture as well, and this need not be more deleterious in agroforestry systems. Interactions may be complementary, as in the case of trees, pasture, and foraging animals, where trees provide shade and/or forage, and airmails provide manure.

Agroforestry systems are designed to produce a range of benefits including food, feed, fuels, often fibers, and usually renewed soil fertility. Agroforestry systems take advantage of trees for many uses: to hold the soil, to increase fertility through nitrogen fixation, or through bringing minerals from deep in the soil and depositing them by leaf-fall, to provide shade, construction materials, foods, and fuel.

Agroforestry systems may be thought of as principal parts of the farm system itself, which contains many other subsystems that together define a way of life.

DEFINITIONS

Alley Cropping	Growing annual crops between rows of trees
Beautification	Planting trees for ornamental purposes
Boundary plantings	Trees planted along boundaries or property lines to mark them well
Dispersed trees	Trees planted alone or in small numbers on pastures or otherwise treeless areas
Earthworks	Constructions made of earth, usually to conserve or control water
Improved fallows	Areas left to grow up in selected trees as part of a trees-crop rotation system
Individual trees	Trees occurring alone, whether spontaneously or planted
Living fences	Fences in which the poles are living trees or in which the entire fence consists of closely spaced trees
Nectar crop	Trees valuable as a source of nectar for honeybees
Terraces	Level areas constructed along the contours of hills, often but not necessarily planted with trees
Vegetative strips	Long, narrow areas of any type of vegetation, usually planted along contours for erosion control; may include trees
Woodlot	An area planted to trees for fuel or timber

SUMMARY OF BENEFITS OF AGROFORESTRY

- Improved year-round production of food and of useful and salable products
- Improved year-round use of labor and resources
- Protection and improvement of soils (especially when legumes are included) and of water sources
- Increased efficiency in use of land
- Short-term food production offsetting cost of establishment of trees
- Furnishing of shade for vegetable or other crops that require it or tolerate it
- Medium and long-term production of fruits
- Long-term production of fuel and timber
- Increase of total production to eat or to sell

COMPONENTS OF AN AGROFORESTRY SYSTEM

Land

Agroforestry is not a system of pots on the balcony, nor is it for the greenhouse. It is a system to manage the agricultural resource—land—for the benefits of the owner and the long-term welfare of society. While this is appropriate for all land, it is especially important in the case of hillside farming where agriculture may lead to rapid loss of soil. Normally land will be what the farmer owns (farmers that rent land may have little interest in the long-term benefits of agroforestry), and thus farmers must think conservatively about how to maintained the land over long periods of time.

Trees

In agroforestry, particular attention is placed on multiple-purpose trees or perennial shrubs. The most important of these trees are the legumes because of their ability to fix nitrogen and thus make it available to other plants. The roles of trees on the small farm may include the following.

- Sources of fruits, nuts, edible leaves, and other foods
- Sources of construction material, posts, lumber, branches for use as wattle (a fabrication of poles interwoven with slender branches, etc.), thatching
- Sources of nonedible materials, including sap, resins, tannins, insecticides, and medicinal compounds
- Sources of fuel
- Beautification
- Shade
- Soil conservation, especially on hillsides
- Improvement of soil fertility

In order to plan for the use of these trees in agroforestry systems, considerable knowledge of their properties is necessary. Desirable information includes the uses described above; the climatic adaptations of the species, including adaptations to various soils and stresses; the size and form of the canopy as well as the root system; and the suitability for various agroforestry practices. Any tree can be used in agroforestry systems, including all trees with edible products. In actual practice, very large trees are usually not used in agroforestry except casually—not by design.

These are some of the most common uses of trees in agroforestry systems.

- Individual trees in home gardens, around houses, paths, and public places
- Dispersed trees in cropland and pastures
- Lines of trees with crops between (alley cropping)
- Strips of vegetation along contours or waterways
- Living fences and borderlines, boundaries
- Windbreaks
- Improved fallows
- Terraces on hills
- Small earthworks

- Erosion control on hillsides, gullies, channels
- Woodlots for the production of fuel and timber

Some very good food-bearing trees for agroforestry are given in Table 7. Table 8 lists some of the best of the nonfood-producing trees used in agroforestry. Some successful uses of trees in isolation are given in Table 9.

Basically all crop plants and animals used by farmers in a region can be part of an agroforestry project. Some examples of the use of trees and crops together are given in Table 10.

TABLE 7

Trees or large shrubs with edible products for agroforestry systems

Species	Common Name	Edibility	Principle Uses in Agroforestry
Anacardium occidentale	Cashew	Flowers, seeds	Garden, fence, pasture
Annona muricata	Soursop	Flowers	Garden, fence, pasture
Borassus aethiopicum	Borassus-	Multiple food uses	Garden, pasture
Cqjanis cajan	Pigeon Pea	Seed, leaves	Hills, nitrogen fixation, fuel, hedgerow
Carica papaya	Papaya	Flowers	Garden, quick shade
Cnidoscolus chayamansa	Chaya	Leaves	Rapid hedge
Cocos nucifera	Coconut	Multiple food uses	Pasture, roadside, construction
Coffea arabica	Coffee	Seeds (bean)	Hedges, hills, fuel
Gliricidia sepium	Mother of Cacao	Flowers	Living fence, feed, fuel
Leucaena leucocephala	Leucaena	Leaves	Hills, alley cropping, nitrogen fixation
Manihot esculenta	Cassava	Roots, leaves	Rapid hedge
*Moringa pterygosperma***	Drumstick	Leaves, flowers, pods	Fence, garden
Theobrama cacao	Cocoa	Pulp, seeds	Understory tree, pasture
Psidium guajava	Guava	Flowers	Pasture, fuel
Sauropus androgynus	Katuk	Leaves	Hedge, alley cropping
Yucca elephantipes	Izota	Flowers	Hedge
Ziszyphus mauritiana	Jujube	Flowers	Erosion control, fuel

**See also Table 8.

TABLE 8

Principal trees for agroforestry systems

Species	Common Name	Principal Uses in Agroforestry
Acacia albida	Apple-ring Acacia	Terraces, dispersed trees, forage, nitrogen fixing
Acacia mearnsii	Black Wattle	Terraces, borderlines, fuel, nitrogen fixing
Bursera simaruba	Gumbo Limbo	Living fences, fuel
Calliandra calothyrus	Calliandra	Vegetation strips, fallows, windbreaks, fuel
Cassia siamea	Siamese Acacia	Terraces, fuel, nitrogen fixing
Erythrina berteroana	Dwarf Machete	Living fences, feed, rapid cover, nitrogen fixing
*Gliricidia sepium***	Mother of Cacao	Living fences, feed, fuel, hardwood
*Leucaena leucocephala***	Leucaena	Alley cropping, soil conservation, food, nitrogen fixing
*Moringa oleifera***	Drumstick	Living fences, rapid cover
Sesbania sesban	Sesban	Planting stakes, quick cover, nitrogen fixing
Sesbania grandiflora	Agati	Rapid cover, feed, nitrogen fixing

**See also Table 7.

TABLE 9
Some examples of successful uses of trees on small farms (not necessarily in combination with other corps and only a very small part of the potential uses)

Location	System	Tree Crop	Benefits	Other Plants
Central America	Living fence	Erythrina, Yucca, Gliricidia	Food, feed	
Tropics	Windbreaks	Casuarina	Fuel	
Central Africa	Dispersed trees	Acacia Albida	Fuel, feed, erosion	
Niger	Improved fallows	Acacia mearnsii, Leucaena, Sesbania	Soil fertility restoral	Grasses
India	Earthworks	Dalbergia, Pongamia, Prosopsis, Others	Food, soil conservation	Rapier grass, mando grass
Tropical Africa	Gully Protection	Tamarix	Food, soil conservation	Grasses

TABLE 10
Some examples of successful agroforestry systems of trees and crops

Location	System	Tree Crop	Benefits	Understory Crops
Costa Rica	Dispersed trees	*Cordia almifolia*	Lumber, shade, nutrients	Coffee
Costa Rica	Dispersed trees	*Erythrina* spp.	Nitrogen, fuel, shade, nutrients	Coffee
Puerto Rico	Dispersed trees	*Inga* spp.	Shade, nitrogen, fuel, wood	Coffee, bananas, root crops
El Salvador	Dispersed trees	*Inga* spp.	Fuel, nitrogen, shade	Coffee, cacao
Central America	Dispersed trees	Leguminous trees, *Acacia albida*	Lumber, fuel, shade special products, nitrogen	Pasture
Malaysia	Dispersed trees	Dwarf Coconut	Food, lumber	Cacao
Tropics	Dispersed trees	Coconut	Food, feed	Pasture
Mexico	Dispersed trees	Brosimim	Food, lumber	Wide variety of crops, pasture
Haiti	Home garden	Mango	Fruit	Rice
Sri Lanka	Mixed perennials	Many fruit trees	Fruit, other products	Spices, vegetables
Philippines	Home garden	Various fruit trees	Fruits, edible leaves	Many vegetables
West Africa	Home garden	Fruit trees	Fruits	Vegetables
Ivory Coast	Mixed perennials	Cacao, bananas	Food	Yams
Puerto Rico	Mixed perennials	Oranges, avocados, bananas	Food, nutrients	Coffee, root crops
Tropics	Alley crop	*Leucaena leucocephala*	Erosion control, fuel, nitrogen,	Annuals, grasses
Nigeria	Alley crop	*Gliricidia sepium*	Erosion control, fuel, nitrogen,	Root crops, grains
Rwanda	Vegetative strips	*Grevillia, Albizzia, Leucaena*		

TABLE 11
Sources of seed of most important tree species

Tree Species	Adaptation*	Possible Source of Seeds
Acacia albida	Hot, dry tropics	CATIE, ILCA, KI, SSC, TSP
Acacia mearnsii	Hot, dry tropics	CSIRO, KFSC, KI, SSC, TSP
Bursera simruba	Hot, dry tropics	FKNN,
Calliandra calothyrsis	Wet tropics	CATIE, TBAIF, TSP
Cassia siamea	Intermediate tropics	CATIE, ILCA, KFSC, KI, SSC
Erythriina berteroana	Intermediate tropics	CATIE
Gliricidia sepium	Intermediate tropics	CATIE, ILCA, KI, TBAIF, TSP
Leucaena leucocephala	Intermediate tropics	CATIE, ECHO, ILCA, KFSC, KI
Moringa oleifera	Intermediate tropics	CATIE, ECHO, ILCA, SSC
Sesbania grandiflora	Intermediate tropics	CATIE, ILCA, KI, SSC, TBAIF
Sesbania sesban	Intermediate tropics	KI, TBAIF, TSP, UH

*Intermediate condition suggests a region of intermediate rainfall.

GETTING STARTED WITH AGROFORESTRY SYSTEMS

The steps in the decision-making process are as follows.

1. Decide whether agroforestry systems are appropriate.
   - Describe family and community needs.
   - Find the limiting constraints in agriculture, including markets and marketing.
   - List the potential benefits of an agroforestry system in the region in question and their relative importance.
   - Then decide if it is worth the effort to develop one.
   - Consider whether the people of the region are willing or capable of adopting a system.
2. Design a system.
   - Select the area.
   - Characterize it (describe it, its strengths, weaknesses) with respect to existing soil, water, and crops.
   - List the needs that could be met with an agroforestry system.
   - Characterize the crops desired by minimum space requirements, water and fertilizer needs, and shade tolerance.
   - Select the trees, shrubs, or grasses to be used.
3. If the system is temporary, do the following.
   - Plan the features of soil erosion control, earthworks, and gully maintenance first.
   - Plan spacing of fruit trees according to final spacing requirements.
   - Plan a succession of annual or short-lived perennials beginning with the most shade tolerant for the final years of intercropping.

   If the system is permanent, do the following.
   - Plan the proportion of the permanent fruit and lumber trees on the basis of relative importance to the farmer.
   - Plan the spacing of long-term trees on the basis of final space requirements times 0.5.
   - Plan succession of annual and perennial understory crops, including crops for soil protection and enrichment.
   - As large permanent trees grow, adjust planting plan to place shade-tolerant crops in most shady areas.

4. In both temporary and permanent systems, do the following.
   - Keep all ground in crops or protective covers at all times.
   - Try the system on a small scale first.
   - Measure the inputs and outputs of the system.
   - Evaluate whether the benefits expected have been achieved.
   - Look for the desired plant materials or for suitable substitutes locally (Table 11).
   - Expand or extend any new system cautiously.

Source Abbreviations

CATIE   Centro Agromomico Tropical de Investigacion y Esperanza, Turrialba, Cartage, COSTA RICA, 56-6431/56-0169. Information & Seeds.

CSIRO   Division of Forestry and Forest Products, PO Box 4008, Queen Victoria Terrace, Canberra, ACT 2600, AUSTRALIA. Seeds.

ECHO   Educational Concerns for Hunger Organization, 17430 Durrance Rd., North Ft. Myers, FL 33917, USA. E-mail: echo@echonet.org. Web site: www.echonet.org.

FKNN   Florida Keys Native Nursery, 102 Mohawk St., Tavernier, FL 33070. Seeds & Forestry

ILCA   International Livestock Center for Africa, PO Box 5689, Addis Ababa, Ethiopia. Information & Seeds.

KFSC   Kenya Forestry Seed Centre, Kari, PO Box 74, Kikuyu, Kenya. Seeds.

KI   Kumar International, Ajitmal 206121, Etawah (UP), India. Seeds.

SSC   Shivalik Seeds Corporation, Panditwari, PO Prem Nagar, Dehra Dun - 248007 (UP), India. Seeds.

TBAIF   The Bharatiya Agro Industries Foundation, "Kamdhenu," Senapati Bapat Marag, Pune-411 016, India. Seeds.

TSP   Tree Seed Program, Ministry of Energy & Regional Development, PO Box 21552, Nairobi, Kenya. Seeds.

UH   University of Hawaii, Department of Agronomy & Soil Science, U. of H. at Manoa, 190 East-West Road, Honolulu, HI 96822, USA. Seeds.

REFERENCES

Folliot, Peter F., and Thames, John L. (1983). *Environmentally Sound Small-Scale Forestry Projects, Guidelines for Planning Volunteers in Technical Assistance* VITA. Arlington, VA.

Forestry/Fuelwood Research and Development Project. (1992). *Growing Multipurpose Trees on Small Farms*. Bangkok, Thailand: Winrock International. To order in the USA, call: 703-351A006 and request book order no. PNABR667.

IIRR (1990). *Agroforestry Technology Information Kit*. The International Institute of Rural Reconstruction, 475 Riverside Drive, Room 1270, New York, NY 10115 ($20.00). Kits are probably available as well from their office in the Philippines: IIRR, Silang, Cavite 4118, Philippines.

MacDicken. K.G. (1988). *Nitrogen Fixing Trees for Wastelands*. FAO Regional Office for Asia and the Pacific, Maliwan Mansion, Phia Atit Road, Bangkok, Thailand.

National Academy of Sciences. (1980). *Firewood Crops, Shrub and Tree Species for Energy Production*, Vol. I, Washington, DC. (Available free to those actively working in development: BOSTID (JH-217D), National Research Council, 2101 Constitution Avenue, Washington, DC 20418, USA.)

Rockeleau, D., F. Weber, and A. Field-Juma. (1988). *Agroforestry in Dryland Africa*. International Centre for Research in Agroforestry (ICRAF). Nairobi, Kenya.

Save the Children/US, Thailand. (1992). *Collection, Storage, and Treatment of Tree Seeds: A Handbook for Small, Farm Tree Planters*. The FAO Regional Wood Energy Development Programme in Asia, Bangkok, Thailand.

USAID (1987). *Windbreak and Shelterbelt Technology for Increasing Agricultural Production*. United States Agency for International Development, Washington, DC.

AQUACULTURE

Michael A. Rice

Aquaculture is the farming or husbandry of aquatic organisms, and in many parts of the developing world, small-scale aquaculture projects provide an inexpensive source of protein-rich food for personal consumption or as a cash crop. The first historical documentation of aquaculture is a treatise on the culture of carp in 475 B.C. by Fan Li, a wealthy Chinese nobleman. However, a number of authors suggest that aquaculture dates back to the very beginnings of agriculture in the third millennium B.C. in the riverine cultures of the Middle East and China. In the west, the culture of oysters was practiced as early as A.D. 100 in Rome and its province of Gaul—now France. The culture of carps was widespread in Eastern Europe by the thirteenth century A.D. (Berdach et al. 1972).

Small-scale artisanal aquaculture projects are a mainstay in Asia where most aquaculture is carried out on small family farms. Although large-scale culture of penaeid shrimp in some Asian countries for international markets has received considerable recent attention for being environmentally unsound (Clay 1997) and causing social problems due to redistribution and concentration of local wealth (Bailey 1997), much of the production is small-scale for family consumption with modest surpluses sold at local markets. By the mid-1990s the top seven aquaculture-producing countries were in Asia, representing about 65 percent of world production (New 1997). The largest of the aquaculture-producing countries is China, which is the world's largest producer of carps in small freshwater ponds and is the only country that produces over half of its domestic food fish supply through aquaculture. Most of the Chinese fish production is the culture of carps in small-scale ponds, although their aquacultural production is diverse in terms of number of species cultured and methods of production. Because of the longevity, success, and simplicity of these small-scale aquaculture projects in Asia, they provide an excellent model for emulation and adaptation in other parts of the world where animal protein is in short supply.

One attractive feature of most forms of aquaculture that are practiced on a small scale is that they can be a highly efficient means of producing high-quality animal protein at low cost in terms of money or feed inputs. For example, using off-bottom techniques for growing oysters in warm estuarine waters in the Philippines, about 2.5 kg of shucked oyster meats can be produced per square meter of farm area with no feed inputs (Rice & Devera 1998). Oysters are filter feeders, deriving their nutrition from filtering phytoplankton and other organic particulates from the waters in which they reside. High productivity in oyster and mussel farms is largely due to the water currents bringing the phytoplankton and other particulate food to the filter-feeding animals from wide, open areas adjacent to the farms. The annual production of some fish like carps that feed on phytoplankton or other plant material have reached yields in excess of 13,000 kg per hectare in Chinese ponds, with no supplemental feed inputs (Liang et al. 1991).

SELECTION OF SPECIES SUITABLE FOR AQUACULTURE

There are a number of characteristics that make an organism desirable as a candidate for aquaculture. For small-scale aquaculture projects focusing on food production, it is important that the selected species is easy to culture. Culturability criteria include ease of reproduction in captivity, hardiness of eggs and larvae, simplicity of feed requirements, adaptability to crowding, and tolerance of a wide range of water quality factors.

Ease of reproduction is an important factor for the culture of most fish species such as carps or tilapias that breed easily in captivity and produce large numbers of eggs per spawn. However, there are a number of fish species that can be cultured by rearing them from larvae or juveniles that have been collected in the wild. For example, in many parts of Southeast Asia much of the culture of milkfish, *Chanos chanos*, in brackish water estuarine ponds is based on the collection of larvae. Although the techniques for hatching and rearing milkfish larvae in hatcheries has been developed (Lee & Liao 1985), most of the culture of this species is based on the collection of larval and juvenile fish from a capture fishery (Villaluz 1986). For the most part, the small-scale cultivation of oysters and mussels is conducted by collection of spat, the collection of the settling larval stages on to some suitable substrate or artificial collector (Grizel 1993).

For aquaculture species that rely on captive spawning for success, the hardiness of eggs and larvae is an important factor. The freshwater carps and tilapias in addition to easy breeding produce eggs and larvae that are very hardy. Most of the cultured tilapias are mouth brooders that effectively use this behavioral strategy to protect their eggs and larvae from predation. On the other hand, carps have a tendency to cannibalize their own eggs in the close-quarters situation of aquaculture ponds, but culture methods include procedures for separating eggs and larvae from parent stock.

There are essentially two approaches to the feeding of aquacultured species. The first approach, which is usually most appropriate for artisanal aquaculture, is to focus on the culture of species that feed low on the food chain—that is, require very little food inputs because they rely on the algae, aquatic plants, or detritus naturally produced in their own ponds. The most basic example of this strategy is the culture of filter-feeding mollusks such as oysters and mussels that feed on phytoplankton and other naturally occurring particulates, but the culture of herbivorous fish is another example because of the reliance on the primary productivity of the ponds. In these simple pond aquaculture systems, the production may be boosted by the supplementary addition of fertilizers to increase primary productivity or adding supplemental feeds that may be simply agricultural wastes. The second approach to feeding is to focus on fish that feed high on the food chain, that require relatively expensive protein-rich diets. This approach, which is often employed by large-scale commercial aquaculture operations such as the pen farming of Atlantic salmon in temperate waters, requires a high market price for the end product (Bettencourt & Anderson 1990). Although the approach of focus on high-value species is most often employed by commercial firms concerned with marketing their product internationally, there are some instances in which this approach has been adopted by artisanal aquaculturists. The techniques for the culture of tropical sea basses, snappers, and groupers in small-scale floating net pens was first developed in Penang, Malaysia in 1973 and spread to a number of other countries in Southeast Asia and Northern Australia (Teng et al. 1978). But if aquaculture systems such as floating cages are adopted that require the feeding of fish, the economic viability of the operation often rests on the ability of the fish to utilize feeds in a cost-effective manner.

Adaptability to crowding is an important criterion to consider when selecting a species for artisanal culture. In the case of fish species, there are those that will tolerate considerable crowding. For example, carps and tilapias are known to tolerate extreme crowding conditions in ponds and indoor culture systems in which water quality is closely monitored and treated. Some species of fish such as the channel catfish of North America are highly territorial in the wild but are adaptable to pond culture at adult stocking densities of 1 fish/m^2. Likewise, there is evidence that some territorial marine species such as seabass and groupers (Rice & Devera 1998) can be adaptable to pen culture. Lack of adaptability to crowding may not be a totally exclusionary criterion for considering a species for aquaculture. If a fish species has a high market value, the stocking densities of territorial fish may possibly be increased using artificial hiding places for fish to seek out (Teng & Chua 1979).

One of the single most important criteria for selecting a species for a small-scale aquaculture project is tolerance of a wide range of water quality factors. Fish with very tightly defined water quality requirements can be cultured successfully, but there is considerable expense incurred to flush ponds with fresh water or to intensively treat the water. Fish that are easily stressed by inadequate water quality are prone to stunted growth, more frequent disease outbreaks, and fish kills. For artisanal aquaculture, especially in small ponds, it is best to select species that are tolerant to fluctuations and extremes of water quality (Swift 1985).

WATER QUALITY

Key water quality factors of concern are temperature, dissolved oxygen, salinity, pH, and dissolved ammonia (Boyd 1990). There are other water quality parameters such as calcium hardness, total alkalinity, and others, but these parameters are rarely of concern in artisanal aquaculture systems. Given the importance of water quality for the success of any kind of aquaculture operation, it is wise to maintain careful records of key water quality parameters on a daily basis.

Temperature

The first water quality parameter of concern is temperature, which can be a limiting factor for fish growth and reproduction. Most warm water fish species like tilapia and carps grow best between 25 and 32°C. Water temperatures in the tropics are within this range year around, but in temperate zones winters are too cool for rapid fish growth. Within their physiological tolerance limits, the metabolism of fish and other poikilotherms—animals with variable body temperatures that conform to their environment—doubles or triples with each 10°C increment in temperature (Roberts 1964). For example, a fish at 30°C will grow two to three times faster and consume two to three times as much oxygen at 20°C. This physiological response to temperature is known by physiologists as the Q_{10} effect.

In relatively deep ponds of 1.5 m depth or more, surface waters are warmed quicker than bottom waters, and warmer waters have less density than cooler waters. The ponds may develop different temperature layers in a condition known as thermal stratification. In most artisanal ponds, thermal stratification is undesirable because there is little mixing of waters across the thermocline. This lack of mixing does not allow minerals from the bottom water (hypolimnion) to reach the surface waters (epilimnion) where most of the primary productivity is occurring. Additionally, the waters of the hypolimnion are prone to oxygen depletion. In the tropics and during the summer in temperate latitudes, epilimnion waters may warm to temperatures greater than 35°C, forcing fish to seek refuge in the cooler oxygen depleted waters below, which can be a stressful situation. Aquaculturists that have intensive deepwater fishponds typically employ some type of expensive aeration devices to break up the stratification and to inject oxygen into the water during the warm periods, but for artisanal ponds it is recommended to construct the ponds with depths of 1.5 m or less. Shallow ponds lessen the effects of thermal stratification by heating up more uniformly during the day and cooling down at night allowing for mixing on a daily basis.

Most species of fish have a poor tolerance to sudden temperature changes, so they should never be removed or stocked into ponds or cages without gradually acclimating them to their new temperature regime. Often a sudden change of temperature as little as 5°C can either kill fish outright or stress the fish, making more prone to disease onset.

Temperature is one of the easiest water quality parameters to monitor using a simple handheld thermometer. The temperature of a pond or cage culture site should be taken at different depths and several times daily to establish a 24-hour temperature profile. This temperature profile at different times of the year will be of great use for making management decisions.

Dissolved Oxygen

Dissolved oxygen is one of the most critical parameters of concern with any kind of aquaculture operation, so the aquaculturist should be familiar with the dynamics of dissolved oxygen concentrations. Measuring dissolved oxygen is fairly straightforward in that there are commercially available oxygen meters or chemical test kits that provide the materials for performing a Winkler titration in the field. The ability to measure dissolved oxygen is essential in deep semi-intensive ponds or in situations of netpen or fishcage aquaculture with fish held at high stocking densities. In these relatively intensive systems, the high stocking densities require the ability to monitor dissolved oxygen and to have methods to aerate the water available. However, with proper care, artisanal methods of aquaculture can be carried out without the need of directly measuring oxygen concentrations in the water. There are a number of simple secondary criteria that can be used to assess the health of the pond and potential for oxygen depletion problems (hypoxia).

Although the atmosphere is 20 percent oxygen, it has a very low solubility in water, and its solubility decreases with increasing temperature and salinity (Table 12). For example, a tropical freshwater fishpond at sea level with a temperature of 35°C will be at 100 percent oxygen saturation, with a dissolved oxygen concentration of 6.949 mg/L. Altitude also affects oxygen solubility. The maximum saturation concentration of dissolved oxygen must be reduced by about 12 percent for each 1,000 m of altitude (Creswell 1993). Based on solubility characteristics alone, there is a greater probability of hypoxia at higher temperatures and salinities. This is compounded by the Q_{10} effect of temperature on organisms in the pond driving up oxygen demand.

Although oxygen will diffuse into surface waters from the atmosphere at rates of 1 to 5 mg/L daily, the primary source of oxygen in most natural water bodies is photosynthesis from phytoplankton and aquatic plants, which ranges from 5 to 20 mg/L daily. Respiratory losses of oxygen include respiration by plankton (5 to 15 mg/L daily), respiration by fish (2 to 6 mg/L daily), respiration by benthic organisms (1 to 3 mg/L daily), and diffusion of oxygen into the air (1 to 5 mg/L daily). Since oxygen is being produced only during the daylight hours, and respiration

TABLE 12

Oxygen solubility (mg/L) in waters of various temperatures and salinities at sea level

Temperature (°C)	Salinity (g/kg)				
	0	10	20	30	40
0	14.621	13.636	12.714	11.854	11.051
5	12.770	11.947	11.175	10.451	9.774
10	11.288	10.590	9.934	9.318	8.739
15	10.084	9.485	8.921	8.389	7.888
20	9.092	8.572	8.081	7.617	7.180
25	8.263	7.807	7.375	6.967	6.581
30	7.558	7.155	6.772	6.410	6.066
35	6.949	6.590	6.248	5.924	5.617
40	6.412	6.090	5.783	5.492	5.215

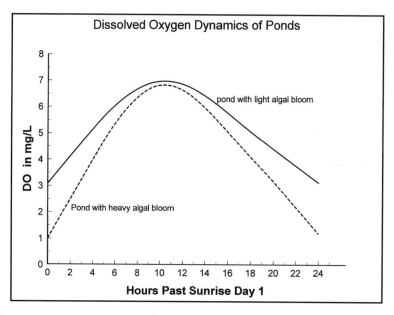

FIGURE 22

The daily dynamics of dissolved oxygen in a typical artisanal fishpond in the tropics. Minimum dissolved oxygen concentrations are usually encountered just prior to sunrise. Throughout the morning hours, oxygen concentrations rise due to net photosynthetic activity of phytoplankton and other aquatic plants, reaching a saturation concentration by midday. The dissolved oxygen concentrations decline at night due to respiratory processes in the pond. Pond fertilization, resulting in more intensive phytoplankton blooms, reduces the overnight oxygen minima by increasing the respiring biomass in the pond.

is occurring on a 24-hour basis, there is a diurnal (day-night) fluctuation in dissolved oxygen concentration in ponds, with minima at sunrise (Figure 22). In a well-managed artisanal pond, most of the respiration is due to the phytoplankton and plants that represent most of the biomass, *not* the fish. During the daytime the plants are respiring, but there is generally a net production of oxygen in the pond during the day.

Fertilizing a pond generally increases the primary productivity of a pond by increasing the density of phytoplankton in the water. This practice increases the amount of food available for zooplankton and grazing fish, but it also increases the probability of hypoxia during the nighttime hours by increasing the respiring biomass. During the daytime the net production of dissolved oxygen proceeds at an accelerated rate, but the final concentration of dissolved oxygen attained is the maximum saturation level based on temperature and salinity (Figure 22). Further details about pond fertilization will be covered in subsequent sections.

Increasing the density of phytoplankton in the water also limits the depth of photosynthesis and thus oxygen production. In ponds with very dense phytoplankton blooms, photosynthesis may be occurring only in the top 10 cm of the epilimnion. In this situation, the fish are forced to the surface and there is risk of an algal die-off that would cause the pond to experience severe hypoxia or anoxia (no oxygen), resulting in a fish kill. Using a Secchi disk one could simply determine relative phytoplankton densities. A limnological Secchi disk is a simple flat, round device that is 25 cm in diameter and painted white and black on opposing quadrants. The Secchi disk is lowered into the pond and the depth at which the disk disappears from sight is noted. Clearer waters result in larger Secchi depths than do turbid waters. It is recommended that for shallow artisanal ponds, the optimum Secchi depths range from 40 to 50 cm. As a general rule, most ponds should contain enough dissolved oxygen to support hardy species of fish at a depth two to three times the Secchi depth. So an artisanal pond 1 meter in depth should have sufficient oxygen levels to support fish all the way to the bottom. If the Secchi depths are less than 40 cm, cease fertilization and partially drain the pond and replace with fresh water.

Fish selected for culture in artisanal ponds should be tolerant to low dissolved oxygen. The most frequently cultured fish in artisanal ponds, carps and tilapias, prefer dissolved oxygen levels to be in excess of 4 mg/L. These fish will tolerate transient overnight hypoxia to 1 mg/L, but continuous exposures to dissolved oxygen concentations less than 4 mg/L is stressful and can increase the likelihood of disease.

pH

Acidity or basicity of water is indicated by pH, which is formally defined as the negative logarithm of the hydrogen ion concentration. In natural waters, the pH measurement scale is logarithmic, ranging from 0 to 14, with the center point pH = 7.0 as neutral. Acidic pH levels are less than 7.0 and basic or alkaline levels are above. Correct regulation of pH in water bodies used for aquaculture is essential for maintenance of fish health. Extremes of pH can be directly harmful to the fish or increase the toxicity of a number of other naturally occurring ions or metabolic waste products such as ammonia.

Determination of pH on a routine basis is essential in intensive or semi-intensive aquaculture, but like dissolved oxygen, proper care can be taken with the establishment of artisanal ponds so that routine monitoring of pH is less critical. The pH of ponds can be determined by the use of commercially available small electronic pH meters. Additionally, commercially available water test kits usually have chemical dye indicator tests for pH included. The normal pH for fish relative to the species one is raising can be determined by sampling from waters in which there are naturally reproducing populations. In general, most fish can survive normally in a pH range between 6.5 to 8.7. Waters with pH extremes of pH = 4 and pH = 11 are deadly to most fish, even the most hardy. When selecting sites for artisanal ponds, it is helpful to keep in mind that acid conditions are often associated with soils with large amounts of decaying organic matter, and alkaline conditions are associated with soils with large amounts of carbonates from limestone or similar bedrock.

The pH of ponds and other natural waters is greatly influenced by the concentration of carbon dioxide in the water. At night when respiration is dominant, carbon dioxide concentration is high and pH is low, while during the daytime the CO_2 scavenging of photosynthesis raises the pH (Figure 23). The chemical relationship between carbon dioxide and pH is described by the carbonic acid-carbonate equilibrium system.

$$CO_2 + H_2O = H_2CO_3 = H^+ + HCO_3^- = H^+ + CO_3^{2-}$$

Excess carbon dioxide drives the equation to the right, producing greater numbers of hydrogen ions that drive the pH down. Conversely, addition of calcium carbonate or removal of carbon dioxide drives the equation to the left, thereby reducing hydrogen ions and raising pH.

In the vast majority of situations, particularly in ponds built in most tropical or good agricultural soils, the daily fluctuations of pH are not great enough to be of concern to the aquaculturist. Typically, pH ranges from 7 or 8 at daybreak to 9 or 10 by late afternoon, the times of pH extremes. Waters that are in the range of pH = 6.7 to 8.0 at daybreak are considered best for fish production. In mountainous areas, or in areas with spring or deep well water, daily pH fluctuations may be greater and a cause for concern. These waters usually have low total hardness

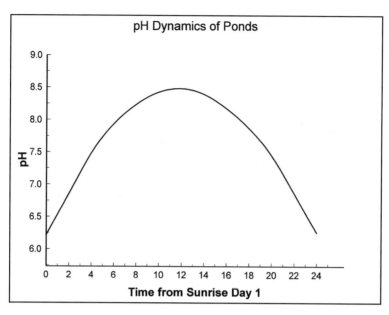

FIGURE 23
The dynamics of pH in a typical artisanal fishpond fluctuates diurnally, or on a day-night cycle. Net respiration in the pond at night results in a greater dissolved carbon dioxide concentration that acidifies the water. During the day, the uptake of carbon dioxide by actively photosynthesizing algae and plants raises the pH.

and alkalinity, which are factors affecting the buffering capacity or the ability of the water to resist rapid changes in pH. Water hardness (expressed in units of mg/L) is the total amount of divalent cations, such as Mg^{++} or Ca^{++} in solution. Total alkalinity (expressed in mg/L) is defined as the total amount of titrable bases expressed as calcium carbonate equivalents. As a general rule, artisanal ponds with total alkalinity and hardness of 20 mg/L each are best for fish production and are adequately buffered against rapid or extreme pH fluctuations. Waters with high alkalinity and low hardness may have dangerously high pH values during periods of rapid phytoplankton growth. If this is the case, the water hardness can be increased by the addition of agricultural gypsum (calcium sulfate). The optimum treatment rate is the amount of gypsum that would increase the total hardness to a level that equals the total alkalinity. The treatment rate may be determined by the following equation.

$$\text{Amount of calcium sulfate (mg/L)} = (\text{total alkalinity} - \text{total hardness}) \times 2.2$$

Conversely if pH is low, the best way to raise pH is to lime the pond with agricultural limestone (calcium carbonate) or other suitable liming agent such as agricultural dolomite ($CaMg(CO_3)_2$), hydrated lime (calcium hydroxide), or quick lime (calcium oxide). These liming agents raise the pH and the total hardness of the water and release carbon for photosynthesis.

If new ponds are to be established in an area without any other ponds for comparison, it would be useful to test for total hardness and alkalinity. Most commercially available freshwater test kits have chemical titration tests for total hardness and alkalinity. Since most individuals may not be able to do chemical analyses on a regular basis, a general rule of lime application may suffice. For new ponds, liming material equivalent to 200 g/m^2 of calcium carbonate should be applied and the total alkalinity determined after a month. If the total alkalinity is below

20 mg/L, another application equal to the first should be made and the total alkalinity should be measured again. This process should be repeated until the alkalinity is up to or exceeding 20 mg/L. Agricultural gypsum and lime can be easily broadcast over bottoms of empty ponds, but application is more difficult when ponds are full of water. Bags of material can be broadcast from a boat, but care must be taken to assure adequate dispersal. In temperate zones, gypsum and lime should be applied in the late summer or fall months to allow reaction with the sediments prior to spring pond fertilization. In tropical ponds, the gypsum and lime applications should be applied one month prior to fertilization. Once ponds are established, gypsum and lime amendments are generally applied at 40–50 g/m^2.

Ammonia

Ammonia is a waste product from fish and is a product of the decomposition of organic matter, including excess supplemental feeds, by bacteria. In water, ammonia occurs in two forms, un-ionized (NH_3) and the ammonium ion (NH_4^+) that are in a pH sensitive equilibrium.

$$NH_3 + H^+ = NH_4^+$$

Un-ionized ammonia is highly toxic to fish, but the ammonium ion is harmless except at extremely high concentrations. Toxic levels of un-ionized ammonia are usually between 0.5 and 2.0 mg/L for most pond fish, but sublethal effects and stress may occur at concentrations as low as 0.1 to 0.3 mg/L. Temperature and pH of water are the key determinants in the relative proportion of the two forms of ammonia. As water becomes more alkaline and temperatures increase, the relative proportion of the un-ionized form increases, thereby increasing toxicity (Table 12). Based on these criteria, warm summer months and tropical regions are more prone to toxic levels of ammonia.

Most commercially available aquaculture water quality test kits include a colorimetric phenol-hypochlorite test for ammonia. The test for ammonia measures the concentration of total ammonia nitrogen in water samples—that is, the sum of un-ionized ammonia and the ammonium ion. Once the total ammonia nitrogen concentration is determined, Table 13 can be used to determine concentration of toxic un-ionized ammonia based on measured values of temperature and pH.

Generally, ammonia is not the greatest killer of fish in artisanal ponds with managed phytoplankton blooms. Phytoplankton growth can remove dissolved ammonia. Additionally other

TABLE 13
Fraction of un-ionized ammonia in freshwater at different pH values and temperatures (To determine the concentration of un-ionized ammonia, multiply the total ammonia nitrogen by the fraction of ammonia that is in the un-ionized form from the table for a specific temperature and pH.)

pH	Temperature (°C)							
	5	10	15	20	25	30	35	40
7.0	0.0012	0.0019	0.0027	0.0039	0.0056	0.0080	0.0111	0.0153
7.5	0.0043	0.0059	0.0081	0.0126	0.0168	0.0254	0.0340	0.0432
8.0	0.0123	0.0182	0.0266	0.0380	0.0537	0.0744	0.1011	0.1345
8.5	0.0419	0.0566	0.0868	0.1132	0.1638	0.2050	0.2729	0.3012
9.0	0.1107	0.1567	0.2144	0.2833	0.3621	0.4455	0.5293	0.6088
9.5	0.3014	0.3714	0.4821	0.5551	0.6561	0.7151	0.8062	0.9123
10.0	0.5535	0.6498	0.7314	0.7983	0.8498	0.8892	0.9167	0.9542

(Calculated from the data of Emerson et al. 1975).

nitrogen cycle processes occurring in the pond can convert ammonia to relatively nontoxic nitrate, or in the sediments it can be converted to nitrogen gas that can eventually diffuse back into the atmosphere. Use of phosphate fertilizers devoid of nitrogen can possibly stimulate phytoplankton growth enough to keep ammonia levels in check. If you suspect ammonia is a problem in ponds, a partial drain down and refilling with water may help in the short term. If you are using supplemental feeds in the ponds, you can control the amount of ammonia in the pond by selecting feeds that provide the maximum growth with the minimum amount of protein. Optimization of feeding strategies will be covered in greater detail in subsequent sections.

Salinity

The last of the major water quality parameters of concern for small-scale aquaculturists is salinity. Salinity is formally defined as the total concentration of dissolved ions in water, and it is typically expressed in units of parts-per-thousand or g/L. Freshwater has less than 1 g/L total salinity, and seawater in tropical areas can be 35 g/L salinity or more. In estuaries, semienclosed bodies of water with freshwater and seawater, salinities usually range between the extremes. Salinity is of interest for a number of reasons. First, as discussed previously, higher salinity greatly influences the solubility of oxygen in water, resulting in decreased dissolved oxygen saturation concentrations. Also, fish are often adapted to specific salinity regimes and become stressed when there is a rapid change in salinity. However, some species of fish can be adapted or acclimatized to grow in salinities well above or below their usual salinity regime. For example, tilapias that are native to freshwater have been successfully reared in floating sea cages in salinity as high as 37 g/L in the Bahamas (Watanabe et al. 1989).

Most fish are greatly stressed when there is a rapid change of salinity. A salinity difference of only 5 g/L can be lethal to some fish, so if you are in doubt it is best to check the salinity of source waters and receiving waters if you are transporting fish from place to place. There are several methods to be used to determine salinity. The simplest method is to measure a known volume of water and evaporate or boil it down until only the salts remain. The weight of all the salts divided by the water volume gives the salinity. There are various other methods for determining salinity. One of the least expensive instruments for determining salinity is a hydro-meter. Glass hydrometers can be commercially purchased, but a simple hydrometer can be made from a lump of modeling clay stuck on the end of a soda straw. The buoyancy of the soda straw hydrometer can be adjusted by adding or removing bits of clay and can be calibrated in glasses of water with varying amounts of table salt. Other more expensive and more precise methods for determining salinity include commercially available handheld refractometers, conductivity meters, and a chemical test for the titration of chloride. In mixtures of seawater and freshwater, the titration for chloride is a good estimate of salinity because the proportion of chloride to salinity of seawater is constant anywhere in the world. If this method is used, the conversion from chloride concentration to salinity is as follows.

$$\text{Salinity (g/L)} = 1.080655 \times [Cl^-] \text{ (g/L)}$$

SITE SELECTION AND CONSTRUCTION OF SMALL-SCALE FISHPONDS

Considerable thought and planning should go into selecting site locations for fish ponds or any other aquaculture system. The costs of construction, the ease and costs of operation, and the overall productivity of the pond can be greatly affected by the site selected. The selection of a good site for aquaculture depends on a number of factors, including (1) the requirements of the species chosen to culture, (2) land and water availability, (3) legal and sociopolitical aspects, and (4) production and marketing infrastructure requirements (Avault 1996).

In a previous section, the criteria for selecting species to culture were discussed. In selecting sites for ponds you must know the biological requirements of the selected culture species for reproduction and growth from larval stages to harvest or market size. Annual fluctuations of temperature at a site may exclude the selection of one species of fish in favor of another. For example, in temperate and some subtropical areas the winter temperature drops below 15°C, the minimum temperature at which most species of tilapia can survive. One strategy to overcome this limitation may be to culture a cool-tolerant species of tilapia, such as the blue tilapia *Oreochromis aureus*, that will survive at temperatures as low as 12°C (Zale & Gregory 1989). Even cool tolerant species of tilapia are unlikely to survive winters with temperatures in the near freezing or freezing range. For this reason, tilapia culture in temperate zones is frequently conducted in greenhouses or other indoor facilities. Species suitable for outdoor ponds that may experience freezing temperatures include the common carp *Cyprinus carpio*, which is easy to culture and has a greater resistance to freezing temperatures (Michaels 1988). In addition to temperature, other factors such as salinity and nutrient content of the water must be matched with the requirements of the species selected to culture.

Water Supply

Ponds must be located in areas where there is an adequate supply of good quality water. Typically water supplies for artisanal ponds include natural surface waters from a nearby lake, stream, or estuary, or from groundwater from a spring or well. For small-scale ponds, it is desirable to be able to have enough water flow to fill the pond in two days or less. This allows enough flow to be able the flush the pond rapidly, and it is usually more than enough flow to compensate for water evaporation and seepage. To estimate the flow rate of a stream, measure the width of the stream using a measured rope; at that same spot, measure the depth of the stream at various points from bank to bank, using a weighted line or measuring stick. Then calculate an average depth based on all of the depth measurements. Determine the current speed of the stream by measuring the time for a leaf or other floating object to traverse a measured distance. The flow rate of the stream is calculated by multiplying its cross-sectional area by the current speed. For example, an irrigation canal with a width of 1.5 meters and an average depth of 0.7 meters has a cross-sectional area of 1.05 m^2. If the water is flowing at 10 cm/sec, then the flow rate is 1.5 m × 0.7 m × 0.1 m/sec = 0.105 m^3/sec = 105 L/sec. In two days (48 h), this irrigation canal would deliver 18,144 m^3 of water, or about enough water for 1.8 hectares of ponds that are 1 meter deep. The flow rate of wells and springs can be estimated by the time that it takes to fill vessels of a known volume.

When using surface water to fill ponds, be mindful that streams and lakes usually have species of wild fish that can get into the pond, creating management problems. In this case, some form of predator exclusion must be employed. Surface waters can also become contaminated with pesticides and other chemicals that could stress or kill fish, so it is wise to investigate water use and discharge practices upstream from your intended site.

Soil Characteristics for Ponds

It is important to keep in mind that a pond is merely a large earthen container for holding water. Subsoil that contains 30 percent clay is best for keeping water seepage to a minimum. Soils that contain a low percentage of clay have high percolation rates and leak if ponds are built, requiring expensive pond liners. Pond linings may include concrete, clay soils brought from outside, or commercially available polyethylene plastic or butyl rubber liners. Test cores of soils should be taken to the depth that the pond is to be built. Agricultural soil laboratories use a combination of sieve analysis and sedimentation rate analysis to classify soils by their grain size (Folk 1968). By sieve analysis, soil particles are classified as gravels, granules, sands, or silts. By sedimentation analysis, or the time it takes for particles to settle in water, larger-grained silt particles are distinguished from clay particles, which are of colloidal dimensions. A practical way to assess the clay content of soils is to roll a moist sample of soil in your hands. Soils with high clay content have a high degree of stickiness or cohesion and maintain shape when molded. When allowed to dry, these clayey soil samples become hard lumps. Soils that pass these simple clay content tests are likely to hold water well enough for the establishment of artisanal ponds.

Land that is suited for rice production in Asia or other areas is usually suited for fishpond development. Soils with high clay content may not have enough drainage for most other terrestrial crops, so these areas may be excellent areas for fishponds. Boggy areas, salt flats behind mangrove forest areas, and areas that are subject to brackish water have all been successfully used as pond sites. It is important to keep in mind that certain riparian or coastal wetlands are critical nursery areas for wild finfish, mollusks, and crustaceans and often support artisanal fisheries that should not be destroyed by the placement of an aquaculture facility (Twilley 1989).

Types of Ponds

Fishponds are arbitrarily classified by their construction methods, location, and by their use. Examples of fishpond types include watershed ponds, also known as rain fed ponds or ravine ponds, and in some coastal areas there are tidal ponds that are flushed daily during the tide cycle. Other pond types include excavated or "dugout" ponds, embankment or "levee" ponds, and raceways. Raceways are long tanks or ponds that are designed to culture fish that grow best in flowing water. They often mimic the flow of rivers and streams. Raceways are a common method for the culture of freshwater trout and other salmonid fish in North America and Europe, but it is a system that is not frequently used in artisanal aquaculture due to prodigious water supply requirements.

Watershed Ponds. In types of terrain that are hilly or mountainous, flat areas for building of fishponds can be scarce and water supplies may be limited. In hilly terrain, pond builders can take advantage of natural ravines or dry watercourses to build ponds inexpensively by simply building a dam structure (Jensen 1989). These dams constructed across valleys form miniature reservoirs that store water during the rainy season. The watershed to water surface area ratios should be large enough so that the pond fills during rainy months and allow enough water retention during dry months. When a watershed is too small to supply enough water, the pond may be supplemented by water brought in from springs or wells, but these means add considerable expense to the project. Generally watershed ponds are designed so that they can be drained by gravity.

Topography will determine the size and shape of a watershed pond. Generally steep sloping valleys require larger volume dams than do gentle sloping hills with wide valleys, so construction costs are higher for valleys with steeper slopes. Ideally for most artisanal watershed ponds, the water depth at the drain should be no more than 2.5 meters, with a dam height of 3 to 3.5 meters. Prior to dam construction, all trees and other vegetation, including roots and topsoil, should be removed from the site. Remaining vegetative material can rot, causing seepage and possible dam failure. The dam location should be staked to determine the construction dimensions and to determine the extent of the impounded water. Leveled lines at the top of the dam

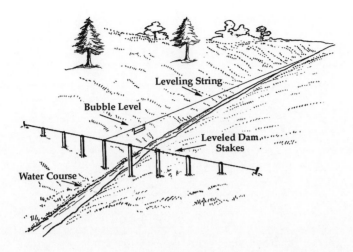

FIGURE 24

Staking the position and height of a watershed dam can be used to estimate the depth and area of pond coverage. Once the height of the dam is determined by leveling the tops of the stakes, the extent of the impounded water can be determined by using a level string leading back up the watercourse from the top of the dam stakes. *Drawing by E. Watkins.*

stakes and a level line reaching back up the watercourse can be used to determine dam height and pond area relationships (Figure 24). To prevent seepage at the dam base, a core trench must be excavated to a depth of 1 meter and backfilled with clay compacted in layers 20 to 30 cm thick. As the dam is constructed, the core of best quality clay should extend from the excavated footer to the top of the dam. The width of the top of the dam should be 3 to 5 meters wide, and the inside and outside slopes of the dam should be constructed with a 3:1 run to rise ratio.

   All watershed dams should be constructed with spillways to allow excess water to safely drain. An overflow pipe connected to a drain or monk drain system should handle small flows of excess water. Unlike simple standpipes and monk drains, a sleeved standpipe drain system allows for the removal of oxygen-poor bottom water during the drainage of excess water. Drain systems for ponds will be covered in greater detail in a subsequent section. All watershed dams should have an emergency spillway to allow removal of large quantities of water during heavy rainfall so that water does not flow over the top of the dam. Emergency spillways vary in width depending on watershed size and slope, the types of vegetation and soils present, and the amount of rainfall expected. Spillway widths range from 3 to 20 meter widths and are generally 0.5 to 1 meter below the top of the dam. Spillways should be planted with grass or riprapped with cobble-sized stones to prevent erosion of the dam when flows occur. The tops of spillways are generally designed to be flat and level to allow for sheet flow of water. Sheet flow of water across the spillway minimizes the formation of erosion channels and acts to retain fish in the pond. Horizontal bar barriers can be placed on the spillways to better assure the retention of fish.

   Once a watershed pond is constructed, it can be used for a variety of purposes. For example, fish can be cultured directly in the pond or in floating cages or netpens. The water can be used for irrigating terrestrial crops or filling other ponds. In some mountainous areas of the Philippines, Indonesia, and China, terraced farming of upland rice using sloping agricultural land technology (SALT) has been developed to a high degree. The upland fields are terraced with retaining walls and levees that follow slope contours (Figure 25). In these countries, some of the upland rice

FIGURE 25

Terraced rice fields on the slopes of the Cordillera Central Mountains of the Benguet Province in the Philippines are typical of upland agricultural practices in much of Asia. In most cases these rainfed, terraced fields can be modified to produce carp or other freshwater fish by simple excavation and reinforcement of the levees and retaining walls. *Photo by the author.*

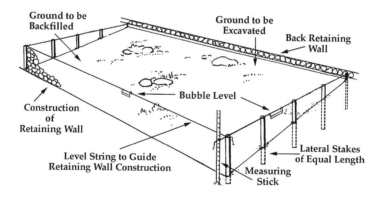

FIGURE 26

Building fishponds on sloping land requires careful attention to site leveling. Lateral stakes should be leveled using a string level. The slope and the desired pond surface area will determine the height of the retaining wall. *Drawing by E. Watkins.*

fields have been converted to fishponds by deepening the ponds and reinforcing the levees for the culture of tilapia or carp. Frequently, these terraced ponds are simply rain-fed watershed impoundments, but often they are serviced by irrigation systems from a watershed reservoir or nearby spring. Building contour ponds requires careful attention to leveling the site using lateral stakes and a string level (Figure 26), as well as excavating and filling the enclosed area of the pond once the outer retaining wall is built (Figure 27).

FIGURE 27
Terraced fishponds and rice fields in Cervantes, Ilocos Sur, Philippines are constructed using sloping agricultural land technology (SALT). After construction of down-slope retaining walls, excavation and backfill level the fields and fishponds. In this location the ponds and fields are adjacent to a river that serves as a water source. Water is diverted to the upper fields and is allowed to cascade down to lower ponds and fields. *Photo by the author.*

Excavated or Dugout Ponds. Excavated ponds are usually the simplest ponds to build on flat terrain. They lie below ground level and are rarely constructed with levees. Frequently excavated ponds result from projects in which the excavated soils are used elsewhere for construction or other purposes. Although excavated ponds are simple, they have some serious drawbacks for fish culture. Excavated ponds generally cannot be drained by gravity or completely drained, especially if they are excavated to the water table. This adds the expense of pumping water to drain and the inability to completely dry the pond bottom is potentially a problem in the transmission of disease pathogens from season to season, requiring modifications to standard pond preparation procedures prior to stocking. Sometimes excavated ponds are used as a location for floating cage culture of fish.

Levee or Embankment Ponds. Levee ponds, sometimes called embankment ponds, are constructed by a combination of excavation and levee construction (Figure 28). Although these ponds cost more than excavated ponds initially, they are usually designed to drain completely by gravity and offer greater control over filling as well, so they are the most frequently constructed type of fishpond in the world. A levee pond may range in size from a simple artisanal or backyard pond with a water surface area of 100 m^2 or less to large commercial ponds tens of hectares in area. Because of the widespread use of levee ponds throughout the world, there are a number of available guidebooks on their construction (Avault 1996; Bard et al. 1976; UNDP 1984).

FIGURE 28
A typical levee fishpond in Mississippi, U.S.A., for the production of North American catfish, *Ictalurus punctatus*. This pond is partially drained, revealing the slope of the pond bottom that aids drainage. Levees are approximately 3 meters wide at the top and are covered with grass for erosion control. *Photo by W. K. Durfee.*

In planning a fishpond, the size and shape of the pond should be given consideration. For artisanal ponds the usual considerations are the amount of land and water available, the amount of time and funds available for construction and what you want to culture. For example, if you decide to build a small pond 100 m^2 in size, you could potentially stock it at 1 fish/m^2, or 100 fish, and depending on the level of management you might harvest anywhere from 30 to 60 kg of fish from the pond in a year. A larger pond can give proportionally higher yields of fish, but are the funds available, and do the costs justify the investment in pond construction?

Prior to construction of the pond, the site must be cleared of all vegetation. Roots must be completely removed, and the outline of the levee foundation should be marked with stakes. The levee staking will consist of two rows of stakes. Inner, parallel rows are the top-width stakes that run along the center of the levee foundation. For example, if the levee top is 1 meter wide, the stakes will be 1 meter apart. The outer rows of stakes are known as the toe stakes, and these delineate the outermost extent of the levee slope. The slope of the levee and its height including freeboard or height above the water surface will determine the distance between the top-width stakes and the toe stakes. For a levee with a height of 1.5 meters and a slope of 2:1, the distance between the top-width stakes and the toe stakes would be 3 meters, and the distance between the two parallel rows of toe stakes would be (3 m + 1 m + 3 m) = 7 meters.

The levee slope depends on a number of factors including size of the pond and the type of soil. Artisanal ponds with an area of 1 ha or less may only require slopes of 2:1, but larger commercial ponds may require slopes of 3:1 or more. Cohesive soil that is rich in clay may only require 2:1 slopes, but if soils are silty or sandy and do not compact well, slopes of 3:1 or even 4:1 may be required. It is important to realize that the greater the levee slope, the greater the amount of earth needs to be moved, thereby increasing construction costs. To save costs while still maintaining levee integrity, some aquaculturists may compromise by constructing a 3:1 slope on the inside of the levee and a 2:1 slope on the outside.

All levees for aquaculture ponds must have some extra height or freeboard above the water level. Freeboard prevents water from waves or excess rainfall from spilling over the levee top. Generally artisanal ponds of a hectare or less require only 60 cm of freeboard. Larger commercial ponds that have a greater exposed surface area for wind wave generation require 90 cm of freeboard, and in some countries with heavy monsoon rains, even more freeboard may be required. Excessive freeboard unnecessarily adds to construction costs.

The most common area for levee seepage or failure is at the base where the levee fill earth meets the grade level. The two most common methods of preventing base seepage are to construct antiseep collars on the inside of the pond or to construct a core trench backfilled with clay (Figure 29). When all the levee staking is completed, the core trench can be dug. Typically the core trench is as wide as the top-width stakes and deep enough to reach clayey or otherwise impervious subsoil. Once the core trench is excavated, it can be backfilled with clay and compacted in layers 20 to 30 cm in thickness. The clay core of the levee should be built to the top of the levee as construction proceeds.

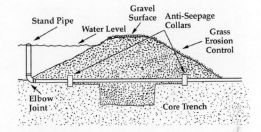

FIGURE 29
A typical pond levee is typically built with a center core trench that is backfilled with soil with high clay content. The most common areas for levee seepage are at the base and around inlet or drain pipes that traverse the levee. A standpipe drain system is one of the simplest to construct and maintain. Concrete antiseep flanges on pipes can reduce the likelihood of seepage along the pipe outer surface. *Drawing by E. Watkins.*

Tidal Ponds. Ponds can be constructed in the intertidal zone using techniques similar to levee ponds, flooded, and then closed off to trap fish and invertebrates carried into the pond by the incoming tide. These simple pond systems are called *tambaks* in Indonesia, where they have been used since the fourteenth century on the island of Java (Shuster 1949). In many parts of Asia and elsewhere, ponds are built in intertidal areas or in coastal wetlands so that they can be flushed by the incoming and outgoing tides, and then stocked with a variety of marine or estuarine species for growth to market size (Figure 30). A key feature of these tidal ponds are specially designed floodgates that can be opened to allow tidal exchange through a predator exclusion weir or closed off to trap water in the pond (Figure 31).

Pond Drain Systems

All ponds, whether they are watershed ponds or levee ponds, require some sort of drain system so that water can be drained completely. There are a number of drain structures that can be employed, which include standpipe drains, sleeved standpipe drains, and monk drains. Watershed ponds all have a natural slope that allows water to flow toward the dam where the drain structure is optimally located. Levee ponds are optimally constructed on land with a 1 to 2 percent slope to facilitate pond drainage. However, the bottoms of levee ponds built on flat land can be backfilled to achieve proper drainage.

FIGURE 30

An aerial photo of intertidal fishponds for milkfish, *Chanos chanos*, in a mangrove estuary in the province of Bulacan, Philippines. These ponds are irregularly shaped due to the location of natural watercourses and their history of ownership and subdivision. The ponds are typically filled during periodic spring tides. *Photo by V. Mancebo.*

FIGURE 31

An intertidal fishpond in the village of Lucap in the province of Pangasinan, Philippines allows free exchange of water throughout a tidal cycle. Fish are retained in the pond, and outside predatory fish are excluded from the pond by a barrier weir on the floodgate. *Photo by the author.*

The most common type of drain is the standpipe, which is a simple device that acts as a drain and maintains water level by virtue of its overall length. Standpipes are usually made from threaded galvanized iron or schedule 40 or 80 PVC material and connected by an elbow connection to a drainpipe that traverses the base of the dam. Simply removing the standpipe from its socket in the elbow connection effects drainage of the pond. Screening material is often put at the top of the standpipe to prevent escapement of fish. The drainpipe is inserted during the construction of the dam and must have antiseep collars that are flanges along the pipe that prevent water from seeping along the outer pipe surface and weakening the integrity of the dam (Figure 29). The diameters of the standpipe and drainpipe depend on the pond size and the amount of water to be drained. For artisanal ponds of a hectare or less, pipe with a diameter of 15 to 20 cm is sufficient, but in larger commercial ponds 20 to 30 cm schedule 80 PVC pipe is frequently used.

A variation on the standpipe drain system is the sleeved standpipe (Figure 32). The sleeved standpipe is identical to the simple standpipe in every particular except for the addition of an outer sleeve fixed to the standpipe. The outer sleeve that extends above the water surface allows for the drainage of bottom water from ponds. This is frequently desirable because bottom water is usually the most oxygen depleted.

Monk drains (Figure 33) are frequently used in artisanal ponds because of their simplicity of construction and design. The monk drain is so called because its design dates back to the thirteenth-century monasteries of Eastern Europe where pond culture of carp flourished. Like the standpipe, the monk drains have a drainpipe that traverses the dam at the lower end of the pond. But since the monk drain does not require an elbow joint, the drainpipe can be constructed from fired clay or concrete, which is easier to obtain in many developing countries or even manufactured on site. The main structure of the monk drain system can be constructed of wood, cement blocks, or formed concrete, depending on what is most available. The water level of ponds serviced by monk drains is maintained by the height of boards stacked on the outer edge of the drain structure.

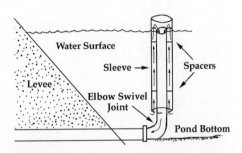

FIGURE 32
A diagram of the sleeved standpipe pond drain system. The water level of the pond is maintained by the height of the inner standpipe, but a fixed outer sleeve allows for draining bottom water. Frequently it is preferable to drain bottom water first because it is frequently the most deficient in oxygen and rich in dissolved carbon dioxide. *Drawing by E. Watkins.*

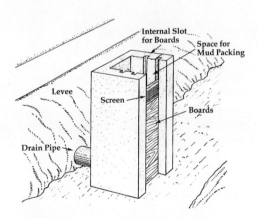

FIGURE 33
Monk drain systems are popular in Europe and in Africa due to their simplicity of design and ability to be constructed from a variety of simple materials. The main framework can be constructed from concrete or wood. The drainpipe can be constructed from concrete or purchased as a prefabricated pipe. The water level of the pond is maintained by staking two rows of boards in the slots provided in the monk frame and packing the median space with mud. A screen can be placed on the monk spillway to prevent fish from escaping. *Drawing by E. Watkins.*

FIGURE 34
Textile sock filter to exclude predator introduction on the water inlet of a fishpond using a surface water supply. *Photo by R. Rudio.*

Pond Water Inlet Systems

Water inlets are generally situated on the side of the pond opposite from the drain. In watershed ponds it is merely the watercourse entering the pond, and in levee ponds it could be an irrigation canal from a watershed reservoir, a watercourse from a spring, or a pipe leading from a well or springbox tank. In general, if surface water is used, it is wise to use a sock filter or similar device to exclude predatory fish (Figure 34). If well or spring water is used, it is free from predators, but it usually is low in dissolved oxygen, so some kind of aeration device to increase its dissolved oxygen concentration as it is added to the pond.

POND MANAGEMENT

The management of aquaculture ponds will largely depend on the species of fish cultured, but there are some general principles that are common to all ponds. One of the important principles for the management of any kind of aquaculture operation is the careful maintenance of complete records. Records of water quality, observations about fish health, amounts of feeds and medications supplied, and any changes in management routine or fish behavior should be noted along with other information that you might think is important. In this section the focus will be on the management of artisanal freshwater fishponds, but many of the principles guiding the proper management of freshwater ponds are equally applicable to most saltwater and brackish water ponds.

Management of Carp Ponds

The freshwater pond culture of carps accounts for a large percentage of the world's freshwater fish culture. Because of this carps are a convenient species to discuss to illustrate a number of management techniques such as breeding, pond fertilization, feeding, and harvest.

Breeding Carps. One of the reasons for the widespread culture of carps is their ease of breeding in captivity. It is common for captive carp broodstock to be many generations away from wild stock. Carps have been selectively bred as meaty, high-yield food fish and as ornamental fish. Indeed, the common goldfish, with a multitude of body forms and colors, and the colorful koi carp have very little superficial resemblance to their wild ancestors with drab coloring.

In tropical southern China and in Southeast Asia, where carp farming is very common, the common carp, *Cyprinus carpio*, breeds year around. In higher latitudes where winter temperatures drop, the breeding is usually during the warmer summer months. In the tropics, both male and female carps reach sexual maturity in a year, but in colder climates maturation may take from two to three years. Broodstock should be maintained on a good quality, relatively high-protein diet to aid egg and sperm development and maturation. Female broodstock exhibit a round, soft, full abdomen when ready to spawn. For both males and females, the best way to test ripeness is to remove them from the pond and apply gentle pressure to the abdomen to extrude eggs or sperm.

In nature, female carps lay several thousand eggs on submerged vegetation, where the males fertilize them. The fecundity or number of eggs produced by female carp is dependent on size and can be expressed by this equation (calculated using data in Bardach et al. 1972).

$$E = a \times L^b$$

when   E is the number of eggs per spawn

L is the overall length of the female carp in cm

a is a constant = 0.148

b is another constant = 3.78

For example, if a female carp is 50 cm in length, it can be expected to produce $[(50)^{3.78}] \times 0.148$, or about 390,000 eggs per spawn. Although the largest fish produce the most eggs, some aquaculturists do not to prefer very large broodstock because there is a decrease in egg viability in older fish.

In nature, the eggs remain somewhat hidden from predation, but in aquaculture ponds where fish densities are high, adult carp will prey upon the developing eggs and larvae. For this reason, established methods of breeding carp involve the separation of parental stock from the developing eggs and larvae. The simplest methods of breeding carp involve preparing special spawning ponds, allowing breeding to proceed, and then removing the breeder fish. One method of this type is the Dubisch method. Grass is planted and allowed to grow on the bottom of a small 0.1 ha pond until it reaches 40–50 cm in length. The pond is then flooded to the top of the grass, and 10 breeding pairs of carp are added (100 breeding pairs/ha). Breeding is allowed to occur. Once egg masses are spotted adhering the submerged grass, the parental stock are then seine netted from the pond (Figure 35), and the eggs are allowed to hatch and develop through their larval and juvenile stages in this hatching pond.

Other methods for breeding carp also focus on the removal or isolation of the eggs from the parental stock. One popular method of breeding carps is the hapa method. A hapa net (Figure 36),

FIGURE 35

Carp breeding requires separation of parental stock from eggs and larvae after spawning in ponds to prevent cannibalism. Seine netting is a frequently used method for removing fish from ponds. Seine nets are wide enough to traverse the width of a pond and have a float line and a weighed sinker line that drags across the pond bottom. *Photo by the author.*

which is a five-sided (four walls with a bottom), 2 to 4 m^2, open top net usually made from 1 mm mesh nylon or polypropylene netting material, is staked by its four corners in a regular carp production pond. Aquatic vegetation such as water lilies, water hyacinths, or bunches of other aquatic plants is placed into the hapa net, and then one or two breeding pairs of carp are introduced. After egg masses appear on the vegetation, one side of the hapa net is lowered to allow the parental stock to return to the general population of fish. The eggs are then allowed to hatch and develop through the larval stages in the hapa net, or alternatively the eggs can be removed for hatching in a specially prepared hatching pond or tank.

In subtropical and temperate zones, the breeding of carps is seasonal and is infrequent and unpredictable. For example, in northern Taiwan (25° N Lat.), the carp breeding season lasts from March to July. The practice of the Taiwanese is to induce spawning by injecting the carp broodstock with three to six pituitary glands removed from the brain tissue of sacrificed adult carp one month prior to the intended spawning time (Chen 1976). This method, called hypophysation, was discovered in Brazil in the 1930s and is the simplest method of inducing breeding of fish by reproductive hormone treatment. Since the 1970s, the Taiwanese have preferred to inject a commercially available product containing a mixture of fish gonadotropins or pituitary extracts and human chorionic gonadotropin (HCG). Although these products may be commercially available from veterinary supply sources, hypophysation remains an excellent tool for artisanal aquaculturists. The only required materials are a scalpel or sharp knife to excise the pituitary tissue from donor carps, a means to macerate the pituitaries in clean water, and a small syringe and needle to inject the mixture into the muscles of the recipient fish. Once the injected fish become ripe, they can be spawned using the methods previously described.

FIGURE 36
Hapa nets a four-sided open top nets that can be placed in a pond to isolate fish. Hapa nets can be seen
in this research pond in the Philippines. *Photo by the author.*

Nursery Culture of Carps. Depending on temperature, carp eggs hatch in 18 to 60 hours. At
21–24°C, the eggs hatch in about 30 hours, and at 27–30°C, it takes about 20 hours. In the first
larval stage after hatching, the carp have a yolk sac from which they derive nourishment, so
they do not require feeding at this stage. After two to six days, the yolk sac is absorbed and the
larvae become free swimming and able to eat particulate food.

The nursery culture of carps and other fish is the critical stage. The larvae are subject to
disease and predation, and they require a steady supply of food in the proper size range, usually
plankton in the water. Mortality rates of larvae may be as high as 50 to 80 percent, which is
much higher than at any other stage of the culture process. To accommodate this critical growth
period, greater care is usually taken with the management of nursery ponds. Nursery ponds are
typically smaller (often 0.1 ha in size) than production ponds for larger fish. Prior to flooding

and stocking, ponds are often treated with plant-derived poisons that are frequently photo-degradable and biodegradable. For example, in the Philippines where tobacco is a common cash crop, dried and shredded tobacco leaves are applied to fishponds at a rate of 1,000 kg/ha and the pond partially flooded. The nicotine in the tobacco acts to kills predators and pests lurking in the pond bottom. After two weeks in the tropical sun, the pond is drained, flushed with fresh water, and then fertilized in preparation for stocking. Commercially available rotenone powder, which is derived from the tropical derris root, can also be used to poison pond bottoms when applied at a rate of 0.5 kg/ha. In most carp nursery operations, pond fertilization prior to flooding for stocking is a critical step to ensure healthy blooms of phytoplankton and zoo-plankton that serve as food sources for the developing fish larvae and juveniles. Stocking densities for carp larvae range from 200,000 to 300,000 in 0.1 ha ponds for the first three months of nursery culture (Chen 1976). During nursery culture, the carp diet can be supple-mented by a host of particulate feeds that are rich in protein to improve growth. Examples of supplemental feeds used successfully for carp are dried insect larvae, dried clam meats, meat powder, dehydrated blood, and fish meal (Bardach et al. 1972). After about three months of culture in nursery ponds, carp can be transferred to production ponds at a stocking rate of about 10,000 fish/ha.

Pond Fertilization

The production of carps and other species such as tilapias in artisanal ponds requires the stimu-lation of phytoplankton production and thus the entire aquatic food web through pond fertiliza-tion. There are a number of ways fertilization of ponds can be carried out. Usually, the initial pond fertilization is part of the preparation of the pond prior to flooding and initial stocking. A number of fertilizers can be used for this purpose including organic fertilizers like composted or fresh animal manures or commercially available inorganic fertilizers. Inorganic fertilizers usually have a N-P-K rating listed on their bag or accompanying literature that indicates the percentages of nitrogen, phosphorus, and potassium the fertilizer contains. It is important to realize that in most freshwater system the most frequent limiting nutrient is phosphorus, so fertilizers with high phosphorus content like ammoniated superphosphate (3-16-0) are often a good choice for pond preparation, but other fertilizer formulations are also acceptable (Swingle et al. 1963). Fertilization can be carried out after stocking of fish by broadcast spreading or other means to maintain phytoplankton blooms (Figure 37). Ammoniated fertilizers are not acceptable for active ponds with fish in them. In brackish or saltwater ponds nitrogen is often the key limiting nutrient, so nitrate-rich fertilizers are often desirable. Calculating the unit cost of desired nutrients in kg/dollar can reduce costs for inorganic fertilizers.

In most artisanal ponds, organic fertilizers are the most frequently used. Most animal manures have a good balance of N-P-K nutrients, but their average nutrient percentages by weight are less than 1 percent. The most common fresh animal manures and their N-P-K percentages are poultry (1.31-0.40-0.54), swine (0.84-0.39-0.32), and cattle (0.43-0.29-0.44) (Avault 1996). Dried or composted manures will contain higher percentages of nutrients by virtue of dehydra-tion. It is useful to calculate the actual nutrient percentages of dried manures based on the average moisture content of fresh manures: poultry (57 percent), swine (74 percent), and cattle (75 percent). An average application rate of manures for pond preparation prior to flooding is 2,000 kg/ha. After fertilizer application on the pond bottom, the pond should be filled and allowed to sit for two to three weeks for the phytoplankton to bloom prior to stocking with fish. Phytoplankton density should be monitored by Secchi disk, and manures can be applied at no more than 80 kg/ha daily to maintain Secchi visibility depths between 40 and 50 cm while

FIGURE 37
One method of pond fertilization is to place bags of fertilizer, usually an inorganic phosphate, directly into the pond. Fertilizer releases slowly as the bag soaks and the fertilizer salts dissolve. In this pond, the farmer has suspended a fertilizer bag in the water from the tripod in the middle of the pond. *Photo by the author.*

keeping the load of organic material low enough to keep oxygen depletion problems from occurring (Avault 1996).

In some areas of China, Southeast Asia, and elsewhere, animal production facilities are constructed in close proximity to fishponds to provide a ready supply of manure. In some cases, animal pens and poultry houses have been constructed directly over fishponds to directly deposit manures in the ponds. To estimate the optimum number of animals to support this type of integrated terrestrial animal-fish culture system, it is useful to calculate numbers of animals required based on average annual manure production of farm animals. Typical annual production of manure per metric ton (1,000 kg) live weight of selected farm animals are cattle (30,000 kg), swine (37,000 kg), and poultry (8,600 kg) (Avault 1996).

Another approach to fertilizing artisanal ponds is to maintain a compost enclosure in the pond to deposit agricultural or household organic wastes (Figure 38). The enclosures can be constructed of split bamboo or similarly flexible material. Typically the enclosures are situated near a levee corner and have a capacity of 2 to 5 m^3. Agricultural wastes such as spoiled foods, fruit peelings, or other organic wastes can be introduced into the compost pit. The compost should be turned with a pitchfork twice weekly to aerate the compost to aid the decomposition. The nutrient value of composted agricultural wastes is highly variable. For example, composted agricultural wastes such as rice straw and cornhusks have low N-P-K values of less than 1 percent each, but some other agricultural by-products from oil presses such as peanut cake or cottonseed cake may have much higher nutrient content.

FIGURE 38

Compost enclosure constructed of split wood basketry materials at the edge of a tilapia pond in Zaire (now Republic of Congo). Household and agricultural wastes, including manures, are placed in the compost enclosure and periodically turned to promote decomposition. *Photo by P. Stiles.*

Fish Polyculture

One of the major ecological concepts that are frequently overlooked by novice aquaculturists is the concept of carrying capacity. If one desire to harvest more fish in a season, there is a natural tendency to push the stocking density of a pond to its limits, usually resulting in overstocking. Overstocking by just a few fish causes stress, resulting in reduced growth and potential losses due to the onset of opportunistic diseases.

One method of stocking ponds for greater overall fish production is to use fish of different species that feed on different food sources. This technique of stocking ponds with two or more different species of fish is called polyculture, which was pioneered by Chinese carp farmers. In classical Chinese carp polyculture, five species of carps are stocked into ponds at a stocking density much greater than could be achieved if one single species of fish were stocked mono-culturally (Chen 1976). The five species of carps often used in polyculture are the silver carp (*Hypopthalmichthys molitrix*) that feed on phytoplankton; the bighead carp (*Aristichthys nobilis*) that feed on zooplankton; the mud carp (*Cirrhinus molitorella*) that feed on detritus, worms, and other bottom invertebrates; the black carp (*Mylopharyngodon piceus*) that feed on freshwater snails and clams; and the grass carp (*Ctenopharyngodon idella*) that feed on grass and other higher plants taking root in the pond bottom. The overall fish-stocking biomass of the pond is increased by virtue of using a number of fish species that do not directly compete for pond space and food resources. The three spatial dimensions of the pond water are effectively used, as are the multiple food resources of the pond (Milstein 1997). Typical stocking densities per hectare for polycultured carp in unaerated Chinese ponds are silver carp 2,000; bighead carp, 600; mud carp, 2,000; black carp, 1,000; and grass carp, 400. In total 6,000 fish are stocked into the typical pond. For comparison, only about 2,000 common carp can be safely stocked into comparable ponds in monoculture. Typically, harvest yields of ponds stocked in this manner can be expected to yield 2,500 to 4,000 kg/ha/year, but some of the highest reported fish yields from unaerated ponds that do not use supplemental feeds are around 9,000 kg/ha/year in polyculture ponds in China (Liang et al. 1991).

In Indonesia and India there are other established methods for the polyculture of carps using other species combinations from the local area. By studying the feeding habits of fish and experimenting with different combinations and stocking densities, it is possible to establish new polyculture combinations and methods just about anywhere.

Pond Culture of Tilapias

Tilapias are a group of fish of the family of cichlids or "mouthbrooders" that are native to Africa and the Near East. As a group they are a highly desirable for aquaculture because of their ease of breeding, rapid growth on inexpensive feeds or natural pond productivity, and a high tolerance of a wide range of environmental factors (Pullin & Lowe-McConnell 1982). They can even be adapted to grow in seawater, but they do not tolerate water temperatures below 15° C, which restricts their outdoor culture to warmer tropical and subtropical regions. Indeed, according to UN-FAO statistics (FAO 1993), the countries producing the largest yields of tilapia are the Philippines, Thailand, China, India, Egypt, Israel, and Nigeria—all warm tropical and subtropical countries. There are a considerable number of tilapia species that are cultured around the world. Among the most commonly cultured species are Nile tilapia (*Oreochromis niloticus*), Mozambique tilapia (*Oreochromis mossambicus*), blue tilapia (*Oreochromis aureus*), and Zill's tilapia (*Tilapia zilli*).

Breeding and Nursery Culture of Tilapias. Behavioral traits are one of the bases by which the tilapias of the genus *Oreochromis* lay eggs in nests in the pond bottom (Figure 39), and the females brood eggs and newly hatched larvae in their mouths. *Tilapia zilli* and other members of the genus *Tilapia* do not mouthbrood, but they are substrate spawners that lay eggs on a flat rock or other surface, then guard the eggs from predation (Fitzsimmons 1997).

Tilapias are known for their ease of spawning. Most species of cultured tilapias reach sexual maturity at a length of 15 cm, which can be reached in 4 to 6 months in warm ponds. Once females have spawned, they can be reconditioned for spawning in 6 to 8 weeks. Females of the genus *Oreochromis* will generally produce 100 to 600 eggs per spawn, depending on size (Peters 1983). Typical spawning ponds for tilapia are about 100 m^2 in area and 1 m deep, and stocked with 12 female and 4 male fish. Tilapia of reproductive age can be sexed by examination of the genitalia. Males have a single urogenital opening on the tip of a genital papilla that is posterior to the anus, whereas females have an oviduct that is separate from the ureter or urinary oriface that is also located posterior to the anus (Bardach et al. 1972). When tilapia are smaller than 10 cm in length, determination of sex by exterior examination is very difficult. A properly stocked 100-m^2 spawning pond can be expected to produce 2,000 to 5,000 fry every three to four months.

The nursery culture of tilapias can be carried out in ponds that are prepared and fertilized similarly to those used for carp culture (Nwachukwu 1997). In monoculture, fertilized nursery ponds for fry and juveniles can be stocked at 20,000 to 30,000 fish/ha for the first three months of growth, after which they are thinned to stocking densities 2,000 to 3,000 fish/ha in ponds for growth to marketable size.

FIGURE 39

Nests of tilapias of the genus *Oreochromis* appear as circular depressions in the pond bottom. In this case the pond is drained and sun dried to kill predators residing in the bottom mud as part of the preparation process prior to seasonal stocking. *Photo by the author.*

Management and Productivity of Tilapia Ponds. One of the key management problems unique to the culture of tilapia is related to their high fecundity. Tilapia breed so often and become sexually mature at such a relatively small size that they breed uncontrollably in production ponds. This breeding causes a large variation in fish sizes and stunting because of eventual overstocking and competition for food resources. Ponds with large number of small fish are not a particular problem in some countries such as the Philippines or Indonesia where fish can be successfully marketed in the 10 to 20 cm size ranges. However, in other countries (like most in tropical Africa), tilapia that are in excess of 25 cm and weighing 600 g or more are the preferred size range (Bardach et al. 1972).

To overcome the problem of excess breeding by tilapia, a number of pond management methods are used. One method is to harvest the largest tilapia from ponds, determine the sex, and isolate these largest fish by sex in separate ponds (Fitzsimmons 1997). The drawback with using this method is that smaller tilapias are very difficult to sex based on external characteristics, and the process can be very tedious and time consuming. A more common method of controlling reproduction is the addition of androgenic steroids to the feed of sexually undifferentiated juvenile tilapia while they are in nursery ponds (Guerrero 1975). Commercially available androgenic steroids such methyltestosterone and androstenedione can be put in the feed at the rate of 30 to 50 mg/kg of feed of first-feeding tilapia fry for 30 days (Guerrero & Guerrero 1997). Using this protocol, androstenedione can induce the development of 80 percent of the fish as males, and methyltestosterone can induce 97 percent of the fish to develop as males. Once fish have completed this sex-reversal protocol, they can be stocked into fishponds as usual with greater growth performance.

Like pond culture of any other species, the production level of tilapia is greatly influenced by the level of management employed. In unfertilized ponds with no active management, tilapia ponds can be expected to yield from 300 to 600 kg/ha/year. By simply fertilizing the ponds with manures to optimum phytoplankton production, the fish yield can be increased to between 3,000 and 5,000 kg/ha/year. In fertilized ponds with supplemental feeding with high-protein agricultural wastes or commercial feeds, the production can be expected to increase to 6,000 kg/ha/year. Finally, fertilized ponds with supplemental feeds and an all-male population have yielded as high as 9,000 kg/ha/year (Swift 1985).

CULTURING FISH IN FLOATING CAGES, NET PENS, AND FISH CORRALS

Raising fish in floating fish cages, floating net pens, or fish corrals set up in a lake or other body of water can be a simple and inexpensive alternative to raising fish in ponds (Beveridge 1987). Floating fish cages and floating net pens are similar, but the containment walls of cages are generally constructed of a more rigid material than simple fish netting (Figure 40). Fish corrals, on the other hand, are enclosures that are constructed in shallow water (Figure 41). Cage culture allows for the holding of fish, but there is free exchange of water between the water body and the fish. Thus, raising fish in enclosures allows an aquaculturist to utilize existing water resources in ponds or common water bodies that may have other uses. For example, fish cages may be

FIGURE 40
Floating fish cages for the estuarine culture of groupers and sea bass, using a four-net floating unit system popular in Malaysia, Singapore, and the Philippines. Each net in the unit is approximately 8 m^3. *Photo by the author.*

FIGURE 41

Fixed net pens or fish corrals directly staked in the estuary bottom is one method to culture fish. These fixed net pens are 70m^3 to 200m^3 in volume and are constructed by suspending nets onto a bamboo framework. They are used to culture milkfish, *Chanos chanos*, in the town of Binmaley in the province of Pangasinan, Philippines. *Photo by the author.*

constructed in estuaries where there are active natural fisheries or used as a waterway for boat transportation.

As with any aquacultural production method, the production of fish in enclosures has its advantages and disadvantages. The advantage of culturing fish in enclosures include the following.

- Ownership of the entire water body may not be necessary, thereby reducing land costs. However, permits from some governmental agency overseeing the use of common water bodies may be necessary.
- There are relatively little investment costs in gear and harvest equipment.
- Harvesting is simple.
- Observing fish behavior and monitoring fish health is simplified.
- It is possible to use the water body in other ways.

Disadvantages of the production of fish in enclosures are mostly related to greater management effort due to large numbers of fish confined in a relatively small space. The following are some specific disadvantages.

- The fish must be fed a complete and nutritious diet. The fish cannot rely on the natural productivity of the water body for their diet.

- Due to the stocking of fish and inputs of feeds into the enclosures, there is the possibility of localized hypoxia or low oxygen stress. Enclosed fish cannot freely swim to areas of better water quality as they do in ponds.
- The incidence of diseased fish may be high, and diseases that appear can spread rapidly.
- Vandalism and poaching are often problems, particularly if the enclosures are placed in common-use water bodies.

In general, most fish that are suitable for culture in artisanal ponds are suitable for culture in floating fish cages or fish corrals. Fish that have been successfully reared in small-scale floating cages or netpens include carps (Masser 1988), tilapia (McGeachin et al. 1987), and North American catfish (Schwedler et al. 1986). For fish like carp, tilapia, and catfish that feed low on the food chain and are tolerant of crowding, typical stocking rates in cages are between 100 and 500 fingerlings per cubic meter of water, depending on fish size and the amount of water flow through the site. They must be thinned as they grow to larger sizes. A fish that feeds low on the food chain and is often cultured in fish corrals in estuaries in Southeast Asia is the milkfish, *Chanos chanos*, at average stocking densities of 36 fish/m^3 in 70-m^3 pens (Rice & Devera 1998).

Although most artisanal cage culture of fishes focuses on species that are tolerant of crowding and can be fed on inexpensive vegetable-based diets, such is not a requirement for successful artisanal aquaculture. For example, several species of very high value fishes can be cultured in floating cages or net pens as a cash crop. Examples of fish that have been successfully cultured using small-scale floating net pens include tropical groupers (Chua & Teng 1982), barramundi or sea bass (Kungvankij et al. 1986), and snappers (Leong & Wong 1987). The culture of these fish that have a very high market value requires considerable attention to feeding a very high-protein diet consisting of fresh "trash fish" or freshly prepared moist pelleted diets with a high percentage of fish or other animal protein (Rice & Devera 1998). Obtaining fry and fingerlings for these fish is often by collection from the wild, but hatchery stock is available for some species in some areas. Typical stocking densities for groupers, seabass, and snappers range from 8 to 32 fingerlings/m^3, which is considerably lower than fish feeding lower on the food chain, indicating a greater sensitivity to crowding (Kungvankij et al. 1986). Seabass cultured in floating cages using this stocking rate can reach a market weight of 500 to 700 g in six to nine months.

Enclosures for fish culture have been constructed from a wide variety of materials. The basic cage materials must be strong, durable, and nontoxic, and the cage must retain the fish while maintaining maximum circulation of water to bring oxygen to the fish and to remove wastes. Possible cage designs are limited only by the imagination of the builder, the requirements of the site location, and the materials available (Milne 1979).

Cage components usually consist of a frame, netting, cover, flotation, and mooring lines. The frame can be made from wood, bamboo, steel, PVC pipes, or any other suitable material. If iron or steel is used in the frame it, should be coated with a water-resistant paint to retard corrosion, especially if they are to be used in seawater or brackish water. Netting materials are most often knotted or knotless fishing nets or plastic polyethylene mesh. In some parts of Southeast Asia, however, a meshwork of split bamboo is often used. For most applications, mesh sizes should be no smaller than 1 cm to allow for good water circulation, and as a general rule it is wise to use the largest mesh possible without allowing the fish to escape. Some net pen operators will have several nets of different mesh sizes to be used as the fish grow. Also, as part of regular net pen management, the nets should be periodically checked for fouling. This is especially critical if the net pens are placed in estuaries or in marine waters. Most cages have lids or covers to restrict predators from reaching the fish. Cage flotation can be provided in a number of ways.

Styrofoam logs, plastic barrels, painted metal drums, and plastic bottles have all been used as flotation devices. Styrofoam is an inexpensive and popular choice for flotation, but it often deteriorates in sunlight. To prevent deterioration and disintegration, it is best to cover raw Styrofoam with a tar or paint.

Mooring lines are an important feature of any floating net pen or cage. There are typically four mooring lines 1 cm in diameter and made from polypropylene, nylon, or similar synthetic material, which is usually adequate to hold the cage in place. Sisal or hemp ropes can be used if they can be obtained inexpensively, but they deteriorate faster than synthetic materials. Mooring anchors can consist of metal boat anchors, cement blocks, rocks, or any other heavy material. Mooring line scope is an important consideration, especially in relatively deep water or in water bodies that are subject to wind and waves during storms. Mooring scope is the ratio of horizontal line run to the depth of the water. If cages are put into a small watershed impoundment pond that is protected from winds, a mooring scope of 2:1 may be all that is required to keep the cages in place. In the extreme case of large net pens in coastal waters, a mooring scope of 10:1 may be required along with steel mooring cables to keep the rig firmly in place; for most artisanal net pens placed in rivers or estuaries, mooring scopes of 3:1 are more typical. If in doubt, more mooring line scope is always better than less even though the cost for materials may be a little higher.

Properly locating cages or net pens is often critical for success. The two main factors influencing the location of net pens and cages are access for maintenance and harvest and water quality considerations. To maximize water flow, cages might be placed in a windswept portion of a pond or, if placed in a river or estuary, in an area that is in a gentle current or is one flushed tidally. Small floating cages are best placed in water depths in excess of 2 meters with a minimum of 0.5 m clearance between the bottom of the cage and the pond or river bottom and away from areas with frequent human activity. To minimize the likelihood of hypoxia or low oxygen around the cages, there should be a buffer zone between cages 4 meters or wider.

FEEDS AND FEEDING

If fish are cultured in cages or net pens, the fish are totally dependent on supplied feed. There are nutritionally complete diets that are commercially available for most popularly cultured species of fish, but sometimes fish farmers opt to formulate their own feeds (Halver 1992). This should be done with care based on observations of what the fish normally eat, what kind of feedstuffs are available, and published feed formulations (New 1987) and their nutritional content (National Research Council 1993). Caged fish should be fed a complete floating diet. A floating diet allows for the observation of the feeding behavior of the fish and allows for less food waste. As a general rule, fish should be fed a greater percentage of their body weight during the juvenile stages than when they become larger adults. For example, most species of warm water fish such as carps, tilapias, catfish, and even some groupers and seabass will consume about 5 percent of their body weight in food daily as a 10-g fingerling. But they would only consume 1.75 percent of body weight daily in food as a 750-g adult (Table 14). Feeding rates are sensitive to water temperature and are species specific. For carps, feeding rates should be reduced when temperatures drop below 15 °C or rise above 32 °C.

As stated earlier, feeds used for finfish culture in cages and net pens must be nutritionally complete. Furthermore, one of the major expenses of fish culture is the cost of feeds. Feeds may not be as critical an expense for fish culture in fertilized ponds where fish are deriving much of their nutrition from the natural productivity of the pond, but in cage culture feeds are the largest expense. For practical applications, the evaluation of production diets can be adequately done

TABLE 14
Estimated daily food consumption by a typical warm
water fish in water temperatures in excess of 20°C

Average Fish Weight (grams)	% Body Weight Consumed Daily
10	5.00
20	4.00
30	3.00
100	2.75
250	2.50
350	2.25
500	2.00
750	1.75

in feeding trials. Different commercial feeds could be compared, or if the aquaculturist is formulating his own diets, different formulations can be tested.

One of the simplest means for an aquaculturist to evaluate feeds is to determine a food conversion ratio (FCR) (Rice et al. 1994). The FCR is the weight of the food supplied divided by the weight gain of the fish during the feeding period and can be expressed by this equation.

$$FCR = F/(W_f - W_o)$$

when F = the weight of the food supplied during the study period
        W_o = the live weight of the fish at the beginning of the study period
        W_f = the live weight of the fish at the end of the study period

For example, if a fish pen operator starts with 500 tilapia fingerlings at an average weight of 100 g each, the aggregate W_o is 50 kg. The fish are fed 3 g food/fish/day for six months, when they are harvested at a final average weight of 450 g each, but there has been 2 percent mortality. The aggregate W_f would be $500 - (500 \times 0.02) = 490$ fish $\times 450$ g = 220.5 kg. The amount of food supplied would be 3 g/fish/day $\times 182$ days $\times 500$ fish = 273 kg. So the FCR would be FCR = 273 kg/(220.5 kg − 50 kg) = 1.60.

A very important factor to remember when FCRs are compared is that they are based on the wet weight of the feed. Different feeds have different moisture levels. For example, a semimoist diet for groupers or sea bass may have a moisture content of over 60 percent, whereas a dry pelleted diet for tilapia may have a moisture content of only 10 percent. Moisture does not contribute to the growth of fish, but it does add bias to FCR values. Thus, if comparisons are made between the performance of two or more diets, it is useful to calculate FCR in a feed dry weight basis. To do this, you must determine the moisture content of your feed by drying and weighing.

Often if you calculate FCR based on supplemental feeds given to fish in fertilized ponds, the FCR values fall well below one. This is indicative that the fish are deriving their nutrition largely from pond production rather than the feed.

High-protein ingredients such as fish meals or fresh fish processing wastes are frequently the most expensive components of artificial diets, so to boost fish farm profitability, a key strategy is to reduce crude protein in diets without sacrificing fish growth. Excess protein in fish diets also results in excess excretory ammonia from the metabolic breakdown of the protein, which may be a stressor to the fish. One means to determine the optimum level of protein in a feed is to compare the protein efficiency ratios (PER) of different feeds fed to the fish (Milne 1979). PER is the weight gain of the fish divided by the dry weight of protein in the feed. Most commercially available diets list crude protein as part of the formulation analysis. PER can be described by this equation:

$$PER = (W_f - W_o)/F \times p$$

when   F = the weight of the feed supplied over the test period
p = the fraction of crude protein in the feed

For example, if the percentage of crude protein in the feed from the previous example were 35 percent, the PER over the 6-month growth period would be

$$PER = (220.5 \text{ kg} - 50 \text{ kg})/(273 \text{ kg} \times 0.35) = 1.78$$

Now, if a feed were chosen or formulated a feed with less fish or fish meal with the reduced protein content 30 percent and the fish growth is the same, the PER would be

$$PER = (220.5 \text{ kg} - 50 \text{ kg})/(273 \text{ kg} \times 0.3) = 2.08$$

The PER values are reduced when protein levels are either insufficient or in excess. The optimum protein content in fish feeds is species specific and occurs when PER is maximized.

ARTISANAL CULTURE OF BIVALVE MOLLUSKS

Many species of bivalve mollusks are suitable for small-scale aquaculture projects. Although there are a few cases of polyculture and harvest of freshwater clams (*Corbicula* spp.) in carp ponds in China (Xu et al. 1987) and Taiwan (Phelps 1994) and research into the polyculture of these freshwater clams in the United States with catfish (Buttner 1986), most artisanal culture of bivalves is carried out with marine or estuarine species. There are a number of reasons why the culture of marine and estuarine bivalves such as oysters, mussels, and clams is attractive. First, many of the methods for raising bivalves are well established in many countries and adaptable to other parts of the world. For example, there have been major reviews and handbooks on the culture of oysters (Korringa 1976; Quayle 1980; Angell 1986), mussels (Vakily 1989), ark clams or blood clams (Broom 1985), and giant clams (Heslinga et al. 1990). Second, bivalve mollusks are filter-feeders that occupy a low position on the food chain and derive all of their nutrition by filtering phytoplankton and other natural particulates from the water, so there is no feeding cost. And finally, the culture of bivalve mollusks can occur in concert with small-scale finfish aquaculture projects (Rice & Devera 1998), and shellfish aquaculture is considered to be relatively benign from an environmental impact standpoint (Kaiser et al. 1998).

Spat Collection

Most small-scale aquaculture of bivalves requires the collection of seed or small juveniles, known as *spat*, as they settle from their planktonic larval stages. All aquacultured marine bivalves spawn seasonally, releasing both eggs and sperm into the water, or alternatively the partially developed larvae into the water where they are dispersed by wind and tidal currents (Loosanoff & Davis 1963). The developmental sequence of bivalves includes the nonfeeding fertilized eggs,

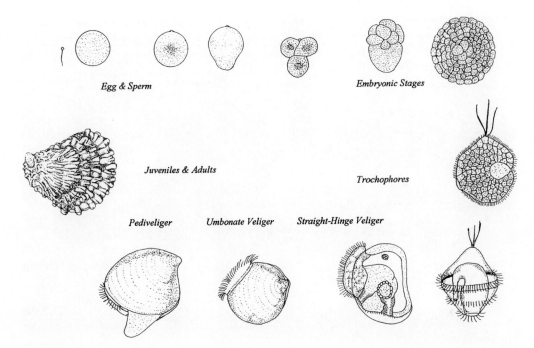

Egg & Sperm

Embryonic Stages

Juveniles & Adults

Trochophores

Pediveliger   *Umbonate Veliger*   *Straight-Hinge Veliger*

FIGURE 42

Identification of the various developmental stages of most commercially important bivalve mollusks from egg and sperm through the larval stages can aid in spatfall prediction. The embryonic and larval stages through the umbonate veliger are planktonic and are dispersed by wind and tidal currents. Pediveligers are also planktonic, but they devlop bottom-seeking behavior and are capable of setting on preferred substrates. *Drawings by V. Encena*

embryos, and trochophore larvae that last for about 24 hours until they develop into feeding straight-hinge larvae, fully developed veliger larvae, and eventually pediveliger larvae that seek out a preferred substrate for settlement and metamorphosis into juveniles (Figure 42).

In nature, settling and metamorphosing juvenile bivalves alight on preferred substrates. For example, reef-building oysters in temperate zones generally settle on oyster shells and rocks, but the mangrove oysters of the tropics primarily settle on mangrove wood, but they also settle on rocks and other hard objects (Angell 1986). Mussels will settle on hard surfaces, but they appear to aggregately settle in areas with other mussels, possibly on the fibrous byssal threads of other mussels (Bayne 1976). Many species of scallops prefer sea grasses as an initial settlement substrate (Marshall 1947), and various species of clams settle and survive in sediments of various consistencies and grain sizes (Pratt 1953).

Most artisanal aquaculture of bivalves relies on either collecting settling spat on some sort of artificial spat collector that mimics the natural preferred settlement substrate or collecting the bivalves from natural seed grounds and moving them to a protected farm site. Spat collectors for various species of bivalves are generally designed from available materials that the shellfish will settle on. For example, a traditional spat collector for European oysters, *Ostrea edulis*, in England has been the coating of curved ceramic roofing tiles with a mixture of cement, lime, and sand and then placing them on the estuary bottom (Walne 1979). Although the traditional tiles have been superseded recently by the use of commercially produced cone-shaped flexible polyethylene spat collectors, the lime, sand, and concrete coating is still used as a settling substrate for oysters. In the Philippines, spat collectors for mangrove oysters, *Crassostrea iredalei*, consist

FIGURE 43

Spat collectors for tropical mangrove oysters can be made from a variety of inexpensive materials, including rubber strips cut from truck or automobile tires (A), waste plastic box strapping material (B), or oyster shells strung onto polypropylene ropes (C). *Photo by the author.*

of oyster shells strung onto polypropylene ropes, rubber strips cut from used automobile and truck tires, or discarded fiberglass and plastic box strapping material (Figure 43). The spat collectors are suspended in the water from a bamboo framework (Figure 44) or floating raft (PHRDC 1991). For the collection of mussels, spat collectors usually consist of a fibrous plant material attached to ropes. In China, spat collectors consisting of palm fibers attached to ropes are suspended from offshore rafts or longlines to catch settling blue mussel (*Mytilus edulis*) larvae (Zhang 1984). Similarly in the Philippines, the fibrous husks or coir of coconuts are strung on polypropylene ropes to serve as spat collectors for the green mussel *Perna viridis* (IIRR 1995). The Japanese use a number of different spat collectors for the collection of spat of the Japanese scallop *Patinopecten yessoensis* (Ito 1991). The spat collector types include bunches of seaweed (*Laminaria*) suspended from ropes, simple stretch nets anchored to the bottom, and mesh bags stuffed with monofilament nylon gill netting. During the 1960s, in the early years of spat collection, the mesh bags were simply discarded onion bags. However, with experience in Japan and elsewhere, it was learned that smaller mesh outer bags with mesh openings of 1 to 1.5 mm or less improved the performance of the spat collector by better excluding predators that feed upon the newly settled scallop spat (Tammi et al. 1997).

Spat collection of infaunal clams has not been as successful as the collection of bivalves that are epifaunal, or residing on top of the substrate surface, like oysters or mussels. However, there are some instances in which there have been successful small-scale clam aquaculture operations based on the collection of seed clams from areas where they naturally set in great numbers. An example of this is culture of the ark clam or blood cockle (*Anadara granosa*) in Peninsular Malaysia and Thailand (Tookwinas 1993), as well as in Indonesia and Taiwan. As with many aquaculture operations that depend solely on the collection of naturally occurring seed, there is the probability that seed supply is the limiting factor for industry expansion. This is true for

FIGURE 44
An oyster farmer in the Salapingao district of Dagupan City, Philippines, tends his oyster spat collectors suspended from a bamboo frame. *Photo by the author.*

blood cockles, which has led to research into methods to spawn and rear seed in hatcheries (Muthiah & Narasimham 1992). This is a useful example of the development of a small-scale shellfish culture industry, first focusing on the development of simple production techniques and market development before moving to the more technically demanding task of establishing shellfish hatcheries.

The timing of the spawning of bivalves is species specific and specific to geographic location. In general, however, bivalves in temperate zones and high latitudes generally spawn in the late spring or early summer months after they become well fed on spring phytoplankton blooms resulting from longer days and warmer water temperatures (Figure 45). Also as a broad generalization, bivalves in tropical regions spawn year round, sometimes with peak spawning periods corresponding to the onset of either wet or dry seasons (PHRDC 1991).

Knowing the timing of the peak spawning period of your bivalve of interest is critical for the success of spat collection. Placing spat collectors in the water too early presents the risk of undesirable fouling organisms like barnacles colonizing the collectors and preventing the attachment of the settling bivalves, and in like fashion, a late placement of collectors results in missing the peak spatfall. In either case, the amount of shellfish seed caught by the collectors is less than optimum. Some oyster and mussel farmers experienced with spat collection advise the placement of spat collectors in the water about one week prior to maximum spatfall. There is evidence that the marine bacterial films that build up on spat collectors are responsible for increased spat set (Tritar et al. 1972). One simple method of determining the time of peak spatfall is to set out a sequential series of test collectors on a biweekly basis. The biweekly setting and retrieval of spat collectors and counting of attached spat can serve as a guide to the full-scale deployment of all collectors (Clarke et al.). Figure 46 shows data of oyster spatfall from a typical river mouth estuary system in the Philippines. Note the year-round spatfall and periodic

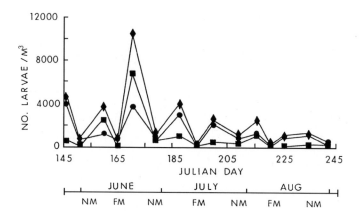

FIGURE 45

Larval bivalves are most abundant in the water during the late spring and summer months in temperate climate. In the Narragansett Bay of Rhode Island, U.S.A., the peak period of bivalve spawning was during June 1995. In the figure, numbers of quahog clam (*Mercenaria mercenaria)* larvae are represented by squares. Circles represent all other bivalve larvae, and diamonds represent the total number of bivalve larvae. Julian Day refers to the numerical date equivalent during the year with January 1 as Julian Day 1 and December 31 as Julian Day 365. The figure indicates the timing of the new moon and full moon by NM and FM, respectively. There is evidence of a periodicity of bivalve spawning that corresponds to the lunar cycle. *Data from N. Butel* (1997).

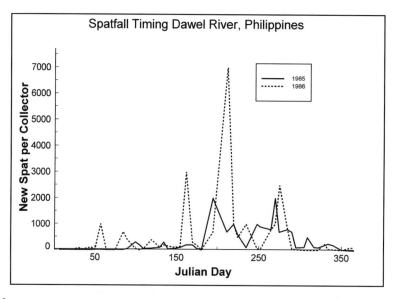

FIGURE 46

Spatfall of oysters occurs in the Dagupan-Agno River estuary system in the Philippines on a year around basis with maximum sets generally occurring July through October. Data represent the average number of juvenile oyster spat appearing on test collectors consisting of 100 oyster shells strung onto polypropylene ropes. *Data are redrawn from PHRDC* (1991).

FIGURE 47
Mr. Joseph Goncalo collecting bivalve larvae by using a 12V battery-operated water pump to pass measured volumes of seawater through a 60 μm mesh plankton net. *Photo by the author.*

spawning events. After several years of experience with the test collectors, the annual cycle of spat collector deployment can be refined. Another method of predicting spatfall requires the use of a microscope and the periodic sampling of water and collecting larval bivalves with a fine mesh (approximately 60-μm mesh) plankton net (Figure 47), then counting the number of veliger larvae observed in daily samples (Quayle 1980). By placing spat collectors in the water when the number of veliger larvae in the water is high, spat collection can be optimized.

Growout or Maturation of Juveniles

Once shellfish attach to spat collectors, they can either be left on the collectors or transferred to other nets or culture apparatus for better protection from predators for growth and maturation to harvest size. In many parts of the world, oysters and mussels are left on the spat collectors

FIGURE 48
In most cases of tropical oyster culture, the shucked oyster meat is the preferred market product. If oysters are allowed to remain on spat collectors, they grow but are small and irregularly shaped. Since labor costs are relatively low for shucking, however, oyster size and appearance are not important. *Photo by the author.*

until they mature to a harvestable market size. In many cases this is a very reasonable way to proceed with the culture operation, especially if the final products are mussels or shucked oyster meats. If the spatfall of oysters is very dense on spat collectors, however, the individual oysters will crowd each other and the shells will be small and irregularly shaped (Figure 48). Similarly, very heavy sets of mussels on spat collectors will have stunted growth because of competition for food (Heasman et al. 1998).

To improve the growth of just about any bivalve that has been caught by spat collection, it is recommended that the juvenile shellfish be separated from the collector and cultured in an apparatus that allows stocking at lower densities (Figure 49). Apparatus for the culture of shellfish after culling from spat collectors include trays (Figure 50), lantern nets suspended from floating lines (Figure 51), and bags in submerged racks (Figure 52) (Rheault & Rice 1995).

It is well known that stocking density of shellfish in nets, trays, or other enclosures will affect growth, and this is related to the competition for food in the filter feeding process (Rheault & Rice 1995). The optimum stocking density for shellfish in any kind of culture system is dependent on the species cultured, the type and size of the culture system used, and the amount of water flow and available food particles in the water (Rice et al. 1994). One simple method of determining optimum stocking density is to experiment with different trays or nets stocked with different numbers of animals. The optimum stocking density will be the one that maximizes shellfish numbers while maintaining rapid growth.

One of the key indicators of food limitation and general health of bivalves is a quantity called condition index (CI). CI is defined as a ratio of meat weight to the internal volume of the shell

FIGURE 49
Dense sets of mangrove oysters, *Crassostrea iredalei*, are culled from bamboo spat collectors and placed in wooden trays. Thinning of oysters from spat collectors and growing them to market size in trays or other kinds of aquaculture gear can improve growth and meat yields. *Photo by the author.*

FIGURE 50
Wooden trays can be used for culturing oysters intertidally. *Photo by the author.*

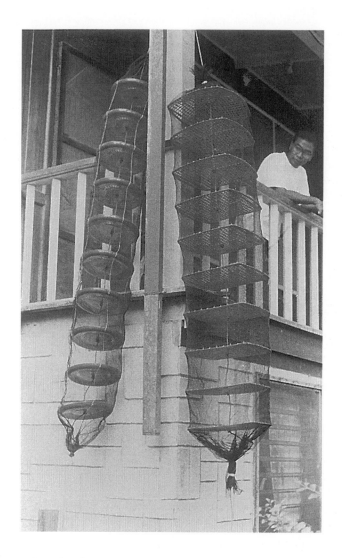

FIGURE 51

Lantern nets are a popular means for culturing a number of shellfish, including oysters, scallops, and pearl oysters. Lantern nets consist of a series of rigid shelves suspended on a rope at various intervals and covered net mesh. The shellfish are stocked onto the shelves and the lantern nets are suspended into the water from floating lines or rafts. *Photo by the author.*

or pallial cavity (Rheault & Rice 1996). CI is most often determined in oysters, but it is directly applicable to most other cultured bivalves. In temperate zones, CI fluctuates throughout the year due to seasonal food availability, temperature fluctuation, and distinct spawning seasons. In tropical areas bivalve CI may fluctuate, but since they spawn year round for the most part and fluctuations in food and temperature are not as great, CI is not as variable throughout the year. To determine CI, one weigh the shellfish whole and freshly from the water so it has trapped water in its pallial cavity. Then shuck the shellfish meat, using an oyster knife or similar instrument, remove the meat, blot it dry with absorbent paper, dry the meat in an oven at 90°C overnight or in the sun, and weigh the dried meat. After weighing the dried meat, weigh the two shell valves to get the shell-only weight. After getting these various weights, the condition

FIGURE 52
Submerged wire mesh cages allow for the culture of shellfish on the bottom, while allowing boat traffic or fishing activities to proceed above. Shellfish seed from spat collectors can be placed in plastic mesh bags that can be placed on the shelves of wire enclosures. These types of enclosures have the disadvantage of requiring periodic cleaning to remove fouling organisms that can impede water flow and food delivery to the enclosed shellfish. *Photo by the author.*

index is determined by this equation.

$$CI = (M \times 1,000)/(W - S)$$

when M = the dry weight of the soft tissues in grams
      W = the weight of the whole animal
      S = the weight of the two shell valves

The volume of the pallial cavity is equivalent to the difference in weight between the whole animal and the shells because the trapped water and bivalve soft tissues have a density of about 1 g/cm^3. The condition index values can be used as an index of bivalve "fatness." For example, Quayle (1980) states that oysters are in good condition if CI is high, up to 150, and in poor condition if CI falls below 75. Low condition index might be corrected by lowering stocking densities in the culture gear.

   Key to the success of a farm for bivalve mollusks is the time it takes to produce a crop. In general, most species of cultured mollusks in tropical regions have rapid crop cycles and are capable of reaching a harvest size between 6 and 18 months. Angell (1986) provides an excellent review of the growth of different species of oysters grown in tropical countries worldwide. Most species of tropical oysters in conditions of adequate food resources and culture conditions can reach a harvest size of about 70 mm in shell height (the maximum linear shell dimension) in 6 to 9 months. Likewise, the culture of tropical mussels *Perna viridis* in Thailand typically take about 8 months from spatfall to harvest (Crosby & Gale 1990). Also, in Thailand and in Malaysia, the crop cycle for ark clams *Anadara granosa* from bed seeding to harvest is about 12 to 15 months (Broom 1985). In the cooler climates of temperate regions, bivalve crop cycles are considerably longer than those in the tropics. For example, under optimum culture conditions it can takes about three years to grow northern quahogs *Mercenaria mercenaria* in New England and the Atlantic Maritime Provinces of Canada. The same species takes only about 18 months to reach harvest size in the warmer waters of Florida (Ansell 1968). Much of the economics of small-scale aquaculture projects—that is, the ability to pay for the gear and other production costs—is dependent on the amount of time it takes to get a crop through the production cycle (Meade 1989). So maintaining complete and careful records on the farm are critical to assessing its success.

Finally, a key factor that needs to be considered is the human health implications of artisanal shellfish farms. Since bivalves are filter feeders deriving their sustenance by the clearance of large volumes of water, they are able to accumulate a variety of microbial pathogens that can be a threat to human health. In most developed nations of the world it is popular to eat raw shellfish, but there are strict governmental water pollution control measures in place to prevent disease outbreaks. However, in most developing countries there are rarely sufficient controls on the disposal of sewage or other wastes in estuaries and coastal waters (Rice 1992). If shellfish are cultured for human consumption in waters of questionable water quality, it is wise to advise consumers to eat molluscan shellfish only if it is well cooked.

ACKNOWLEDGMENTS

This is publication 3704 of the Rhode Island Agricultural Experiment Station and the College of the Environment and Life Sciences, University of Rhode Island. Partial support for this work was provided in a 1996–97 Senior Fulbright Research and Teaching Fellowship to the author through the Philippine-American Educational Foundation, Manila, and the Rhode Island Agricultural Experiment Station Project H-886. I thank the following aquaculture students at the University of Rhode Island who performed literature searches in support of this review: Charles Gedraitis, Jeffrey Gibula, Jeffrey Kosiorek, and Ethan Lucas.

REFERENCES

Angell, C. L. (1986). *The Biology and Culture of Tropical Oysters.* ICLARM Studies and Reviews 13, International Center for Living Aquatic Resources Management, Manila.

Ansell, A. D. (1968). "The Rate of Growth of the Hard Clam, *Mercenaria mercenaria.* Throughout the Geographical Range." *Journal Conseil International pour l'Exploration de la Mer* 31: 346–409.

Avault, J. W. (1996). "Chapter 5. Site Selection and Culture Systems." In *Fundamentals of Aquaculture: A Step-by-Step Guide to Commercial Aquaculture.* Baton Rouge, Lousiana: AVA Publishing Company.

Bailey, C. (1997) "Aquaculture and Basic Human Needs." *World Aquaculture* 28(3): 28–31.

Bard, J., P. deKimpe, J. Lazard, J. LeMasson, and P. Lessent. (1976). *Handbook of Tropical Fish Culture.* Centre Technique Forestier Tropical, Nogent-sur-Marne, France.

Bardach, J. E., J. H. Ryther, and W. O. McLarney. (1972). *Aquaculture: The Farming and Husbandry of Freshwater and Marine Organisms.* New York; Wiley Interscience.

Bayne, B. L. *Marine Mussels: Their Ecology and Physiology.* Cambridge, England: Cambridge University Press.

Bettencourt, S. U. and J. L. Anderson. (1990). *Pen-Reared Salmonid Industry in the Northeastern United States.* Northeastern Regional Aquaculture Center and Rhode Island Cooperative Extension Service, University of Rhode Island, Kingston.

Beveridge, M. C. M. (1987). *Cage Aquaculture.* Farnham, Surrey, England: Fishing News Books.

Boyd, C. E. (1990). *Water Quality in Ponds for Aquaculture.* Alabama Agricultural Experiment Station, Auburn University, Alabama.

Broom, M. J. (1985). *The Biology and Culture of Marine Bivalve Molluscs of the Genus Anadara.* ICLARM Studies and Reviews 12, International Center for Living Aquatic Resources Management, Manila.

Buttner, J. K. (1986). "*Corbicula* as a Biological Filter and Polyculture Organism in Catfish Rearing Ponds." *Progressive Fish-Culturist* 48: 136–139.

Butet, N. (1997). "Distribution of Quahog Larvae Along a North-South Transect in Narragansett Bay." Masters thesis, University of Rhode Island, kingston.

Chen, T. P. (1976). *Aquaculture Practices in Taiwan.* Farnham, Surrey, England: Fishing News Books.

Cherlamwat, K., and R. A. Lutz. (1989). "Farming the Green Mussel in Thailand." *World Aquaculture* 20(4): 41–46.

Chua, T. E., and S. K. Teng. (1982). "Effects of Food Ration on Growth, Condition Factor, Food Conversion Efficiency, and Net Yield of Estuary Grouper, *Epinephelus salmoides* Maxwell, Cultured in Floating Net-Cages. *Aquaculture* 27: 273–283.

Clarke, T. M., S. Little-Saunders, and G. F. Newkirk. "Spatfall Monitoring and Spat Collection of Oysters in Jamaica." In G. F. Newkirk and B. A. Field (Eds.). *Oyster Culture in the Caribbean.* Mollusc Culture Network, Biology Dept., Dalhousie University, Halifax, NS, Canada.

Clay, J. W. (1997). "Toward Sustainable Shrimp Aquaculture." *World Aquaculture* 28(3): 32–37.

Creswell, R. L. (1993). *Aquaculture Desk Reference.* New York: AVI/Van Nostrand Reinhold.

Crosby, M. P. and L. D. Gale. (1990). "A Review and Evaluation of Bivalve Condition Index Methodologies with a Suggested Standard Method." *Journal of Shellfish Research* 9: 233–237.

Emerson, K., R. C. Russo, R. E. Lund, and R. V. Thurston. (1975). "Aqueous Ammonia Equilibrium Calculations: Effect of pH and Temperature." *Journal of the Fisheries Research Board of Canada* 32: 379–388.

FAO. (1993). *FAO Fisheries Circular 815, Revision 5.* United Nations Food and Agriculture Organization, Rome.

Fitzsimmons, K. (1997). "Introduction to Tilapia Reproduction." In K. Fitzsimmons (Ed.), *Tilapia Aquaculture: Proceedings from the Fourth International Symposium on Tilapia in Aquaculture.* Northeast Regional Agricultural Engineering Service, Cornell University, Ithaca, New York.

Fitzsimmons, K. (1997). "Introduction to Tilapia Sex-Determination and Sex-Reversal." In K. Fitzsimmons (Ed.), *Tilapia Aquaculture: Proceedings from the Fourth International symposium on Tilapia in Aquaculture.* Northeast Regional Agricultural Engineering Service, Cornell University, Ithaca, New York.

Folk, R. L. (1968). *Petrology of Sedimentary Rocks.* University of Texas, Austin.

Grizel, H. (1993). "World Bivalve Culture." *World Aquaculture* 24(2): 18–23.

Guerrero, R. D., and L. A. Guerrero. (1997). "Effects of Androstenedione and Methyltestosterone on *Oreochromis niloticus* Fry Treated for Sex Reversal in Outdoor Net Enclosures." In K. Fitzsimmons (Ed.), *Tilapia Aquaculture: Proceedings from the Fourth International Symposium on Tilapia in Aquaculture.* Northeast Regional Agricultural Engineering Service, Cornell University, Ithaca, New York.

Guerrero, R. D. (1975). "Use of Androgens for the Production of all Male *Tilapia aurea* (Steindachner)." *Transactions of the American Fisheries Society* 104: 342–348.

Halver, J. E. (1992). *Fish Nutrition.* San Diego, CA: Academic Press.

Heasman, K. G., G. C. Pitcher, C. D. McQuaid, and T. Hecht. (1998). "Shellfish Mariculture in the Benguela System: Raft Culture of *Mytilus galloprovincialis* and the Effect of Rope Spacing on Food Extraction, Growth Rate, and Condition of Mussels. *Journal of Shellfish Research* 17: 33–39.

Heslinga, G. A., T. C. Watson, and T. Isamu. (1990). *Giant Clam Farming.* Pacific Fisheries Development Foundation (NMFS/NOAA), Honolulu, Hawaii.

IIRR. (1995). *Livelihood Options for Coastal Communities.* International Institute of Rural Reconstruction, Silang, Cavite, Philippines.

Ito, H. (1991). "Japan." In S. E. Shumway (Ed.), *Scallops: Biology, Ecology and Aquaculture.* Amsterdam: Elsevier Press.

Jensen, J. W. (1989). *Watershed Fish Production in Ponds: Site Selection and Construction.* SRAC Publication No. 102, Southeastern Regional Aquaculture Center, Mississippi State University, Stoneville.

Kaiser, M. J., I. Laing, S. D. Utting, and G. M. Burnell. (1998). "Environmental Impacts of Bivalve Mariculture." *Journal of Shellfish Research,* 59–66.

Korringa, P. (1976). *Farming of the Cupped Oysters of the Genus Crassostrea.* Amsterdam: Elsevier Publishing.

Kungvankij, P., L. B. Tiro, B. J. Pudadera, and I. O. Potestas. (1986). *Biology and Culture of Sea Bass, Lates calcarifer.* Aquaculture Extension Manual No. 11. Aquaculture Department, Southeast Asian Fisheries Development Center, Tigbauan, Iloilo, Philippines.

Lee, C-S., and I-C. Liao (Eds.). (1985). *Reproduction and Culture of Milkfish.* Hawaii: The Oceanic Institute Waimanalo.

Leong, T. S., and Wong, S. Y. (1987). "Parasites of Wild and Diseased Juvenile Golden Snapper, *Lutjanus johni* (Bloch), in Floating Cages in Penang, Malaysia." *Asian Fisheries Science* 1: 83–90.

Liang, Y., J. M. Melack, and J. Wang. (1991). "Primary Production and Fish Yields in Chinese Ponds and Lakes." *Transactions of the American Fisheries Society* 110: 346–350.

Loosanoff, V. L., and H. C. Davis. (1963). "Rearing of Bivalve Mollusks." *Advances in Marine Biology* 1: 1–136.

Marshall, N. (1947). "Abundance of Bay Scallops in the Absence of Eelgrass." *Ecology* 28: 321–322.

Masser, M. P. (1988). *Cage Culture: Species Suitable for Cage Culture*. SRAC Publication No. 163, Southern Regional Aquaculture Center, Mississippi State University, Stoneville.

McGeachin, R. B, R. I. Wicklund, B. L. Olla, and J. Winton. (1987). "Growth of *Tilapia aurea* (Steindachner) in Seawater Cages." *Journal of the World Aquaculture Society* 18: 31–34.

Meade, J. W. (1989). *Aquaculture Management*. AVI-Van Nostrand Reinhold, New York.

Michaels, V. K. (1988). *Carp Farming*. Farnham, Surrey, England: Fishing News Books.

Milne, P. H. (1979). *Fish and Shellfish Farming in Coastal Waters*. Farnham, Surrey, England: Fishing News Books.

Milstein, A. (1997). Do Management Procedures Affect the Ecology of Warm Water Polyculture Ponds?" *World Aquaculture* 28(3): 12–19.

Muthiah, P., and K. A. Narasimham. (1992). "Larval Rearing, Spat Production and Juvenile Growth of the Blood Clam *Anadara granosa*." *Journal of the Marine Biological Association of India* 34: 138–143.

National Research Council. (1993). *Nutrient Requirements of Fish*. National Academy of Sciences, Washington, DC.

New, M. B. (1987). *Feeds and Feeding of Fish and Shrimp: A Manual on the Preparation and Presentation of Compound Feeds for Shrimp and Fish in Aquaculture*. Aquaculture Development and coordination Proggramme, FAO. Rome.

New, M. B. (1997). "Aquaculture and the Capture Fisheries." *World Aquaculture* 28(2): 11–30.

Nwachukwu, V. N. (1997). "Tilapia Nutrition Through Substrate Enhancement in Ponds: A cheap, Sustainable, and Environmentally Friendly Feeding Method." In K. Fitzsimmons (Ed.), *Tilapia Aquaculture: Proceedings from the Fourth International symposium on Tilapia in Aquaculture*. Northeast Regional Agricultural Engineering Service, Cornell University, Ithaca, New York.

Peters, H. M. (1983). "Fecundity, Egg Weight and Oocyte Development in Tilapias (Chichlidae, Teleostei)." *ICLARM Translations 2*. International Center for Living Aquatic Resources Management, Manila.

Phelps, H. L. (1994). "Potential for *Corbicula* in Aquaculture." *Journal of Shellfish Research* 13: 319.

PHRDC. (1991). *The Science and Business of Growing Oysters. Seafarming Research and Development Center*. Philippine Human Resources Development Center, Bonuan Binloc, Dagupan City, Philippines.

Pratt, D. M. (1953). Abundance and Growth of *Venus mercenaria* and *Callocardium morrhuana* in Relation to the Character of Bottom Sediments." *Journal of Marine Research* 12: 60–74.

Pullin, R. S. V., and R. H. Lowe-McConnell. (1982). *The Biology and Culture of Tilapias*. International Center for Living Aquatic Resources Management, Manila.

Quayle, D. B. (1980). *Tropical Oysters: Culture and Methods*. Report No. IRDC-TS17e, International Development Research Centre, Ottawa, Canada.

Rice, M. A. (1992). Bivalve Aquaculture in Warm Tropical and Subtropical Waters with reference to Sanitary Water Quality, Monitoring and Post-Harvest Disinfection." *Tropical Science* 32: 179–201.

Rice, M. A., and A. Z. DeVera. (1998). "Aquaculture in Dagupan City, Philippines." *World Aquaculture* 29(1): 18–24.

Rice, M. A., D. A. Bengtson, and C. Jaworski. (1994). *Evaluation of Artificial Diets for Cultured Fish*. NRAC Fact Sheet 222, Northeastern Regional Aquaculture Center, University of Massachusetts, Dartmouth.

Rice, M. A., R. B., Rheault, M. S. Perez, and V. S. Perez. (1994). "Experimental Culture and Particle Filtration by Asian Moon Scallops, *Amusium pleuronectes*." *Asian Fisheries Science* 7: 179–185.

Rheault, R. B. and M. A. Rice. (1996). Food-Limited Growth and Condition Index in the Eastern Oyster, *Crassostrea virginica* (Gmelin 1791), and the Bay Scallop *Argopecten irradians irradians* (Lamarck 1819)." *Journal of Shellfish Research* 15: 271–283.

Roberts, J. L. (1964). "Metabolic Responses of Fresh-Water Sunfish to Seasonal Photoperiods and Temperatures. *Helgolander Wiss. Meeresunters*. 9: 459–473.

Shuster, W. H. (1949). *Fish Culture in Saltwater Ponds on Java*. Department van Landbouw en Visserij Publicate No. 2 van de Onderafdeling Binnenvisserij, Amsterdam.

Schwedler, T. E., M. L. Berry, and D. R. King. (1986). *Raising Catfish in a Cage*. Cooperative Extension Service, Clemson University, Clemson, South Carolina, USA.

Swift, D. R. (1985). *Aquaculture Training Manual*. Farnham, Surrey, England: Fishing News Books.

Swingle, H. S., B. C. Gooch, and H. R. Rabanal. (1963). "Phosphate Fertilization of Ponds." *Proceedings of the Southeastern Association of Game and Fish Commissioners* 17: 213–218.

Tammi, K. A., E. Buhle, W. H. Turner, and V. Satkin. (1997). "Making the Perfect Spat Bag for Collection of the Bay Scallop, *Argopecten irradians*." *Journal of Shellfish Research* 16: 295.

Teng, S.-K., and Chua,T.-E. (1979). "Use of Artificial Hides to Increase the Stocking Density and Production of Estuary Grouper, *Epinephelus salmoides* Maxwell, Reared in Floating Net Cages." *Aquaculture* 16: 219–232.

Teng, S.-K., Chua, T.-E., and Lim, P.-E. (1978). "Preliminary Observation on the Dietary Protein Requirement of Estuary Grouper, *Epinephelus salmoides* Maxwell, Cultured in Floating Net-Cages." *Aquaculture* 15: 257–271.

Tookwinas, S. (1993). *Commercial Cockle Farming in Southern Thailand.* (Translated from Thai by E.W. McCoy.) ICLARM Studies and Translations, No. 7, International Center for Living Aquatic Resources Management, Manila.

Tritar, S., D. Prieur, and R. Weiner. (1992). "Effects of Bacterial Films on the Settlement of Oysters, *Crassostrea gigas* (Thunberg, 1793) and *Ostrea edulis,* Linnaeus, 1750 and the Scallop *Pecten maximus* (Linnaeus, 1758)." *Journal of Shellfish Research* 11: 325–330.

Twilley, R. R. (1989). "Impacts of Shrimp Mariculture Practices on the Ecology of Coastal Ecosystems in Ecuador." In S. Olsen and L. Arriaga (Eds.), *A Sustainable Shrimp Mariculture Industry for Ecuador.* Coastal Resources Center, University of Rhode Island, Narragansett.

UNDP. (1984). *Inland Aquaculture Engineering.* United Nations Development Programme, Food and Agriculture Organization of the United Nations, Rome.

Vakily, J. M. (1989). *The Biology and Culture of Tropical Mussels of the Genus Perna.* ICLARM Studies and Reviews 17, International Center for Living Aquatic Resources Management, Manila.

Villaluz, A. C. (1986). "Fry and Fingerling Collection and Handling." In C-S. Lee, M. S. Gordon, and W. O. Watanabe (Eds.), *Aquaculture of Milkfish* (Chanos chanos): *State of the Art.* Waimanalo, Hawaii: Oceanic Institute.

Walne, P. R. (1979). *Culture of Bivalve Mollusks: 50 Years Experience at Conwy.* Farnham, Surrey, England: Fishing News Books.

Watanabe, W. O., K. E. French, D. H. Ernst, B. L. Olla, and R. I. Wicklund. (1989). "Salinity During Early Development Influences Growth and Survival of Florida Red Tilapia in Brackish and Seawater." *Journal of the World Aquaculture Society* 20: 134–142.

Xu, X., L. Qian, L. Zhang, and Z. Yu. (1987). "The Reproductive Cycle of *Corbicula fluminea* (Mueller) in Dianshan Lake, Shanghai." *Journal of Fisheries of China/Shuichan Xuebao, Shanghai* 11(2): 135–142.

Zale, A., and R. W. Gregory. (1989). "Effect of Salinity on Cold Tolerance of Juvenile Blue Tilapias." *Transactions of the American Fisheries Society* 118: 718–720.

Zhang, F. (1984). "Mussel Culture in China." *Aquaculture* 39: 1–10.

BEEKEEPING

Bees are not difficult to raise, although one needs special equipment: hives and protective gear for handling. Bees find their own food. One usually begins by buying a swarm of bees, and once the hive is set up, little care is needed. People even raise bees successfully in urban areas.

GETTING STARTED

To get started one needs bees, a hive, protective clothing, and a smoker. A tool to assist with manipulating the hive is helpful. Equipment is also required to extract the honey from the cells, although in Third World markets the entire comb—honey, wax and often some dead bees—is sold. In the United States extraction equipment can be rented, at least until one decides whether he or she will definitely pursue beekeeping. One could probably set up a small beekeeping operation for less than $500.

In the United States one can purchase a package of bees. When doing so, attempt to get bees certified free of diseases and mites. Try to find a beekeeper who is willing to sell a swarm of bees complete with a queen. It is probably best to start the hive in spring, so one should start negotiating for the swarm several months earlier—in January in the northern hemisphere.

You will also need a hive. It is probably best to start with a single hive and eventually expand to four or five hives. A standard hive is shown in Figure 53. The base is about 10 inches by 15 inches, and the height is about 20 inches. The "supers" hold vertical wax frames, with a hexagonal pattern matching the honey cells to be built by the bees. The useable honey is taken from these frames. The "queen excluder" is a screen that allows the worker bees to reach the supers but

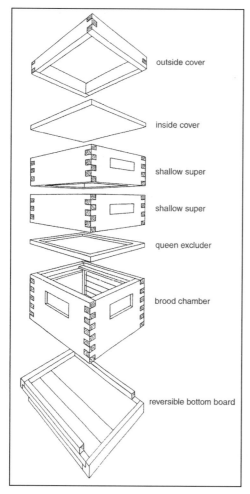

FIGURE 53

A hive. (*More Other Homes and Garbage*, Jim Leckie, Gill Masters, Harry Whitehouse, Lily Young, copyright Sierra Club Books)

excludes the larger queen—to keep eggs out of the honey. The brood chamber also holds frames, but these are for rearing the young and storing honey and pollen for short-term and winter use. It is best, especially for a beginner, to purchase a hive that has already had bees in it for a year or more because installing bees into a new hive stresses them and usually reduces the honey crop.

The minimum personal equipment that is needed includes a veil, hive tool, and bee smoker. The bee veil keeps bees away from your head and face. A wide-brimmed hat is needed to go with the veil. This need not be a special hat, but a wide brim on all sides is essential. High-top boots with your pants legs tucked in should also be worn. A bee suit and bee gloves are useful but not essential. Light-colored, smooth clothing is recommended.

The bee smoker allows the beekeeper to keep the bees from becoming agitated. A small puff of smoke usually calms the bees so the owner can extract honey or check on the hive. A hive tool is specially designed to assist the beekeeper in removing the frames in the supers. If one is not available, a screwdriver will do.

CARE

Bees should have access to water. They do need 15 to 20 pounds of reserve honey during the year. To get this, three frames of honey should be left in the hive as surplus. If medication is available, the hive should be medicated every two weeks in the spring until a month or so before honey will be harvested. At the time the medication is given inspect the hive to ensure that the queen is present and laying eggs, that there is no sign of disease, and that the colony has sufficient stores to last until the first nectar and pollen become available from plants. When most of the frames are filled with honey, another super can be added to the hive.

EXTRACTION

The first step is to blow smoke into the hive to pacify the bees. Chemicals can also be used. The chemical is placed on the underside of the cover, and the smell drives the bees out of the honey super. An extensive operation could use a big blower to blow the bees right off the frames, out of the super, and onto the ground in front of the hive.

The next step, especially when chemicals or a blower were not used, is to separate the combs of honey from the bees (pulling the honey). This can be as simple as using a bee brush to sweep the bees from each frame. With just a few hives a bee brush is adequate and inexpensive but a fair bit of work.

Once the honey is pulled, you can extract it yourself, or perhaps an established beekeeper will extract it for you. First the wax cappings from the honeycomb must be removed. If electricity is available, an electrically heated knife simplifies the task. Honey extractors range in size from two-frame, hand-powered devices to motor-driven machines that can handle 100 or more frames. An extractor spins the frames, and centrifugal force throws the honey against the walls.

Once the honey is extracted, it should be strained (cheesecloth or nylon works well) and then stored in a warm place in a tall tank or container to allow the fine impurities to rise to the top. At this point it helps to have a proper tank with an outlet at the bottom so that the clean, warm honey can be drawn from the bottom directly into the honey containers.

OVERWINTERING

If you wish to winter your bees, then the colony should be examined to ensure the queen is still viable and that no disease has developed. Medicate the colony at this time. To overwinter a colony in cold climates, 20 pounds of honey or supplementary sugarwater should be provided. Before the first frost the bee colony can be moved indoors for winter or wrapped to protect the bees from the elements. If the bees are outside, there is little that needs to be done to assist them. They will survive even if they are completely covered by snow for a while. If the bees are indoors, ensure that the temperature stays low and constant (about 50°C) and ventilation is maintained. It is normal for bees to leave the hive during the winter and die.

STINGS

If you keep bees, you will be stung by them, and most likely members of your family will get stung eventually. For most people a bee sting hurts, and a brief period of discomfort follows. For others (about .4 percent of the population) there is a danger of death from anaphylactic shock brought on by the bee sting. Frequently there are indications that a person is becoming highly allergic to bee stings, but occasionally the problem occurs unannounced. (Swelling at the site of the sting is normal; hives over the body, itching in areas of the body remote from the sting, and shortness of breath are abnormal and cause for concern.)

REFERENCES

www.agric.gov.ab.ca/agdex/600/616-23.html. Alberta Food and Rural Development.
Graham, Joe M. (Ed.). (1992). *The Hive and the Honey Bee*. Oradant & Sons, Inc. Hamilton, Illinois.

ABC & XYZ of Bee Culture, 40th Edition, A.I. Root, Medina, Ohio.

The New Starting Right with Bees, A.I. Root, Medina, Ohio.

The Honey Bee: A Guide for Beekeepers, by V.R. Vickery, Particle Press, Quebec.

Beekeeping in the Tropics, Compiled by P. Segeren. 1991, AGROMISA, P.O. Box 41, 6700 AA Wageningen, The Netherlands, Fax: +31 317 419178, agromisa@worldaccess.nl

RAISING CATTLE

Based on Better Farming Series 11. Cattle Breeding, A.J. Henderson: Food and Agriculture Organization of the United Nations, Rome, 1977, ISBN 92-5-100151-0

Cattle breeding can create wealth. In traditional breeding little effort is taken with the herd, and production is minimal. To make the herd produce more, different methods are required. A modern farmer must do the following.

- Feed the animals well
- House the herd sufficiently
- Take good care of ill animals
- Make good choices in breeding animals
- Sell the animals at a good price

It is very important to feed cattle well. An badly fed animal will have stunted growth, will not produce much meat, will not produce good calves, and will frequently get sick. All the animals in the herd do not use their food in the same way. For example, one ox in the herd may gain weight faster than the others; this ox makes better use of its food. Pregnant cows and working oxen also need more food.

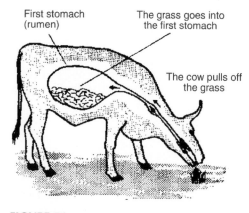

First stomach (rumen)

The grass goes into the first stomach

The cow pulls off the grass

FIGURE 54
A cow eating.

HOW CATTLE EAT

When a cow bites grass, it does not chew it but swallows it whole. The grass goes into the first stomach (or rumen)—see Figure 54.

A cow can eat a lot of grass: There is room for up to 15 kilograms of grass in its first stomach, depending on the size of the breed. But a cow needs a lot of time to feed, to fill up its first stomach. So you must give a cow, and especially working oxen, at least eight hours a day to feed off pasture. When a cow has finished filling its first stomach, it often lies down, but it continues to move its jaws. This is called ruminating. The cow brings up a little grass from its first stomach into its mouth. It chews this grass for a long time with its molars. When the grass is well chewed and broken down, the cow swallows it again. This time, however, the grass does not go into the first stomach but into the second one. A cow needs several hours to ruminate.

A cow can ruminate well when it is quiet and above all when it is lying down. If you make a shelter, the cow will be protected from rain, wind, and sun. That will give it a quiet place where it can rest and ruminate well. Young calves do not ruminate because their first stomach is not yet developed, so they need different food.

HOW TO FEED CATTLE

Cattle eat grass, which supplies them what they need to build their bodies and become strong. However, it is often necessary to give them a feed supplement. In the rainy season there is plenty of grass that grows quickly, so it is easy to feed cattle then. In the dry season it is difficult to feed the animals adequately because grass is scarce. The grass is tough and dry, and the animals will not eat it. As a result, they get thin and sometimes die. If the cattle were better fed, especially in the dry season, it would take less time for them to reach selling weight. A modern cattle breeder who hasn't got enough food for his animals during the dry season should sell some of the animals at the end of the rainy season.

In traditional animal husbandry, farmers used to move the herd from place to place. When the grass is dry and deep, not only will the cattle not eat it but it is difficult for them to walk around, too. In this case, the grass can be burned. After the fire the grass grows again and is better for the animals to eat. Brush fires can damage soil and destroy useful plants, however, which cannot stand burning as well as grass.

In the modern way of feeding cattle, farmers should do the following.

- Improve their pastures
- Make new pastures and grow fodder crops
- Store green fodder as silage and hay
- Give their animals feed supplements
- Give their animals enough water

IMPROVING PASTURE

A pasture is the field where cattle find grass to eat. In order to have tender, young grass, divide the pasture into four parts. Each week put the cattle in one part and let the grass grow in the other parts. At the end of four weeks go back to the first part. After the herd has been through the pasture, cut the weeds before they seed so they won't multiply.

STORING GRASS

During the rainy season, grass grows a lot, and the cattle do not eat all of it. Grass can be stored in the form of silage or hay. Dig a pit 1.50 to 2 meters deep and 1.50 to 2 meters wide. This pit is called a silo. It has to be made rather long so that all the cut grass can fit in it. At the bottom of the silo put some large stones. On top of these stones, put the grass to be stored.

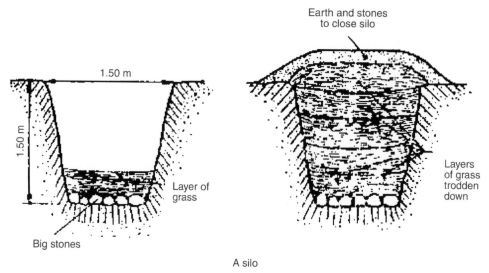

FIGURE 55
A silo.

Tread down the grass well by trampling on it. On top of the full silo, on the pressed-down grass, put earth and stones. The silo must be well closed so that air and rain cannot get in and the grass will not rot. A silo is shown in Figure 55. Grass kept this way stays fresh for several months. It shouldn't take more than two days to fill, tread down, and close the silo to make sure the grass stays in good condition.

HAY

You can also dry grass, called hay. Cut the grass when it is green and let it dry. Many farmers keep the dried stalks and leaves of groundnuts in order to feed them to animals. This is groundnut hay. Hay is nearly as good as green grass. For hay to be good food, you must cut the grass when it is still green before it starts going to seed and before it becomes too tough. If you wait too long before cutting the grass, you will get not hay but straw. Animals do eat straw, but it is not easy to digest. Straw is used for making manure.

To make hay, cut the grass with a machete or a scythe, which will get the work done quicker. When the grass is cut, let it dry in the sun. Then turn it over and leave exposed the parts that are not yet dry. This is done with a fork. When all the grass is very dry, make it into a big heap next to the animal shed. Then you can give the animals food during the dry season. Sun is needed to dry grass, so you must wait for the end of the rainy season before you make hay.

FEED SUPPLEMENTS

When there is not enough fodder, give the cattle a feed supplement. When oxen are working, when cows are about to calve, or when cows are giving milk, they need a feed supplement. You can give them oil cake made from groundnuts, copra, or cottonseed (see the "Oil Press" article). You can also buy cattle meal. Some manufacturers make a feed in which every 100 kilograms contain 50 kg maize meal, 10 kg copra oil cake, 38 kg groundnut oil cake, and 2 kg mineral salts (dicalcium phosphate and salt).

For example, a cow that weighs 300 kg and gives 3 liters of milk should be given every day 15 kg of pasture grass and 1 kg of palm kernel or copra oil cake.

MINERAL SUPPLEMENTS

A mineral supplement supplies mineral salts. If animals do not get enough mineral salts, their bones will not grow well. You can also give them mineral salts by putting salt in their water or hay or by giving native soda or a mineral lick (licking stone). A licking stone weighing 1 kg contains salt, calcium, phosphorus, and other mineral salts.

WATERING CATTLE

Animals need water. Animals can lose weight in the dry season because they are not well fed but also because they do not drink enough. An ox can drink 30 to 40 liters of water a day, or even more in the dry season, if it is very hot and the grass is very dry. Oxen do not need to drink as much if it is not very hot and if the food contains plenty of water, such as green grass or silage. Animals drink at the cattle shed, from a river or stream, or at a well. If the animals drink at a stream, you must be careful because the water is often dirty and may give the animals some disease. Do not let the animals remain in the water after they have drunk. It is best to let the animals drink two or three times a day. It is good to add a little salt to the water.

HOW TO FEED CALVES

At the beginning, use mother's milk. The first stomach of a young calf is not fully developed, so it cannot ruminate. If it is fed grass, it cannot digest it. Milk is digested without ruminating. However, too often the cows do not have much milk and the calves cannot drink enough. Never feed two calves with the milk of only one cow. The cow gives too little milk to feed two calves. A two-month-old calf fed with its mother's milk needs 4 to 6 liters of milk a day. During the first two months, leave all the mother's milk for the young calf. During this period do not milk the cow.

Later, feed the calf grass. At the age of three weeks, it can be given a little green grass. Its stomach develops and it begins to ruminate, so by at three months it can digest grass. At this point, it is considered to be weaned. It no longer needs all its mother's milk, so the cow can be milked. After weaning, the calf stops drinking its mother's milk and feeds on grass. Weaning is often the time when calves die. It is difficult for calves to change from one food to another. To help a calf at weaning, give it a feed supplement as well as grass. If you mix this feed supplement with water, the calf will digest it better. Do not forget to give calves a mineral supplement.

FEED SUPPLEMENTS AT WEANING

You may give the calf any of the following.

- Cereals—Millet, sorghum, maize, and rice are good feed for calves. Crush these cereals so that they are well digested. One kg of crushed millet feeds a calf as well as 2 kg of whole millet grain. These feeds can be expensive. They are food for people, but you can give animals grain that is broken or damaged by insects and the part that people do not eat: the bran of rice, maize, or millet.

- Oil cake—This is the name for what remains when the oil has been taken from groundnuts, copra, oil palm kernels, or cottonseed. Oil cake is good food, rich in protein.
- Meal for calves—Dealers sometimes sell meal for animals. This meal is a mixture of crushed grain and oil cake.

For instance, to make 100 kg of meal, mix 62 kg of crushed sorghum, 35 kg of groundnut oil cake, and 3 kg of mineral supplement.

The 3 kg of mineral supplement contain 0.6 kg lime, 0.3 kg salt, and 2.1 kg bone ash.

WATCHING ANIMALS

A farmer who doesn't watch her or his animals and leaves them to roam freely may not have to do much work but should be aware of the following.

- The cattle will not make good use of the grass. They will eat the good grass first and leave the poor growth. The good grass is eaten before it goes to seed, and so it cannot multiply. On the other hand, the poor grass that is not eaten can grow well and make seed. So it multiplies, and the pasture becomes poor.
- The cattle may have accidents and get diseases. They may go near streams, where they can be bitten by tsetse flies and catch sleeping sickness. If an animal is bitten by a snake or has an accident, nobody knows about it, and nobody looks after the animal. The oxen can also be stolen more easily.
- The cattle can damage crops. To prevent this, fields must be surrounded by fences, or fields a long way from the village must be farmed. Then the farmer wastes a lot of time getting to his fields.

HOUSING ANIMALS

Shelter is needed to protect the animals from wild beasts, wind, sun, rain, and diseases. In a traditional enclosure there are often too many animals. The cattle stand on a mixture of earth, excrement, urine, and water. They can't lie down. They can't ruminate well, and they do not make good use of their food. They are very dirty, and when animals are dirty, they get more diseases and their wounds do not heal well, especially those of the feet. The calves are in danger. Parasites and diseases attack them more easily. Good manure cannot be made. Instead you have only a mixture of earth and excrement. This mixture is not as good for the fields as real manure. The traditional enclosure must be improved by making a shed and a manure heap.

COW SHED AND MANURE HEAP

Choose a dry place. If you put up the shed in a hollow, the rainwater will collect there and will not run off. You can greatly improve the animals' housing without spending too much money by using wood, earth, and straw that you can find on the spot. Animals must be protected from wind. Build a wall on the side from which the wind usually blows. Animals must be protected from sun and rain, so put up a straw roof.

When the shed is built, spread straw on the ground. This straw, mixed with excrement and urine, rots and makes manure. When the straw is rotted, put clean, dry straw on top of it so the animals are always on clean straw. When there is a lot of manure, remove it. You can either take it straight to the field or mix it at once with the soil by ploughing it in, or you can make a manure heap near the shed and take the manure to the fields when you are ready to plow.

The animals must not be too crowded in the shed. They must have room to lie down. A cow needs 5 to 6 square meters of space (3×2 meters). For example, there is room for six cows in a cow shed 5 meters wide by 7 meters long. The shed should be disinfected once a month to kill germs. Position the shed so that the wind will carry the smell away from the house.

Next to the shed, make a paddock where the animals can walk about. Surround it with a strong fence made out of posts, branches, or thorns. Leave a few trees to give shade. Inside the paddock, put feed troughs where you can give the animals their feed supplement and watering troughs where the animals can drink. The feed and watering troughs can be made with hollowed tree trunks or barrels cut in half. The gates of the shed and paddock must be big enough for a cart to enter.

HEALTH

An animal can be in bad health due to diseases, injuries, or parasites.

Diseases

Nowadays rinderpest and other serious diseases are much less common. All the same, there are still many diseases to treat. Good ways of controlling diseases are vaccination, proper feeding, and adequate housing. A veterinary surgeon should be consulted about a sick animal. You must keep the sick animal away from the others because of the danger of infecting other animals. The chief diseases of cattle are Rinderpest, *Pleuropneumonia*, Anthrax, Black-quarter, and *Trypanosomiasis* (sleeping sickness). Vaccines exist for all of these diseases.

Do not eat the meat of animals that have died from certain contagious diseases, such as tuberculosis. This disease can be passed on from animals to people. Do not mix your herd with herds passing through, especially if they come from far away. Passing herds may bring diseases with them. Do not mix with your herd an animal you have just bought or that comes from somewhere else unless you are sure it has been vaccinated. If an animal dies of a contagious disease, burn the body or bury it 2 meters deep with quicklime to kill the germs.

Care of Wounds

Wounds need to be attended to carefully. If you see an ox or a cow that has difficulty walking (limps), bleeds after a fight with another animal, or has hurt itself, lose no time in looking after it. An infected wound does not heal quickly. It may prevent the animal from walking, going to the pasture, working, and giving milk. A cow in pain gives less milk. Find out how the animal got hurt. Has it a thorn in its foot? Has a piece of wood or iron torn the skin? Has the rope, the collar, or the yoke rubbed too much, or is it too tight? Is there a vicious animal in the herd? Do not work a sick animal; it is better to lose a few days' work than to lose the animal.

Clean the wound with hot water with disinfectant added that will prevent the wound from becoming more infected. A wound that is always kept clean heals quickly, so wash the wound often.

Parasites

Parasites are little animals that live on the skin or in the bodies of other animals. Chief among the parasites that live on the skin of cattle are ticks. Ticks stick to the skin of the animals and suck blood. If an animal has many ticks, it can lose up to half a liter of blood a day. After a time it may become very weak. Ticks wound animals. Often you can see an animal's ears or udders damaged by ticks or an animal walking with difficulty. The cows will be difficult to milk, and they will not let their calves suck. Ticks may also bring serious diseases. They spread fevers, typhus, *brucellosis*, and *piroplasmosis*. Ticks can be killed with a pesticide such as toxophene or with paraffin oil. Soak a piece of cloth in paraffin oil and rub the areas where there are ticks. Veterinary services can tell you what pesticides to use and help you to apply the treatment. This must be done over and over again.

Some parasites live *in* the body. Generally, parasites live in the digestive tract. Many are worms: tapeworms, roundworms, and pinworms. Sometimes they live in the muscles or the lungs, like *strongyles*. They injure the digestive tract, and the animals cannot digest properly. Animals that have worms lose weight and sometimes die. To kill these parasites, the animals are given medicine such as phenothiazine. There are traditional medicines that can also be used.

A good way to control parasites is to let pastures rest. The eggs of the parasites fall on the pasture with the animals' excrement. They grow in the grass, and then they can attach themselves to the skin of the animals, or the animals may eat them together with grass. If you let the pasture rest long enough, the parasites cannot feed on the skin or in the bodies of the animals, and they will die.

CATTLE REPRODUCTION

A cow has two ovaries. Every three weeks the ovaries produce an ovum (egg). If the cow mates with a bull at this time, the ovum is fertilized, and it develops and becomes a calf.

Pregnancy lasts about nine months. If the cow has already had a calf, she must not feed this calf more than five or six months after the new fertilization. The calf she is carrying needs more food. The cow cannot feed the calf in the womb and give milk at the same time.

Some days before the birth, the cow's udders will swell. During birth, part of the water-filled membranes that cover the calf in the womb will come out. Next either the calf's two forelegs or two hind legs emerge. Sometimes it is necessary to pull downward on the calf's legs to help the birth. Once the calf is out, if the umbilical cord is still connected to the mother, cut it and clean it well with a little iodized alcohol. After the birth, the rest of the membranes will come out. (If all the membranes do not come out, they can rot inside the cow and kill her.) When the calf is born, the mother licks it all over. At this time the cow is often thirsty, so give her a drink. During the first few days after the birth, the mother's milk is thick and yellow. The calf must drink this milk, which will clean its digestive tract.

Calves are delicate and can easily catch parasites, so give them good care. To protect them, give them a medicine to get rid of internal worms at the age of three weeks and again at ten weeks. They easily catch diseases, so have them vaccinated. At three weeks, the calf begins to eat grass with a little cooked cassava.

AGE OF BREEDING ANIMALS

Heifers are young female cows. The ovaries begin to produce ova when a heifer is nine or ten months old. From that time, heifers can be fertilized. Do not have a heifer serviced by a bull when she is too young. The heifer cannot go on growing herself and feed the calf she is carrying, too.

In fact, you may have accidents when the calf is born. So wait until the heifer is big and strong enough—about two years old—before having her mated.

The testicles of young bulls begin to produce semen when the bulls are about eight months old. Do not have cows serviced by too young a bull. The bull will get tired, not grow well, not eat well, and become a poor breeding animal. Do not have the bull service cows before it is 18 months old. Keep cows younger than two years and bulls younger than 18 months away from each other.

CASTRATING BULLS

A herd of 25 cows needs only one good bull, a good breeding animal. The other males in the herd must be castrated. A castrated male is called an ox or bullock. To castrate a bull, either remove the testicles or crush the ducts that connect the testicles to the penis. The animal husbandry service and the livestock assistants have instruments for castrating bulls. After castration bulls are quieter, they are not vicious, and it is easier to harness them. They put on weight more quickly, and the meat is better. They cannot fertilize cows; in this way you prevent poor breeding animals from reproducing and can leave them in the herd.

Bulls should be castrated at about 10 months if you want to sell them to the butcher and at about 18 months if you want to make working oxen. If you wait until 18 months, the ox is stronger for work, but in that case it must be kept away from the herd so that it cannot come into contact with cows.

CHOOSING BREEDING ANIMALS

Bulls and cows must be carefully chosen because the calves take after their parents. Choose breeding animals with the following qualities.

- They are well formed. Sell all poorly developed animals. Keep animals with plenty of muscle, especially of the back and rump, because they give the best meat.
- They gain weight quickly.
- They are resistant to disease.
- They give plenty of milk.

It is easier to improve the herd by a good choice of bull. A cow passes on her good qualities to only one calf each year. A bull passes on his qualities to all the calves of the herd. Animals that are not used for breeding should be sold for slaughter. All cattle do not yield the same amount of meat or the same quality of meat. Usually you can earn more by selling a few high-quality animals than many lower-quality ones.

HOW TO KNOW YOUR HERD

Modern farmers keep a herd book. Give a number to each animal in the herd. This number is the animal's name. Mark the number on the animal's rump, for instance, by branding. Write in the book everything you need to know about your animals. For a female record year of birth, number of sire, number of dam, when serviced and by whom, number of offspring, number of deaths before weaning, production (milk, weight), vaccinations and disease, and remarks.

For a male record year of birth, number of sire, number of dam, number of female serviced, and date of each service.

MILK

Milk is formed in the cow's udder and comes out through the teats. Squeezing the teat makes the milk come out. The milk is produced by the blood that circulates in each quarter of the udder. If plenty of blood circulates in the udder, plenty of milk is produced. Emptying the cow's udder of milk may take five to ten minutes. For good milking, the cow must be calm; if you strike her or she is frightened, she will not let herself be milked easily. Make sure you always empty the udder. If all the milk is taken away, the udder can develop, and a well-developed udder can give more milk. Often a cow gives more milk after her third calving than after the first.

Milking must be done every day at the same time—for example, in the morning before going to the pasture. The cow gets into the habit of giving her milk at the same time every day. A cow with large blood vessels can have a lot of milk. Milk production varies greatly according to breed, the cow's health, the cow's age, recent birth, and grass intake. If the cow calves in the rainy season when there is plenty of good grass, she gives a lot of milk. A well-fed cow gives more milk than a badly fed cow. A cow in milk needs a feed supplement and plenty of water.

ORGANIZING SALES

To earn money, it is not enough to work well—you must also sell well. You should sell sterile cows that do not produce calves. You should feed them well for several months and sell them when they are really fat. You should sell a cow before it is too old. If you wait too long before selling, maybe you will get one more calf, but the cow will be too old to fetch a good price. By keeping a cow that is too old, you lose more money than you can earn from the calf she may produce. You should sell oxen as soon as they no longer gain weight. It is useless to keep them for five or six years. Sell oxen at four years. If you keep them longer, the ox eats food that would have enabled you to raise another animal. If a farmer has too many young calves and not enough grass to feed the animals, he or she can sell some calves to another farmer who will fatten them.

Sell some animals at the end of the rainy season. Then you will be able to feed the herd well in the dry season. You know when meat is bought at a high price—for example, at festivals and at the end of the dry season. Organize your breeding so you have animals for sale at that time.

MILKING COWS

Many Third World countries are tropical, which makes good hygiene difficult. More than 10 percent of the milk produced in India, for example, is lost due to spoilage. The essential problem is to prevent contamination right from the moment the milk leaves the udder. Likewise, cooling should start as soon as possible. If ice-making facilities are available at the milk collection center, hygiene can be greatly improved.

Hand milking makes sense for a small farmer. Cows should be given a shady place to wait to be milked. Good stimulation by rubbing the udders and squeezing the teats before milking is needed. Unless adequate washing water with chlorine in it is available, it is best not to try to wash more than the teats and lower parts of the udder. Full hand milking is essential. Milking of both buffalos and cows should be carried out with the teats as dry as possible. If a lubricant is thought to be essential, the use of coconut oil in small quantities is helpful. Coconut oil may be added to soap made from this oil to make an udder wash. A small quantity of the creamy mix is rubbed onto the udder surface and teats and washed away with a final squeezing away of residual water. A farmer with more than ten milk cows will probably want to invest in a milking machine.

The milker's bucket should have a partial cover to minimize dirt falling in during milking. The receptacle into which the milk is poured from the milker's bucket must be covered and provision made for cooling when possible. A simple immersion cooler is very helpful. Ideally, the milk should be refrigerated.

All vessels used for milk must be thoroughly scrubbed with a detergent or soap and must be rinsed with chlorine solution. The solution is about 2.5 percent chlorine and is diluted to 200 parts per million for use.

REFERENCES

Heifer Project International. "List of Publications," 1015 S. Louisiana, PO Box 808, Little Rock, AR 72203. Tel.: 501-376-6836.

Hibbs, John W., and W. G. Whittlestone. (1985). "Understanding Dairy Production in Developing Countries." Volunteers in Technical Assistance, 1600 Wilson Boulevard, Suite 710, Arlington, Virginia 22209 USA. Tel.: 703-276-1800. Fax: 703-243-1865. vita@vita.org.http://idh.vita.org/pubs/docs/uhn.html

RAISING CHICKENS

Based on Better Farming Series 13. Keeping Chickens, A.J. Henderson: Food and Agriculture Organization of the United Nations, Rome, 1992

A male is called a cock. A mother bird is called a hen. Their little ones are called chicks. When the chicks have grown big enough to be sold, they are called chickens (see Figure 56). All the birds in the poultry house (cocks, hens, chicks, and chickens) are called poultry. To succeed in keeping poultry, you must choose hens of a good breed, feed them well, house them well, and protect them well against diseases. A family poultry unit should not cost much, but it should yield a lot.

Chickens from improved breeds usually make better use of their feed than local hens do. They grow quickly, become fat, yield a lot of meat, and produce many big eggs. A local hen grows fat very slowly, uses a lot of feed, and uses it badly. If a farmer decides to feed his local hens well, he or she will not earn much money and will not get much meat. On the other hand, a chicken of improved breed needs only 3 to 5 kilograms of feed from its birth to the day when it can be sold or eaten.

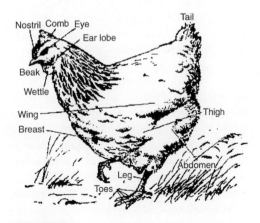

FIGURE 56
A chicken.

Crosses between various breeds create valuable chickens. Here are some of the more successful crosses. In naming the crossed breeds, the breed of the male is always given first.

- Rhode Island-Sussex cross: The hen is a good layer. In ten months—that is, about 300 days of laying—it will lay 165 to 180 eggs.
- Sussex-Rhode Island: The hen is also quite a good layer. A three-month-old pullet will weigh from 1.5 to 1.7 kilograms.
- Rhode Island-Wyandotte cross: This bird is also called P-60. The hen is a good layer, but it dislikes the damp. The adult hen weighs between 1.7 and 2 kilograms.
- New Hampshire-Leghorn and Rhode Island-Leghorn crosses: These crosses often produce white birds and sometimes birds with mahogany feathers. They produce white eggs.

Good chicks must be selected. You can buy either day-old chicks or three-month-old pullets. Day-old chicks cost less, but you have to know how to raise them. You have to be able to house them well, for they are very delicate and can die easily. Three-month-old pullets cost much more, but they require less looking after. They have been vaccinated, and they are more resistant to diseases. When you begin modern poultry keeping, it is better to buy three-month-old pullets because they are easier to raise than day-old chicks.

FEEDING CHICKENS

It is no use selecting cocks, hens, and chicks of a modern breed—unless you feed them well. The cocks, hens, and chicks bought from the animal husbandry centers have been well fed, and you must go on feeding them well. If you do not, they will not get fat, they will catch diseases, and they can die. In a well-run poultry farm with 100 laying birds, not more than 5 or 10 birds should die during a year. Feeding poultry well is important—and difficult. A hen is not like a goat that can feed on grass only. Poultry need feed rich in proteins and calcium to produce eggs. Poultry need a certain quantity of each type of food. If a hen has too much protein and not enough calcium, the protein cannot replace the calcium. You must give poultry the exact quantity of each feed that is needed.

It is the muscles of poultry that yield meat. Poultry are good if they yield a lot of meat in a short time, if the meat is not hard, and is white. If poultry eat plenty of protein feeds, they develop good muscles. The protein feeds are also part of bodybuilding feeds.

If poultry have well-formed bones, they can walk well. Poultry bones are long, thin, and light but hard. In order to have strong hard bones, poultry must eat mineral salts, which are part of the body-building foods.

Energy food is the most important ingredient in poultry feed, and often the farmer can grow these feeds. The main energy feeds are maize, sorghum, millet, and rice. These can be given in the form of grain or of meal. Chickens really like maize, and they can eat a lot of it without harm. Cassava is also a useful food. It can be given as meal or boiled, but don't give them too much. In 10 kilograms of feed there should not be more than 2 kilograms of cassava. Rice bran is another energy feed, but in 10 kilograms of feed, there should not be more than 1 kilogram of rice bran. Palm-kernel oil cake—what is left when the oil has been taken from palm kernels—is both an energy feed, which can replace maize, and a bodybuilding feed, which can replace groundnut oil cake. In 10 kilograms of feed there should not be more than 1.5 kilograms of palm-kernel oil cake.

Bodybuilding feeds are rich in proteins. Poultry need proteins that come from both animals and plants. If you give poultry 2 kilograms of feed containing proteins, there should be 1.5 kilograms of vegetable proteins and 0.5 kilograms of animal proteins. Oil cake contains a lot of proteins, and poultry digest it well. In 10 kilograms of feed, 1.5 kilograms can be groundnut oil cake. Poultry cannot digest a large amount of cotton oil cake. In 10 kilograms of feed, not more than 0.5 kilogram should be cotton oil cake. Palm-kernel oil cake can take the place of maize and groundnut oil cake, but you should not give more than 1.5 kilograms of it in 10 kilograms of feed. Proteins that come from animals include boiled blood, meat meal, milk powder, or fishmeal. You should not give poultry too much animal protein. It costs a lot, and if you give too much, it can make the poultry ill. Energy feeds and body-building feeds are not well used by poultry unless you give them at the same time.

If a farmer buys poultry feed, he or she may buy all the feed—that is, a meal containing all the foods that poultry need to live and grow. No other food need be given. This meal is costly, and the seller's instructions should be given, including the necessary quantities. Or a farmer can buy only part of the feed—the concentrates. These are the kinds of meal that contain chiefly proteins, mineral salts, and vitamins. If the farmer buys concentrates, crushed grain or oil cake should be given. Poultry feed means less work to do but costs a lot of money.

MINERAL SALTS

In 10 kilograms of feed, there should be 200 grams of mineral salts. Bones, oyster and snail shells, and eggshells are rich in mineral salts. If mineral salts are not given, poultry cannot grow well; their bones will be small and badly formed. Vitamins should be mixed in poultry feed if the birds are kept in a yard. Otherwise the birds get vitamins by eating grass. Vitamins are given in very small quantities. A farmer cannot produce them, they must be bought from the store or animal husbandry centers.

EGGS

A hen has only one ovary. The ovary produces the ovule, which consists of a germ and reserves. It is these reserves that make the egg yolk. The ovule passes into the oviduct, where the white and the shell of the egg are formed. The egg passes into the cloaca and out of the hen. The eggshell is made of mineral salts, especially calcium. The egg white contains a lot of water, protein substances, and mineral salts. The egg yolk contains a little water, a lot of protein substances, fat, and vitamins. A hen begins to lay from the age of five months and can produce an egg nearly every day from the age of seven months.

If you want to have chicks, a cock must fertilize the hen. The cock makes a lot of sperm, so it also needs to be well fed. A cock eats more than a hen. One cock can fertilize ten hens. To improve the quality of the chicks you may want to take part in a cock distribution scheme if one is available in your area.

CLEAN WATER

It is very important to give poultry plenty of clean water. Poultry do not make good use of their food if they do not drink enough water. One hen can drink a quarter-liter of water a day. If the weather is very hot, the birds will drink more—about half a liter of water every day. Put the water in fairly big drinking troughs so that several hens can drink at the same time. Put the drinking troughs in the shade, near the feeding troughs. Never let drinking troughs be empty. The water must always be clean. Dirty water gives poultry a lot of diseases.

SPECIAL NEEDS OF CHICKS, LAYING HENS, AND TABLE POULTRY

From birth to eight weeks, chicks need above all body-building foods. You should give them water and a feed containing 7 kg of crushed maize and/or other grains; 2 kg of groundnut oil cake; 1 kg of a mixture consisting of remains of meat, fish, or blood; oil or vegetables; bones and crushed shells; and termites per every 10 kg of meal.

From 8 to 14 weeks, the feed should be more plentiful. There should be, per 10 kg of meal, 8 kg of crushed maize and/or other grains; 1.5 kg of groundnut oil cake; .5 kg of a mixture consisting of remains of meat, fish, blood; grass; termites; and vegetables.

After 14 weeks, give only maize grain or a mixture of maize with other grains. Remember, if the poultry are in a yard, you must also give them feeds that are rich in vitamins—grass and vegetables—and rich in proteins—termites, meat scraps, or fish scraps.

Laying hens need plenty of mineral salts to make the shells of their eggs. To make the reserves in the egg, laying hens need plenty of proteins. In every 10 kg of feed, there should be 8 kg of crushed grains (maize, sorghum); 1.5 kg of groundnut oil cake; .5 kg of a mixture consisting of meat or fish; boiled blood; grass and vegetables; and especially 300–400 grams of crushed bones, oyster or snail shells, or eggshells.

During the first three months of life a bird eats about 5 kilograms of feed. The food must be well mixed and not prepared too far in advance of giving it to the poultry, or the food may go bad. Feeding and drinking troughs and the drinking founts must always be very clean.

HEALTH CONCERNS

Do not put too many birds together, or they may wound or even kill each other. The stronger ones peck the weaker. The grass in the run is soon eaten up by the birds and cannot grow again. Diseases are passed more easily from one bird to another. For 50 laying hens, you need an area of about 25×20 meters. Hens should not be put together with ducks, guinea fowls, and turkeys, since the diseases of these birds can be passed to the hens.

Take any sick birds out of the run. Do not eat sick poultry. Kill them and burn them so that the microbes are not left in the ground to be passed on to the other birds. It is also better to remove from the run hens that are thin and not growing anymore. These birds do not resist diseases well and can give them to the healthy poultry. When a bird is sick or dead, take it to the animal husbandry service or the nearest veterinary assistant and do what is necessary so that the disease is not passed on to all the poultry in the village.

All poultry must be vaccinated when they are very young, before they have begun to lay eggs. Young birds that have not been vaccinated do not resist diseases and can die. If you have to vaccinate a hen that is laying, it will not lay any more eggs. Vaccination is generally used against fowl pox, cholera, and Newcastle disease. There are two chief ways of vaccinating: mixing the vaccine with the drinking water or injections.

MAIN DISEASES OF POULTRY

There are many poultry diseases. Some of them are difficult to recognize. The following are the most prominent.

- Bone disease: The birds walk with difficulty; they limp. The leg bones are badly formed. This disease is chiefly caused by lack of vitamins and mineral salts. These birds must be given food that contains more vitamins and mineral salts, such as vegetables and crushed bones and shells.

- *Pullorum* disease: The chicks are listless and walk with difficulty. They have a very big belly and drag their wings. Their excrement is liquid and turns white. Many of the birds die at the age of eight days. The disease is transmitted by the hens' eggs. A hen that has had *pullorum*, even if it has been cured, always produces infected eggs. All its chicks will be diseased. Such hens can be kept for food or eggs. To prevent poultry from catching this disease, do not buy chicks from unknown sources.
- Fowl pest (Newcastle disease): Fowl pest is a very common disease and very dangerous. It can kill a large number of poultry very quickly. The birds breathe heavily and with difficulty. They digest their food badly. When they have this disease, they cannot be treated. Fowl disease can be prevented by vaccination and by not mixing chickens of local breeds with the chickens bought from the animal husbandry service.
- *Coccidiosis*: Parasites living in the digestive system are the cause of this disease. Blood is seen in the excrement of chicks between ten days and three months old. If the chick is not dead in 30 days, it will always remain thin and will be very late in laying. To cure diseased poultry, add coccidiostats to their water. To prevent poultry from catching this disease, do not put too many birds together, and be very careful about the cleanliness of drinking troughs and poultry houses. There is no vaccine against *coccidiosis*.

PECKING

Chickens peck each other. They pull out feathers and make the skin bleed. Then the birds become more and more vicious. If there are too many poultry in a run, if their house is not shaded from light, if the drinking and feeding troughs are not big enough, the birds are quick to fight and may even kill one another. If this happens, do the following.

- Remove wounded and vicious birds from the run.
- Treat wounds with a bad-smelling medicament.
- Sometimes it may be necessary to cut off the tip of the beak.

You can also hang bundles of vegetables or green grass from the roof of the poultry house. The poultry get tired from reaching for this food and become less vicious.

POULTRY HOUSE

A poultry house is shown in Figure 57. You can build a poultry house like this without spending much money. The walls are made of earth with wooden posts or clay bricks, and the roof can be made with straw or big leaves, or even with old sheet iron. The ground of the poultry house floor must be well firmed down, or you can cover it with concrete. If you want to raise chicks, put in brooders. If you want to raise laying hens, put in nests. Do not make solid walls more than half a meter high. Close the remainder of the space with wire netting or bamboo laths.

For 50 hens, the poultry house can be 4 meters wide × 4 meters long × 2 meters high.

For 100 hens, the poultry house would be 6 meters wide × 6 meters long × 2 metres high.

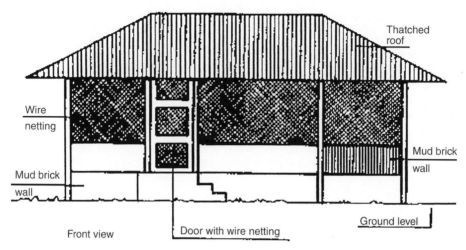

FIGURE 57
Poultry house.

The poultry house must be built near the farmer's own house because he or she will have to visit it several times a day. The poultry house must be built on dry ground because damp ground is dangerous for poultry, which can get diseases as a result. If the ground slopes, dig a ditch all round the poultry house to get rid of rainwater. The poultry house must be sheltered from sun and wind. You must build the poultry house so that it is sheltered from the sun during the hottest time of the day. To allow the sun to shine in during the evening, orient the ridge from east to west. During the rainy season put mats and branches on the sides of the poultry house to prevent the rain and wind from getting in.

NESTS

In the poultry house put wooden boxes or baskets lined with straw for the hens to lay eggs in. You need one nest for every five hens. Collect the eggs three times a day: every morning, at noon, and in the evening.

POULTRY RUN

These are necessary so the poultry can walk around and find green grass, insects, and worms. Put a fence around the run so the poultry do not roam too far and to protect them from animals. Try to leave trees for shade. Divide the run into two parts: one that is in use and the other to let grass grow back. For 50 hens you need a run about 25 meters wide × 20 meters long.

FEEDING TROUGHS

Feeding troughs should be sufficient in number and long enough for each bird to have its place when it wants to eat. For 100 birds the length of feeding troughs should be 2 meters for young chicks and 7 meters for older chickens. Feeding troughs should not be too wide, so that the birds cannot leave their droppings in them. Hollowed-out bamboo trunks can be used.

A feeding trough is shown in Figure 58. It can be made in the village. Nail the bottom of the trough to two planks. To the right and left of the trough nail two perches. The sides of the trough are made of thinner boards. They are high if the trough is for hens and low if the trough is for chicks. To prevent the hens from leaving their droppings in the troughs, add a wooden bar that turns (a roller). If the trough is outside the poultry house, make a little roof of sheet iron to shelter the food from the rain. Take off the roof to fill the trough.

DRINKING TROUGHS AND FOUNTAINS

For 100 birds two drinking troughs are needed with a capacity of 4 liters for young chicks and 10 liters for older chickens. Drinking troughs must not let the chicks fall into the water. You can use bowls or buckets put on a stand or set into the ground and partly covered by netting. For chicks put the water in a shallow bowl or can where the chick can drink easily. Fill a bottle with water, turn it upside down, and put the neck in the bowl; lean the bottle against a wall or make a support to hold the bottle. An ordinary 10- or 15-liter bucket serves very well, too. Sink it into ground so that only about 10 centimeters are above ground. Be sure to change the water frequently.

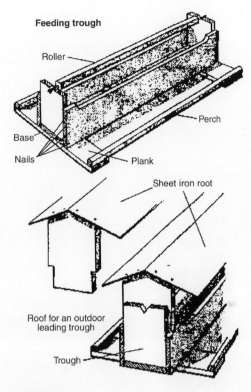

FIGURE 58
Feeding trough.

BREEDING CHICKENS

Choose hens of a good breed that do not have *pullorum* disease, and mate them with cocks of good breed that also do not have *pullorum*. To determine which hens produce a lot of big eggs, use nests that close after the hen has gone in. Choose hens that sit on the eggs. These hens should be big, in good health, and have plenty of feathers. Separate them from the other hens. Put them in a corner of the poultry house surrounded by netting, with a feeding trough and drinking trough with fount. Feed them very well, and protect them well against parasites. Kill any parasites with wood ashes or with products that are sold for the purpose.

If you buy day-old chicks, you must protect them from cold and from animals—rats, snakes, cats. Put the chicks in the brooder for three or four weeks. To protect the chicks from predatory animals, put them in a big wooden case or in a big basket. Put netting over the top. To feed the chicks, put a feeding trough and a drinking fount inside the brooder.

To protect the chicks from cold, put a storm lantern in the middle. Surround the lantern with netting so the chicks do not burn themselves. A storm lantern gives enough warmth for 20 to 40 chicks. Adjust the warmth by observing the chicks' movements. The chicks are too warm when they move away from the lantern. The chicks are too cold when they crowd up to each other.

REFERENCES

Bird, H. R. (1984). "Understanding Poultry, Meat, and Egg Production." Volunteers in Technical Assistance, 1600 Wilson Boulevard, Suite 710, Arlington, Virginia 22209 USA. Tel.: 703-276-1800. Fax: (703) 243-1865. vita@vita.org. http://idh.vita.org/pubs/docs/uhn.html.

Bishop, John. (1995). "Chickens: Improving Small-Scale Production." Echo Technical Note, ECHO, 17391 Durrance Rd., North Ft. Myers, FL 33917. Tel.: 914 543-3246. echo@echonet.org. http://www.echonet.org.

Klein, G. (1947). *Starting Right with Poultry.* Charlotte, VT: Vermont Garden Way Publishing.

RAISING DUCKS

Based on Better Farming Series 39. Raising Ducks 1, Tom Laughlin: Food and Agriculture Organization of the United Nations, Rome, 1990, ISBN 92-5-102939-3

Ducks lay eggs and supply meat. They are good to eat and very good for you. Ducks are strong and hardy and easy to raise. They need less care than chickens. They do not get sick easily. With very little time and work you can raise a small flock of ducks.

What do you need to raise ducks? You will need (1) a good place near your house to keep a flock of ducks, (2) a simple shelter for your ducks to protect them in cold, hot, or wet weather and to keep them safe from their enemies or from people who may steal them, (3) enough food and water for your ducks to eat and drink, and (4) enough strong and healthy, fully grown or young ducks to start your flock.

WHAT KINDS OF DUCKS CAN YOU RAISE?

There are several kinds of ducks you can use. You must find out what kinds of ducks you can get where you live. Usually, you can get local ducks. Local ducks are strong and hardy and used to living in your area, so they grow very well and do not get sick easily. However, local ducks are often small, give little meat, and lay few eggs. You might be able to buy ducks at the local village market, or there may be another farmer near where you live who will sell you some ducks from his own flock. Perhaps you can buy ducks from a nearby duck hatchery or from a government farm. Sometimes you can get improved ducks that have been brought from another place. Improved ducks have been carefully mated over many years, and they are bigger and better than local ducks. Improved ducks cost more money, but they give more meat and more eggs. A suitable duck is shown in Figure 59. When you are choosing ducks to start your first flock, try to get the kind of ducks that are good for both eggs and meat.

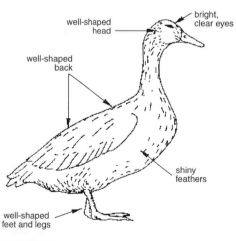

well-shaped head

bright, clear eyes

well-shaped back

shiny feathers

well-shaped feet and legs

FIGURE 59
A good duck to buy.

WHAT CAN DUCKS EAT?

Ducks eat just about everything, so it is not hard for them to find enough food for themselves, even if you feed them nothing. Ducks eat insects, worms, slugs, snails, frogs, grass, weeds, roots, most water plants, seeds, grain, plant materials, materials left on the ground after harvest, damaged or overripe fruits, and vegetables. You can also give ducks stale bread and food scraps.

Ducks can also find food to eat in home gardens and farm fields. However, keep the ducks out of your gardens and fields when the plants are young and tender or the ducks may eat them. After most plants are big, you can let your ducks feed between the rows. When your crops are fully grown and you have harvested them, be sure to let your ducks look for food there. After you have harvested your gardens or fields, your ducks will find a lot of food that is very good for them to eat and would otherwise be wasted. However, when ducks are in your home garden, it is a good idea to watch them. Ducks may eat snails, slugs, worms, and other things on the ground, such as seeds or fallen leaves. You must, however, be very careful if you have low-growing berries or fruits because the ducks will eat them, too.

Always remember that ducks must have fresh water with their food. If you are raising your ducks where there are no ponds or streams, make sure they always have plenty of fresh drinking water.

HOW MANY DUCKS SHOULD YOU RAISE?

If you are going to raise ducks that will live by themselves and find their own food, with very little help from you, you can raise up to 24 ducks. However, until you know more about ducks and how to take care of them, it is best to start with a small flock—for example, five female ducks and one male duck. From a flock of six ducks you will get five to ten eggs each week. In addition, with a male in your flock, you will get fertilized eggs. This means that you can raise your own baby ducks.

HOW TO KEEP DUCKS SAFE AND WELL

Although ducks can live outside by themselves, they will live much better if they have a shelter. The main reason for building a shelter for your ducks is to keep them safe from enemies at night when they are sleeping. Ducks, especially young ducks, have many enemies: dogs, cats, foxes, rats, snakes, predator birds, and thieves.

There are also other reasons for building a shelter. Ducks sleep on the ground, and if the ground is cold or wet or dirty, they can get sick. A shelter can be kept dry and clean so the ducks will stay healthy. Strong sunlight is bad for ducks—protection from the sun is needed in very hot weather.

Ducks lay eggs mostly at night or early in the morning. A shelter with nests where the ducks are kept inside at night makes collection easier and causes fewer broken eggs.

CHOOSING A PLACE TO RAISE DUCKS

It is best to keep your ducks where you can watch them easily, so choose some place close to where you live. Look for a place on your land that has enough of the kinds of food that ducks like to eat, shade for your ducks on hot days, and has a protected place where they can escape wind, cold, or rain. A good place is near a pond or stream where the ducks can easily get to water

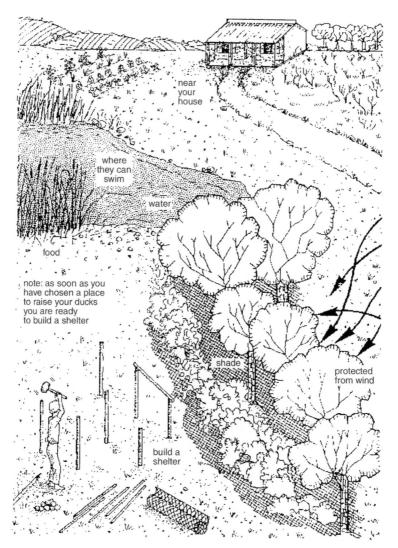

FIGURE 60
Location for raising ducks.

and swim. Raising a flock of ducks is only one use for land, so be careful not to choose a place for them that could be better used for something else, such as planting crops or growing a vegetable garden. Remember that ducks can live just about anywhere outside as long as they can find enough to eat and drink. Figure 60 shows what is important in choosing a location.

BUILDING A SHELTER FOR YOUR DUCKS

You can build either a pen or a house to shelter your ducks and keep them safe at night. A pen should have about one square meter of space for three ducks. So if you begin with a flock of six, you will need a pen of two square meters. A house where your ducks can sleep should have about one square meter for a flock of six. A completed pen and house is shown in Figure 61. Never try to put too many ducks in either a pen or a house. If a shelter is too crowded, it will quickly become wet and dirty, and your ducks may get sick.

A pen or a house can be built using local materials such as bamboo, used wood, palm leaves, or grass. A mat of woven bamboo, palm leaves, or strong grass makes a good pen cover. It must be strong enough to keep out the enemies of the ducks but fine enough to keep in small ducks. Wire mesh is suitable but expensive.

Build the fence around the pen using posts at least 1.20 meters high. The posts should be about one meter apart and about one-half meter in the ground. To keep animals from digging under the fence of a pen made of bamboo or woven material, put a row of stones along the bottom on the outside of the fence. This row of stones is shown around the

FIGURE 61
Duck pen and shelter.

shelter in Figure 61. To keep animals from digging under the fence of a pen made of wire mesh, bury the wire mesh about one-third to one-fourth meter in the ground. If there are meat-eating birds nearby, cover the pen as well—the same material as the fence is suitable.

Try to find a place for your pen on high ground so water will run off and the pen will stay dry. Have the pen built and ready for your ducks before you get them. That way they can become used to their new home from the beginning.

THE FLOOR OF A DUCK SHELTER

You must be very careful to keep the floor of the shelter as dry and clean as you can to keep the ducks healthy. A shelter that is on high ground can have a dirt floor. However, if the shelter is built in a low, wet place, you will need to keep it dry by covering the floor with some kind of dry material. Cover the floor with sand or fine gravel, pine needles, wood chips, or leaves, and give your ducks a bed of straw or cut grass to sleep on as well. Change the floor material when it becomes wet or dirty and especially if it becomes moldy. Mold will make your ducks sick. With a flock of six ducks, change the floor covering at least once a month or sooner if it becomes wet, dirty, or moldy. Pick a day each month to change the floor covering.

BUILDING NESTS

As soon as you have finished building a pen or a house, nests should be built. One nest for is needed for every three female ducks. So a starting flock of six ducks (five females and one male) will need two nests. Ducks like small nests with just enough room to get in, turn around, and sit down. Give your female ducks a cozy nest of just the right size.

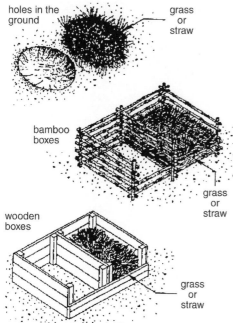

A good duck nest is about 30 centimeters wide, 38 centimeters deep, and, if it has a roof, from 30 to 35 centimeters high. Ducks live and sleep on the ground, and they will lay their eggs on the ground as well, so put the nests on the ground. A nest can be a hole in the ground or a box made of bamboo or wood that is lined with clean material such as cut grass or dry straw. Ducks prefer dark, quiet places to lay their eggs, so it is best to cover the nest with a roof of thatch or wood to make it dark and quiet. Some nests are shown in Figure 62. As soon as you have finished building the nests, put them in the shelter. That way your ducks can become used to their nests as soon as you bring them home. Be sure to change the grass or the straw in each nest as soon as it gets dirty.

FIGURE 62
Nests for ducks.

HOW TO TELL THE DIFFERENCE BETWEEN FEMALE AND MALE DUCKS

When you first begin with a flock of six ducks, you must have one male duck to be able to raise your own baby ducks. Therefore, it is very important to be able to tell the difference between female and male ducks. You can tell the difference by listening to the quack made by the duck or by looking at the feathers near the tail of the duck. If you are going to begin with ducks of eight weeks or older, you can tell females by listening to them quack. When ducks have reached this age, the quack made by a female is very different from the quack made by a male. Gently hold the duck by the tail until it quacks. A female duck will make a hard, loud quack. A male duck will make a soft, rough quack.

If you are going to begin with ducks of four months or over, you can tell female from male ducks by the feathers on their tails. When ducks have reached this age, male ducks have curled feathers on their backs near the tail and female ducks do not.

HOW TO HANDLE YOUR DUCKS

The legs or wings of a duck can easily be hurt or even broken. Never grab a duck by the legs or the wings. To catch a duck, grasp it firmly but gently at the base of the neck. You can also catch a duck by holding its wings against its sides with one hand on each side of its body and a thumb over each wing. After you have caught a duck, slide one hand under its body and hold its legs firmly. Then rest the body of the duck on the lower part of your arm for easy carrying.

If you have to move a duck from place to place, carry it in a basket or a crate with a cover. First, tie the legs of the duck together. Then put it gently inside and put on the cover to keep it in.

When you first bring new ducks home, put them in their shelter, close the door, and go away. That way they can settle down, become calm, and get used to their new home. Later the same day, just before the sun goes down, give them some food and water, but give it to them inside the shelter. You can give them food that is left over from your last meal or some chopped green plants. If you see your ducks eat all the food you give them, give them a little more.

Keep your ducks in their shelter for the first two or three days. However, be sure to give them some leftover food and chopped greens each night just before dark, and make sure they have water. When you see that your ducks are calm and used to their new home, you can let them out for the day. Let your ducks out a few hours after the sun is up so they will lay their eggs inside. During the day your ducks will wander about looking for insects, worms, grass, and roots and other things they like to eat. Then, each night just before dark give them the leftovers. However, this time give them food in front of the shelter, not inside. That way it will stay clean inside. By giving your ducks food in front of their shelter each night, they will become used to coming back to eat at that time. When your ducks have eaten, you can close them safely inside.

During the first few weeks, check on your ducks from time to time during the day to see where they are. However, soon they will learn to go out in the morning and come back by themselves at night. You will have to do very little for your ducks.

GROWING YOUR OWN BABY DUCKS

If you would like to have more ducks, the easiest way to get them is to raise your own baby ducks. You must, of course, have fertile eggs. If your small flock has a male duck, the eggs you get will be fertile, and you can begin to raise baby ducks.

A baby duck grows inside the shell of a fertile egg when it is kept warm. Eggs are kept warm when a female duck sits on them, called setting. After 28 days the baby duck is ready to hatch out of its shell. (The eggs of most kinds of ducks that you are likely to find where you live take 28 days to hatch. However, Muscovy duck eggs take 35 days to hatch.)

SETTING THE FEMALE DUCK

When a female duck tries to hide or sits on a nest more often, she is probably ready to set. Make her a setting nest in a quiet, dark place well sheltered from rain, sun, and wind. The nest can be a simple hole in the ground or a box lined with clean, dry grass or straw. A female duck can cover 10 to 12 eggs, so try to collect this many eggs from your flock. However, when you collect the eggs, handle them very carefully. If you shake eggs too much, they may not hatch. Eggs that are very small or very large may not hatch—choose medium-sized eggs. Eggs to be used for setting should be no more than ten days old. The eggs should be clean. If they are dirty, clean them with a soft, damp cloth. When you have the right number of eggs, put them in a nest in a quiet place, and the female will begin to set.

When one of your female ducks is setting on a nest, make sure that she has enough food and water nearby. If she has to go too far to find food and water, the eggs may get cold and not hatch.

On or a little before the twentieth day your baby ducks will begin to hatch. They will begin breaking out of their shells little by little. It may take as long as two days until all the baby ducks are hatched. As the baby ducks hatch, take away the pieces of broken shell from the nest.

After all the baby ducks have hatched, disturb them as little as possible. The female duck will take good care of her babies. However, your baby ducks will need some special care for the first four weeks. You should be sure to (1) keep them warm, (2) shelter them in bad weather, (3) keep them apart from the rest of the flock, and (4) feed them well.

For the first four weeks give the baby ducks all of the leftover food you usually give to the rest of the flock.

PUTTING YOUR NEW BABY DUCKS WITH THE FLOCK

When your baby ducks are four weeks old, they can begin to live with the rest of the flock and eat the same food. If you see that the baby ducks are bitten by the older ducks and that they are not able to get enough food to eat, give them their food away from the rest of the flock.

At first you can leave all of your younger male ducks with the flock. As the young males grow older, it is best to eat or sell them because the young male ducks may be from the same family as your female ducks, and inbreeding weakens the flock. However, as your flock grows larger you will need more male ducks. Buy them at the market or from another duck farmer, or trade one or two of your young males for new ones. Remember, with a larger flock you must be sure to have the right number of male ducks: one male for every five females.

WHEN TO USE OR SELL THE MEAT

As your ducks grow older, you can take them eat or sell as soon as you have enough young ducks to keep your flock the size you want. However, take the male ducks first. Ducks are old enough to eat or to sell at ten weeks. When your ducks reach two years of age, replace them with young ducks.

TAKING CARE OF YOUR DUCKS

You can raise as many as 60 ducks in much the same way as you raise a smaller flock. You will have to learn a little more about ducks and work a little harder. However, with 60 ducks you will get many more eggs and much more meat to eat and to sell at the market.

REFERENCE

Bauer, F. (1983). "Muscovy Ducks." Echo Technical Note, ECHO, 17391 Durrance Rd., North Ft. Myers, FL 33917. Tel.: 914-543-3246. echo@echonet.org. http://www.echonet.org.

OSTRICH PRODUCTION

Based on Ostrich Production, *Part of the Agricultural Alternatives Series, Penn State, College of Agricultural Sciences, Cooperative Extension, 1994*

An ostrich produces three marketable products: the skin, which is soft and durable; the meat, which is similar to beef in color, taste, and texture (but is lower in fat and cholesterol); and the feathers. Needless to say, before starting an ostrich-raising business, one should make sure a market exists for at least one of these products. In many tourist lodges in Africa ostrich meat is popular.

OBTAINING THE BIRDS

There are four ways to get started.

1. Incubate eggs. It may be possible to purchase fertile eggs. Using commercially made incubators is probably best for beginners. Incubators are used to keep eggs at a constant temperature, to maintain constant humidity, to provide good air circulation, and to turn the eggs on a regular basis. The incubation period for ostrich eggs is 42 to 43 days. Management practices include candling the eggs to monitor embryo development, removing eggs that aren't fertile, and ensuring that conditions are as sanitary as possible.
2. Purchase chicks more than three months old—mortality is highest in the first three months.
3. Purchase yearlings or young adults. Yearlings should be productive within two years.
4. Purchase proven breeders.

PENS

Ostriches require high-tensile-strength mesh fences that will not allow them to get their heads or legs caught. They can be quite aggressive and will bite or nip. They will reach over, through, and under any type of fence if possible, and they can injure their necks or legs if caught. Ostriches can grow to more than eight feet tall, so the fencing should five to six feet in height around a running area of one to three acres. Separate grow-out pens may be needed to separate chicks into groups of similar size.

Ostriches need shelter from the extreme cold of winter and heat of summer. Buildings can be inexpensive pole structures or a variety of renovated farm buildings, like barns and sheds. An area of 20 feet by 20 feet is adequate for a breeder pair. The shelter should have adequate space, good lighting, and good ventilation. Wide doors, a heat source, and proper flooring are required for both breeding pairs and feeder chicks. Chicks, breeding pairs, and feeder birds should be kept separate. Flooring and ground cover are important management issues. Producers need to ensure that the flooring and the ground in ostrich facilities don't contribute to leg problems in their birds. This requires removing all sharp objects; providing sand, alfalfa, or some other material as ground cover; and maintaining sanitary conditions.

CARE

Make sure the birds get enough exercise, which helps to prevent leg problems. Avoid as much as possible moving birds from one location to another. Any object that can be picked up and swallowed should be removed from the pen.

The most common health problems are stress, stomach impaction, and diarrhea. Chicks are the most vulnerable to these diseases, and the threat of death is highest from the time of hatching to three months of age. Managers need to practice good sanitation, disease monitoring, and disease prevention. Mixing grit with feed may help to decrease the incidence of impactation.

NUTRITION

Generally, adult ostriches consume three to four pounds of feed per day. New entrants to ostrich production should utilize the expertise of feed companies already producing ostrich feed. These companies can determine the most appropriate form of feed and the proper level of nutrition to feed ostriches at the different stages of development.

Feeds for growing chicks should be designed to achieve even growth without having to push the birds too hard. After the first three weeks, feed all they can consume in two 20-minute feeding periods daily. Gaining weight too rapidly can cause leg problems in young chicks. Chicks are fed a starter ration for three months. Then they are fed a grower ration until 10 to 14 months of age (at which time they are ready for slaughter). Breeding stock is generally kept on a breeding ration for six months of the year and a maintenance ration for the other six months.

Always have clean water available to all birds at all stages of development. Water containers should be rinsed daily and scrubbed with soap and water every three days. Grit is essential to an ostrich diet. Small stones or commercial grit is best.

BREEDING

Ostriches are normally paired off for breeding. Hens will start laying anywhere from 24 to 36 months of age. Hens tend to mature earlier than males. The female can lay up to 20 eggs per year. The egg incubation process takes approximately six weeks.

REFERENCE

www.agric.gov.ab.ca. A Web site focused on raising ostriches commercially in North America.
http://aqalternatives.cvs.psu.edu/

RAISING PIGS

Based on Guidelines and References: Livestock training component (Small Animal Husbandry) Volume IV: Livestock
From: Agricultural Development Workers Training Manual
Printed By: Peace Corps, Information Collection and Exchange November 1985

THE FREE-RANGE SYSTEM

In Latin America and most of Africa where pigs are raised by subsistence farmers the pigs are allowed to free-range and scavenge off whatever feeds and garbage they find. In Southeast Asia pigs are more commonly penned and fed rather than being allowed to free-range. Neither of these two systems is inherently "better" than the other. Both are systems that have developed over time as a response to local culture and conditions. Most importantly, they are systems of husbandry that work and are appropriate to the local conditions. Listed below are some of the advantages and disadvantages of the free-ranging system of swine husbandry.

Advantages

1. Feed, scraps, and garbage available locally gets consumed and produces meat.
2. The feed is free, and therefore production costs are nearly nonexistent.
3. People do not have to spend much time caring for the pigs.
4. There is no need for expensive housing and feeding equipment.
5. The pigs are not competing with humans for scarce cereal grains.

Disadvantages

1. The production of meat tends to be very low.
2. The pigs harbor diseases and parasites that are transmittable to humans and generally lower the standards of public health.
3. They burn off a lot of energy looking for food that could be used to produce meat or fat.
4. They are subject to greater predator losses than if they were penned.
5. There is no control over breeding or genetics in a free-ranging environment.

For the generalist volunteer working with subsistence farmers who produce a limited amount of cereal grains, it is my contention that housing of pigs is not a major issue. Of the five components of livestock production (nutrition, management, diseases and parasites, genetics, and housing), I consider housing to be the least important in terms of increasing production. If the goal of the farmer and the volunteer is to increase the production of meat from pigs, then an investment of money and time should be given to the other four components before housing, because they will yield better results. It is totally inappropriate to invest in sophisticated housing and labor-saving devices for pigs when working with local or creole breeds while not having solved all the problems relating to their feeds and nutrition. Unless there is a real excess of cereal grains being produced in your area, it is probably not appropriate to attempt to move farmers from their free-ranging system of husbandry to one of raising pigs in confinement. When pigs are confined, the farmer and the volunteer have committed themselves to a higher level of production that involves feeding the pigs daily, worrying about production costs and the scarcity of grains, marketing pigs for profit, and improving the levels of husbandry as well as attempting to improve the genetic strains or breeds of hogs that you are working with. To go from a free-ranging or survival level of production to a moderate level is a major step and not one that volunteers should promote unless they are very sure that all the resources needed to do so are available.

SHELTER

I offer the following points to be considered in your decision concerning the type of shelter you might provide to pigs.

1. Visit other farmers who are already raising pigs in confinement and see what designs and materials they are using.
2. Use materials that are available locally (either free or at minimal cost) such as bamboo.
3. Do not invest much money in housing for pigs.
4. Make sure that the pigs that are confined have protection (shade) from the sun. This can be provided by a thatched roof or trees.
5. Provide a water source or a mud "wallow" for the pig to cool in as a protection from the heat.
6. Design the pen in such a manner as to protect the pigs from predators.
7. The pen should be kept near the farmer and the source of feed. This creates the possibility of increased access and (hopefully) better care for the pig.
8. Consider the following space requirements.
   - Mature sows, gilts, and boars require 15 to 20 square feet of space each.
   - Gilts with piglets and mature sows with piglets require at least 50 to 65 square feet of space.

- Weaner pigs to pigs of 75 lbs. require 5 to 6 square feet.
- Pigs of 75 to 125 lb. need 6 to 7 square feet.
- Pigs of 125 to 200 lbs. need 8 to 10 square feet each.

Disinfecting

All swine housing and equipment (feeders, waterers, and farrowing crates) need to be disinfected from time to time. If you are rotating pastures or pens, the pens and all equipment should be disinfected. If you have a buildup in parasites or a disease outbreak, you also need to disinfect. Scrubbing with soap and water, followed by washing with a disinfectant (such as chlorine or iodine), and then sun drying for 48 hours is effective in controlling most diseases and parasites.

FARROWING CRATE

The purpose of a farrowing crate is to provide an area where the sow can remain while farrowing, which affords protection to the piglets and allows the farmer to tend to the sow during the farrowing. (Farrowing is giving birth.) Newborn piglets are often not quick enough to avoid being laid upon by a sow as she rolls. When this happens the piglet can be squashed or may suffocate. Farrowing crates or guardrails along the wall of the farrowing pen create a space where the piglet can lie down while nursing and still be protected from the weight of the sow. Each farrowing crate may be from $4\frac{1}{2}$ to 5 feet wide and 8 feet long. These widths allow from $1\frac{1}{4}$ to $1\frac{1}{2}$ feet of space on both sides of the 2-foot-wide sow stall for the piglets. The bottom guardrail should be about 15 to 18 inches off the ground. The size of the far rowing crate also depends on the size of the sow and local breeds. The crate can be built out of a variety of materials such as thick bamboo, wood, or pipe. The important thing to remember is that the crate be built out of as inexpensive (yet functional) material as is available.

FEEDERS

One of the best feeders is the trough. This type of feeder is good for either dry feed or a wet mash. They can be built cheaply out of wood. They should be built so that they form a V, which causes the feed to fall to the bottom, prevents the pigs from standing in it, and may reduce waste. The trough should also have slats spread along the top of it to prevent the pigs from lying in the feeder. The trough should be supported by wide end pieces that prevent it from being turned over and the feed wasted. Creep feeders are designed so that the piglets can be fed in an area where the sow cannot enter and feed. (Creep-feeding is the practice of feeding concentrates to young pigs in a separate enclosure away from their dams.) Pigs that feed directly off the ground, rather than from a trough feeder, are more subject to infection by parasites. One to two feet of linear space per pig on a feeder is needed (depending on the size of the pig).

WATERERS

Ideas for building waterers can be gained from local farmers and their creative uses of bamboo and wood. One to two linear feet, depending on the size of the pig, is needed for watering space. It is important to make them easy to clean and heavy enough so they are not constantly tipped over by pigs. Also, pigs should not be able lie down in the waterer to cool themselves, thereby contaminating their drinking water.

SWINE NUTRITION

Swine in many countries of the world are considered a "dirty" animal because in free-range environments they eat primarily garbage and feces. Any attempts to alter this free-range environment would involve confining the swine and feeding them. Since 70 to 90 percent of the cost of raising pork is feed, volunteers must work to provide as inexpensive a ration as is possible and appropriate to the farmer's production level. Furthermore, good feeds can reduce the stress level on a pig and improve its health. The nutrient needs of swine are influenced by age, function, disease level, nutrient interaction, and environment.

Nutrients

Energy is supplied mainly by two types of carbohydrates for swine.

1. Nitrogen free extract (NFE). This includes the soluble carbohydrates such as sugars and starches and is highly digestible. Examples include corn, sorghum, maniac, and tarot.
2. Crude fiber such as lignin and cellulose. These are highly indigestible for swine. Examples include overripe hay, straw, and grain hulls.

Protein. Although it is common practice to refer to "percent protein" in a ration, this term has little significance in swine nutrition unless there is information about the amino acids present. For swine, quality is just as important as quantity. It is possible for pigs to perform better on a 12 percent protein ration, well balanced for amino acids, than on a 16 percent ration with a poor amino acid balance. Symptoms of protein (amino acid) deficiency are reduced feed intake, stunted growth, poor hair and skin condition, and lowered production.

Minerals. Of all common farm animals, the pig is most likely to suffer from mineral deficiencies. This is due to the following peculiarities of swine husbandry.

1. Hogs are fed cereal grains and their by-products (as well as garbage and feces), all of which are relatively low in mineral matter, particularly calcium.
2. The skeleton of the pig supports greater weight in proportion to its size than that of any other farm animal.
3. Hogs do not normally consume great amounts of roughage (pasture or dry forage), which would tend to balance the mineral deficiencies of grains.
4. Hogs reproduce at a younger age than other classes of livestock.

Water

The daily water requirements of swine vary from .5 to 1.5 gallons per 100 pounds of live weight. The higher requirements are for young pigs and lactating sows. Also, the higher the temperature, the greater the water consumption. Remember, too, that many of the fecally spread, water-borne parasites that affect humans have swine for their intermediate host. Therefore, humans and swine should not share the source of water.

FEEDING PRACTICES FOR DIFFERENT CLASSES OF SWINE

Breeding Gilts. Prospective breeding gilts should be kept from getting too fat. This can be a problem in moderate-level production where the ration of the animal may be too high in carbohydrates and too low in protein. Meat-type hogs can usually be left on a high-energy ration until they reach 175 to 200 pounds without becoming too fat. It is neither desirable nor necessary that females intended for breeding purposes carry the same degree of finish as market animals. Breeding gilts should be fed as follows.

- Five pounds per head per day through their second heat period.
- Flush—full feed—after the second heat period until breeding in the third heat period.

After breeding, limit the feed intake to three to five pounds daily. Overfeeding during gestation can cause embryonic death and thus decrease litter size.

Breeding Boars. In moderate to high production operations where a good supply of feed is available, feed 120- to 150-pound boars six to nine pounds daily. Mature breeding boars should be limited to five to seven pounds per day to keep them from becoming fat (which reduces their libido and fertility).

Brood Sows. The nutrition of brood sows is critical because it may materially affect conception, reproduction, and lactation. Proper feeding of sows should begin with replacement gilts and continue through each stage of the breeding cycle—flushing, gestation, farrowing, and lactation.

Gestation Sows.

- Limit feed to four pounds per day.
- Approximately two-thirds of the growth of the fetus is made during the last month of the gestation period. Therefore, the demands resulting from pregnancy are greatest during the latter third of the gestation period. You may wish to increase the feed to the sow by 20 percent during the last month of gestation. It is important for the sow not to be too fat or too thin as she approaches farrowing. Leguminous pastures are good for gestating sows, and if the pasture is of good quality, the sow's grain feeds can be reduced by one-half during this period.

"Flushing" Sows. The practice of conditioning or having the sows gain weight just prior to breeding is known as flushing. The purpose of flushing is to increase the number of ova shed during estrus. About 10 to 14 days prior to expected breeding, the sow should be fed a ration that will make for gains of 1 to $1\frac{1}{4}$ pounds per day. Generally 6 to 8 lbs. per head per day of a high-energy 14 to 16 percent protein feed that is well balanced in minerals and vitamins is adequate. This is a particularly good technique where you have a limited amount of good feed and seek large litter sizes. Under such circumstances flushing can provide good results. Immediately after breeding, the sows should be put back on limited feeding. Continuation of a high level of feeding after breeding will result in higher embryo mortality.

Farrowing Sows.

- They should be given all the water they will drink.

- It is good, since sows are prone to constipation during farrowing, to reduce their feed by one-half 48 hours before farrowing. Remove all feed the day the sow farrows. The day following farrowing, provide the amount of feed you were giving during gestation.

Lactating Sows.

- This is time of highest stress on the sow in production.
- She needs good quality concentrates rich in high-quality protein, calcium, phosphorus, and vitamins. If you have a limited amount of protein feeds available for your swine, this would be a good time to use what you have because the survival of the litter depends on the sow's nutrition.
- A good sow produces one gallon of milk daily. The growth of the litter (as well as the survival) is directly controlled by the sow's nutrition.
- A good rule of thumb in feeding lactating sows is one pound of feed per piglet daily (minimum of five pounds).
- Give at least $1\frac{1}{2}$ gallons of water per 100 pounds of body weight daily to the sow if the sow does not receive an adequate supply of milk.
- If possible it is advisable to increase the crude protein percentage to 16 percent during the lactation of the sow.

SWINE DISEASES

Volunteers who work with swine projects (be it with farmers, crops schools, or clubs) should place more emphasis on disease prevention than on treatment. Working to improve the diet of the herd, improving sanitation, and management practices are all good ways to prevent disease. Remember that prevention is generally less expensive than treatment and, therefore, more profitable to the project. Consider the following points.

1. Morbidity disease decreases the productivity of swine (that is, the feed to gain ratio drops) and usually is more damaging to a swine operation than mortality disease.
2. Reproductive diseases are often the most difficult to diagnose.
3. Disease diagnosis is often difficult for generalist volunteers. Many times there are no labs available to do diagnostic work.
4. There will probably not be veterinarians to assist in diagnosis or treatment of animals. In order to have a project that faces these limitations of resources and is still capable of being profitable, it is important first to emphasize prevention of disease. Furthermore, the volunteer should be competent with postmortem procedures in order to be able to assist in diagnosis of disease(s) and parasite(s) through identification of clinical symptoms.

The following herd management practices will help reduce disease.

1. Closely observing the herd for abnormal or diseased behavior.
2. Keeping precise feed consumption, breeding, and production records.
3. Maintaining a regular vaccination and parasite control schedule for your herd against diseases known to be in your area.

4. Quarantining all new stock for at least 30 days.
5. When buying new stock, having them tested first for *brucellosis*.
6. If your pigs are penned, preventing contact with free-ranging pigs.
7. Lessening stress by providing shade—especially during the hot season.
8. Providing adequate water for drinking and cleaning the pen regularly.
9. If you have a disease outbreak, following proper quarantine and cleanup procedures.
10. Assisting the sow during farrowing.

Finally, since pigs are common intermediate hosts for diseases and parasites that affect people (this transfer between species is called *zoonoses*), be careful with your own health. This is especially true if you have an outbreak of an undiagnosed disease in your herd.

- The normal rectal temperature of swine is 102.6°.
- The normal pulse rate is 60 to 80 beats per minute.
- The normal breathing rate is 8 to 18 per minute.

PARASITES

Pigs are susceptible to many different internal and external parasites. You can be sure that in a free-ranging environment where the pigs eat feces and vegetable waste they will be heavily infested with parasites. In attempting to improve production one must rid the pigs of parasites or they will not be able to utilize the feed efficiently. Feed is always too expensive to give to parasites. Treatment for parasites is expensive, potentially harmful to the pig, and does not prevent reinfection. Prevention and control of parasites by breaking their life cycle is far more important than treatment. With both internal and external parasites it is important for you to know the various life stages and habits so you can plan to disrupt their life cycles and destroy them. Pigs are also the intermediate host for many parasites that affect humans. Therefore, eliminating parasites in swine will serve to eliminate them from people as well.

PRODUCTION

As mentioned previously, the majority of volunteers who find themselves working to improve the production of swine are going to be working in survival production environments where the pigs are free-ranging, of poor genetic stock (creole), and receiving minimal care. Introducing an exotic breed of swine into this management environment will not necessarily improve production. There is no "magic" in exotic breeds. They have been bred for increased production through selection of genetic traits that yield the results desired (large hams, quick growth, etc.) by the breeder. However, for the pig to reach its genetic potential, it must have high-quality feed, a disease-free environment, good housing, and so on. Without these resources, the new breed will not perform as expected and may not produce any better than the local breed of pig would under similar conditions. Remember that in many ways production is the opposite of survival genetically. The local breed of pig has been selected genetically for survival, and the exotic breed has been selected genetically for production. To clarify this point, let us look at one genetic characteristic: body size.

Large body size is desirable for commercial farmers in the United States, and pigs are bred accordingly. However, large body size requires more than mere genetic selection; it also requires good feeds. The Third World farmer who does not have the good feeds or the market for such large animals has bred for small body size because the energy and water requirements for such a pig are reduced. This makes the pig a better survivor in the face of drought and a lack of feed. Generally, small Third World farmers are not concerned with production but with survival. Volunteers should be sensitive to local farmers' needs concerning production and not assume that more is always better. Volunteers should remember that lowering production costs (and thereby increasing profit margins) is as important a concern for many farmers as is increasing overall production. Improved breeds will not necessarily lower production costs.

When selecting new, exotic breeding stock, it is important to consider whether you want a lard, bacon, or meat-type hog. Lard hogs would be favored in countries that use animal fat for cooking or for eating. The back fat of hogs is a favorite dish in Central America. These animals tend to be smaller in size, thick, compactly built, and have very short legs. This type of hog is prone to small litter size. Bacon-type hogs are more common where the available feeds consist of dairy by-products, peas, wheat, barley, oats, rye, and root crops. They are not common in the tropics. Compared with corn, such feeds are not fattening. Thus, instead of producing a great amount of lard, they build sufficient muscle for desirable bacon. They are more common in Northern Europe. Meat-type hogs are intermediate between lard and bacon hogs. Meat hogs have a large body size and high-energy requirements.

PURCHASE OF BREEDING STOCK

Volunteers and farmers who seek to improve production through crossbreeding of exotics with local breeds are faced with the choice of whether to buy a sow or a boar. If you have limited money to work with, buy the boar. A boar can have many more offspring during a given season or a lifetime than a sow. He is, from a hereditary standpoint, a more important individual than any one sow as far as the whole herd is concerned. Therefore, because of their wider use, boars are usually culled more rigidly than sows, and the farmer can well afford to pay more for a good boar than for an equally good sow.

If you mate your boar with sows of other farms (or from the streets) in an attempt to improve the genetic stock, you may also be exposing your boar to diseases such as *brucellosis*. Be cautious!

REFERENCES

Pinkston, Herman, and Eugene Snyder (1989). "Understanding Swine Production." Volunteers in Technical Assistance, 1600 Wilson Boulevard, Suite 710, Arlington, Virginia 22209 USA. Tel.: 703-276-1800. Fax: 703-243-1865. vita@vita.org.http://idh.vita.org/pubs/docs/uhn.html.

RAISING RABBITS

Based on Better Farming Series 36, 37. Raising Rabbits 1,2,
Food and Agriculture Organization of the United Nations,
Rome, 1988, ISBN 92-5-102583-5 and ISBN 92-5-102584-3

Rabbit meat is good for you, and it tastes good. Rabbits are easy to raise. They are clean, quiet, and small. You can keep many rabbits in a very small place. You can easily find enough food to feed a few rabbits. You can even grow some of the food yourself. Since one rabbit makes a good meal, and since you can eat them one at a time when you need meat, you will not have to worry about rabbit meat going bad. Rabbits have babies often. Starting with only a few full-grown rabbits, you can raise many baby rabbits, which will give you a lot of meat.

To raise rabbits you will need the following.

- Healthy, full-grown rabbits
- The kinds of food that rabbits eat and plenty of fresh water
- A pen for each full-grown rabbit and pens for the young rabbits
- Time to take care of the rabbits

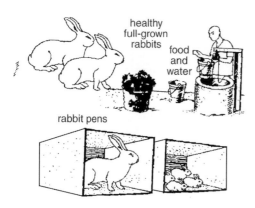

FIGURE 63
Rabbits, food, and water, and rabbit pens.

Rabbits are usually very healthy and hardy animals. If you give them a good home and take care of them well they will be less likely to get sick. It is much easier to keep rabbits from getting sick than it is to make them well after they have become sick. Rabbits must have a quiet place to live where they will feel safe and well protected. Rabbits are very quiet animals and are easily frightened. So you must give them a home away from people and noise and animals that may frighten or even harm them, such as foxes, rats, snakes, dogs, or cats. If you decide to raise rabbits, the same person should take care of them all of the time so the rabbits will become used to their keeper and will not be frightened (see Figure 63).

Rabbits may not grow well if they get too much sunshine, so they must have a shady place. Rabbits can get sick or die if it is too windy or if they get wet. Rabbits should *never* be wet. They must have a place that is protected from wind and rain, but they need fresh air, too. So they need a place to live that is closed enough to protect them from the sun, the wind, and the rain and yet open enough to allow for plenty of fresh air.

Rabbits should have clean, dry homes. You should clean their pens every day. Rabbit droppings should never be allowed to collect on the floor of the pen.

Rabbits should have clean fresh food and fresh water at all times. Rabbits should be given food twice each day, once early in the morning and once in the evening, after sunset, and water as often as they need it.

KINDS OF RABBITS

There are many kinds of rabbits, all with different sizes, shapes, and colors, and with different kinds of fur. In most places, you can find several kinds of rabbits. Usually, you can get local rabbits, which are strong and hardy and used to living in your area. They grow well and do not get sick easily. However, local rabbits are often small and give little meat.

Sometimes, you can get improved rabbits that have been brought from another place. Improved rabbits have been carefully mated over many years and are larger and give more meat. However, they are not as hardy as local rabbits. So if you get improved rabbits, you will have to take better care of them. Sometimes local rabbits are mated with improved rabbits. When the babies are born, they have some of the qualities of both kinds of rabbits. Three recommended improved rabbits are New Zealand rabbit (white or red), Californian rabbit, and Blue Vienna Rabbit.

WHAT RABBITS EAT

Rabbits eat almost any kind of plant that is not sour or spoiled. Much of the food that they eat is not used for anything else. Rabbits mostly eat green plant materials such as leaves or stems, or parts of green plants that you do not use. They also eat many kinds of weeds. If you have nothing else, you can feed your rabbits only green plants and weeds.

If you want rabbits to grow very well and fast, you will also have to feed them some richer foods. In many places, you can buy good, rich foods made especially for rabbits. These are called rabbit pellets and cost a lot of money. There are many other rich foods that you can give your rabbits that are cheaper or that you can grow yourself, such as wheat, barley, beans, maize, sorghum, millet, or soybeans.

Here is a list of foods that rabbits like to eat.

- Fresh plants
- Nearly all green plants and especially rich plants such as bean plants and alfalfa
- Many kinds of grass and weeds
- The outside leaves and the tops of vegetables
- Lettuce, endive, and chicory
- Tender banana, cane and bamboo leaves
- Cut-up pieces of the stalks of plants such as maize or banana
- Roots such as cassava, yams, carrots, beets, and turnips

Note: The leaves of the potato or the tomato plant may make your rabbits sick.

Rabbits also eat nearly all dried plants when they are green, including grass and weeds.

Rich foods for rabbits are barley, maize, wheat, oats, field beans, rye, buckwheat millet, sorghum, cottonseed, groundnut, coconut, linseed, and sesame cakes.

If you have some food that you think your rabbits might like, give them a little of it. If they eat it, you can feed them that. However, a sudden change of food may make your rabbits sick. So when you give them something new, give it to them little by little for a week or more until they become used to it.

HOW MANY RABBITS SHOULD YOU RAISE?

You can raise many rabbits if you have enough for them to eat, the materials to build enough pens, and the time to take care of them. However, until you know more about how rabbits live and grow and what you must do to take care of them well, it is best to start with only a few. A good number to start with is two full-grown female rabbits and one full-grown male rabbit. With two females, you will get as many as 40 to 50 baby rabbits each year. Later, if you find that you could eat or sell more rabbits if you had them, you can begin to raise more.

HOW TO BEGIN

The first thing that you will have to do is to build pens for your rabbits. To begin you will need one pen for each full-grown rabbit. You will also need one pen for the young rabbits of each female. If you decide to start with two females and one male rabbit, you will need five pens.

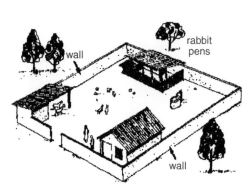

FIGURE 64
Rabbit pens inside the walls.

The best place for rabbit pens is near your home. That way you can watch your rabbits, take care of them easily, and protect them from animals. Try to choose a place that is quiet and peaceful, away from animals, people, and noise. The location should be protected from the wind, perhaps by the walls of your house, a fence, some trees, or a hill. If it rains a lot where you live, put your pens in a place that is sheltered from the rain. You can also build a shelter over the pens to protect your rabbits and keep them dry. If you have a wall around your house and farm buildings, put your rabbit pens inside the wall, as shown in Figure 64.

MATERIALS FOR BUILDING RABBIT PENS

Wire mesh is the best material for building rabbit pens. It lasts a long time and is easy to keep clean, but it costs a lot of money. You can also build good pens using materials that cost little money such as used wood, old boxes, crates, poles from straight tree branches, or bamboo. Build your pens on legs above the ground. That way, the rabbit droppings will fall to the ground and not collect on the floor. In addition, a pen on legs will help to keep the rabbits safe from other animals. Be sure to fit your pens with metal guards to keep rats and other small animals from getting to your rabbits. These metal guards are made of metal sheets, such as flattened tin cans cut into a circular shape and rolled into a cone with an open top.

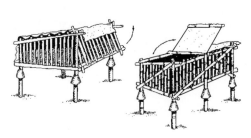

FIGURE 65
Rabbit pens.

A good size for a rabbit pen is 70 centimeters wide by 90 centimeters long and 50 to 60 centimeters high. This size of pen is good both for full-grown and for young rabbits, so you will need to build only one size pen. Make the inside of each pen as smooth as you can to keep your rabbits from hurting themselves. The pens can be improved in several ways. If there are small animals that might harm your rabbits, build a fence around your pens. If it rains a lot where you live, add a plastic sheet that can be rolled up and down. It will also help protect them from wind. There are many ways to build pens. Figure 65 shows some simple rabbit pens that are easy to build.

For a grain feeder you can use a hollow piece of bamboo, a dish, or other container. Attach the feeder to a piece of wood to keep the rabbits from spilling their food. You can also make a grain feeder using tin or metal (see Figure 66).

Build a nest box for each female rabbit using wood about 1.5 cm thick. A simple nest box is shown in Figure 67.

CHOOSING RABBITS

Once you have a pen it is time to buy your full-grown rabbits. You will have to learn how to choose healthy rabbits. It is best to buy your rabbits from a place where you can go to see them before you buy them. Here are some things you should look for when you are choosing your rabbits.

- Never buy a rabbit that moves slowly or looks dull and sleepy.
- Never buy a rabbit with a runny nose or with sores in its ears or feet.
- Look at its fur. The fur of a healthy rabbit is smooth and clean. Never buy a rabbit with fur that is rough or dirty or grows in patches.
- Look at its teeth. The front grinding teeth of a rabbit should be in line. Teeth that are out of line grow and grow until the rabbit cannot eat.
- Try to get rabbits from females that have five or six babies at a time.
- Try to get rabbits that grew well and were healthy when they were young and that weighed $1\frac{1}{2}$ to 2 kilograms when they were three to four months old.

Rabbits that come from big families and that grow well and weigh a lot when they are young are more likely to have big families and big, healthy baby rabbits. When you buy your rabbits, make sure that your female and your male rabbits are not siblings. The female rabbits that you choose should have eight teats so that they can feed eight babies.

With most kinds of rabbits, females are ready to mate when they are 4 to $4\frac{1}{2}$ months old. Male rabbits are ready a little later, when they are 5 to $5\frac{1}{2}$ months old. When you first begin, try to get rabbits that are old enough to mate so that you can begin to raise your own baby rabbits as soon as you can. When you have decided what rabbits to buy, find out what kinds of food they have been eating. This will help you to know what to feed them when you get them home. You may even be able to take home

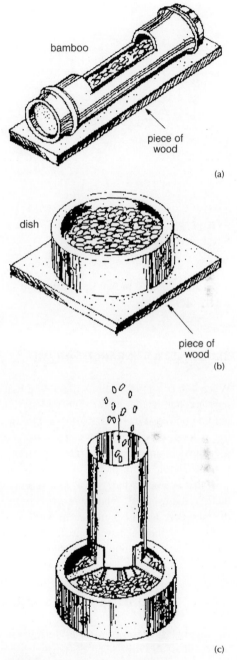

FIGURE 66

Three kinds of grain feeders. (a) Bamboo feeder, (b) Dish feeder, (c) Metal feeder.

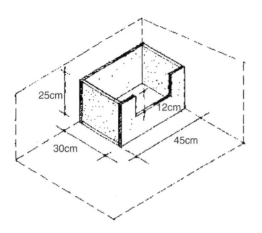

FIGURE 67
Simple nest box.

some of this food to feed them for the first few days.

HANDLING RABBITS

Rabbits must be handled with great care in order not to hurt them. When you pick up a rabbit, hold it firmly but gently. Make no sudden movements or you may frighten the rabbit, and it will begin to struggle or scratch you. Remember, never pick up a rabbit by the legs or the ears or you may hurt it. When a rabbit is full grown, the skin over the shoulders at the back of the neck is very loose. Gently grasp this loose skin with one hand to hold the rabbit and put the other hand under the rabbit to pick it up. When you lift the rabbit, the hand under the animal should carry nearly all of the weight.

BRINGING YOUR RABBITS HOME

When you bring your rabbits home you will have to be very careful not to frighten them or they may get sick or die. Move them quietly and gently. Do not give any food to rabbits the night before you move them. It is not good to move a rabbit when its stomach is full. However, if the trip is long, they should be given water from time to time. It is best to move each rabbit in a separate container such as a crate or a basket with a lid that can be closed. However, the containers you use should let in plenty of fresh air during the trip.

If the weather is hot, move your rabbits early in the morning. If the weather is cooler, move them in the evening. When you reach home, put the containers down gently near the pens. Let the rabbits rest and become calm while you prepare their pens.

Put fresh food and water in the pens. This will help your rabbits to feel at home and settle down more quickly. Stay away from them as much as you can until they become used to the pen.

WHEN TO FEED YOUR RABBITS AND HOW MUCH TO FEED THEM

Rabbits eat at night as well as during the day. So you must be sure that they have enough to eat all of the time. The best time to feed rabbits is once early in the morning and once in the evening, before it is dark. Each time you feed your rabbits, put fresh or dry plant materials in the plant feeder, and put some of the rich food that you are using, such as crushed maize, in the grain feeder. If you are feeding your rabbits any larger foods, such as sections of banana stalk or whole carrots, beets, or yams, you can put them on the floor of the pen.

Rabbits need different amounts of food at different times in their lives. Female rabbits, when they are not having babies, and male rabbits, when they are not mating, need much less food.

They must be given enough food to keep them strong and healthy but not so much that they get fat. Female rabbits that are fat do not have babies easily. Male rabbits that are fat are lazy and do not want to mate. Female rabbits that are going to have babies need more food, and after their babies are born they need a lot more food. Once you know that a female is going to have babies, give her as much food as she can eat. After the babies are born, she will have to be able to produce a lot of rich milk for them.

You will have to learn by your own experience just how much to give them. If you see that your rabbits do not eat all of their food, give them a little less. If you see that your rabbits eat all of their food, give them a little more (unless they are fat).

Here are some other things that you should know about feeding your rabbits.

- The food you feed your rabbits must be very clean or it may make them sick. Never gather green food from places made dirty by other animals.
- Never give your rabbits food that is moldy or dusty. Even food that was once clean, such as dry hay, grass, or grain, can become moldy or dusty after only a short time.
- Do not to feed your rabbits food that is sour or spoiled.
- Never give your rabbits fresh green food that has been standing in piles. Fresh green food left standing becomes spoiled very quickly. If you are going to keep green food even for a very few hours after it has been cut, spread it out and turn it over from time to time so that it will not become warm, sour, or spoiled.

When you give your rabbits new food, uneaten food from the last feeding may be left if the food is clean and the feeders and containers are clean. But if the old food or the feeders are dirty, take away the old food and clean the feeders before you put in new food. In addition, if the water or the water containers become dirty, empty out the water, clean the containers, and put in new water.

PENS

It is very important to keep your rabbit pens clean. Your rabbits may get sick if their pens become dirty. Sometimes the droppings may not fall out of the pen. So watch carefully to see that no rabbit droppings collect on the floor or in the corners of your pen. If they do, brush them out. Clean the pens of your full-grown rabbits with soap and water at least twice each year and more often if they become dirty. When your baby rabbits are big enough and no longer need their nest boxes, clean the boxes with soap and water and put them away for the next time. When all of your young rabbits have been eaten or sold, clean their pens with soap and water before you put in more young rabbits.

Note: If a rabbit begins to ruin its pen by chewing the wood of the walls or the floor, give it a piece of wood to chew.

TAKING CARE OF YOUR RABBITS

Watch your rabbits carefully to see that they are not sick. A rabbit may be sick or getting sick if it does the following.

- It does not eat its food.
- It loses weight.

- It dirties the fur around its tail.
- It sits in strange positions or cannot move about easily.
- It has rough, dry fur.

If one of your rabbits becomes sick, take it out of its pen, wash the pen with soap and water, and when the pen is dry, put the rabbit back. If a rabbit becomes very sick, take it out of its pen to keep the sickness from spreading to your rabbits. This is especially important when there are many rabbits in the same pen. If a sick rabbit dies, burn it at once to keep the sickness from spreading.
   Here are some other things that you should watch for and what to do about it.

- Sometimes rabbits have sore ears. This may be caused by very small mites under the skin inside the ear. If this happens to any of your rabbits, wash out their ears using a clean cloth and vegetable oil.
- Sometimes rabbits have sore or runny eyes. This may be caused by flies, or they may have scratched their eyes. If this happens to any of your rabbits, wash out their eyes using a clean cloth and clean water.
- If any of your rabbits sneeze or rub their nose or have a runny nose, they may have a cold. Make sure that they are dry and protected from wind and rain.
- Make sure that their food is clean and free from dust.
- If any of your rabbits have sore or bleeding feet, it may be caused by a rough place or a wet, dirty floor in their pen. First, smooth out rough places and clean and dry the pen. Wash their feet in warm, soapy water and rinse and dry them well. Then rub the bottoms of their feet with vegetable oil.

Rabbits that live in pens often grow very long claws. If their claws become too long, they may get caught in the pen floor, and the rabbits can injure their feet. If this happens, trim the claws carefully. However, avoid the red center of the claw (the vein).

RAISING YOUR OWN BABY RABBITS

Mating

Remember that female rabbits are ready to be mated when they are 4 to $4\frac{1}{2}$ months old and male rabbits when they are 5 to $5\frac{1}{2}$ months old. So once your rabbits are old enough (and if they are not sick), you can begin to mate them for the first time. The best time for mating is early in the morning or in the evening when it is cool.
   When you mate your rabbits, always put the female into the pen of the male. Usually they will mate quickly. If the male mounts the female and in a short time falls off to one side, mating has taken place. As soon as this has happened, put the female back in her pen and write down the date of mating so you will know when to expect the baby rabbits to be born. If the rabbits do not mate after about five minutes, put the female back in her cage and try again the next day.

Baby Rabbits

Baby rabbits are usually born about one month after mating—possibly a few days earlier or later. About five or six days before you expect the babies to be born, put one of the nest boxes that you have built in the pen with the female. Cover the bottom of the box with sawdust or wood chips. This will help to keep the box dry after the baby rabbits are born. Put a little dry grass or hay in the box. The female rabbit will then make a nest in the box to protect the baby rabbits, using some of her fur mixed with the dry grass or hay.

The nest box is very important because baby rabbits are weak and helpless when they are born. Baby rabbits have no fur, they cannot see, and they cannot walk. Soon after the female rabbit has finished the nest, you can expect the baby rabbits to be born. Stay away from the pen as much as you can during this time, and do not bother the female. This is especially true while she is having the babies. After the baby rabbits are born, look at them carefully to see that they are all well.

Check the babies for the following.

- If they are lying close together or far apart
- If they are warm and well protected in the nest
- If they are alive and well
- How many baby rabbits there are

Never touch baby rabbits unless absolutely necessary. If you must, make sure your hands are clean.

All of the baby rabbits should be lying close together. If they are lying far apart, carefully move them together. A female rabbit gives milk to her babies only once a day. If the babies are not close together, she may not feed them all. All the baby rabbits should be warm and well covered in their nest. If they are not, put the fur and hay in the nest all around them. If any of the baby rabbits are dead or deformed, take them away and destroy them.

If there are too many babies for the female to feed, take some of them away. A female rabbit with eight teats can feed only eight babies. If you have another female rabbit with too few babies, you can give some to her to feed. However, the baby rabbits that you give to another female should be no more than two days younger or two days older than her own babies. Gently rub the baby rabbits you are going to move with some of the grass or hay of the new nest. This will give them the same smell as the new nest so that the new female rabbit will be more likely to accept the new babies as her own.

Sometimes female rabbits will kill their babies, and sometimes they will not feed them. This can happen if a female rabbit does not have enough milk. It can also happen because she does not know how to take care of them. If she kills her babies or will not feed them the next time she is mated, do not use her for mating again. Replace her with a new female.

At first, your new little rabbits will sleep most of the time and move very little. They will take milk once a day. When they are about two weeks old their fur will begin to grow. They will begin to see, and they will start to move about. At about three weeks, the little rabbits will come out of their nest box, and they will begin to eat food in addition to the milk they drink. From this point on, the little rabbits will eat more and more food. When that happens make sure that there is enough food for all the rabbits to eat. By the time they are six weeks old, the little rabbits will no longer take milk and they will eat all of the foods that full-grown rabbits eat. When the rabbits are eight weeks old, it is safe to move them. Then you can take them away from the female rabbit and put them in their own pen.

Raising Young Rabbits

You can put as many as six to eight young rabbits in a pen. It is best to keep all the young rabbits from the same female rabbit together in the same pen. Once they are in their own pen, you can begin to fatten them to eat or to sell. After four months of age, rabbits begin to eat a lot more food. You should try to eat or sell all your rabbits by this time. If you keep rabbits longer than this, the male rabbits may begin to fight.

THE DIFFERENCE BETWEEN FEMALE AND MALE RABBITS

It is not too hard to tell the difference between female and male rabbits after they are eight weeks old. (The easiest way to learn is to ask someone who already knows how to help you.) Hold the rabbit in your arms or put it on its back on a table. You will see two openings just behind the tail. The opening nearer to the tail is where the rabbit droppings come out. This opening looks much the same in all rabbits. However, the second opening of a female looks quite different from the second opening of a male. Push down gently with your thumbs on each side of the openings. You will see that they are red and moist inside. Look carefully at the second opening. If you see a slit, the rabbit is a female. If you see a circle, the rabbit is a male.

MATING FEMALE RABBITS

As you have already learned, rabbits can give birth frequently. Female rabbits can have babies six or seven times each year. However, a female rabbit must be strong and healthy and be fed good food to produce healthy babies that many times. When you first begin to raise your own rabbits, you should mate your females only four or five times each year. After you have been raising rabbits for some time and you see that your female rabbits are strong and healthy, you may be able to mate them more often. However, if you do mate your female rabbits more often, make sure that you feed them enough so that they do not lose weight.

Note: Remember, if you mate all of your female rabbits at about the same time, all of your baby rabbits will be born at about the same time. That way you can easily move them from one female to another if you need to. Babies will be born four times a year if you mate your female rabbits when their babies are eight weeks old.

REPLACING FULL-GROWN FEMALE RABBITS

If any of your full-grown female rabbits become sick or do not have healthy baby rabbits, do not mate them. Replace them with new full-grown females. As long as your full-grown female rabbits are healthy and have healthy baby rabbits, you can continue to mate them until they are three years old. However, after three years you should replace them with other full-grown females.

REPLACING FULL-GROWN MALE RABBITS

Male rabbits can be used for mating until they are about three years old and even longer if they are healthy and the females with which they mate continue to have healthy babies. However, you should not mate rabbits from the same family or your baby rabbits may not be strong and healthy. So after you begin to use your own females for mating, you should replace your male rabbit about once every year. That way you will be sure not to mate a father to a daughter. You can either buy a new male or exchange your old male rabbit for a new male rabbit. However, be very careful that your new male rabbit does not come from the same family as your old male rabbit.

Note: Remember to choose rabbits that come from big families of five to six babies that weighed at least $1\frac{1}{2}$ to 2 kilograms when they were three to four months old. In addition, remember that new females should have eight teats.

EATING OR SELLING RABBITS

Do not give your rabbits any food to eat the night before you are going to eat them or take them to the market. It is not good to kill a rabbit or move a rabbit from place to place when its stomach is full. However, during this time be sure that it has water to drink. Remember, you can kill rabbits one at a time when you need meat. If you are going to sell rabbits at the market, you should move them when it is cool. If it is too hot, they may die before you get to the market.

HOW TO KILL A RABBIT

1. You can kill a rabbit quickly and easily by hitting it on the back of the neck.
2. You can also kill a rabbit by holding its back feet and pulling its head down and out to break its neck.
3. When the rabbit is dead, tie it up by the back feet, cut off its head and front feet, and let the blood drip out.

HOW TO CLEAN A RABBIT

1. Slit the skin around both back feet and make a cut from one leg to the other.
2. Pull off the skin from both back legs and cut off the tail.
3. Continue to pull the skin until it is completely off.
4. Cut the rabbit up the middle and take out everything except the kidneys, liver, and heart (which are good to eat).

Cut the rabbit into seven pieces: two back legs, two rump pieces, one ribs piece, and two shoulders and front legs.

REFERENCES

College of Agricultural Sciences. (1994). "Rabbit Production." Pennsylvania State University, University Park, PA 16802-2801. Tel.: 814-863-0471.
Price, Martin L., and Fremont Regnier. (1982). "Rabbit Production in the Tropics." Echo Technical Note, ECHO, 17391 Durrance Rd., North Ft. Myers, FL 33917. Tel.: 914-543-3246. echo@echonet.org. http://www.echonet.org.

RAISING SHEEP AND GOATS

Based on Better Farming Series 12. Raising Sheep and Goats,
A. J. Henderson: Food and Agriculture Organization
of the United Nations, Rome, 1977,
ISBN 92-5-100152-9

TRADITIONAL SHEEP AND GOAT BREEDING

The traditional way of breeding sheep and goats does not take much work, but it also does not produce much. It takes little work because the animals are not looked after. They are not fed, they are not given water to drink, and they are not given any shelter. This method of breeding, however, produces little. The animals are small, they are often ill, and their young ones often die. If a farmer breeds sheep and goats in the modern way, he or she can earn more money.

A few terms should be defined: A *flock* of sheep consists of 1 male, called the *ram*; 20 females, called *ewes*; and young ones called *lambs*.

A flock or herd of goats consists of 1 male, called the *buck* or *billy goat*; 30 females, called *she-goats*; and the young ones, called *kids*.

BREEDS OF SHEEP AND GOATS

Sheep

In Africa there are many well-adapted breeds of sheep. Among them are a breed of wool sheep, the Macina (Mali). These are rather big sheep weighing 30 to 40 kilograms. It is a breed that produces little meat and milk. These sheep are raised for their coarse wool, which is made into blankets. Other sheep include the Nar variety—a big sheep that also produces little meat. They are used for crossing with Astrakhan sheep to produce fur. The Touabir variety are very big black and white sheep that produce a lot of meat. Peulh sheep are the most common breed in West Africa. Djallonquee variety are small black and white sheep that weigh 20 to 25 kilograms. They have short legs and are good meat producers. Toronquee variety are bigger sheep that weigh 40 to 50 kilograms and are white and brown in color. They produce a lot of meat, and the ewes give a lot of milk. Targui sheep have long legs, and can walk a long time in search of grass. The big Targui sheep have mottled white hair. The small Targui sheep have longer hair, brownish grey in color. The Targui are good mutton sheep because they produce a great deal of meat. Attempts are now being made to cross them with European breeds, but foreign sheep adapt very badly.

Goats

There are many breeds well adapted to the climate of their regions. The chief breeds are Northern goats of the savanna country and Southern goats of the wetter, forested regions in the south. Figure 68 shows these goats. Northern or Savanna goats are very big goats weighing between 25 and 30 kilograms. The buck has very big horns. The she-goats are white with black spots. They can produce two kids in a litter and produce a lot of meat and milk. These are small animals weighing 18 to 20 kilograms. The body is short and fat, and they are brown in color, with the tip of the tail and legs black or white. They are bred for their meat and are resistant to sleeping sickness in the very wet regions.

DIGESTIVE SYSTEM

When you look into the mouth of a sheep or goat, you see two jaws and a tongue. Toward the back of the mouth you can see large teeth with which the animal chews grass. These are called molars. The upper jaw has no front teeth, but the lower jaw has eight front teeth. The older the animal is, the more these teeth are worn. You can tell the age of a sheep or goat by looking at its front teeth.

When a sheep or a goat eats, it grips the grass between the upper jaw and the teeth of the lower jaw. It jerks its head to pull off the grass. It does not chew the grass but swallows it at once.

Northern goat   Forest goat

FIGURE 68
Southern or forest goats.

The grass goes into the first stomach (or rumen). When sheep and goats have filled the first stomach, they often lie down. But they go on moving their jaws—ruminating.

They then bring up a little grass from the first stomach into the mouth. They chew the grass for a long time with their molars. When the grass is well chewed, they swallow it again. This time the grass does not go into the first stomach but into the other parts of the stomach. Sheep and goats can ruminate well when they are quiet and lying down.

FEEDING SHEEP AND GOATS

If an animal does not get enough food, it does not put on weight. In the dry season there is often not enough food, and animals lose weight. They must be given rich food. Ruminants eat grass, from which they get what is needed to build their bodies. In addition, they can be given very rich foods, called feed supplements. A sheep or a goat raised for meat should be grown quickly. A pregnant ewe or a she-goat needs good food to nourish the young in her womb, who will later drink her milk.

In order to give animals enough food all year round, the flock is moved from place to place. In the dry season sheep and goats can feed more easily than cattle. They make better use of the grass because the sheep cut the grass closer to the ground, and the goats pull up the grass. Sheep can be fed on pasture where cattle have already fed because sheep eat short grass, but they leave nothing behind. Sheep or goats must not be allowed to feed in very wet places because they catch diseases of the feet and body.

A good shepherd knows how to move the animals (a good dog can help). Then the flock is well fed, it does not catch diseases, and the little ones seldom die. During the rainy season it is easy to feed animals well. Grass grows quickly, there is a lot of it, and it is tender and nourishing. During the dry season, animals are badly fed. The grass is hard and scarce, the stems are tall, and the leaves are dry. The animals won't eat this grass, so they get thin and sometimes die. During the dry season it is necessary to give the animals a feed supplement.

BALANCED RATIONS FOR ANIMALS

Rations for Lambs Five Months and Over and for Breeding Males

In the rainy season an animal eats about 2.5 kilograms of grass a day. In the dry season, give them the following in this order.

First ration: 1 kg of hay and 500 grams of silage

Second ration: 1 kg of hay and 100 grams of cooked cassava

Third ration: 1 kg of silage and 200 grams of rice bran

Fourth ration: 1 kg of hay and 100 grams of rice bran

Fifth ration: 1.5 kg of silage and 150 grams of cooked cassava

If you want to fatten an animal for sale or for eating, add 350 grams of oil cake cottonseed, copra, or oil palm kernel. Oil cake is costly, but it makes animals put on weight and fatten quickly.

Do not give the same rations to females and their young ones: Their needs are different. Instead, in the rainy season, give the following rations.

Pregnant ewe or she-goat weighing 30 kg
- 2 kg of grass
- 100 g of rice bran
- 300 g of oil cake

Ewe or she-goat suckling young of 0 to 4 weeks
- 2 kg of grass
- 400 g of cooked cassava
- 400 g of rice bran
- 600 g of oil cake

Ewe or she-goat suckling young of 5 to 10 weeks
- 2 kg of grass
- 200 g of cooked cassava
- 400 g of rice bran
- 600 g of oil cake

Ewe or she-goat suckling two young ones of 0 to 4 weeks
- 2 kg of grass
- 900 g of cooked cassava
- 500 g of rice bran
- 600 g of oil cake

Ewe or she-goat suckling two young ones of 5 to 10 weeks
- 2 kg of grass
- 700 g of cooked cassava
- 500 g of rice bran
- 600 g of oil cake

Food Supplements and Mineral Salts

When food is scarce, when the grass is tough, or when the females are pregnant or giving milk, they must be given a feed supplement. A farmer can, for instance, buy meal for sheep and goats, sold commercially but expensive. A farmer must also give mineral salts, such as a licking stone.

One kilogram contains 400 g of salt, 150 g of calcium, 80 g of phosphorus, as well as other mineral salts. You can also give native soda. Put the salt in the water, hay, and silage. Mineral salts are needed to form the animals' bones.

WATER

Sheep and goats get thin during the dry season because they are not well fed but also because they do not drink enough. A sheep can drink 5 to 6 liters of water a day. If ruminants do not drink enough, they cannot digest grass. Animals can drink in their shelter, from a hollowed-out tree trunk, a barrel cut in half, a concrete trough, or at streams or rivers. Make sure that the water is clean and clear; there must be no mud in it. Sheep and goats easily catch diseases from water. It is important not to let the sheep and goats go into the water. They can catch diseases from it.

CARE AND HOUSING OF ANIMALS

To be well fed, animals must be watched over. If animals are left alone, they do not make good use of the grass. They eat the good grasses and leave the poor ones. The good grasses are always eaten before they make seeds, and so they cannot multiply. On the other hand, the weeds are not eaten, they grow and make plenty of seed, and multiply. Unwatched animals go into plantations and destroy the harvest. Farmers have to make their fields a long way from the village, so farmers lose a lot of time going to work. The animals may hurt themselves or get diseases. They go to the streams and catch parasites or diseases.

Animals need a paddock because it protects the animals from wild beasts and thieves: from wind, sun, and rain; and from diseases. In the village animals are kept in a traditional enclosure. There are often too many animals and the ground is dirty and wet, so the animals catch diseases. They cannot lie down to ruminate, and they make poor use of their food. Their wounds heal badly, especially those of the feet. Diseases increase. The little ones are often ill. You cannot make good manure, though the dung of sheep and goats is good for manure. In a traditional enclosure there is only a mixture of earth and droppings.

Pastures should be fenced, or the animals can get into plantations and destroy them. A field 100 meters on each side is needed to feed about eight adult animals. Do not leave the animals too long in the field, or the grass will not grow again. Divide the field into seven parts. Every five days, or when you see that the grass has been well eaten, move the flock to another part. When the last part (the seventh) is finished, go back to the first part where the grass has grown up again. The animals manure the soil of the field with their droppings. You can make fences by planting little trees very close together, by planting two rows of sisal, or by planting thorns. Leave a gate 2 meters wide. Before dividing the field into seven parts, pay attention to where the trees are in the field so they can serve as shelter the animals from rain and sun. Each part should have a tree and should not be far from the path.

To keep the animals in one part, you need movable fences made of wood. Make them 2 to 3 meters long and 80 centimeters high. Move the fences when you put the animals in another place. Making the fences requires money and especially work. It is useless to do a lot of work and spend the money if you do not also improve the animals' food, housing, and care.

To give the animals good housing, make a building next to the paddock. (A modern building of concrete and sheet iron is too expensive.) You can improve the housing of sheep and goats without spending too much money by using local materials. Choose a dry place on a little rise. If you build the shed in a low-lying place, rainwater and urine cannot flow away. Put a layer of

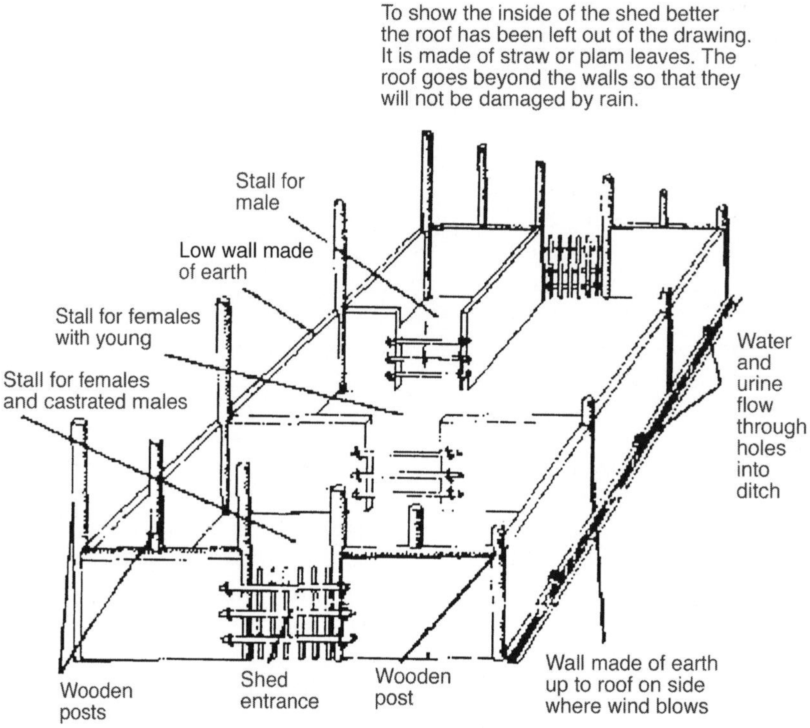

To show the inside of the shed better
the roof has been left out of the drawing.
It is made of straw or plam leaves. The
roof goes beyond the walls so that they
will not be damaged by rain.

Stall for
male

Low wall made
of earth

Stall for females
with young

Stall for females
and castrated males

Water
and
urine
flow
through
holes
into
ditch

Wooden
posts

Shed
entrance

Wooden
post

Wall made of earth
up to roof on side
where wind blows

FIGURE 69
A sheep shed.

concrete (cement and gravel) on the ground. Build the shed where the wind will take the smell away from the house. To protect the animals from the wind, build a wall of earth up to the roof on the side where the wind usually blows. To protect the animals from sun and rain, make a roof of straw or palm leaves. Put a gutter on the lower side of the roof to drain rainwater away.

When the shed is finished, make three stalls inside. One small stall is to isolate the male or males from the herd so they can't mate with the ewes. Two other large stalls are for females with young and for females without young and castrated males (see Figure 69).

Put straw on the ground, which, when mixed with droppings and urine, will rot and make manure. When the straw is partly rotted, put clean, dry straw on top of it. See that the animals are always on clean straw. When there is a lot of manure in the shed, take it out. You can take it out to the field and plow it into the ground at once, or you can make a heap by the side of the shed and use it as you need it. Sheep or goat dung makes good manure, adding much organic matter and mineral salts to the fields.

The animals must not be crowded in the shed. If they are crowded, they do not have enough room to lie down, they ruminate badly, they hurt themselves, and they get ill. Two adult animals need a space of 1.5 square meters. For example, put six adult animals in a shed 3 meters by 3 meters. The doors of the shed must be wide—2 meters wide so the animals will not be crowded when going through and will not get hurt. Disinfect the shed every two weeks with water and potassium chloride or water and creosote. Alongside the shed make paddocks where the animals can walk around. Make a small paddock for the males next to their door and a big paddock for the females and their young. The young ones are left with the rest of the flock when they are between one and two weeks old (see Figure 70).

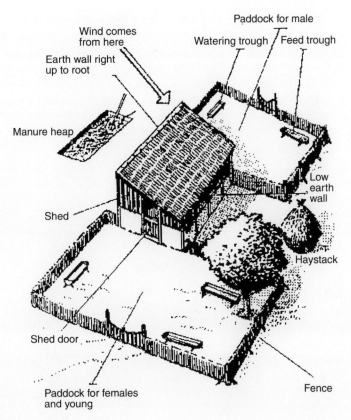

Wind comes from here

Paddock for male

Watering trough / Feed trough

Earth wall right up to root

Manure heap

Low earth wall

Shed

Haystack

Shed door

Paddock for females and young

Fence

FIGURE 70
Shed and paddocks.

In the paddocks put (1) feed troughs to give the animals their feed supplement and (2) watering troughs from which they can drink. Feed troughs and watering troughs are made of hollowed tree trunks or drums cut in half.

DISEASES OF SHEEP AND GOATS

Vaccinate animals *before* they are ill. Vaccination protects an animal from disease. Vaccination tires animals a little, but it is not dangerous if they are well fed and well housed.

Parasites are the worst problem in sheep and goat breeding. Protect animals from parasites. Give them clean water to drink. Do not leave the flock near streams. Do not keep more than 20 animals together. Parasites and contagious diseases multiply in big flocks. Take special care with the young ones and mothers; they are the most delicate.

PARASITES

- Mange or scab: The animal scratches, its hair comes out, and scabs form. Wash the animals with warm water and soap. Soak a piece of cloth in mineral oil and rub the animal. Repeat every day.

- Ticks: They stick to the animal's skin and suck the blood. Wash the animal with water and a pesticide such as toxophene. Rub the animals regularly every week.
- Parasites in the body: Usually they live in the digestive system, in the lungs, or in the nerves. The parasite eggs are left by flies in the pasture. These eggs develop in the grass and are eaten with the grass by the animals. Then they develop in the animal's body. Parasites living in the lungs, such as lungworms, are controlled by the use of aerosols or with phenothiazine. Parasites living in the digestive system, such as strongyles, are controlled with phenothiazine *before* the animal is ill. Ask the animal husbandry service for advice on treating liver rot (liver fluke infestation), coccidiosis, and tapeworm.
- Parasites living in the brain cause gid (or sturdy). Animals stagger as if drunk. They must be slaughtered before they die.

When parasites have got into a pasture, do not take animals there for a long time. The parasite eggs hatch out, but since there are no animals, the parasites cannot attack them and have nothing to eat, so they die. Animals should not feed in wet pastures because that is where parasites live.

INFECTIOUS DISEASES

- Enterotoxaemia: This is a serious disease of lambs. They should be vaccinated to protect them against it.
- Anthrax: This disease infects all animals and people. Vaccination beforehand is needed; otherwise all the animals die. Animals that die of this disease must be burned. The blood of the dead animal is black. People must not eat the meat of animals with anthrax because they can catch the disease and die.
- Foot rot: The horny parts of the foot are destroyed, and the animals limp. To cure it, dig a little ditch at the door of the shed and fill the ditch with water and antiseptic. Then make the animals walk through the water. Their feet will heal.
- Foot-and-mouth disease and sheep-pox: The animals will not catch these two diseases if they are vaccinated *beforehand*.

PREGNANCY AND BIRTH

Pregnancy takes about five months. The newborn animals are called a litter. If the she-goat or ewe has had one litter in the year, she should not have another before the following year. The she-goat or ewe cannot both feed the young one(s) in her womb and suckle those already born.

Usually the birth takes place without difficulty. There is nothing for the farmer to do except in the case of a female that is giving birth for the first time. In this case one can help her by pulling downward on the legs of the young one. Females that are about to give birth stay in one corner and the udders swell and harden. At birth, part of the membranes that cover the young one(s) comes out, and the water in these membranes should flow out. Next you see the legs of the young animal coming out, either the two forelegs or the two hind legs.

After the young animal has come out, if it is still joined to the mother by the umbilical cord, cut the cord and tie knots at both ends and clean it well. After the birth, the rest of the membranes come out. They must all come out. Otherwise they may rot inside the womb and cause the mother to die.

When the young animals are born the mother rubs them with her tongue; she licks them. You must let her do this. At this time the mother is often thirsty, so give her water to drink.

LOOKING AFTER THE YOUNG ONES

Take great care of the newborn animals. They are very delicate. They easily catch diseases and parasites. To protect them, have them vaccinated.

After birth the mother suckles her offspring for about four months. But from the third week the lamb or kid can take other food besides the mother's milk. At six months the lambs or kids no longer suck and they are said to be weaned. If the mother still has milk, she is milked. After one or two months have her served by a male.

The mother may refuse her young. This often happens when the mother gives birth for the first time, or when she has two young ones. In that case put the mother and young ones together in a stall to get them used to each other. If the mother is dead, suckle the young one with the milk of another female, or give pure cow's milk in a feeding bottle. Give five to seven feeds a day in small amounts.

If there are 20 females in a flock there should be 20 litters a year. If there are only 10 litters a year, the flock is not producing enough. A female who produces no offspring during the year should be sold or eaten. If a female has no young five months after being served by the male, she is sterile. But a female can be sterile because she is badly fed; because she is ill; because she is too old. Give her plenty to eat for four or five months to fatten; then you can sell her for a good price.

The females must be served five months before the beginning of the rainy season when there will be plenty of grass at the time of birth. A male should not serve his daughter; the offspring will be malformed.

A flock of 30 to 40 ewes or she-goats needs two rams or bucks. The other males of the sheep flock are castrated. In a goat flock, the bucks are sold or eaten while they are young and before they can breed so there is no need to castrate them.

To castrate a male, either the testicles are removed, or better, the ducts joining the testicles to the penis are crushed. The animal husbandry service has special instruments for this, and the livestock assistants are experienced. If the males are not properly castrated their wounds will not heal up well and they may die. The males who are not wanted for breeding should be castrated at the age of two months. Castrate old rams and old bucks, because then they fatten more quickly, the meat has less taste, and you can sell them better and faster.

CHOOSING BREEDING ANIMALS

When selecting a male choose the son of a good female and a good, wellbuilt male, with a mother that gives a lot of milk. The young male should be well built and in good health. If you can buy a young male at a breeding station he will improve the flock. This costs less money than buying several females from the breeding station. The male should be lively and strong, and should be well fed, especially for two months before service. The male should have a flat back with broad loins. Rams do not serve before the age of 15 months. Otherwise they remain small and do not give the ewes good litters. Bucks should not serve before the age of 18 months.

When selecting a female, choose daughters of good, well-built milk types and of fine males. They should be well developed and in good health. Their hind legs should be well spread but straight, with broad loins. Their bellies should be well developed and muscular. Ewes who have two lambs in the first litter almost always have two in other litters. The ewe lambs should not be fertilized before the age of 18 months. Otherwise they will remain small and will produce small lambs and will have little milk to feed their young ones. The female kids should not be fertilized before the age of one year.

FLOCK MAINTENANCE

Modern farmers put a mark on each animal in the flock. Give each animal a number, which serves as its name. Mark the number on the back of the animal by cutting the wool with a pair of shears. For example, mark A on the male, on females mark 1, 2, 3, and so on. On the young animals mark A1 if, for example, the sire is A and the dam is 1. You can tell from which litter of dam 1 the offspring comes by marking a second number. For example, if A1 is from the third litter, it is marked A13, and so on.

Keep a herd book, with a double page for each animal. Write on it all you need to know about each animal in the flock. List in the herd book deaths among the stock. This will make it easy to see if a dam has many deaths among her offspring. If the dam has deaths in each litter, she should be fattened and sold. Have the veterinary assistant writes in the herd book what was done and what the farmer should do. The farmer should write down approximately how much food is given per day (not counting grass) so the amount of food needed in a year can be determined. Notes on each animal should be made to determine which are the best animals.

SELLING AND BUYING

Choose the animals to be sold or bought. All you have to do is put your hand over the animal's loins and pinch its ribs. If you can feel the bones, the animal has not got much meat. If you do not feel the bones, the animal is good for selling or buying. Also, look at the age. Ewes are fattened for sale or eating when they are five or six years old. After that age they cannot be fattened any more, they produce meat of poor quality, and they do not fetch a good price. She-goats are fattened for sale or eating after five or six births—that is, at the age of six or seven years. After the age of four years the males are weaker and produce poor offspring. The males should be castrated at the age of four years and fattened for sale or eating.

Animals are often sold alive at the time of important festivals. So you must always have fine animals for the chief festivals. Plan to fatten animals so they will be ready for sale at a time when the price is high. You might also want to sell some animals at the beginning of the dry season if you have not stored enough fodder.

GOAT MILK

In much of the Third World goats are not used for milking purposes, though it is not clear why. Goats require little care and use a small area—much smaller than a cow pasture. A good goat will give 2 liters a day, enough for a family. Goat milk is similar to cow milk in its basic composition.

Goat milk may have a peculiar taste. This can be caused by the presence of a buck at milking time or a low-grade udder infection (subclinical mastitis). Another reason is the goat's diet—left to its own devices, a goat will eat many different things, which will affect the taste of the milk.

MILKING

Noncommercial herds use mostly hand milking, which requires few facilities and little equipment. Routine, once-daily milking is not recommended. The doe's udder produces milk throughout the day and night, but production is slowed as milk accumulates. During the height of lactation, heavy producers can be milked three times a day at eight-hour intervals to relieve pressure in the udder.

Milking equipment should include a strip cup, a seamless milking pail, and milk strainer with a filter that is thrown away after each milking. Goats should be milked in an environment free of dust, odors, dogs, and disturbing noises. To produce clean milk it is necessary to have clean equipment, a clean area for milking, healthy goats, clean clothes, and clean hands. The udder can be washed with a clean cloth, but both the udder and hands should be dried before milking.

The first stream or two of milk should be directed through a fine wire mesh, such as a tea strainer, into a separate strip cup, so the presence of flaky milk, which is often an indication of mastitis, can be detected. Dairy goats should be milked dry at each milking.

As soon as the milk has been collected from the doe, it should be poured through a single-use filter. The milk should be cooled promptly and rapidly (to as near 0°C as possible) to ensure good flavor and retard the growth of bacteria. After cooling, the container of milk should be taken promptly to the consumer, stored in a refrigerator, or immersed in ice water. Unnecessary temperature changes can cause a bad flavor.

All milking equipment should be rinsed in warm water immediately after use and then washed in hot water to which a mild chlorine solution and detergent are added. Finally, the utensils should be rinsed in clean—preferably boiling—water and kept in a dust-free place to dry.

REFERENCES

Attfield, Harlan H. D. and George F. W. Haenlin. (1990). "Understanding Dairy Goat Production." VOLUNTEERS IN TECHNICAL ASSISTANCE, 1600 Wilson Boulevard, Suite 710, Arlington, VA 22209 USA. Tel.: 703-276-1800. Fax: 703-243-1865. vita@vita.org. http://idh.vita.org/pubs/docs/uhn.html.

Ingham, Claudia S. "Understanding Sheep Production." VOLUNTEERS IN TECHNICAL ASSISTANCE, 1600 Wilson Boulevard, Suite 710, Arlington, VA 22209 USA. Tel.: 703-276-1800. Fax: 703-243-1865. vita@vita.org. http://idh.vita.org/pubs/docs/uhn.html.

NUTRITION

Nancie H. Herbold

HEALTHY EATING

Following the Food Guide Pyramid (see Figure 71) is an easy way to assist in planning a diet that contains the essential nutrients for good health. The pyramid allows you to select foods from various food groups while not overconsuming calories, fat, and sugar. The pyramid contains a range of serving recommendations. The lower serving number would be appropriate for sedentary women and some older adults. The moderate range of servings is recommended for children, adolescent girls, active women, and sedentary men. The highest number of servings is recommended for adolescent boys, active men, and very active women (see Tables 15 and 16).

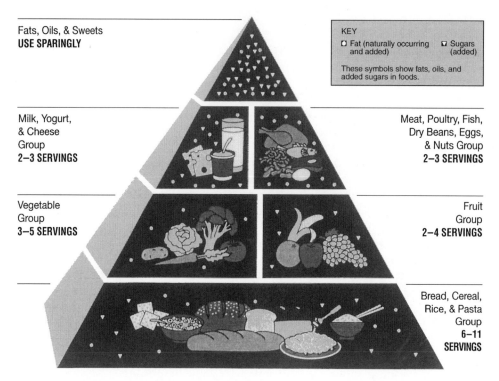

FIGURE 71
Food guide pyramid—a guide to daily food choices.

TABLE 15
Sample diets for a day at three calorie levels

	Calorie Levels		
	1,600	2,200	2,800
Bread Group Servings	6	9	11
Vegetable Group Servings	3	4	5
Fruit Group Servings	2	3	4
Milk Group Servings	2–3*	2–3*	2–3*
Meat Group (Total Ounces)	5 oz.	6 oz.	7 oz.

*Pregnant and breastfeeding women, teenagers, and young adults to age 24 need three servings.
Source: The Food Guide Pyramid, United States Department of Agriculture Home and Garden Bulletin Number 252, August 1992.

NUTRIENTS PROVIDED BY THE FOOD GROUPS

Bread, Cereal, Rice, and Pasta Group

Foods that are in this group provide the following.

- Complex carbohydrates (starches)
- Protein
- Fiber
- Iron
- Magnesium

- B vitamins:
- Niacin
- Riboflavin
- Thiamin

Some examples of these foods from other countries include the following.

- Rice all kinds
- Tortillas
- Noodles: rice, wheat, cellophane (bean thread), soba
- Chapati (Indian whole wheat bread)
- Pouri (Indian fried bread—high fat)
- Couscous
- Cracked wheat (tabouli uses cracked wheat, tomatoes, cucumbers, and lemon)
- Kasha

- Maize
- Millets
- Sorghum
- Oats
- Rye
- Barley
- Tef
- Quinoa

TABLE 16
Food group serving sizes

Bread, Cereal, Rice and Pasta: One Serving
  1 slice of bread
  1 ounce ready to eat cereal
  1 tortilla (7")
  $\frac{1}{2}$ cup cooked cereal, rice, pasta
  $\frac{1}{2}$ English muffin, small bagel, hamburger roll, pita (6")

Vegetable: One Serving
  1 cup raw leafy vegetables
  $\frac{1}{2}$ cup cooked or chopped raw vegetables
  $\frac{3}{4}$ cup vegetable juice

Fruit: One Serving
  1 medium apple, orange, banana
  $\frac{1}{2}$ cup chopped, cooked, or canned fruit
  $\frac{3}{4}$ cup fruit juice
  $\frac{1}{4}$ cup dried fruit

Milk, Yogurt, and Cheese: One Serving
  1 cup milk or yogurt
  $1\frac{1}{2}$ ounces natural cheese
  2 ounces processed cheese
  1 cup fortified soymilk
  $1\frac{1}{2}$ ounces fortified soy cheese

Meat, Poultry, Fish, Dry Beans, Eggs, and Nuts: One Serving
  2–3 ounces lean meat, fish, poultry
  2–3 ounces tofu
  $\frac{1}{2}$ cup cooked dry beans, 1 egg, 2 tablespoons peanut butter, $\frac{1}{4}$ cup nuts or seeds count as 1 ounce of meat

---

Vegetable Group

Vegetables provide the following.

- Vitamin C
- Vitamin A
- Folate
- Potassium
- Magnesium
- Iron
- Fiber

A general rule of thumb is the darker the color of the vegetable, the more nutrients it has. For example, winter squash has a greater amount of vitamins than summer squash; spinach has more nutrients than iceberg lettuce. Try to include both dark leafy green vegetables and orange-yellow vegetables in your diet.

Vegetables from other countries that are good sources of vitamin A or C include the following.

- Peppers
- Bok choy
- Broccoli
- Pea pods
- Mustard and turnip greens
- Avocado
- Kale
- Watercress
- Cajion, bayan, and amaranthus (dark green leaves)
- Pumpkins
- Cabbage
- Yams and sweet potatoes
- Nopales (cactus leaves)
- Jicama
- Chayote (squash)
- Starchy vegetables: plantain, cassava, potato

Fruit Group

Fruits provide vitamins A and C, potassium, and fiber.

Fruits from other countries that provide good sources of these nutrients include the following.

- Pineapple
- Mangoes
- Papaya
- Apricots
- Kumquat
- Pawpaw
- Guavas
- Rosehips
- Zapote (sweet yellowish fruit)
- Ackee (Use only mature fruit; discard bruised or damage fruit—it can contain a natural toxin.

Milk, Yogurt, and Cheese Group

Milk, yogurt, and cheese provide the following.

- Protein
- Calcium
- Riboflavin

- B$_{12}$
- Vitamins A and D

Foods from other countries that provide these nutrients include the following.

- Queso blanco—white cheese
- Soft curd cheeses
- Yogurt and yogurt drink (yogurt, water, and sugar)
- Milk: goat, buffalo, sheep
- Condensed and evaporated milk (can be made from skim milk—may not be fortified with vitamins A and D)
- Soured milk
- Soy milk (select if possible brand fortified with vitamins A and D)

Meat, Poultry, Fish, Dry Beans, Nuts Group

This group provides the following.

- Protein
- Iron
- B$_6$
- Niacin
- B$_{12}$*
- Thiamin
- Zinc
- Magnesium
- Phosphorous

Foods from other countries that provide these nutrients include the following.

- Soybeans
- Beans—curd (tofu), pastes, cheese beans of all kinds—cowpeas, pigeon peas
- Peanuts (Avoid moldy nuts and legumes; they can contain aflatoxins, a natural toxin.)
- Lentils—all colors
- Fish of all kinds (may be dried or canned)
- Game

Fats, Oils, and Sweets Group

This group provides calories but not significant sources of nutrients.

- Fats and oils
- Sugars
- Olive oil
- Sugarcane
- Sesame oil
- Syrups
- Coconut oil
- Shaved ice with syrup
- Soybean oil
- Soft drinks
- Red palm oil (increased vitamin A content)
- Candy
- Ghee
- Lard

*The only practical food sources are in animal products or fortified foods such as soymilk.

TABLE 17
Dietary guidelines for Americans

AIM FOR FITNESS...
Aim for a healthy weight.
Be physically active each day.

BUILD A HEALTHY BASE...
Let the Pyramid guide your food choices.
Choose a variety of grains daily, especially whole grains.
Choose a variety of fruits and vegetables daily.
Keep food safe to eat.

CHOOSE SENSIBLY...
Choose a diet that is low in saturated fat and cholesterol and moderate in total fat.
Choose beverages and foods to moderate your intake of sugars.
Choose and prepare foods with less salt.
If you drink alcoholic beverages, do so in moderation.

Source: *Nutrition and Your Health: Dietary Guidelines for Americans*. USDA, Home and Garden Bulletin No. 232, Washington, D. C., 2000.

DIETARY GUIDELINES

U.S. Dietary Guidelines encourage people to eat less saturated fat, sugar, and salt and to increase the amount of fiber in the diet (see Table 17). A food rule of thumb to help accomplish this is "The less processed the food, the less salt, sugar, and fat it generally contains." For example, canned foods such as fruits and vegetables have more sugar and salt than fresh fruits and vegetables. The majority of salt in the U.S. diet is not from the salt added at the table but the salt found in the processed food we consume. There are exceptions to this rule of thumb. Some processed foods are preferred for safety reasons over unprocessed foods—for example, pasteurized milk versus raw milk.

Fats in the diet come in a variety of types, and all fat in food is a mixture of these three.

- Saturated fat—found mostly in animal products but also present in coconut, palm, and palm kernel oils.
- Monounsaturated fat—found mostly in olive, peanut, and canola oils.
- Polyunsaturated fat—found mainly in corn, safflower, soybean, cottonseed oil, and some fish.

The Dietary Guidelines recommend moderating the amount of fat and limiting saturated fat and cholesterol. (Cholesterol is a fatlike substance found in animal products.) The current recommendation is to consume 30 percent or less of daily calories as fat and 10 percent or less as saturated fat. Saturated fat can cause blood cholesterol levels to rise, increasing the risk for heart disease. Trans fatty aids can also raise blood cholesterol. Trans fatty acids are found in partially hydrogenated vegetable oils, which are ingredients of hard margarines, bakery items, and some crackers. By selecting the lowest fat choices from each food group in the pyramid and adding no fat to your food in cooking or at the table, you will only get up to half of the 30 percent.

VEGETARIAN DIETS

Vegetarian diets are nutritionally sound when they include a wide variety of foods. Here are few points to remember when following a vegetarian diet.

Protein. All the essential amino acids can be found in any one protein food of animal origin, such as milk, meat, and eggs. Vegetarians, however, need to combine foods that complement one another to obtain all the necessary amino acids. This need not be difficult. Many native cultural food dishes do this already—for example, rice and beans. The essential amino acids that are lacking in the beans are found in the rice, thereby providing all the essential amino acids when the two foods are combined. It was previously thought that these complementary protein foods had to be consumed at the same meal. This is not the case: One food can be eaten at the next meal, and the benefits will still be derived. See Table 18 for examples of foods that complement one another.

Vitamin B_{12}. This vitamin is found only in animal foods in reliable amounts. Seaweed, fermented products, and yeast have been used as B_{12} sources. These are not reliable sources when confirmed by laboratory analysis. Therefore, vegans (vegetarians who do not consume cow's milk, eggs, or cheese) should drink B_{12}-fortified soymilk.

Calcium and Vitamin D. The use of a calcium and vitamin D–fortified soymilk is suggested. Vitamin D is also supplied by the action of sunlight on the skin. Tofu that is preserved in a calcium precipitate is recommended. Dark green vegetables such as collards, turnip and mustard greens, and kale are examples of calcium-containing vegetables.

Iron and Zinc. High amounts of iron and zinc are found in meat. Vegetarians can obtain these nutrients from whole-grain cereals (some cereals are fortified with these nutrients), legumes, nuts, and soy products. Consuming a high–vitamin C food, such as orange juice, with these suggested foods will increase the iron absorption from the food.

TABLE 18
Protein combinations for obtaining essential amino acids

Grains		Legumes
Rice		Lentils
Bulgur	and	Beans
Barley		Dried peas
Pasta		Beans
Legumes		Seeds and nuts
Chickpeas	and	Sesame seeds
Soybeans		Sunflower seeds
Lentils		Walnuts

HYDRATION

When working, especially in warm climates, it is important to stay well hydrated. The average fluid requirement is 2.5 liters (McCardle 1996). This requirement increases as activity and temperature increase. Individuals obtain water from three sources: liquid, food, and products of metabolism.

Individuals lose water in urine, feces, perspiration, and respiration.

Increased physical activity, hot weather, fever, vomiting, and diarrhea increase fluid requirements. Fluid balance is maintained when fluid intake is equal to fluid output. Liquids should be consumed before, during, and after physical activity to maintain hydration equilibrium. Thirst is not always a reliable indicator of hydration status. By the time an individual becomes thirsty, he

TABLE 19
Foods to avoid if diarrhea develops

	Avoid	Choose
Bread, Cereal, Grain Group	Whole-grain breads, cereals	White rice, pita bread, French bread, pasta
Fruit	Fresh fruit Fruit with seeds/skins	Ripe banana, applesauce, fruit juice (no pulp) or nectar
Vegetables	Raw vegetables, salad, Strong vegetables: broccoli, cauliflower, cabbage, onions Vegetables with seeds/skins	Cooked mild vegetables, potatoes
Meat, Chicken, Fish, Legumes	Fatty meats, legumes, nuts, beans, hotdogs, fried meat, chicken, fish	Lean broiled or baked meat, fish, chicken
Milk, Cheese, Yogurt	Milk may bother some	Lowfat milk, yogurt, cheese
Other	Gravy, spicy foods, fried foods, caffeine	Softdrinks, mild soup, broth

Notes: Generally, minimal or no treatment is necessary for mild attacks of diarrhea with the exception of fluid replacement. Replace fluid with bottled water, fruit juice, fruit beverages, softdrinks, or broth. Progress to mild, bland foods as listed.

or she is already mildly dehydrated. Older adults and young children may especially be at risk for dehydration because they do not necessarily pay close attention to their thirst. If you are performing strenuous physical work, you can lose several pounds of weight. For each pound of weight lost, a person needs to drink the equivalent in ounces of fluid. For example, if a person loses 1 pound (16 ounces), they must drink 16 ounces of water. You can determine if you are well hydrated by the frequency and color of your urine. Your first urine of the day upon rising is generally a dark color because it is concentrated. If you are urinating frequently and your urine is a pale color, you are probably hydrated sufficiently. If you are vomiting or develop diarrhea, replacing electrolytes as well as water is important. The use of an oral rehydration solution may be necessary. (See the Health chapter for an article on diarrhea and oral rehydration.) Foods to eat and avoid if diarrhea develops are listed in Table 19.

FOOD-BORNE ILLNESS

To avoid food-borne illnesses and their consequences, some precautions are in order. The World Health Organization developed The WHO Golden Rules for Safe Food.

1. Choose foods processed for safety. Drink pasteurized milk rather than raw milk.
2. Cook food thoroughly. Cook food to 70°C all the way through.
3. Eat cooked foods immediately. Do not give microorganisms the time or the temperature to grow.
4. Store cooked foods carefully. Hot foods should be stored at 60°C or greater. Cool foods should be stored at 10°C or lower. Do not store warm food in very large quantities in the

TABLE 20
Foodborne illnesses

Organism	Incubation	Symptoms	Duration	Food Involved
Salmonella	6–72 hours	Cramps, diarrhea, fever, headache, nausea, vomiting	2–6 days	Meat, poultry, eggs, milk
Staphyloccous	1–6 hours	Nausea, vomiting	1–2 days	Poultry, eggs, ham, custards, cream-filled pastry, potato salad
Shigellosis	1–7 days	Diarrhea, fever	1–2 days	Potato, tuna, chicken, shrimp, macaroni salads, lettuce
Clotridum perfringens	8–24 hours	Diarrhea	24 hours	Vegetables, slow-cooked meat at low temperature, raw meat
Campylobacter	1–7 days	Diarrhea, nausea, headache	Up to 7 days	Raw meat, poultry, raw milk
Bacillus Cereus	1/2–1 hour	Nausea, vomiting, diarrhea	6–24 hours	Spices, cereals, rice, dry foods mixes
Listeria	1 day–3 weeks	Nausea, vomiting, meningitis	Depends on treatment	Unpasteurized milk and cheese, vegetables, meat, seafood, poultry
E-coli	2–4 days	Bloody diarrhea, vomiting	10 days	Ground beef, meat, poultry, raw milk
Botulinium	12–36 hours	Vertigo, double vision, difficulty swallowing and breathing	Up to 10 days	Improperly canned low acid foods such as vegetables, reheated baked potatoes, sautéed onions, garlic in oil
Trichinosis	1–2 days; 7–15 days; larval invasion	Nausea, vomiting, fever, diarrhea, swelling of eyelids muscle stiffness/pain	2–4 weeks	Pork products, wild game
Hepatitis A Virus	2–6 weeks	Nausea and vomiting, fever, loss of appetite, dark urine	2–4 weeks	Contaminated shellfish
Prion Protein (bovine spongiform encephalopathy-BSE) Mad Cow Disease	Months to Years	Neurological	6 Months to Years	Beef

References
Bovine Spongiform Encephalopathy and New Variant Creutzfeldt-Jacob Disease, http://www.cdc.gov/ncidod/diseases/cjd/bse_cjd.htm. National Center for Infectious Disease, 2001.
Questions and Answers Regarding Bovine Spongiform Encephalopathy (BSE) and Creutzfeldt-Jakob disease (CID), http://www.cdc.gov/ncidod/diseases/cjd/bse_cjd_qa.html. National Center for Infectious Disease, 2001.

refrigerator. The refrigerator may not be able to handle the load, and the food will not cool in sufficient time to avoid microbe growth. Avoid storing infant food.

5. Reheat cooked foods thoroughly. Internal temperature should reach 70°C throughout the food.

6. Avoid contact between raw foods and cooked foods. Do not put cooked poultry on the same plate that was used for the raw poultry without washing first.

7. Wash hands repeatedly. Wash hands before preparing food. Wash hands after using the toilet or changing a diaper. Wash hands after touching pets. Wash hands after handling raw meat, fish, or poultry before touching other foods. Make sure all cuts and infections on hands are bandaged properly.

8. Keep all kitchen surfaces meticulously clean. Launder all dishcloths in boiling water.

9. Protect foods from insects, rodents, and other animals. Store food in covered containers.

10. Use safe water. If water safety is questionable, boil water before adding to food.

Infants, pregnant women, immuno-compromised people, and the elderly are at increased risk. By practicing safe food handling procedures food-borne illnesses can be decreased. The following are some other prevention measures to follow to avoid food-borne illnesses and diarrhea.

- Drink only bottled water.
- Use only ice that has been made with bottled water, and use bottled water when brushing your teeth.
- Avoid uncooked foods unless they can be peeled or shelled, such as fruits and vegetables.

Most harmful organisms will be on the outside of the food. Some fish species can contain toxins when caught in tropical reefs rather than open ocean: red snapper, amberjack, grouper, and sea bass. It is not recommended to consume barracuda or puffer fish, since they can also be toxic. Ask local health workers about safe fish and shellfish. Street vendors provide interesting cultural foods, but the risk for food-borne illnesses is high. Table 20 provides a summary of the more common food-borne illnesses.

REFERENCES

Bennion, M. (1995). *Introductory Foods*, 10th Edition. New Jersey: Simon and Schuster.

The Educational Foundation of the National Restaurant Association. (1992). *Applied Food Sanitation: A Certification Course Book*, 4th edition.

Food and Water Precautions and Travelers' Diarrhea. (1996). www.cdc.gov/travel/foodwatr.htm.

A Guide on Safe Food for Travelers. www.who.int/fsf/trvl.htm.

Johnson, P. K., and Haddad, E. H. (1996). "Vegetarian and Other Dietary Practices." In V. I. Rickert (Ed.), *Adolescent Nutrition Assessment and Management*. New York: Chapman and Hall.

King, F. S., and Burgess, A. (1996). *Nutrition for Developing Countries*, 2nd Edition. Oxford: Oxford University Press.

McCardle, W. D., F. J. Katch, and V. L. Katch. (1996). *Exercise Physiology Energy, Nutrition and Human Performance*. Baltimore, MD: Williams and Wilkins.

UNICEF. (1998). *The State of the World's Children*. Oxford: Oxford University Press.

USDA. (1990). *Nutrition and Your Health: Dietary Guidelines for Americans*. Home and Garden Bulletin No 232, Washington, DC.

CHAPTER 4

Health

Edited by STEVEN G. McCLOY
Brown University

HEALTH OVERVIEW

Steven G. McCloy

WHY THERE IS A CHAPTER ON HEALTH IN THIS BOOK

Initially, it might seem puzzling that a portion of this book is devoted to aspects of health and health care. Medicine and engineering are very different disciplines. Some would say they are disparate disciplines. Engineers focus on the concrete in what is viewed as a hard science. Medicine likes to talk about both its "art" and its "science," but a well-designed bridge, building, or diode is seen by some as artful, too. And isn't screwing back together a fractured femur rather mechanical? It may be our fields are not so far apart after all.

One reason for this association that immediately comes to my mind is the historical parallel between the development of medicine and engineering as disciplines. Surgery is not possible without metallurgy and the design of instruments (See Figure 1).

FIGURE 1
Removing a spear tip on the battlefield.

A second reason lies in the essential contributions of the field of engineering to the formation and maintenance of a healthy environment and society. Humankind learned to group together, to master fire and heat their caves, to gather goods and foods, and to domesticate beasts. They left the cave and learned to build their own shelter. Perhaps the first engineer was the man or woman who figured out how to erect a lean-to or keep a roof from falling down.

Beyond safety and shelter, engineers brought water to make possible the *civitas* (city). A city cannot exist without a source of water. To preserve the health of the inhabitants, the water must be clean or must be cleaned. This is the role of the engineer.

FIGURE 2
Roman Aqueduct Near Nimes, France, Fourteenth Century C.E.

The third reason closely follows: Engineered solutions may carry with them profoundly ill effects for society. Our example of the Roman aqueduct (see Figure 2) points rather quickly to the growing attribution of chronic lead poisoning as a cause of the decline of the Roman Empire. Water flowed from those same aqueducts directly to the baths and to the homes of Roman citizens in pipes formed of sheet lead. This material was far from inert and quite probably caused chronic lead poisoning in the populace. The great convenience of indoor plumbing carried with it a high price in human health.

The engineer's solution, like the physician's surgery or prescription, can have both good and ill effects. A dam in Egypt or Ghana will bring water for drinking and irrigation into lands. By providing larger expanses of open, still water, a dam will also bring an explosion of diseases like shistosomiasis or malaria.

In these articles the authors apprise readers of some of the essential underpinnings of health and how the engineer contributes so greatly to them. The problem of health in less developed communities is examined. The articles also consider the health of the person (perhaps you, the reader) going to developing countries from the outside.

When the traveler gets sick (which is inevitable), what kind of health resources might there be? Who are the providers of these resources? Are the doctors familiar to us from the developed world? Healers? Indigenous health workers? Shaman? The "Community Health Workers" article will acquaint you with the possibilities, while the "Health Care Delivery System" article focuses on the particular problems of women.

In their treatment of you, will they provide a prescription of a substance from a recognized pharmaceutical company? More likely they will provide a natural remedy or herb, like those discussed in several articles on traditional medicine.

In the Third World, it is water as much as bread that makes the staff of life. The article "Water and Health" reviews what makes water that staff to support the drinker or impale her.

Diarrhea, malaria, and AIDS—particularly serious problems in the Third World—are each discussed in separate articles. Smoke from an open fire pollutes the air inside traditional homes—the amount of pollution is described.

An article on dentistry is included because much can be done with simple tools.

Will your efforts bring better health to the area in which you work? What are the health needs of the citizens of the Third World? These are the topics addressed in the article "Third World Health Care Needs." This article sets a context and is recommended to all readers.

To be effective you must be in good health, and one article gives particular suggestions on keeping healthy. All the authors of articles in this section offer it to benefit and promote your good health and the health of those you serve. Not all subjects related to health would fit in this book so we urge the volunteer to take a copy of *Where There Is No Doctor*, which can be ordered from the Hesperian Foundation, PO Box 11577, Berkeley, California 94712-2577, USA.

WORDS TO THE VILLAGE HEALTH WORKER

This article is a collection of pieces of advice. Much of it was based on *Where There Is No Doctor*, by The Hesperian Foundation. It is strongly recommended that a volunteer to a Third World country bring a copy of that book along, since it contains worthwhile suggestions and explanations about health care. The present volume simply did not have space to include such material.

WORKING WITH LOCAL PEOPLE

Here are some suggestions that may help you serve people's human needs as well as health needs.

1. *Be kind.* A friendly word, a smile, a hand on the shoulder, or some other sign of caring often means more than anything else you can do. Treat others as your equals.

2. *Share your knowledge.* As a health worker, your first joy is to teach. This means helping people learn more about how to keep from getting sick. It also means helping people learn how to recognize and manage their illnesses—including the sensible use of traditional remedies and common medicines.

3. *Respect your people's traditions and ideas.* If you can use what is best in modern medicine, together with what is best in traditional healing, the combination may be better than either one alone. In this way, you will be adding to your people's culture, not taking away. So go slowly—and always keep a deep respect for your people, their traditions, and their human dignity. Help them build on the knowledge and skills they already have. Work with traditional healers and midwives—not against them. Learn from them and encourage them to learn from you.

4. *Know your own limits.* Do what you know how to do. Do not try things you have not learned about or have not had enough experience doing, especially if they might harm or endanger someone. But use your judgment. Often, what you decide to do or not to do depends on how far you have to go to get more expert help.

5. *Keep learning.* Use every chance you have to learn more. Always be ready to ask questions of doctors, sanitation officers, agriculture experts, or anyone else you can learn from.

6. *Practice what you teach.* Good leaders do not tell people what to do. They set the example. Before you ask people to make latrines, be sure your own family has one. Also, if you help organize a work group—for example, to dig a garbage hole—be sure you work and sweat as hard as everyone else.

7. *Work for the joy of it.* If you want other people to take part in improving their village and caring for their health, you must enjoy such activity yourself. If not, who will want to follow your example? Try to make community work fun as a "work festival"—perhaps with refreshments and music. The job will be done quickly and can be fun.

8. *Look ahead—and help others to look ahead.* Many sicknesses can be prevented. Your job is to help your people understand the causes of their health problems and do something about them. Most health problems have many causes, one leading to another. To correct the problem in a lasting way, you must look for and deal with the underlying causes. You must get to the root of the problem.

PREVENTION: HOW TO AVOID MANY SICKNESSES

Many common infections of the gut are spread from one person to another because of poor hygiene and poor sanitation. Germs and worms (or their eggs) are passed by the thousands in the feces of infected persons. These are carried from the feces of one person to the mouth of another by dirty fingers or contaminated food or water. Diseases that are spread or transmitted from feces-to-mouth in this way, include the following.

- Diarrhea and dysentery (caused by amebas and bacteria)
- Intestinal worms (several types)
- Hepatitis, typhoid fever, and cholera
- Certain other diseases, like polio

The way these infections are transmitted can be very direct. For example, a child who has worms and forgets to wash her hands after her last bowel movement offers her friend a cracker. Her fingers, still dirty with her own stool, are covered with hundreds of tiny worm eggs (so small they cannot be seen). Some of these worm eggs stick to the cracker. When her friend eats the cracker, she swallows the worm eggs, too, so the friend will also set worms.

Many times pigs, dogs, chickens, and other animals spread intestinal disease and worm eggs. For example, a woman with diarrhea or worms has a bowel movement behind her house. A pig eats her stool, dirtying its nose and feet. The pig goes into the house where a child is playing on the floor. This way some of the woman's stool gets on the child, too. Later the child starts to cry, and the mother picks her up. Then the mother prepares food, forgetting to wash her hands after handling the child. The family eats the food, and soon the whole family has diarrhea or worms.

If there are many cases of diarrhea, worms, and other intestinal parasites in your village, people are not being careful enough about cleanliness. If many children die from diarrhea, it is likely that poor nutrition is also part of the problem. To prevent death from diarrhea, both cleanliness and good nutrition are important.

BASIC GUIDELINES OF CLEANLINESS

Personal Cleanliness (Hygiene)

1. Always wash your hands with soap when you get up in the morning, after having a bowel movement, and before eating.
2. Bathe often—every day when the weather is hot. Bathe after working hard or sweating. Frequent bathing helps prevent skin infections, dandruff, pimples, itching, and rashes. Sick persons, including babies, should be bathed daily.
3. In areas where hookworm is common, do not go barefoot or allow children to do so. Hookworm infection causes severe anemia. These worms enter the body through the soles of the feet.
4. Brush your teeth every day and after each time you eat sweets.

Cleanliness in the Home

1. Do not let pigs or other animals come into the house or places where children play.
2. Do not let dogs lick children or climb up on beds. Dogs, too, can spread disease.

3. If children or animals have a bowel movement near the house, clean it up at once. Teach children to use a latrine or at least to go farther from the house.

4. Hang or spread sheets and blankets in the sun often. If there are bedbugs, pour boiling water on the cots and wash the sheets and blankets—all on the same day.

5. De-louse the whole family often. Lice and fleas carry many diseases. Dogs and other animals that carry fleas should not come into the house.

6. Do not spit on the floor. Spit can spread disease. When you cough or sneeze, cover your mouth with your hand, a cloth, or a handkerchief.

7. Clean house often. Sweep and wash the floors, walls, and beneath furniture. Fill in cracks and holes in the floor or walls where roaches, bedbugs, and scorpions can hide.

Cleanliness in Eating and Drinking

1. Ideally all water that does not come from a pure water system should be boiled, filtered, or purified before drinking. This is especially important for small children and at times when there is a lot of diarrhea or cases of typhoid, hepatitis, or cholera. However, to prevent disease, having enough water is more important than having pure water. Also, asking poor families to use a lot of time or money for firewood to boil drinking water may do more harm than good, especially if it means less food for the children or more destruction of forests.

   A good, low-cost way to purify water is to put it in a clear plastic bag or clear bottle and leave it in direct sunlight for a few hours. This will kill most germs in the water.

2. Do not let flies and other insects land or crawl on food. These insects carry germs and spread disease. Do not leave food scraps or dirty dishes lying around, as these attract flies and breed germs. Protect food by keeping it covered or in boxes or cabinets with wire screens.

3. Before eating fruit that has fallen to the ground, wash it well. Do not let children pick up and eat food that has been dropped—wash it first.

4. Only eat meat and fish that is well cooked. Be careful that roasted meat, especially pork and fish, do not have raw parts inside. Raw pork carries dangerous diseases.

5. Chickens carry germs that can cause diarrhea. Wash your hands after preparing chicken before you touch other foods.

6. Do not eat food that is old or smells bad. It may be poisonous. Do not eat canned food if the can is swollen or squirts when opened. Be especially careful with canned fish. Also, be careful with chicken that has passed several hours since it was cooked. Before eating leftover, cooked foods, reheat it until it is very hot. If possible, give only foods that have been freshly prepared, especially to children, elderly people, and very sick people.

7. People with tuberculosis, flu, colds, or other infectious diseases should eat separately from others. Plates and utensils used by sick people should be boiled before being used by others.

HOW TO PROTECT CHILDREN'S HEALTH

1. A sick child should sleep apart from children who are well. Sick children or children with sores, itchy skin, or lice should always sleep separately from those who are well. Children with infectious diseases like whooping cough, measles, or the common cold should sleep in separate rooms, if possible, and should not be allowed near babies or small children.

2. Protect children from tuberculosis. People with long-term coughing or other signs of tuberculosis should cover their mouths whenever they cough. They should never sleep in the same room with children. They should see a health worker and be treated as soon as possible. Children living with a person who has tuberculosis should be vaccinated against TB (B.C.G. Vaccine).

3. Bathe children, change their clothes, and cut their fingernails often. Germs and worm eggs often hide beneath long fingernails.

4. Treat children who have infectious diseases as soon as possible so that the diseases are not spread to others.

5. Teach children to follow these guidelines, and explain why they are important. Encourage children to help with projects that make the home or village a healthier place to live.

6. Be sure children get enough good food. Good nutrition helps protect the body against many infections. A well-nourished child will usually resist or fight off infections that can kill a poorly nourished child.

Public Cleanliness (Sanitation)

1. Keep wells and public waterholes clean. Do not let animals go near where people get drinking water. If necessary, put a fence around the place to keep animals out. Do not defecate or throw garbage near the water hole. Take special care to keep rivers and streams clean upstream from any place where drinking water is taken.

2. Burn all garbage that can be burned. Garbage that cannot be burned should be buried in a special pit or a place far away from houses and places where people get drinking water.

BREASTFEEDING

Breastfeed rather than bottle-feed babies. Give only breast milk for the first four to six months. Breast milk helps babies resist the infections that cause diarrhea. If it is not possible to breastfeed a baby, feed her with a cup and spoon. Do not use a baby bottle because it is harder to keep clean and more likely to cause an infection. When a baby is given new or solid food, start by giving her just a little, mashing it well, and mixing it with a little breast milk. The baby has to learn how to digest new foods. If she starts with too much at one time, she may get diarrhea. Do not stop giving breast milk suddenly. Start with other foods while the baby is still breastfeeding.

Mothers normally have a choice between breastfeeding and the use of a commercial infant formula. The benefits of breastfeeding are straightforward. It is much healthier for the baby because the milk provides all the essential nutrients and is not contaminated, while the water used in preparing formula may be. Infant formula is expensive, around $200–300 per year, which is perhaps 30 percent of the monthly salary for an average government worker in an African country. Breastfeeding protects the child against other diseases besides diarrhea, since the mother's natural immunity (antibodies) is passed to the infant through the milk. Finally, breastfeeding helps prevent pregnancy. Women who do not breastfeed have a postnatal infertility period of 3 months, while women who breastfeed are usually postnatally infertile for a period of 13 months (but this usual period of infertility is not reliable enough to be depended on).

One problem with formula is that clean water, sterile bottles, and nipples are hard to obtain in many villages. Furthermore, once the formula is made, it will spoil unless kept cool, and refrigeration is often not available. A third problem with formula is that the correct proportion of mix to water must be used, but not all parents are able to read the mixing instructions.

The reason that mothers do not breastfeed as much as they could seems to stem back to well-intentioned but misguided aid programs in the 1960s. Well-regarded medical research indicates that a 6-month-old baby needs more protein than a mother's milk can provide, so some kind of supplement is helpful until the infant is entirely weaned at 24 months or so. Locally made supplements of cow's milk, thin rice gruel, or the best commercial formulas are good supplements to mother's milk for an older infant. Formula actually contains slightly more nutrients than mother's milk but no antibodies. In any case, formula is not needed before the baby is 6 months old, in normal situations, and formula was originally intended as a supplement. Formula does tend to be more convenient, since feedings are required less often on a routine schedule. So for a variety of reasons, some mothers have come to favor formula over mother's milk. The need exists to educate people about the relative advantages of both mother's milk and formula. Recent reports seem to indicate that breastfeeding is now being done more often than it used to be.

REFERENCES

Interagency Group on Breastfeeding Monitoring. (1997). "Cracking the Code, Monitoring the International Code of Marketing of Breast-Milk Substitutes." London.
Where There Is No Doctor. The Hesperian Foundation, PO Box 11577, Berkeley, California 94712-2577, USA.

COMMUNITY HEALTH WORKERS

David J. Bermon

Most of the world's people survive from day to day without access to basic health services. To fill this void, dedicated inhabitants of various villages, towns, and cities work to improve the health of their communities. These individuals have been called community health workers (CHW), and they may represent the best hope for making appropriate health care accessible to everyone, especially the rural poor. I describe two programs in two different hemispheres and their impact on the populations served. I discuss similarities and differences and the potential application of these models to other parts of the developing world.

WHAT IS A CHW?

CHWs are people: men, women, young, and old. Often, they are selected by their communities because they are good leaders, responsible, wise, just, or known to be especially kind. Beyond the roles of providing curative medical care in the community, some have also been agents for social change. CHWs may work to raise community consciousness and organize against unjust societal structures and practices that influence the health of the community. Four qualities characterize the potential benefits that the CHW brings to the delivery of health services.

Availability. A CHW provides services and technologic information that may never before have been possible in a community.

Accessibility. The CHW makes health care accessible to everyone.

Acceptability. The CHW is of the community and shares a common awareness of local beliefs and practices. These qualities help the CHW to provide the community with acceptable health services on the community's terms.

Affordability. The CHW represents a cost-effective means of offering health services and activities where resources available to the community are limited.

THE MODEL

Most CHW models state that health and illness are the result of complicated interactions between biological, physical, social, economic, and political elements and that health work must address *all* of these elements. Consider a child suffering from hookworm. The infection is as much a result of poor environmental sanitation as it is of behavioral practices like walking barefoot. The CHW fails if he or she only administers worming medicine and fails to engage the community level in building latrines and promoting shoe wearing.

BAREFOOT DOCTORS

An attempt by the People's Republic of China to bring health care to its underserved rural population led to the development of one of the largest community health worker programs the world has ever seen. It remains the backbone of China's national health system. The shift to communist rule in the country ushered in a revolutionary vision for national health care. Under Chairman Mao Tse-tung, China's new national health policy was based on four principles.

1. The health system must serve the people.
2. Preventive medicine must be emphasized over curative medicine.
3. Traditional Chinese and modern medicine must be integrated.
4. Health work must be integrated with mass campaigns.

Mao Tse-tung supported experiments in rural health care that eventually led to the creation of over a million and a half community health workers, agriculturists who served as part-time health workers. They came to be known around the world as *barefoot doctors*, a name that alludes to the workers who toiled barefoot in the rice paddies around Shanghai and emphasized the barefoot doctors' identification as peasant farmers.

During the 1950s the People's Republic of China was reorganized into a decentralized administrative system based on agricultural production units. The smallest unit was the production team, which was composed of 200 to 700 people. Several production teams made up a brigade, and several brigades formed a commune.

In the rural communes surrounding Shanghai, efforts were made to train health workers who would remain dedicated to agricultural work while providing health services to their production teams. The idea was promoted during the Cultural Revolution of the 1960s. Some estimate as many as 1.8 million barefoot doctors have been trained.

Closely tied to the Chinese administrative structure, they were accountable to their brigades and compensated for both their agricultural work and health services. Selection of barefoot doctors was also a brigade responsibility. Criteria included at least six years of elementary school and a strong dedication to serve the people.

Training is highly varied from region to region, but usually consisted of three to six months' instruction at a commune or county hospital. Theory and practice were included in training on topics of both preventive and curative medicine. Aspects of traditional Chinese medicine were also taught, including the use of acupuncture and herbal medicines. Commune health personnel provided training, technical supervision, and support to the barefoot doctors.

Barefoot doctors' responsibilities included the following.

- The promotion of preventive health activities and health education
- Management of environmental sanitation
- Organization of health campaigns
- Immunizations and first aid
- Treatment of simple and common illnesses
- Attention to maternal and child care and family planning services
- Collection and report of health information
- The training and organization of volunteer health aides

Some barefoot doctors were even trained to perform types of surgery. Technical supervision and subsequent training sessions were used to improve the barefoot doctor's knowledge and skills.

The Chinese barefoot doctors were based on an ideology that emphasized equity, prevention, participation, and appropriate technology in health care. They radically changed the form of the Chinese health system in an effort to provide accessible health care services to the country's unimaginably huge rural population.

CHIMALTENANGO GUATEMALA DEVELOPMENT PROGRAM

In 1966, the Chimaltenango Development Program began training community health workers in the highlands of Guatemala. The program, founded by Dr. Carroll Behrhorst, aimed at forming a self-reliant and self-determined health program to serve the country's predominantly indigenous rural poor. In doing so, the program created a new type of health worker for Guatemala. These new *health promoters* were trained in curative and preventive medicine as well as community development. Across the years the program grew to include over 70 health promoters working in 50 villages.

The promoters were village members with an average of three years of elementary school education. They were selected by and accountable to their communities through the formation and functioning of community health committees within the villages.

Teaching focused on symptom description allowing the promoters to recognize and treat the majority of ailments afflicting their communities. Health education was continuous in weekly sessions in the Behrhorst clinic seeing the cases that came in through the clinic's doors. In periodic exams, the promoters were asked to describe a patient's condition, suggest treatment, and explain appropriate preventive measures. Senior health promoters visited the trainee's village to assess local use of skills outside of the formal clinic setting.

After attending the weekly sessions for at least a year, the promoters were allowed to begin treating patients in their communities. They received financial compensation for their services through the sale of medicines at prices monitored by the community health committees. An understanding of individual limits was included in the educational program, and promoters were trained to recognize and refer cases beyond their abilities to the Behrhorst clinic or another area health facility.

Community development and preventive medicine were as important in promoters' education as curative medicine. Work responsibilities included immunizations, tuberculosis control and treatment, water projects, literacy programs, family planning, and agricultural and animal husbandry programs.

The Chimaltenango Development Program provided the villages it served with a type of community health worker who could respond to the community's urgent medical needs and work as a catalyst for development and social change to secure its future.

The training of health personnel in a community-based, comprehensive approach like Behrhorst's has been the goal of other CHW initiatives around the world. The village health workers of the Bondolfi Mission in Zimbabwe and the health promoters of Project Piaxtla in western Mexico are examples of nongovernmental organizations' effort. In addition, this model has been borrowed by national health ministries across Africa, Asia, and Latin America.

INTERNATIONAL CONFERENCE ON PRIMARY HEALTH CARE IN ALMA-ATA

In 1978, ideas about health and health services shifted dramatically at the International Conference on Primary Health Care in Alma-Ata. The World Health Organization and its member states redefined health as *a state of complete physical, mental, and social well-being and not merely an absence of disease or infirmity.*

This concept integrates social, economic, and political development with health work. A new primary health care model based on the principles of equity, prevention, community involvement, self-reliance and the use of appropriate technology was advocated as the best strategy to address the world's unmet health needs. The conference also recommended the creation of national CHW programs as an important part of achieving health for all.

Programs that have both influenced and followed these new developments in health policy have advocated a CHW model that places particular emphasis on the health worker's relationship to the community that selected her and where she lives. The CHW also serves as a bridge between the community and the formal health sector. Most importantly, the CHW is a catalyst for change through preventive and promotive health activities, health education, development work, and the introduction of appropriate technologies into the community.

CRITICISMS AND FAILURES

Over the years, CHW programs have received criticism in countries such as Nepal, Honduras, India, Zimbabwe, and Nicaragua. Programs fail when the need is vastly greater than resources. The formal health sector may fail to support the CHW. Unsuitable selection criteria and poor design of a program's training component may lead to failure. The nuclear structure of the traditional village is absent in some countries. In others, a lack of awareness and consideration by the federal health ministry of the prevailing health beliefs and practices of the rural population was a powerful impediment to ministry initiatives.

The "classic" programs like the Chimaltenango Development Program and the Chinese Barefoot Doctors have encountered difficulties in more recent years. In the early part of the 1980s, the health promoter program of the Behrhorst clinic was devastated in the horrific violence of Guatemala's civil war. The organizational skills of health promoters made them targets of the Army units and secret death squads. Societal restructuring in China has removed the critical underpinnings of the barefoot doctors' relationship with their communities. In recent years, these conditions, exacerbated by new popular attitudes favoring specialized services and moves away from traditional Chinese medicine, have compromised the barefoot doctor's fulfillment of initial program goals.

Nonetheless CHW's continue to have a profound impact on the health of communities around the world. Shortcomings serve as a guide for improving current programs and the creation of future ones.

Frequently, shortcomings in CHW programs evidence a lack of dedication and commitment by government and the formal health sector to the underlying principles of primary care. CHW programs are only part of a much larger process of health sector reform. They cannot be the entire solution.

CONCLUSION

These brief descriptions illustrate the variety of experiences found in community health efforts around the world. While they do not provide a universal definition for the community health worker, they do offer insight into their potential to dramatically influence the health of their communities.

A new understanding of health has combined curative and preventive medicine with social, economic, and political development. Emphasis has been placed on the principles of equity, accessibility, participation, self-determination, and self-reliance in health care. The community health worker addresses all of these diverse elements. If the hope of health for all is to be a reality, CHWs will remain on the forefront of providers.

HEALTH CARE DELIVERY SYSTEM

Float Auma Kidha

In many developing countries health facilities are so far away from the families that a mother would plan to be away from home for the whole day to get her child immunized or have other maternal and child health services for herself, such as antenatal care, postnatal care, family planning, and so on. In many countries these services are segmented in that mothers have to make multiple trips to the health facilities—for example, visit on Monday for her child's immunization, on Wednesday for Nutrition Clinic for another malnourished child, and yet again on Friday for her antenatal care. Many times it is the same woman who must fetch water some three kilometers away, gather firewood in a forest two kilometers away, gather vegetables, go to the market, and the grinding mill before preparing food and feeding the family. It is too much to expect her to walk ten kilometers to the health facility and back unless her child is very, very ill. More often than not, by the time she makes the long trip to the health facility, she has tried the village herbalist's medicine or the witch doctor's charms. The health services in rural areas still are used more for emergency curative care than for preventive purposes.

IMPLICATIONS FOR WOMEN

To raise the health standards of the family ways must be found to lighten the woman's load in one or more ways. Water, firewood, and health services must be within easy reach. Even the earthen pots for carrying water are too heavy. Women must be allowed to make decisions regarding their own fertility so that they can independently decide not only on how many children they want to have but also when they will have them. In many African cultures for example, it is the mother-in-law who decides on the number and spacing of children for the couple.

SCHEDULING

Health services should be arranged in such a way that all services—immunization, antenatal care, child spacing or family planning, postnatal care and curative services—are provided every day. In this way, a woman can plan to go to the health facility for multiple services on a single day, which will leave her with more days at home to attend to those chores that are necessary for the health and general living standards of her family. More days at home of itself would go a long way in preventing many deaths among children, which are usually due to communicable diseases and malnutrition. An integrated system of health care delivery, especially in the rural areas of developing countries, is a technology that takes into consideration a wider perspective than health alone. It embraces the total way of life of a woman!

DRUGS

Medicinal drugs are an important aspect of medical technology. The list of drugs needed for primary health care is those drugs required in specific local circumstances.

The most common diseases for which drugs are needed in villages are malaria, diarrhea, respiratory tract infections, and skin conditions. Drugs for malaria, like chloroquine, as well as pain relievers like aspirins are usually kept by the local shopkeeper, from whom people buy after clinic hours or when the clinic has run out of chloroquine—a situation that occurs more often than not. Because of lack of knowledge about chloroquine dosage, people buy two or more tablets, which are taken in one dose rather than the needed series. There are several reasons for single dosages. One is that people do not know the proper dosage of chloroquine, even if they had enough money. The second reason is that the shopkeeper is most interested in selling her or his stock and not advising people on dosages of the drugs sold over the counter (of course, often the shopkeeper does not know the dosage). The third reason is that drug manufacturers usually write the instructions in English, so even if the illiterates were literate, they may not understand. In some communities chloroquine and quinine have been used by young pregnant girls to procure abortions, so pregnant women with malaria may not wish to take chloroquine for fear of miscarrying. The result of all these is that people do not take the right treatment for malaria. Another result of low dosages is that the parasite becomes resistant to the drugs available (a big problem in many Third World countries at the moment).

The problem could be overcome by training local shopkeepers on the right dosages of the simple drugs they sell. The shopkeepers should insist that a total course of chloroquine be purchased. The shopkeeper must know the side effects of every drug sold, to educate the customers. The drug stock should be appropriate for local health needs.

Traditional midwives and healers should be taught the dosages of the common drugs, and a supply of the basic drugs should be left with them by the local clinic. After all, more patients go to traditional medicine people, who are in every village, than to government clinics, which serve several villages. Places where the health care system has integrated traditional healers into their activities have succeeded in reducing the incidence of common conditions such as malaria and diarrhea.

CULTURE, HEALTH, AND THE ENVIRONMENT

Sarah Scheiderich

Many of the countries of the developing world are rich in plant life. Alarmingly, the widespread exploitation of this resource progresses at a rapid rate. For example, Guatemala is one of the 25 most plant-rich countries in the world. Its original forests have been reduced by 65 percent in the past 30 years, and more than 130 species of animal are threatened with extinction.

Native peoples have of necessity used local plants and herbs for their medicinal qualities since prehistory. Pollens and herbs found in Ice Age burial sites are both preservatives and presumed gifts for the dead. Modern-day usage of plant materials may also serve to guide our knowledge of indigenous conceptions of health. They are tied to an understanding of how nature interacts with humans and in turn how humans affect nature.

Anthropological research performed in summer 1998 in San Lucas Tolimán, Guatemala, Central America revealed the presence of a medicinal plant garden that sustained 58 species of plants. Some were common to that area, some had been collected by health promoters from surrounding communities, and others had been imported by visitors from the developed world. One of the men who tend this garden has written a *botanica*, a book about the plants and their use. His information has come from many sources, among them the oral history gathered from the various *comites de los ancianos* (old people's groups) in and around this community.

Health promoters not familiar with herbal remedies may have access to *curanderas*, lay healers recognized by their communities to have special knowledge about a wide variety of folk remedies. In other cultures, this role may be played by a wide variety to people going by names like shaman, witch, or healer, without necessarily the association with the occult that these terms may carry in western cultures.

People in developing countries often have interesting insights into and explanations for illness. In San Lucas Tolimán illness is understood as an interaction between the physical being, the emotional state, and the external environment. This "primitive" idea is not at all different from the so-called "holistic" approach to health utilized by sophisticated citizens of the developed countries. Further, Maya perceive this connection of the external and internal environments by the sensitivity of both to changes in temperature, pollution, and the movement of substances through them.

The blood mediates this interaction of the body with the environment by its sensitivity to external stimuli that can alter health for better or worse. An illness may be described as an attack of "*frio*" or cold caused by such things as wearing wet clothes or being immersed in cold air. At the same time blood that is too hot can produce fevers and sweats. Using the *temescal* (sweat bath) also may be a means of cleansing the body's internal environment of pollutants.

Foods may bring toxins to the blood either because of toxins contained in the food itself or substances that have been placed on the foods (fertilizers, pesticides). Blood may be viewed as "weak" because of its inherent qualities of the way it flows through the body. Blood may be weak because of deficiencies of vitamins and minerals (where these substances are known by the indigenous population) or other undefined deficiencies. Blood can be too thick and cease to flow properly.

The view of health in this perspective reveals an understanding of health as more than a biological phenomenon. In cultures that have not been exposed to "western" medicine and the Galenic inheritance, there seems often to be a closer and more intuitive association between the ill person and her or his environment. Further, the medicinal plants used to cure such illnesses are specifically adapted to local needs and speak to local conceptions of health.

Herbal healers frequently refer to using plants to restore a balance. A "cold' illness requires a "hot" plant, for example. In other cases, specific plants will have a specific indication for use. An example in San Lucas Tolimán is the use of the plant *Amor seco* (literally "dry love") for women who are sad and have experience "dolor de corazón"(heartache). Plants can be used alone. Often several are used at one time in the form of a tea. Healers observed by the author "prescribed" combinations of herbs that they claimed were unique for that patient and that patient's complaint.

THE INDIVIDUAL AND THE ENVIRONMENT

The patient cannot be passive in the curing. Instead, he or she must actively alter physical and mental habits to maintain an environment with the body and the mind. During the duration of the treatment with herbs (which can be for months) the individual alters the diet (perhaps avoiding coffee, cigarettes, and alcohol) and exercises a strong faith in God(s) for the medicine to have effect. This recognition of the importance of the spiritual in health and illness seems easier for the citizens of the developing countries. They do not have to pay for personal trainers to rediscover New Age truths. God's blessing is often invoked when medicinal plants are harvested. These plants also are never subjected to pesticide spraying.

Few cultures are totally isolated from civilization and from the availability of modern pharmaceuticals. Guatemalan *campesinos* differentiate between *quimicos* ("chemicals," commercial pharmaceuticals) and natural plants. They do not express a relationship to the former. In the developing world, most people feed themselves with the food they raise on their own land with their own effort. They have a close relationship to the soil and its produce. They witness the process of planting, growth, harvest, and consumption. What does it mean to ingest a pill? Illness is an intensely personal, even intimate, event. How can a foreign-made tablet or injection "understand" or intervene? What does one make of the often wracking side effects of modern pharmaceuticals?

CONCLUSION

Rocheleau (1996) notes that our first environment begins with our own bodies. Traditional medicine in San Lucas encourages healthy use and integration of one's environmental resources in order to alter or enhance the body. Lines are blurred between the internal physical and emotional being and the external environment in that most intimate context of sickness and health. Traditional medicine provides a model for living in a more culturally and economically autonomous fashion.

The Third World faces huge threats of environmental and health degradation. Workers at the level of the community will succeed at the grassroots to the extent that they understand and integrate issues of health and the environment with an appreciation for local cultural conceptions and belief.

Finally, the longer one works in the field, the more likely one is to become ill. Awareness of the availability of local herbs and people familiar with their usage offers the advantages of effectiveness, familiarity, and availability.

REFERENCES

Berger, Susan. (1997). "Environmentalism in Guatemala: When Fish Have Ears." *Latin American Research Review* 32: 99–117.
Rocheleau, D., B. Thomas-Slayter, and E. Wangari. (1996). *Feminist Political Ecology: Global Issues and Local Experiences*. New York: Routledge Publishing.
Tradepoint Guatemala Network. (1998). http://tradepoint.Org.gt/inguin/enareaspro13.html.

TROPICAL RURAL HEALTH CARE: PHARMACEUTICALS VERSUS MEDICINAL PLANTS

Thomas J. S. Carlson

CONTRIBUTIONS OF MEDICINAL PLANTS TO TROPICAL RURAL HEALTH CARE

Most people living in tropical rural areas do not have access to or cannot afford modern pharmaceuticals. These rural cultures do, however, use locally available plants for medicines. The World Health Organization (WHO) estimated that 80 percent of people in the world utilize medicinal plants as their primary care medicines (Farnsworth et al. 1985). The knowledge system on how to use locally available tropical plants for medicines is embodied within the local indigenous cultures (Johns 1990; Berlin 1992; Etkin 1996; Carlson & King 1997). Research and development for herbal medicines has already been conducted over thousands of years of cultural evaluation. The importance of traditional medicine has been recognized by the WHO and this organization has established the Traditional Medicine Programme to evaluate the safety and efficacy of medicinal plants (Bannerman 1983; WHO 1991; Akerele 1992).

The most biologically diverse ecosystems and culturally diverse traditional medicine systems are present in tropical countries (Durning 1992). As biological ecosystems around the world are destroyed, vascular plant species become extinct, resulting in a loss of potential new medicines (Abelson 1990). This loss is not only of plant-derived pharmaceuticals but also of traditional herbal medicines used by indigenous peoples that live in these ecosystems. Farnsworth et al. (1985) reported that 25 percent of modern medical drug prescriptions written between 1959 and 1980 in the United States were 119 pharmaceuticals derived from 90 different botanical species.

Pharmacological evaluation has demonstrated that many tropical botanical medicines are safe and effective. Fifty-six percent (28 of 50) of plants used ethnomedically in tropical countries to treat Type 2 diabetes mellitus in adults demonstrated antidiabetic activity in mice with diabetes (Carlson et al. 1997a). Publications describe detailed experimental studies demonstrating antidiabetic activity in rodents and humans in tropical plant species (Bierer et al. 1998; Luo et al. 1998a;

Luo et al. 1998b). Studies characterize tropical medicinal plants with anti-viral activity (Vlietnick and Van den Berge 1991; Carlson et al. 1997a) and anti-HIV activity (Balick 1990; Balick 1994).

Tropical medicinal plants have demonstrated pharmacological activity for the treatment of common tropical infectious diseases including malaria, infant pneumonia, and diarrhea. For example, experimental studies have shown anti-malarial activity in numerous different medicinal plants (Carvalho & Krettli 1991). A common medicinal plant from Asia used to treat malaria is *Artemesia annua* L. A synthetic derivative of a compound isolated from this plant, arthemether, has shown compelling activity against *Plasmodium falciparum* malaria in human studies (Hien et al. 1996; Van Hensbroek et al. 1996).

Pneumonia and bronchiolitis caused by respiratory syncytial virus is a major cause of morbidity and mortality in infants in tropical countries. Numerous tropical medicinal plants used to treat this infectious disease have shown both in-vivo and in-vitro activity against this viral pathogen (Carlson et al. 1997a; Kernan et al. 1998; Chen et al. 1998).

The most common cause of death in the world is infant diarrhea with dehydration. Anti-diarrheal activity in animals has been demonstrated in experimental studies of guava (*Psidium guajava* L.) leaf tea, a commonly used traditional medicine (Maikere-Fanino 1989; Lutterodt 1992). The red bark latex of a common medicinal tree, *Croton lechleri.*, is taken orally in South America to treat different types of diarrhea including cholera (Carlson, 2001). Antidiarrheal activity in in vitro and rodent studies was demonstrated by a compound isolated from this bark latex (Gabriel et al. 1999). A double-blind placebo controlled human clinical trial in AIDS patients with chronic diarrhea showed that this compound produced a statistically significant reduction in diarrhea.

There are examples of the use of the same plant species to treat the same disease in different continents, countries, and cultures: leaf sap from *Senna alata* (L.) Roxb. is applied topically to treat fungal skin infections in Africa (Crockett et al. 1992), Asia (Ibrahim & Osman 1995), and Latin America (Caceres et al. 1991); leaf tea from *Psidium guaja,va* L. is taken orally to treat diarrhea in Africa (Maikere-Faniyo et al. 1989), Asia (Grosvenor et al. 1995), and Latin America (Caceres et al. 1993); a decoction of the fruit of *Momordica charantia* L. is taken orally to treat Type 2 diabetes mellitus in adults in the Middle East (Mossa 1985), Asia (Lotlikar & Rajarama Rao 1966), in Latin America (Zamora-Martinez 1992). These studies highlight the important contributions to health care provision made by locally available medicinal plants. The ethnolinguistic groups around the world have evaluated locally available plants for medicinal qualities through their own traditional scientific methods to produce safe and effective botanical medicines. While this is different from the modern pharmaceutical evaluation, it should still be recognized as a legitimate method of assessing the safety and therapeutic efficacy of plants.

NONSUSTAINABILITY OF PHARMACEUTICALS IN TROPICAL RURAL SETTINGS

The research and development on a single compound pharmaceutical from a herbal medicine requires inputs of hundreds of millions of dollars and typically takes 10 to 15 years to complete. The production, packaging, distribution, and refrigeration of pharmaceuticals also require large resource inputs and energy consumption. These capital inputs result in the relatively high cost of pharmaceuticals, making them unaffordable to most tropical rural communities. If pharmaceuticals are donated to rural communities, there is typically not a continuous supply over a longitudinal period of time. Also, many of these drugs require refrigeration, which is seldom available in these settings. If refrigeration is available, it requires a relatively high capital input and consumption of energy. An additional problem with donated pharmaceuticals is that they are often for ailments that are rare or not present in the community (WHO 1992). It is also known that many of the donated pharmaceuticals have expired and can no longer be sold in Europe and the United States (Berkmans et al. 1997).

Perishability of pharmaceuticals is a major problem in hot tropical village settings where refrigeration is typically not available. When pharmaceuticals are used by rural populations, they are often administered inappropriately (wrong dose and/or for wrong disease) because a modern medical professional is seldom present to correctly administer these medicines. Anti-bacterial pharmaceuticals, such as penicillin, donated to tropical villages are used to treat viral infections or, worse yet, are seen as a panacea to treat a broad spectrum of diseases including headaches and backaches. It is especially dangerous when these medicines are injected because the needles and syringes are often reused without proper sterilization, resulting in the potential transmission of infections like HIV (Garrett 1994) and hepatitis B. By comparison, the use of botanical medicines is typically more therapeutic and safe because the source is locally harvested and knowledge of its medicinal use is known by the local culture. Traditional botanical medicines are inexpensive or free and locally available to tropical rural peoples whereas modern medicines are much more expensive and typically only intermittently available to these populations.

Traditional botanical medicines may also be a commodity under indigenous control that is harvested by local rural communities and sold to urban areas and/or northern countries resulting in capital flow from urban to rural areas and from north to south (King 1994; Iwu 1996; Carlson et al. 1997b; Carlson et al. 2001). These local communities may become suppliers of dried plant material used as herbal medicine or as a source for extraction of medicinal phytochemicals. This provides communities with economic benefits from sustainable harvesting of locally available biological resources.

CONCLUSION

Traditional botanical medicines have different energy and control characteristics when compared to pharmaceuticals. Pharmaceuticals are under external urban and/or northern control resulting in capital flow from rural to urban and south to north. The research and development, packaging, refrigeration, and distribution of these pharmaceuticals make them both capital and energy intensive forms of medicine. By comparison, tropical rural traditional medicine systems use locally available plants that do not require the capital and energy inputs needed for pharmaceuticals. These capital and energy inputs make the cost of pharmaceuticals too high and the access too low for tropical rural communities. Tropical botanical medicines under local rural indigenous control are more affordable, available, and sustainable for these communities.

While the efficacy of pharmaceuticals has been verified by western scientific evaluation, the tropical botanical medicines have gone through thousands of years of cultural evaluation. As discussed above in this paper, the therapeutic activity of many of these tropical medicinal plants has been confirmed in scientific laboratories. Medicinal plants are locally available, and the knowledge on the therapeutic uses of these plants is known by the local indigenous cultures. This allows these communities to use the traditional botanical medicines in a safe and effective fashion. To the contrary, pharmaceuticals are typically not therapeutic when taken in these settings because a medical trained professional is typically not present. This results in the drugs being given at the wrong doses for inappropriate diseases.

Pharmaceuticals are also quite perishable in hot tropical villages because of lack of refrigeration and because many drugs arrive in these settings already beyond their date of expiration. If refrigeration is available, it is expensive and energy intensive. Botanical medicines gathered locally in rural communities are typically used soon after harvesting and do not require refrigeration. These botanical medicines may also be cultivated or harvested from local ecosystems by tropical rural communities and sold to urban areas and to the north as herbal medicine or for the extraction of pharmaceuticals; this results in capital flow from urban to rural and from north to south. Hence, these botanical medicines typically provide the most reliable source of health care for the local rural communities, and in some cases economic benefits as well. The multiple benefits of botanical medicines establish incentives for rural tropical peoples to conserve their ecosystems, biological species, and languages to nurture and further develop their local indigenous traditional medical systems.

Many of the common diseases that afflict tropical rural communities can be treated by botanical medicines in a safe, effective, affordable, and sustainable fashion. It is unfortunate when the more energy intensive, capital intensive, expensive, and less sustainable modern medicines are introduced to treat diseases already well managed by locally available traditional botanical medicines. The use of these modern pharmaceuticals at the local rural level should only be reserved to treat those diseases not well managed by the local botanical medicines. When pharmaceuticals are used for specific ailments, there should be careful monitoring of the treatment to make sure the correct dose is given and the appropriate disease is being treated. If modern medical public health programs are introduced into rural communities, they should work to complement rather than replace the local traditional medical systems. The local traditional healers and their botanical medicines should be included as integral components of these health care projects.

BIBLIOGRAPHY

Abelson, P. H. (1990). "Medicine from Plants." *Science* 247: 513.

Akerele, Olayiwola. (1992). "WHO Guidelines for the Assessment of Herbal Medicines." *Fitotherapia* 43 (2): 99–110.

Balick, Michael J. (1990). "Ethnobotany and the Identification of Therapeutic Agents from the Rainforest." In Derek J. Chadwick and Joan Marsh (Eds.), *Bioactive Compounds from Plants*. CIBA Foundation Symposium 154: 22–39.

Balick, M. J. (1994). Ethnobotany, Drug Development and Biodiversity Conservation: Exploring the Linkages. In Derek J. Chadwick and Joan Marsh (Eds), *Ethnobotany and the Search for New Drugs*, CIBA Foundation Symposium 185, pp. 4–24.

Bannerman, Robert H. (1983). The Role of Traditional Medicine in Primary Health Care. In Bannerman, Robert H., Burton, John, & Wen-Chieh, Ch'en (Eds.), *Traditional Medicine and Health Care Coverage: A Reader for Health Administrators and Practitioners*, pp. 318–27. World Health Organization, Geneva.

Berkmans, P., V. Dawans, G. Schimets, and D. Vandenbergh. (1997). Inappropriate Drug-Donation Practices in Bosnia and Herzegovina, 1992 to 1996. *The New England Journal of Medicine,* 337(25): 1842–1845.

Berlin, B. (1992). *Ethnobiological Classification: Principles of Categorization of Plants and Animals in traditional Societies*. Princeton: Princeton University Press.

Bierer, D. E., D. M. Fort, L. G. Dubenko, C. Mendez, J. Luo, M. Reed, P. Peterli-Roth, R. E. Gerber, J. Litvak, N. Waldeck, T. J. Carlson, S. R. King, R. C. Bruening, R. F. Hector, and G. R. Reaven. (1998). Ethnobotanical-Directed Discovery of Cryptolepine from *Cryptolepis sanguinolenta*: Its Isolation, Synthesis and Antihyperglycemic Activity. *Journal of Medicinal Chemistry,* 41, 894–901.

Caceres, A., L. Fletes, L. Aguilar, O. Ramirez, L. Figueroa, A. M. Taracena, and B. Samayoa. (1993). Plants used in Guatemala for the Treatment of Gastrointestinal Disorders. 3. Confirmation of Activity Against Enterobacteria of 16 Plants. *Journal of Ethnopharmacology,* 38(1): 31–38.

Carlson, Thomas J. S. (2001). Medical Ethnobotany research as a method to identify bioactive plants to treat infectious diseases. In Iwu, Maurice M. & Wootton, Jaqueline C. (Eds.), *Ethnomedicine and Drug Discovery,* Elsevier Science B.V., pp. 45–53.

Carlson, T. J., and S. R. King. (1997). Ethnomedical Field Research Methods to Assess Medicinal Plants. In M. M. Iwu, E. N. Sokomba, C. O. Okunji, C Obijiofor, & I. P. Akubue, (Eds.), *Commercial Production of Indigenous Plants as Phytomedicines and Cosmetics.* BDCP Press, pp. 152–165.

Carlson, T. J., M. M. Iwu, S. R. King, C. Obiolar, and A. Ozioko. (1997a). Medicinal Plant Research in Nigeria: An Approach to Compliance with the Convention on Biological Diversity. *Diversity*, 13(1): 29–33.

Carlson, T. J., R. Cooper, S. R. King, and E. J. Rozhon. (1997b). Modern Science and Traditional Healing. Special Publication, *Royal Society of Chem.*, 200 (*Phytochemical Diversity*), pp. 84–95.

Carlson, T., B. M. Foula, J. A. Chinnock, S. R. King, G. Abdourahmaue, B. M. Sannoussy, A. Bah, S. A. Cisse, Camara, Mohamed 54, Richter, and K. Rowena. (2001). Case Study on Medicinal Plant Research in Guinea: Prior Informed Consent, Focused Benefit Sharing, and Compliance with the Convention on Biological Diversity. *Economic Botany*, 55(3).

Carvalho, L. H., and A. U. Krettli. (1991). Antimalarial Chemotherapy with Natural Products and Chemically Defined Molecules. *Memorias do Instituto Oswaldo Cruz,* Rio de Janeiro, Vol. 86, Suppl. 11, pp. 181–189.

Chen, Jian Lu, Michael Kernan, Philippe Blanc, Cheryl Stoddart, Mark Bogan, Nigel Parkinson, Zhi-jun Ye, Raymond Cooper, Michael Balick, Weerachai Nanakorn, and Edward Rozhon. (1998). "New Iridiods with Activity Against Respiratory Syncytial Virus from the Medicinal Plant *Barleria prioniti*." *Journal of Natural Products* 61: 1295–1297.

Crockett, C. O., F. Guede-Guina, D. Pugh, M. Vangah-Manda, T. J. Robinson, J. O. Qlubadewo, and R. F. Ochillo. (1992). "*Cassia alata* and the Preclinical Search for Therapeutic Agents for the Treatment of Opportunistic Infections in AIDS Patients." *Cellular Molecular Biology,* 38 (5): 505–511,

Durning, Alan Thein. (1992). *Guardians of the Land: Indigenous Peoples and the Health of the Earth.* Ed Ayres (Ed.). Worldwatch Paper 112. Worldwatch, Washington, DC.

Etkin, N. L. (1996). "Medicinal Cuisines: Diet and Ethnopharmacology." *International Journal of Pharmacognosy* 34 (5): 313–326.

Farnsworth, Norman R., Olayiwola Akerele, Audrey S. Bingel, Diaja D. Soejarto, and Zhengang Guo. (1985). "Medicinal Plants in Therapy." *Bulletin of the World Health Organization* 63 (6): 965–981.

Gabriel, S. E., S. E. Davenport, R. J. Steagall, V. Vimal, T. J. Carlson, and E. J. Rozhon, (1999). "A Novel Plant-Derived Inhibitor of cAMP-Mediated Fluid and Chloride Secretion." *American Journal of Physiology* (GI Section), 276 (39): G58–G63.

Garrett, Laurie. (1994). *The Coming Plague: Newly Emerging Diseases in a World Out of Balance.* New York, Penguin Books.

Grosvenor, P. W., P. K. Gothard, N. C. Me William, A. Supriono, and Do Gray. (1995), "Medicinal Plants from Riau Province, Sumatra, Indonesia. Part 1: Uses." *Journal of Ethnopharmacology* 45 (2): 75–95.

Hien, T. T., N. P. J. Day, N. H. Phu, and H. T. H. Mat. (1996). "A Controlled Trial of Artemether or Quinine in Vietnamese Adults with Severe Falciparum Malaria." *The New England Journal of Medicine* (335) 2: 76–83.

Ibrahim, D., and H. Osman. (1995). "Antimicrobial Activity of *Cassia alata* from Malaysia." *Journal of Ethnopharmacology* 45 (3): 151–156.

Iwu, M. M. (1996). "Biodiversity Prospecting in Nigeria: Seeking Equity and Reciprocity in Intellectual Property Rights through Partnership Arrangements and Capacity Building." *Journal of Ethnopharmacology* 51: 209–219.

Johns, T. (1990). *With Bitter Herbs They Shall Eat It: Chemical Ecology and the Origins of Human Diet and Medicine.* Tucson: The University of Arizona Press.

Kernan, Michael, Ambrose Amarquaye, Jian Lu Chen, Jody Chan, David F. Sesin, Nigel Parkinson, Zhi-jun Ye, Marilyn Barrett, Cheryl Bales, Cheryl Stoddart, Barbara Sloan, Phillipe Blanc, Charles Limbach, Salehe Mrisho, and Edward Rozhon (1998). "Antiviral Phenylpropanoid Glycosides from the Medicinal Plant *Markamea lutea*." *Journal of Natural Products* 61: 564–570.

King, S. K. (1994). "Establishing Reciprocity: Biodiversity, Conservation and New Models for Cooperation Between Forest Dwelling People and the Pharmaceutical Industry." In Tom Greaves (Ed.), *Intellectual Property Rights for Indigenous Peoples: A Source Book.* Oklahoma City: The Society for Applied Anthropology.

Lotlikar, M. M., and M. Rajarama Rao. (1966). "Pharmacology of a Hypoglycemic Principle Isolated from the Fruits of Momordica charantia, *Indian*." *Pharmacy* 28: 129.

Luo, Jian, Tom Chuang, Jeanne Cheung, Josephine Quan, Joyce Tsai, Cynthia Sullivan, Richard Hector, Michael J. Reed, Karl Meszaros, Steven R. King, Thomas J. Carlson, and Gerald M. Reaven, (1998a). "Masoprocol: A New Antihyperglycemic Agent Isolated from the Creosote Bush (*Larrea tridentate*)." *European Journal of Pharmacology* 346: 77–79.

Lou, Jian, Diana Fort, Thomas J. Carlson, Ben K. Noamesi, D. nii-Amon-Kotei, Steven R. King, Joyce Tsai, Josephine Quan, Cristina Hobensack, Pricilla Lapresca, Nancy Waldeck, Christopher D. Mendez, S. D. Jolad, Donald E. Bierer, and Gerald M. Reaven, (1998b). "Cryptolepine, a Potentially Useful New Antihyperglycemic Agent Isolated from *Crytolepis sanguinolenta:* An Example of the Ethnobotanical Approach to Drug Discovery," *Diabetic Medicine* 15: 367–374.

Lutterodt, G. D. (1992). "Inhibition of Microlax-Induced Experimental Diarrhoea with Narcotic-Like Extracts of *Psidium guajava* Leaf in Rats." *Journal of Ethnopharmacology* 37 (2): 151–157.

Maikere-Faniyo, R., L. Van Puyvelde, A. Mutwewingabo, and F. X. Habiyaremye. (1989). "Study of Rwandese Medicinal Plants Used in the Treatment of Diarrhoea I." *Journal of Ethnopharmacology* 26 (2): 101–109.

Mossa, J. S. (1985). "A Study on the Crude Antidiabetic Drugs Used in Arabian Folk Medicine" *Int. J. Crude Drug Res.* 23 (3): 137–145.

Van Hensbroek, M. B., S. Onyiorah, S. Jaffar, G. Schneider, et al. (1996). "A trial of Artemether or Quinine in Children with Cerebral Malaria." *The New England Journal of Medicine* (335) 2: 69–75.

Vlietnick, A. J. and D. A. Vanden Berghe (1991), "Can Ethnopharmacology Contribute to the Development of Antiviral Agents?" *Journal of Ethnopharmacology* 32 (1-3): 141–154.

WHO. (1991). "Guidelines for the Assessment of Herbal Medicines." Finalized at WHO consultation in Munich, Germany (June 19–21, 1991) and presented to *6th ICDRA in Ottawa,* Canada, 1991.

WHO. (1992). "The Use of Essential Drugs: Model List of Essential Drugs: Fifth Report of the WHO Expert Committee." *World Health Organization Technical Report Series, 1992,* 825: 1–75.

Zamora-Martinez, M. C. and C. N. P. Pola. (1992). "Medicinal Plants Used in Some Rural Populations of Oaxaca, Puebia and Veracruz, Mexico." *Journal of Ethnopharmacology* 35 (3): 229–257.

TRADITIONAL HEALTH PRACTITIONERS

Gwinyai Dziwa

In the context of social development, indigenous traditional health practitioners across the African continent play a significant role in health care education, community leadership, and provision of preventive and curative remedies. Traditional health practitioners are well known for their advocacy for environmental protection. Their approach to health poses little environmental degradation, since they obtain their herbs and wait for the bushes to regenerate. Lately, they have been recruiting local communities in reforestation programs.

Traditional health practitioners come in the form of spirit mediums, herbalists, and birth attendants. The spirit mediums treat psychosomatic illnesses and play a significant role in helping families with victims of mental illness. They are held in high esteem in their local communities, and their participation in environmental and health education programs is often very effective. The herbalists prescribe herbal remedies for a wide array of ailments and function like the modern pharmacist. However, the spirit mediums also prescribe herbal remedies. The traditional birth attendants help women in labor and give women birth relaxants and early pediatric concoctions.

APPLICATIONS OF TRADITIONAL HEALTH MEDICINE

African traditional health practice is still operating effectively in African communities despite the absence of institutionalized regulatory structures. However, in some countries such as Zimbabwe, organizations have been set up to promote the activities of traditional healers. Although there is scarce documentation of African traditional medicine, most of the knowledge and skills have been passed orally over several generations. This knowledge is still useful in prevention of tropical diseases such as malaria or bilharzia. For instance, traditional healers have herbs that contain mosquito repellants, but that knowledge is not as ubiquitous as one may wish. The methods used by traditional health practitioners are normally environmentally friendly. One good example is their use of mosquito repellants that are void of nonbiodegradable material. Although traditional health practitioners cannot explain their procedures in scientific terms, their methods have been found to be effective for generations.

Besides prevention, traditional practitioners provide herbs that have analgesic components. This helps to reduce pain in cases of headaches, backaches, and other traumatic injuries. Traditional herbalists also keep herbs that neutralize the effects of snake venoms such as the cobra or black mamba. Although the traditional healer does not label his herbal concoctions in the herbarium, he knows exactly where to pick a laxative or an emetic. Traditional birth attendants know different herbs that can help women in labor. One thing that traditional herbalists are known for is their ability in treating a wide range of sexually transmitted infections.

TRADITIONAL HEALTH TECHNOLOGY

Government and nongovernmental agencies can cooperate with traditional medical practitioners on various levels. They can be used to disseminate information on preventive measures, environmental issues, and civic development. Western technocrats should understand the beliefs of local people before they engage in developmental activities. In fact, Western institutions should realize that fitting use of locally appropriate, available, and affordable technology is an important element.

Fitting use of traditional medical practitioners suits the definition of primary health care as outlined by the World Health Organization. Primary health care was defined as "essentially health care based on practical, scientifically sound, and socially acceptable methods and technology made universally accessible to individuals and families in the community through their full participation and at a cost that the community and country can afford to maintain at every stage of their development in the spirit of self-reliance and self-determination."

One observer of the African health situation observed that it should be obvious that the local community must participate in selecting technology considered appropriate with necessary advice from all supporting partners. Technology is appropriate when it takes into account local expertise and resources and when it enables people to cope better, resulting in self-esteem and encouraging self-responsibility and self-reliance. Professional rigidity stifles the identification, development and use of appropriate technology and tends to encourage "quick-fix" technical interventions with lip-service to people's "empowerment."

Another observer states that the greatest health care challenges in the developing world are around nutritional and communicable diseases, and health service infrastructure—particularly the need to make services accessible to rural communities.

Therefore, traditional health practitioners can be used to incorporate both western and local technologies as deemed fit. For instance, traditional healers can be asked to distribute condoms and brochures on HIV/AIDS, while at the same time they continue to prescribe their herbal remedies to those already afflicted with sexually transmitted infections.

THE ESSENCE OF TRADITIONAL PRACTITIONERS

Traditional health practitioners are useful in local communities to disseminate information on appropriate technology because they are familiar with local people. There is considerable mutual trust between the healer and his clients such that methods fit more comfortably with local beliefs than do unfamiliar foreign attitudes. In most cases, traditional healers speak the same language as the client and communication is usually effective. The healer does not normally interfere with cultural beliefs and taboos. Since healers are normally in the nearby villages, travel is made easier.

REFERENCES

Allied Health and Health Education: Multidisciplinary Approaches. (1992), Piggs Peak: W. K. Kellogg Foundation Report. Battle Creek, Michigan.

THE TRADITIONAL MEDICAL PRACTITIONER IN ZIMBABWE

Her Principles of Practice and Pharmacopoeia

(Adapted from M. Gelfand, S. Mavi, R. B. Drummond, and B. Ndemera Mambo Press, Harare, 1985)

INTRODUCTION

It is difficult to separate religion from medicine in the faith of the Shona, since they are closely linked to each other. The *n'anga is* not only a minister of religion but also a diagnostician and healer. She achieves this skill, it is claimed, by being spiritually endowed. She is able to contact the spiritual world and so learn which of the ancestral spirits in a family is responsible for the illness or death or, if it should be an evil person, who caused it and what measures should be taken to remove this influence. Once the *n'anga* learns the reason for the illness, she proceeds to find out what are the requirements or offerings that have to be made in order to propitiate the offended spirit or, in the case of a witch, what action should be taken to eradicate the evil already perpetrated.

FIGURE 3
Traditional healer in Zimbabwe.

THE *N'ANGA* OR GENERAL MEDICAL PRACTITIONER

The traditional folk practitioner in Zimbabwe is the *n'anga* or *chiremba*. Most *n'anga* are spiritually endowed and have the gift of healing and divining. It is generally believed that these special powers are given to the individual concerned either by a *mudzimu* (spirit of a departed relative) or by a *shave* spirit (the spirit of someone unrelated who had a talent for healing). Figure 3 shows a traditional healer from the Matopos area of Zimbabwe, taken in 1997.

THE PRINCIPLES AND PRACTICE OF THE *N'ANGA*

The *n'anga* follows a medical protocol that has its own special characteristics. She treats disease. Like those who practice scientific or Hippocrene medicine, her aims are to find the cause of the illness and to remove its effects with her treatment. However, the principles involved in the diagnosis and treatment of disease by the *n'anga* are different from those of a European doctor.

In contrast to the European doctor's sensory perception of the causation of disease, that of the *n'anga is* based on a mystical basis, rational though her argument may be. Thus, she claims that disease or death may be due to the upset of one's mudzimu (ancestral spirit) or to the evil practices of a person (muroyi). By her special diagnostic procedure of divination with her bones or through her healing spirit that speaks through her, the *n'anga* learns not only the cause of the illness but also what must be done to propitiate the offended spirit or to overcome the evil of the witch. Only then does she enquire about the symptoms and prescribe medicines to cure them. She does not take a history relative to the body structure and function, nor does she make a thorough physical examination of the patient. Thus, it is in the realm of the causality of disease that the practice of a *n'anga* and that of a Western or scientific doctor becomes difficult to understand, as with a Western Christian Scientist, spiritualist, or faith-healer.

Not only is the *n'anga* a diagnostician but she is also a therapist, employing herbal medicines along lines similar to those used in ancient Egypt. The idea behind this is probably the belief that for every human illness there is a plant that possesses the property of neutralizing its effect. Whilst the *n'anga* occasionally employs parts of the body of an animal or bird, her remedies are more often composed of at least one plant, sometimes two or more compounded together, producing a mixture that is consumed in a liquid form. Again, one or more roots may be ground into a fine powder, burnt to charcoal in a piece of broken clay pot (*chayenga*), and stored in horns or tins. When required it is rubbed with the finger into incisions (*nyora*), usually cut over the painful part of the body as for instance over rheumatic joints, on the back, over painful chest walls, or on the site of an aching head.

CLOTHING AND EQUIPMENT

The *gona* or *nyanga,* as it is alternatively termed, is either a horn or a calabash containing special medicine. For most *n'anga* they are essential items. The *n'anga* usually employs a calabash or a large horn into which she dips a wooden rod. The rod is usually simply a twig of *Gardenia volkensii* or *G. ternifolia*. The *gona is* filled with oil, generally castor-oil from the seed of *Ricinus communis* (*mupfuta*). A little honey or sugar and small pieces of roots and twigs, often of various species of *Loranthus* (*gomarara*), are added to its contents.

Another prized possession of the *n'anga is* the *muswe* (whisk), which is the tail of a zebra, ox, lorie, wildebeest, or other animal. It is one of the items most usually passed on from an *n'anga* to her successor. Its principal use is to drive out evil spirits.

Other pieces of equipment important to a *n'anga* are her *hakata* (bones), her wooden plate from which she eats and into which a patient puts the payment for her treatment, the rattle (*hosho*), *mbira* and drums played to appease the spirits, her walking stick (*tsvimbo*) and axe (*gano*), which are insignia of her profession rather than put to any practical use, and her snuff container (*nhekwe*), which used to be made from the neck of a calabash or is more often fashioned from wood and may be ornate.

ORGANIZATION OF THE PRACTICE

An *n'anga* has a fixed place from where she practices, usually her house. In rural areas there is no sign to indicate where this may be, but in town it is becoming common for them to advertise, at least with a board labeled "Chiremba" outside their place of practice. Usually the patient herself comes to the *n'anga* or sometimes only her relatives, leaving the patient at home. It is only rarely that an *n'anga will* visit a patient. If treatment is likely to be prolonged, an *n'anga* may occasionally keep patients at her house.

In order to ensure that she gets maximum benefit from her visit to the *n'anga*, it is customary for the potential client to inform her spirit of her intent. She should then put some money in a plate and clap her hands, saying, "Sekuru tichamboona chiremba titungamireyi" ("Grandfather, we are going to visit a *n'anga*, can you direct us?"). It is believed that her spirit will then direct her to the right *n'anga*.

The patient usually visits the *n'anga* with one or more of her close relations, such as her father, mother, or brother, to make certain that all questions connected with her family, which are asked by the *n'anga*, can be answered. If the patient is too ill to come, the relatives visit the *n'anga* to be told the nature and cause of the malady. She then visits the patient at her home and prescribes the medicine, or she may request that the patient be brought to her.

Usually the patient waits outside the house until called in by the *makumbi*. When she has entered the house, the *makumbi* asks her to put some money in a wooden plate; the usual amount ranges from 50 cents to one dollar. At this stage of the proceedings the *n'anga* is ready to divine the nature and cause of the illness; this she may do by becoming possessed by her healing spirit and going into a trance or by throwing the *hakata* (bones).

After three days, should there be no improvement in the patient, the *n'anga* tries an alternative, but should there still be no improvement, she may refer her to another *n'anga* or hospital. Usually she suggests the name of an *n'anga*, whom the patient should consult, but she does not hold a consultation with the other *n'anga*. There is no such procedure as consultation with physicians or surgeons as takes place in Western medicine. When referring her patient to another *n'anga*, she may throw her *hakata* to be certain that she has chosen the correct person.

DIAGNOSIS AND METHODS OF TREATMENT

In some instances the *n'anga*, will merely treat the patient with herbs without divining. The decision whether to divine or not depends to some extent on whether the patient considers her complaint to be normal or abnormal: in other words whether the disease is natural or caused by some external agency, such as a witch or angered spirits. A patient may suspect that her complaint is abnormal if it is prolonged and does not respond to the treatment being applied. So it is the patient who usually requests that divination be performed.

The *n'anga* has three basic and important principles of treatment: prevention, finding the cause, and curing the illness. These fundamental principles are precisely the same as those followed by scientific medical practitioners who look for an objective cause, such as a bacteria, virus, or growth that is responsible for the disease found in the body. The *n'anga*, on the other hand, believes that the cause of illness may frequently be of spiritual origin. The cause, be it bacteria, virus, or other agency, can, according to their belief, be introduced into an individual by an evil spirit that may be an avenging one (*ngozi*) or a witch (*muroyi*). The spirit able to enter the person when the family spirit (*mudzimu*) is for some reason not affording its usual protection. Another type of spirit that is believed to cause illness is a *shave*, the spirit of a stranger who died nearby and was not buried in the traditional way.

There are three kinds of divining lots or dice, all three of which are commonly referred to as *hakata*: wooden bones, *mungomo* seeds, and *shoro* (bones of animals). Which type of dice a *n'anga*, uses depends on her place of origin and training. For instance, *mungomo* are more popular in Matabeleland, *shoro* in the eastern parts of Zimbabwe, and *hakata* in central Mashonaland. In the main centers all three types are used.

TREATMENT

Once the spiritual cause of the illness has been established, the next stage in the treatment of the patient is the prescription of medicines. Medicines are commonly given in the form of powders, decoctions, infusions, or ointments.

Fine powder to be taken orally is generally prepared by stamping the roots in a *duri* (stamping block). The roots are first dried in the sun for several days and then crushed into small pieces in the *duri*. Ointments are prepared by mixing the prepared potions with oil from the seeds of the castor oil plant (*Ricinus communis*) (*mupfuta*). To prepare a decoction, the roots, bark, or leaves are boiled in water for a short while and the resultant liquid may be used immediately or stored in bottles. Similarly an infusion may be prepared by steeping the required portions of the plants in water for a short period or for several hours.

Medicine may also be given intravaginally or as an enema. When used in the vagina its purpose may be to prevent or induce abortion, to dilate or constrict the vagina, or to increase fertility. The medicine is left in the vagina for two to three days. When used as an enema, the medicine is normally intended to cure constipation or fever.

Inhalations are commonly used for respiratory disorders such as asthma and to drive away bad spirits. The aromatic ingredients are usually made into a form of powder (*mbanda*), which is added to water already boiling. The patient's head and neck are covered by a sheet, towel, or blanket, and the rising steam is directed under the covering so it is inhaled while still hot. A variation of this method is for some very hot stones to be dropped into the boiling water just before the patient commences inhaling. Another method is for the powdered ingredient to be put directly on the fire and the patient inhales the fumes arising from the burning embers. Sometimes the patient is given some of the powder to use as snuff, or she may be required to smoke it in the form of a cigarette.

When it is wished to rid a hut of evil spirits, the aromatic preparation is put in a clay pot with water, and the *n'anga* dips her *muswe* (tail) into the liquid and sprinkles it over the room and its occupants. *Mbanda* may also be burned in a piece of clay pot (*chayenga*), and the smoke emanating from it fumigates the hut.

A popular method of treatment is scarification, in which linear incisions (*nyora*) are made with a razor blade or sharp metal instrument (chisvo). These *nyora* are often made in pairs close to each other and are about 1 cm in length. An irritant powder is rubbed into the incisions, leading to an increased blood supply to the affected parts. Usually *nyora* are made at the site of the pain, on the joints for rheumatism, on the temple for headaches, and perhaps on the back, abdomen, chest, and elsewhere. *Nyora* are also made in the treatment of suspected witchcraft or to prevent it. In this case, they may cover a large proportion of the body.

Application of a *murumiko* to a painful or swollen part is a common method of treatment. Traditionally a *murumiko* is a small horn from which the tip has been removed and the opening thus formed is covered with beeswax. Another type of *murumiko* consists of a half of a tennis ball. The *murumiko* is applied especially in the treatment of rheumatism, pleurisy, painful abdominal conditions, or other disorders in which severe and continuous pain is experienced. The principle is to create a vacuum by the air being withdrawn over the affected part. The underlying skin is drawn up or sucked into the space created. This causes an increased flow of blood, dilatation of the vessels, and at least an initial swelling and redness of the part. The *n'anga* pierces the wax with a pin and then applies the opposite open end to the skin of the affected area. She sucks vigorously through the pinhole until she has created sufficient vacuum. When the *murumiko* becomes firmly attached, she releases her lips, closing the passage in the wax and leaves the instrument attached for some 20 minutes. The *murumiko is* said to suck out or remove impure blood from the infected parts or, in the case of *chitsinga* or *chipotswa*, foreign bodies like insects, bones, and snakes.

CASE STUDY OF TREATMENT BY *N'ANGA*

When a patient arrives at the consulting room of the *n'anga*, she is asked to wait outside until invited to enter by the *makumbi* (assistant of the *n'anga*) or by the outgoing patient. She is asked to remove her shoes before going in as a mark of respect. She may be accompanied by anyone she wishes. No smoking is permitted.

Some *n'anga* use divining sets (*hakata*); others divine without them through the guidance of their healing spirit. When an *n'anga* divines with *hakata*, she asks the patient to make the first throw, and after noting how they lie, she proceeds to throw them himself. Whether she divines with the help of *hakata* or without them, the *n'anga* does not ask the patient many questions about her symptoms. The questions asked are mainly whether the patient has come for divining alone, herbal remedies for her complaint, or for both.

The *n'anga* then tells the patient the spiritual cause of her illness, which spirit is annoyed, whether a witch is responsible, or if the illness is due to natural causes. She then inquires about the patient's reaction to her findings. The patient may now ask questions or argue about the pronouncements of the *n'anga*, who then answers questions or settles her doubts. Usually, however, the patient agrees with what has transpired. While divining, some *n'anga* mentions the patient's actual symptoms before being told what they are. When the divining session is over, the *n'anga* asks the patient whether she wants medicine for her ailment. If she does the assistant is told which preparation to give her, but in particular cases the *n'anga* herself dispenses them.

Another group of *n'anga is* that of pure "herbalists" who do not divine. Patients come to them, tell their symptoms, and are given the necessary medicines. Sometimes patients are referred to them by *n'anga* who divine. *N'anga* who divine with the aid of their healing spirits without *hakata* are by far the largest group and are followed by those who use divining devices. The smallest group is composed of *n'anga,* who provide medicines only and do not divine. A very small number divine for a patient in the presence of other patients; out of the ten *n'anga*, we encountered only one of this type. Whatever the patient's symptoms, divination is performed so that the healing spirit (*mudzimu* or *shave*) of the *n'anga* reveals their cause, but as yet the *n'anga* does not enquire about the complaints. In most cases, both a witch and the *mudzimu* are responsible for the illness, since no witch or evildoer alone can harm an individual whose family spirit protects her. But if it is annoyed with her or one of her family, it withdraws its protection, allowing the evil person or witch to harm her.

An *n'anga* divines primarily to find out why the patient's is stricken with illness, not what her symptoms are. Therefore, these are not asked for until after the spiritual causes have been determined. She is always careful not to mention the name of a witch, giving a vague description of the type of person she is and leaving it to the patient or her family to decide which of her suspects fits into the description.

The 42 patients who were interviewed agreed that most *n'anga* are not completely accurate in their diagnosis, and most of them said they did not altogether rely on the findings of the first *n'anga* and that they usually seek confirmation from at least two others. However, they all agreed that a small percentage of *n'anga* are able to make accurate diagnoses.

MEDICINAL PLANTS

Fred V. Jackson

In many regions of Central and South America, people use plants for a wide variety of needs, from construction of roofs for their homes to weaving baskets and bags, and for reasons as commonplace as food for sacred rituals. In this article, however, our interests lie in the medical realm. There has been renewed awareness in the use of medicinal plants as an alternative to conventional medicines, so scientists have been exploring the rainforests in search of new plant discoveries. Though the forest has existed for years, we have only recently begun to recognize the need to explore its vast wealth of resources.

This article presents a list of medicinal plants that have been used by the local indigenous people in Peru along the Amazon. It should serve as a guide to motivate you to find other sources of plants native to the specific country you will be visiting. Many of these plants have been used for years by the people who share a habitat with the plants. As you look through the list, note the many uses, in addition to medicinal purposes, of each plant.

MEDICINAL PLANTS

Annona squamosa, "Custard Apple," "Sweetsop." Flesh of this tropical fruit tastes like sweet custard. Some medicinal uses include boils, common colds, diarrhea, fever, stomachache, rheumatism, insect and parasite repellant. Leaves are used to make cough syrup in some South American cultures.

Artocarpus heterophyllus, "Jackfruit." Cultivated edible fruit pulp eaten for its sweet flavor. It has an unpleasant odor. Enormous fruit weighing 15 to 20 kilograms are sometimes used to feed livestock. Medicinal uses are for astringent, laxative, diarrhea, sores, stomach problems, toothaches, tumors, and, in some cases, leprosy.

Brunfelsia grandiflora, "Moca pari." Used for ornamental, medicinal, and ritual purposes. In many S.A. cultures used for arthritis, as a diuretic, and for rheumatism, fever, snakebite, and yellow fever. According to Jim Duke, a noted ethnobotanist, the plant is also used to cure impotence.

Cassia reticulata. Beautiful yellow-flowering ornamental tree used by many tribes for various stomach disorders. Leaves of the Cassia can be used as an insect repellant by rubbing them on clothes when your DEET has run out.

Carica papaya, "Papaya." One of the tastiest of tropical fruits, it is also the best aid in digestion problems. This is especially true for people who have a hard time digesting certain foods like red meat and chicken. Other curative properties of papaya are as an antibiotic, diuretic, and laxative. Corns, diarrhea, fever, flu, gas, toothache, warts, and scorpion stings can also be alleviated by papaya.

Cymbopogon citratus, "Lemongrass." Many people use it for making carbonated drinks and use its fragrant leaves in cooking. Crushed leaves are used as an insect repellant. In Peru, shamans use crushed leaves mixed with other herbs to wash their hair before taking the ritual drink "ayahuasca." This ceremonial drink is administered by the tribal shaman to anyone who wants to seek a spiritual awareness, visions, and closeness to the natural surroundings of the rainforest. Additionally, lemongrass has medicinal uses for colds, coughs, fever, flu, headaches, digestive problems, mouth sores, and toothache pain.

Croton lecheri, "Sangra de drago," "Dragon's blood." A relative of the Poinsettia family is one of the most interesting curative plants. The shaman Don Antonio Montero took a few drops of the red latex from a tree, rubbed it into the palm of his hand, and a white powder formed. This powder is used mainly for healing wounds and cuts.

Ficus insipida, "Oje." The latex or sap of this large tree is used internally for removing parasitic worms by some cultures of Panama and Peru. According to Jim Duke's *Amazonian Ethnobotanical Dictionary*, people who take this medicine must stay out of direct sunlight, avoid eating greasy and salty foods, and not talk to strangers for a prescribed number of days. If these instructions are not followed, the patient's skin will turn white.

Hymenaea courbaril, "Stinking Toe." A large tree with interesting hooflike foliage. Although it has many medicinal uses, the most popular uses are eating the pulp from the seedpods or using the pulp to make a refreshing beverage. Many Central American and South American cultures use this tree for constipation, respiratory ailments, diarrhea, headaches, malaria, sores, bruises, and foot fungal infections.

Ixora coccinea, "Flame-of-the-woods." Cultivated tropical ornamental shrub noted for its glossy foliage and deep red flowers has medicinal value in the rainforest. It is used for various stomach disorders including diarrhea and dysentery.

Lantana camera, "Yellow Sage." Anyone who has seen a formal garden will undoubtedly recognize this cultivated beauty. Fragrant foliage with clusters of pink or yellow and red or orange flowers can be found in a number of places in the world. This plant has many medicinal uses such as treating colds, coughs, chicken pox, fever, flu, cuts, inflammation, itch, measles, malaria (quinine substitute), sores, tetanus, toothache, and yellow fever,

Mimosa pubic, "Sensitive plant" or "Shame plant." An interesting houseplant that is found in many parts of the world including Central America. When the leaves are touched, they fold curiously inward. Medicinally, there are a number of uses as a sedative, for diarrhea and dysentery, headaches, and insomnia.

Psidium quajava, "Guava." Edible tropical fruit used medicinally to control common cold, coughs, diarrhea, indigestion, hemorrhoids, scabies, sores, sprains, stomachaches, toothaches, wounds, and the dreaded evil eye.

Zea mays, "Maize," "Corn." Widely grown throughout the New World, corn has other uses beside the sweet kernels of the cob. Beverages, animal feed, and baked goods can be made from corn. There are many medicinal uses, but the best is as a diuretic. Other uses include treatment for bladder, kidney, and prostate ailments. As a liquid, corn can be used to soak swelling feet of pregnant women.

These plants are just a few that are used throughout the New World for their medicinal properties. This list is meant to inform the traveler that in certain instances when conventional resources are depleted, there are other local resources available. Many books and Web sites exist to further inform the reader of the world of botanical medicine and ethnobotany before venturing out to parts unknown.

REFERENCES

Duke, James A., and Rodolfo Vasquez. (1994). *Amazonian Ethnobotanical Dictionary*. Boca Raton: CRC Press.
Duke, James A. (1997). *The Green Pharmacy*. Emmaus, PA: Rodale Press.
Gelfand M. et al. (1985). *The Traditional Medical Practioner in Zimbabwe*. Zimbabwe, Africa: Mambo Press.
Graf, Alfred Byrd. (1978). *Exotic Plant Manual*, 5th Edition. New Jersey: Roehrs Company.

Johnson, Timothy. (1999). *CRC Ethnobotany Desk Reference*. Boca Raton: CRC Press.
Werner, David. (1992). *Where There Is No Doctor*. Berkeley: The Hesperian Foundation.
Web Sites
  Dr. Duke's Phytochemical and Ethnobotanical Databases
  http://www.ars-grin.gov/cgi-bin/duke/ethnobot.pl
  The WWW Ethnobotany Resource Directory
  http://hammock.ifas.ufl.edu/~michael/EB/list.htrnl

WATER AND HEALTH

Rachel Salguero

*(Ms. Salguero has studied water supplies in San Lucas Tolimán, Guatemala, C. A.
She developed original teaching programs on water ecology for school children
in the school system there.)*

Water comprises over half (57 percent) of the human body. Safe water promotes good health,
while contaminated water can lead to illness and even death. There are numerous contaminants
that can threaten the water supply. Articles related to this one are in the water supply chapter,
particularly the sanitation article, the water supply article, the sand filters article, the VIP latrine
article, and the composting toilet article.

WATER CONTAMINANTS

Microbiological

Drinking water is never pure in the chemical sense. Ideally, drinking water should not contain
any microorganisms known to be pathogenic. It should also be free from bacteria indicative of
excremental pollution. The coliform group of organisms is the primary indicators of bacterial
contamination because they are easy to detect even after considerable dilution. The following
pathogenic organisms can be transmitted via a contaminated water supply.

Amebas. These parasites are transmitted through the stools of infected people and through
drinking water or food. They can cause bloody diarrhea as well as abscesses in the liver.

Giardia. This parasite lives in the gut and causes intermittent/chronic diarrhea that is yellow
and frothy.

Helminths. These infective stages of many parasitic roundworms and flatworms can be transmitted to humans through drinking water by a single mature larva or fertilized egg. The guinea
worm (see Figure 4) and the human schistosomes are hazards that are primarily encountered in
unpiped water supplies. The guinea worm is transmitted by freshwater crustaceans, that may be
collected in plankton nets and examined for parasitic larvae under a microscope. Schistosomes
are worms that migrate to the human bloodstream, causing schistosomiasis. They are spread by
water snails.

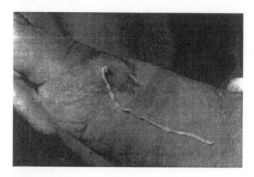

FIGURE 4
Female guinea worm emerging from foot.

Legionella. These can grow in water heating and cooling systems. Legionella is the cause of Legionnaire's disease, a respiratory infection that is potentially fatal.

Shigella. These bacteria that cause bloody diarrhea and fever.

Typhoid Fever. This is an infection of the gut spread by contaminated food and water. It begins like a cold or flu and then proceeds to fever, trembling, and delirium.

Chemical and Physical

Contamination from industries such as mining and smelting or from agricultural practices can lead to increased chemical and physical hazards in the drinking water supply.

Organic Compounds. These compounds, derived from pesticides, herbicides, and various household and industrial chemicals, can pose a threat to humans who consume them via drinking water. Organic compounds carried in water that have been implicated in various cancers are benzene, carbon tetrachloride, p-dichlorobenzene, 1,2-dichloroethane, 1,1-dichloroethylene, simazine, tetrachloroethylene, toxaphene, trichloroethylene, and vinyl chloride. In addition, elevated levels of 1,1,1-trichloroethylene have been shown to have adverse effects on the liver and nervous system.

Inorganic Compounds. Inorganic compounds such as asbestos, barium, cadmium, and chromium have also been shown to have adverse effects on human health when ingested from the water supply. Asbestos has been implicated in cancer; barium has been shown to have circulatory effects; cadmium has shown effects on the kidney; and chromium can lead to liver, kidney, respiratory, and circulatory disorders. These are three particularly common inorganic poisons.

- *Lead and Copper.* Lead from natural/industrial deposits, plumbing, solder, and brass alloy faucets can lead to kidney and nervous system damage. Copper from natural/industrial deposits, wood preservatives, and plumbing can cause gastrointestinal irritation.
- *Nitrates.* These compounds can cause methemoglobinemia ("baby blue" syndrome) in which hemoglobin binds carbon dioxide rather than oxygen. Nitrates can enter the water from fertilizers, sewage and septic systems, animal waste, and natural deposits.
- *Fluoride.* Too much can cause skeletal and dental flourosis (a brownish discoloration of the teeth).

HOW CONTAMINANTS GET INTO THE WATER SUPPLY

Contaminants of any kind that are found within a particular watershed will eventually drain to a common outlet, such as a particular lake, underground aquifer, or reservoir. At any point in the watershed, precipitation runs off the land surface and collects in natural and manmade drainage pathways. Some precipitation seeps through the ground and soil to emerge in another water body, and some seeps deeper and goes into the groundwater supply.

Groundwater is often connected to surface water bodies such as lakes and reservoirs. Lakes that seem to have no major inlet may be fed primarily by groundwater. Contaminants that seep into the groundwater supply can eventually enter a drinking water supply.

SPECIAL CONCERNS

Clean water for drinking, preparing and cooking food, washing the body, and washing clothes is essential for community health. Extreme poverty or low socioeconomic conditions deny these standards to much of the developing world. Indigenous citizens who survive are adapted in part to their water supply. The visitor or traveler is much more at risk than is the local. In general, the following measures can be helpful.

1. Keep drinking water safe. Protect wells and springs from dirt and animals by putting fences or walls around them. Use cement or rock to provide good drainage around the well or spring so that rain or spilled water runs away from it.
2. Keep outhouses (latrines) at least 20 meters from the source of water. If there are no outhouses, go far away from where people bathe or get drinking water to defecate.
3. A clean container must be used to transport water from the source to its site of consumption. When not in use, the container should be kept covered with a clean cloth or lid. Clean cups only should be dipped into the water.
4. If water for drinking or cooking is collected from a rainwater tank, the water is safest if it enters the tank through a screen or a filter to keep out leaves, dirt and insects. The tank should be kept covered and emptied and cleaned at the beginning of each rainy season. Take water from it either through a tap or with a hand pump, not by dipping unclean containers. In general, a barrel of still water is an excellent breeding ground for vectors of disease, such as the mosquito.
5. Water used for drinking or preparing foods (especially for infants and children) should be boiled or filtered. Filtration does not disinfect as well as boiling, but it does remove some disease-causing germs and the eggs of some worms. If there is a shortage of fuel for boiling water, a low-cost purification method is to put the water in a clear plastic bag or bottle and leave it in direct sunlight for a few hours to kill most microbes. Iodine tablets are effective for pathogenic organisms. Filtration (except with ion-exchange resins) does not remove organic or inorganic chemicals, although slow sand filtration will remove pathogens.
6. To prevent schistosomiasis, which is endemic to much of Africa, avoid contaminated water. The snails that are hosts to the infective forms of the parasitic worm live in slow-flowing or stagnant water. Water should not be drawn from shallow or vegetated parts of a pond or steam. If contaminated water does contact the skin, rapid drying within five minutes or rubbing with alcohol may prevent infection. The disease is also called Bilharzia.

The prevention of enteric infection and disease will result in improved nutrition, since infections tend to decrease food intake and increase metabolic losses. Diarrhea also produces malabsorption of nutrients as well as chronic, subclinical enteric disease, associated with impaired tests of intestinal function. Another article in this chapter deals with infant diarrhea.

To decrease the incidence of enteric infection, water supply and environmental hygiene must be improved, coupled with behavioral changes through education. Once these changes can be effected, better nutrition and therefore greater resistance to disease in general will result.

WATER AS MEDICINE

In the developing world, one of the most common causes of death in children is severe dehydration resulting from diarrhea. By giving water with a pinch of salt and a scoop of sugar, dehydration can often be prevented or corrected, allowing the victim to fight off the intestinal infection—more details are in the infant diarrhea article.

In general, in areas where medicine is not readily available, water itself can serve as medicine. For example, hot soaks or compresses can be used for infected wounds. Consuming liquids can help reduce fever. Drinking plenty of water can also help treat minor urinary infections. Drinking water and breathing hot water vapors can help ease cough, asthma, bronchitis, pneumonia, and even whooping cough. Finally, gargling with warm saltwater can help alleviate sore throat or tonsillitis.

ASSESSING WATER QUALITY

Several factors go into the assessment of water quality.

1. *Dissolved oxygen (DO)*—90 percent or more DO is considered healthy for a body of water. The oxygen in water is derived from the atmosphere or from photosynthesis. Buildup of organic wastes can lead to high bacterial consumption of oxygen during the decomposition of this organic matter. In such a case, species that cannot tolerate low levels of DO would be replaced by pollution-tolerant organisms.
2. *Fecal coliform (FC)*—A high FC count (over 200 colonies/100 mL of water) indicates a greater chance that pathogenic organisms are present. For drinking water, 1 colony/100 mL water is the highest acceptable amount.
3. *pH*—In the United States, the acceptable H ion concentration is 6.5 to 8.5 (on a scale of 0–14). Increased amounts of nitrous oxides and sulfur dioxides, primarily from auto and coal-fired power plant emissions, are converted to nitric acid and sulfuric acid in the atmosphere and fall as acid rain. Most organisms have adapted to life in water of a specific pH and may die if it changes even slightly. Also, very acidic waters can cause heavy metals to be released into the water, which can cause deformities in the young of the exposed organisms.
4. *Biochemical oxygen demand (BOD)*—A measure of the quantity of oxygen used by microorganisms in the aerobic oxidation of organic matter. In rivers with high BOD levels, much of the available DO is consumed by aerobic bacteria, taking it from other organisms that need it.
5. *Temperature*—should be within 13 to 20°C. Temperature has an influence on the amount of oxygen that can be dissolved in water, the rate of photosynthesis by algae and larger aquatic plants, the metabolic rates of aquatic organisms, and the sensitivity of organisms to toxic wastes, parasites, and disease. Humans influence water temperature by contributing to erosion, cutting down trees that shade rivers, and thermal pollution.
6. *Total phosphate (TP)*—TP concentrations in water should be less than 0.1 mg/L. Plants need phosphorus to live and grow, but when there is too much, a large amount of *eutrophication* (enrichment of water with nutrients, in this case, phosphorus) occurs. Overenriched waters result in extensive algal growth ("blooms"), depleting the supply of DO.
7. *Nitrates*—Levels should not exceed 1 mg/L. Nitrogen in the form of ammonia and nitrates acts as a plant nutrient and therefore causes overeutrophication. Nitrates come from sewage, fertilizers, and runoff from feedlots, dairies, and so forth.

8. *Turbidity*—Lower turbidity (cloudiness) is better for water quality. Turbidity can be raised because of soil erosion, waste discharge, urban runoff, or algal growth. Raised turbidity can cause water to become warmer as suspended particles absorb heat from sunlight, causing DO levels to fall. Photosynthesis will also decrease in the case of high turbidity due to lack of light.

9. *Total solids (TS)*—High TS concentrations lower water quality, whereas low concentrations may limit the growth of aquatic life. TS includes those that are dissolved and those that are suspended and can include urban area runoff, wastewater treatment plants, leaves and plant materials, and decayed plant and animal matter.

Besides regular testing, it is important to note color, taste, and odor of water. Consumers may turn to alternative, perhaps unsafe, sources if their drinking water is not aesthetically pleasing.

REFERENCES

Joubert, L., and A. Gold, (1990). "What Is a watershed?" *Natural Resources Facts.* Fact sheet #90–20.

Lave, Lestor B., and A. Upton. (1987). *Toxic Chemicals, Health, and the Environment.* Baltimore: John Hopkins Press.

Rice, R. (1985). *Safe Drinking Water.* Michigan: Lewis Publishers, Inc.

Torun, B. (1983). "Environmental and Educational Interventions Against Diarrhea in Guatemala." In Lincoln C. Chen (Ed.), *Diarrhea and Malnutrition: Interactions, Mechanisms, and Interventions.* New York: Plenum Press.

US EPA Office of Water. (1994). *Is Your Drinking Water Safe?* EPA 810-F-94-002.

Werner, David. (1996). *Where There Is No Doctor.* California: The Hesperian Foundation.

WHO. (1987). "Limits on Releases into the Air and Water." In H. W. de Koning (Ed.), *Setting Environmental Standards: Guidelines for Decision Making.* Geneva WHO, 37–47.

CONSTRUCTING A SMALL RURAL CLINIC

Steven G. McCloy and Barrett Hazeltine

PURPOSE OF A SMALL RURAL CLINIC

A small rural clinic is in most cases part of a more complex health care system. It provides first aid and deals with a few common diseases, such as malaria and diarrhea. Minor surgeries may be performed if trained staff is available. Medicines are often dispensed. Immunizations may be given. Young children may be brought to such clinics for check-ups. This sort of clinic is especially valuable for treating young children who need immediate care. The clinic would refer serious cases to an area hospital and arrange for transportation. A rural clinic may also include a birthing facility. In some cases it will be the center for health education and for community health workers who visit homes. Other names for such clinics are dispensaries or health posts. A typical clinic provides care for an area containing about 4000 people.

A typical rural clinic could be staffed by one or two medical auxiliaries, sometimes called physician's assistants. If a birthing facility is present, a trained midwife and perhaps a traditional birth attendant would also be on the staff. In some cases a physician will make regular visits.

PLANNING THE CONSTRUCTION OF A SMALL RURAL CLINIC

The first issue is the site and position. An important consideration is the necessity to leave space for future additions—more treatment rooms and a larger reception area. In many cultures it is expected that relatives of patients will stay at the hospital to care for the patients and perhaps bring them meals. Space for these relatives and for cooking should be provided. Space should be reserved for housing of the clinic staff—not all will be residents of the village. Of course a source of water, probably a well, is needed as well as a latrine. Latrines are vital for community health. A latrine should be close enough to the school, clinic, or home to be reached easily, but far enough away (30 meters is recommended by many experts) to keep the main building free of odors and potential contamination. A latrine must not contaminate ground or surface water that may enter springs, wells, or fields. A Peace Corps manual comments: "In many areas, community acceptance of latrines as an integral part of a clinic project may be more important than any other construction idea." Wells and Latrines are discussed in the Water Supply Chapter of this book.

Other considerations in choosing a site include locating it centrally for the convenience of patients, close to a roadway suitable for vehicles, away from standing water harboring mosquitoes or other sources of disease. It should be away from woodlands, which reduce breezes, can be susceptible to fire, and harbor pests like rats and ants. A building should be sited open to prevailing breezes. The soil should be firm enough to build on and the site well drained. A special concern is that no standing water be close to the clinic, to reduce the risk of malaria and other parasites.

FLOOR PLANS

A building for a small health clinic should contain a treatment room, a reception area, an office area, and a sleeping room to accommodate patients that may have to spend the night. Most patients will not spend the night, however. Very sick patients would be transported to a regional hospital, and the less sick patients would go home. A possible floor plan is shown in Figure 5. The building includes a covered porch to serve as a waiting area for patients and families.

As illustrated in Figure 5, doors should be positioned (especially in the sleeping room) not only to allow ventilation, but also to maintain as much privacy as possible. Windows should be

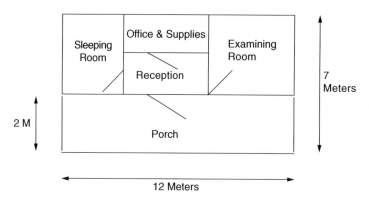

FIGURE 5
Small rural clinic (not to scale).

placed to promote cross ventilation. Ventilation and screening are very important. When electricity is not available for artificial lighting or air circulation, large windows can provide natural sunlight and ventilation for patient examinations. Traditional forms of construction are appropriate: baked clay bricks, earth bricks, bamboo, and so forth. Construction methods are described in Chapter 5 of this book. A separate building is usually built for maternity patients, which would typically include an office for the midwife, a delivery room, and a sleeping room with several beds. A separate maternity building is necessary not only for privacy, but also to prevent the spread of infection to mothers and babies.

Most small rural clinics do not contain a laboratory because the diagnoses are made on the basis of symptoms. Specimens needing tests would be sent to a regional hospital. Two situations where onsite testing may be necessary are the presence of drug-resistant malaria and giving of blood transfusions.

An important concern for any medical facility is the proper disposal of waste. Syringes and needles need especial attention because they are attractive to children. Used x-ray film is also an attractive nuisance. Blood containers and discarded blood, as well as other body fluids, should not be allowed to circulate. Incineration and deep burial of ashes is a possible disposal mechanism.

OUTFITTING A SMALL RURAL CLINIC

A major determinant of the type of equipment to be installed is whether electric power is available. If the clinic cannot be connected to the national electric power grid then installation of photovoltaic panels can significantly improve the operation of the clinic. (Photovoltaic panels are described in the Energy chapter of this book.) Photovoltaic panels produce much less electrical energy than can be taken from the power grid—it would be difficult to power an x-ray machine, for example, from a feasible photovoltaic installation. However, the small amount of power available from solar panels can make a significant difference. A particularly important use for electricity is emergency lighting—the alternative might be an open flame close to volatile chemicals. Photovoltaic panels can power a small refrigerator to store pharmaceuticals in a manner that does not lose their effectiveness.

Equipment

Equipment and supplies will vary with the level of practitioner skills in the rural clinic. Some health promoters or *feldshers* "barefoot physicians" are well trained and able to dispense relatively sophisticated medications and provide advanced care. Others have less training and experience. Equipment and supplies for a rural clinic may include the following.

- Thermometers
- Battery-operated otoscope
- Stethoscope
- Snakebite kit
- Clean bandages (if sterile not available)
- Bucket or tray for washing, soaking and sterilizing syringes and needles
- Scales, either commercial or homemade for weighing babies
- Soap and chlorine bleach

- Cooking pan (for field sterilizing)
- Razor blades, needles and thread, scissors, sharp knife
- Pencils, paper or index cards
- Source of water
- Oral rehydration salt kits or raw materials
- Rubbing alcohol
- Lantern or other reliable light source
- Urine dipsticks for testing diabetes, infection, etc.

Medications

A list of possible medications could be endless. The list below is offered as very basic medications that should serve well in most parts of the world. An excellent resource is *Where There Is No Doctor*. Though not exhaustive, a list of stocked medications could include the following.

- Antibiotics (ciprofloxacin, amoxicillin, penicillin, doxicycline)
- Antiparasitics (albendazol, metronidazol, antimalarial, and others appropriate to the location)
- Antifungals (clotrimazol, griseofulvin)
- Oral rehydration supplies
- Medication for cough and cold symptoms (acetaminophen, antihistamine, cough drops)
- Antiseptics (bacitracin ointment, soap)
- Antacids

Careful consideration of local religious, ethical and cultural concerns should be made before deciding to dispense, display or discuss family planning, condoms, and other contraceptives. Posters and other materials about AIDS, malnutrition, infant diarrhea and other illnesses are excellent for wall hangings in the clinical area and waiting areas. A good source of information about family planning is the John Snow Incorporated Center for Women's Health; 44 Farnsworth Street; Boston, MA 02210; USA; Phone: 617-482-9485.

REGIONAL HEALTH CENTERS

These centers may be in part subsidized by the government and may have more equipment and supplies. Ill and injured may be treated there, or triaged to an area hospital for surgery or other care. In many developing countries, regional health centers are simply empty buildings without staff, supplies or equipment. The sophistication of a regional center directly bears on the skills of its staff. If there is no skilled staff, scarce resources are better used to support the local health promoter or to facilitate transfer to an area hospital.

Medical equipment is expensive but can sometimes be received as donations. One such source is the American Medical Resources Foundation (AMRF). Their web site states, "The American Medical Resources Foundation donates used, but functional medical equipment to hospitals serving the poor worldwide. AMRF also develops and provides training programs for medical equipment repair technicians and hospital managers responsible for maintenance, repair and calibration of medical equipment." The AMRF may be reached at: American Medical Resources Foundation; P.O. Box 3609; 56 Oak Hill Way; Brockton, MA 02404-3609; USA; Phone: 508-580-3301; Fax: 508-580-3306; E-mail: info@amrf.com.

REFERENCES

James E. Herrington, Jr. *Understanding Primary Health Care for a Rural Population.* Published by Volunteers in Technical Assistance 1600 Wilson Boulevard, Suite 710, Arlington, Virginia 22209 USA Telephone: (703) 276-1800, 1985.

Médecins Sans Frontières. *Refugee Health: An Approach to Emergency Situations.* Macmillan, 1997.

David Werner. *Where There is No Doctor.* Hesperian Foundation. PO Box 1692, Palo Alto, CA 94302 USA, 1998. The Hesperian Foundation publishes a unique series of books and pamphlets to aid health care in the developing world.

DIARRHEA

In rural villages in Africa, infant mortality is high; perhaps 150 out of 1,000 babies die before reaching the age of one. Many of these children die from diarrhea dehydration. A closer look at a village may help us understand why. Most of the people are subsistence farmers with a diet restricted to a few vegetables and grains and a little meat and fish. Their resistance to illness is low. The lack of running water in homes makes cleaning difficult, although in some African societies each meal begins with passing a bowl of clean water for rinsing hands. The emphasis in this article is on infant diarrhea but some notes on adult cases are given at the end.

Three beliefs prevalent in village society are relevant to infant diarrhea. The first is that people do not realize human wastes carry disease. The general awareness of the cause of disease may be low, and the concept of germs may not be known at all. The second belief, present in many cultures, is that diarrhea is best treated by withholding fluid intake. Such treatment, of course, increases the severity of dehydration. The third belief, that commercial infant formula is more effective or convenient than breastfeeding, has created a trend away from breastfeeding.

Besides diet, lack of running water, and these beliefs, the proximity of a health center to the village is an important determinant of mortality from infant diarrhea. Infant diarrhea can kill quickly, so timely treatment is important. Health centers can be a long walk from homes. Mothers take care of infants and also work in the fields and do household chores. Thus, it may take some time to realize that a baby is very sick and get her or him to the health center.

The result of all these factors is that dehydration due to acute diarrhea is a significant factor in infant mortality. It is the most prevalent childhood disease. In 1980, 5 million children—approximately 10 children per minute—worldwide—under the age of five died from acute diarrheal dehydration. In many less industrialized countries diarrhea disease is one of the two most common reasons for visiting health clinics, pharmacies, and hospitals.

DIARRHEAL DEHYDRATION

Now let us consider diarrheal dehydration in more detail. Acute diarrhea is an attack of sudden onset, which lasts usually 3 to 7 days but may last up to 14 days. Death is usually caused by two factors: dehydration and malnutrition. Dehydration is directly linked to the fluid loss, as well as to the common practice of withholding fluids from the patient. Malnutrition is due to the victim's loss of appetite and the body's inability to absorb food properly.

Dehydration results when the body loses more liquid than it takes in. This can happen with severe diarrhea, especially when there is also vomiting. It can also happen in very serious illness when a person is too sick to take much food or liquid. People of any age can become dehydrated,

If you lift up the skin and you can still see the fold after you let go, the child is dehydrated.

How to prevent dehydration in diarrhoea

Many of the herbal teas and soups that mothers give to children with diarrhoea do a lot of good because they help get water back into the child. Breastfeeding provides both water and food and should always be continued.

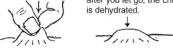

KEEP BREAST FEEDING
A BABY WITH DIARRHOEA

Demonstrations help children understand how important it is to give a child with diarrhoea as much water as he is losing. For example:

Water level

As long as just as much water is put back as that which is lost, the water level will not go down (so the child will not get dehydrated.)

A child with diarrhoea needs 1 glass of liquid for each time he has a loose stool.

Giving lots of liquid to a child with diarrhoea may at first increase the amount of diarrhoea. This is all right. *THE DIRTY WATER MUST COME OUT.* The important thing is to *be sure that the child drinks as much liquid as he loses.*

The Special Drink
How to mix it

The Special Drink, made from SUGAR, SALT and WATER is especially good for children (or adults) with diarrhoea.
It is simple to make:

MIX: SUGAR   +   SALT   +   WATER

one level teaspoonful of sugar   +   a little at the end of a spoon   +   one glass of water

OR   a scoop of sugar   +   a pinch of salt   +   one glass of water

A special plastic spoon will be available. Use this to teach everyone the right amount of sugar and salt for each glass or cup.

BEFORE GIVING THE DRINK TASTE IT - IT SHOULD BE NO MORE SALTY THAN TEARS

How to give the Special Drink

Start giving the Special Drink AS SOON AS diarrhoea begins. A child should drink one glass for each stool he passes.

If the child vomits the drink, keep giving him more. A little of it will stay in his stomach. Give it in sips every 2 or 3 minutes. If the child does not want to drink, gently insist or coax him to do so.

CHILD

One glass each stool

ADULT

Two glasses each stool

Keep giving the drink every 2 or 3 minutes, day and night, until the child urinates normally (every 2 or 3 hours.) Older children and their mother can take turns through the night.

Warning signs

Take child with diarrhoea to the health centre if he:
1 SHOWS ANY OF THE SIGNS OF DEHYDRATION
2 Cannot or will not drink
3 Vomits so much he cannot drink
4 Makes no urine for 6 hours (time from dawn to noon, or noon to dark)
5 Has diarrhoea so often he cannot drink one glass per stool
6 Has blood in his stool
7 Diarrhoea lasts more than 2 days

Figuring out how well the children are doing

Encourage the children to discuss how they can find out such things as:
- how much they have learned;
- whether other people in the community have learned some of the same information;
- how many of the children have put their new knowledge (about diarrhoea) to work;
- if fewer babies and children suffer and die from diarrhoea as a result of this activity.

If this discussion takes place when they are beginning the activity they will be able to gather the information they need to make comparisons as they go along.

Counts can be made each month (or after six months or a year) to see, for example:
- how many children (or their mothers) have made the 'special drink' for those with diarrhoea?
- how many cases of diarrhoea have there been?
- how many children have died?

Ask any child who has used the Special Drink for a brother or sister with diarrhoea to tell the story to the school, explaining how he (or his mother) made and used it and if it seemed to help.

Ref: AS3E   CHILD-to-child January 1979

FIGURE 6
Instructions for preparing ORT mixture.

but dehydration develops more quickly and is most dangerous in small children. Very severe dehydration may cause rapid, weak pulse; fast, deep breathing; fever; or fits (convulsions). When a person has watery diarrhea or diarrhea and vomiting, quick action is needed. Figure 6 shows how one can test for dehydration by lifting the skin.

Diarrhea is caused by some 25 different pathogenic parasites, virus and bacterial, which cause diarrhea by various means—usually by damaging the lining of the walls of the intestine, thereby causing the malabsorption of food and thus diarrhea and vomiting. These pathogenic organisms are transmitted in many ways. The most common methods are either by contamination of drinking water or through contaminated foods. Food contamination results from flies moving from animal or human feces to the food.

The most likely source of contamination in a village is an open latrine because it can spread disease in different ways. (Latrines are discussed in another article in this chapter.) Improperly built latrines may contaminate groundwater supplies; such as shallow wells, if the pit is not properly lined with concrete or some other appropriate material. Open latrines give flies access to the feces, and then the flies can go on to uncovered food. Another source of contamination are feces from livestock roaming around the village.

ORAL REHYDRATION THERAPY

A treatment for diarrhea is oral rehydration therapy (ORT), which basically consists of giving the victim a large amount of water mixed with salt and sugar. Parents and children can make the mixture very easily. People involved can understand what has to be done and can act rather than just watching the victim die.

Oral rehydration therapy, although simple and inexpensive, is effective. Most patients (90 to 95 percent) can be treated with ORT alone regardless of the cause of the diarrhea or the age of the patient. The average cost of treating one patient with ORT is 50 cents, while intravenous therapy costs approximately $5. In areas where ORT programs have been implemented, there have been substantial decreases in infant mortality.

The principle of ORT consists of the replacement of fluid and essential salts that are lost during acute diarrhea. A dehydrated patient is treated in two phases. The *rehydration phase* consists of the replacement of fluid and the essential salts lost by vomitus and stool production. The *maintenance phase* follows, in which one compensates for continuing abnormal losses due to diarrhea and vomiting as well as losses due to normal activity—sweating, breathing, and urination. It is important, therefore, to continue giving the mixture until the diarrhea is passes. One guide is to continue to give it until the patient urinates normally, every two or three hours.

Of course, ORT is not a cure for acute diarrhea. In most case, diarrhea only lasts for a limited time, and the ORT helps the patient survive the effects of the illness.

THE ORT MIXTURE

The ORT mixture is simply salt, sugar, and water and can be prepared at home. A page from a manual for health workers in the Third World is shown in Figure 6. It shows how the mixture is made and administered. It also suggests how a health care worker can explain the treatment. An advantage of having those who need the mixture learn how to prepare it is that powerful members of the village cannot withhold it from those less powerful. Here are two more recipes.

Two More Ways to Make "Home Mix" Rehydration Drink

1. Making the drink with sugar and salt (raw sugar or molasses can be used instead of sugar): In one liter of clean water put $1/2$ level teaspoon of salt and 8 level teaspoons of sugar. Caution: Before adding the sugar, taste the drink and be sure it is less salty than tears (see Figure 7).

WITH SUGAR AND SALT (Raw sugar or molasses can be used instead of sugar.)

In 1 liter   put   half of a   and
of clean   level   8 level
WATER   teaspoon   teaspoons
                     of **SALT** of **SUGAR**.

CAUTION: Before adding the sugar, taste the drink and be sure it is less salty than tears.

To either Drink add half a cup of fruit juice, coconut water, or mashed ripe banana, if available. This provides potassium which may help the child accept more food and drink.

FIGURE 7
Making the "Home Mix" rehydration drink.

WITH POWDERED CEREAL AND SALT
(Powdered rice is best. Or use finely ground maize, wheat flour, sorghum, or cooked and mashed potatoes.)

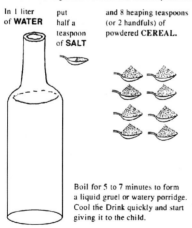

In 1 liter   put   and 8 heaping teaspoons
of **WATER** half a   (or 2 handfuls) of
              teaspoon   powdered **CEREAL.**
              of **SALT**

Boil for 5 to 7 minutes to form a liquid gruel or watery porridge. Cool the Drink quickly and start giving it to the child.

CAUTION: Taste the Drink each time before you give it to be sure it is not spoiled. Cereal drinks can spoil in a few hours in hot weather.

FIGURE 8
Another way.

2. Making the drink with powdered cereal and salt (powdered rice is best, or used finely ground maize, wheat flour, sorghum, or cooked and mashed potatoes): In one liter of water put $1/2$ teaspoon of salt and 8 heaping teaspoons (or two handfuls) of powdered cereal. Boil for 5 to 7 minutes to form a liquid gruel or watery porridge. Cool the drink quickly and start giving it to the child (see Figure 8).

The drink should be tasted each time before it is given to be sure it is not spoiled. Cereal drinks can spoil in a few hours in hot weather.

To either drink, add cup of fruit juice, coconut water, or mashed ripe banana, if available. This provides potassium that may help the child accept more food and drink.

Give the dehydrated person sips of this drink every 5 minutes, day and night, until he or she begins to urinate normally. A large person needs 3 or more liters a day. A small child usually needs at least 1 liter a day, or 1 glass for each watery stool. Keep giving the drink often in small sips, even if the person vomits. Not all of the drink will be vomited.

In some countries packets of oral rehydration salts (ORS) are available for mixing with water; such packets are useful in hospitals. These contain a simple sugar, salt, soda, and potassium. However, homemade drinks—especially cereal drinks—when correctly prepared are often cheaper, safer, and more effective than ORS packets.

CARE OF BABIES WITH DIARRHEA

Diarrhea is especially dangerous in babies and small children. Often no medicine is needed, but special care must be taken because a baby can die very quickly of dehydration. Continue breastfeeding, and also give sips of rehydration drink. If vomiting is a problem, give breast milk often but only a little at a time. Also give rehydration drink in small sips every 5 to 10 minutes. If there is no breast milk, try giving frequent small feedings of some other milk or milk substitute (like milk made from soybeans) mixed to half normal strength with boiled water. If milk seems to make the diarrhea worse, give some other protein (mashed chicken, eggs, lean meat, or skinned mashed beans mixed with sugar or well-cooked rice or another carbohydrate, and boiled water).

CHOLERA

"Rice-water" stools in very large quantities may be a sign of cholera. In countries where this dangerous disease occurs, cholera often comes in epidemics and is usually worse in older children and adults. Severe dehydration can develop quickly, especially it there is also vomiting. Treat the dehydration continuously, and seek medical help. Cholera should be reported to the health authorities.

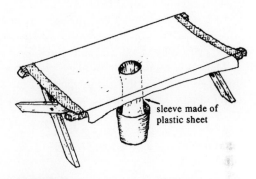

A cholera bed, shown in Figure 9, can be made for persons with very severe diarrhea. Watch how much liquid the person is losing, and be sure he or she drinks larger amounts of rehydration drink. Give the drink almost continuously, and have the person drink as much as he or she can.

FIGURE 9
Cholera bed.

WHEN TO SEEK MEDICAL HELP IN CASES OF DIARRHEA

Diarrhea and dysentery can be very dangerous, especially in small children. In the following situations you should get medical help.

- If diarrhea lasts more than four days and is not getting better, or more than one day in a small child with severe diarrhea
- If the person shows signs of dehydration and is getting worse
- If the child vomits everything he or she drinks, or drinks nothing, or if frequent vomiting continues for more than three hours after beginning rehydration drink
- If the child begins to have fits, or if the feet and face swell
- If the person was very sick, weak, or malnourished before the diarrhea began (especially a little child or a very old person)
- If there is much blood in the stools. This can be dangerous even if there is only very little diarrhea

REFERENCES

Where There Is No Doctor. The Hesperian Foundation, PO Box 11577, Berkeley, California 94712-2577, USA.
Hirschhorn, Norbert, and William B. Greenough III. (1991). "Progress in Oral Rehydration Therapy." *Scientific American,* Vol. 264, No. 5, 50–56.

MALARIA: RESURGENCE OF A DEADLY DISEASE

(Based on "Resurgence of a Deadly Disease," Ellen Ruppel Shell,
Atlantic Monthly, *Vol. 280, No. 2, August 1997, pp 45–60)*

Malaria kills roughly twice as many people worldwide as AIDS. Drugs no longer work against some strains, and mosquitoes in diverse parts of the United States now carry the disease. Why aren't we doing more to fight it?

Malaria has returned in full force to North Africa, India, Southeast Asia, China, South America, and the Caribbean. Worldwide incidence of the disease has quadrupled in the past five years, and resistance to available drugs for prevention and treatment is growing rapidly.

Malaria is transferable in blood. Temperature and humidity may well be among the most important factors in the rate of spread of the disease, yet we have only a vague notion of what effect, if any, climate change will have on malaria transmission.

DEADLY, CAGEY, AND RESILIENT

It is not unheard of for an African child to go to school in the morning and die of malaria in the afternoon. Understanding this adds perspective to the public obsession with other fast-acting microbes, such as so-called "flesh-eating" bacteria and the Ebola virus. The 1995 Ebola outbreak in Zaire that inspired Hollywood and transfixed the world caused approximately 250 deaths over a period of six months. More than 20 times that many Africans die every day of malaria.

The parasites are transmitted to human beings through the saliva of the female mosquito. Once injected, the parasites quickly retreat to the liver, where they mature and multiply. It is not until they reemerge in the bloodstream and invade the blood cells that symptoms appear. By this time the parasites have reproduced thousands of times. They thrash about, popping blood cells, clogging blood vessels, debilitating their host, and in some cases killing within hours.

We can be infected repeatedly, or carry them for any amount of time, without developing a full resistance to them. Malaria leaves a population alive but debilitated—with victims to be taken care of.

Malaria inspires such fear that 40 years ago powerful, long-lived insecticides were sprayed into the interior walls of homes. Workers tramped through villages in North Africa and Asia with spray guns loaded with DDT. Malaria rates went down, and hopes for public health soared. But the optimism was short-lived. It soon became clear that spraying was most effective in areas that were only marginally malarious—areas such as Egypt and southern Europe, where the parasite had only a slippery hold. Meanwhile, for complex reasons, mosquitoes where malaria was solidly endemic started showing resistance to the insecticides.

A SOLUTION THAT ISN'T

Shell visited several counties in Africa to learn of developments in the war against malaria. Mosquito bed nets impregnated with insecticide are widely touted as a cheap and easy way to prevent malaria in the tropics. "Nets might help, but they will not cure the problem," a medical officer in Senegal explains. "Bed nets reduce exposure, but they cannot stop it. Not all mosquitoes bite people in bed." The officer himself says he doesn't use a net. "The West brought the nets. But people here do not believe that they work, and most of us do not use them. Nets keep out the breeze, and Senegal is a very hot country."

Jean-Francois Trape, a French physician and malaria expert, argues that providing people with bed nets to ward off malaria in areas where it is highly endemic may in fact increase their risk of dying from the disease by reducing but not eliminating their exposure. The *gambiae* mosquito can transmit the malaria parasite in a single bite. If the transmission rate is reduced to fewer than two attacks a year, a person may lose his or her partial resistance to the disease between bouts, therefore, in areas where malaria transmission is reduced but not eliminated, adults get sicker and older children die at only a slightly later age than do their counterparts where the rate of transmission is higher. "I have two children, and I can tell you, as they get older, you invest in them more," Trape said. "In African culture people prepare for the death of an infant but not a ten-year-old. The nets in many cases simply delay the inevitable."

BAD AIR

In the tropics mosquitoes breed not just in swamps and ponds but everywhere—in upturned soda bottles, discarded automobile tires, and animal footprints. Even if African nations had the money to drain swamps, doing so would not be enough. Large-scale spraying for mosquitoes is equally impractical in most areas, and even small-scale spraying is problematic. DDT is one of the more benign pesticides known. It is certainly among the cheapest. But it is banned or heavily restricted in most African nations, as in the United States, and the alternatives, pyrethroid insecticides, are expensive.

Even if insecticides were magically made available free to all, they would probably fail to solve the global malaria problem. The same chemicals that are used in small amounts to combat malaria are used in large amounts to keep bugs away from crops, a practice encouraged by the demands of agribusiness. Resistance to insecticides is growing rapidly because insects have tremendous exposure to the chemicals and hence ample opportunity to develop protective tactics.

MALARIA FIGHTS BACK

Equally troubling is the ability of the malaria parasite to develop resistance to drugs. Malaria has for centuries been a treatable disease. Quinine, an antimalaria compound that is extracted from the bark of the South American cinchona tree, is one of the oldest effective pharmaceuticals in existence. This is still useful in the treatment of acute falciparum malaria. But it is expensive and short-acting, has side effects ranging from dizziness to deafness, and fails to prevent relapses. Chloroquine, developed during World War II, killed malaria parasites and had none of the drawbacks of quinine, but resistance to it has spread throughout the world. A relatively new drug, mefloquine, a synthetic analogue of quinine, is expensive and is losing its effectiveness in many regions. U.S. doctors prescribe these drugs and others to prevent and treat malaria of various kinds, but their continued use is too expensive for most local people.

Short-term visitors to malaria-infested areas of the world should seek advice about the most effective malaria prophylaxis in their area. Malaria pills should first be taken two weeks before entering the infested area and their use continued for several weeks after leaving.

Qinghaosu, a drug made from *Artemisia annua*, a cousin of wormwood that grows wild in fields and thickets across China and is a favorite of natural healers the world over, is showing the strain of overuse as well. Practitioners of traditional Chinese medicine have used qinghaosu to treat fevers for something like 2,000 years, although whether in its natural form it actually kills the malaria parasite is unclear. Like quinine, qinghaosu fails to prevent relapses of the disease and may be neurotoxic.

The Thai-Cambodian border region harbors a variety of malaria that responds to no known drug. This multidrug-resistant malaria has spread from the border deep into Thailand and Cambodia and as far as India, Bangladesh, and Nepal. Brazil is another cradle of drug resistance.

Trials with experimental vaccines have not been encouraging, partly because the number of strains of malaria is so large.

WHAT TO DO

The Senegalese medical officer advocates a more nuanced approach, one that would contain malaria region by region until it retreats into the background of everyday life in the tropics. "Malaria is our environment; it is part of what we are," he says. "To control it we must take care not to disturb the equilibrium, to respect the local ecology and customs, to work with human behavior as well as with science." Education would do much to stem the tide of the disease—as would public health measures such as the drainage of standing water and the judicious use of insecticides and drugs.

Pedro Alonso, a physician and an epidemiologist at the University of Barcelona who oversaw the first vaccine trials in Africa, explains how malaria was wiped out in Spain with just such a low-key approach. The Spanish Civil War in 1936–1939 somehow prompted a resurgence of malaria in his country, and it hung on stubbornly into the 1950s. Pesticides were tried, and they helped—but not enough. The Spanish stood back and reassessed the situation—why this resurgence of malaria after war? What had been disrupted? They couldn't be sure what, but something had increased the mosquito population. The trick was to reverse the process. They stocked their ponds and lakes with gambusia, a fish that eats mosquito larvae. "Now Spain has an enormous number of gambusia," Alonso said, laughing, "but almost no malaria."

China's Hunan province, too, has beaten back malaria. Its caseload fell from nearly a million in 1985 to 68,500 in 1993. No new drugs or vaccines were used, but rather a mixture of strategies was employed. Swamps were selectively drained, and malaria cases were quickly treated with both traditional and Western medicines.

MORE INFORMATION CAN BE OBTAINED FROM:

Division of Public Information, World Health Organization, Geneva.
Malaria Division, Public Health Service Center for Infectious Diseases, Chamblee, GA 30333.
Tropical Disease Office, Pan American Health Organization, 525 23rd Street, Washington, DC.

PALLIATIVE CARE FOR AIDS PATIENTS

(Based on Handbook on AIDS Home Care,
*World Health Organization, Regional office for South-East Asia
New Delhi, India 1996)*

At some point in the disease process of AIDS, there is nothing more that can be done to effectively treat the opportunistic infections or completely relieve the symptoms that they cause. The infections or illnesses have progressed beyond what medicines can cure. At this point, the goal of all care (medical, nursing, religious, and psychological) is to keep the person as comfortable as possible and to maintain his or her dignity. In some places this is called palliative care.

WHEN DOES THIS TIME BEGIN?

It is often difficult to decide when the focus on medical treatment should stop and care for the dying should begin. The change in care may begin, for example, with the following.

- When medical treatment is not available or is no longer effective
- When the person says he or she is ready to die and really does appear to be very sick. This is clearly different from someone who is depressed for a time and who only needs encouragement
- When the body's vital organs begin to fail

WHERE CAN YOU PROVIDE CARE FOR SOMEONE WHO IS DYING?

Care for the dying can be provided in a hospital or in the home. Most people prefer, or are forced by circumstances, to remain at home. However, some people may not want to actually die in the home. They may want to stay at home until the last moment but, either because of their own or the family's wishes, may want to go to the hospital to die. If this is the case, a plan for transporting them will need to be thought out.

WHAT ARE THE GOALS OF CARING FOR SOMEONE WHO IS DYING?

- Keeping them comfortable and protecting them from problems that can make them feel worse
- Helping them to be as independent as possible
- Assisting them in grieving for, and coping with, the continuing losses they experience
- Helping them and their families prepare for death; this may include making a will, tending to relationships in the family or the community, and arranging for the transfer of responsibilities
- Keeping them within the community and family groups for as long as possible. Family members can bring them into this part of their lives even when it seems they are too ill to enjoy or understand what is going on.

WHAT CAN YOU DO TO MEET THESE GOALS?

Give Comfort

- If the person is in constant pain, make sure that pain medication is available in regular doses. It should not be taken just when the pain is really bad.
- Use relaxation techniques, such as encouraging deep breathing or giving back rubs or body massages.
- Continue basic physical care to keep the person clean and dry and to prevent skin problems and stiffness of joints.

- Encourage communication within the family and community. People with AIDS and those they love need to feel that they are not outside the love and life of their community. Help them use this time as a chance to heal old wounds and to make peace with each other. This will help to increase the comfort and acceptance of the whole family.
- Provide physical contact by touching, holding hands, and hugging. Provide or arrange for counseling if desired—for example, from religious representatives. They can be very helpful for spiritual counseling.

Allow the Sick Person Independence

- Accept the person's own decisions, such as a refusal to eat or get up (or even a *demand* to get up when you think that resting would be better for them.)
- Respect requests—for example, not wanting to see visitors.
- Ask them what they are feeling. Listen and allow the person to talk about how they feel.
- Accept the person's feelings of anger, fear, grief, and other emotions.

Prepare for Death

Talk about death if the person wishes to. Many people feel that it is not good to talk about the fact that someone is going to die, as if mentioning death is a wish for death. But discussing death openly can help the dying person prepare for death. It may take great courage to talk about it, but it can be a big help for the person to feel that their concerns are heard, that their wishes will be followed, and that they are not alone.

One of the most common worries is for the future of the children in a family. People may fear that their children will be hungry or lack money for school fees after they have died. Begin planning with relatives, friends, or orphan programs for the future of the children. It will ease such worries if the person knows that suitable arrangements have already been made.

The person may be worried about being in pain as he or she nears death. The fear can be lessened by knowing what it will be like. If the person asks, describe what might happen, such as difficulty in breathing or passing in and out of consciousness. If pain medications are available, reassure the person that they will be used to prevent unnecessary pain.

The person may be worried about what will happen after they die. The anxiety can be lessened by helping them to write a will, by planning and writing down details such as funeral arrangements, and by discussing spiritual beliefs, perhaps with a representative of the person's religion.

PRECAUTIONS WHEN SOMEONE HAS DIED OF AIDS

Immediately after death, you need to follow the same rules in dealing with the body as you did when helping the person through her or his illness. Hands should be protected when cleaning and laying out the body, particularly if there are body fluids such as diarrhea or blood, and then washed with soap and water afterwards. Wounds on hands or arms should be covered with a Band-Aid or bandage.

Shortly after the person has died, the virus will also die. HIV can only live and reproduce inside a living person. Therefore, you do not need to worry about special precautions during the funeral itself. The person can be either buried or cremated, according to local custom.

EXPOSURE TO INDOOR AIR POLLUTION FROM BIOFUEL STOVES IN RURAL KENYA

Daniel M. Kammen
Majid Ezzati
B. M. Mbinda
Burton H. Singer

THE DETERMINANTS OF EXPOSURE TO INDOOR AIR POLLUTION

In a study of five towns in Kenya it was found that more than 80 percent of low-income households and approximately 30 percent of medium-income ones use wood for cooking. The numbers are likely to be even higher in rural areas with no access to gas and electricity. Exposure to indoor air pollution, especially to particulate matter, has been implicated as a causal agent of acute respiratory infection (ARI) and eye infection. "Improved" (high-efficiency, low emission) cook stoves have been a celebrated means of reducing exposure to indoor air pollution. On average the improved (ceramic) wood stoves reduce the level of suspended particulate emission by approximately 50 percent from that of traditional open fire (see Table 1). An article on improved wood stoves is in The Energy chapter.

The largest reduction in pollution is associated with transition from firewood to charcoal (and then to kerosene). In comparing the pollution level for different charcoal stoves, a relationship similar to that of wood stoves can be seen: The improved stoves, in average, reduce emission, but this reduction could have been achieved using the metal jiko under its best operating conditions.

Women have higher exposure than men. We attribute this to the extra time spent inside the house by female members of the household. This result suggests the need for reduction of emissions from stoves and modifications to the structure of the cooking area or the time spent near the stove.

TABLE 1
Stove-fuel combinations in the study group

Stove Name	Material		Fuel	Price (US$ Equiv.)	Efficiency (%)
	Body	Liner			
3-stone	N/A	N/A	Firewood	0	10–15
Kuni Mbili	Metal	Ceramic	Firewood	$8	25–40
Metal	Metal	N/A	Firewood and charcoal	$2	10–18
Kenya Ceramic Jiko (KCJ)	Metal	Ceramic	Charcoal	$6	25–40
Loketto	Metal	Metal	Charcoal	$8	

The improved stoves used in the region of study reduce the emissions of suspended particulate matter, although the same reductions could have been achieved using the best-scenario use of traditional stoves. The largest reduction in pollution is achieved by switching from firewood to charcoal. Exposure to indoor pollution is determined by both the level of stove emissions and behavioral factors, with infants and female household members at the highest risk.

THE HEALTH IMPACTS OF EXPOSURE TO INDOOR AIR POLLUTION

Exposure to indoor air pollution, and especially to particulate matter, from combustion of biomass fuels has been widely associated with the incidence of acute respiratory infections (ARI) and eye infections. This association, coupled with the high level of prevalence of respiratory and eye infections in developing countries, has put preventive measures to reduce exposure to indoor air pollution high on the agenda of development and public health organizations. Acute and chronic respiratory infections together account for approximately 15 percent of the total burden of disease in developing countries. In 1996 and 1997, for example, the leading cause of death from infectious diseases was acute lower respiratory infections (3.9 and 3.7 million deaths worldwide for the two years, respectively). For efficient and successful design and dissemination of preventive measures, a fundamental question must be answered: What is the quantitative relationship between the incidence of disease and exposure to indoor air pollution?

We studied the relationship between exposure to suspended particulate matter and the incidence of acute respiratory infection (ARI), eye infection, and headache. For adults, we find significantly higher incidence of ARI, eye infection, and headache among females. Illness incidence rate, especially among adults, is significantly related to average exposure. Cooking duties are performed entirely by women who are exposed to high concentrations within short periods of time. This intensity of exposure is likely to be as important as average exposure in determining health risk associated with indoor air pollution.

We find a positive relationship between exposure to suspended particulate matter and incidence of ARI, eye infection, and headache. An increase in average total suspended particulate concentration would result in an increase in the number of illness cases among adults and a greater increase among children. We also find that women, due to exposure to high intensity of total suspended particulate, are in higher risk than men. From a policy standpoint, this result indicates that a reduction in the average pollution level should be accompanied by a reduction in peak values for effective mitigation in health impacts of indoor air pollution.

REFERENCES

Ballard-Tremeer, G., and K. Jawurek. (1995). "Comparison of Five Rural, Woodburning Cooking Devices: Efficiencies and Emissions." Submitted to *Biomass and Bioenergy.*

Kammen, D. M. (1995). "Cookstoves for the Developing World." *Scientific American,* 273, 63–67.

Kammen, D. M., M. Ezzati, and B. M. Mbinda. (1999). "Determinants of Exposure to Indoor Air Pollution from Biofuel Stoves." *The Proceeding of Indoor Air 99: The 8th International Conference on Indoor Quality and Climate, Edinburgh, Scotland.*

McCracken, J. P., and K. R. Smith. (1998). "Emissions and Efficiency of Improved Woodburning Cookstoves in Highland Guatemala." *Environment International,* 24, 739–747.

Westhoff, B., and D. Germann. (1995). *Stove Images: A Documentation of Improved and Traditional Stoves in Africa, Asia and Latin America.* Brussels: Commission of the European Communities.

DENTAL HEALTH CARE

(Based on What to Do When There Is No Dentist
Murray Dickson
Hesperion Publishing Company, San Francisco, 1996)

PREVENTION

Eat Only Good Healthy Foods

The best food is food that you grow or raise yourself. Traditional food is usually good food. Sweet food, especially the kind you buy from the store, can mix with germs and make cavities—holes in the teeth. Soft food sticks to the teeth easily, and it, too, can make a coating of germs and food on the teeth that starts an infection in the gums—gum disease. Breastfeed to help a child's teeth grow and stay strong.

Clean Your Teeth Every Day

If you do not clean properly, the food that is left on your teeth can hurt the teeth as well as the gums near them. Bits of food stay longer in grooves and "hiding places." This is where both tooth and gum problems start. To prevent problems you must take special care to keep these protected places clean. It is better to clean your teeth carefully once every day than to clean poorly many times a day. There are three places where problems start: grooves on top, between the teeth, and near the gums (see Figure 10).

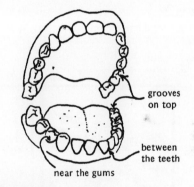

FIGURE 10
Where problems start.

Use a soft brush to clean your teeth. Buy one from the store (be sure it says "soft" on the package), or make a brush yourself. Figure 11 shows how to make a brush.

You can also twist the fiber from inside a coconut husk into a kind of brush: First rub it and shake away the loose bits; then use the end to clean your teeth (see Figure 12).

Whatever kind of brush you use, be sure to clean your back teeth as well as your front teeth. Scrub the tops and sides where the grooves are, and then push the hairs between the teeth and scrub.

Toothpaste is not necessary. Charcoal or even just water is enough. When your teeth are clean, rinse away the loose pieces of food.

Examination and Diagnosis

You can prevent much suffering and serious sickness when you notice and treat problems early. Whenever you hold a health clinic, try to find out how healthy each person's mouth is (see Figure 13).

1. Use a small branch, young bamboo, strong grass or the skin from sugar cane or betel nut.

2. Cut a piece that is still green and soft.

3. Chew one end to make it stringy like a brush.

4. Sharpen the other end so it can clean between the teeth.

FIGURE 11
Making a toothbrush.

FIGURE 12
Making a brush from coconut husk fiber.

During examination, look for the following problems (see Figure 14).

- Cavities, which should be filled when they are still small. Tell the person what is happening and how to prevent it from getting worse or affecting other teeth.
- A tooth that is dark—it is dead. Infection from its root can go into the bone. This can make a sore on the gums.

CAVITIES, TOOTHACHES, AND ABSCESSES

Cavities are holes in the teeth made by the infection called tooth decay. A black spot on a tooth may be a cavity. The tooth may hurt when you eat, drink, or breathe. If you do not fill a cavity, it grows bigger and deeper (see Figure 15).

When decay touches the nerve inside, the tooth aches, even when you try to sleep (see Figure 16).

When infection reaches the inside of a tooth, it is called a tooth abscess (see Figure 17).

A tooth with an abscess needs treatment at once, before the infection can go into the bone (see Figure 18). In most cases the tooth must be taken out. If it is not possible to do this right away, you can stop the problem from getting any worse by doing the following.

1. Wash the inside of the mouth with warm water. This removes any bits of food caught inside the cavity.
2. Take aspirin for pain. Every 6 hours, adults can take 2 (300 mg) tablets, children 8–12 years old can take 1 tablet, and children 3–7 can take $^{1}/_{2}$ tablet. Children 1–2 years old should only take $^{1}/_{4}$ of a 500 mg tablet of acetaminophen (paracetamol) 4 times a day.
3. Reduce the swelling.
   - Hold warm water inside your mouth near the bad tooth.
   - Wet a cloth with hot water and hold it against your face. Do not use water hot enough to burn yourself!

FIGURE 13
Dentist examining child mouth.

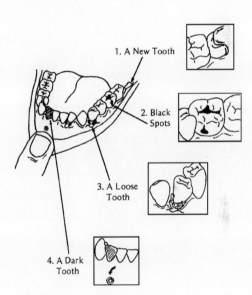

FIGURE 14
Types of tooth problems.

FIGURE 15
A cavity.

FIGURE 16
Decay touches a nerve.

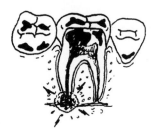

FIGURE 17
Tooth abscess.

A tooth abscess can cause swelling like this.

FIGURE 18
Swelling from a tooth abscess.

Sores

Look for sores under the smooth skin on the inside of the lips and cheeks. Look also under the tongue and along its sides. A sore on the gums may be from an infected tooth. Sores on the inside of the lip or cheek may be from a virus. Sores on the lips or tongue may be cancer.

WHERE TO EXAMINE

Examine people in a light and bright place. It is dark inside a person's mouth, so you need light to see the teeth and gums. Use the sun. Examine either outside or inside a room facing the window. With sunlight alone you will be able to see most places in the mouth well enough. If you cannot, set up a lamp or have someone hold a lamp for you. Reflect the light off a small mouth mirror onto the tooth or gum.

If you have a low chair, lift up the person's chin so that you do not have to bend over as far when you look into the mouth. An even better way is to have the person sit on some books. The person's head can lean back on a piece of cloth. A headrest can be put on a chair. Use an old chair with a strong back. Attach two flat sticks vertically to the chair back, and then tie a strip of clean cloth to the sticks above the back of the chair. Tie it strong enough to support the head but loose enough to let the head lean back.

Instruments

The following three instruments are really enough.

1. A wooden tongue blade to hold back the cheek, lips, and tongue.
2. A small mirror to let you look more closely at a tooth and the gums around it.
3. A sharp probe to feel for cavities and to check for tartar under the gum.

If you have many people to examine, it is helpful to have more than one of each instrument, but be sure they are clean. Dirty instruments easily can pass infection from one person to another. After you finish an examination, clean your instruments in soap and water and then put them in a germ-killing solution.

A GOOD DIAGNOSIS

People have some problems more often at certain ages. When a person first comes in to see you, notice the person's age. Then before you ask the person to open his mouth, look at his or her face for a sore or swollen area.

Causes of Swelling
Child

- Mumps
- Infection in the spit gland
- Tooth abscess

Young Person

- New tooth growing in
- Broken jaw
- Tooth abscess

Adult

- Tumor
- Tooth abscess

Causes of a Sore
Child

- Impetigo
- Vincent's Infection

Young Person

- Fever blisters
- Tooth abscess

Adult

- Tooth abscess
- Bone infection (osteomyelitis)

Examine inside the mouth. Look at the teeth: Is a new one growing in? Is a tooth loose? Is there a dark (dead) tooth? Look at the gums: Are they red? Is there any swelling? Do they bleed? Are the gums eaten away between the teeth?

Touch the sore place. Touching is a good way to find out how serious the problem is. This will help you decide which treatment to give. Push gently against each tooth in the area of pain to see if a tooth is loose. Rock the loose tooth back and forth between your fingers to see if it hurts when you move it. Using the end of your mirror, tap against several teeth, including the suspect. There is probably an abscess on a tooth that hurts when you tap it. Press against the gums with cotton gauze. Wait a moment, and then look closely to see if they start bleeding. Then use your probe gently to feel under the gum for tartar. Carefully scrape some away. Wait and look again to see if the gums bleed. When gums bleed, it is a sign of gum disease.

GUMS

Healthy gums fit tightly around the teeth. Gums are infected if they are loose, sore, and red, and if they bleed when the teeth are cleaned. Infection in the gums is called gum disease. Gum disease, like tooth decay, happens when acid touches the teeth and gums. This acid is made when sweet and soft foods mix with germs (see Figure 19).

   A bubble on the gums below the tooth is a clear sign that the person has an abscess. The abscess may be from the tooth or the gums. To decide, look carefully at both the tooth and the gum around it. A bubble beside a healthy tooth is a sign of infected gums. A bubble beside a decayed tooth is a sign of a tooth abscess. A sore on the gums from a badly decayed tooth appears when a gum bubble breaks open and lets out the pus from inside.

   Infection from gum disease can spread into the root fibers and bone. To stop gum disease and prevent it from coming back, clean your teeth better and strengthen your gums. To strengthen your gums eat more fresh fruits and green leafy vegetables, and rinse your mouth with warm saltwater. Do this every day, even after your gums feel better (see Figure 20).

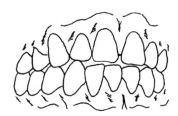
HEALTHY TEETH AND GUMS

CAVITIES AND GUM DISEASE

FIGURE 19
Healthy and diseased gums.

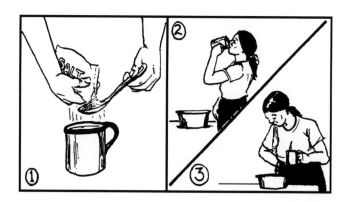

FIGURE 20
Mouth rinse.

MORE SERIOUS GUM DISEASE

Painful gums that bleed at the slightest touch need special treatment. At home, you can do some things to help.

1. Clean your teeth near the gums with a soft brush. Gently push the brush between the tooth and the gum. It may bleed at first, but as the gums toughen, the bleeding will stop.
2. Make your food soft, so it is easier to chew. Pounded yam and soup are good examples.
3. Eat plenty of fresh fruits and vegetables. If it is difficult to bite into a fruit, squeeze it and drink the juice.
4. Rinse your mouth with a mixture of hydrogen peroxide and water. The strength of hydrogen peroxide is important Use a 3 percent solution mixed evenly with water—that is, $^1/_2$ cup of hydrogen peroxide with $^1/_2$ cup of water. Take some into your mouth and hold it there for about two minutes. Then spit it out and repeat. Do this every hour you are awake. Use hydrogen peroxide for only three days; then change and start rinsing with saltwater.

Warning: Read the label to be sure the solution is 3 percent. A mixture with more than 3 percent hydrogen peroxide can burn the mouth.

If you take good care of them, your teeth will last a lifetime.

THIRD WORLD HEALTH CARE NEEDS: A PERSPECTIVE FROM A SMALL TOWN IN GUATEMALA

Elaine L. Bearer

INTRODUCTION

The needs of the Third World in the area of health are protean. In this article, I provide a limited analysis based on my own experiences practicing medicine in Guatemala over the past seven years. My personal contact with health care in the Third World has been in a small town, San Lucas Toliman, in the Western Highlands of Guatemala.

Some of the principles operative in Guatemala can be extrapolated to many, if not all other Third World nations, while other principles concern specific problems unique to the history and ethnic combination in this Guatemalan town. The purpose of this article is to provide a basic description of the problems surrounding health care and disease in this Third World country, with the goal of inspiring engineers and others to participate in their resolution.

In writing this article, I have become even more acutely aware that it is almost impossible to assess the statistical significance of the health care needs even in this small town. Since the beginning of my involvement there, I have tried to perform a needs-assessment analysis. To this end, I and my other American physician colleagues have recruited the assistance of a large number of students, both from the undergraduate and medical curriculums. Despite this effort, we have gained only an anecdotal understanding of the health care needs in this village. Some of the local problems that frustrate our efforts to obtain a needs assessment are (1) difficulty with communication in Mayan languages; (2) lack of a government census, data on births, and causes of deaths, absence of local medical records; (3) indigenous and religious superstitions and beliefs concerning health and sickness; (4) differences between the Mayan cultural traditions and our own, which has resulted in our own cultural assumptions blinding us to health issues; and (5) the complex family network system which channels our work within extended families and away from communities as a whole.

Communication in Guatemala is difficult for many reasons, both logistical and sociological. Telephones, radios, newspapers, and the Internet are not dependable. When I first traveled to San Lucas, there were very few phones, and these were often not in service. The only newspaper was La Prensa, delivered from Guatemala City irregularly. Radio and television reception were poor if at all. Most news travels by mouth; is often full of misinformation, gossip, and opinion; and is only sporadically available. The lack of an educational system results in widespread illiteracy. Word of mouth is the primary method of information dissemination. Verbal information is further distorted, since it often must be translated from one Mayan language to another. Since the peace accords, the phone system has been more dependable, although the post office is virtually never open, and when it is, there are no stamps and usually no pickup or delivery services. Efforts to establish an Internet link are in progress. I witnessed Ethernet cables being installed along the Pan-American Highway last summer (1999) and heard rumors in San Lucas of a Cyber café being equipped with computers. It is not clear what the effect of entry into the Information Age will be on San Lucas. It is unlikely that the rumor network will be replaced in the near future.

Sociologically, communication between Guatemalans living in San Lucas is affected by networks of family ties and allegiances that channel verbally transmitted information within groups. The various Mayan languages are different enough that adjacent peoples cannot understand each other's speech. In San Lucas, there are at least three different Mayan languages: Quiche, Tzutzujil, and Kakchikel, in addition to the national language, Spanish. Many of the women, who receive even less schooling than the men, speak only their own Mayan dialect. Thought process is very different among these people than for us Americans. Their languages not only have a different vocabulary and pronunciation from European languages but the process of describing, analyzing, and relating events is very different. This difference profoundly affects the way a Mayan will describe her or his illness. The concept of time is also quite different from ours, and there are likely to be many other fundamental ways that life and reality are experienced, remembered, and described that differ between Mayan thought and our Western European ways. These differences affect our ability to acquire information about individual illness or to assemble large compendiums of health needs. Any technical advance that would improve communication would be a major contribution for health care.

My experience of health and disease in the San Lucas area has been primarily gained through offering outpatient clinics in the small towns (population 70–250 families) and fincas (plantations) in the surrounding countryside. These clinics ("consultas") are usually conducted in the manner of instruction or advice-giving sessions. The patients are typically mothers and their children. I very rarely see men, who are often working in the fields during the daylight hours when the clinics are most often held. Because of the Mayan dialect, I often had an interpreter, usually a villager, who translated the Mayan into Spanish for me. I kept detailed notes of the complaints, the symptoms and signs of each patient, and the treatment I recommended. The following essay relies on these notes for my impression regarding the prevalence of the illnesses I personally diagnosed and treated. In these outpatient clinics, I did not witness the more serious life-threatening diseases of acute onset. However, I also occasionally attended in the mission clinic, the central health care service for San Lucas, where urgent cases were routinely seen. There I saw a number of other diseases that were either more serious or had reached a critical stage. Although this experience did not provide a basis for statistical assessment, it does allow me some sense of the types of acute-onset diseases present in the area.

BACKGROUND

San Lucas is a market town on the southeastern shore of Lake Atitlan. Its location on the lake in relation to roads and other towns contributes to some of the health problems it faces. San Lucas is reached by one main road that passes up from the coast, around the lake, and on to connect at a distant mountain pass, La Quintla, with the Pan-American Highway. Reached by boat from San Lucas across the lake are a number of other villages, including the larger tourist center, Panajacel, which counts prostitutes of both sexes and IV drug users among its inhabitants, as well as a booming tourist business that brings visitors from all over the world. During the 1970s, American "hippies" established a presence around the lake, bringing with them a drug culture. Some of the smaller lake villages, such as San Antonio Palopoj, do not even have latrines, and human waste washes directly down stony slopes into the lake.

The population of San Lucas is primarily Mayan, although there is a Catholic mission run by four American priests and visited by dozens of Americans over the year, and several Evangelical missions in the town. When I first began working in San Lucas, most of the indigenous peoples worked on any of several large plantations (fincas) or on their own land. Since the peace accords were signed, that has changed. Several of the fincas have changed over from housing the workers and their families year round on the land to hiring migrant workers only for planting and harvest times. This has resulted in great upheaval for the "fin-queros," people who had lived for generations on the plantations. These people were thrown out of their finca-owned housing and off the land. They are now struggling to survive on tiny plots bought for them by the Catholic mission. The peace accords also brought improvements to the roads. This allows workers who live in the San Lucas area to travel to the coast daily on cattle cars to work on the sugar plantations. They bring with them coastal diseases not previously seen in San Lucas, including malaria. The crowded conditions on the cattle cars, where workers are packed in, standing, for the two- to three-hr trip, half of it at humid tropical temperatures, favors transmission of air-borne infectious illness, the most serious of which is tuberculosis.

The Mayan dialect spoke in San Lucas is Kakchiqel, although just north of the town, Tzutzuhil is spoken. There are about 27 different Mayan dialects, and most Mayans cannot understand any dialect but their own. Although Spanish is the national language, many do not speak it. In the outreach clinics I hold every year, more than half the women cannot speak Spanish, although the majority of the men have learned it. This could be because the men are more often responsible for the bartering at market and other family business that involves the Latino business world or because male children are often afforded more years of schooling. In the elementary schools I have had a chance to visit, the vast majority of the students were male. The language difficulties impact health care tremendously, since it is difficult to educate the mother, who is often responsible for preparation of food and maintenance of the household environment, about nutrition, the microbial basis of disease, hygiene, and preventive health care.

The ancient Mayan belief system profoundly affects health care prevention and delivery. Originally suspicious of Western medicine, the Maya have become increasingly dependent on it. There were 13 Shaman in the San Lucas area in 1972, but since 1994 I have not found a single one. Now there are native healers, including herbalists, called "curanderas"; witch doctors of both the good and evil sort, called "bruja/o"; and midwives, "comedrones." In addition to these more traditional native healers, I have also witnessed a number of charlatans, con artists who peddle "Western medicine" without medical training or knowledge. Pharmacists in the town and "doctors" with questionable training and apparently little government licensing validation prescribe Western medicine and give injections for which they receive remuneration. Desperately ill people take the wrong medicine at the wrong dosage, or they take an appropriate antibiotic but for too short a period. This practice is likely to produce antibiotic-resistant organisms and could result in an infectious disease nightmare in the town.

The ready availability of antibiotics and other drugs will contribute to a future problem. San Lucas hosts a large number of pharmacies that sell drugs that could only be obtained in the States with a prescription. These drugs can be purchased in whatever amount the buyer wishes and can afford. The buyer uses her or his own discretion on what drugs to buy. The drug choice is made on the basis of a diagnosis for the illness that can be entirely arbitrary.

The disappearance of the Shaman does not mean that the old belief system is gone. In fact, in a recent study of how mothers' beliefs affect the way they care for their sick children, it was found that most mothers still believe that the evil eye can cause disease. In addition, there are "illnesses" that the Maya perceive that are not included in our Western health/disease concept. Finally, preventive medicine often involves lifestyle changes, particularly in an environment so heavily burdened with infectious organisms as this one. But lifestyle in traditional cultures is often prescribed by the beliefs of good and evil, of the supernatural world and how to please its powerful deities. Furthermore, efforts to change people's behaviors toward healthier lifestyles can cause direct conflicts between beliefs and health. We are establishing a new medical student project to examine belief systems about health and disease in San Lucas.

The major causes of disease in San Lucas are sociological. Poverty with its attendant poor nutrition, lack of sanitation, and poor personal hygiene combine with high birth rates and overcrowding to produce a situation where infectious disease flourishes. In addition to infectious disease, environmental factors also cause illness, including smoke inhalation from wood-burning fires maintained without ventilation in the hovels, and exposure to herbicides and pesticides, which would be banned in the United States. The tropical climate also leads to insect-borne diseases. Finally, trauma from accidents during manual farm labor—including machete wounds, injuries from vehicles, particularly from riding standing on the back of a pickup truck, and osteoarthritis of the spine from the heavy load many carry on their backs daily—are common-place. Although the Guatemalan government claims that cervical cancer is the leading cause of death, I have seen only one case.

The main illnesses that I have treated affecting the people in the San Lucas parish are infections by virus, bacteria, protozoa, and multicellular parasites. Some of these organisms cause infection after ingestion of contaminated soil or water, others infect via an insect vector, and others can penetrate the skin on their own. The two most common illnesses presenting in clinic are infections of the gastrointestinal tract or of the upper airways and lungs.

DIARRHEA

The most common illness I observed in these outpatient clinics was diarrhea of varying severity and associated symptoms. Despite this, given the huge load of pathogenic organisms in the environment, it always astounds me that there are in fact so few of the population affected.

Even mild forms of diarrhea can be life threatening in the undernourished victim, while the severe forms involving vomiting and dehydration are acutely life threatening. In addition, digestive system infections in children that cause pain, vomiting, and diarrhea can leave them with a persistent inability to eat. When a vegetarian diet low in fat is all that is available to them, their digestive tract can shrink, leaving them without the capacity to consume the volumes necessary to maintain adequate caloric intake. I have seen a number of children who have no current GI complaint but whose parent reports a stomach upset in the past and a subsequent refusal to eat enough. These children are often emaciated and near death from starvation, even though they have survived the actual infectious process. The addition of a tablespoon of oil to their meal of rice or cornmeal can reverse the process of starvation.

To help a volunteer to consider how to intervene in these health hazards, I give a brief synopsis of the symptoms, diagnosis, and treatment for the major forms of diarrheal disease I have seen in San Lucas.

Infectious diarrhea can be caused by virus (typically rotavirus), a number of different bacteria, and protozoa, including ameba and giardia. The symptoms of the GI tract upset can often help to diagnose the organism causing it. Viral diarrhea is usually accompanied by nausea and vomiting and a low fever that lasts 6 to 12 hours and resolves quickly and spontaneously. Viral illness usually passes so quickly that it is rare in Guatemala for a person so affected to come to an outpatient clinic such as mine.

In contrast, the sudden onset of frequent, brown, watery stools often with cramping is typical of bacterial infections, including cholera. If the diarrhea progresses to "rice-water" stools and significant water loss, or includes vomiting and the inability to retain fluids, it can rapidly become life threatening, and the victim requires intravenous fluid support. Cholera produces several toxins that interact with the host cells to excite them to secrete water into the bowel lumen. The bacteria itself does not penetrate the bowel wall. Fortunately, these diseases do not typically produce vomiting, and the patient can be managed with oral rehydration using a formula that replaces the water, salt, sugar, and bicarbonate. (Oral rehydration is discussed in the "Diarrhea" article in this chapter.) Antibiotic therapy is usually not necessary or even advisable for these illnesses. In Guatemala, I frequently treat these diseases with oral rehydration and bismuth, which is thought to coat the stomach, allowing the organism to pass through on its own.

Frequent, bloody diarrhea with a fever is typical of the more aggressive intracellular bacteria, *Salmonella, Shigella, enteropathogenic Escherichia coli* (EPEC), and *Listeria monocytogenes*. These bacteria have the ability to invade host cells and replicate within them, thereby bypassing the immune system. In addition to causing diarrhea, these bacteria can invade the bowel wall and cause sepsis and disseminated infection. Listeria is a serious infection in the pregnant woman, since it usually infects the fetus, causing intrauterine demise. Again, the volume of fecal material and clinical symptoms or signs of dehydration determine whether oral rehydration is necessary. There are mixed opinions on whether these infections should be treated more aggressively with antibiotics than those that do not produce watery stools. In Guatemala, Bactrin (Trimethoprin-sulfamethoxyzol) has so far been usually effective.

There are numerous types of protozoa that can infect the GI tract and cause illness. My personal belief is that we have not yet identified most of them. The two we have become the most familiar with are *Giardia lamblia* and *Entamoeba histolytica*. Giardia typically causes a green, frothy diarrhea with upper abdominal pain. Because Giardia attaches to the wall of the duodenum, it induces a fat malabsorption. There is typically no blood in the stool and no fever. In contrast, *Entamoeba h.* erodes the bowel wall by secretion of lytic factors. As a consequence, diarrhea caused by this organism is bloody, although there is usually no significant fever. Both of these organisms are eukaryotic, and thus antibiotics that are effective against them are also toxic to humans. As yet there is only one effective treatment for both: metronidazol (Flagyl). This is an imidazole derivative containing azide. In another form (antabuse) it is used to curb alcoholism because it causes severe nausea and vomiting when taken in combination with alcoholic beverages.

While diagnosis by symptomatology is often accurate, as evidenced by response to specific therapy, these diseases are definitively diagnosed by microscopic examination of the stool. Phase microscopy can determine numbers of bacteria, although stool always contains bacteria, and without any special tests, microscopy alone is not sufficient to diagnose infection with pathogenic bacteria.

Giardia, ameba, and all the worms can be identified in the microscope, with some caveats. Low-grade infections with Giardia can cause significant symptoms without the organism itself being present in the stool. In the United States, endoscopic biopsy of the duodenum can be required for diagnosis. There are two types of ameba, only one of which is pathogenic, but both are very similar in appearance in the microscope. Other simple, more definitive tests that could be performed in the field would be very useful in diagnosing these pathogens such that appropriate therapy could be selected.

The exact route of infection for any of the organisms is not known. All of them are passed by fecal contamination of water, food, or any other material, including clothing or fingers, that is put into the mouth. Those children who suck their thumbs or bite their nails are highly at risk, since these pathogens appear to be ubiquitously present in the environment. If the person preparing the meals, typically the mother of the family unit, touches a child who has been crawling on the dirt floor and then touches food or dishes, this can lead to contamination. The lack of adequate latrines has produced an environment where almost every surface carries some burden of pathogenic organisms.

Of significant interest to me is the fact that many of the local people drink this contaminated water and do not become ill, although I have not found any of our American tourists or volunteers to be so resistant. When the stool of healthy local people is examined by microscope, they do not have a high burden of bacteria and are not carriers of protozoa for the most part. Medical science does not know how they remain uninfected.

Where do these organisms come from? The widespread contamination of the water source is a principal source of infectious disease in the area. The lake serves as the major source of water for San Lucas and many of the tiny pueblos that dot the surrounding countryside. Water is pumped to these villages from the San Lucas town cistern. Although there is running water in most of the homes in San Lucas, this water comes directly from the lake and is not filtered or treated in any other way for purification. When we first wished to analyze the water, we thought we would need to pass gallons of water through a filter to isolate infectious organisms. Instead, we found large numbers of microbes in a single tiny drop of water sampled directly from a faucet in one of the San Lucas houses and examined by phase microscopy. Few of these organisms could be identified, given the crude laboratory equipment available. However, simply by their morphology and motility, it was easy to find ameba, giardia, and trichomonads together with numerous bacterial forms.

Unfortunately, there is no sewage treatment plant in the area. The lake itself serves as the only sewage-treatment center, since it receives the runoff from the streets, including latrine overflow in the rainy season. Houses, businesses, schools, and health centers all have either latrines or more complex cesspools. The cleaning, recycling, and waste removal from these facilities is not known to me.

The water as it leaves the lake is already contaminated and becomes more so on its transit to the central holding tanks and through the city water pipes to the homes. The tanks are virtually never cleaned, and the underground pipes have blind alleys where stagnant water allows proliferation of microbes.

Once in the home, water could be purified. However, because of its altitude (5,000 feet above sea level), purification by boiling is difficult. Water boils at a lower temperature at high altitude, which requires that it be boiled longer to kill the contaminating organisms. Fuel to boil water is scarce, and the smoke and heat from burning cause a host of other health problems. In Quixaya, a smaller village within the San Lucas purview, a project to build solar stoves has been started but is slow to catch on. Although San Lucas is very close to the equator, the skies are not always sunny, and during the rainy season (three to four months of the year), there is often not enough light to boil water to sterilization. Furthermore, the social system of the family is disrupted when there is no hearth with a wood fire burning.

Pathogenic organisms also come from the soil. The life cycle of each pathogen is uniquely suited to maximize its infectivity in the sort of primitive housing conditions found in the San Lucas area. Principal is the unavoidable contact with contaminated soil. Amebas, bacteria, and parasites all have life forms that can withstand desiccation or inundation. These "cysts" can survive for years in the soil, only to become activated when ingested and exposed to stomach acid or to bile.

Some of the parasites infect domesticated animals and are spread to people by the animals. Giardia is commonly carried by dogs, and dogs are a ubiquitous presence in San Lucas. When I began explaining this to the people of San Lucas, a new program to poison stray, semiwild dogs was put into effect, and the following year I saw many fewer cases of Giardia in clinic. Hookworm is spread through swine. Pig feces in the soil produce larvae that invade the human host by passing through the skin of a naked foot. Many children go barefoot and become infected.

Other pathogens only infect humans and are passed from human to human only through human excrement. In villages with latrines, the prevalence of these diseases, particularly amebic dysentery, puzzled me. My patients would insist that they boiled their water, washed the produce with boiled water, washed their hands, and used latrines. Finally, in my sixth year, I observed a two-year-old child squat and defecate in the yard outside the house where I was holding clinic. Surprised, I questioned the mother. "Oh, she is too small to sit on the latrine!" she exclaimed, as if this were entirely appropriate. Now I wish I had a large truck to transport potty seats to San Lucas for all the two- to five-year-olds, who have the highest death rate from diarrhea and who contaminate the environment for everyone else.

Even without direct contact of contaminated soil or water with food, ameba can be spread. Flies carry infective cysts from feces to food or dishes. Thus, even in a village with latrines, if the latrines are not kept covered and the fly population controlled, amebas are spread invisibly.

In some villages where there are no latrines, the people use the fields. These fields have become contaminated to such an extent that it could take years for the soil to recover. The water runoff from these fields carries the organisms on to other villages lower down the mountain. In one such village we visited, a dentist accompanied us. We were glad to have him with us, since tooth decay and consequent abscesses in the jaw are major health problems. The dentist brought with him new toothbrushes, which he proudly presented to all his patients as a gift. His assistant carefully explained the importance of brushing the teeth but not the importance of using clean water to do it. On our way out of this village, we waved at two laborers who were bent over a ditch beside the fields. They straightened up to wave back at us, each with a toothbrush proudly displayed between his teeth. They had been brushing their teeth with the runoff water in the ditch.

The soil does not wait to be touched, either. In the dry season, it rises up. The rich fecal mud turns to dust and is blown about, coating everything. The dust problem is exacerbated by the lack of paving and the use of packed earth that surrounds housing compounds and the floors of dwellings, dirt roads, playing fields, and footpaths. The passage of vehicles on dirt tracks kicks up huge dust clouds. These roads often also act as the main sewage conduit during the rainy season. Dust blows into your mouth and nose, and the pathogenic organisms it carries are swallowed. During an overnight visit to one of the outlying villages, I was guided around town by a troop of young boys. I found an orange lying on the road and picked it up. One of the boys snatched it quickly from my hands—don't eat it, he warned, cholera, cholera! I realized that the road was the best drainage track for overflow from the village latrines, inevitable during the torrential rains. These latrines were positioned along the road just above where I was standing on the side of the volcano.

RESPIRATORY DISEASES

This contamination of the environment by infectious organisms that cause diarrhea is equally relevant for infectious respiratory disease.

Poor nutrition—a consequence of poverty—intestinal parasites, and chronic diarrhea lead to a lowered immune system and increased susceptibility to respiratory infections. Immunizations are also lacking. There has been inconsistent delivery of vaccines in the area, and at one finca we were told that many young babies and children had died recently from "tos farina," whooping cough, which has been virtually eliminated in this country by vaccination. Because of the lack of refrigeration vaccines may be denatured during transport and no longer effective when delivered. We have seen outbreaks of measles and polio in villages where "vaccination" has been performed.

Respiratory illness is also exacerbated by indoor smoke. Few families have chimneys, so cooking and heating fires are allowed to smolder in closed rooms without ventilation. Mothers and the younger children are often the most affected, with both respiratory reactions to smoke inhalation, including both asthma and irritative pneumonitis. Eye irritation from smoke also occurs, and the children will rub their eyes with dirt-contaminated fingers, leading to serious cases of conjunctivitis (bacterial eye infections).

The most seriously ill children I have seen are those infected with intestinal worms, who have had repeating bouts of diarrhea and are presented at clinic because of bronchitis or pneumonia. I believe that without the intervention of Western medicine in the form of appropriate antibiotic and anti-Helminthics, many of these young children would die.

NUTRITION, REPRODUCTION, AND SANITATION

Inadequate dietary intake, intestinal parasites, and chronic diarrhea combine to produce poor nutrition. The reasons for inadequate intake are many. Those I have identified include (1) insufficient land to produce enough food to sustain the family; (2) no refrigeration and only limited techniques for food preservation and storage; (3) the rigidity of traditional lifestyles that determine food types, cooking styles, and farming techniques and that obstruct change; and (4) large families and rapid population growth that overburden existing resources.

Insufficient land is most commonly cited as the reason for the poverty of the Mayan people. While it is true that the vast majority of the land is owned by Westerners, there are other reasons often overlooked that also contribute to the failure to produce enough food to sustain their large families. Much of the land owned by the indigenous people is at high altitudes where cash crops do not flourish. The altitude compromises the types of food crops that can be grown. Although the volcanic soil appears fertile, water is not always available. In the rainy season, torrential rains wash away topsoil without penetrating, and many areas have no natural, year-round water sources for irrigation in the dry season when no rains fall for months at a stretch.

In addition, there appears to be virtually no concept of family planning, and family size is not influenced by access to sufficient resources. In fact, I have held some preliminary meetings with herbalists and women's groups where many women who already had several young children did not know that their menses was a monthly event. They had married at menarche, bearing their first child within the year, and continued either lactating or pregnant ever since, with only a few menstrual cycles between each pregnancy. These meetings were overcrowded with women eager to hear how they might choose the number of children they bear. The increase in infant survival has not been matched with a commensurate decrease in birth rate. The high birthrate also compromises the women's health in numerous ways, including increasing her nutritional needs for vitamins and minerals that are difficult for her to obtain in that culture, such as iron and B_{12}.

The lack of refrigeration and food storage techniques means that there are cycles of feast and famine. Many people eat meat—beef, pork, or chicken—only on market day, when the animals are freshly slaughtered. Whole villages seem to travel in the bed of the head man's pickup truck down the mountain to the nearest market village, where they sell their vegetable produce and purchase meat, which they eat on the spot. I have seen dried beans, rice, cornmeal, and seeds. I believe they have an ancient traditional method for preserving corn, but otherwise, I have not witnessed any food storage method in process.

Traditional methods of farming, cooking, and behavior concerning meals and eating are integral parts of this culture. It is very difficult for any outsider to understand all the social implications of a traditional behavior. Changing the types of food that are eaten can impact the life of the family on a daily, possibly even hourly, basis. Furthermore, despite our advanced medical science, we probably don't know how the current diet is balanced to provide all the essential nutrients. Only recently was it discovered that the stone they use to grind their corn contains calcium, which sets into the cornmeal, thus supplying this dietary requirement. There are likely to be many other ways that the traditional methods contribute nutrients to the diet that we are not aware of and that would be lost if the diet were changed.

DEATH

Since we have not found any organized death records, my impression about what people die of are, again, anecdotal. I have attended one funeral and one wake and visited dying elderly people in their homes. On one occasion, Dr. Steve McCloy and I returned to a small, independent mountain village a few days after we had held a consultas there. The headman casually mentioned to us that a woman had died that night. We were surprised, since none of the patients we had seen two days earlier in the clinic were that sick. When we asked who it was, he laughed and said, "Oh, we didn't bring her to see you because, you see, it was her time to go." We attended her wake, where both Catholic and Mayan prayers were said in two languages.

I have questioned patients presenting to me in clinic about their families, asking detailed questions about those who have died. I have not been able to extract specific information. The most common causes of death seem to be fever in the young people and old age in the adults. But the underlying, specific etiology of the fever cannot be determined without medical studies unavailable in San Lucas.

Death seems to be more accepted in this culture, viewed as a normal process of life. Attitudes about death and ideas about afterlife probably affect emotional and behavioral responses to death.

CONCLUSION

Third World health needs as exemplified by this village in Guatemala are many, and they involve all levels of the socioeconomic structure, including belief systems, traditional lifestyles, access to resources, water purification and distribution, and the availability of trained health professionals, pharmaceuticals, and surgical services. Poor nutritional status, together with lack of sanitation, overcrowding, and high reproductive rates combine to produce increased incidence of a wide range of infectious disease. Communication gaps between groups speaking different tribal languages are augmented by high illiteracy rates and unreliable news sources. Health is tightly woven into a broader complex web of cultural behaviors, thought processes, and lifestyle. Any plan for direct intervention by outsiders should take into consideration the impact and repercussions on the individual, the extended family, and the community. No plan should be put into action without in-depth analysis of its consequences on this fragile and precious social order.

REFERENCES

Werner, David. (1994). *Where There Is No Doctor: A Village Health Care Handbook,* The Hesperian Foundation, Palo Alto, CA.

Woods, Clyde M., and Theodore D. Graves. (1973). "The Process of Medical Change in a Highland Guatemalan Town." In J. Wilbert (Ed.), *Latin American Studies,* Vol 21. Latin American Center, University of California, Los Angeles, CA.

HEALTH GUIDELINES FOR TRAVELERS

Fred V. Jackson

Every year many people travel to undeveloped countries seeking knowledge in a particular region, staying from six months to a year. Traveling to different parts of the world can be rewarding and memorable. Yet, when a person is not prepared for certain hazards, the trip can be horrible. Certain precautionary health measures taken before you leave, such as knowledge of where you are going, can make for a pleasant stay. I would like to share my thoughts and experiences when I visited Central and South America. A companion article, "Medicinal Plants," describes particular plants that can be helpful.

The first step when venturing out of the country is to contact your local traveler's clinic to find what necessary shots and pills are needed. Each area has different outbreaks of malaria, yellow fever, hepatitis, and so on, with some regions being worse than others. Your health clinic will have the necessary information about each country and will prescribe to you what should be administered. At the clinic, ask where health services are located in the country you will be traveling to, in case an emergency arises. The best plan is to carry medicinal items with you before departure. Common items in your kit bag should include aspirin, diarrhea pills, bandages, bacteria ointment, Pepto-Bismol tablets, insect repellent, medicated powder for rashes, rehydration powder, powdered Gatorade (for tasteless distilled water), and whatever prescription drugs you need. Remember, drugs in other countries are not as stringently regulated as in the United States, so dosages given in foreign pharmacies will vary.

Many times international travelers end up in remote sites for field work, miles away from local health facilities. In such a case, it's probably a good idea to know who in the area is knowledgeable about home remedies. The necessity of such knowledge will quickly become apparent when ailments such as infections, parasites, diarrhea, and headaches occur and the bag of common medicines that you have so diligently brought with you is now empty. Each village usually has a designated person who is knowledgeable about plants that have medicinal properties. Sometimes the person is called healer. Maybe it is the local shaman who can help in times of need. The shaman can be the administrator of healing with medicinal herbs. Other times the shaman is a spiritual person who is not easily accessible to people in the village.

It is the women of the household, however, who are more likely to have dried medicinal herbs and home remedies. They possess medicinal knowledge that as been passed down through generations of plant collecting and have, in their homes, extracts of plants to assist with minor discomforts. It is probably a good idea for the local people to collect the plants themselves because they are the most experienced, and problems could occur if the wrong plant is taken. Use commonsense judgment when taking these remedies, and be sure the people that are offering them to you are reliable, respected people of the village. A local guide of the area you are visiting could also serve as a resource to help introduce you to a medicinal plant person.

CHAPTER 5

Construction

Edited by RICHARD F. HERBOLD
PB Consulting

BUILDING

Richard F. Herbold

OVERVIEW

Engineering, architecture, construction management, construction, and—in general—"built environment projects" have become very complicated over the last 50 years. Rules, regulations, inspections, registrations, licenses, and so on are commonplace in the business. In Europe and North America another layer has been placed on top of the basic built environment project system, one that includes lawyers, delay claims, finance experts, the stock market, return on investments, lawsuits, and so on.

In this chapter, "Bibliography," a companion article to this one, gives references relevant to the following material. The Transportation and Water Supply chapters also have articles that discuss construction.

Built environment work in the Third and Fourth Worlds is moving toward the complexity of project work in the developed world. But there are still pockets of civilization where the rules are extremely basic. A nonengineer/architect/construction professional should know a few things about the more complex built environment world but keep a focused eye on the following simple issues.

- Drinking and cooking water must be pure and clean—free of disease.
- Sewage must be moved away from the populated area and treated if possible.
- Built environment projects (houses, schools, community facilities, work enclosures, medical complexes) in the Third and Fourth Worlds must be built to last. Poorly constructed facilities will require vast amounts of resources to maintain over the useful life of the same. Resources cannot be wasted in this environment.

Here is some general advice.

- Use simple concrete mixes and designs.
- Use simple masonry and mortar mixes.
- Use available wood for structures.
- Use simple systems.
- Keep rainwater out of the built environment project.
- Install simple air conditioning in very hot climates.
- Install simple heating systems in very cold climates.
- Always compact the soils under a structure.
- Always build on suitable materials.
- Stay with simple one- or two-story structures.
- Electricity is a basic need anywhere—learn to work with it; install wires and devices.
- Keep an eye on fire protection items—fire is both a friend and an enemy.
- Learn about simple plumbing systems—plumbing is needed to bring in pure water and take away sewage.
- Do not attempt to deal with complex systems such as structural steel and joists, elevators, complicated concrete mix designs, and advanced water and sewage treatment plants. If encountered in the field, find an expert to help out.
- Learn about cell phones and PCs with Internet connections to the world—this is the future for built environment projects—for example, Web sites list thousands of built environment products that can be purchased off the Web and delivered to anywhere in the world—*anywhere*!
- Learn to work with plastic pipe for water and sewage. Avoid metal piping; plastic is the way to bring water to a facility and take sewage away in Third and Fourth World environments.

IMPROVEMENT OF EXISTING DWELLINGS

In many cases, improvements can be made to existing houses at little or no cost. For example, separating the animals from the dwelling and installing a well-designed latrine should improve sanitary conditions. Developing a nearby water supply of adequate quantity and good quality will make women's lives easier. A mud stove will save firewood and contribute to forest resource conservation; however, the waste heat from a traditional fireplace may be needed for warming the home in cool climates.

Another improvement desirable in many rural homes is additional backfilling to raise the floor level to 10 to 15 cm above the outside ground level. Unfortunately, this will sometimes make ceiling and door heights undesirably low. Cut-off drains will also help to prevent surface water from entering the home. Although it may be difficult to install in an existing house, a waterproof foundation will be helpful in preventing moisture from penetrating the floor and lower walls.

CONTEMPORARY FARM DWELLINGS

Two sets of expandable farm dwellings are shown in Figure 1. For these structures a number of local materials are suitable. A foundation of stone and brick masonry or concrete is desirable, but adobe blocks, mud and poles, or stabilized soil blocks are suitable for the walls. While corrugated steel makes a clean, leakproof, and durable roof, where it is available, thatch is less

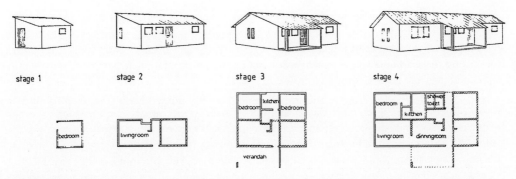

FIGURE 1
Improved farm dwelling design. (From *Farm Structures in Tropical Climates*, FAO-SIDA, Rome.)

expensive and perfectly satisfactory. Thatch will require a roof slope of approximately 45 degrees, and the frame should be built high enough to allow the eaves to be a minimum of 2 m above the ground. An overhang in the verandah areas will require support, as shown in Figure 1.

When resources allow, the same designs may be built with concrete foundations and floors, along with durable masonry walls of brick, concrete blocks, and other available material. The temperature extremes typical of corrugated roofs can be reduced with the installation of insulated ceilings. The final result will be a secure, easily cleaned, and durable home. Although considerably more expensive than dwellings made completely of local materials, this type of construction should be feasible for the emerging farmer who is producing some crops and animals for the commercial market.

TEAMWORK

For thousands of years the built environment was designed, constructed, and operated by master builders. A sole project manager (master builder) acted as the programmer, the architect, the structural engineer, the mechanical engineer, the soils expert, the construction manager, the electrical engineer, the general contractor, the facilities manager, and so on. The great Mayan temples, Roman roads, Egyptian pyramids, and Peruvian mountain meeting palaces were all built under this master builder system. Fast forward a few thousand years, and you come to our current system of designing, building, and operating the built environment: strategic teams of experts that come together for a project and produce a built environment product (a road, a bridge, a building, a dam, a water treatment plant, a sewage treatment plant, and others). The basic cast of characters for a project includes developer, finance facility, architect, structural engineer, mechanical engineer, electrical engineer, civil engineer, geotechnical engineer, interior designer, landscape designer, fire protection engineer, construction manager, general contractor, subcontractors, material suppliers, equipment vendors, scheduling engineer, budget/cost engineer, safety engineer, quality control expert, testing firms, and furnishings and fitout consultant. Some teams are larger, some smaller; there is an expert for every component of a project. Such is the modern-day built environment delivery system: a vast number of experts coming together at a point in time with a sole purpose—to transform natural products into something built by hand.

This is the delivery system for the developed world; team members are readily available and easy to find in Boston, New York City, London, Sydney, Tokyo, and Paris. The construction of a built environment product in the developing world sometimes follows this system for large-scale projects such as dams, major roads, and major water and sewage projects; smaller projects are built using a different set of rules. The process for a small Third World project (schools, religious buildings, medical centers, homes, water supply to a village, sewage removal from a village, a basic road between village and town) involves going back thousands of years to the time of the master builder—one sole person acting as the programmer, designer, constructor, and operator of the built environment project.

Unfortunately, the world is a complicated place, and built environment projects have moved from a structurally simple stone Mayan temple to extremely complex facilities such as water treatment plants, medical centers, and World Wide Web–ready schools. But one sole master builder can still organize and manage the required resources (labor, materials, equipment, money, subcontractors, subconsultants) and build a small-scale road, school, home, or medical facility. All that is needed is a little knowledge—some rules of thumb given above and later and the desire to assist in the human scale built environment development of a small village or town.

THE PROJECT SCHEDULE

Large-scale built environment projects normally have full-blown critical path method schedules and a staff to monitor project activities and time frames. Small-scale developing country projects must have a schedule. It should be noted that most field installation activities are interrelated and rely on many things to get done prior to their own completion. (Related articles are in the Planning and Implementation chapter of this book.)

For example, foundation concrete cannot be placed until the land is cleared, a ditch is dug, reinforcing steel is placed, and forms are built. A simple hand-written schedule can be used. This chart shows the steps that should be followed.

Schedule for Creating the Actual Fieldwork.

Day	1	2	3	4	5	6	7	8	9	10	11	12	13	14	15	16	17	18	19	20	21	22
Clear the land	x	x		x																		
Dig the ditch					x		x		x		x											
Build the forms						x	x	x														
Place the rebar									x		x	x			x		x					
Place the concrete																x		x				
Wait for the concrete to dry																x		x	x	x	x	
Remove the forms																			x			
Fill in the ditch with dirt																			x			x

This schedule shows that on days 1, 2, and 4 the land is cleared. On day 7 the ditch is being dug and the forms built.

Back off each activity—the rebar is needed on day 10—and create a schedule for preinstallation work.

- Place the rebar order.
- Deliver the rebar.
- Fabricate the rebar.
- Approve the submittal.
- Submit data on the rebar.
- Award a contract.
- Bid a contract.

- Design the rebar.
- Formulate a program.

Basically, any field installation activity must be preceded by the following series of office or paperwork activities.

- Formulate a program—what are you building?
- Do a design—put it on paper!
- Bid the work, organize the resources—labor, materials, equipment, tools.
- Award a contract—put it in writing.
- Get submittals—in writing from your resource.
- Approve the submittals—make sure you will get what you want!
- Fabricate the item—make it or take it off the shelf.
- Deliver it to the site. Watch out! This *may* take some time, using boats, planes, trucks, horses, camels.
- Install the item in the field.
- Maintain the item over time. Watch out! Make sure that what you designed and installed is appropriate for the project and for the country. (For example, if you design, purchase, and install a great looking light fixture with a special bulb, and in two months, the bulb breaks and it is not available locally, the fixture is worthless for the project.)

THE BUDGET

All large-scale built environment projects have a budget—a line item by line item breakdown of project costs. Small developing country projects must also have a budget. A budget with a slight twist—available labor and training—must be factored into the overall time frame. The budget should show labor, training, tradeable goods and services, and nontradeable goods and services (that is, what is available in the country and what must be purchased outside of the country, what is available locally and what is miles away).

Example

Item: Concrete footings for one-story medical facility.

Clear the land

- Local labor   200 hours @ $2/day U.S.
- Train local labor   20 hours @ $10/day U.S.
- Heavy equipment and operator   100 hours @ $4/day U.S.
- Train operator   not needed

Design the footings

- Completed before moving to the site cost @ $500 U.S. using in-country engineer

Dig the ditch (all labor)

- Local labor   100 hours @ $2/day U.S.

and so forth

FINANCING A MAJOR PROJECT

International agencies or large banks will probably support large projects. These will do a financial analysis of the project using net present value/discounted cash flow techniques. (These techniques are described in the Planning and Implementation chapter.) Project support may be in the form of loans. Often two different loans are used over the course of the project—a construction loan and a long-term loan. The first, with the money outstanding for a year or two, pays expenses associated with the construction. Once the building is done, the long-term loan is used to pay off the construction loan. The reason for two loans is that different lenders specialize in different duration loans.

Basically, as a project moves forward, resources, including money, are spent. Bills are sent to the lender/finance facility (a firm, individual, country, multilateral bank), and the facility sends the builder a check to cover bills and labor costs. The finance facility looks at the payment as a loan, a savings account that will make interest. The interest is merely a book entry until the project is complete, and a long-term loan is put in place. The long-term loan pays the finance facility all moneys lent and interest. The long-term loan is a form of mortgage that is paid off over time.

SURVEYING—LINE AND GRADE WORK

The world of project layout, surveying, and line and grade work has reached an extremely complicated level over the last ten years. Computer-controlled survey equipment, costing thousands of dollars, is now the norm. Satellite links are being used more and more. For the small size developing country built environment project five pieces of survey gear are needed: the level, the transit, the rod, the plumb bob, and the tape measure. Project managers can easily learn to use this equipment, and usage can be taught to in-country assistants within a few days. (The *Standard Handbook for Civil Engineers*, listed in the "Bibliography" article, describes surveying.)

A purpose of surveying is to lay out the boundaries of the building—to decide where the foundations will be. Another purpose is to ensure that the floors are level and the walls vertical. A level, together with a rod, determines the slope of the ground, so the excavations can compensate for this slope and ensure that the floors are level. The transit is used to measure angles so the corners of a building will be square or another desired angle. The plumb bob is used to verify that a wall is built vertically—that it does not lean in or out. A tape measure determines lengths of foundations and walls.

CIVIL AND SITE ENGINEERING

Sewage

Gravity Sewers: In many small developing country villages, sewerage is a minor problem. A family of eight living on a farm of, say, 40 acres does not create much of a sewerage. As density increases, sewage problems become greater. The basic requirement of any sewage system is to move the raw sewage from the populated area to a treatment or disposal facility; treatment is referable. Gravity sewers are the simplest way of accomplishing this activity. Pumps and pumping stations (high-cost and high-maintenance items) are not required; gravity sewers will last for many years. (*Note*: Archaeologists have unearthed sewer systems hundreds of years old.) The "Sanitation" article in the Water Supply chapter describes sewer systems in more detail.

Asbestos cement, clay, reinforced concrete, cast iron, and plastic pipe are the common types of piping used for gravity sewer systems. Reinforced concrete manholes are the common underground structures between piping runs.

For large-scale projects, experts such as engineers and contractors are needed. For small-scale projects in developing countries, the following rule of thumb built environment information can be used by the project manager.

- Size of pipe: 6 inch *minimum*, plastic
- Flow velocity: 2 feet per second
- Infiltration (groundwater allowed into system): 200 gallons per day per inch of pipe diameter per mile of pipe or less
- Slope: for a 6" pipe, flow of 2 feet per second, use .0060 slope
- Wastewater flow: at 2 feet per second a 6-inch pipe will handle 0.26 million gallons per day

Pressure Sewers: Here experts—engineers and contractors—are needed. The systems are extremely complicated and hard to maintain. Pumps, pressure lines, pressure manholes, and electrical services are required. Pressure sewers are not east to design, build, and maintain.

Sewage treatment systems for densely populated areas are extremely complicated and require experts in the design, construction, and operations of the same. Small scale developing country built environment project managers (SSDCBEPMs) can design and build some simple systems: plastic pipe from the populated areas and lagoon, natural-type treatment systems.

Aquaculture—Water Hyacinth and Wetlands Systems

Many wastewater treatment facilities using aquaculture are for secondary or tertiary treatment only. The systems may be used in developing countries as a primary means of treatment for small-scale waste production village. (Aquaculture is described in an article of the same name in the Food chapter.)

Design and construction details vary widely; the following can be used as a rule of thumb.

- Depth of pond or ditch: 4 to 5 feet
- Land requirements: 8 acres per million gallons per day

Septic Systems

In developing countries step number one in sewerage management is to remove the sewerage product from the populated areas. Step number two is to provide some form of treatment; a ditch or lagoon with some form of biotreatment is step number 2A; an alternative, step number 2B, is providing a septic system.

The basic septic system consists of a sewer pipe to the septic tank, a septic tank, and a drain field—see Figure 2.

Design of the septic system is straightforward. Wastewater enters the septic tank near the top and drains off near the bottom—on the other side of the tank. Baffles in the tank are useful to slow the passage of waste through the tank. Perforated pipe is used in the leach field so the effluent drains evenly into the field. The required minimum size septic tank can be estimated by multiplying the number of people in the family using the septic system by 60—that is, a tank for a family of 10 people should be at least 600 gallons. A family of 10 in the Third World uses about 100 gallons per day—much less than in the United States (the septic tank should

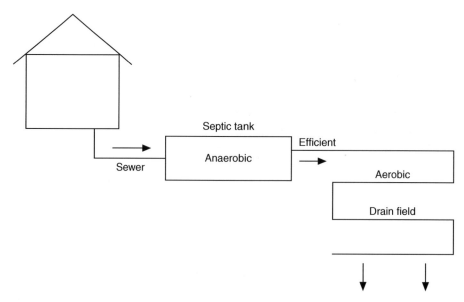

FIGURE 2
The parts of a septic system.

hold the waste water for about 6 days). If a well is onsite, the drainage field should be at least 100 feet away from the well system. The size of the aerobic drain field, also called the leach field, depends on how quickly water drains though the soil. Sandy or gravely soil drains best. 1 square foot of field is needed per gallon per day—that is, a leach field for a family of 10 should be about 100 square feet in sandy soil. In soil composed of clay and sand the field should be three times as large, and for soils with much clay—poor drainage—the field should be six times as large. (Septic systems are described in the Sanitation article in the Water Supply chapter.)

Wells

Wells can be constructed with hand tools by digging. It is feasible to dig down as far as 50 feet. A minimum diameter of 3 feet is required to give space for working. While digging or after, a lining—brick, timber, or other material—should be constructed to prevent the walls from collapsing and to protect the water source. Expert advice is helpful in deciding where to dig. The water table will be close to the surface near lakes, rivers, and marshes. Professional well diggers use drilling machines that can go down several hundred, even thousands, of feet. (An article on well construction is in the Water Supply chapter.)

Well water will need disinfection if contaminated. The usual source of contamination is feces, either human or animal. The location of the well relative to latrines, leach fields, and farmyards is important. Mines and factories can also poison groundwater. Contamination from human and animal waste can be removed by disinfection or by boiling. Industrial pollutants are very difficult to eliminate—if present, another water source should be chosen. Chlorine-based disinfection chemicals are commercially available. Filtering through a column of sand perhaps 4 feet long can clean water that is cloudy or has a bad taste or odor. A layer of charcoal will augment the removal of tastes and odors. Filtering removes suspended particles but cannot be counted on to remove dangerous bacteria. (An article on water filtering is in the Water Supply chapter.)

SOIL AND FOUNDATION ENGINEERING

Foundation engineering is an extremely complicated area of practice. Expert advice is needed for large structures. An example of a failure is the Shrine of Guadeloupe in Mexico City. The building has settled 5 feet into the surrounding plaza. The purpose of a foundation is to give a firm base for the walls and the floor. Without a foundation the soil at the bottom of the walls or under the floor might wash out and the building collapse. Buildings in earthquake-prone areas need special designs—an expert should be consulted or, at least, a proven design used.

When designing and building a simple facility, the following rules of thumb may be used when dealing with soils.

- Never build near or on a slope or incline. If possible build on the flat only.
- Keep the building site dry—remove all standing water and divert storm water runoff around the site.
- Never deal with pile or caisson foundations—leave this for the engineers.
- Keep excavations to 5 feet or less.
- Always backfill and compact the backfilled soil in 4- to 6-inch layers.

A foundation for a small structure can either be "pads" of concrete or rock underneath the main supports or a solid piece of concrete under the entire house. One approach to choosing between pads or a slab is to look at the soil under the house. Muddy soil can hold pressure of 30 KiloNewtons per square meter. The designer should calculate the weight of the structure, adding the weights of the walls, roof, and any load supported by the walls. The weight of each part of the building is found by looking up the density of the relevant material and multiplying by the volume of the part.

As an illustration we will determine the foundation area of a particular brick wall. The wall is 5 meters long, 2.5 meters high, and 0.1 meters thick. It has a volume of 1.25 meters, and the density of brick is 1,500 kilograms per cubic meter, so the mass of the wall is 1875 kgs. The force is found by multiplying by the gravitational constant—9.8. The force is 18,375 Newtons or 18.375 KiloNewtons. To keep the force on the soil less than 30 KiloNewtons/m^2 the foundation under the wall must have an area of 0.6125 m^2—only slightly larger than the base of the wall. If a roofed structure were being designed, the weight of the roof would have to be added. The foundation under a floor is designed the same way, allowing for the weight of the people, animals, or equipment, using the floor. The thickness of the foundation is made to be about 0.20 meters— sufficient so bending is not a problem.

Reinforced Concrete

Concrete is one of civilization's oldest building materials. The modern-day name for the material is "cast-in-place" concrete or "reinforced cast-in-place" concrete. On a large-scale project, engineers, architects, testing firms, sample takers, cylinder breakers, mill test reporters, and a host of other experts are involved in the concrete production and placement activity. Small-scale placement is possible on a simple scale.

Concrete is produced by mixing water, Portland cement, and aggregate—gravel. Other additives may be added to the mix (such as form release agents, accelerators, curing compounds, air-entraining agents, water reducing agents, and corrosion inhibitors). The proportions of cement, sand, and gravel in most cases should 1:2:4—less cement for foundations, less gravel for floors. Only as much water as is necessary to create a workable mixture should be used. The weight of the water used should be about 50 percent of the weight of the cement. Concrete will set in three days but reaches maximum strength after seven days. It should be kept wet during this curing time.

Reinforcing steel should be used with all concrete placements. Usually reinforcing is done with steel rods called "rebar." Such bars should be placed deep enough in the concrete so they are covered at least 3 diameters. Standard diameters for reinforcing bars range from 0.275 inches to 1.270 inches. If commercial bars are not available, wire mesh or sisal fibers can be used. The reason for using reinforcing is that concrete, although strong in resisting compression, is weak under tension—the steel gives tensile strength. Tensile strength is important because when a load is applied to one place on top of a concrete beam or foundation, the beam deforms so the top compresses and the bottom stretches. An article on using bamboo to reinforce concrete is in this chapter.

Formwork is needed to hold the wet concrete in place until it solidifies. Forms should be very sturdy and should be able to hold the wet concrete in and take the weight and push/pull of the staff placing the material. The weight of a cubic foot of concrete is 150 pounds.

Several manufacturers make bagged concrete mixes that contain cement, sand, and aggregate; all that is needed is the addition of water. The bagged concrete is easy to use and can be transported to a project site in small bag weights (40-pound bags). The material is ideal for foundations, stairs, concrete floors or slabs, column bases, and water runoff systems.

Typical foundation wall and slab on grade construction includes continuous footings, drainage piping, base courses under the slab on grade, fabric on the outside of the walls, damp proofing, and concrete with reinforcing steel and/or wire mesh. Some projects require concrete foundations for wood posts and columns, concrete sidewalks, and water runoff ways, where reinforcing steel is not required.

Some advice about concrete work: Stay with a simple mix; make sure that reinforcing steel is in the concrete; make sure that all formwork is sturdy enough to hold the wet concrete and the weight of those placing the concrete; and do not work on any building project over two stories high.

Masonry, Brick, Concrete Blocks

Brick and concrete blocks have been used for many years in developing countries. Most brick and concrete block systems are structural in such projects; they support floor and roof loads. If used on a small-scale project, the project manager should limit the height of a brick or block wall system to one floor, or 9 to 10 feet maximum. The placement of reinforcing steel within the brick or block wall is highly recommended. Blocks should be laid in a so-called "bond" so cracks do not run directly from top to bottom—see Figure 3. A typical mortar holding the bricks together is composed of 1 part cement, .5 parts hydrated lime, and 4.5 parts sand. Brick walls can fail by overturning—buttresses and

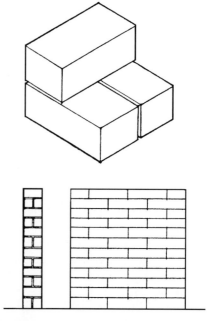

FIGURE 3
Placement of bricks in a wall.

bracing may be necessary. A waterproof plastic film may be needed between the lowest course and the foundation to the bricks from acting as a wick and bringing water up.

Bricks are often made by small-scale builders. The article on "Stabilized Soil Blocks" in this chapter describes one process. Another process uses a charcoal kiln to fire bricks made of clay. Here bricks are formed from clay, dried in the sun for three days, stacked for a week, and finally heated to 900° for six hours in a charcoal kiln. Local artisans will know the properties of local soil and fuel. They will understand how to build the fire and to judge when the bricks have been sufficiently fired.

Wood Structures

Roofs or floors can be supported by wood. Wooden joists 2 inches by 8 inches spaced 16 inches apart can span a distance of 11 feet. The joists are positioned upright—with the long dimension vertical. To span 18 feet, 2 inch by 12 inch joists are needed. In the United States a common form of construction for a house is a wood frame covered by shingles, bricks, stucco, or other kinds of siding. Usual practice is to use 2" × 4" studs—uprights—spaced 16 inches apart, actually 16 inches between centers. A two-story structure is made by framing one story, laying joists across it with double joists at the edges, and then erecting second-story framing on top of the first story.

Sloping roofs can be made from rafters—like joists but positioned as an inverted "V"—or supported by trusses (see Figure 4). A longer span can be obtained from a given sized board (joist) if it is used to support a sloping roof rather than a flat floor. Trusses made from laminated boards can be purchased in many places. Metal flashing needs to be installed where parts of the roof are joined, such as where chimneys protrude or dormer windows are used. Gutters and downspouts are useful to collect rainwater and to lead it away from the foundation. A major problem with roofs can be wind. Wind can push down with a force of 30 KiloNewtons per square meter. Wind flowing over a structure can also create lift, so roofs need to be fastened down securely.

When building with wood, always consider the dead load of the wood and framing system, wind loads, snow loads, and of course, live loads. A good rule of thumb for safety is to multiply the live load (what the structure needs to carry) by three in designing a frame. If you build a test floor for a simple structure and feel that it must support 4 people, test load it for 12 people prior to building.

Wood structures can and will be destroyed by mold, stain, and decay. An article on protecting timbers is in this chapter. Rule number one is to never allow a wood frame system to contact

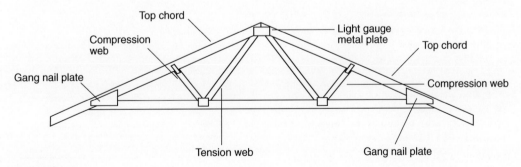

FIGURE 4
Roof trusses.

the ground. Always build wood structures on concrete, stone, or masonry foundations. Contact with the ground will lead to termites and general decay damage. Here are some other rules of thumb when building with wood.

- Always drain the construction site properly; do not allow standing water at or near the wood construction.
- Always separate the wood from moisture sources. Never allow wood to be in contact with a moisture or water source.
- Insects, termites, flying ants, and wood eating bugs in general live in moist environments. They will destroy a wood building/facility quickly. Other insects to watch for are wood-boring beetles, wood wasps, powder-post beetles, carpenter ants, marine borers, and shipworms.
- Always ventilate enclosed spaces. Do not allow moist air to remain within a built structure.

OTHER ASPECTS OF BUILDING

Use of arches and domes allows the construction of fairly elaborate structures by local builders, as described in an article in this chapter. Buildings can be built effectively of straw bales, also described in an article in this chapter. Concrete roofing tiles are fairly easily made by untrained crafts persons and are also described in an article in this chapter. Another article in the chapter deals with making stonewalls.

REFERENCE

Bengtsson, Lennart P., and James H. Whitaker (Eds.). (1988). *Farm Structures in Tropical Climates.* FAO/SIDA Cooperative Programme, Food and Agriculture Organizations of the United Nations, Rome.

BUILDING BIBLIOGRAPHY

1. Aith, Max, Charlotte Max, and Duncan S. Blackwell. *Wells and Septic Systems,* 2nd Ed. 1992. Tab Books, McGraw-Hill, Inc., New York, NY.
2. Ambrose, James. *Design of Building Foundations,* 2nd Ed. 1988. John Wiley & Sons, New York, NY.
3. American Institute of Timber Construction. *Timber Construction Manual,* 3rd Ed. 1985. John Wiley and Sons, Inc., New York, NY.
4. Behrens, W., and P. M. Hawranek. *Manual for the Development and Preparation of Industrial Feasibility Studies.* 1991. The United Nations Industrial Development Organization (UNDO), New York, NY.
5. Brungibber, P. E., and L. Robert. *Timber Construction and Design,* 2nd Ed. 1987. Professional Publications, San Carlos, CA.
6. Ching, Francis D. K. *Building Construction.* 1991. Van Nostrand Reinhold, New York, NY.
7. Clark, David B. *Roofing and Siding.* 1981. Sunset Books, Lane Publishing Company, Menlo Park, CA.
8. Coleman, P. B., Robert A. *Structural Systems Design.* 1983. Prentice Hall, Inc., Englewood Cliffs, New Jersey.
9. Harberger, Arnold C. *Project Evaluation.* 1976. The University of Chicago Press, Chicago, Illinois.
10. Harris, Richard. *Construction and Development Financing.* 1983. Warren, Gorham & Lamont, Boston, MA.
11. Huang, Yang II. *Stability Analysis of Earth Slopes.* 1983. Van Nostrand Reinhold Company, New York, NY.
12. Horsley, F., and William Means. *Scheduling Manual—Critical Path Method Scheduling.* 1981. KS. Means Company, Kingston, MA.
13. Jones, Jr., J. P. Rear Admiral, US Navy. "Economic Analysis Handbook." June 1986. NAVFAC P442. The United States Navy, Washington, DC.

14. Kerzner, Harold. *Project Management—A Systems Approach to Scheduling, Planning and Controlling*, 1979. Van Nostrand Reinhold Company, New York, NY.
15. Leet, Kenneth P. E. *Reinforced Concrete Design*. 1982. McGraw-Hill Book Company, New York, NY.
16. McQuiston, Faye C., and Jerald D. Parker. *Heating, Ventilating and Air Conditioning—Analysis and Design*, 2nd Ed. 1982. John Wiley and Sons, New York, NY.
17. Merritt, Frederick S. *Standard Handbook for Civil Engineers*, 3rd Ed. 1983. McGraw-Hill Book Company, New York, NY.
18. Mullin, Ray C., and Robert L. Smith. *Electrical Wiring—Commercial*. 1981. Van Nostrand Reinhold Company, New York, NY.
19. Ray, Anandarup. "Cost-Benefit Analysis—Issues and Methodologies." 1986. The International Bank for Reconstruction and Development. The World Bank & John Hopkins Press, Washington, DC.
20. Schwartz, Max. *Basic Engineering for Builders*. 1993. Craftsman Book Company, Carlsbad, CA.
21. Scott, Ronald F., and John J. Schuster. *Soil Mechanics and Engineering*. 1968. McGraw-Hill Company, New York, NY.
22. Souder, William B. *Project Selection and Economic Selection*. 1984. Van Nostrand Reinhold Company, New York, NY.
23. Staff of Ovid W. Eshbach. *The Eshbach Handbook of Engineering Fundamentals*, 2nd Ed. 1980. John Wiley and Sons, Inc., New York, London, Sydney.
24. Stein, Benjamin, and John S. Reynolds. *Mechanical and Electrical Equipment for Buildings*, 8th Ed. 1992. John Wiley & Sons, New York, NY.
25. The Commonwealth of Massachusetts. "The State Building Code—780 CMR—6th Edition." The Commonwealth of Massachusetts, Boston, MA.
26. The Institute of Civil Engineers. *Engineering Economics and Appraisal*. 1969. William Clowes and Sons, London, England.
27. Tietenberg, Tom. *Environmental and Natural Resources Economics*. 1988. Scott, Foresman and Company, Glenview, IL.
28. Ward, William, Bany J. Deren, and Emmanuel D'Silva II. 'The Economics of Project Analysis." 1991. The International Bank for Reconstruction and Development. The World Bank, Washington, DC.
29. Vernick, Arnold S. and Elwood C. Walker. "*Handbook of Wastewater Treatment Processes*." 1981. Decker, New York, NY.

BUILDING WITH ARCHES, VAULTS, AND CUPOLAS

Condensed from Basics of Building with Arches, Vaults, and Cupolas, *A SKAT publication, Swiss Center for Development Cooperation in Technology and Management, 1994, Thierry Joffroy, assisted by Hubert Guillaud, architect researchers, CRATerre-EAG*

The numerous types of arches, vaults, and cupolas allow a great variety of architectural models (see Figure 5). As a result, the technology can adapt to the most varied climatic conditions: zones that arid or rainy, cold or hot. Although arches, vaults, and cupolas are traditionally used to cover limited spaces, they are perfectly well suited to build much larger spaces, up

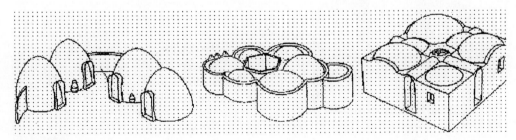

FIGURE 5
Dome construction.

to tens of meters. Thus, they can meet the needs of any building program, public or private, low-cost or quality housing, granaries, warehouses, shops, schools, public, religious buildings, and so on.

ADVANTAGES OF ARCHES, VAULTS, AND CUPOLAS

The materials used for the construction of arches, vaults, and cupolas can be the same as those used for walls and can be found or produced locally. Construction is therefore less expensive, creates jobs, and allows foreign currency savings. There is no use of wood, which also totally eliminates the risk of fire. The massive nature of these structures provides good heat storage capacity and delay in heat transmission, meeting comfort requirements, especially in dry climate regions. This mass also gives good sound insulation.

The cost of buildings using arches, vaults, and cupolas varies according to the materials employed, the complexity of the design and of the construction technique, the size of the structures, and the surface protection used. However, buildings employing sun-dried earth bricks (adobe) and protected with an earth render can be built at a lower cost than traditional buildings of a similar standard. In general, arches, vaults, and cupolas made from water-resistant materials and protected by durable waterproofing cost more than simple buildings covered with roof sheeting but remain less expensive than buildings with reinforced concrete slab roofs.

Significant foreign currency savings are possible. The basic materials are generally available on site, and little power is required for processing them. Transport costs are also reduced. The investment in site equipment, centering, compass, and scaffolding is generally very small, and most of these can be manufactured locally without any major difficulty.

Building with arches, vaults, and cupolas is highly labor intensive. Therefore, the technology has great job creation potential, not only with regard to construction but also with regard to building materials production.

The many possibilities of forms, sizes, and combinations of different elements and types of finishing enable highly attractive spaces to be created. Thus, arches, vaults, and cupolas can be used not only for low-cost housing programmes but also for high-quality, luxury dwellings (see Figures 6 and 7).

FIGURE 6
Section of house in Tunisia.

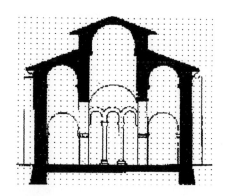

FIGURE 7
Section of church in Spain.

LIMITATIONS OF USE

Some local conditions can limit the benefits of building with arches, vaults, and cupolas.

- Rejection by the inhabitants for cultural reasons
- High cost of labor
- High cost of suitable building materials
- Use in earthquake areas requires special care
- Lack of building norms

The technology must be adjusted to a given context, and the techniques must be mastered to warrant the advantages of using arches, vaults, and cupolas.

SPATIAL CONSIDERATIONS

The curved forms generated by arches, vaults, and cupolas define free variable heights that increase with the span and the rise. Depending on requirements, one should first determine the curvature of the element and the height of its spring-point. The ability to move between one space and another also makes it important to take account of the ability to open out between the interior and the exterior of a vault or of a cupola. One can distinguish between open, semiopen, and closed types.

Building systems using arch, vault, and cupola structures can range from the simplest to the most complex. The main distinction to be made is between modules that are independent from one another (whether connected or not) and sophisticated systems that respect very precise geometrical traces. The sophisticated systems get an optimum use of the material and perform very well from the point of view of costs. On the other hand, their replicability can be tricky, since a high degree of precision at the laying out stage is required. As a general rule, the more the shapes of the vaults or the cupolas are visible, the cheaper the building. Certain models permit a simple and inexpensive transition to sloped roofs, as shown in Figure 8.

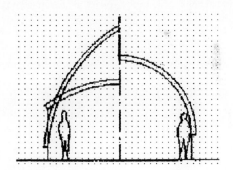

FIGURE 8
Three vault shapes.

ACOUSTICS

The rounded shapes, especially of cupolas, can result in a significant degree of reverberation that is an asset or a disadvantage depending on the intended use of the space. Simple measures, such as corbelled elements or suspended cloth, can be used to reduce reverberation.

STABILITY OF ARCHES, VAULTS AND CUPOLAS

To ensure stability, the form should be as close as possible to the line of force. In the case of vaults, for evenly distributed loads, the best form resembles that of an inverted suspended chain (catenary). This applies both to flatter and to slender forms. See Figure 9.

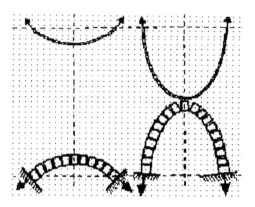

FIGURE 9
Arches and catenaries.

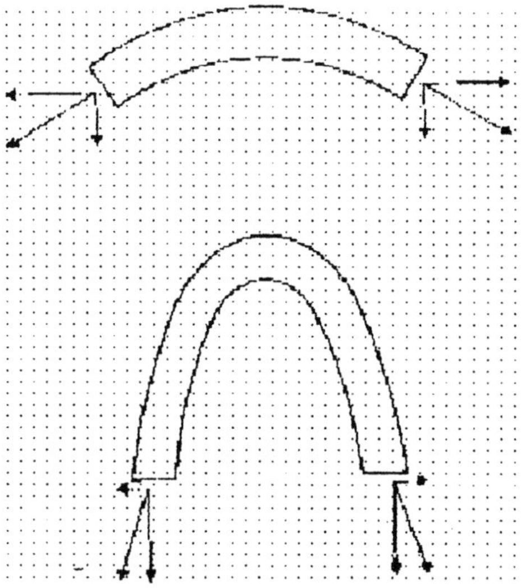

Arches, vaults and cupolas are heavy roofing elements and they exert thrusts that increase with the degree of flatness, the span and the weight of the structure. These thrusts may be concentrated or dispersed. They have a tendency to push out the supporting bases (walls, piers and foundations) and must be taken into consideration–Figure 10. The most common solutions are: canceling them out by juxtaposition of arches, widening the wall, buttresses, and using tie-beams or ring-beams. Juxtaposition is shown in Figure 11 but note that segmental arches exert greater thrusts than slender arches.

BUILDING MATERIALS

The building materials required for the masonry of arches, vaults and cupolas can be divided into two broad groups: solid components (adobe, stone, etc.) and binder components (mortar). The solid components ensure compressive strength. The binder components ensure that compressive forces are transmitted and the cohesion of the whole. To avoid the risk of shrinkage during drying out after building work, a minimum amount of binder component should be used. The various solid and binder components are rarely incompatible but it is preferable to use components of similar strengths.

FIGURE 10
Forces from an arch.

SOLID COMPONENTS

Stone is often of very irregular dimensions, which forces one to use very expensive preparation and implementation techniques. Soft, light stone offers interesting possibilities, as does layered stones, which can be standardized by simply chiseling them to the required shape.

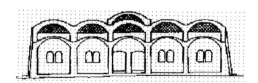

FIGURE 11
Segmented or juxtaposed arches.

Adobe block is simply prepared by moulding plastic soil and leaving it to dry, the adobe block is always inexpensive. Its shape and size can be easily adapted to the various methods of construction, so that this material is the most widely used. However, its rather low mechanical performances require the construction of fairly massive structures.

Stabilized compressed earth block are insensitive to water and have good mechanical strength–this material provides an alternative to the more costly-fired brick. Most of the presses used to manufacture compressed earth blocks can be adapted to produce special blocks. Brick Presses are described in the "Soil Stabilized Bricks" article in this chapter.

Plain molded cement blocks made from sand-cement mortar have been used for very diverse buildings as many shapes and sizes of both simple elements, and also of special parts, can be prepared.

Fired solid brick is more expensive but generally performs well, which allows for either lighter buildings or larger dimensions. Special shapes and sizes can be obtained, depending on the means of production.

Fired hollow brick enables special application techniques to be used, and very light buildings to be realized. Materials produced using craft pottery methods can be used in a similar way.

BINDER COMPONENTS

Earth mortar is used to bond adobe, fired brick or stone. Its good sticking properties enables one to build vaults and cupolas without shuttering–using a screen to hold the plaster as it dries. Plastic earth mortar can be used to make hand-shaped cupolas. To avoid shrinkage, straw or sand is added to soils with an excessive clay content.

Stabilized earth mortar–cement, lime, plaster or bitumen improve the characteristics of earth mortar. Such mortars are very suitable for binding compressed earth blocks or fired bricks.

Lime-sand mortar displays average characteristics. Its main advantage is that it stays malleable for quite a long time and sets very slowly. It can be improved by adding crushed brick or pozzolanas. Lime-based mortars are very frequently used in conjunction with fired brick and stone.

Cement-sand mortar is suitable for bonding good-quality fired earth and full sand-cement blocks. Given the high cost of cement, the amount used must be carefully calculated according to the mechanical strength required. Because this mortar has little adherence, it is not well suited to shutterless building methods, apart from corbelling. Medium-span vaults have been made directly using a sand-cement mortar. Small size elements can be prefabricated.

Gypsum mortar sets very quickly, which enables adhesive application techniques to be used. Small span vaults can be made with bricks or even prefabricated directly using this mortar. The stability of water-soluble gypsum can be improved by combining it with lime.

SURFACE PROTECTION

To avoid infiltrations and therefore risks of deterioration, vaults and cupolas must be protected from rainwater, which must be channeled away from the building using waterspouts or downpipes.

An earth render—wall coating—is the least expensive but requires periodical maintenance whose frequency depends on the nature and the quantity of rainfall. A sand-cement mortar is too rigid and will always crack, which allows water infiltration to take place. It should not be used without a complementary membrane. Lime-sand and gypsum-lime-sand renders are better suited because of their greater pliability, but a certain amount of maintenance might still be needed.

Details and finishings can influence the choice of structure. Attention must be paid to the bonding pattern(s)—particularly for visible masonry—to window frames, and to channeling and removal of rainwater.

BUILDING ARCHES

With formwork: For building an arch, a formwork of the same shape of its intrados needs to be used. This can be made of wood or metal and should be used in conjunction with a system allowing its easy removal. For most arch forms, the formwork can be removed immediately after construction. The formwork can therefore be reused straightaway. This means that if there are a great many identical arches to be built, clearly it makes sense to prepare a precise and robust formwork to ensure that it can be reused and that the work will be of high standard and carried out quickly and economically. Building up bricks or blocks with dry joints and shaping the curve with a layer of mortar is a very economical way of making a formwork.

Without formwork: Corbelled arches are built using courses that each jut out farther than the previous one. Cut-out arches and arches shaped with wooden reinforcement are also built without formwork.

BUILDING VAULTS

Vaults can be built in a similar way to arches. With formwork, building can be done on a full formwork setup in place. This is mainly used for floors over basements. It is possible to use an integrated formwork. Being very heavy, this has to be assembled and dismantled in place. This technique is economical only if the vaults—and therefore the formwork—are light.

Sliding formwork can be used. Building takes place in stages, with the formwork being repositioned as often as necessary. This is one of the most interesting methods but applies only to barrel vaults.

Without formwork, Corbel vaults—made from blocks in each row projecting into the space to be covered—can be built to cross small spans. Building vaults in so-called "slices," without formwork, is the most economical method. This exploits the properties of certain mortars, which enable the bricks to be laid face on layers leaning at a steep angle to the vertical. To improve adhesion, small and fairly thin elements of regular dimensions should be used. Using this principle, and by varying the shape and the size of the courses, one can build barrel, groined, dominical, trough, squinch, and boat vaults. Traditionally, all these vaults are built "by eye," but light tools can be used to guide the masons so that they build the required shape correctly.

BUILDING CUPOLAS

Building techniques without formwork are always preferred as a formwork would be too complicated to manufacture. Corbelling enables large areas to be covered but results in conical forms that are very high relative to the span. Horizontal rings that gradually decrease in diameter can be used to create all types of cupolas. This method uses the same brick-laying technique as for vaults built in "slices." Simple rotating guides, which describe the form in the air, can be used to obtain correct, regular forms. By hand shaping a straw-reinforced earth mortar, it is possible to build fairly large sized cupolas. These are built up in successive layers.

REFERENCES

Joffroy, Thierry, and Hubert Guillaud. (1994). *Basics of Building with Arches, Vaults and Cupolas.* A SKAT Publication. Swiss Center for Development Cooperation in Technology and Management.

Norton, John. (1997). *Building with Earth, A Handbook.* Intermediate Technology Publication, 103/105 Southampton Row, London WC1B 4HH.

COB STRUCTURES

Jessica Wurwarg

Cob is an ancient building material composed of clay, sand, and straw. It has been used on the British coast since before the 1400s, and some cob structures built in the 1400s are still standing and inhabited today. These old homes are usually stuccoed and whitewashed on the outside and are made of cob underneath. Cob was used then and is used now because it is durable, inexpensive, and easy to use. People have discovered some new things about cob over the years that have inspired a resurgence of its use in building homes.

Cob is nontoxic and completely recyclable, a welcome relief to the headaches, heartaches, and stomachaches that some recently utilized building materials, like asbestos and lead paint, have caused. Cob is environmentally sound; it makes no contributions to deforestation, pollution, or mining. No power tools or manufactured materials are needed to make a cob structure. The thermal mass of cob helps to keep the structure a comfortable temperature.

Cob is a combination of clay, sand/small stones, straw, and water. These ingredients are mashed together, as shown in Figure 12. The clay, sand, and straw are combined in different proportions, depending on the soil conditions and contents of the area. Cob, if one were to make it out of pure ingredients, would be 70 percent sand, 20 percent clay, and 10 percent water. There are also methods of making cob by using soil instead of clay. Courses are offered in cob building through different organizations in the United States and abroad. These courses, which are generally only a week or so long, can give people enough education in cob building and the preparation of cob that they would be able to build a structure; it is that simple. Before building and making the cob mixture it is necessary to test the clay/soil and water. One aspect of cob that makes it so unique and special is that it can be made anywhere, as long as the components can be shipped or trucked or flown into the area. It has been used in high latitudes and extremely windy and rainy climates throughout Western Europe.

When cob is dries, if it is at least 8 inches thick, its hardness is comparable to that of cement, and it is fully load bearing—which means it does not need supports or a frame for a roof to be added to it. Cob is extremely weather resistant, and it can withstand long periods of rain without weakening. The old cob houses along the shore of England have withstood hundreds of years of heavy rains and strong winds due to cob's porous nature. Wide roof eaves protect the walls from overexposure to rough weather.

The word *cob* is rooted in Old English, coming from a word meaning "a lump or rounded mass." Cob structures are built by hand, with "lumps or rounded masses" of cob shaped into any form the builder desires. Cob structures require no frame. This unique quality allows the builder an infinite amount of freedom and creativity in the design of the structure. Bricks aren't even formed from cob; only handfuls or "lumps or rounded masses" are used, so organic sculptural shapes can be formed, and curved walls, arches, and vaults can be made in the structure. Furniture, such as shelves and benches can be built into the walls of the cob house.

FIGURE 12
Making cob (Photo by Jan Sturman from www.into-solutions.com/cob/).

Building with cob is similar to sculpting with clay. Cob can be added on to, cut out of, or reshaped even after the cob has dried.

This adding on process, however, can be very difficult. Someone once told me the story of an old cob wall that someone was trying to chop down with a pickax. The wall probably was originally about eight inches thick. When the man finally chopped it until it was only one inch thick, his friend on the other side of the wall still could not see the wall vibrating when the pickax struck it. One good way to add on to cob structures is to plan ahead when first building the structure. Leaving some twigs sticking out in a part of a wall would provide a suitable spot to make additions later on. Any jagged areas on a wall are ideal for making additions; even jagged rusty nails can be used. Cob is fireproof, so fireplaces, stoves, and chimneys can be shaped with it. Cob beds and benches can be placed near the stoves and fireplaces to be warmed.

Usually the walls of a cob home are whitewashed or stuccoed over. This undoubtedly helps to preserve them, protecting especially against harsh winds. Russell Holvinger at the Cob Cottage Company has told me about cob buildings whose walls have not been covered and protected, and they have lasted hundreds of years. Cob does act like clay in its properties of allowing itself to be reshaped. Reshaping cob can prove to be very difficult, but it is possible. It simply must be soaked and wet repeatedly for long period of time. Sometimes the cob accepts the water and allows itself to be reshaped or allows another piece of cob to be joined with it, but other times cob is not so cooperative. Due to the solid nature of cob—it is one monolithic unit—it stands up relatively well against earthquakes.

Cob is a good soundproofer and has good acoustical properties. It also has been known to preserve food for long periods of time; perhaps this is due to the breathable walls of cob structures.

Cob is not an engineered process. The "buildability," strength, and composition of cob varies from site to site. It is not uniform. For this reason engineers cannot regulate it. There are no codes specifying what is safe and what is not safe. Building codes and laws vary from place to place. Most people just build without permits. The Cob Cottage Company says, "Codes today protect the industrial manufacturers of building components better than they protect homeowners. Not surprisingly, there is no code for cob, though nowhere is earthen building prohibited. Many cob builders choose not to involve building officials and have had no problems. Legally permitted cob buildings are beginning to appear; there's considerable expense and paperwork involved, as with any permit."

Because cob composition varies from site to site, the builder must test little clumps of cob before he or she begins building. Clumps from a few different cob mixes should be left to dry and their consistency and hardness examined.

"It is an intuitive process," said Russell Holvinger of Cob Cottage Company. It is straight-forward and simple; no specialists are needed to build the actual structure. If someone is building one's own home, they can get permission to install their own plumbing and electricity. Otherwise, specialists are needed to install plumbing and electricity. Pipes and wires can be laid directly into the cob, so there is no need for sheetrocking, taping, sanding, and so on. Most of the cob houses constructed through Cob Cottage Company don't utilize lights during the day, and because of the thermal mass of cob, they need very minimal heating and cooling systems. The concept of cob as a thermal mass means that cob acts as a battery. It absorbs heat energy from the sun during the day, stores it, and then releases it at night. Water, salt, sand, and stones are each excellent thermal masses. Interestingly these are the components of cob. Cob also maintains relatively cool temperatures in the summer and warm temperatures in the winter. One cob house in New Mexico had no heating or cooling systems, and the temperature inside the house never went above 90°F or below 65°F. For cob's thermal mass and insulation to be effective, cob walls should be one to two feet thick.

Both the composition of cob and the positioning of the house/cob structure affect the electricity consumed in the structure. It is more energy efficient to position the house so that it faces the south (in the northern hemisphere) and north (in the southern hemisphere), with big windows in this direction. This way the big south/north facing windows and solar panels, if employed, can capture sunlight and then the thermal mass of the cob can absorb and make use of the solar energy. Planting trees and ivy near this big window can be very helpful for temperature control. When the trees are in bloom, in the summer, the shade provided by the trees will help keep the house cool; all the heat from the sun won't be absorbed by the thermal mass. But in the winter, when the trees and ivy do not have leaves and no shade is provided, the sun's heat will be absorbed and the cob's thermal mass will store up the heat and warm up the house. This is one great way cob building makes use of and works with its environment. Further notes on house design are in the solar heating article in the Energy chapter.

Cob building is so inexpensive, it provides a way to own a home right away without having a mortgage. If the builder can provide his or her own labor, use the dirt from the foundation to build the walls, and use mostly recycled materials like wood and stones, a house can be built for as little as $500.

The best foundation for cob houses, or at least the favorite among those at the Cob Cottage Company, is a dry rock foundation. This way as little cement as possible is used.

Cob housing, especially building one's own homes, helps in "deconsumerizing." Building a home oneself, using friends to help in the labor, utilizing natural resources without wasting them reduces the flow of cash. Building and designing one's own house creates one well suited to the builder. Such houses are built with nature. The direction they face, the way they are shaped, and the materials they are made of utilize solar energies and emphasize the character and climate of their site. An old Native American adage comes to mind when considering the philosophy of the cob house: "Our world was not given to us by our parents but loaned to us by our children." We must keep in mind that we need to give back what we take, and the clay, rock, straw, and wood that are used in cob houses can be given back.

People interested in cob construction can contact The Cob Cottage Company, PO Box 123, Cottage Grove, OR 97424. Tel.: 541-942-2005. Web site: http://www.deatech.com/cobcottage/cob-contact.html.

STRAW-BALE CONSTRUCTION

Based on Stephen O. MacDonald, A Visual Primer to Straw-Bale Construction, *Appropriate Development Project Upper Gila Watershed Alliance*
PO Box 383, Gila, NM 88038

Straw is an annually renewing, relatively abundant building material. In most grain-producing regions, straw—the long, hollow stems of cereal crops such as wheat, oats, rice, barley—is considered waste that is discarded, often by burning. Built correctly, straw-bale shelters are very energy—and cost—efficient and can cut down heating needs through high levels of insulation.

DESIGN BASICS

Simple, energy-efficient shelters made from bales of straw have taken on many forms. These have ranged from quick, temporary structures, like those used for disaster relief and as wrap around existing structures, to more permanent structures for long-term residence.

Keep in mind these basic rules of thumb for designing appropriate straw-bale shelters.

- Keep the size of the building modest in size. How much space is really needed? How much can people realistically afford?
- Keep the design simple so that it is not only easier to build but uses less material as well. Simple, straightforward designs allow people with reasonable common sense, some building skills, and minimal tools to be able to build their own homes.
- Insulate the living space of the building completely.
- Maximize solar energy with passive-solar designs. Face most of the windows south to capture daytime sunlight. (The amount of heat you want to retain will depend, of course, on the climate in which you live.) Use insulated curtains or shutters to keep out the cold at night.
- Use local materials, preferably ones minimally processed like poles, slab wood, and mud. This not only makes economic sense but ecological sense as well.

SUPERINSULATION BASICS

Straw as Superinsulation

To work well, your straw-bale "thermos" must follow these two basic rules.

1. Provide a continuous, surrounding layer of insulation! Leave no heat-leaking gaps in the straw-bales wrapping your building. Have at least as much insulation in your ceiling as there is in the walls. Insulate your floor well. Windows and their openings should be of modest size, of good quality, mostly facing the sun, and covered at night with insulated curtains or shutters.

2. Keep the insulation dry. Insulation won't work if it's wet (and straw as insulation can also rot if it stays wet). Protect it on the outside with good roof and wall coverings. Protect it on the inside (particularly in cold climates) with a leak-proof vapor barrier to prohibit the flow of water vapor into the wall, floor, and ceiling insulation. Adequate ventilation is essential for dissipating the inevitable movement of some moisture into your straw insulation.

Fresh, Clean Air

A potentially serious problem with a tightly sealed building is maintaining enough clean, oxygen-rich air inside. One solution is to provide your heating stove with its own oxygen supply piped in and preheated under the floor. Another is a mechanical air/heat exchanger. Sunspaces and attached greenhouses can be sources for adequate preheated supplies of fresh, clean air to replace stale, polluted air.

STRAW-BALE BASICS

What Is a Good Bale?

A building-quality bale has the following characteristics.

- Dry throughout. In hand, a dry bale feels relatively light when lifted and has no dampness inside when probed deeply with your fingers.
- Stems are strong and flexible. Bend and pull individual stems to check for strength and brittleness. Mature, brightly colored, harvested-dry straw (and kept that way) makes the best bales.
- Well compacted. A compact bale deforms little when lifted by its tie strings. The ties—plastic twine or wire are best—are tight enough around the bale such that you cannot get more than two or three fingers inserted between them and the straw.
- Relatively uniform in size. Bale sizes vary, especially in their length. Two-tie bales measure about 46 cm (18 in.) wide, 35 cm (14 in.) high, and between 86 and 102 cm (34 to 40 in.) long.

Check the size and quality of your bales before you buy.

Handling and Storage

A fundamental rule of straw-bale construction is "keep your bales dry from field to finished building." Always handle them with care, protect them from getting wet during transport, and store them well, preferably in a secure warehouse building or under good tarps. Plan out their purchase and proper storage well in advance.

Enemies of Straw

While straw is durable, stay mindful of its three basic enemies.

- Rot—Rot, caused by enzymatic actions of fungi, occurs in wet bales with a moisture content greater than 20 percent.
- Fire—While well-compacted bales are quite fire resistant, loose straw can be a fire hazard. Be particularly mindful of loose straw in attics and around chimneys. Always keep your building site clean and prohibit smoking. Keep a fire extinguisher handy at the job site.
- Pests—Rodents, birds, and some insects may also find straw hospitable. Seal off or wire screen all potential access points. A few pellets of rodent poison strategically distributed before closure may sometimes be necessary. Keep all curious, hungry livestock away from your straw.

SITE PREPARATION

Select a Site

A good building site incorporates the following.

- It does the least destruction to the area and its habitats.
- It is large enough and relatively flat, reducing the need for excessive earthmoving and foundation work.
- It gives the building a firm, permafrost-free substrate to sit on.
- It allows good drainage away from the building.
- It permits an unobstructed southern exposure for passive-solar gain.
- It protects from cold winds and other climatic factors.
- It allows reasonable access and traffic flow.

Prepare the Site

These three steps are required in preparing your site.

1. Take corner positions using 3-4-5 squaring techniques. An efficient way to ensure a square layout of smaller buildings is to use two tape measures and basic Pythagorean geometry.
2. After marking out the outline of the building on the ground, replace unwanted soil with an elevated pad of gravel.
3. Check corner stakes for accuracy; then secure leveled strings between them to guide the placement of the foundation form.

Plan Ahead

A successful project requires careful planning well in advance of any construction. Consider the following before you build.

- Have detailed and accurate blueprints that include wall elevations showing bale layout. Let the average measurements of your bales help dictate the length (using average bale length from ten or more bales sampled) and width of your building's foundation, window openings, and the height of its walls. This helps you avoid having to create many custom-length bales in each course and having these shortened bales break up the "running bond" (where each bale overlaps the two bales below it by nearly equal measures).
- As determined by a complete materials list, have all your building materials purchased and on site prior to construction time. Plan ahead for their safe transport and storage.
- Begin fabricating all window and doorframes, roof trusses, metal roofing, and so on as early as possible, so they are ready when you need them.
- Keep your onsite stack of bales well covered and protected.

FOUNDATION AND FLOOR

Foundation Fundamentals

A good foundation creates a stable, durable base, while protecting bale walls from water below. In nonload-bearing buildings, such as the post-and-beam structure illustrated in Figure 13, all live-and-dead loads of the building concentrate on the foundation at each post location. Between these high-load points, the metal-reinforced foundation functions primarily as a way to hold the building together in one integral unit (very important in earthquake- and frost-prone areas) and as an elevated platform for the relatively light plastered bale walls.

Protruding metal or wooden "spikes" are sometimes incorporated along the top surface of the foundation to impale and thus help secure the first course of bales. In nonload-bearing structures, however, where the bales serve basically as insulative infill rather than critical structural elements, these bale "impalers" have little function; they are also dangerous to work around.

Creating a wider foundation "footprint" at each post location helps support the concentrated loadings. In Mongolia, we inserted 10×10 cm [4×4 in.] wooden posts into 25-cm [10-in.] deep "pockets" created during the concrete pour.

Consider placement of all reinforcement rod, metal post support straps, other imbedded attachments, and passageways before pouring concrete into the foundation formwork. Having an elevated (about 5-cm [2-in.]) foundation "toe-up" above the floor assures extra bale protection from interior flooding.

A Straw Bale Floor

Straw bales have been used in cold climates to insulate a wooden floor or concrete slab. This technique is relatively new and its longevity unknown; further testing and evaluation are needed. With this system of floor insulation, it is critical that the bales are sitting on an adequate (10 + cm [4 in.]) and well-draining layer of gravel. Also critical is having good drainage away from the perimeter of the building.

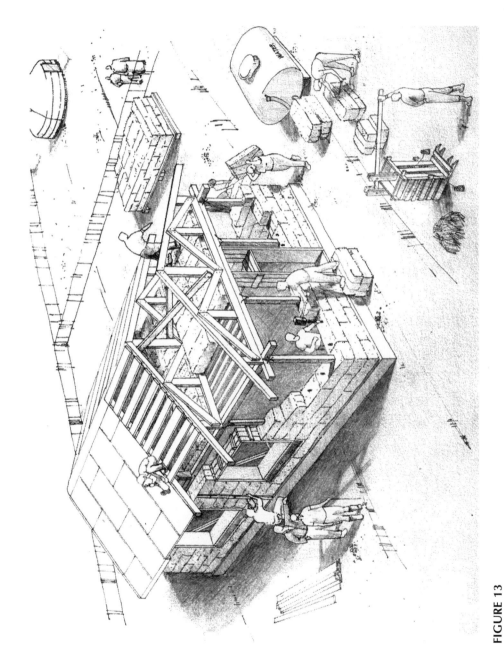

FIGURE 13
Straw bale building. (From *A Visual Primer to Straw-Bale Construction.*)

For both concrete-slab and wooden flooring systems, continuous rows of bales are spaced about 5 to 10 cm [2 to 4 in.] apart, with the space filled either with well-tamped gravel or concrete. For making a slab floor, concrete is poured directly over the bales (about 10 cm [4 in.] in thickness) and then screeded level, using the help of temporary guides. In wood flooring systems, wooden "floor joists" are imbedded level between bale rows to attach the floor decking, preferably tongue and grooved boards.

Once the building is complete, it is wise to insulate the outer perimeter of the building so that cold will not pass through the concrete foundation. A technique used in Mongolia is to pile fly ash around the foundation's perimeter and cover it with a sloping "skirt" of concrete.

FRAMEWORKS AND BRACING

Raising the Framework

A rigid, load-bearing framework carries the total loads of the roof system and transfers them to the foundation. A well-engineered framework makes multiple-storied straw-bale structures possible. Many variations for frameworks are possible. In our example, 10×10 cm [4×4 in.] timbers are used for both vertical posts and horizontal beams.

Posts can be notched into the bale, left unnotched with the gap stuffed with loose straw, or positioned on the outside or inside of the bale wall, which has the advantages of running the bale wall without interruption and requiring less work.

The horizontal beams must be securely attached and braced to the posts. Tracing "staples," fashioned from a reinforcement rod sharpened for easy driving, were used in Mongolia as well as metal brackets and wooden braces.

It is good building practice to provide additional cross bracing to the building's framework, particularly in areas prone to earthquakes and wind wear. Crossties of heavy wire or cables have been used. To be effective, these must be firmly attached to both the foundation (preinstalled loops of metal rod securely attached to the foundation's imbedded reinforcement system works well) and to the upper frame/roof truss structures.

Adding the Roof

A straw-bale structure is only as good as its sheltering roof. While it may comprise the structure's most costly element, don't skimp. Get the best roof you can afford.

Trussed roof systems have many advantages, including efficient use of materials, and flexibility of exterior and interior design. The raised truss allows enough space at the eaves for full bale insulation and adequate airflow. (*Caution*: Don't close off ventilation here by overstuffing it with loose straw.)

The roof structure needs to be firmly attached to the building's framework, especially in situations where long, overhanging eaves and high winds can pose problems. Heavy metal fasteners or several wraps of stout wires give added protection beyond simply toenailing. Always diagonally cross-brace between trusses below the ridgeline once they are installed (see Figure 13).

Caution: Roof trusses can be heavy and dangerous to install. Use ropes, inclines, helping friends, and calm to do this job safely.

Many roofing options are available. Metal roofing is preferred by many for its durability and ability to harvest clean rainwater for drinking. Whatever surfacing is used, however, make sure it doesn't leak, especially if straw bales are being used as ceiling insulation.

Roofs on straw-bale buildings have taken on many shapes, including gabled, hipped, shed, gambrel, offset gabled, vault, dome, and even cone shapes.

Straw Bale Wall Raising

Raising the bale walls is the fun and fast part of building. Consider doing the following.

- Have all the doorframes securely in place and the window frames built and ready to install.
- Install corner guides—upright posts at the inside and outside of each corner. When using wooded lath boards to reinforce the plaster, these guides are permanent. Make sure they stay plumb throughout the wall raising. Never force bales into position; that's how you get out-of-plumb. It is better that the bales be a little too short for the space rather than too long. Stuff the gap with a little loose straw, or look around the stack and find just the right sized bale to do the job.
- Start laying bales at corners and on both sides of door/window frames. Make sure each course evenly overlaps the course below it with a "running bond." Stuff the gaps and dips between bales as you go up.
- If the first course of bales is in any danger of getting wet from below, cover or coat the foundation with a waterproof barrier first. Some builders might even drape the outside of the first course with a waterproof membrane such as tarpaper. Place a waterproof drape under each window frame just before they are set in place.
- It may be useful to pin each course of bales, two per bale, with wooden stakes (5 cm × 1 m [2 in. × 3.25 ft] sharpened poles work well), starting from the third course, then each course thereafter. (If your roof is in the way, don't worry if you can't stake the last course or two.)

Making Custom-Sized Bales

Ideally, you have designed your building to match the average length and height of your bales so that few need customizing. However, some resizing of bales—actually shortening—will be necessary, regardless. To do this make several "bale needles" by pounding flat and filing sharp the end of a round metal rod (about 5 cm × 1 m long [1/4 in. × 3 ft]). Drill a 5-mm [1/4-in.] hole at the flattened point so that baling twine easily threads through it. Now thread twine through the existing bales to wrap a new bale of the proper size. When the wrapping is done, cut away the extra straw from the original bale, leaving a wrapped, shortened bale.

Too loose bales may need more compression and longer length to make them building quality. Make a simple press or hand baler (shown in Figure 13) to accomplish this. To accommodate notched-in posts, the bale can be easily cut using a handsaw, but be careful not to cut the twine.

Insulating the Ceiling

Bales have been used as insulation in the attic. Tightly baled straw is more fireproof than loose straw (use for stuffing gaps only). If you use bales, do the following.

- Insulate around chimneys with something other than straw—fiberglass or fly ash work well here.
- Coat the top and bottom of ceiling bales with something fire-resistant, like plaster.
- If you're in a cold climate, make sure you have a good and continuous vapor barrier between the bales and the interior living spaces.
- Ventilate the attic space well for good airflow. Screen off all vents openings.

PLASTERING THE WALLS

Some builders have plaster-coated their straw-balewalls without the benefit of some kind of reinforcement. This works well enough with sticky, clay-based plasters. With less sticky, cement-based plasters, reinforcement, usually metal netting, is recommended, especially in damage-prone places such as near corners and around window openings (note poured-concrete window sills.)

A system of reinforcing plaster developed in Mongolia (where metal netting is unavailable) uses thin wood strips (about 1 cm × 3 cm × 2 m long [1/4 in. × 1 in. × 6.5 ft]) nailed horizontally or cross-diagonally (about every 3 or 4 cm [1 or 1+ in.]) to vertical 5 × 5 cm [2 × 2 in.] evenly spaced wooden strips. These upright strips are notched flush with the surface of the bales (digging with the claw end of a hammer works well) and firmly wired to inner posts or matching uprights on every other course of bales with the help of a baling needle and a friend. Upright corner guides are secured similarly.

Partition Walls and Ceilings

An effective partitioning technique, also developed in Mongolia, uses the same wooden lattice system, outlined above (or metal netting, where available), nailed on both sides of 5 × 5 cm [2 × 2 in.] frameworks. Loose straw is gently stuffed in the spaces between the latticework and then plastered.

Plastered ceilings provide good fire protection to attics insulated with straw-bales. Wire netting rather than wooden lath works best in this situation. A coat of waterproof enamel or latex paint over the plaster should provide an adequate vapor barrier.

Plastering

Plaster is applied by hand or with trowels, and to do this well takes practice. There are many techniques, and most will probably work on straw-bale walls. Cement-based plasters are the most durable but are not as easy to apply as lime, gypsum, and clay-based "mud" plasters. In many areas, cement may not be readily available—or affordable.

This is commonly used recipe for cement-based plaster.

1 part Portland cement

1 part lime (type S, if available)

8 parts clean, finely sifted sand

In the mixer put the sand and lime and add 80 percent (of the entire volume) clean water and mix for 1 minute. Then add the cement and more water as needed. Mix for at least 10 minutes to make a sticky consistency. Usually two or three coats are applied (decreasing sand to 6 parts in the final coat to some desired finished texture). Keep cement plasters damp for several days to ensure proper curing.

Clay-based "mud" plasters are some pretested mixture of fine clay soil with enough finely sifted sand added—perhaps with some chopped straw—to prevent cracking (at times in a ratio of 3–5 sand to 1 clay soil).

Let Your Walls "Breath"

In cold climates, a good vapor barrier paint should be applied to inside wall and ceiling surfaces. Do not seal off your outside plastered walls so thoroughly that they cannot breath away water vapor that inevitably gets inside the walls from time to time.

MAINTENANCE AND REPAIRS

Construction work is never done—it just goes on into maintenance and repairs. Here is a partial checklist of things to monitor.

- Roof—leaks? corrosion? paint?
- Wall coverings—excessive plaster breaks and cracks? need repainting?
- Straw in walls and attics—dry? pests?
- Windows, doors—in good repair? need repainting? air leaks? still fit?
- Attic—complete insulation coverage? any water damage? adequately vented?
- Heating/plumbing system—in good repair? safe?
- Electric system—good, safe condition?
- Floor—in good repair? straw in floor in good condition?
- Perimeter drainage—still adequate?

SOIL STABILIZED BRICKS

Based on Building with Stabilized Soil Blocks—A Training Manual,
P. Tawodzera, IT Zimbabwe, Harare, Zimbabwe, 1998

A stabilized soil block is a rectangular block formed in a press by compacting soil with a predetermined percentage of stabilizer—cement or lime. These blocks can be used for building homes or community centers. Stabilization increases the block's strength, reduces volume changes due to water absorption, and improves resistance to erosion by wind or rain.

ADVANTAGES OF SOIL STABILIZED BRICKS

- The major raw meterial—soil—is available in large quantities almost everywhere.
- The technology is simple and affordable.
- The product is low cost, has high insulation properties, and a smooth finish.
- Production can be done close to the building, reducing transportation costs.
- Large amounts of water and fuel are not required—pollutants are not released.

SOIL

Subsoil is used for the production of soil stabilized bricks. The organic material in topsoil interferes with the settling and hardening of the brick. The top .5 meter of soil is usually topsoil and unsuitable. Sandy soil with a low proportion of clay particles makes the best blocks. Studying the soil and crumbling it in the hand will give a preliminary indication of whether the soil is suitable. It should not contain organic matter and should feel smooth. Rolling a sample of soil and determining if it holds together will indicate whether sufficient clay is present. A rough analysis of a soil's composition, a sedimentation test, can be made by filling a glass bottle half full of soil and half full of water. The bottle should be shaken, the mixture allowed to settle for an hour, then shaken again and allowed to settle for 45 minutes. The soil will settle in layers, gravel at the bottom, then sand, clay, and silt, with water and organic material above—see Figure 14. The clay and silt should be approximately 20 percent of the solid material. Another test, using the shrinkage of a block, is to make a block of wet soil, filling a box. The block is allowed to dry for three to seven days. A suitable soil will shrink from 2.5 percent to 10 percent—the more clay, the greater the shrinkage.

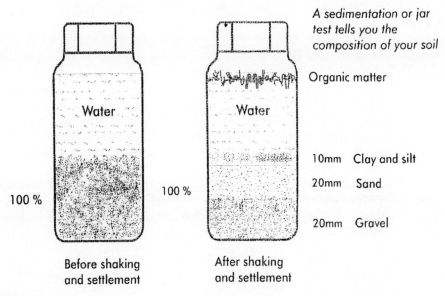

FIGURE 14
Sedimentation test.

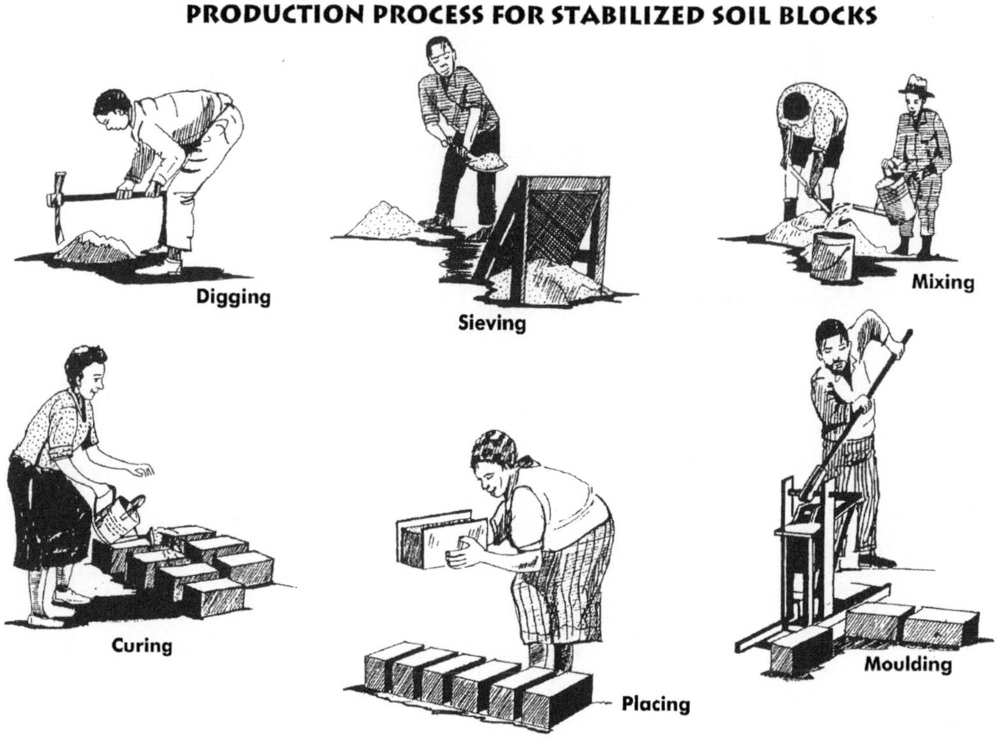

PRODUCTION PROCESS FOR STABILIZED SOIL BLOCKS

Digging

Sieving

Mixing

Curing

Placing

Moulding

FIGURE 15
Production process.

BRICK PREPARATION

This process is shown in Figure 15. Sieving is done to remove soil particles coarser than 6 mm. Cement and water are mixed with the soil. The ratio, by volume, of soil to cement should be about 16:1. Soils with more clay require a higher percentage of cement, up to about 12:1. Water is added to make a semidry mixture. To obtain uniform mixing, it is convenient to apply the water from a watering can. The amount of water is optimum if a fist-sized ball, dropped about 1.2 m, breaks into several large pieces. If the ball flattens without disintegrating, the moisture content is too high. If it breaks into numerous small pieces, then the soil is too dry. The soil should be used within one hour of mixing, before the cement sets.

A typical block press is shown in Figure 16. This is the Aptech Block Press, manufactured in Zimbabwe. Its operation is straightforward. Care should be taken to fill the block mold completely, checking the corners for spaces. On the other hand, the press can be damaged if the block mold is overfilled and the lever forced down. As shown in Figure 15, the block is taken from the press between two boards and placed on a flat area for curing. A team of six people, two operating the press, can make between 500 and 800 blocks a day. One bag of cement is sufficient for about 85 blocks.

Curing takes place in two stages: humid curing and air curing. Humid curing requires that the blocks be kept in a humid or moist state by sprinkling water on top of them at least twice a day for 5 to 7 days. The moisture in the blocks must come out slowly and evenly to ensure

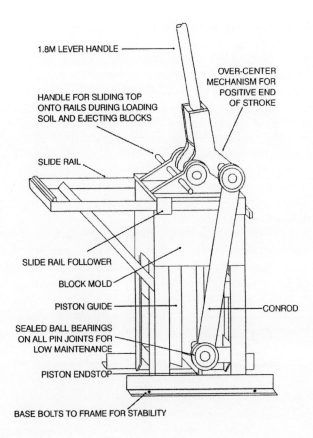

1.8M LEVER HANDLE

OVER-CENTER
MECHANISM FOR
POSITIVE END
OF STROKE

HANDLE FOR SLIDING TOP
ONTO RAILS DURING LOADING
SOIL AND EJECTING BLOCKS

SLIDE RAIL

SLIDE RAIL FOLLOWER

BLOCK MOLD

PISTON GUIDE

CONROD

SEALED BALL BEARINGS
ON ALL PIN JOINTS FOR
LOW MAINTENANCE

PISTON ENDSTOP

BASE BOLTS TO FRAME FOR STABILITY

FIGURE 16
Block press.

the setting or hydration process of the stabilizer. During the humid curing stage the blocks should be kept under shade and covered with grass or plastic sheets to protect them from direct sunlight—causing excess evaporation—and from excessive rain. After at least two days of humid curing, blocks can be stacked on top of each other to a height of about five blocks to create more space for the fresh blocks. After 7 days of humid curing, the blocks are left to air cure for a further 21 days. Thus, after 28 days the blocks are ready for use in construction.

TESTING THE BLOCKS

Some simple tests should be performed to ensure that the blocks will last in use. The blocks should be examined and those with broken edges and more than three cracks discarded. The dimensions should be checked. The blocks should be given a wetting/drying cycle test. Five blocks should be immersed completely in water for 12 hours, then allowed to dry for 12 hours. The procedure should be repeated for seven days. Defective bricks will fall to pieces, even burst. As part of this test the wet and dry bricks should be weighed—a good brick will not absorb more than 15 percent of water by weight. Two strength tests should be done. The bricks should be dropped from a height of 1.5 m. A test brick should be supported at either end by two other blocks, 150 mm apart. An average weight person should stand on the supported block over the gap. The bricks are satisfactory if they do not break in either test.

BUILDING WITH STABILIZED SOIL BRICKS

Foundations should be laid on firm ground that is least susceptible to moisture movements, settlements, and subsidence. A strong foundation with sufficient base course height—also called plinth height—is required. Foundation walls should be built with concrete blocks, stone, or well-fired clay bricks whenever possible, particularly in areas with high rainfall or bad drainage. In areas with low rainfall and good drainage, stabilized soil bricks with high cement content can be used. In this case, the foundation wall should be at least one block length in thickness. The base course should be .40 m high where the rainfall is average and .60 m where the rainfall is heavy. These base walls should go up to a damp proof layer. Plastic sheeting or well-fired bricks can be used for the damp proof layer, which is essential to arrest rising dampness by capillary action. Two foundations are shown in Figure 17. Suggestions about building are also given in the "Building" article in this chapter.

Runoff water should be drained from the building through provision of an apron and drainage courses. This ensures that the lower parts of the walls are well protected from chronic dampness that could easily lead to erosion of the stabilized soil blocks. Rainwater gutters and downpipes fitted to the building should be kept clear from the walls so the wall will remain dry in the event of leaks.

It is advisable to provide generous roof overhangs, especially in high rainfall areas, to shield the walls as much as possible from erosive effects of rain. The overhang should be about .7 m to 1 m measured horizontally from the walls to the eaves—see Figure 18.

Walls are built from stabilized soil bricks in the same way as from other kinds of bricks (see the "Building" article). Both load-bearing and nonload-bearing walls can be built. Soil stabilized bricks, like other bricks, function well in compression but have low tensile, bending,

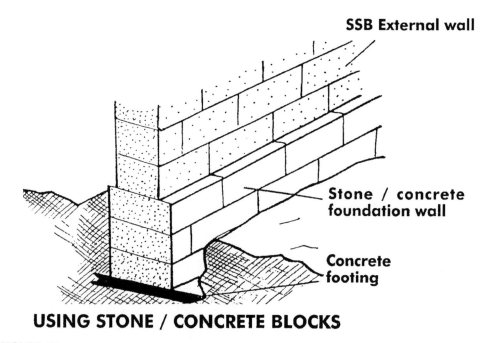

SSB External wall

Stone / concrete foundation wall

Concrete footing

USING STONE / CONCRETE BLOCKS

FIGURE 17
Two foundations.

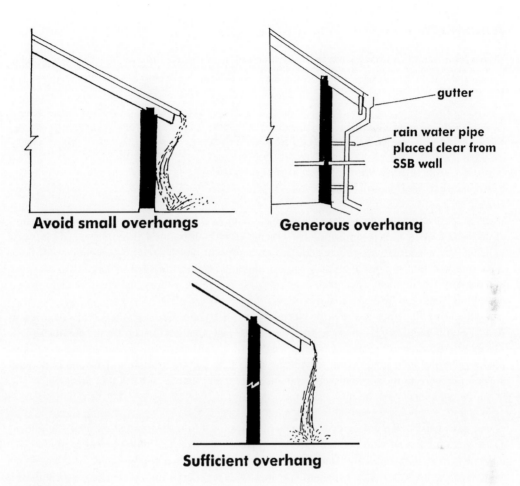

Avoid small overhangs

Generous overhang

gutter

rain water pipe placed clear from SSB wall

Sufficient overhang

FIGURE 18
Roof overhang.

and sheer strength. Off-center or point load should be avoided, as well as bending and possible bulging. Thin and long walls should be buttressed at about 3 m to 6 m center to center—see Figure 19. Over an opening for a door or window a long piece of wood or concrete, called a lintel, is used. The lintel should extend at least 25 cm beyond the opening at both ends. Doors and windows weaken the structure—they should be kept far apart and away from corners. Corners are another potentially weak place—use of half-sized bricks to fill in should be avoided.

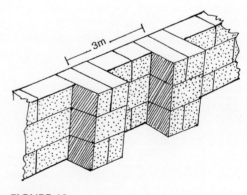

FIGURE 19
Buttresses.

REFERENCE

Thomas, David W. (1990). "Understanding Small Scale Brickmaking." VITA Technical Note 70. Volunteers in Technical Assistance, 1600 Wilson Blvd., Arlington, VA 22209, USA. vita@www.vita.org

MICRO-CONCRETE ROOFING TILE PRODUCTION

Based on an article by the same name by Otto Ruskulis, IT Technical Enquiry Service, in Appropriate Technology, *Vol. 23, No. 1, June 1996*

Over the past 25 years much attention has been paid to developing the small-scale production of concrete roofing tiles as an affordable alternative to both traditional roofing materials, such as thatch, and modern, mass-produced, often inappropriate, galvanized iron sheeting or asbestos cement. These tiles are relatively low in cost, durable (with a life span of more than 20 years in most areas), aesthetically acceptable, able to offer adequate security and comfort, and provide protection from both heavy rain and hot sun.

Concrete roofing tiles are now produced by small businesses in a number of countries in Africa, South and Central America, Asia and Southeast Asia, and in the former Soviet Union. The key to the success of this technology was the development of equipment and techniques to produce the tiles on a small scale. It typically costs U.S. $5,000 (excluding land and buildings) to set up a concrete roofing tile workshop and can be less than U.S. $1,000 in areas where the vibration equipment and the molds are made locally.

When the technology was first developed, it was decided to make large roofing sheets similar in size and shape to the corrugated asbestos or galvanized iron sheets used on many buildings. These were reinforced with natural fibers such as sisal or coir. The fiber-cement mortar (FCR) mix was simply spread out by hand on a flexible plastic sheet in a large mold. Afterwards the sides of the mold were taken away and the sheet with the mortar on top was gently pulled over onto a corrugated mold where it took its shape.

Problems were experienced with decay of the fibers and cracking of the sheets after only a few years, and so the production of fiber-reinforced concrete roofing sheets has been abandoned in many countries.

A more recent development has been to make concrete tiles without any fiber at all. These are the so-called micro-concrete roofing (MCR) tiles. Greater care needs to be taken with MCR tile production compared with FCR if the number of damaged or substandard tiles is to be kept low. MCR tiles are also more brittle than FCR tiles and can be damaged if dropped or handled carelessly when transporting or fixing them to the roof.

QUALITY CONTROL

To produce a good MCR tile, care needs to be taken in the quality of the sand used to make the morter. The sand should have regular grain-size grading—without too much material of one size and, particularly, without too much fine, silty material. If the sand from one source contains too much material of any one size, it should be mixed with a sand of different grading from another source. In addition, the batching of the quantities of sand, cement, and water needs to be done accurately—to ensure that there is enough cement and that the mix is not too wet. The tile maker needs to mold the tile with care and skill, and it is important that the tiles are properly cured.

With FCR tile production there is some capacity for these quality aspects to be less rigorously exercised, but with MCR production there is no margin to be lax on quality control to avoid large numbers of damaged or substandard tiles. If the potential producer cannot ensure good quality control at all stages of production, then it probably is not a good idea to produce MCR tiles.

EQUIPMENT AND MATERIALS

The equipment and materials needed to produce MCR tiles are the same as for FCR tiles, except no fiber is used, and the sand used needs to be of good quality, noted above.

Essential equipment includes the following.

- Tile vibrator
- Molds
- Plastic sheets
- Batching boxes
- Water curing tank
- Table to work on

The use of a vibrator is essential for MCR tile production. Vibration helps to consolidate the mortar mix and remove air bubbles that would otherwise cause weak spots in the hardened tile. The vibrator unit itself consists of a flat metal plate that is suspended on dampers and to which is attached a rotating eccentric cam. It is the rotation of this cam that translates into the up-and-down motion of the plate. A hinged metal cover fits onto the plate. This defines the sides of the tile. The vibrator may be driven manually, electrically with a standard 12-volt truck battery, or on mains electricity.

Because cement mortar sets slowly and the tiles need to be left on the molds at least overnight before they can be removed, the producer will need enough molds to cover a whole day's production. A single person should comfortably be able to make at least 200 tiles per day and probably considerably more. (*Note:* The cost of 200 or more molds should not be underestimated—they could cost more in total than the vibrating unit itself.) Because it is important that MCR tiles cure in a damp environment, the enveloping type of mold (Figure 20) must be used. These molds are stacked one on top of the other and hence cover the curing tiles and prevent them from drying out too quickly. Although cheap concrete molds have been used for FCR tiles, which then cure open to the air, more rigorous standards are needed to cure MCR tiles. The stack-up type of mold is made of plastic. Note also that out of a batch of 200 molds, at least 10 should be ridge molds for making the specially shaped tiles for the ridge of the roof.

Tiles are molded on top of a plastic sheet on the tile vibrator. After vibration the tile is removed carefully from the vibrator, still on its plastic sheet, and positioned on the mold, which forms its shape. The same number of plastic sheets as molds will be needed, but some additional sheets should be kept in stock to replace those that wear out.

Accurate batching of sand and cement is essential. For this reason it is usual to have two batching boxes to measure out exactly the right amount of sand and cement needed for one tile. A measuring jug for water, to ensure that similar amounts of water are added for each tile, would also be useful.

A water tank is needed for curing the tiles. Since the tiles are cured in warm water for at least five days, the tank should be large enough to hold 1,000 or more tiles. A single tank to accommodate this number of tiles would need to be about 8 meters long, .8 meters wide, and at least .6 meters high, although it probably would be more convenient to use a number of smaller tanks.

These materials are needed to make MCR tiles.

- Sand
- Cement (ordinary Portland cement)
- Water
- Pigments (optional)

The need for sand with a suitable grading has already been noted; more specifically, the following guidelines are used for sand grading for MCR tiles.

Tile Thickness	6 mm	8 mm	10 mm
Maximum grain size	4 mm	5.5 mm	7 mm
Above 2 mm	25–45%	30–50%	35–55%
.5 to 2 mm	20–50%	15–40%	15–40%
Below .5 mm	15–45%	15–45%	15–45%

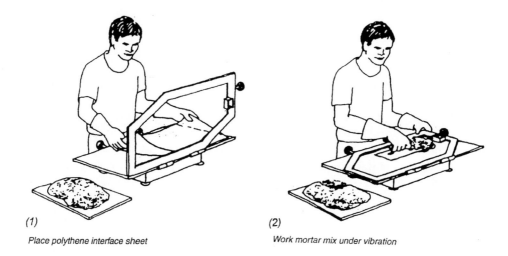

(1)

Place polythene interface sheet

(2)

Work mortar mix under vibration

(3)

Smooth under vibration

FIGURE 20
Making concrete roofing tiles. (continued on next page)

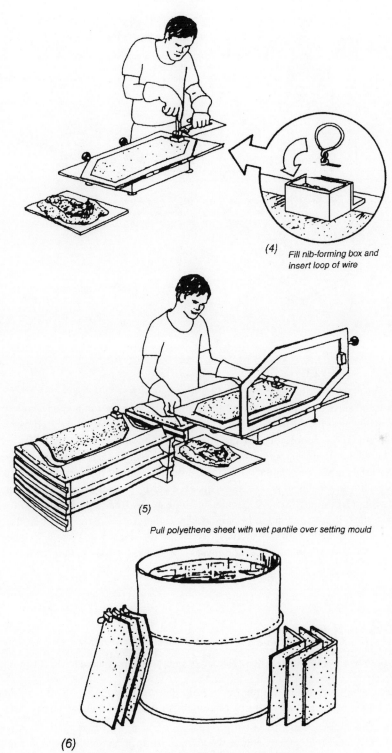

(4) Fill nib-forming box and insert loop of wire

(5)

Pull polyethene sheet with wet pantile over setting mould

(6)

Cure in a water tank, in this case an old oil drum

FIGURE 20
(Continued)

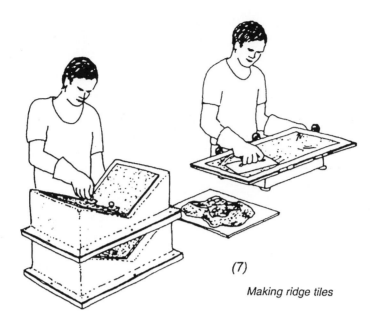

(7)

Making ridge tiles

FIGURE 20
(Continued)

In addition, the clay and silt content should be less than 4% in all cases.

A set of three sieves with openings of approximately the sizes indicated above would be a very useful acquisition for the serious MCR tile producer. These could be used for determining the sand size grading and, possibly, for making sand heaps of different sand sizes, which could afterwards be mixed in appropriate proportions. A clean, sharp sand is the best to use.

The water used should be clean and free from significant quantities of dissolved salts, particularly sulphates. If water that is of good enough quality to drink is available, then this should be used.

Pigments for coloring the tiles are popular in some areas. Red is probably the most common color used for tiles. Pigments tend to be imported and therefore expensive. Their use makes the cost of the tile significantly more expensive, but in some areas a market does exist for more expensive colored tiles. Pigments add nothing to the strength of a tile and may even reduce it slightly.

THE PRODUCTION SEQUENCE

Production is shown in Figure 20. The sequence of operations to make an MCR tile is as follows.

1. Fill the cement and sand batching boxes fully to the top and level off. It is normal practice to use three volumes of sand to one of cement so the sand batching box is three times the size of the cement box.

2. Tip out the sand and cement onto a wooden, plastic (polythene is used in Figure 20), or metal mixing board placed on a table. Mix thoroughly for up to a minute until all the meterial is a uniform color.

3. Add water to the mix gradually, turning the mix with a trowel constantly until it becomes wet enough to be workable. Add the water slowly to ensure that too much is not added. It is best to use a measuring jug and add a measured amount of water each time. A few trial mixes can be made to find out how much water would normally be needed. If a set of scales is available, a standard water-to-cement ratio (by weight) can be determined. First, find out how much cement is used per tile by weighing the cement batching box empty, then full to obtain the weight of the cement used. Mix the sand and cement and add a mesured volume of water until a mix of the required consistency is obtained. The volume of water in milliliters is equal to its weight in grams, and from the weight the water-to-cement ratio can be calculated. A good mix will have added to it a weight of water equal to half the weight of cement (that is, the water-to-cement ratio is .5). One milliliter of water weighs one gram, so the weight of water in grams is equal to its volume in milliliters. If after adding your calculated volume of water the mix is too stiff, add a little more water, but make a note of how much extra is added. If a water-to-cement ratio of more than .65 is needed to make the mix workable, then the resulting tiles will be of low strength and quality, so stop and reexamine your materials before proceeding.

4. Place the plastic sheet onto the vibrator unit and clamp down the sides.

5. Transfer the mortar mix onto the vibrator, spreading it out with a trowel.

6. Switch on the vibrator unit and continue to spread the mix with the trowel. A vibration time of 30 seconds will usually be sufficient. Vibration for more than a minute is not recommended—it can cause the mix to separate.

7. Add the tile nib to be used to fix the tile to the roof manually.

8. Remove the sides of the unit and carefully slide the green tile, still on its plastic sheet, onto the mold so that it takes the corrugated or ridge shape of the mold.

9. Stack the molds and leave overnight.

10. The next day, remove the tiles from the molds and place them in the water-curing tank, leaning the tiles against each other. Leave the tiles in the tank for at least five days.

11. Remove the tiles from the tank and leave them to cure in the air for at least 20 days in a cool, shaded place. Sprinkle with water at least twice a day.

FURTHER INFORMATION

Appropriate Building Materials, 3rd Ed. Roland Stulz and Kiran Mukeiji. SKAT, IT and GATE copublication, 1993.
Production Guide: Fibre and Micro Concrete Roofing Tiles. FCR/MCR Toolkit Element 22. SKATIILO Publications, St. Gallen and Geneva, 1992.
Quality Control Guidelines. FCRIMCR Toolkit Element 23, SKAT, St. Gallen, Switzerland, 1991.
The Basics of Concrete Roofing Elements. Fundamental Information on the Micro-Concrete Roofing (MCR) and Fibre Concrete Roofing (FCR): Technology for Newcomers, Decision-makers, Technicians, Fieldworkers, and All Those Who Want to Know More About MCR and FCR. SKAT, St. Gallen, Switzerland, 1989.
An Introduction to FCRIMCR Production. A BASIN Video. IT/GTZ-GATE, Eschborn, Germany, 1990.

FURTHER ADVICE

SKAT in Switzerland can offer further advice on MCR production and MCR technology. SKAT is a member of the Building Advisory Service and Information Network (BASIN). Contact: The Roofing Advisory Service, SKAT, Vadianstrasse 42, CH-9000 St. Gallen, Switzerland. Tel.: +41 71 228-54-54. Fax: +41 71 228-54-55. E-mail: 100270.2647@compuserve

Drawings from *Appropriate Building Materials*, Published by SKAT.

THATCH

Based on Farm Structures in Tropical Climates, *edited by Leonard Bengtsson and James H. Whitaker, FAO/SIDA Cooperative Programme, United Nations*

Thatch is a very common roofing material in rural areas. It has good thermal insulating qualities and helps to maintain rather uniform temperatures within the building, even when outside temperatures vary considerably. The level of noise from rain splashing on the roof is low, but during long, heavy rains some leakage may occur. Although thatch is easy to maintain, it may also harbor insects, pests, and snakes.

A number of different plant materials such as grass, reeds, papyrus, palm leaves, and banana leaves are suitable and inexpensive when locally available. Although the materials are cheap, thatching is rather labor intensive and requires some skill.

The durability of thatch is relatively low. In the case of grass, a major repair will be required every 2 to 3 years, but when well laid by a specialist and maintained, it can last for 20 to 30 years or longer. The supporting structure of wooden poles or bamboo, although simple, must be strong enough to carry the weight of wet thatch. The use of thatch is limited to rather narrow buildings, since the supporting structure would otherwise be complicated and expensive and the rise of the roof very high due to the necessity of a very steep slope. Palm leaves should have a slope of at least 1:1.5, but preferably 1:1 and grass thatch. Increasing the slope will improve the durability and reduce the risk of leakage. The risk of fire is extremely high but may be reduced by treatment with a fire retardant, as described later.

GRASS THATCH

Grass for thatching should have the following characteristics.

- It should be hard, fibrous, and tough, with a high content of silicates and oils and low content of easily digestible nutrients like carbohydrates, starches, and proteins.
- It should be free of seeds and harvested at the right time.
- It should be straight and have thin leaves and be at least 1 meter long.

Proper thatching procedure requires the following.

- Stems that are parallel, densely packed, with the cut side pointing outward.
- A steeply sloping roof frame or more—a minimum of 1:1 but preferably 1:0.6.
- The eaves should be low to offer protection for the walls.
- For best results the roof shape should be conical, pyramidical, or hipped in shape, rather than double pitched, where the verges present weak points.

HOW TO THATCH

For easy handling the grass is tied into bundles—see Figure 21. The thatching is started from the eaves in widths of about 1 m. A number of grass bundles are put next to each other on the roof, with the base of the stems to the bottom (Figure 22). The grass is tied to the purlins with bark fiber or preferably tarred sisal cord. In subsequent layers the bundles are laid to overlap the layer underneath by half to two-thirds of their length, which means there will be two to three layers in the finished thatch.

A long needle is used to push the string through and tie the bundles of grass onto the roof-laths. Then the bundles themselves are untied and with the hands the grass is pushed into the right position, giving a smooth surface to the roof. Then the string is pulled, and this fixes the grass securely in place. Another method leaves the bundles of grass as they are, which gives the roof a stepped surface. The thickness of the new thatch layers varies between 15 and 20 cm, but later on this will become somewhat thinner because of settling.

Figure 23 shows three types of ridge caps that can be used on thatch roofs.

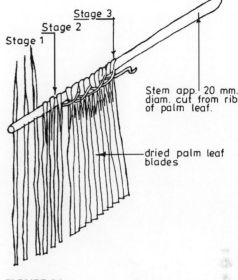

FIGURE 21
Tying bundles.

SEWN THATCHING

The grass or straw is bound in bundles to the battens forming thatch boards. These boards are manufactured on the ground and bound to the rafters beginning at the eaves and continuing to the ridge. Each board covers with its free ends the board underneath.

Stitch at bottom of first thatch on lowest batten. The second layer must overlay the stitching of the first row and include the top section of the underneath layer in the actual stitch. It is better to have each layer held by three rows of stitching. The stitching of every row must be completely covered by the free ends of the next layer above it.

FIGURE 22
Thatching with grass.

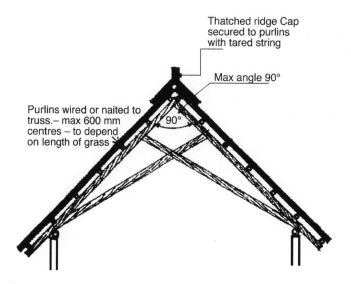

Thatched ridge Cap
secured to purlins
with tared string

Max angle 90°

Purlins wired or naited to
truss.– max 600 mm
centres – to depend
on length of grass

90°

Cement reinforced ridge
Cap on expanded metal
Secured to purlins with
galvanized wire

Galvanized iron sheet tied to
purlins with galvanized wire

FIGURE 23
Roof caps.

FIRE RETARDANT FOR BAMBOO AND THATCH

Fire retardant paints are available as oil-borne or water-borne finishes. They retard ignition and the spread of flame over surfaces. Some are intumescent—that is, they swell when heated, forming a porous insulating coating.

A cheap fire-retardant solution can be prepared from fertilizer-grade diammonium phosphate and ammonium sulphate. The solution is made by mixing 5 kg of diammonium phosphate and 2.5 kg of ammonium sulphate with 50 kg of water. The principal disadvantage is that it is rendered less effective by leaching with rain. Therefore, the fire retardant impregnation must be covered with a water-repellent paint. The entire roof construction—bamboo trusses, strings, wooden parts, and thatch—should be treated with the fire retardant. The following procedure is recommended.

IMPREGNATION OF THATCH

1. Dry the thatching materials such as reeds, palmyra, leaves, bamboos, ropes, and so on by spreading out in the sun.
2. Prepare the solution of fertilizer-grade diammonium phosphate, ammonium sulphate, and water as recommended.
3. Immerse the material in the chemical solution and let soak 10 to 12 hours. A chemical loading of 10 to 14 percent by weight of the thatch (dry basis) is adequate.
4. Take out the material, drain excess solution, and again dry in the sun.
5. Prepare thatch roof in the conventional manner, using the impregnated material and similarly treated framing material.

Such roofs do not catch fire easily, and any fire spreads very slowly.

PALM LEAF THATCH (MAKUTI)

Palm leaves are often tied into makuti mats that are used for the roof covering. They consist of palm leaves tied to a rib (part of the stem of the palm leaf), using the dried fiber of Doum palm leaves or sisal twine (see Figure 24).

The mats are laid on the rafters (round poles) and the stems tied to the rafters with sisal twine. The mats are usually produced to a standard size of 600 mm × 600 mm and laid with a 100 mm side lap, thus requiring a rafter spacing of 400 mm. For a good quality mat 600 mm wide, an average of 75 blades will be required. Spacing up the roof slope that is, the distance between the ribs of the makuti mats—is usually 150 mm to 100 mm, thus forming a four- to six-layer coverage 5 cm to 8 cm thick.

Detail showing tying of blades

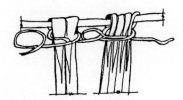

FIGURE 24
Assembling makuti mats.

PAPYRUS THATCH

First a papyrus mat is placed on top of the purlins; then a layer of black polyethene and another one or two papyrus mats complete the roof. These materials are fixed to the purlins with nails and iron wire. Nails are fixed to the purlin at 15 cm to 20 cm spacing, and the iron wire is then stretched over the top of the papyrus mat and secured to the nails. The papyrus has a life span of about three years, but that can be extended by treating the papyrus with a water-repellent paint.

GYPSUM

Otto Ruskulis IT Technical Enquiries Service (Based on ATBrief No 19 "Gypsum Processing and Use," Appropriate Technology, *Vol. 23, No. 4)*

Gypsum plaster is a very useful building material that can be produced with simple processing equipment. It is made from deposits of gypsum rock or sand. Raw gypsum is heated to eliminate a portion of the chemically combined water. The resulting gypsum plaster, when recombined with water, sets like cement. It can be used both as a plaster and for making building blocks. It is suitable for indoor use and in drier regions may be used outdoors as well.

In addition to its use as a building material, gypsum plaster is used to make casts for broken limbs; to make molds for pottery or dentistry; to condition the soil in agriculture; to make writing chalk; as an additive in certain foods, medicines, and cosmetics; and in both paper making and sculpture.

Small-scale production of gypsum plaster is possible if there is access to gypsum rock or gypsum sand of satisfactory quality and sufficient quantity. Production of as little as 500 kg of plaster per day may be economically viable in some areas. In addition, at the most basic production level only a small number of relatively low-cost equipment items would be required, and these could probably be fabricated or sourced locally.

The main advantage of gypsum plaster over some other binders is that temperatures of only 120 to 160°C are required during production, so high-temperature kilns are not needed. The main disadvantage of gypsum plaster when used in building is that it is not durable in prolonged wet or damp conditions, becoming soft and soggy.

RAW MATERIAL

Gypsum rock (calcium sulphate dihydrate—$CaSO_4 \cdot 2H_2O$) was formed in geological times through the evaporation of seawater. It is often laid down in beds, ranging in thickness from a few centimeters to several tens of meters.

Because gypsum rock is slightly soluble in water, it is not usually found above ground in wet or damp areas, but it may be found underground, where is it not affected by the water table. In dry regions it may also be found on the surface, sometimes in the form of gypsum sand. The presence of gypsum on or just below the surface of the ground is often indicated by changes in vegetation. Some plants thrive on gypsum-rich (alkaline) soil, whereas others are not at all tolerant of those conditions. Gypsum rock is usually white or colorless, although it may sometimes have gray, yellow, pink, or brown hues. Gypsum is much softer than minerals of similar color, such as calcite or quartz, and is the only one that can be scratched with a fingernail. If a piece of gypsum is held over a flame, it will turn cloudy and opaque and give off water.

Some gypsum sand deposits contain only about 60 percent gypsum, and these are not very suitable for producing a plaster. Those containing more than 80 percent would be most suitable.

DEHYDRATION: ROCK INTO PLASTER

Gypsum rock is converted into gypsum plaster by evaporating some of the chemically combined water. Heating gypsum at 120°C for one hour results in hemi-hydrate ($CaSO_4 \cdot 1/2H_2O$), with three quarters of the water removed. Gypsum hemi-hydrate is also known as Plaster of Paris. Prolonged heating over several hours results in the formation of anhydrite—practically none of the chemically combined water is left. Anhydrite sets more slowly and is a slightly stronger plaster than hemi-hydrate, but with the drawback of added production cost might not justify the improvement achieved. In practice, a simple production system would most likely give a mixture of the hemi-hydrate or anhydrite phases. Nearly all commercial plaster produced today is plaster of paris. Gypsum plaster sets by chemically combining with water to form solid calcium sulphate dihydrate.

PROPERTIES OF GYPSUM PLASTER

In general, gypsum plaster sets very rapidly when mixed with water, so only small quantities should be mixed at a time and used almost immediately. Gypsum plaster will begin to stiffen in about 5 minutes and become completely rigid in under 20. The addition of retarders can delay this process by as much as half an hour. Lemon juice, borax, sugar, molasses, fish liquor, and keratin (made from animal hoofs and horns) can all be used, and the actions of some of these are enhanced by a small quantity of building lime.

Once hardened, gypsum plaster is a hard and durable material, suitable for many building applications. In a wet environment, however, it will soften irreversibly and eventually dissolve. Portland cement or lime is often preferred as cement for outdoor use, since there is no easy way to improve significantly the water resistance of gypsum.

PRODUCTION PROCESSES

Gypsum processing plants vary widely in scale and level of technology. They range from plants producing one or two tonnes per day, using low-cost manual technologies, to plants of a thousand tonnes per day that are highly mechanized and capable of producing different types and grades of gypsum plaster. There are five basic stages in gypsum processing.

Excavation is sometimes carried out by digging out an area of ground where the gypsum is located using opencast techniques. To reach deeper deposits, drift or shaft mines may be needed.

Crushing the gypsum rock is advisable before processing further, especially if subsequent heating is to be done in a pan rather than a shaft kiln. Crushing will ensure a product that is more uniform and requires less energy to heat. Crushing can be done manually with a hammer or handheld roller, but mechanical crushing is faster and less laborious. Most clay-crushing equipment, such as that used for brick making and pottery, would be suitable. Crushing should

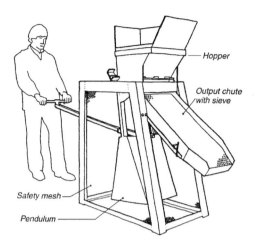

FIGURE 25
Crushing the gypsum. (Clay crusher design by J P
M Parry and Associates Ltd.)

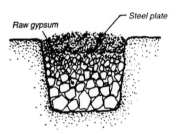

FIGURE 26
A pit kiln.

reduce the gypsum to grains of less than a
few millimeters across. Gypsum sand does
not need crushing (see Figure 25).

Screening with a sieve (manual or motor-
driven) will remove large grains that have not
been crushed properly and that may contain
impurities.

Grinding—for example, in a ball, rod, or
hammer mill—is necessary if the gypsum is
to be used for high-quality plasterwork or for
molding, medical, or industrial applications.
Unlike with other cements, such as lime and
ordinary Portland cement, special mills for
mineral grinding may not be required, and
the relatively soft gypsum could be pulverized in agricultural mills, which are generally
widely available.

Heating may be done in a number of
ways involving a range of technology levels
and costs. The simplest method is to mix the
gypsum stone and fuel in a mound or in a shallow pit in the ground and burn it (Figure 26).
Medium-scale batch production might be
carried out in an excavated hillside kiln, a
shaft kiln with alternate fuel and stone layers, or a permanent walled kiln. An alternative method is to heat the gypsum in large
pans or on a flat metal plate positioned
above a kiln (Figure 27). Industrial production may be carried out in a purpose-built
enclosed batch kiln, a continuously fed vertical shaft kiln, a specially designed large

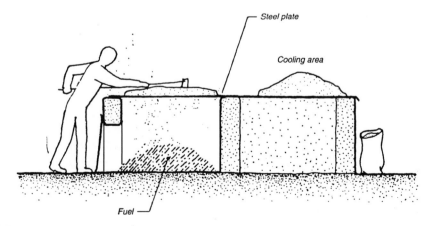

FIGURE 27
Heating the gypsum on a flat metal plate.

kettle, or in rotary kilns. Kiln-based systems are more efficient than burning the gypsum in mounds or pits, and even a small kiln may use less than half of the fuel of a pit or mound (Figure 28).

It is easiest to judge when enough heat has been applied if an indirectly heated pan or metal plate is used. When the temperature of the surface is increased steam will be produced and the material seems to boil. The temperature is maintained until this "first boil" is completed, which removes all but a quarter of the water and leaves hemi-hydrate plaster. If the temperature is allowed to rise further, this will start to convert to anhydrous plaster. Allowing the material to cool naturally after firing will help to remove some residual water left in the mass. It is also important to stir the material continuously to help the steam escape. If the material is heated in a kiln, then it is more difficult to know when the removal of water has taken place, and fuel usage and loading need to be judged from experience.

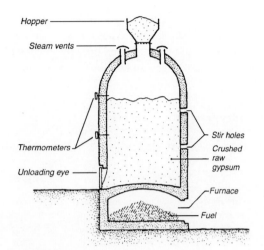

FIGURE 28
Hopper fed furnace heated kiln.

As a rough guide, a field kiln might require .2 tonnes of wood for every tonne of raw gypsum burned, or 70 liters of fuel oil, although kiln efficiencies will vary widely and so would the optimum amount of fuel to use.

In some small batch kilns gypsum rock is burned in the form of lumps, with the larger lumps nearest the fire supporting the smaller stones farther away. After burning, the gypsum would still need to be crushed, screened, and, possibly, ground down further in a mill.

USING GYPSUM PLASTER

Plaster of paris is mixed with sand and water to produce a mix suitable for plastering walls and ceilings, for external rendering in dry climates, for mortar, and for making building blocks. It can also be used as a soil stabilizer in stabilized soil blocks. If mixed into lime-based plasters and mortars, it will give a fast initial set compared with the much slower setting lime.

Gypsum-based mortars and plasters are typically made up of one volume of gypsum to two or three of sand, with enough water added to make the mix sufficiently workable. Because the mix may start to stiffen only five minutes after adding the water, it is important to make up only small quantities of plaster and to apply these quickly, unless retarders are used.

A typical mix for wall blocks would be one volume of gypsum plaster to one volume of sand, two volumes of gravel, and one volume of water. Such blocks would be hard enough to demold after 10 to 15 minutes, before being left to dry for several days. The blocks should satisfy the standard requirements for building blocks for general internal structural purposes but not in wet or damp conditions.

Gypsum plaster can also be reinforced with fibers and cast in molds to producce precast decorative panels. Certain fibers, such as sisal and conventional glass fiber, will not deteriorate appreciably in a gypsum medium, although they do deteriorate over time in a more alkaline ordinary Portland cement mix.

Note that because of its affinity for water, gypsum plaster should be stored in a dry place and never outdoors if there is a risk of rain. If storing gypsum in bags or in bulk for long periods, then it should not be stored directly in contact with the ground and preferably raised on pallets or slats. Prolonged storage in damp or humid conditions is not advisable.

REFERENCES

Coburn, Andrew, Eric Dudley, and Robin Spence. (1989). *Gypsum Plaster: Its Manufacture and Use*. London: IT Publications.

Bureau of Indian Standards. (1989). *Low Grade Gypsum; Use in Building; Code of Practice*. IS 12654. Bureau of Indian Standards, New Delhi.

Gypsum Today. (1989). Gypsum Products Development Association. London.

Smith, Ray. (DATE ?). "Small-Scale Production of Gypsum Plaster for Building in the Cape Verde Islands." *Appropriate Technology*, Vol. 18, No. 4, 4–6. London: IT Publications.

BAMBOO-REINFORCED CONCRETE

Based on Simon Perry, Professor of Civil Engineering and Head of Department, Trinity College, University of Dublin, Dublin 2, Republic of Ireland, "Is Bamboo-Reinforced Concrete Really Useful," Appropriate Technology, *Vol. 14, No. 4, March 1988*

Concrete is a material that is strong in compression and weak in tension. However, the load-carrying capacity of concrete is improved considerably when it is reinforced with a material that is strong in tension. The most commonly used reinforcement is steel, usually in the form of round bars (from 6 mm to 40 mm in diameter) or welded mesh (typically 3 mm to 6 mm diameter wire on a 50 mm grid). However, the shortage and high cost of steel in most Third World countries have prompted several studies into the practicability of using natural fibers, notably bamboo, as substitutes for steel in the reinforcing of concrete.

In the Savannah region of Africa, where low seasonal rainfall minimizes erosion, a tradition of highly sophisticated bamboo-reinforced mud building has developed over several hundred years. With proper maintenance, which for these buildings involves no more than periodic (every three to five years) resealing of the weathered external face with a skim of mud, the bamboo itself can remain effective for at least 100 years.

CHARACTERISTICS OF BAMBOO

Bamboo is a perennial grass that occurs in the natural vegetation of many parts of the tropical, subtropical, and mild temperate regions, with about 1,250 species identified throughout the world. New bamboo plants may be produced by means of seeds (which are not easily obtainable) or by vegetative fractions or layers (the more usual method with some species). Bamboos have jointed and hollow or, rarely, solid culms or stems.

Each culm is divided into nodes (the joints), where solid partition walls occur, and internodes, which are the lengths between the nodes. The culms can vary from 15 to 30 meters in height, 2.5 to 25 centimeters in external diameter, and 30 cm to 50 cm in internodal length.

The epidermis of the bamboo is overlaid with a waxy covering (cutin), which restricts the loss of water from the culms. Bundles of subepidermal cells (sclerenchyma), which constitute the skeletal framework of the bamboo, have great ultimate strength as well as a high elastic limit. They have similar elongation to steel before the limit of elasticity is reached. It takes about three years for the sclerenchyma to mature. Reinforcement is prepared by cutting strips, called splints, from the culm; typically, splints are 5 mm thick, 20 mm wide, and 1.5 m long.

Although as a whole bamboos are versatile, not all of them are suitable for engineering applications. For example, of the three species that are available in Ghana, namely, *Oxytenanthera Abyssinica, Dendrocalamas Strictus*, and *Bambusa Vulgaris*, the first two are hard and relatively strong in tension, while the third species is soft and of low tensile strength. The principal disadvantages of a bamboo otherwise suitable for use as reinforcement when it is in natural untreated form are as follows.

- Poor bond with the concrete
- Low modulus of elasticity
- High water-absorption tendencies
- Variations in mechanical properties
- Variable dimensions
- Extreme flexibility
- Low durability
- Poor resistance to fire

Some of these disadvantages, however, can be significantly offset by subjecting the bamboo to appropriate treatment.

DESIGN AND CONSTRUCTION GUIDE

The following recommendations are based on the experimental work of the authors and others. A bamboo-reinforced beam is shown in Figure 29. Culms at least three years old should be stacked and allowed to air-season in the shade for at least 28 days. Splints are produced by first splitting the culm horizontally. Splint width should not exceed 20 mm; otherwise, difficulty may arise in the placing of the concrete.

An adequate bond between the bamboo splints and the concrete is essential for the success of bamboo as a conventional reinforcement. Bond strength, which may vary from as low as 0.2 N/mm^2 for unseasoned, untreated splints to as high as 3.0 N/mm^2 for carefully seasoned and waterproofed splints, may be improved by different methods.

- Increasing the seasoning period of the bamboo.
- Arranging for nodes to lie in the anchorage length; the node protrusions provide extra mechanical resistance to augment the bond and should always be left intact.
- Friction introduced by light filing of the smooth epidermal skins of the splints.
- The application of two coats of a suitable bituminous paint at intervals of 24 hours, dusting with sand immediately after the second coat, and leaving to dry for a further 24 hours before use.
- Higher-strength concrete (though a practical upper strength may be 40 N/mm^2, even only 30 N/mm^2, with most Third World ordinary Portland cements).

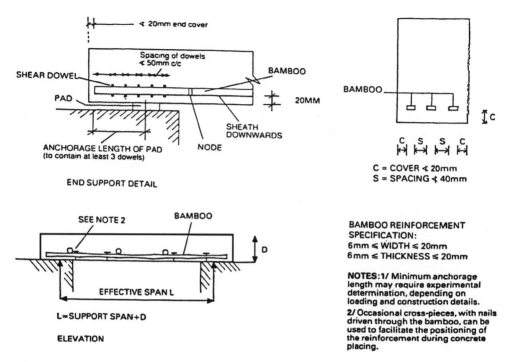

FIGURE 29
Bamboo-reinforced beam.

The waterproofing paint is to prevent swelling of the splints due to the absorption of moisture from the wet concrete and subsequent shrinkage due to moisture leaving the bamboo after the concrete has hardened; the sand adds an extra roughness to the surface. In most developing countries, purchase of the paint will probably require foreign exchange. Clearly, this will need to be offset against the savings in steel cost. Another reason why the use of green, unseasoned bamboo is not recommended is that it tends to season and shrink while embedded in the concrete, leading to a decrease in bond strength.

Long lengths of bamboo splints tend to be rather flexible and are not always completely straight. Therefore, in order to facilitate the maintenance of the specified cover to the reinforcement, a form of construction that employs long-span in situ concrete slabs should be avoided.

CRACKING, COVERING, AND SPACING

Cracking is an important serviceability limit in bamboo-reinforced concrete, especially in structures that are not temporary. If a crack is wide enough to expose the bamboo splints to termites or other forms of attack, then the structure may eventually collapse. Also, in some situations, the exposed bamboo splint may expand due to the absorption of rainwater. This can lead to an enlargement of the crack and eventual collapse of the structure. Thus, as with steel reinforcement, it is essential that steps are taken to limit cracking. Although further research is needed to define crack width limits precisely, the usual recommendations for steel reinforcement appear satisfactory.

Covering of the bamboo reinforcement, necessary both to ensure proper structural behavior and to protect the bamboo, should not be less than 20 mm in thickness.

In practice, spacing limitations, rather than the risk of overreinforcement, govern the bamboo ratio (the ratio of the cross-sectional area of the bamboo tension reinforcement to that of the concrete section); this is normally not greater than 4 percent. The splints should be arranged and placed with their nodes staggered and their outer sheaths facing the zones of tensile stress (that is, facing downward in simply supported beams and slabs). In order to avoid excessive deflection and cracking, the limiting span/total ratio for simply supported beams and squat slabs should not exceed 13 and 23, respectively.

Estimates of the load-carrying capacities of concrete elements reinforced with bamboo may be obtained by using methods used for steel reinforcement. Where shear reinforcement is required, bamboo *can* be used, but it is likely to be very much less effective than steel; it should be ignored in calculating a beam's shear strength of a beam.

When using steel, hooks or bends are provided at the ends of beams in order to provide the anchorage necessary to avoid reinforcement "pullout" failure. In order to reduce the (often) excessive length that would otherwise be required to develop the full tensile capacity of the bamboo on the support side of a beam, positive anchorage in addition to that provided by the nodes may be created by piercing the reinforcement with dowels.

This must be undertaken with care. Although more research into this technique is being done at present, it is suggested that tight-fitting (but not force-fit, which might split the bamboo) bamboo or wooden dowels, about 20 mm to 40 mm in length, be fitted through drilled holes. With a diameter not exceeding 25 percent of the reinforcement width, the holes should not be within either 50 mm of each other or of a node.

Bamboo reinforcement may be used in lightly loaded structural concrete elements such as door lintels, sun-shades, stairs, fence posts, simple roofs, manhole covers and pavement slabs not subjected to vehicle loading, and irrigation channel lining slabs. Such concrete elements may be either precast or cast in situ.

A CAUTIONARY TALE

In theoretical calculations, it is customary to calculate the moment of resistance using the nodal strength, which may be as low as 50 N/mm^2. However, in practical slabs and beams with staggered nodes the actual ultimate moment can be much higher than calculated, since it will be governed by the internodal strength, which can be as high as 400 N/mm^2. Since these slabs and beams would be able to carry much higher loads than the calculated maxima, this would not normally pose a problem in use.

Such concrete elements, however, are actually overreinforced and would fail suddenly if accidentally overloaded. This could be serious in situations where the safety of users requires a ductile type of failure with adequate visual warning. Designers should, therefore, be aware of the danger of the brittle and explosive type of failure associated with overreinforced beams and slabs.

Also, it must be emphasized that, although bamboo-reinforced mud buildings are known to have survived for over 100 years, the bamboo in such buildings was used primarily as a former in compression members. Its use as tension reinforcement in structural concrete is much more recent, and, since it is an organic material, much research into its long-term behavior, especially its response to humidity and alkalinity, is required.

CAUTIOUS CONCLUSIONS

For reasonably complex applications, such as multistory (perhaps even only two-story) buildings, neither the engineering viability nor the long-term material suitability of bamboo is proven. Indeed, even if bamboo is used only as main reinforcement in a more heavily loaded structural member, some steel is likely to be essential—for instance, as shear links in beams. Certainly, present knowledge would suggest that the benefit of a reasonably good quality cement can only be fully realized if it is used in conjunction with steel reinforcement but that a suitable bamboo, when used with care, is a viable but structurally less efficient alternative to steel as tensile reinforcement in simple concrete structures.

REFERENCES

1. Kankam, J. A., M. Ben-George, and S. H. Perry. (1986). "Bamboo-Reinforced Concrete Two-Way Slabs Subjected to Concentrated Central Loading." *The Structural Engineer,* Vol. 64B, No. 4.
2. Kankam, J. A., S. H. Perry, and M. Ben-George. (1986). "Bamboo-Reinforced Concrete One-Way Slabs Subjected to Line Loading," *Int. Journal for Development Technology,* Vol. 4.
3. Perry, S. H., J. S. Kankam, and M. Ben-George. (1986). "The Scope for Bamboo-Reinforcement Concrete," *Proc. Tenth Quadrennial FJP Congress,* Vol. 2. New Delhi.

NONPOISONOUS TIMBER PROTECTION

Based on "ATBrief No. 1: Non-poisonous Timber Protection," Appropriate Technology, *September 1992, Vol. 19, No. 2*

Timber is a common building material. It is convenient to use and is a renewable resource. It is strong, versatile, and often offers the best economic option for the construction of a variety of buildings. The benefits of building with wood are greatly limited, however, if the timber is vulnerable.

As an organic material, unprotected wood is affected by the following climatic, biological, and human factors.

Climatic—weathering (rain, blown sand, or grit), moisture (condensation), and heat (solar radiation)

Biological—insects (termites, beetles, wood wasps) and fungi (molds, stains, rots)

Human—fire

In recent years, many people have recommended chemical preservative treatments for timber protection. There is no doubt that these methods are effective, but a number of health hazards and environmental problems have been found to result. In some cases (such as in situations with a high risk of fire or fungal and insect attack) chemical treatment may be unavoidable, but if other protective measures are employed as well, the preservatives used can be of low toxicity and therefore less harmful. In many instances (such as in situations with a low risk of fire or fungal and insect attack) nonpoisonous timber protection practices can be employed to avoid *any* chemical treatment of wood.

This article outlines (1) the disadvantages of treating timber chemically and (2) means of protecting timber, including examples of protecting timber by proper felling and seasoning, by good design and construction practices, and through the use of natural preservatives.

DISADVANTAGES OF CHEMICAL TREATMENT

The chemical treatment of wood is designed to prevent wood-destroying organisms such as insects and fungi from attacking and weakening the timber or to prevent the further decay of timber that is already affected. Usually, this is achieved by using highly toxic insecticides and fungicides, but fungicides and insecticides have to be sufficiently toxic to be effective. They cannot differentiate between harmful and harmless organisms, which are both destroyed.

The chemicals affect animals and humans by skin contact, inhalation, or through contaminated food, causing a variety of health problems ranging from headaches, nausea, dizziness, depressions, and rashes to diseases of the lungs, heart, liver, kidneys and other organs, paralysis, and even cancer.

The production, application, and disposal of chemical treatments all contribute to serious environmental pollution. Toxic chemicals can enter the food chain and accumulate in increasing concentrations in the bodies of all living organisms and, most of all, in human beings. Factors such as solar radiation, high temperatures, humidity, and atmospheric pollutants can transform certain preservatives into other even more dangerous substances.

MEANS OF PROTECTING TIMBER

Timber can be protected by the following.

- The reduction of its moisture content to less than 20 percent, below which timber is usually immune from fungal attack.
- The avoidance of distortion and splitting, which weakens the structure. Splitting also provides access for insects.
- The protection of vulnerable surfaces. These surfaces include the ends of beams, concealed surfaces, or those exposed to climatic elements.
- The exclusions or quick removal of water. This prevents deterioration due to excessive expansion and contraction as a result of frequent wetting and drying.
- The use of natural preservatives, which minimize the development and spreading of fire, fungi, and insects.

These methods should be used together to protect timber effectively, although the relative importance of each will depend on local circumstances.

REDUCTION OF MOISTURE CONTENT

Allowing the felled tree to lie for some time with all its branches and leaves accelerates the drying process. Seasoning timber is the process of drying sawn timber under a roof for protection from the rain. The saturated air around the timber is removed by fresh air, which is circulated naturally or forced through the gaps between the pieces of timber. Seasoning should be done to a moisture content in equilibrium with the atmosphere in which the timber will be used in order to avoid excessive moisture movement after construction. Kiln seasoning is faster than air seasoning and provides timbers of uniform moisture contents by controlling the air circulation, humidity, and temperature within the kiln (Figure 30). The interruption of the drying process after sunset enables

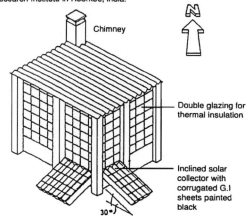

The solar kiln works without fans on the principle of thermal air circulation. The solar collectors supply the kiln with heated fresh air and replaces the humid air which escapes through the chimney. This prototype was developed by the Central Building Research Institute in Roorkee, India.

FIGURE 30
A solar kiln.

the moisture from the center of the wood to come to the surface, thus preventing excessive tension between the surface and the interior, which can cause distortions and splitting.

AVOIDANCE OF DISTORTION AND SPLITTING

Distortion and splitting are caused by wood shrinking. This varies according to the direction of the shrinkage. *Longitudinal* shrinkage is negligible—about .1 to .2 percent. *Radial* shrinkage is about 8 percent from the "green" to the dry state. The corresponding *tangential* shrinkage is about 14 to 16 percent (Figure 31). The stacking of sawn timber for seasoning and for storage afterwards requires careful horizontal placement on equally spaced battens to avoid sagging and differential stresses.

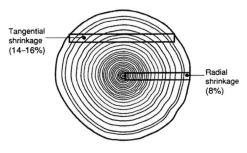

FIGURE 31
Timber shrinkages.

PROTECTION OF VULNERABLE SURFACES

Surfaces of timber beams and columns that are cut across the grain are extremely vulnerable to insect and moisture penetration. Problems are avoided by fixing protective wooden boards or metal plates at the end of beams (Figure 32).

Wooden surfaces in contact with the ground or with other materials that are wet (such as concrete) are in constant danger of moisture and insect penetration. Splashing rainwater and floods are additional hazards to wood components close to the ground. To avoid this, the following steps should be taken.

- Timber components should be installed at a minimum of 30 cm above the ground.
- Timber columns should be fixed on metal supports and never embedded in concrete or masonry footings.
- If necessary, adequate resistance to insect penetration should be provided using metal plates projecting at least 5 cm. See Figure 33.

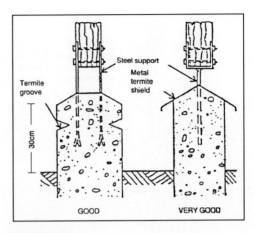

FIGURE 32
Simple ways of protecting end grain.

EXCLUSION OR QUICK REMOVAL OF WATER

- Wide overchanging roofs are essential where rain occurs frequently. Eaves gutters help to avoid the splashing of rainwater.
- Timber wall elements with the grain running vertically drain water fastest. The lower edges should be shaped so that dripping water is discharged outwards and not encouraged to find its way into joints and openings (Figure 34).
- To hasten the removal of any moisture (from rain or condensed water) that penetrates timber components, air must be allowed to circulate on all sides. Provision for ventilation gaps should therefore be made at the design stage of a timber construction.

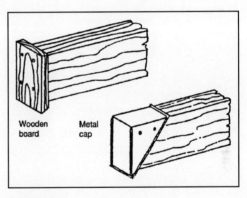

FIGURE 33
Timber posts on concrete footings.

LOW TOXICITY PRESERVATIVES

Chemical treatment may still be necessary in high-risk situations. If the protective measures described previously are employed, the preservatives used can be "natural" or of low toxicity, thus reducing, if not eliminating, health and environmental hazards.

- Borax is a fire retardant. It is also effective as an insecticide and a fungicide. Dissolved in water, it can be applied by brushing. Adding a binder such as a natural resin reduces leaching.
- Soda (sodium carbonate) is boiled in water and applied by brushing. As it dries, certain substances in wood, which attract harmful organisms, are decomposed.

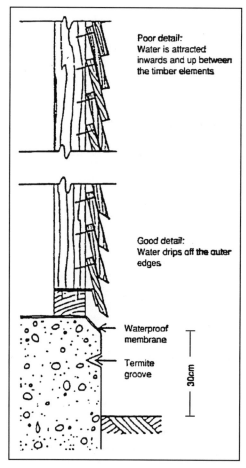

FIGURE 34
Timber wall cladding.

- Potash is a traditional preservative for outdoor applications, made by boiling wood ash in water and diluting the solution. It contains the same protective substances such as potassium carbonate and sodium that are present in the bark of trees.
- Linseed oil mixed with turpentine is applied by brushing and provides a tough, water-repellant surface.
- Beeswax is an ancient wood preservative for indoor use. Heated and applied thinly with a soft cloth, it seals cracks and pores. Its resistance to water is low, but this can be improved by adding a natural resin or oil.

A NOTE ABOUT PAINT

The surface of wood can be protected by water-repellant paints: Paint can only be applied on fully seasoned timber. If paint is applied on wood that is not seasoned, then blistering, flaking, peeling, or mold growth will result. The complete sealing of well-seasoned timber can be advantageous but requires regular renewal, since insects and moisture can attack damaged portions. Sharp edges should be sandpapered to ensure a uniform application of paint.

This ATBrief is based on the ESCAP *Building Technology Series No. 9: Non-poisonous timber protection*, prepared by Karen Maker on behalf of the ESCAP/UNCHS joint Unit on Human Settlements, Bangkok, Thailand, 1987. Adapted for IT Publications by Rod Shaw.

DRY STONE WALLING: APPLICATIONS AND TECHNIQUES

Based on article of same name by Richard Tufnell Brae House, Balmaclellan, Castle Douglas, Dumfries, SG7QE, Scotland, in Appropriate Technology, *Vol. 18, No. 1, June 1991*

Dry stone walling is an ancient skill with its origins in Neolithic times. This article is a guide to a widespread craft with enormous potential to solve one of the world's most pressing needs that is, surprisingly, often passed over.

Dry stone walling is the technique of assembling stone structures without the use of any form of binding material, although the term is sometimes stretched to include stone-earth combinations.

The name is something of a misnomer in the sense that houses, barns, wells, traffic-carrying bridges, wolf traps, temples, and even entire towns complete with infrastructure, have been built using stone alone. The largest "dry stone" domestic dwelling was built by Native Americans near Arizona; its 800 rooms housed up to 5,000 souls. Sadly, the craft is declining, partly because of rural depopulation, often coupled with social unrest, and partly as a result of "efficient" and "status enhancing" methods and materials that rarely offer any advantage over stone.

Another of its key applications—terracing—enables steep mountainsides to be productive. The valley floors are often irrigated with mud-lined dry stone channels, irrigating crops and carrying water to drive mills and small hydro plants. On arid flatlands low stone walls are a weapon against flash flooding, and soil erosion can also be countered on steep terrains. Retaining walls are used to prevent roads from eroding and disappearing down precipices, and young tree plantations in the Himalayas have been stock-proofed using dry stone walled enclosures.

DURABILITY

Dry stone walling has astonishing advantages over other forms of fencing and construction. It is extremely durable, compacting as its settles over time, thereby decreasing the effects of seismic activity. Skillfully constructed, dry stone will outlast mortared work because the water has no medium to attack. Dry stone work is self-draining, fireproof, and may be used where water and/or cement is unavailable. It can be used on waste ground where fence posts cannot be driven. It clears the ground of its burden of waste stone, and, most importantly, the prime material is essentially free. Shelter is given to both the lee and windward sides, providing a haven for plants and animals and giving a longer growing season to crops. This effect is enhanced because the stone acts as a heat sink, gently releasing the stored warmth during the night. There are no adverse environmental effects, using as it does natural local materials. The benefits include the pleasing aesthetics of craftsman's work on the landscape and the provision of habitats for many species of flora, lichens, mosses, and ferns, as well as small birds and mammals.

Not least among these numerous benefits is the possibility of providing local rural employment because the basic tools are simple, available universally, and inexpensive. The methods are not difficult to understand and can be mastered anyone. Experience has shown that a short course under expert tuition is imperative, however, if the would-be craftsman is to obtain the grounding essential in developing hand/eye coordination.

The freestanding dry stone wall described here is 1.4 m high, which is suitable for stock-proofing against grazing animals, except for goats. A wall may be built to virtually any height, provided that the height remains approximately twice the base width.

SAFETY

Serious injury in dry stone in dry stone walling is very rare, but the following points should be noted. When lifting heavy stones, bend at the knees with back straight. The greatest risk comes from stone chips flying off the stone when it is struck with a hammer. Shut your eyes or turn away at the moment of impact, or wear protective glasses.

TOOLS AND EQUIPMENT

- A ruler or measuring stick
- A spade for preparing the foundations
- A club-type hammer, about 1.5 kilograms
- Crowbar. Can be useful particularly with very heavy stones
- Buildings lines. Two are required, at least 10 meters long
- Walling frame. These are templates for each particular wall. They are constructed as shown (Figure 35) out of any wood available that has a straight outer edge; 75 mm × 25 mm is a good size.

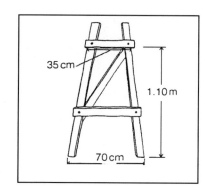

FIGURE 35
A walling frame.

Two are needed and give the wall its characteristic "A" shape or "batter" as the guiding strings are moved up the frame during construction

BUILDING THE WALL

Walls may be built in a number of different styles; a wall built of slate will look very different from one built of granite boulders, but the principles are the same. The wall described here is built with a large proportion of small stones.

Much time and frustration can be saved if a certain amount of stone sorting is carried out before operations commence. There are essentially five categories of stone.

- *Topstones* are placed at the top of the wall to protect it. They should be regular in shape, at least 20 cm high, and seven or eight are needed for each meter length of wall.
- *Building stones*, often called "double," are used in the face of the wall.
- *Through stones* are long stones placed about halfway up the wall that go right through binding the two sides together. Laid at 1-meter intervals, they are important. Do not break them up for use elsewhere.
- *Foundation stones* are large, flat stones used for the bottom layer.
- *Filling*, also known as "hearing," is waste stone used to pack the middle of the wall between the two faces.

THE FOUNDATIONS

A foundation course, wider than the rest of the wall, is laid to allow slow, even settlement, except when building on rock. Mark out the line of the wall with the strings attached to pegs (Figure 36). Ensure that the strings are parallel to each other and 85 cm apart. Remove soil and vegetation to a depth of about 6 cm or until there is a firm base; then move the strings slightly inwards to 80 cm apart.

The foundation stones should now be laid, following the strings carefully, one after the other. Try to match the stones for height where possible to make subsequent building operations easier. Always make sure that each stone laid touches the one beside it. Lay the stones so that they are

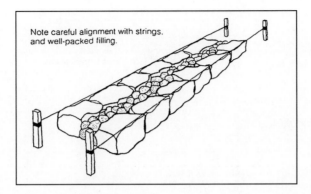

FIGURE 36
The foundations.

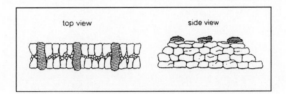

FIGURE 37
Through stones in position.

level by lifting the rear of the stone and slipping wedge-shaped stones beneath as necessary. Always pack the center of the wall carefully with waste stones as work proceeds, taking great care not to allow the formation of empty spaces. Work as closely as possible to the strings without actually touching them.

THE LOWER COURSES

With the foundation completed, the "A" frames are placed 5 m to 10 m apart, taking care that they are level. Support them with stakes driven into the ground or with several large stones placed carefully around the legs. Attach the strings some 5 cm above the foundations, and begin laying in the lower courses. Level off at 55 cm, ready for the through stones. These long stones bind the wall together and are laid at 1-m intervals (Figure 37). They should project some 5 cm either side but not much more; otherwise, animals will rub up against them, and they could hurt passersby. Do not bang them down or slide them across; either action can damage the wall built thus far. Once in position, they must be stabilized with slivers of stone wedged underneath until they are immovable.

SECOND LIFT

Now the strings are repositioned higher up the "A" frames once again, and the courses are continued for a further 55 cm or so. Level off accurately to the strings, having ensured as always that the wall is tightly packed in the middle.

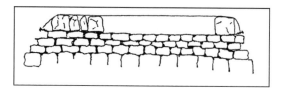

FIGURE 38
Using string to set the topstones.

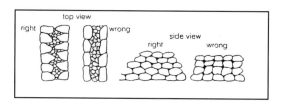

FIGURE 39
The right and wrong way to lay stone.

THE TOPSTONES

The wall will now be about 35 cm wide and will be ready for the topstones. Take two of the regular-shaped stones set aside for this purpose, and place one at each end of the stretch being worked on. Stretch one of the building lines tightly between them; then place the topstones vertically, one after the other (Figure 38). Some will be too short, so raise them until they almost touch the string by inserting small flat stones under one or both sides. Fill the holes between the stones with pins (pointed stones that prevent the topstones moving and twisting).

The last task is to lock the topstones securely. This is done by taking narrow, wedge-shaped stones (these can be made by hitting the edge of a large stone with a hammer) and driving them hard in between the topstones. The wall should now be able to be walked on without any movement.

Observing a few simple rules will ensure a strong, good-looking wall. First, build up course by course. When the strings are reached, raise them by a further 15 cm and so on. Ensure that each stone is firmly in place before moving on to the next. Use the larger stones available at the bottom, as there may not be room for them in the higher courses as the wall narrows. Always lay the stones so that the longer side runs *into* the wall, and *always* break the joints. See Figure 39. The Golden Rule is "One stone on two, two stones on one."

Once this basic technique has been mastered, the possibilities for application are limited only by the imagination. It is very difficult to begin the learning process without an element of expert instruction in the first stages, either through a short course or through apprenticeship to a craftsman. Those unused to handling stone may well find the initial process is very confusing, and guidance in mastering the basic principles is really essential. Once this is obtained, true skill develops purely through practice.

LESSONS FROM A WORKSHOP IN NEPAL

We learned that four factors are necessary for successful technology transfer.

1. The participants should have had at least some experience in handling stone.
2. There must be an NGO or similar institution to provide initial support and organization.

3. An area should be chosen where enclosure walls or other dry stonework is planned as part of an economic or environmental strategy.
4. It is important that existing masons do not suffer any loss of earnings by adopting new methods.

The Dry Stone Walling Association of Great Britain is the only body in the world devoted solely to the craft. It handles enquiries from around the world, both from individuals and ins-titutions, and will give help wherever possible, including a list of relevant publications. Send a stamped, self-addressed envelope to The DSWA, YFC Centre, The National Agricultural Centre, Kenilworth, Warks, CV8 2LG, UK.

CHAPTER 6

Transportation

Edited by Barrett Hazeltine
Brown University

TRANSPORTATION

Based on A Review of Technologies for the Provision of Basic Infrastructure in Low-Income Settlements United Nations Centre for Human Ssttlements (Habitat) PO Box 30030 Nairobi, Kenya, Telephone 621234, Cable Unhabitat; Fax (254)-2-624266/624267; Telex: 22996 Unhabke, No date given

This chapter focuses on two issues: vehicles and roads. Bicycles and related vehicles, such as rickshaws, offer much in the Third World, and ingenious artisans can adapt them to special uses. In Africa and other places a variety of handmade carts are in general use, as well as other simple transport devices. Oxen and donkeys are the usual source of power. Not all transportation is on land, and improvements have been made on boat design—with efforts focused on improved fishing. The second issue deals with roads, including low-technology roads; bridges and other ways of crossing streams; and ropeways used in mountains.

Transportation opportunities affect the lives of people significantly. Transportation problems involve both roads and vehicles. The pattern of transport services available to residents of urban and rural settlements depends mainly on the spatial relationship between the settlement and centers of services and employment. The problems faced by rural inhabitants of developing countries in moving produce and materials, inside or outside places of agricultural activity, are exacerbated by a shortage of appropriate devices and vehicles. Public passenger transport facilities are very limited in the rural areas, and carrying heavy loads on foot remains the predominant form of traffic for the rural poor. When feeder roads have been provided to rural settlements, passenger travel has tended to increase markedly. In some areas roads become so muddy during the rainy season that villages are effectively cut off from the rest of the country—where one is early in the rainy season is where one will be at the end of the rainy season.

Low-income urban dwellers try to locate themselves in places close to potential employment and, thus, minimize transport expenditure, but frequently, the cost of shelter or lack of space forces people to locate away from the centers of employment and services. Walking to places of work can sometimes involve considerable distances—five kilometers or more. Low-income groups prefer walking to spending their scarce resources on fares that can sometimes range from 10 to 30 percent of the family income.

The network of roads, paths, and walkways in low-income urban settlements develops gradually as the need for circulation and access increases. Pedestrian facilities are usually of poor quality, and they are often encroached on by new building. On main roads, where heavy pedestrian traffic usually concentrates, walkers must compete with vehicles or use improvised tracks parallel to the carriageway. Most of the roads, lanes, and paths in squatter settlements have surfaces that become dusty during dry seasons and muddy when it rains. (An article in the Health Care chapter mentions the transmission of disease by windblown dust because children and animals defecate on dust roads.) Inner-city slums normally have an established circulation pattern, but, even here, the building of dwellings sometimes precedes the planning of streets and paths. Another transport problem the urban poor face is in relation to access to their own dwellings and services. Poor access for water carts, refuse vehicles, firefighting vehicles, and other emergency services contributes to insecure living conditions.

Transportation needs, like other needs, are not the same for women and men. Fetching water is perhaps the most arduous and time-consuming women's task. Women generally have less access to means of transport and hence have no alternative to head loading. Only in the tasks at present performed by both men and women (farm-to-field transport of both cash and subsistence crops, firewood transport) will women in fact benefit from the use of intermediate means of transport. In many cases, the task of transporting water can only be alleviated by reducing the distance between the home and the water supply.

Bicycles are an important means of transportation, and the developement of a local bicycle manufacturing industry looks appealing. There is a lesson, however, from Central America: Any country (even in the throes of economic collapse) can start a bike industry, but it should start small and assemble imported bicycles. Only gradually, it appears, can a country substitute imports with "home-grown" frames and parts. There are models to be avoided, such as Tanzania's large-scale, tariff-protected bike industry. Brazilian and Taiwanese export-oriented bike industries offer models for large-scale bicycle production in the context of heavily industrializing economies, with substantial available capital investment.

Road standards and maintenance are discussed in the "Earth Roads" article. Road construction standards should be reduced and more labor-intensive, lighter-weight motorized and nonmotorized vehicles should be favored over very heavy vehicles. Exceptions to this would be for certain types of raw material and bulk commodities transport on selected routes, and in some cases this traffic might be well served by rail. These strategies can create jobs and substantially reduce the costs of roads.

A lesson from Kenya is that the maintenance system established for completed roads can employ local individuals, usually former roads department casual laborers, each of whom is made responsible for a two-kilometer section of the road. The cost involved in this system of maintenance has proved to be about 120 Kenyan pounds per kilometer per year compared with an average of 460 pounds for more highly mechanized methods used on conventional roads.

RICKSHAWS AND BICYCLES

Based on "The Rickshaw: Appropriate Transport for a New Age," by C. O. Bhatmagar, Appropriate Technology, *Vol. 24, No. 3, December 1997*

RICKSHAWS

The lowly rickshaw, shown in Figure 1, is often thought to be an anachronism in the age of development, where high-speed motorized transport is king. Yet, the rickshaw continues to survive, thanks to the extensive patronage extended to it by society's burgeoning middle class. In fact, for short-haul transport (say, 5–10 km) of passengers and goods, it simply has no peer. The rickshaw is popular all over the Indian subcontinent, China, and Southeast Asia, since it provides the puller with an element of dignity in comparison to the hackney cart. On the Indian subcontinent and China, the pull-type driver-front design (with varied designs for the body) is popular, whereas in Southeast Asia, the push-type driver-rear or the types that use a sidecar along with the bicycle are more common. Using readily available (and often surplus) human power, the rickshaw transports passengers and goods right to their

FIGURE 1
Rickshaw. (From *Appropriate Technology.*)

doorstep, and it is used, among other things, for delivery, vending, hawking, transporting schoolchildren, and collecting garbage.

Rickshaws are tricycles, with the driver in front and the passengers, as many as three, seated over the rear wheels. Bicycle parts—wheels, hubs, handlebars, sprockets and so on, are often used, with a longer chain. The awning gives some protection from sun and rain, but pieces of waterproof cloth or plastic sheeting are often carried to be wrapped around passengers in heavy rain. Cycle-rickshaw can also be used to carry freight. Traditional rickshaws powered by a person running have been essentially replaced by these cycle-rickshaws, perhaps because the latter go faster, perhaps because they are considered less degrading for the operator.

Conventional Design

The usual cycle-rickshaw suffers from two major defects: First, they are hard to stop, and second, they are hard to start, especially on a hill. The stopping problem results from using only front-wheel brakes—it is difficult to modify standard bicycle parts to get rear-wheel braking. The starting problem results also from use of standard bicycle parts—the gear ratio is not proper for the much heavier loads in a cycle-rickshaw. Conventional rickshaw designs also suffer from inefficient use of road space, inadequate structural strength, high dead weight, poor ergonomics, poor vehicle and steering stability, and an absence of lights and reflectors for night protection. All these shortcomings lead to an uncomfortable ride for passengers, frequent accidents, high maintenance, and adverse effects on the puller's health.

Improved Design

One new design, made at the Center for Technology and Development in New Delhi, has the following new features.

- Sturdy welded frame and fork, using oversized steel pipes
- A convertible two-in-one design incorporating both passenger and load applications with the use of foldable seats
- An all-steel body design that totally eliminates the use of wood
- A low floor and rear exit that makes it easy for passengers, especially the elderly and young, to alight
- Space for luggage and overflow passengers and children
- A canopy with rain covers that also protects the puller from sun and rain

- Smaller but sturdier wheels
- Rubber vibration cushions
- The possibility of incorporating three-speed gears and modern 50–75 cc automatic gear engine

Different designs have been developed for the varied applications demanded by various sectors, such as for passengers or for loads: schools, industry, tourism, entertainment, vending (vegetables, ice cream), and municipal use (garbage collection).

Another improved design is the "Oxtrike," shown in Figure 2, developed at Oxford University, with support from Oxfam. Its design includes both front and rear brakes, the latter a novel design actuated by a separate pedal. A standard three-speed gearbox is used and is located midway between the pedals and the rear wheels, so two chains are necessary, but they are standard. The frame is not made from thin-walled tubing, as in most bicycles, but standard steel sheets, folded and welded. Local metal workers can do this kind of construction, whereas thin wall tubing requires special tools for cutting and welding. Various kinds of bodies, for carrying people or baggage, can be mounted on the frame.

FIGURE 2
The oxtrike. (From *Appropriate Technology*.)

BICYCLES

In most areas of the world the bicycle is the transportation of choice. Autos are too expensive, and public transportation is overcrowded and underfunded.

Bicycle Technology

How is it that a person on a 30-pound bicycle can go faster and farther in a day than a person walking, even though the walker's load is 30 pounds less? One reason is that rolling friction, predominant in a bicycle, is less than the sliding friction predominant in walking. Another reason is that the force and speed used in propelling a bicycle matches the force and speed at which human muscles are most effective. A final reason is that the frame and handlebars help support the cyclist's body and additional loads being carried, so more of the bicyclist's energy goes to producing motion, compared to the walker.

Bicycles evolved to their present form about 100 years ago, and most present-day designs are fairly standard, but it is instructive to consider some of the design decisions. In designing the frame the tradeoff is between strength and weight. Ordinary bicycles in China and India, where big loads—200 pounds or so—are carried on bumpy roads, might weigh 40 pounds. A 24-speed bicycle used for pleasure in the United States might weigh 20 pounds. The frame can be improved by using expensive metal alloys, which are stronger for the same weight than the steel normally used in bicycles. Metal failure in a bicycle frame is usually produced by "fatigue"—repeated small bending rather than a one-time overload. So for durability, one would like a frame that would not bend, but a somewhat flexible frame is more comfortable than a rigid one. Frame design is in part a tradeoff between durability and comfort.

The main source of friction, which makes pedaling difficult, is the tires against the ground. Another source is internal to the bicycle—at the wheel hubs. The friction from this source is small because ball bearings are used in the hubs. Early bicycles did not have these ball bearings, and the resulting friction made starting difficult. Tire friction is increased if the tire pressure is low—force must be exerted constantly as the wheel turns to compress different parts of the tire; in a sense, one is always pedaling uphill. Although rubber tires create more friction than uncompressible ones, they soften the ride by absorbing bumps. Without rubber tires the ride would be uncomfortable and the frame would break sooner. Once the bicycle is moving normally the greatest source of friction, and the one that limits the top speed, is air resistance—pushing the air ahead of the bicycle out of the way. Really high-speed bicycles have a streamlined body around them to reduce air resistance. One bicycle so equipped attained a speed of 65.48 miles per hour, but this bicycle was hardly an intermediate technology device—it also had special tires and a lightweight frame.

Bicycles generally have gears and a chain. The purpose of the gears is to allow the cyclist's legs to move at their optimum rate, 50 to 100 revolutions per minute, no matter how fast the bicycle wheels are moving. The chain, of course, transfers the power from the pedal to the wheels. A well-oiled, clean chain is about 98.5 percent efficient—that is, only about 1.5 percent of the power from the pedals is used to move the chain itself.

Brakes are, for obvious reasons, essential. On some racing bicycles one pushes backwards on the pedals to brake, but the work required is greater than that of rubbing something against the wheel—the usual type of brake. The major problem with conventional brakes—rubber pieces pushed against the wheel rim—is that they lose effectiveness when wet. The braking action is produced by the friction of the rubber brake shoe against the metal rim, and this friction nearly disappears when a thin film of water is between the rubber and the metal. Stopping a bicycle on a rainy day can be a perilous thing! So-called "coaster brakes" work by forcing metal rings inside the rear hub against each other. Coaster brakes are enclosed, so they are not bothered by weather, but they work only on the rear wheel. Enclosed drum brakes for front wheels have been developed, but they require a heavier wheel and more force from the bicyclist.

Maximum Speed, Power, Capacity of a Bicycle

As was mentioned, a sophisticated bicycle in special conditions has achieved a speed over 65 miles per hour for 200 meters. A more typical speed for an average bicyclist is 12 miles per hour. This corresponds to about 1/10 of a horsepower or 75 watts and can be sustained for at least an hour. The load capacity of a heavily built bicycle seems to be limited more by the skill of a cyclist at balancing than by the strength of the bicycle—300 or 400 pounds can be carried without damaging a strong bicycle.

Uses for a Bicycle

Uses can be divided into two categories: transportation and stationary power. Because the technology of the bicycle itself is so mature, it seems that fruitful approaches to using bicycles will be mostly either in the supporting infrastructure—making bicycle transportation more convenient—or in imaginative ways of applying their power—driving different kinds of tools.

Transportation Uses

Bicycles certainly have many advantages when used for transportation. In crowded cities, including those in the United States, a bicycle tends to be a faster vehicle for commuting than an automobile, at least up to distances of 5 miles, because of traffic. A bicycle goes nearly door-to-door and is easier to park. It does not pollute and can use narrower roads. It does not use gasoline. It is cheaper than an automobile to purchase and much cheaper to run. More people can manage bicycle maintenance than auto maintenance. Bicycle riders tend to enjoy riding, and their health is probably improved by the exercise.

What are the drawbacks of a transportation system based on bicycles? The following list is from Whittin and Wilson's book.

- No protection from rain or snow
- Poor braking in wet weather
- Difficulties in seeing a moving bicycle at night
- Difficult to carry large packages or children
- Exertion from long or steep hills
- Commuters resisting cycling more than five miles to work
- Fairly frequent failure of key parts, particularly tires, brake, and gear cables. (It is worth noting the well-known Chinese bicycle, Flying Pigeon, uses metal rods instead of brake cables.)

Some of these problems could be solved by covered, illuminated, and redesigned roads, which would certainly be cheaper and more environmentally benign than superhighways.

Stationary Power

Power generated by stationary pedaling can be used for many things—electric generators; water pumps; and small tools such as grinders, drills, or kitchen appliances. The major design question is whether one uses an ordinary bicycle temporarily mounted to run a machine or builds one from scratch using bicycle parts. In a sense, the bicycle exercise machine used in many homes is an example of the latter, but the power is wasted.

An example of a machine that uses an ordinary bicycle is the grain mill, shown in Figure 3. The grain is actually milled in the box in the right side of the picture.

Adapters allowing conventional bicycles to power machines are fairly easy to build. They need to be strong and rigid as the forces are considerable. In these adapters power is usually taken from the bicycle by a rubber tired wheel, a roller, or a pulley firmly pushed against the rear tire. Electric power can be generated using an automobile generator, perhaps connected by a belt to the bicycle wheel. One could use also the small generators sold to power bicycle lights but these can produce much less current.

FIGURE 3
Bicycle-powered grain mill. (From *Pedal Power.*)

REFERENCES

Hoda, M.M. (1984). "Appropriate Methods for Improvement of Cycle-Rickshaws." *GATE*, Vol. 84, No. 4, 34–36.
McCullagh, James C. (1997). *Pedal Power.* Emmaus, PA: Rodale Press.

Whittin, Frank Rowland, and David Gordan, Wilson. (1982). *Bicycling Science*, 2nd Ed. Cambridge, MA: The MIT Press.

Wilson, Stuart S. (1977). "The Oxtrike," *Appropriate Technology*, Vol. 3, No. 4, 21–22.

USING OXEN FOR FARM WORK AND TRANSPORTATION

Drew Conroy

Oxen have been an important part of human development for centuries. Globally oxen greatly outnumber horses and donkeys as beasts of burden. The debate of the ox as a backward way of farming continues, as it has for centuries. Horses are faster; tractors are more efficient. The poor farmer who is unable to have either of these technologies, however, continues to use oxen to pull plows, carts, and nearly everything else on the farm. Plugging along at a slow pace, often in crude and uncomfortable yokes with poor training, the ox continues to till more soil than any other beast on earth. Often the farmers that could benefit from oxen do not understand this technology. Others who have seen the animals remain skeptical about their use. The use of oxen in agricultural development has many faces. It also has many different degrees of adoption, use, and interest. There is no one simple solution to agricultural development all over the world. Oxen are just one cog in the wheel of developing agricultural systems.

Agricultural development is the ability to develop some "better" system of agricultural production. For many rural poor the world over, human labor is a major constraint to greater agricultural production. Labor constraints during the plowing, weeding, or harvesting seasons are where oxen can most easily fill the gap between what people can do and what they would like to do. Greater efficiency and timeliness are easily accomplished if oxen are employed in all agricultural operations. Many farmers plow with oxen and then leave them idle for the remainder of the year. Employing the animals in labor-saving and profitable ways year round takes a great deal of creativity. However, this creativity can pay great dividends when greater harvests are possible. The implements needed for plowing, weeding, and transportation are often a larger constraint for many farmers than the acquisition, training, and employment of animals.

INTRODUCING OXEN

Oxen are not appropriate for all farmers. Potential users of oxen must understand the animal's limitations. Cattle require feed, water, security from theft and large predators, and shelter from extremes in the weather. The investment in an ox is a substantial investment for poor farmers. To lose an animal to disease or theft is a tremendous financial loss. Many farmers prefer hand labor rather than risking what little resources they have on a technology they do not understand. If the animals are employed just for plowing and planting and greater acreage is planted, there may be a severe labor bottleneck during weeding. Simply increasing acreage has proven time and again to be of little value if there is not additional labor or the employment of animals in the weeding and harvesting of the crops (Figure 4).

Ideally, oxen should generate a greater profit for any farmer using them. Oxen are appropriate where people are genuinely committed to using them. Cultures unfamiliar with cattle often fail in training and using the animals without strong educational, moral, and/or technical support. Even with such support, local capacity to maintain the technology must be encouraged from the beginning.

FIGURE 4
An ox pair.

People must be motivated in order to help each other train, work, and use the animals. Simply introducing this exciting "new" technology will result in failure. Most successful technologies are passed down from generation to generation. The adoption of any new technology is slow and not universal the first time it is introduced. Cultural bias, a history without draft animals, and the added burden of caring for livestock is often more than enough to halt the spread of draft animals. Cattle can be a drain on small farms with limited land for grazing or money for medical supplies for the animals. Finally, oxen must also be readily available and affordable. If animals must be trucked in from long distances, without ample replacements, the use of oxen in the long term is not going to be successful.

OXEN IN AGRICULTURAL DEVELOPMENT

If oxen are going to be encouraged, it is important to first understand the animal. Before investing in expensive equipment, elaborate harnesses, and farm implements, learning about the animals must be the first step. Oxen can help increase farm productivity, but this is not without a serious commitment to putting the animals to work. There is no magic to training or using oxen. Their training and their comfort must be top priorities. There is no need for imported animals and fancy equipment. Local people should use their indigenous breeds of cattle for oxen. Local artisans should be trained with their tools in the manufacture of yokes and other equipment. The technology will spread only if people have a reason to adopt it, if they are already raising cattle, and if they have the infrastructure to keep the animals and equipment functional.

OX TRAINING

Ox training is a sequence of events designed to create animals that are trustworthy in the yoke and willing and able to do the work at hand. Rushing through training and beating the animals rarely produce animals that are eager to work. Using a simple series of exercises for ox training, which begins by handling the animals and progresses to leading and teaching them to start and stop, is preferred. This progresses to wearing the yoke and slowly building them up to heavy pulling or plowing over a period of weeks. With mature animals this may take weeks. With young animals it may take days. Training is made easier by yoking a younger team behind an older team. While this type of step-by-step training may seem like a waste of time to some farmers, the result will ultimately be better than an animal that was beaten into submission.

Mature oxen that are trained to plow in one day might be considered amazing by the person using them—allowing more villages to be visited in a week, a month, or a year. Sometimes this works, provided the oxen have had previous handling. However, the next time the oxen are put into the yoke, their attitudes may be altogether different, especially if they are sore from plowing the day before. If the animals have to be beaten so severely that they refuse to work or lie down in the furrow, this will not encourage others to use the technology.

OX YOKES IN AGRICULTURAL DEVELOPMENT

(Yokes are also discussed and illustrated in the "Pack Animals" article.) Ox yokes come in many designs. It is easy to be very critical of the yokes the oxen are wearing, but it does not matter what type of a yoke they are wearing. The more important aspect is that the animals should simply never be suffering in any yoke that is poorly designed. An extra couple of hours spent carving a comfortable yoke will pay huge dividends in preventing animal breakdowns and injuries over time. Ox teamsters must understand that oxen with sores on their necks are begging for an improved yoke design. Yokes that are designed so poorly that the animals are always uncomfortable are not a function of poverty but of ignorance. The poorest person will go without shoes before wearing shoes that cause sores that will make her or him unable to walk. Just as the athlete cannot run in shoes that do not fit, the ox should not have to wear a yoke that wounds him as he works.

Simple improvements, such as carving the yoke beam very smooth and increasing its surface area to accommodate the animal's neck or withers, can pay great dividends. Making the yoke strong enough to withstand the jobs it is designed for not only protects the animal from injury but also saves time during critical periods like plowing season. If a head yoke is used, it must be comfortably fitted to the animal's head and horns and strapped in a manner that doesn't injure the animal or its head. Wide straps work better than ropes, and a head pad on the forehead protects the head from chafing due to the straps. The yoke should be designed so that it allows the animal to carry its head in a position slightly lower than its back. Animals with their heads held up high or near the ground are uncomfortable and cannot perform properly in the yoke.

Farmers already using oxen will also benefit from improvements that can be made in animal training and yoking. Initially they may be the first to resist the new ideas. However, once they see the advantages of more comfortable yokes and better-trained animals, they are usually the first to quietly implement improved techniques. Time saved in the field by having animals that can accomplish a greater variety of tasks in a shorter period of time is not something many farmers will ignore.

PRIMARY TILLAGE

Primary tillage is one of the foremost functions of the ox. Plowing is the most common form of primary tillage. Plowing with oxen takes many forms (see Figure 5). In some parts of the world the plow is the only implement on the farm, and it may be the only function of the ox. Plowing is accomplished in some areas by simply scraping the surface with a simple pointed plow and providing a place for soil-seed contact. The moldboard plow was a great invention and has been improved and perfected over its many centuries of use in Europe and America. The ease with which a well-scoured and adjusted moldboard plow can turn a furrow is an amazing feat, yet plowing is the most difficult task an ox performs. Logging and stump pulling can also be difficult, but with proper planning, they can be easier than plowing. The physical challenge of plowing is made more difficult by the many variables that affect its movement through the soil.

Before animals are used for plowing they should be "hardened" or physically conditioned for the job. If the animals are not physically prepared for the task, they will have great difficulty following commands, staying in the furrow, and walking a straight line. Animals that are worked year round will be physically prepared for the job. Animals that are idle for much of the year will benefit from three to four weeks of lighter work before attempting to plow. Plowing is very different from pulling a cart. Therefore, the prior conditioning should involve work like logging or pulling a sled loaded with stones or other heavy objects.

Training a team to plow is like training the animals for any other task. The animals should be first physically acclimated to the task, slowly introduced to the job, and methodically trained to follow a furrow. A pair of oxen can be trained to follow a furrow and controlled by one teamster.

FIGURE 5
Plowing with oxen.

FIGURE 6
An ox plow.

However, this is not before the animals are acclimated to just what it is they have to do. Maintain high expectations, be patient, and offer proper instruction to the animals. The animals will soon do what it is that is asked of them. Give them the opportunity to learn what it is they have to do and plenty of time to practice. Cattle will soon learn to follow a furrow and respond to the commands of a driver behind them as easily as they follow a path back from the pasture.

Besides conditioning animals, it is also important that one be familiar with the plow itself—how to maintain it and how to adjust it for the existing soil conditions. There are very few parts to a walking plow, but each part plays an important function. Know the parts of a plow and how to identify broken and worn parts. Walking plows are the most common plows used with oxen and are relatively simple to adjust and maintain (Figure 6).

TRANSPORT

(Some of the following material is also discussed in the "Low-Cost Transport Device" article in this section.) There is no limit to the ways oxen can be used for transport. Transport can be as simple as hauling manure to the fields on a sled or drawing a cart with the season's harvest. There are many examples of how oxen have pulled everything from cannons to battle and boats across land. Carts or wagons of all sizes and shapes are easily adapted to oxen. Wheeled transport is easier than sledges but generally is more costly and challenging to build. Sleds work well in winter in snow-covered areas and give idle or young oxen a good workout in cooler climates.

The two-wheeled cart has been the traditional mode of transportation when oxen are used (see Figure 7). It has remained important in many developing nations around the world. It is simple in design, cheaper to build, and easier to maintain than a four-wheeled cart. This is not to say that it is the best system of using the ox for transport. However, for small farmers with

FIGURE 7
Pulling a cart.

limited resources, the ox and the two-wheeled cart seem to have always been used together. The best ox cart is one that is well balanced with only a slight amount of weight on the tongue, so as not to allow the cart to lift the oxen if it is loaded too heavily in the rear. The ox cart is often improved with a dump body, which allows easy unloading of manure, gravel, or crops that have been harvested. It is likely the most important improvement to the simple ox cart's design.

Wagons—four-wheeled carts—of many designs offer greater capacity in moving almost anything with oxen compared to the two-wheeled cart. The weight of the tongue is also reduced, decreasing the burden on the animal. The wagon is harder to maneuver, requiring more room for turns and backing. However, any disadvantage is usually outweighed by its capacity to haul almost anything.

Today, rubber inflatable tires on wagons and carts offer an improved ride and ease of use on both crude paths and improved roads. There are many manufactured designs that are appropriate for oxen. For someone interested in welding and cart construction, car axles, tires, and rims are often readily available at auto junkyards. Using additional scrap steel and wood construction, there are numerous possibilities for custom-designed ox-powered wheeled transport.

A major concern is controlling the wagon, cart, or sled when going downhill. Horned oxen wearing a neck yoke will learn to hold a small load with their horns. A brake is certainly a possible aid in checking the speed or stopping a load on a hill, especially in oxen that have no horns. It is imperative that the wagon tongue be equipped with a stop, which will prevent the tongue from riding through the yoke ring and the wagon overtaking the team.

Wagons or carts should use a braking mechanism of some sort. This might be as simple as a chain wrapped around the wheels, which prevents them from rolling and overtaking the animals. More innovative ox teamsters often design more improved brakes that in some way slow the wheeled vehicle, thereby preventing it from overtaking or overpowering the oxen.

The size of the load in either a sled or wagon has a considerable effect on how much effort must be used to control its downhill speed. Huge loads of gravel, logs, or stones must be braked in some manner. Head yokes are traditionally used in many hilly nations, which rigidly attaches the animals to whatever it is they are pulling. This has the advantage of controlling the lightweight wagon or sled, but it can also create problems if the load is too heavy or builds up too much momentum if the animals run downhill.

SUMMARY

There are many possibilities for using oxen in agriculture. As the most important draft animal in the world today, as well as the primary beast that "paved" the way for other draft animals in this country, the ox certainly has a place in agriculture. For the small farmer who understands both the attributes and shortcomings of the ox, they certainly have a place on some farms. They offer a great deal of flexibility and readily available animal power. The key is to use the ox properly: Give it good care, work it regularly, and employ it in every possible task. The dividends will be great. Ignore its nature, training, potential on the farm, and level of efficiency, and the ox will be overlooked by those who need it most.

REFERENCES

Ashburner, John, and Paul Starkey. (1994). *Draught Animal Power Manual*. United Nations Food and Agriculture Organization, Rome, Italy.

Cannon, Arthur. (1985). *The Bullock Drivers Handbook*. Night Owl Publishers, PO Box 764, Shepparton 3630, Australia.

Conroy, Drew, and Dwight Barney. (1986). *The Oxen Handbook*. Butler Publishing and Tools, PO Box 1390, LaPorte, CO 80535, USA.

Conroy, Drew. (1987). "Basic Training of Oxen," Video, 43 minutes, Rural Heritage, Gainesboro, TN 38562-5039, USA. Tel.: 931-268-0655.

Conroy, Drew. (1988). "Advanced Training of Oxen." Video, 46 minutes, Butler Publishing and Tools, PO Box 1390, LaPorte, CO 80535, USA.

Conroy, Drew. (1988). "The Traditional Ox Team and Its Yoke." *The Tillers Report*, 8:1. Tillers International, 5239 South 24th Street, Kalamazoo, MI 49002, USA.

Conroy, Drew. (1999). *Oxen, A Teamster's Guide*. Rural Heritage, Gainesboro, TN 38562-5039, USA. Tel.: 931-268-0655. Web site: http://www.ruralheritage.com.

Crossley, Peter, and John Kilgour. (1983). *Small Farm Mechanization for Developing Countries*. New York: John Wiley and Sons.

Food and Agriculture Organization of the United Nations (FAO) (1968). *The Employment of Draught Animals in Agriculture*. Rome, Italy.

Kramer, Dave, and Drew Conroy. (1998). "Ox Yokes I: Carving a Yoke." Video produced by Rural Heritage, Gainesboro, TN 38562-5039, USA. Tel.: 931-268-0655. Web site: www.ruralheritage.com.

Pingali, Prabhu, Yves Bigot, and Hans P. Binswanger. (1987). *Agricultural Mechanization and the Evolution of Farming Systems in Sub Saharan Africa*. Baltimore, MD: Johns Hopkins University Press.

Rosenberg, Richard. (1992). "A Neck Yoke Design and Fit: Ideas from Dropped Hitch Point Traditions." *Tillers TechGuide*. Tillers International, 5239 South 24th Street, Kalamazoo, MI 49002, USA.

Starkey, Paul. (1989). *Harnessing and Implements for Animal Traction*. Available for $31 through IT Publications 103-105 Southampton Row, London, WC1B 4HH, ENGLAND. Tel.: 011-44-171-436-9761. Fax: 011-44-171-436-2013. E-mail: orders@itpubs.org.uk.

Starkey, Paul, Sirak Teklu, and Michael R. Goe. (1991). "Animal Traction: An Annotated Bibliographic Database." International Livestock Research Center for Africa (now called ILRI) in Addis Ababa, Ethiopia.

PACK ANIMALS

Based on An Animal Traction Resource Book for Africa, *Paul Starkey, A Publication of the Deutsches Zentrum für Entwicklungstechnologien—GATE, a Division of the Deutsche Gesellschaft für, Technische Zusammenarbeit (GTZ) GmbH—1989*

Donkeys and mules are the main pack animals in most regions of the world. Mules are produced by crossing a female horse with a male donkey. Mules are larger and stronger than donkeys, but donkeys are cheaper to buy and to maintain. The reliability of donkeys is legendary. Once trained, donkeys can follow particular routes with minimal supervision; they will wait patiently for several hours, and they can often be trusted to return "home" unattended. Horses can be fast and efficient pack animals, although they are not as hardy as donkeys. Being more expensive to purchase and maintain than donkeys, horses are used mainly for high-value or strategic operations. Camels are excellent pack animals, unrivaled in their ability to cope with severe desert conditions, but they also are more costly than donkeys. Llamas and yaks are locally used in the foothills of the Andes and Himalayas. It is rare for cattle to be used as pack animals.

SADDLE BASKETS

When donkeys are used for pack work, it is normal to place some form of protective padding over their backs. This may be sheepskin, sacking, or discarded cloth. Soft loads such as sand, fertilizers, canvas water containers, and straw are placed symmetrically over the back and held in place by one (or more) leather or rubber straps around the girth or belly and under the base of the tail. Hard loads such as firewood, stones, or rigid containers are generally supported on simple wooden symmetrical saddle frames sitting on light padding and held in place with tail and girth straps. Simple pannier baskets may also be used (Figure 8). Pannier baskets with opening

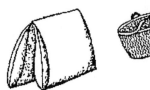  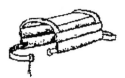 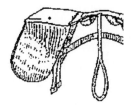

FIGURE 8
Some packsaddle designs. (From *Harnessing and Implements for Animal Traction*, GATE-GTZ.)

bottoms that allow loads to be shed easily have been used in Western Samoa. In Ethiopia, donkeys are widely used as pack animals and animals averaging 100–110 kg body weight regularly carry loads of 25–50 kg over distances of up to 20 km.

The distribution of donkeys in Africa is restricted by several ecological factors, notably the disease trypanosomiasis. With cattle being much more readily available, there has been some interest in the potential of cattle as pack animals. While cattle do not readily take loads on their back, they can certainly be trained to do so. In parts of Mali and Chad cattle may be ridden for personal transport by farmers, and some pastoralists in Sudan and Somalia use cattle to transport their effects when moving between sites. As animals can pull greater loads than they can carry, in most areas work relating to oxcarts will probably be more productive than trying to develop systems of using cattle as pack animals.

YOKES FOR OXEN

Withers—shoulder yokes—are numerically the most important system of harnessing in the world. They are almost universally used in Asia and Ethiopia, and are widely used in parts of western, eastern, and southern Africa and areas of Europe and the Americas. They are almost always made of wood, although a few projects in Africa and Asia have made yokes from steel pipe. In their simplest form they are just wooden poles with small descending pegs ("sticks") to restrict lateral movement. The wooden yokes may be shaped into double bows to more closely match the shape of the withers, thus giving a greater surface area of contact (Figure 9). Such simple shaping may well be the simplest and most cost-effective means of increasing the comfort and therefore the effectiveness of a wooden yoke. Withers yokes can be lightly padded, and in Ethiopia the traditional yoke is padded with sheepskin or cloth covered with cowhide.

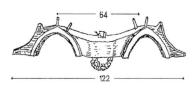

FIGURE 9
Withers yoke used in Zimbabwe. (From *Harnessing and Implements for Animal Traction*, GATE-GTZ.)

YOKES FOR DONKEYS AND HORSES

The breast band is the simple and cheap system for donkeys, mules, and horses. The work force is primarily taken from a broad band of leather, rubber, or strong canvas material across the animal's chest. Attached to either end of the breast band are the traces (ropes or chains) or shafts that pass back to the implement or swingle tree—see Figure 10. The breast band is held in position by one or more straps. Usually there is a neck strap crossing the withers and a back strap across the middle of the back. These straps not only maintain the position of the breast band, they also transmit the vertical component of the work, and they are often padded on the back and referred to as "saddles." The back straps may be adjustable or made to size. While leather is the traditional material for breast band and straps,

FIGURE 10
Breast band on a donkey. (From *Harnessing and Implements for Animal Traction*, GATE-GTZ.)

rubber carefully cut from old truck tires is increasingly used, and pieces are sewn together with wire. A study of several donkey harnesses in Botswana concluded that carefully made and padded breast harnesses made from either tire rubber or from webbing could be both cheap and effective. The use of breast band harnesses made from padded rope has also been reported and in Senegal some breast bands are made from nylon rope surrounded by cloth, contained within an old innertube.

REFERENCE

IT Zimbabwe and Institute of Agricultural Engineering, *User's Guide: to Low Cost Transport Devices*, IT Zimbabwe, Harare, Zimbabwe and ILO/ASIST, Harare, Zimbabwe, (1997)
Starkey, Paul, (1985), *Harnessing and Implements for Animal Traction* GATE-GTZ, Eschborn

LOW-COST TRANSPORT DEVICES

Based on An Animal Traction Resource Book for Africa, *Paul Starkey A Publication of the Deutsches Zentrum für Entwicklungstechnologien—GATE, a Division of the Deutsche Gesellschaft für, Technische Zusammenarbeit (GTZ) GmbH—1989*

SLEDGES

Wooden sledges are quite widely used in certain areas of eastern and southern Africa, Madagascar, and parts of Asia and Latin America. In southern Africa simple sledges are made by joining two wooden beams in the form of a V or by selecting a naturally occurring fork in the branch or trunk of a tree, perhaps 150 mm in diameter. A traction chain is attached to the single end of the "V" or "Y" (Figure 11). The load is supported by the two arms onto which a simple platform can be built, and sides can be fitted if required. More expensive sledges can be made using separate wooden or steel runners onto which can be mounted a variety of bodies. Such refined sledges have been evaluated for transport work on oil palm plantations in Malaysia (Kehoe and Chan 1987).

The advantages of sledges are that they are cheap and simple to make and maintain. They have a low center of gravity, and they are narrow, enabling them to be used on tracks too narrow or steep for carts. They can often be used in sandy, muddy, or rutted conditions where

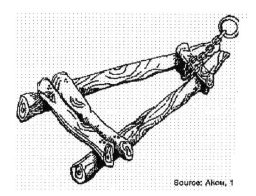

Source: Akou, 1

FIGURE 11
Simple wooden sledge used in Uganda and southern Africa.

a cart might become stuck. However, these advantages are offset by many disadvantages. In most conditions they require more effort to pull than does a cart. They have limited clearance and can be stopped dead by projecting stumps. Most importantly they tend to accelerate erosion by leaving rutted tracks, often only passable by other sledges, which become watercourses during heavy rains. In several areas of southern Africa, including Lesotho and Zimbabwe, the dangers caused to the environment by sledges have led them to be officially discouraged and even banned.

CARTS

A design widely used in West Africa and elsewhere is a simple steel cart frame with a relatively high-level load-bed bolted onto the axle and a wooden or steel platform fitted to this (Figue 12). While such designs are not particularly cheap, they are usually long lasting, with the only regular problem being tire punctures. Roller bearings are also used in carts made from old car axles. The bed may be either wood or metal—often a flattened-out oil drum or recycled roof sheeting is used. In Zimbabwe, scrap automobile body parts are common. Although some carts are made with scrap axles recovered from old cars, many more have much lighter, purpose-made axles with only the bearings made of steel. This type of cart is relatively lightweight, typically weighing around 125 kg, but sturdy enough to take heavy loads on rough ground, and in firm conditions are easily maneuverable by a single donkey. A cart of this type can carry a load of up to 700 kg, the equivalent of about 30 adult headloads, yet can be easily driven by a child. Donkeys are easy to train and will learn to follow a route. (It is not unusual to see a donkey with a loaded cart on a long journey, such as collecting firewood for sale, walking unguided while the driver sleeps.)

Most carts are designed to withstand loads of up to one tonne. The ability of animals to pull such loads will depend on the road surfaces and the inclines. An easy load to pull on a tarred road surface may be impossible to pull on a track with steeply sided holes or muddy ruts. Single donkeys can generally pull loads of 500 kg, single horses can pull 700–1,000 kg, and pairs of oxen can pull 1 tonne or more. Pairs of oxen of large Indian draft breeds are reported to be able to pull 1.5 tonne loads over 60 km of rough roads in a day. Balancing the load on two-wheeled carts is important, since any imbalance will cause upward or downward forces on the animal's harnessing systems. A heavy load shifting backwards during use can cause a donkey to be literally lifted off its feet, with disastrous consequences.

The fitting of brakes on carts is not common in flat areas, but it may be desirable. Brakes are important to save the animals from discomfort where steep slopes are encountered. Such slopes may be major hills or simply the steep sides of a road embankment. Even on flat ground, a loaded cart pulled at normal speed has a considerable momentum, and absorbing this through the harnessing system on a downward slope can be very uncomfortable for the animals. The choice of harnessing system can influence the efficiency with which animals can brake carts with their own bodies. Basic wheel brakes can be made from concave wooden blocks (or even

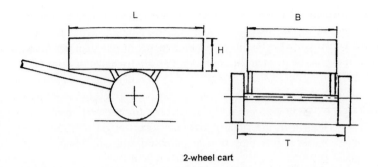

2-wheel cart

FIGURE 12
Cart design used in West Africa. (From R.A. Dennis, *Guidelines for Design, Production and Testing of Animal-Drawn Carts.*)

just logs) that are pushed against the wheel or tire surface. In the simplest case no fixings are necessary, although a lever mechanism can be arranged. Some manufactured wheels for carts come with internal brake shoes. Old car brakes can be quite easily adapted if mechanical parking brake linkages (not simply hydraulic mechanisms) are available.

Assembled carts are very expensive to transport over long distances, due to their great volume. For this reason, and to facilitate local construction and repair services, carts should be made, or assembled, as close to the point of use as practicable. Several African countries, including Burkina Faso, Mozambique, and Togo, have adopted the system of supplying basic cart kits to rural centers. Simple kits may comprise two wheels, an axle, and the struts that fix this axle to a wooden platform. Others may contain a complete steel frame in component form and even a steel drawbar. Some components may have to be imported (several countries import complete axle and wheel assemblies), while others may have been made in local workshops. Artisans, traders, and/or small workshops assemble the kits and build on wooden platforms, and perhaps removable sides, for sale to the end users.

WHEEL OPTIONS FOR CARTS

(See also the article "Wheel Manufacturing Technology" in this chapter.) Large-diameter wheels are better able to negotiate ruts and holes than small-diameter wheels. The ability of a wheel to accept poor conditions is dependent not only on wheel diameter but also on the width and type of tread and the strength, weight, and elasticity of the tires or wheel rims. The use of small wheels (400–600 mm diameter) allows cart platforms to extend over the wheels in a manner that is impracticable with large wheels (800–1800 mm). Such a design provides a wide but not too high loading area and easy access from the sides, and thus greater convenience.

Large wooden wheels with wooden spokes were standard in most parts of the world before the development of pneumatic tires, and such designs are still widely used in Asia and Latin America. Wooden-spoked wheels have for many years been made and used in Egypt, North Africa, and the islands of Madagascar and Mauritius. One recent project initiative in Zaire, where timber is plentiful, found that each wooden-spoked wheel required well-seasoned wood and about one month's skilled labor. With large fluctuations in the ambient humidity between seasons, any inferior work or poorly seasoned timber became apparent as wheels buckled and disintegrated. It was concluded that steel wheels of similar diameter might be more durable.

Wheels can be constructed from wood even if the technical refinement and complexity of spokes is neglected. In several parts of Asia and Latin America long-standing designs of such "solid" wooden wheels are seen, but they are much less common than wheels with spokes. Solid wheels are heavier, relative to their strength, than spoked wheels, and so large-diameter solid wheels are rare. In Africa several designs of "solid" wooden wheel have been evaluated. Some designs are made by cutting a circle from parallel timbers, glued or nailed into position. These are then supported by other timbers or by a second circle made from boards aligned in a different direction. The wheels are usually given a rubber tread cut from an old tire. The main advantage of such wheels it that they do not puncture and can be made mainly from local materials by village artisans. However they are heavy and they are not considered fashionable or prestigious (in one country they have earned the name "Flintstone" carts, after the famous cartoon characters).

Steel-spoked wheels generally are lighter than solid wheels, and they are easier to manufacture and maintain than wooden-spoked wheels. They are usually of larger diameter than wheels fitted with pneumatic tires and thus may be preferred for use on rough tracks where their larger diameter is advantageous for negotiating ruts and holes. However, steel wheels are much less resilient than wheels fitted with pneumatic tires, and so they tend to transmit unabsorbed shock loads to the wheel bearings, cart body, passengers, and animals. Their lack of resilience also makes steel wheels more likely to damage roads and tracks. Steel wheels are relatively cheap to make and easy to maintain. One problem is that shock loads and stresses imposed on steel-spoked wheels can cause fatigue in the welds joining the spokes and the rim. If weld failures are not noticed and repaired, the whole wheel may distort or even collapse. However, farmers adopting carts with steel wheels are much more likely to have problems with the wheel bearing than with the wheels themselves.

In recent years small wheels fitted with pneumatic tires have become the accepted standard for animal-drawn carts in many African countries. The adoption of common automobile tire sizes on carts allows farmers the option of making use of old vehicle tires. In practice, farmers have often found that the problems caused by punctures make wornout tires a false economy. Since the specifications of new car tires are unnecessarily high for slow-moving carts, special lower-cost animal-drawn vehicle tires have been produced in India. However, the development of these large-diameter tires was based on the potentially enormous Indian domestic market (with around 15 million carts), and similar investment in special cart tires seems unlikely in African countries. An alternative approach, widely used in West Africa, is to purchase at considerable discount the reject tires from large factories. Low-grade reject tires are dangerous if put on cars, but they can be safely used with animal-drawn carts. In a few countries the use of standard car tires on carts may be seen as a disadvantage because during shortages of car spares, compatible cart tires become targets for theft.

In many countries, a proportion of carts in use have been made from old car axles or from the entire rear section of light pickup trucks. These are generally heavier than carts with purpose-built axles, but where the necessary scrap vehicles and skills are available, such carts can be very effective. The increasing popularity of front-wheel-drive cars means that lightweight dif-ferential-type axles are rare, but some pickups have suitable axles. The independent stub axles from the front or rear of a car can be welded onto a steel beam or attached to a wooden frame, but the necessary dismantling, refitting, and correct alignment is not easy. In general the construction of carts based on old axles can be regarded as useful, small-scale initiatives for entrepreneurs or small organizations. For larger organizations, particularly those in areas of high demand for carts, the restricted availability of scrap parts, their heavier weight, and the quite modest cost savings suggest that car axles and pickup bodies should be regarded as supplementary rather than primary sources of animal-drawn carts.

CART AXLES AND BEARINGS

Simple bush bearings made of cylinders of cast iron, hard wood, or steel tube can be very effective, provided they are well prepared, appropriately lubricated, and regularly maintained. The majority of the world's carts still use simple bush bearings. Many traditional wooden carts are based on a large wooden hub rotating around a greased steel axle. In the center of the hub may be inserted a replaceable bush bearing, with cast iron often being preferred to hard wood or steel tube. Such bearings are commonly associated with large-diameter wheels on which the hub rotates relatively slowly. Furthermore, traditional wooden wheels have big hubs, allowing long bearings with a large surface area that, if well made and maintained, can last a long time (even if they do impose significant frictional loads). Carts with wooden block bearings are generally heavier to pull than other designs, due to the inherent friction and the weight of the cart. Bearing blocks have to be kept tightly clamped together, and the relative simplicity of the design should not disguise the fact that axles will only run freely and truly if the bearing tolerances are correct.

Many metal wheels are also designed to rotate around a fixed steel axle, and the search for suitable bush-bearing materials has occupied the staff of many projects in Africa (see Figure 13). Metal wheels are often of medium diameter so that the speed of rotation of the hub is faster than that of large, traditional cartwheels, and consequently the rate of wear of bearings is greater. In Tanzania some projects, such as that at Iringa, have tried to use oil-soaked wooden bushes as replaceable bearings. It was assumed that wooden bushes would be cheap and very easy to replace. In practice in the early years, both new and replacement bushes rapidly disintegrated, leaving very wobbly wheels. Furthermore, the wooden bushes were not sufficiently uniform to be fitted easily into the wheel hubs. As a consequence, farmers tended to tolerate worn bushes longer than they should, until the steel hubs of the wobbling wheels started wearing themselves. PVC bushes were evaluated, but these were expensive and wore rapidly. Bronze bearings (made from locally mined copper) have also been tried in Zambia, and these have been found more durable than hardwood or PVC. Mild steel does not have very good bearing characteristics, but it is readily available and easy to work. Although bush-bearings are a major source of frustration to projects and farmers, with regular repair and maintenance they can be kept going for many years In Ethiopia horse-pulled light carts dating back several decades are still in regular use, even though the original bearings have long since been replaced by wheel centers made from steel pipes and bushes (where present) made from a range of local materials, including rags.

A different approach to bearings that has also been tried in Tanzania, Zambia, and elsewhere is the use of "live" (rotating) stub axles made of water pipe or old half-shafts from pickups and trucks. The axles are held in place by two bearings, each made of two oil-soaked blocks of wood,

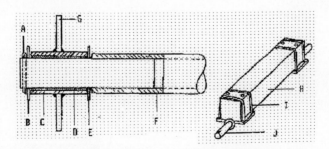

FIGURE 13
Fixed axle with bush bearing tested in Zambia. A—Split pin; B—Washer; C—Wheel hub; D—Bearing (bronze, nylon, or PVC); E—Washer; F—Stub axle; G—Spoke; H—Wooden beam; I—U-bolts; J—Stub axle.

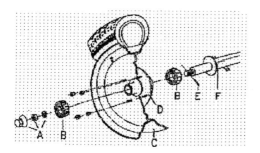

FIGURE 14

Wheel and axle unit with roller bearing and pneu-matic tires. A—Dust cap, lock nut, and washer; B—Tapered roller bearing; C—Wheel rim; D—Wheel hub; E—Split pin; F—Axle shaft.

hollowed out to the shape of the axle and bolted. Thrust washers are welded onto the axle to restrict lateral movement. The bearing blocks are bolted onto the wooden chassis (see Figure 13).

In conclusion, development projects are often faced with the choice between expensive, high-technology roller bearings (Figure 14) or various "appropriate technology" options. Many projects have spent a great deal of time and endured much frustration trying to perfect the simpler technology, but long-term mainte-nance problems have often been serious and adoption rates disappointing. Since transport is often very profitable, the higher cost of roller bearings that allows carts to be used very frequently, yet with little maintenance, may well be justified in the long term. With the benefit of hindsight it is apparent that several projects in Africa might have had more impact if they had provided credit to allow farmers to purchase higher-cost products, rather than employing people to try to develop low-cost alternatives.

FOUR-WHEEL TRAILERS

Four-wheeled carts, or trailers, are used for urban transport in many towns in Asia and some in Africa and Latin America. They are also used on some estates and plantations. The four wheels support the whole load, so animal power is only needed for forward movement. This allows heavy loads to be pulled, particularly if the road surface is smooth. Four-wheeled trailers can be left with loads in place even when the animals are not present (two-wheeled carts tip up when left, although it is a useful practice to always carry pieces of wood to support the front and rear of the cart to prevent such tipping). While two-wheeled carts can pivot around the wheels during sharp turns, four-wheeled trailers need some form of articulation to ensure maneuverability, which makes the design of trailers significantly more complex than just adding a set of wheels to a two-wheeled cart. While two-wheeled carts are likely to increase rapidly in rural Africa, it is unlikely that four-wheeled trailers will become common.

CART ADAPTATIONS

The standard two-wheeled cart can be modified for particular applications. In Figure 15 a tipping cart is shown—such is useful when transporting bulk loads such as grain, sand, soil, or manure. A bicycle-driven cart is feasible. It can be built from lightweight steel and outfitted with bicycle tires or built from heavier material. One extension is an ambulance cart, used to carry a sick person. Such can be hitched to a bicycle where the roads are good or pushed/pulled by hand in other places. A cloth or canvas canopy can be added to protect the sick person from rain or sun. Carts can be modified for agricultural use—to carry tools or water. The water cart is basically a wider two-wheel cart with a removable body; the frame is built to hold two water drums.

Human-powered carts are also useful and can be built by village artisans. A smaller version of the two-wheel cart can carry one water drum, firewood, building material, or agricultural produce.

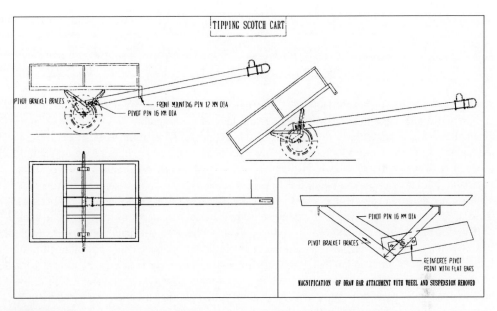

FIGURE 15

Tipping scotch cart. (From *User's Guide to Low Cost Transport Devices.*)

A donkey can be hitched to this small cart, making it more suited to the traditional tasks of rural women than a large ox cart needing tow oxen. It may be equipped with a stand in front of the wheels so it remains upright when at rest. A small model with 13-inch diameter wheels is lighter and better balanced than a wheelbarrow; a larger 18-inch model is easier to push on rough ground Wheelbarrows are feasible for village artisans—special models can be made adapted to carrying water tanks with the frame for the tanks on either side of the wheel rather than above the wheel. A sack barrow is shown in Figure 16. It is a basic L frame trolley for carrying sacks and crates. Wheels of different sizes can be fitted, depending on road conditions. A metal or wooden body can also be added, converting the trolley to a two-wheel barrow.

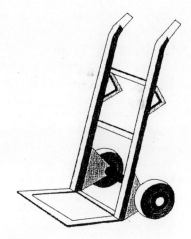

FIGURE 16

Sack barrow. (From *User's Guide to Low Cost Transport Devices.*)

REFERENCE

Dennis, R.A. (1996). *Guide Lines for Design, Production and Testing of Animal-Drawn Carts*, Intermediate Technology Publications, London.

IT Zimbabwe and Institute of Agricultural Engineering. (1997). *User's Guide to Low Cost Transport Devices*, IT Zimbabwe, Harare, Zimbabwe and ILO/ASIST, Harare, Zimbabwe.

A WHEEL-MANUFACTURING TECHNOLOGY FOR RURAL WORKSHOPS

Ron Dennis, IT Transport Ltd. The Old Power Station, Ardington, Oxon OX12 8PH, UK
(Based on an article of the same name in Appropriate Technology, *Vol. 17, No. 3, Dec. 1990)*

A major constraint on the production of affordable, nonmotorized vehicles such as handcarts and animal-drawn carts in rural workshops is the lack of good quality, low-cost wheel and axle assemblies. This is particularly the case in African countries. Most rural workshops do not have the equipment needed to produce good quality wheels and bearings, while setting up centralized production of conventional wheels involves a level of technology and capital investment that is not viable in many developing countries. Imported versions, even those that are taken from scrap vehicles, are general in short supply and very costly. Because of these problems is not unusual for the cost of the wheel/axle system to comprise 50 percent to 70 percent of the total vehicle cost.

The skills and materials required to produce wooden wheels of consistent quality are available in some Asian countries but in few African countries. Also, these wheels are rather limited in their range of application and in their compatibility with different types of tires. Steel wheels made from commonly available steel sections such flat and round bars and tube and angle sections could be produced quite readily in most rural workshops, provided that the means of producing good quality rims was made available. Also, these wheels could be produced in a range of sizes to suit various types of tires.

Two basic designs of steel wheels have been developed.

- A split-rim wheel that can accept most car or truck tires. This wheel is mainly for use on animal-drawn carts but it could also be used on tractor-drawn trailers.
- Rigid-spoked wheels, which can accept bicycle or moped tires, rubber tires, and steel tires. These wheels can be used on a range of vehicles, including wheelbarrows, handcarts, bicycle trailers, rickshaws, and animal-drawn carts.

The manufacture of these wheels is based on a simple hand-operated rim-bending device, which is able to bend a variety of steel sections into good quality wheel rims to suit a range of different tire types. The device is simple to construct and can readily be made in workshops equipped with welding equipment. Operation of the device is quite straightforward, and experience in a number of developing countries has shown that it can be successfully introduced and used in workshops to produce a range of good quality wheels. The versatility of the rim bender is shown in Table 1, which lists the maximum sizes of different steel sections that can be bent in the device.

The wheel manufacturing equipment also includes assembly jigs, which ensure that the wheels are set up accurately for welding and that a consistent quality of manufacture is maintained.

WHEEL-RIM BENDING DEVICE

The device is illustrated in Figure 17. It comprises a mainframe on which are mounted two lower fixed rollers and a lever arm that supports a central upper roller or forming tool. The section to be formed rests on the two fixed rollers and is bent by the forming tool, which is forced downwards by the lever arm. Prior to bending, the section is marked out into equal increments of 25–50 mm, and during bending the section is fed over the lower rollers in increments. At each increment the lever arm is depressed to a preset stop to produce a short length of the desired curvature in the rim.This method of incremental bending produces a formed circle in which the variation in radius is of the order of 2 mm. The amount of bending, and hence the diameter of

TABLE 1
Capacity of wheel-rim bending device

Type of Section	Maximum Size That Can Readily Be Formed	Minimum Diameter That Can Be Formed
Roller in outer position (290 mm centers)		
Flat	100×12 mm	450–500
Angle	$40 \times 40 \times 6$ mm	400–450
Round	25 mm diameter	~400
Tube	$^3/_4$-inch water pipe	~400
Roller in inner position (150 mm centers)		
Flat	100×6 mm	~300
Angle	$25 \times 25 \times 5$ mm	~300
Round	16 mm diameter	~300
Tube	$^1/_2$-inch water pipe	~300

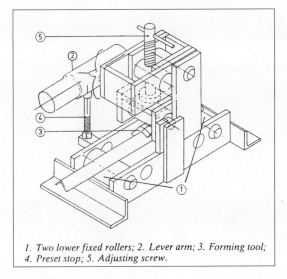

1. Two lower fixed rollers; 2. Lever arm; 3. Forming tool; 4. Preset stop; 5. Adjusting screw.

FIGURE 17
The wheel-rim bending device.

rim produced, is set by the adjusting screw. The section only needs to be passed through the device once to obtain the fully formed rim, so it is possible to produce a rim in minutes.

Three forming tools are needed to cater for the range of sections listed in Table 1. Additional tools can easily be made if other sections such as channel are to be formed. Two positions are provided for the lower rollers. For the outer positions, less force is needed to bend the rim, and this position is used for forming heavier sections and larger-diameter rims. The inner position is used to bend smaller-diameter rims.

The bending device has two major advantages over conventional rolling machines: first, the simplicity of its construction: no rotating shafts, bearings, or drive mechanisms are needed; second, the small degree of sliding movement at the rollers during bending means that unhardened rollers can be used without risk of significant wear.

ASSEMBLY JIGS

The most important function of the assembly jig is to ensure that the rim is aligned concentrically with the wheel hub and axle. If this is not achieved, the wheel will run eccentrically and/or wobble from side to side, reducing the efficiency of the vehicle and increasing the wear of both tires and bearings.

The design of assembly jigs may be quite complicated in order to cope with changes in wheel diameter and the various materials that may be used in different workshops. For instance, an assembly jig for split-rim wheels needs to be adjustable to cater for rim diameters to suit tire sizes from 13-inch to 16-inch. Also, a great deal of care needs to be taken to ensure that the jig is constructed as accurately as possible; otherwise, the quality of wheels made on the jig will suffer. The use of a well-made assembly jig greatly simplifies the construction of wheels and ensures a consistent quality of manufacture.

An assembly jig used for the construction of bicycle-type wheels is shown in Figure 18. In this case the brackets supporting the adjustable stops may be bolted to the frame at different positions to cater for wheel sizes from 20 inches to 28 inches.

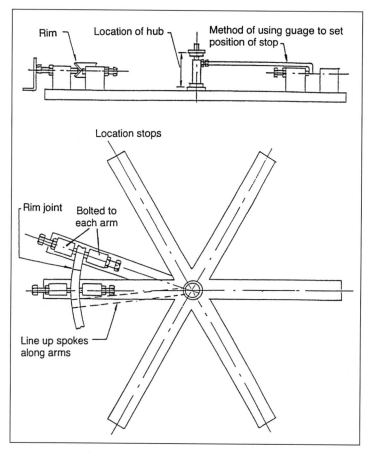

FIGURE 18
The assembly jig used in the construction of bicycle-type wheels.

WHEEL DESIGNS

Figure 19 shows a basic range of wheels that can be manufactured using the bending device. The following features are noteworthy.

- The novel concept of fitting a standard bicycle tire and tube into a rim made from 25 × 25 × 3 mm angle section. This produces a low-cost wheel that is significantly more robust and durable than a standard bicycle wheel. Heavier duty wheels may be made by fitting moped tires into rims made from larger-size angel.
- In the split-rim wheel the two parts of the wheel are fitted to either side of the tire and then bolted together to hold the tire in position. It is therefore very simple to assemble and disassemble the tire by hand. A similar type of wheel can be produced to accept motorcycle tires.
- A variety of rims can be produced to suit different types of tires, all from materials that are commonly available in developing countries, such as angle, flat, and round bar.

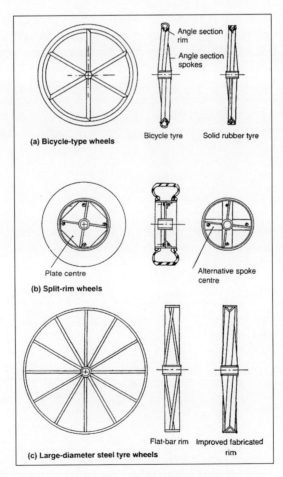

FIGURE 19
A selection of wheels that can be produced using the bending device.

BOATS

Based on Chapter 1, "Boat Design, Construction, and Propulsion" in Fisheries Technologies in Developing Countries, *National Academies Press, Washington, DC, 1988*

The most common use for boats in the Third World is for fishing, but the same boats can be used to take people across rivers or to and from islands. Alternative ways to cross rivers are fords and culverts or pontoons—basically a strong barge—attached to two cables. One cable keeps the barge on course; the other cable is pulled or winched to propel the boat. Diesel engines are also used. (Fords and Ropeways are discussed in other articles in this chapter.)

TRADITIONAL BOATS

Traditional boats have passed the test of time. Nevertheless, traditional craft are not without their problems. They often have a very limited range of operation and are not able to go beyond nearshore areas. Many will sink if swamped, providing no reserve of safety. Customary building materials are often unavailable. Deforestation in many coastal areas has created a scarcity of quality wood for dugout canoes and larger craft. Traditional boats can be improved, often without radically altering the basic design; a respect for tradition will increase their acceptance. In areas without harbors, beachable craft are a priority. Improved designs should also help ensure the safety of the crew by including a second means of propulsion and sufficient buoyancy so that the vessel remains afloat when flooded. This article covers some design considerations, examines some new boat construction methods and materials, and describes a few propulsion techniques.

DESIGN

A boat may be described as a floating platform used to transport crew and passengers and, in some cases, to support the crew and equipment. Some of the major factors that affect the design of this platform include the following·

- Available funds
- Available materials
- Skills for building and maintenance
- Size limitations dictated by water depth or requirement for beaching
- Distance to be traveled
- Fuel costs
- Vessel speed requirements
- Safety features

Small craft design should be based on the traditions of a given region. Vessel sizes and designs that have evolved in an area are usually well adapted to the local fishing gear and methods, the range of operations, construction materials, the winds, and local water conditions. A radical departure from the traditional hull design may not gain local acceptance.

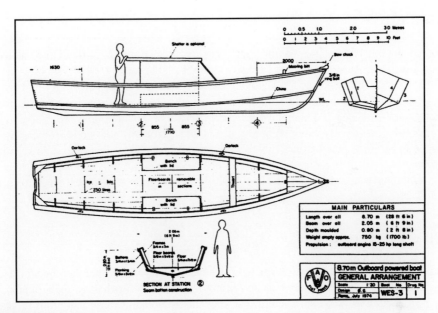

FIGURE 20
FAO-developed 8.7-m boat.

Rafts are keelless vessels that are common in many areas of Asia. They may be constructed of bamboo, logs, or plastic cylinders, lashed or fastened together. These vessels are beach landing craft, well suited for heavy surf conditions that would exclude many other boat types. The kattumaran of South India is a wooden log raft that ranges from 3 to 9 m long. Each log is individually shaped with a definite fore and aft curvature. Longer logs are placed inboard and shorter ones outboard, and all are lashed together. Planking is then nailed over the logs to provide a smooth working surface.

Single-hulled vessels are most commonly used in small-scale fisheries and often for transport. Designs with a high length to beam width ratio and a low displacement (displacement is the weight or volume of water displaced by a boat) to length ratio have less resistance per unit of displacement than do fat, heavy hull forms. Therefore, narrowing the beam, lightening the draft (draft is the depth of water that a boat displaces), and decreasing the displacement length ratio will result in less fuel consumption at a given speed. A number of FAO-designed (Food and Agriculture Organization of the United Nations) hulls based on these principles have been adopted in the South Pacific. The FAO 8.7-m boat (Figure 20) has been designed as an easily propelled, narrow beam, light displacement craft suitable for village fishery operations. An outboard-powered model of this craft has been built in Western Samoa for U.S. $1,250. With a crew of four and a 200-kg catch, the vessel can achieve a speed of 10 knots with a 20-hp outboard motor.

Multihulled vessels, such as catamarans and trimarans, have traditionally been used as fishing boats in the Pacific Islands. They show promise as fishing boats in other areas, especially where fishermen use one or two outriggers and are accustomed to the idea of multihulls. Multihulled boats have a number of positive features for small-scale fisheries. Their hulls have low displacement to length ratios and high length to beam ratios (long and narrow) and therefore offer minimum resistance and are easily propelled. Moreover, the stability of multihulls makes them ideal candidates for sail power. Small catamarans are lightweight and can be beached and carried with relative ease.

CONSTRUCTION

Traditional dugout canoes and bamboo rafts are common throughout the Third World. The construction materials are usually inexpensive and available locally. However, both materials severely limit the hull shape and are relatively short lived. Wooden logs are heavy and can result in high fuel consumption. While bamboo has the advantage of being lightweight, it is not especially durable. Wood and bamboo will remain important boat building materials in villages where they are readily available. Where there is a scarcity of good wood, there may be no alternative to adopting new materials. Newer materials and methods can offer many advantages that compensate for their increased cost. The choice of material will depend on a number of factors, including cost, availability, longevity, ease of repair, strength, and resistance to corrosion and rot.

WOOD CONSTRUCTION

Timber planked hulls have been constructed for hundreds of years throughout the world, and in many areas they are still very popular and highly regarded. Nevertheless, their importance is clearly diminishing as new construction materials are accepted. Several variations of planking are commonly used. In carvel planking, the outside planking is laid edge to edge, giving the hull a smooth surface. If the planks are very narrow (2.5–4 cm wide) and wedged together with the edges fastened, the method is called strip planking. Marine glue or caulking is used to keep the seams watertight. In clinker planking, each plank overlaps the upper edge of the plank below and is attached to it by nails driven from the outside. This variation is strong and flexible and is ideal for such small craft as dinghies.

Wood can be a very satisfactory boat-building material: It has good resistance to chafe, gives thermal and acoustic insulation, and allows great variation in hull shape. If good timber is available locally and is economical, it is a logical choice. However, in many tropical regions, suitable boat-building timber is scarce and expensive. Another disadvantage is the high degree of skill required to build a wooden boat. With only hand tools, construction can be very time-consuming. The hulls produced are of medium weight and, as they become increasingly water-logged with age, consume large amounts of fuel. Many woods are also subject to rot and attack from marine borers.

In Tahiti, V-bottom boniiers are built of imported redwood planking with local timber used for the frames. Hot dipped, galvanized carver nails are used for the fastenings. These boats are reported to last well, in spite of being stressed when they are run at high speeds.

Plywood is a sandwich of wood veneers and filler material held together by adhesives. There are many grades of plywood, but generally, marine plywood made with a waterproof adhesive is required for boat building. Lower-grade plywoods can sometimes be upgraded for marine use if they are coated with a polyester resin.

Plywood is very adaptable to small boat-building operations. It is light, can be cut to any shape, and is easily bent. Since sections are cut from large plywood sheets, there are fewer seams than in planked boats. Plywood construction involves building a framework for the hull from planks and then attaching sections of marine plywood to this frame. The plywood hull is held together by nails; marine glue is used to seal the seams.

Plywood boat building can be quick, inexpensive, and easy. As long as the surface, and especially the edges, of the plywood are treated with epoxy resin or another sealer, the boat will have a long life. However, the use of plywood does restrict the hull to hard chine shapes, such as flat or V-bottomed boats. Moreover, its resistance to chafe is not high.

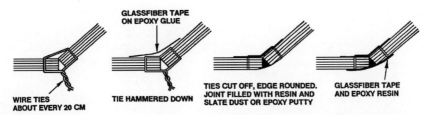

FIGURE 21
Stitch and glue.

There are many successful examples of plywood boats built and used throughout the world. Some 250 plywood versions of the Alia, an 8.5-m fishing catamaran, were built in Western Samoa in the 1970s and have survived almost a decade without hull rot or delamination. These vessels have an emergency sail but rely on outboard motors as their principal method of propulsion. Fishermen generally employ these catamarans for trolling and handlining. In Fiji, more than 130 V-bottom fishing boats (8.6 m) have been constructed of plywood. They are equipped with inboard diesel motors and are also used primarily for handlining and trolling.

A plywood single outrigger canoe was designed by FAO in 1985 specifically for the waters of Papua New Guinea. This 7-m canoe is sail-assisted and is designed to use an 8-hp outboard motor. The outrigger is filled with foam and helps support the weight of two or three persons in the canoe. In sea trials it was shown that this new vessel equipped with an 8-hp outboard engine was faster than a traditional dugout powered with a 25-hp engine and could travel about twice as far on the same amount of fuel. Similar plywood outrigger canoes (proas) have proved their worthiness throughout the South Pacific, where they can replace canoes made from timber.

Plywood skiffs have wide acceptance throughout the world as inexpensive, rugged workboats. In southern New England (United States), plywood skiffs are extremely common. With good waterproof adhesives, these skiffs can have a 15-year service life.

Marine plywood is also used in the stitch-and-glue technique (Figure 21). Precut sections of plywood are wired together with galvanized wire; the seams are then sealed with epoxy resin. Bonding the epoxy resin glue with glass fiber makes the final connection. Once the resin has set, the wires can be cut and a finish applied. The product can be a strong, light boat with a life expectancy at least as good as traditional timber vessels.

Boat construction by this technique is easy and fast. Precut sections of marine plywood may be assembled in a village workshop without sophisticated equipment. Skilled carpenters are not required, but it may be necessary to import the epoxy resin and glass fiber.

Double-hulled boats have been constructed by stitch-and-glue methods. They can be landed on the beach and offer stability and a large platform for fishing. One small version, the 4.8-m Sandakipper, was also introduced into South India (Figure 22). It can carry half a ton of gill net and an additional ton of catch.

A plywood houri has also been designed as a replacement for the dugouts and planked houris of the Indian Ocean (Figure 23). Built from only four sheets of plywood, it can be rowed, paddled, or powered with a 4-hp motor.

The boat-construction technique known as cold molding uses veneers or thin plywood strips to build up a laminated hull. The veneers are applied in diagonally opposed layers. The thickness of the veneers varies in proportion to the hull size, but typically they are from about 2 mm to about 10 mm thick. These thin boards can be produced by a plywood mill or with a bandsaw or circular saw. One cold-molding method involves fabricating a mold that provides surfaces on which the planking is stapled. The veneers must be carved to a shape that will fit with their neighbors.

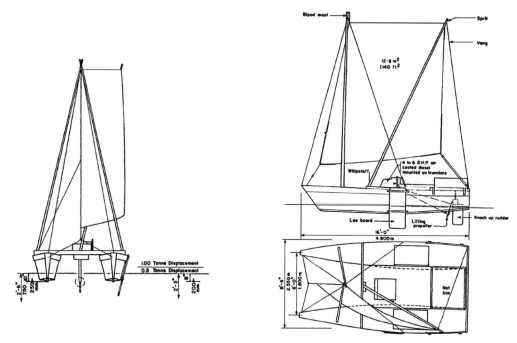

FIGURE 22
Double-hulled plywood boat.

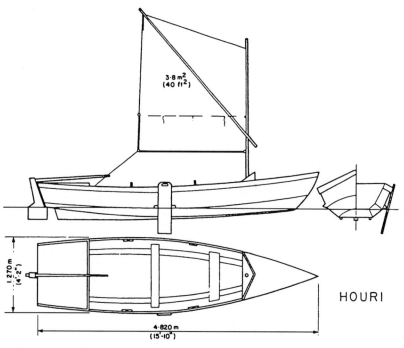

FIGURE 23
Plywood boat.

The first layer of veneer is stapled longitudinally to the frame. Epoxy adhesive or another gap-filling glue is applied to this first layer, and a second layer of veneer is stapled diagonally over the wet, uncured glue. A third layer of veneer may be placed diagonally to the second.

After lamination has been completed, the frame can be removed and the staples clipped. The gunwale and keel are then attached. If necessary, a fiberglass-epoxy resin coating can be applied inside and outside the hull. A water-repellent preservative or paint will protect the wood satisfactorily.

The cold-molding technique creates a very light and strong hull, resulting in low fuel consumption. These relatively thin hulls are not highly resistant to puncture, but this can be improved by increasing the fiberglass-resin layer. Although in most areas it is probably easier to obtain veneers than good timber, the adhesives may have to be imported.

"Constant Camber" is an improvement on this lamination technique. It requires a reusable mold shaped like a curved trellis. The hull geometry is such that the veneers can be precut and can be easily mass-produced. Each veneer strip does not have to be hand carved to fit perfectly with neighboring pieces. Another great advantage is that one mold can produce various hull sizes and types.

The mold is best suited to hull forms that have a relatively constant amount of curvature throughout, such as the long narrow hulls of multihulled vessels. However, wide-body hulls can also be produced, and craft as long as 19 m have been fabricated.

Using the Constant Camber process, the veneers are bent diagonally across the mold and stapled, as in cold molding. Additional layers are held by epoxy resin and can be applied immediately. No screws or nails are required in the process. The staples can be left in and later cut and sanded down.

Alternatively, a process called vacuum bagging can be used to eliminate the need for staples. The defects in the wood are filled with glue, and even imperfect wood can be substantially strengthened. The reusable equipment for vacuum bagging costs about $500.

The resulting veneer-epoxy composite is stronger than the original wood itself. The hulls are strong, light, waterproof, and rot-resistant, and have a predicted life of 20 years.

A 35-foot panel can be laminated by several people in a matter of hours. Two half-hull panels are then sewn or glued together to form the hull. Plywood or veneers of fast-growing woods could be obtained locally in many Third World villages and the molding technique learned by village craftsmen. Liabilities are the lack of expertise in using this relatively sophisticated method.

The Constant Camber technique has been used to construct a fleet of 100 paddle-powered catamarans used by Burundi fishermen on Lake Tanganyika. These boats are especially energy efficient because they are easily paddled. A local wood was used for the veneers, but most of the equipment and adhesives as well as the expertise had to be imported.

In Tuvalu in the South Pacific, several Constant Camber catamarans transport people and cargo around the atoll lagoons. These boats were originally financed by the Save-the-Children Federation but are now self-supporting. Over 100 smaller wood-epoxy boats have recently been constructed there.

NONWOOD CONSTRUCTION

Ferrocement is the term used to describe a steel-and-mortar composite material. It differs from conventional reinforced concrete in that its reinforcement consists of closely spaced, multiple layers of steel mesh completely impregnated with cement mortar. Ferrocement can be formed into sections of less than an inch thick. Ferrocement reinforcing can be assembled over a light framework into the final desired shape and mortared directly in place. Ferrocement boats are usually constructed close to the water's edge because of their weight. The building site should be chosen with the size of the craft, its draft, and its launching in mind.

There are five fundamental steps in ferrocement boat construction.

1. The shape is outlined by a framing system.
2. Layers of wire mesh and reinforcing rod are laid over the framing system and tightly bound together.
3. The mortar is plastered into the layers of mesh and rod.
4. The structure is kept damp during the cure.
5. The framing system is removed (unless it has been designed to remain as part of the internal support).

There are several ways to form the shape of the boat. A rough wooden boat can be constructed as a matrix or an existing, perhaps derelict, boat can be used. Pipes or steel rods may be used to frame the shape of the hull. In the construction of Chinese sampans, a series of welded steel frames and precast ferrocement bulkheads are erected. Layers of wire mesh are then attached to this framework and mortar applied. The steel frames and ferrocement bulkheads are left in place as part of the boat structure.

Using these and similar techniques, ferrocement boats from 8 to 20 m long have been constructed. Above and below this size range there has not been enough experience to recommend this type of construction. Ferrocement hulls less than about 8 m are usually heavier than comparably sized hulls in wood, steel, or fiberglass. This characteristic also prohibits ferrocement use in multihulled vessels.

Problems with chafing, penetration by sharp objects, and saltwater corrosion of the steel mesh have also been reported. Perhaps the most positive aspect of ferrocement as a construction material is the very low cost of materials. A high percentage of materials can usually be obtained locally. Construction is straightforward and rapid.

Any desired hull shape can be produced in ferrocement. Because the hull is homogenous, there are no seams to leak. Damage from impact simply requires chipping away the broken concrete, reshaping the mesh support, and applying new cement. The repair process is easier and cheaper than repairs for many other materials.

Ferrocement boats have been constructed and operate in Southeast Asia, South Asia, the South Pacific, and Africa. Many of these boats have been pilot projects, but in some cases, ferrocement has become a leading boat-building material.

Plastic tubing is durable and inexpensive, resistant to marine borers and rot, and does not react with saltwater or become waterlogged. Rafts in Taiwan have been traditionally made of bamboo; although very strong and light, this wood is also short lived. The bamboo is being replaced by sealed plastic (PVC) tubes that are 15 cm in diameter. Nevertheless, the vessel design is very restricted.

From 6 to 20 4-m-long pieces of plastic tubing are fastened together to construct the raft. The first meter of tubing at the bow is curved upward at 45 degrees to minimize resistance to the water. Inboard diesel motors are generally used to propel the plastic rafts. Sail power seems to have fallen into disuse with this vessel.

Fiberglass-reinforced plastic (FRP) has gained increasing acceptability as a structural material for boats since the 1950s. This material was first used for pleasure craft and is now increasingly used to construct fishing boats in the Third World. FRP is a composite material made of fiberglass and a polyester resin. The fiberglass provides the material's strength, and the resin, which is absorbed by the fiberglass, allows the material to be easily shaped.

After a prototype has been chosen, a female mold is manufactured. A polyester resin gel coat is sprayed onto the mold's surface, and then fiberglass and more resin are used to laminate the hull. After transom and keel reinforcements have been installed, the hull is removed from the mold.

FRP is an outstanding construction material for boats. Virtually any complex hull shape can be created. Because of the one-piece hull structure, leakage is practically impossible. The material is highly resistant to scratching and does not rot, rust, or corrode. Thus, less maintenance time is required, and durability is good. FRP shells have a much higher strength-to-weight ratio than similar wooden shells and are also lighter. The actual boat construction does not require high skills or special tools.

The major disadvantage of FRP is the cost of materials. Fiberglass and polyester resin often must be imported at high cost. The development of the female mold required for production is an additional expense. Repair of the hull in remote areas may also be a problem. The resin presents some difficulties for the tropics because it must be stored in an air-conditioned room and replenished every six months. The fibers and resin also can be hazardous to the health of the workers.

A boat evolution has occurred in Sri Lanka through FAO's Bay of Bengal Program. The traditional oru is a Pacific proa-type vessel with a single outrigger. Built of jak timber, it is seen in sizes from 15 to 40 feet. Because of the shortage of large jak timber, the FAO program designed an FRP oru that involves a modification of the hull but retains the traditional rudders and rigging. In some locations, such as the eastern Caribbean, FRP is also used to sheath traditional wooden vessels to extend their lifetime. The Bay of Bengal Program has proposed to protect the logs of South India's traditional kattumaran with FRP sheathing.

C-Flex is a fiberglass planking that can be used to build boats without the standard mold required for fiberglass-reinforced plastic boat construction. C-Flex is composed of parallel rods of fiberglass and reinforced polyester resin alternating with bundles of continuous fiberglass rovings. This structure is held together by two layers of lightweight, open-weave fiberglass cloth. Each plank is 112 cm wide. The planks are laid over plywood frames, tacked in place, and covered with resin. Fiberglass mats are then applied at right angles to the C-Flex. Sanding and a final finishing complete the process.

Construction of small aluminum vessels involves standard metalworking techniques. Aluminum plates are cut and bent to fit the frame of the hull. Welding and riveting are then used to seal the seams and fasten the plates. Aluminum alloys are excellent materials for small vessels. They can be shaped to almost any hull form and produce a greater variety of shapes than glued wood can. Aluminum is also light, which is another advantage, because it reduces the displacement and results in low fuel consumption. In addition, aluminum shows a high resistance to chafe, has an excellent strength-to-weight ratio, and holds up well under bending stress. Aluminum oxide forms in a thin coating on the alloy and provides protection against corrosion. Thus, boats constructed of this material can have great longevity.

The disadvantages of aluminum are significant. The cost of aluminum alloys suitable for boat building is very high, and the alloys may be difficult to purchase in small quantities. Although dents may be easily hammered out, punctures may require welding equipment, which is not likely to be available in coastal fishing villages. Moreover, aluminum is far more difficult to weld than steel and requires the high temperatures of arc welding.

PROPULSION

New technologies in propulsion include alternative fuels, alternative engines, and unconventional wind-based methods. Alternative fuels include biomass-derived gasoline and diesel-fuel substitutes. Alternative engines include units powered by steam and producer gas. Unusual types of sails and wind-powered rotors complete this section.

ALTERNATIVE FUELS

Both alcohol (ethanol) and vegetable oils have been examined as potential alternative fuels for small island communities. It was proposed, for example, that it would be possible to produce alcohol from cassava on one of the smaller islands in Fiji. Using a simple fermentation unit and distillation column, ethanol of 95 percent purity could be manufactured and used in modified outboard engines.

Coconut oil and other vegetable oils have been examined for use in diesel engines. There have been three general approaches in the testing of vegetable oils as diesel substitutes. First, the oils can be used as 100 percent substitutes for diesel oil. In many short-term performance tests, vegetable oils have proved almost equal to diesel fuel. The use of pure vegetable oils in longer-term endurance tests has rarely been satisfactory, however. Problems arise with coking and clogging of the injector ports and with fouling of the crankcase oil. Various blends of vegetable oils and diesel oil have also been tested. The use of 80:20 (or higher) blends of diesel oil to vegetable oil has generally proved satisfactory in both short-term and long-term tests. In the Philippines, however, when there was a national program to include 5 percent coconut oil in the diesel fuel, there were significant problems with clogging of fuel filters.

The most promising approach in the use of vegetable oils as diesel fuels involves their chemical transformation. Through the reaction of vegetable oil glycerides with alcohols (such as methanol or ethanol), the original high molecular weight glycerides are converted to methyl or ethyl esters, much closer in molecular size and shape to diesel oil. Performance tests with the esters derived from many vegetable oils have demonstrated good results in both short- and long-term testing.

ALTERNATIVE ENGINES

Both steam- and producer-gas-powered engines have a special appeal for developing countries—that of fuel diversity. A wide variety of forest and agricultural products and wastes can be used as fuel in these systems. Using coconut-shell-derived charcoal as fuel, producer-gas-powered fishing boats have been tested in the Philippines. The Intermediate Technology Development Group (ITDG) in London has begun development and testing of a small steam engine specifically for use in developing countries.

WIND POWER

Despite the presence of favorable winds in many areas, sailing as a means of propulsion for fishing craft in the developing world has declined in recent years. Wind patterns in the tropics are generally stable and predictable; large regions benefit from regular trade winds. In some areas, such as the northeast Indian Ocean, the China Sea, and Malaysia, fishermen continue to use their sailing skills. Large parts of Africa and Central and South America have not developed sail craft because they lack information, suitable materials, or incentive. Retrofitting sails to existing vessels can also be troublesome: hulls may not be suitably designed or sufficiently strong to accommodate masts or the strain imposed by sailing.

Natural or synthetic fabrics are most commonly used for sails. Dacron has proved to be one of the most durable and efficient materials for sails, but for most developing countries, local materials will be more practical and less expensive. Depending on wind strength and sail configuration, a sail area ranging between 1.9 and 6.5 m^2 (20–70 ft^2) is equivalent to 1.0 hp.

HUMAN POWER

Arm- and leg-powered devices, including oars, paddles, and pedal-driven propellers, have all been used for boat propulsion. A highly efficient shell requires about 230 watts (.3 hp) effective power to attain a speed of 4 m per second (9 mph). Maximum instantaneous human output is about 1,500 watts (2 hp), but for a one-hour period, this decreases to about 500 watts (.67 hp) and for 24 hours to about 370 watts (.5 hp).

Since humans can probably produce more power by pedaling than any other endeavor, much could be done with pedal-powered propellers. Rowing is a relatively inefficient way to use human power for boat propulsion. Sculling, the use of a rear-mounted oar fixed on a fulcrum, is significantly more efficient.

LIMITATIONS

To gain ready acceptance of users, changes or improvements in boat design and construction should not depart radically from traditional designs. This concern will be automatically satisfied if the local users play an important role in deciding the changes they would like to see in their boats. What works well in one area will not necessarily work well in another. If new construction materials are used, they must be economical and, if possible, available locally. Local facilities must also exist for the repair and maintenance of the vessels. Before its introduction, a new vessel must first be carefully evaluated and modified as a prototype. Improvements should be recommended and adopted only when it can be clearly proved that they will be economically justifiable.

REFERENCES

Design

Food and Agriculture Organization of the United Nations (FAO). (1984). *Manual of Fishing Vessel Design*. FAO, Rome, Italy.

Fyson, J. (1986). *Design of Small Fishing Vessel*. Fishing News Books Ltd, Surrey, UK.

Reinhart, J. M. (1979). *Small Boat Design*. ICLARM Conference Proceedings, No. 1, ICLARM, Manila, Philippines.

Todd, J., and L. G. Lepiz. (1986). "An Integrated Approach to Development of the Small-Scale Fisheries of the Talamanca Coast of Costa Rica." In F. Williams (Ed.), *Proceedings of the 37th Annual Gulf and Caribbean Fisheries Institute*. GCFI, Miami, FL.

Traung, J. O. (1967). *Fishing Boats of the World*. Surrey, UK: Fishing News Books Ltd.

Construction

Sleight, S. (1985). *Modern Boat building, Materials and Method*. Camden, ME: International Marine Publishing Company.

Wood

Steward, R. M. (1980). *Boat Building Manual*. Camden, ME: International Marine Publishing Company.
Harper, E. (1980). *Wood Vessel Layup*. Institute of Fisheries and Marine Technologies, St. John's, Newfoundland, Canada.

Plywood

Kurien, John. (1986). "South India Fishermen Helped by Introduction of New Boats." *Appropriate Technology*, Vol. 22, No. 2.

Ferrocement

Harper, E. (1981). *Ferrocement Boat Building*. Institute of Fisheries and Marine Technology, St. John's, Newfoundland.
MacAlister, G. (1980). "Ferrocement and the Development of Small Boats." *Journal of Ferrocement*, 10: 47–50.
National Academy of Sciences. (1973). *Ferrocement Applications in Developing Countries*. Washington, DC.
Sharma, P. C., and V. S. Gopalaratnam. (1980). *A Ferrocement Canoe*. Asian Institute of Technology, Bangkok, Thailand.

Fiberglass-Reinforced Plastic

De Schutter, J. (1985). "Glassfibre Reinforced Polyester: Its Application in a Boat Building Project in Lombok, Indonesia." *Vraagbaak* 13(3): 56–62.
Vaitses, A. (1984). *Boat Building One-Off in Fiberglass*. Camden, ME: International Marine Publishing Company.

C-Flex

Kennedy, K. (1977). *C-Flex Construction Manual*. Camden, ME: International Marine Publishing Company.
Taylor, M. (1982). "C/Flex—Fantastic Wood Sheathing." *National Fisherman*, November.

Propulsion

Mitchell, R. M. (1982). *The Steam Launch*. Camden, ME: International Marine Publishing Company.
National Academy of Sciences. (1980). *Alternative Fuels for Maritime Use*. National Academy Press, Washington, DC.

EARTH ROAD CONSTRUCTION

Based on Earth Roads A Practical Guide to Earth Road Construction and Maintenance, *by* Jack Hindson, revised by John Howe and Gordon Hathway IT Publications, London, 1996

The density and quality of roads is of great importance in the lives of rural people. Without dependable roads, farm produce cannot be brought to market. Where adequate roads do not exist people cannot move during the rainy season between urban areas and rural areas. Families are separated, and stocks of food, clothing, and medical supplies cannot be replenished. A graduate engineer and mechanized equipment is need to design and construct paved roads, but satisfactory earth roads can be laid out and built by nonprofessionals with shovels, hoes, and basic measuring tools.

EROSION

The major problem in designing earth roads is to minimize erosion, especially from water (wind erosion is typically much less serious). Water erosion is acerbated when the pattern of rainfall is short, heavy storms concentrated into a few months. Erosion on the carriageways (where the traffic runs) or in the side drain of the road results from too much water being allowed to accumulate there.

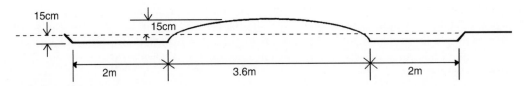

FIGURE 24
Section through an earth road.

ROAD CONSTRUCTION

Earth roads must be raised above ground level to allow water to drain off (see Figure 24). Earth for the road is dug from the side drain until the road surface is 15 cm—about 6 inches—above the former ground level. The road surface should be about 3.6 meters—4 yards—wide. The side drains should be about 15 cm below the former ground level so the road surface is 30 cm above the road drains. The width of each side drain is about 2 meters—about 2.2 yards. A wide drain will allow drainage water to flow slowly and thus permit grass to grow in the drain, minimizing erosion.

Using earth from the side drains to create the raised road surface saves the effort of transporting fill material. If the earth from the side drains were not put on the road surface, it would probably end up near the drain, making a barrier for water to flow out of the drain onto the neighboring field.

DIVERSION BANKS

Long bumps, located diagonally across the road, should be used to divert water off the road. A typical bank is 30 cm—about 1 foot—above the road surface. The slope of the bank should be about 1 in 20, so the length of each side of the bank would be about 6 m—20 feet. These banks do slow vehicles, but they protect the road from erosion. A vehicle going 40 kph or 24 mph could negotiate such a bump without difficulty. In any case, it is best if the speed of vehicles along earth roads were slowed. A diversion bank is shown in Figure 25. Diversion banks should run into diversion drains to lead the water away from the road. Diversion banks should be located close to where water would tend

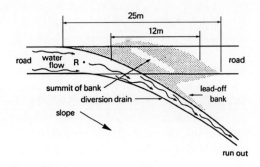

FIGURE 25
Diversionary bank. (From *Earth Roads.*)

to run along a road—for example, where a gully crosses the road or above a downward slope in the road. A footpath crossing a road will often bring water onto the road, so diversionary banks should be built along the footpath to divert the water before it reaches the road.

ROAD LOCATION

Roads should be located, if possible, along a watershed—the line of high land where water flows away on both sides. To find the watershed, work back uphill from the lowest point on the projected road. When working downhill, shoulders between small tributaries may be confused with the main watershed. When the watershed cannot be followed expect to construct diversionary banks and drains. Steep gradients should be avoided—1 in 12 is probably the maximum gradient tolerable; 1 in 18 is better if trucks are expected to use the road. Sharp turns should be avoided partly to improve visibility, partly because erosion is likely where vehicles turn.

Laying out a road requires exploration of the terrain to find the best use of the watershed, to minimize gradients, and to make as straight a road as possible. Once the general path is found, then specific points on the road should be marked and compass bearings between points made. In laying out a road between two points it is best to start in between the points and work back to either end, doing small sections alternately. In this way, the endpoints to which one is working will be closer than if one started at one end and had to make major corrections when reaching the other end.

CONSTRUCTION

The first step in constructing the road is to insert tall pegs along the line that will be the center of the road. These pegs should be about 1.8 meters high and be placed 15 m to 20 m apart. A 2.1-m-long stick should then be used to lay out the road boundaries. The stick is placed on the ground at each peg, perpendicular to the direction of the road, and the end of the stick marked with a peg. These pegs will indicate the inner boundary of the side drains. As shown in Figure 24, the road surface extends 1.8 m on either side of the centerline. Thus, 30 cm remains between the road surface and the drain.

All trees within 1 meter from the pegs indicating the edge of the side drains should be removed, including their stumps. When the stumps are removed—usually a major task requiring animal power or a tractor—as much of the side roots as possible should also be removed to reduce sprouting. Tree branches that extend to within this 1-meter distance should also be removed. Rocks in the road should be removed when possible. If it is not possible, then they should be broken down. Building a hot fire on top of the rock, removing the hot ashes, and then immediately dousing with cold water should split a big rock, making removal easier. Holes left by rocks and stumps should be filled in and bumps in the road leveled off.

Once the road surface has been formed, then narrow—20 cm, the width of a hoe—side drains should be dug out. The excavated sod is used to fill in low places on the road. Once these narrow side drains have been dug, then the entire road position can be checked to see if changes are necessary.

Next, the side drains can be widened to 80 cm, from the original 20 cm. The earth removed is used to build up the road surface. The center pegs can now be removed and vehicles driven along the road to compact the earth. Depression places will appear, some where the stumps were removed. High points should be leveled, with the soil going to the depressions. Now the width of the side drains is extended another 60 cm, building up the roadway with the earth, sod, and grass, and beginning to create the rounded crown. Some trees may occur in this second 60-cm strip, and they can be left standing, but bushes are discarded.

Some months later, after the road surface has consolidated, the drainage ditch is widened another 60 cm. After the months' use, depressions and bumps in the road surface will undoubtedly become evident. The soil from the widening is used to fill in the depressions, and the bumps are leveled. The resurfacing process is done in the several stages just described so the surface has time to consolidate.

When the road traverses damp areas, wider drainage ditches—3.3 m or so—may be necessary to keep the water level in the ditches 30 cm below the road level. If the area is very damp, the road surface may have to be built up higher.

When the road goes across the side of a hill so water flows directly across the road, then the uphill drainage ditch should be wide and the downhill ditch can be narrow. The height of the water in these two drains should be the same—see Figure 26.

Initial consolidation of the road surface can be done by tractor or jeep. A loaded rubber-tired cart pulled by cattle will also work. Some sort of scraper will help to level the surface. A long, rigid roller does not compact small depressions at all, but cattle hoof action and wheeled traffic work well. Even deep depressions in the surface can be eliminated by filling in, even leaving a slight mound at first, and periodically refilling.

Excessively large gradients are removed by filling in low places and removing soil from high places, as shown in Figure 27.

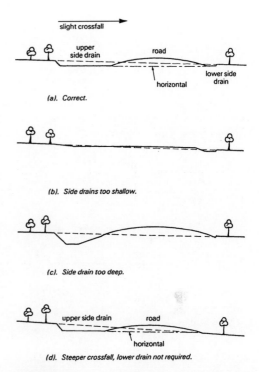

(a). Correct.

(b). Side drains too shallow.

(c). Side drain too deep.

(d). Steeper crossfall, lower drain not required.

FIGURE 26
Widening drains when water flows across. (From *Earth Roads.*)

MAINTENANCE

The basic task in maintenance is to ensure that every drain is working properly. Drainage ditches may silt up. Ridges may develop along the carriage way as vehicle wheels push soil away or improper grading moves soil from the center of the carriage way to the sides. Some soil will be lost every year as dust blows off the road in the dry season and mud flows off in the wet season. In general, low places are filled in, using earth from mounds that are blocking water flow from the road or earth from drainage ditches. Erosion in the area above the road that causes sand to be deposited in the drainage ditches should be attended to.

If roadwork must be done during the rainy season, use the driest soil that can be found. Allow soil taken from drainage ditches to dry before using it on roads. Bail standing water out of holes before refilling the holes.

REFERENCES

Barcomb, Joe, and David K. Blythe. (1986). *Understanding Low-Cost Road Building.* Technical Paper #45. VOLUNTEERS IN TECHNICAL ASSISTANCE, 1600 Wilson Boulevard, Suite 710, Arlington, Virginia 22209, USA. Tel.: 703-276-1800. Fax: 703-243-1865. vita@www.vita.org.

de Veen, J.J. (1980). *The Rural Access Roads Programme: Appropriate Technology in Kenya.* Geneva, Switzerland: International Labor Office.

Edmonds, G. A., and J. D. F. G. Howe. (1980). *Roads and Resources: Appropriate Technology in Road Construction in Developing Countries.* London: Intermediate Technology Development Group.

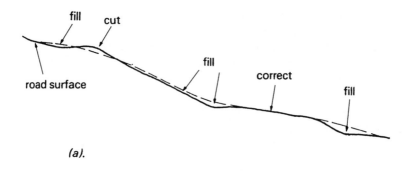

(a).

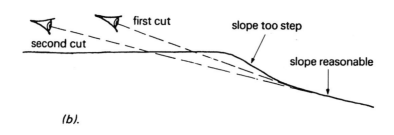

(b).

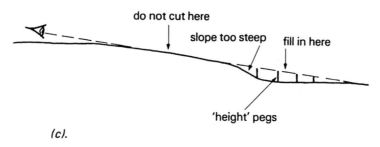

(c).

FIGURE 27
Reducing gradients. (From *Earth Roads.*)

International Labor Office. (1981). *Guide to Tools and Equipment for Labour-Based Road Construction.* Geneva, Switzerland: International Labor Office.

Jackson, Ian. (1955). *Handbook of Fundamentals of Low-Cost Road Construction.* Awgu, Nigeria: Community Development Training Center.

Weigle, Weldon K. (1960). *Designing Coal-Haul Roads for Good Drainage.* Berea, Kentucky: U.S. Forest Service, Experimental Station.

SMALL-SCALE BRIDGES

Bridges are needed, of course, to span rivers or gorges. In the United States they are commonly used to carry one road over another, but such applications are hardly small-scale technology. An alternative to a bridge for crossing a river is a ford—a hard surface under the water that can be negotiated on foot or by vehicles. Fords are discussed in another article in this chapter. Ropeways are a type of bridge used in mountain areas, and some bridges are constructed from arches. There are articles on ropeways and arches in the Construction chapter.

Wood, steel, and concrete are used for bridges. Concrete usually lasts much longer than wood, and it is cheaper than steel. Information on making concrete is provided in an article in the Construction chapter.

TYPES OF BRIDGES

The four basic types of freestanding bridges are beam, arch, truss, and suspension (see Figure 28). Pontoon bridges built on floats are used in some situations. Suspension bridges large enough to carry vehicles require sophisticated design, but simple suspension bridges to carry pedestrians across rivers and chasms are feasible projects. They are similar to ropeways.

A basic problem in bridge design is that a bridge must support its own weight. One cannot make a bridge stronger by simply making it from thicker material because the thicker material increases the weight.

When a board or concrete piece bends because of a load placed on its center, as in Figure 29, the top part of the board or slab is compressed and the bottom part is extended, so it is under

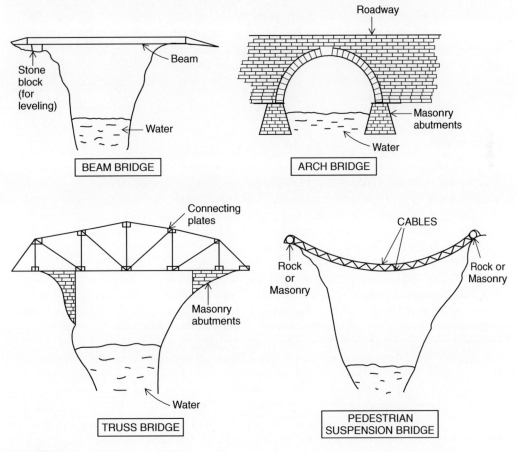

FIGURE 28
Four types of bridges.

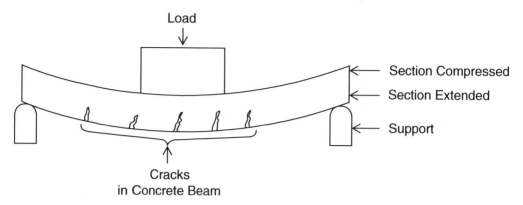

FIGURE 29
A bridge beam under load.

tension. Rocks or concrete pieces resist compression well but crack under tension. Steel beams behave differently; they resist tension well but tend to buckle when compressed. A common design solution, then, is to place steel reinforcing rods in concrete before it has set. In this way, the steel resists the tensile force and the concrete the compression force. It is most efficient to put the steel where one expects the tensile force to be—at the bottom of the concrete beam in Figure 29.

The strength of a reinforced beam can be increased further by "pre-stressing"—stretching the reinforcing rod before the concrete sets. When the concrete hardens, the stretched rod exerts a compressive force on the beam, so the beam remains under compression even when the load would create tension. Pre-stressing is probably too complicated for a small-scale builder, but the builder should know about it.

Beam Bridge

The beam bridge uses one or more beams crossing the river or obstacle. (A fallen tree across a stream is an example). Wooden beam bridges can cover spans up to 15 meters, concrete made from a flat slab up to 10 meters, concrete made with a beam on the undersurface up to 15 meters, and steel up to 25 meters. Construction is not difficult. Material—wood, concrete and reinforcing steel for concrete bridges, or structural steel—should be chosen appropriately. Beam bridges are usually the least costly for spans less than 12 meters.

A top view of a wooden beam bridges is shown in Figure 30. The dotted horizontal lines represent long wooden beams under the ropeway. The beams should be placed about 1 meter apart. The beams themselves are approximately .5 meter deep and .1 meter thick. The flooring is made from planks .05 meter thick. Diagonal bracing is used between beams. Railings are shown at each side of the bridge. People with ordinary carpentry skills and tools can build such a bridge, although a team would probably be needed to put the heavy beams in place.

Instead of wood, steel beams can be used, and these will probably last longer, but there must be regular inspections for corrosion and cracks. Steel beams should be 2 meters apart, and steel flooring about .02 meter thick. Wood or concrete flooring can also be used. If concrete is used, it is best to pour it in place, and a strong wooden form must be built to hold the concrete as it is curing. Steel beams come in standard sizes, and the supplier can suggest the appropriate size for a particular application—that is, for a particular length span and expected load.

Concrete is another material used for beam bridges, but reinforcing rods need to be placed in the beams. The beams should be about 1.2 meters apart, .2 meter deep, and about .05 meter

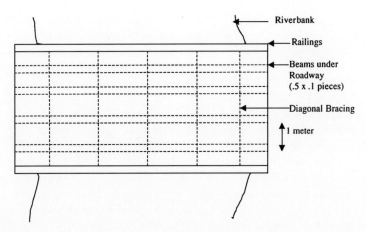

FIGURE 30
Wooden beam bridge.

thick. When the bridge is being constructed, wooden forms need to be built strong enough to hold the concrete until it sets. Care should be taken not to remove these forms too early. They should be allowed to set for at least a week.

Arch Bridge

Arch bridges are particularly suitable for building in stone because each of the stones in the arch will be under compression. Arch bridges are also built of concrete and steel. Small-scale arch bridges can cover spans up to 10 meters, although the ancient Romans built masonry arch bridges three times as long. The design requires some experience. If the arch is too flat, the lateral force on the sides will be large, and the arch will separate. (A simple vector diagram can be used to show this). If the arch is too high, the lateral force will be small, and the walls will be unstable. A semicircular shape is a good initial shape. A wooden framework is useful in building an arch bridge, since the structure is not stable until all the stones are in place. Construction of an arch bridge will be much easier with experienced stonemasons and with a quarry nearby.

Truss Bridge

The truss bridge is composed of both tension and compression members—the various members share the load. Many of the covered bridges of the world are truss bridges. Steel truss bridges are used for intermediate spans (12 to 30 meters). An advantage of truss bridges is that they require shorter pieces of steel than beam bridges. Of course, skill is required in cutting and fastening the steel. Riveting has been a common fastening technique, but welding may be more convenient these days. Different configurations of the truss members are possible; a common one is shown in Figure 28.

Suspension Bridge

A suspension bridge is built around long cables extending over the river or gorge. The bridge consists of these cables plus a walkway or roadway. In most small-scale suspension bridges the walkway is fastened directly to the cables. Side rails, using some of the cables, are also needed. In the pedestrian suspension bridge shown in Figure 28 the walkway could be boards across the bottom cables; the upper cables would form the side rails. These bridges can be used to span longer distances—50 meters or so—than other small-scale bridges. A problem is that they tend to sway in the wind and to shake when being crossed. A simple suspension bridge is entirely appropriate for pedestrians—although possibly frightening in a serious wind. A suspension bridge suitable for vehicles is a fairly elaborate structure as a framework to keep the driveway stable must be included.

The design is straightforward. The cables should dip down—sag—as shown in Figure 28. The tension on nearly horizontal cables would be excessive—again, a vector diagram can show the forces. A reasonable amount of sag is 1:6—the span is 6 times the sag. Deeper sags require more cable; shallower sags create undue force on the cable. If the sag puts the walkway too close to the river, then the cables should be raised on pillars. Large suspension bridges in the United States and Europe have tall towers so the roadway, suspended from the cables, is at ground level at either end. The number of cables needed depends on the cable strength, the span, and the load expected; the cable supplier should have specifications that can be used to estimate the number of cables. A generous safety factor is appropriate. At least six cables should be used. The cables must be strongly anchored at both ends. Masonry structures are often used as anchors. If trees are chosen, they should be periodically examined to ensure their firmness. Metal chains or heavy ropes can be used instead of steel cables. The construction process starts by attaching a string to a rock and throwing it across the gap. The string is used to pull a rope across, and the rope can then pull the cable. An alternative to throwing is flying a kite across.

Pontoon Bridge

A common pontoon bridge consists of a roadway supported by floats. A pontoon bridge is similar to the floats used in swimming areas in lakes. The floats may be sealed empty barrels, inflated animal skins, or bags of polystyrene packing—anything lighter than water. These floats can be anchored to the bottom of the river or fastened only to the roadway. It is probably easiest to build a large bridge out of several structures and fasten them together than to build one big, unwieldy structure. A pontoon bridge can tip and sway as people and, especially, vehicles cross it. Design of a stable pontoon bridge to carry vehicles is a major project, but one for pedestrian traffic is certainly feasible. Pontoon bridges prevent river vessels from going past, and such bridges will probably be destroyed in a flood—unless brought to shore in time.

An alternative to a pontoon bridge is a barge and a cable extending across the river. The barge travels along the cable so it does not need to be steered and is not pushed downstream by the current. Usually a second cable is used for propulsion; the people on the barge pull the cable to move the barge across the river. Handles that ratchet along the cable make the pulling easier. More elaborate systems winch the barge along this second cable. When the barge is not moving, the cables rest on the river bottom so river traffic can pass. A sufficiently sturdy barge can carry an automobile or small truck, certainly an animal drawn cart.

BRIDGE DIMENSIONS

A bridge used for pedestrians will need a width of 2 or 3 meters. Vehicles, however, require at least one lane of 3 to 4 meters, plus an additional width for pedestrians. If both pedestrians and vehicles will use the bridge, a raised sidewalk or curbing should be used.

A bridge crossing a river should be above the flood level. The flood level can usually be determined by examining the riverbank and asking local people the highest water they have observed. The bridge must also provide an adequate underclearance for boat passage.

MAINTENANCE

Wood structures require regular application of wood preservative. Steel structures require periodic painting. Concrete structures require patching of flaked or chipped areas with cement grout. Worn roadway surfaces should be given a suitable wearing and paving coat for protection. Cracks should be sealed with a commercial sealer.

PIERS AND ABUTMENTS

Piers are the intermediate supports for multispan structures, and abutments are the end supports. For example, in the arch bridge in Figure 28, the abutment would be a pier if the bridge consisted of several arches. Piers and abutments are supported by either spread footings—masonry blocks with a wider area than the pier—or piles—poles driven into the river bottom. The important forces in designing the foundation are the vertical loads from the superstructure and the ice pressures in deep rivers of cold-climate areas. If ice pressure is not a concern, the pier is a compression member and can be built of masonry or unreinforced concrete.

Maintenance of substructure units consists of patching of concrete or masonry. Major maintenance is needed only if erosion undermines abutments or piers. In this case, filling in the eroded area and placing rock protection to prevent further erosion are required. Substructure units should be inspected yearly for erosion damage or immediately after unusual runoff.

REFERENCES

Cobb, Newton H., and Andrew Zanella. (1993). *Discovery, Innovations, and Risk.* Cambridge, MA: The MIT Press.

Commins, Robert J. (1990). "Understanding Small-Scale Bridge Building." VITA, 1815 North Lynn Street, Suite 200, Arlington, VA 22209.

GPO. (1972). *Small Footbridges: Design and Construction.* London.

Comp, T. Allan, and Donald Jackson. (1977). *Bridge Truss Types.* Nashville, TN: American Association for State and Local History.

EMERGENCY RIVER CROSSINGS: CULVERTS, FORDS, AND DRIFTS

Based on ATBrief No. 30: "Emergency River Crossings: Culverts, Fords, and Drifts, Appropriate Technology, Vol. 26, No.3, December 1999

In the aftermath of a disaster, whether flood, earthquake, hurricane, or civil war, engineers are often called on to provide emergency infrastructure to service large numbers of people. One common problem is enabling supply vehicles, often large trucks, to reach the people affected. This article deals with methods of crossing gaps or obstacles such as drainage channels, streams, or rivers. It includes the simplest methods—culverts, drifts, and fords—rather than more complex bridge building.

The choice of crossing will depend on the time, skills, materials, and equipment available for construction and the time period for which a lack of access can be tolerated. Culverts can carry a natural watercourse or convey water under the road from a side ditch on the uphill side of the road. Culverts require materials of sufficient strength to support the road above, and they can take time to put in place. In an emergency, fords, drifts, and simple dips in the road are often a better first-stage option. Culverts can be constructed at a later date if necessary. Fords, drifts, and dips (submersible crossings) are often adequate where only short floods are expected.

In selecting a crossing over a stream or river, a critical parameter is the level to which the water will rise at peak flow. This is related to the discharge of water and to the profile of the stream or riverbed in the vicinity of the crossing. The discharge will depend on the catchment area, the intensity and duration of the rainfall, the nature of the soils, and the terrain. The most reliable method of estimating high water levels is to look for indicators of flood levels. Look at the riverbanks, note changes in vegetation, and ask local people what the patterns are.

CULVERTS

Culverts can be made of timber, concrete, or steel pipes. If timber is available, a culvert can be made from sawn logs, as shown in Figure 31.

Corrugated steel culverts are lighter than concrete and therefore easier to handle and to transport to the site. The choice of materials will probably be decided by availability. A typical pipe culvert installation is shown in Figure 32.

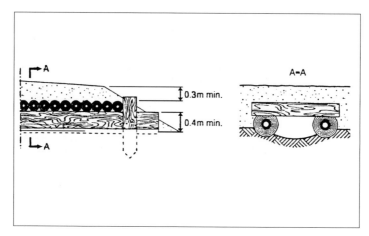

FIGURE 31
A log culvert.

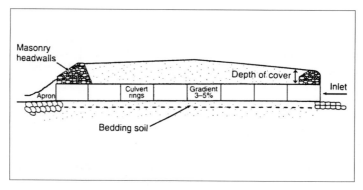

FIGURE 32
A typical pipe culvert (from Antoniou et al. 1990).

TABLE 2
Depth of cover for culvert pipes

Type of Pipe	Diameter: mm	Depth of Cover: mm
Spun concrete	600	300
	900	450
Concrete cast on site	600	450
	900	700
Corrugated steel	600	300

Source: Antoniou et al. 1990.

Ensure that culverts have sufficient soil cover to spread the traffic load (see Table 2). Ensure that erosion near the culvert is controlled, particularly on the discharge side.

FORDS AND DRIFTS

There is little time in an emergency to construct anything but the simplest of bridges. Even pre-fabricated modular bridges take time to purchase, transport to site, and erect. An alternative to a bridge is a submersible crossing. There are two basic types.

- Bed-level fords and drifts
- Vented fords and submersible bridges raised above bed level

Submersible crossings are generally simpler and quicker to construct than bridges and are often adequate for small crossing points that experience intermittent, short-term flooding. Always consider a submersible crossing before opting for a bridge.

The siting of a submersible crossing differs from the siting of a bridge, since a wide crossing often provides slower, shallower water and a gentler approach. If the banks are steep (more than 10 percent), reduce the gradient by approaching at an angle. Try to ensure that, on the approach, drivers have a clear view of the crossing.

Where crossings disrupt the natural flow of the river, scour may be induced, which, if left uncontrolled, can rapidly damage the crossing. Take appropriate measures to control scouring.

FORDS

Fords, as defined here, are unpaved crossings. The simplest ford is a crossing at a shallow river section over a naturally supportive streambed. A natural ford can be improved by reducing the approach gradient and strengthening the crossing's surface.

Large stones placed across the ford to retain a gravel surface suffer two problems. Scour may occur downstream if the stones are too large or placed too high. Conversely, if the stones are not large enough, they will be swept away under flood conditions.

Figure 33 shows three methods of retaining natural river gravel, or imported fill, and a method of scour prevention. To ford a stream using gabions (Figure 33a), place a line of gabion baskets in a trench, .2–.3 m deep, dug across the stream so that they protrude no more than .3 m above the level of the streambed. Fill the middle basket to act as an anchor and tension the baskets with a rope before filling, as shown in Figure 34. (Gabions are discussed later in this article.)

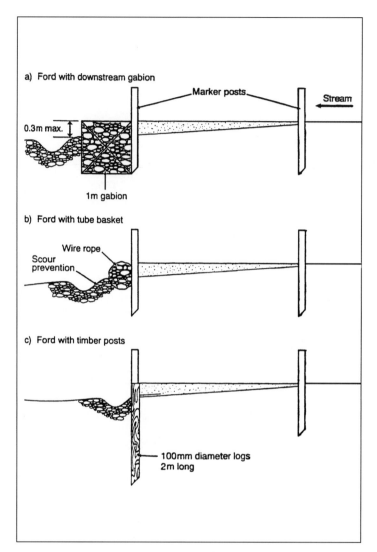

FIGURE 33
Three methods of improving fords.

If gabions are not available, make a wire tube by rolling up chain link fencing mesh (Figure 33b). After filling, tie the tube at the top and tension with a wire rope anchored at each end.

Figure 33c shows how timber piles can be used if suitable timber is available and piles can be driven into the streambed.

Drifts

Drifts, as defined here, are paved fords. A drift may also be known as a bed-level causeway, paved dip, or ford. Figure 35 shows a number of alternative designs. Curtain walls and aprons protect the vehicle running surface from scour. If you are using concrete, remember that it requires time to gain strength and, being inflexible, it is liable to cracking. Gabions provide an

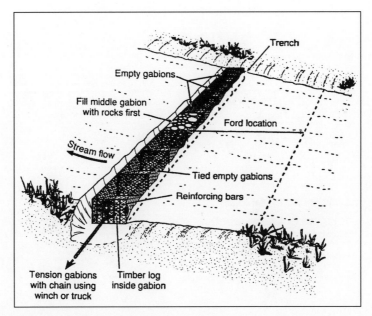

FIGURE 34
Placing gabions to improve a ford.

instant and flexible structure and may be appropriate for hurriedly constructed crossings or for sandy crossings subject to intense flash flooding.

Vented Fords

A vented ford (also known as a vented causeway) raises the vehicle running surface above the level of the streambed. It is passable in normal flood conditions but submerged and impassable under high flood conditions. The resistance offered by a vented ford to the natural flow of a stream or river means that it must be built to withstand both the water pressure and the impact of debris carried in the flow. Construction takes considerably longer than a simple drift and should be carried out during periods of low stream flow. A vented ford may be an appropriate alternative to a bridge where access for food and relief supplies can be improved in the dry season in anticipation of the wet season.

Figure 36 shows a temporary vented earth ford. Protect the earth fill from erosion with stone and/or gabions. If time and materials are available, a more permanent structure can be constructed in concrete or cemented masonry. The vents are standard concrete or corrugated steel culvert pipes. Pipes spaced too far apart can induce water flow parallel to the road, which could result in scour. Set the vents level with the streambed and at the same gradient to avoid siltation.

GABIONS AND MATTRESSES

Gabions and mattresses (also called Reno mattresses) are boxes formed from hexagonal woven steel mesh. They are placed in position and filled with stone. Adjacent units are tied together with wire lacing to give a stable but flexible wall or base structure. Typical examples are shown in Figure 37.

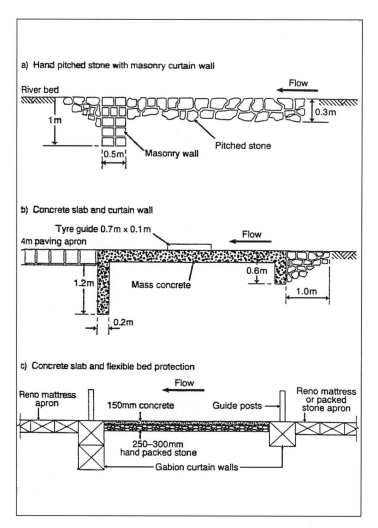

FIGURE 35
Alternative drift designs.

The advantages of using gabions are the following.

- Flexibility—a gabion structure can tolerate movement due to settlement, floodwaters, and shifting streambeds.
- It requires simple foundation preparation, since the ground only needs to be level.
- Placement by hand rather than machine is possible, although a machine may be useful.
- Construction with gabions is suited to an unskilled workforce.
- They are convenient to transport because gabions are supplied folded flat in compressed bundles and are opened on site.
- Local materials can be used. Gabions are filled with stone that can be obtained locally in many cases.

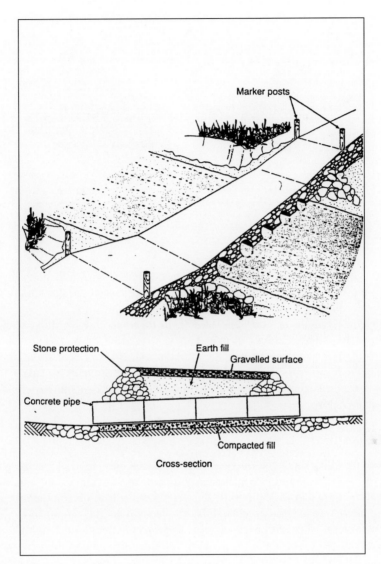

FIGURE 36
A vented earth ford.

A gabion structure is permeable although, if required, it can be sealed with a mix of suitable clays and plastic sheeting to form a low impermeable wall or dam.

Installation

Foundations: The basic requirement for a foundation is that it is reasonably flat. Flat stones can be put in place if this makes working easier.

Assembly: Unfold, stretch out, and flatten each gabion on a hard, flat surface. Fold the sides to make a box. Boxes and mattresses may have internal diaphragm walls, which are also folded vertically. The edges are strengthened with thicker selvedge wire that protrudes from each side.

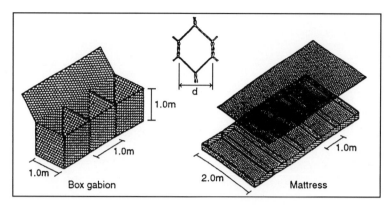

FIGURE 37
A box gabion and mattress.

Use the selvedge wire to tie the corners. Tie galvanized binding wire at the top corners and with a continuous piece of loops about every 100 mm. A pair of 8-inch (20 mm) long-nose pliers is recommended for binding.

Place empty gabions in position and lace them together. Before filling with stone it is important to stretch empty gabions in line. This can be achieved by filling a central or end gabion to provide an anchor against which to pull the rest. Use pieces of strong timber or steel rods to spread the load across the central or end gabion when stretching, using a "Tirfor" type of winch or a team of laborers. Multilayered gabion structures should be wired to adjacent layers above and below.

Filling with Stone

The stone used for filling should be hard and durable. Woven mesh is sized according to the width of opening "d" shown in Figure 37. Standard mesh sizes are 80 mm and 60 mm. Stone should be graded between 100 mm to 150 mm. If this size is difficult to obtain, then the core can be filled with smaller-sized material, provided that the facing material can retain it.

Bracing wires must be fixed to exposed faces to prevent bulging. Fill 1-m-high gabions one-third full and then place bracing wires between the sides. Continue filling and place another row of bracing wire when two-thirds full. Half-meter-high gabions can be braced at the midpoint, if required. Overfill each gabion by 30 mm to allow for settlement. Fold the lid down and lever into position for lacing to the front, sides, and diaphragm walls.

The installation procedure is similar for mattresses. On slopes, mattresses may need to be pegged into position during filling.

Standard Sizes

Gabions and mattresses come in a variety of unit sizes, mesh sizes, wire diameter, and numbers of partitions (diaphragm walls). See Table 3. The following are typical.

Mesh sizes: 60 mm and 80 mm
Wire diameter: 2.5 mm, 2.7 mm, 3.0 mm

Note: Wire can be coated in PVC to provide improved resistance to corrosion.

TABLE 3
Typical Gabion and Mattress Sizes

Box Gabions		Mattresses	
Unit Size (m)	No. of Partitions	Unit Size (m)	No. of Partitions
1 × 1 × 1	0	3 × 2 × .5	2
1.5 × 1 × 1	0	6 × 2 × .5	5
2 × 1 × .5	1	3 × 2 × .3	2
2 × 1 × 1	1	6 × 2 × .3	5
3 × 1 × .5	2	3 × 2 × .17	2
3 × 1 × 1	2	6 × 2 × .17	5

Bridges

For larger rivers, a prefabricated modular bridge (such as a Bailey bridge) or a temporary floating causeway may be appropriate. To erect these, an engineer without significant bridge experience should, if at all possible, request outside help. This can often be provided by phone and fax. If such help is required, contact RedR (see address in the References) or the supplier of the bridging system. If a major bridge has to be constructed, a bridge specialist should be called in.

REFERENCES

Antoniou, J. et al. (1980). *Building Roads by Hand: An Introduction to Labour-Based Road Construction*. London: Longmans/ ILO.

Beenhaker, H. L. et al. (1987). *Rural Transport Services: A Guide To Planning And Implementation*. London: IT Publications.

Davis, Jan, and Robert Lambert. (1995). *Engineering in Emergencies: A Practical Guide for Relief Workers*. London: IT Publications.

TRRL. 1992. *A Design Manual for Small Bridges, Overseas Road Note 9*. Crowthorne: TRRL.

RedR can be contacted at 1-7 Great George Street, London, SW1P 3AA, UK. Fax: +44 02072220564. E-mail: <bobby@redr cemon.co.uk>

ROPEWAYS IN THE HINDU KUSH-HIMALAYAS: HELPING TO SOLVE THE PROBLEMS

David Sowerwine and Greta Rana

THE PROBLEMS

In a world in which you have to cover a vast, diverse terrain of almost insurmountable altitudes and in which the only way to get from here to there is by walking, river gorges and very difficult ground conditions make travel slow and hazardous and, when transporting goods, extremely costly. It also means that it is very difficult to get to a hospital. The school across the river may look close by, but it is really half a day's walk away, and in case of disaster, there is no quick way round the river. In Nepal, during Tij, the principal festival for women in the Hindu calender, eight women returning home across the river from a celebration held in honor of their husbands lost their lives while crossing. Something that seems normal commerce in the level world is life threatening in the Hindu Kush-Himalayas. Transport in mountain areas is all about overcoming obstacles and lessening the risks posed by hazardous terrain.

Ropeways are not a new thing. Mountain dwellers have used some kind of means to swing across rivers for several generations. Yet, these have never been without their problems. Before we discuss the types of ropeway used and their problems, environmentally and otherwise, let us examine the developement responses to inaccessibility in the mountains and their drawbacks and plus points.

The standard response to inaccessibility in general has been to build roads. An extremely costly undertaking, high maintenance costs, and—in terms of positive impact—benefits seem to accrue unevenly to those who have access to transport-to wheels. A road doesn't really solve the problem of the poorest of the poor, the mountain porters, and those who cannot afford bus fare to get their goods to market. To some extent, feet are free goods; they come with the human package. Wheels have to be acquired—not without great expense, in grass-root terms. Roads, too, tend to bypass villages that once thrived because of the footpaths. Roads bypass many people who make their living carrying goods on their backs. Roads should never be built without attention being paid to every aspect of the impacts they might have, and that means negative impacts, too. Mountain terrain means that problems are different and greater than those encountered in the plains. Unfortunately, this rarely happens. Yet, roads have the possibility of linking major towns in the mountains, although not without great expense. They often are poorly maintained because they are not commercially viable to the whole community.

When it comes to crossing rivers, the Swiss Development Co-operation and Helvetas have been successful in building suspension and suspended bridges. These provide access to goods and people. The costs, however, compared to ropeways are high (a ropeway costs 20 percent of the cost of these bridges). They are excellent but need subsidies. The Swiss subsidize by providing building materials and training to build bridges locally through self-help. In the long run, the fact that subsidies are needed could render them unsustainable. These bridges are not as environmentally threatening as roads, particularly roads that are built without taking mountain characteristics into consideration. Some roads have literally been blasted through the mountains, destabilizing an already unstable terrain.

There are three kinds of ropeway in Nepal: The first, of which there is only one, is for transporting passengers. It is a powered ropeway built from the highway to Manakamana, a popular pilgrimage site. This ropeway was very expensive and it is impractical. The local community has misgivings about it, too, because it deprives them of wages earned through portering for pilgrims. The second type is a long-distance monorail; its Nepali name means "a wire road." The design is based on an ancient system of transportation used in banana plantations, and there are about half a million kilometers of wire road worldwide. Nobody translated it into use for people, and this is an important plus point for the ones being constructed in Nepal. The wire road is environmentally positive, since it doesn't involve using the earth as you go over the top. For goods only there are four types of wire road: namely, a simple wire cable (*gheling*) to transport goods from one side of a river to the other. The fact is they have often been used to transport wood from the forests and have, thus, accelerated deforestation. The fact that people are tempted to use them and that they are not built for people adds danger to their bad impact. The second type of ropeway for goods in Nepal is that from Hetauda to Kathmandu, built over 40 years ago with U.S. assistance. Commercially it could have run well, but the government took it over and it eventually fell into disrepair. The third ropeway is one built by the European Union (EU) just outside Kathmandu at a place called Batidanda. It was intended for transporting milk from the hills. The reasoning behind it was that by transporting

FIGURE 38
Ropeways.

milk people would not have to cut wood to boil the milk on the spot (traditionally the farmers had boiled the milk to make *panir*, a kind of curd cake, in order to preserve the product over the distance to Kathmandu). The problem with this one was that the people could not ride on it, so they lobbied for a road to run alongside it. So the road was built at great expense.

People want to travel and donors see people's travel as a luxury, not a need, so they ignore it. Ignoring felt needs is the swiftest way to failure. Another example of people who wanted to travel comes from Barphak in Gorkha district. Barphak is renowned for good local community leadership. They wanted a ropeway, and they had electricity generated from a locally built and run mini-hydro plant. They approached the Swiss, who offered them a powered ropeway built by the Swiss army. The British government provided the money to purchase this ropeway. The ropeway was built to carry goods, but people should only have been carried on it in medical emergencies. Instead, people rode it all the time. There were no inbuilt safety factors for this type of use. The end result was that the ropeway only lasted a year; it wasn't maintained properly (lubrication of bearings and cables). During a storm, four people were on it when the bottom slid into the river and the whole structure was lost. Finally, the one category specifically for getting across rivers, just from here to there, is the traditional *gheling*. It consists of a large twisted cable tied to a tree on both sides, a pulley with a bucket, and perhaps a separate rope with which you can inch your way across. The passenger has to pull. It is extremely dangerous and, if one does not fall into the river or even if the cable is maintained at full strength, the system of pulling oneself across the river can lead, occasionally, to missing fingers.

Of the two ropeways that carry passengers—the Manakamana system and the wire roads now being built by Ecosystems' Private Ltd. (see Figure 38)—the Manakamana system is economically unfeasible for carrying the amounts of goods available from individual farmers footing the bill as individuals. Farmers have to compete with tourists, many of them foreign, and this they cannot do. For the Ecosystem's model local communities have to provide half of the costs of materials and services and supply the labor. NGOs and other organizations often volunteer to pay the other half, especially if the wire road is helping a community transport commercially viable goods. Ecosystems gives training for maintenance and carries out the monitoring, evaluation, and overhaul; this is all comprehended in the initial cost. In use of their model, Ecosystems is trying to address the problems that struck the other types of ropeway, to ensure that their ropeways satisfy a need, and that there is a body in place that has the incentive to make sure the system works and that the cost is affordable by the community.

To ensure all of these, one has to spend time getting to know the users and to come up with the solutions technically, financially, and operationally. Only if there is sufficient incentive to maintain a wire road—economic returns—will they do the work intended. We cite these examples only to point out that when people with good intentions come to help, they should be aware of certain things. How to construct a decent ropeway for the mountains is really not something you learn from an article like this one or through the Internet. You learn it by being on the spot, examining the terrain, and watching what works. The one thing that is certain is that the mountain community must have a vested interest in the ropeway. It must mean something to them in terms of earning from their produce and freedom of movement. Both give them mobility: Earnings give them economic mobility and life choices and physical movement social mobility, a physical freedom they have not had. You need to know how the ropeway will be used, how it will be maintained, who will pay for it, what problem is it intended to solve, and will the ropeway solve that problem.

Ropeway are not the only venture in the Hindu Kush-Himalayas. Other INGOs are looking at appropriate technologies. An American INGO, International Development Enterprises (IDE),is investigating appropriate technological developement for water pumps, treadle pumps, and drip irrigation by indentifying areas of need and putting people in position to transfer technologies. The Intermediate Technology Group U.K. also works in the area of transferring technologies suitable for a mountain terrain. One lesson learned from the Nepal Himalayas, however, is that using the market mechanism brings sound returns on the investment on technologies. The mountain dwellers of Nepal will work hard to make ropeways work and keep on working if they can see the benefits translated into improved living standards.

CHAPTER 7

Household Technologies

Edited by BARRETT HAZELTINE
Brown University

HOUSEHOLD TECHNOLOGIES

Many useful items in everyday life can be made at home. In some cases these same articles—baskets are a particular example—can be sold either in local markets or to wholesalers who will sell them in urban areas or even export them. The line between objects made for the home and those made for distant markets is indistinct. When access to city shops—especially shops catering to visitors—is possible, many domestic articles can be sold for cash. The proverbial person in the Land Rover who visits a village every month or so, collecting baskets and curios, can bring appreciable cash into a village. Whatever the distribution channel, it should be able to transmit suggestions to village artisans about demand for particular products. The purpose of this section is to review the technology of making useful household items so the volunteer can evaluate the potential of establishing a project. It is probably essential for a volunteer to start by first learning what has gone on in the past—what expertise is available in the village. It may be very difficult to transfer an entirely new technology into a village, but it is feasible to nurture knowledge that has waned or is presently overlooked.

This section, in general, focuses on utilitarian objects. The first few articles deal with textiles: spinning, weaving, and dying. In many countries imported cloth and garments are very cheap. Also, in many countries textile factories can make high-quality fabrics inexpensively, so village-made cloth finds only specialized markets—often high-quality material for ceremonial use. Inexpensive fabric, however, gives an opportunity for handmade clothing to be competitive in price. Such clothing is often made on foot-powered treadle sewing machines. Such machines last a long time but do need to be serviced, as explained in an article. The next several articles deal with making simple items: soap, candles, and chalk. Pottery making, discussed in one article, is an important and traditional village craft. Basket making, another article, has an equally long tradition, and baskets are attractive to tourists. Useful objects—chairs, toys, baskets, and

FIGURE 1
Kerosene lamps.

so forth—can be made from scrap paper and starch (maize) paste. The technology has been effective in Zimbabwe and other countries and is worth considering. Ropes and knots are an integral part of many devices and industries. The next three articles deal with somewhat less traditional technologies: reclaiming rubber (partly to make sandals), paint production, and paper making.

Useful items are made by local artisans other than those discussed in this section. Kerosene (called "paraffin" in parts of the world) lamps are an alternative to candles. The simplest lamp is a wick in a bowl. A craftsperson—called a "fundi" in parts of Africa—can do much more. Two lamps are shown in Figure 1: one in the shape of an airplane, the other made from a discarded lightbulb. Buckets, cooking pots, dustpans, and similar items made from scrap sheet metal are sold in many Third World markets. These can be made with hand tools or something more elaborate (see the Tools chapter). Blades for axes, hoes, and adzes, made by local blacksmiths, are also common in such markets. Wooden serving and eating dishes are another common product. Enameled plates and bowls, commercially made, are inexpensive and widely available, so wooden pieces compete because of aesthetic qualities. A simple wooden lathe, as shown in Figure 2, can be used to produce attractive turned bowls, as well as other items. Woven articles, such as fish traps, mousetraps, and hot pads, are also seen frequently. The "Basketry" article discusses another woven item—mats. High-fashion, tailored clothing, based on traditional designs, are often desired by foreign visitors but often the problem is gaining access to the market. Many tourists will purchase handmade musical instruments: thumb pianos, xylophones, drums, whistles, and similar objects. A final handmade item is beadwork.

FIGURE 2
Handmade lathe and turned bowls. (From *Inventors of Zambia*.)

SPINNING AND WEAVING

Kylee Hitz

Spinning and weaving are intimately related to each other in the production of both utilitarian and decorative textiles. However, while spinning and weaving are often processes utilized hand-in-hand with each other, they are indeed two very different technologies, each requiring very different skills and equipment. Spinning is the process of taking a raw fiber, such as wool, flax, or cotton, and drawing it out into a yarn by introducing a twist using either a spinning wheel or a hand spindle. The spun yarn is the final product of spinning and is ready to be used in a variety of textile techniques. Some of these include knitting, crocheting, netting, braiding, plaiting, bobbin lace-making, and of course, weaving. Weaving involves taking the spun yarns and organizing them to interlace at right angles in some sort of over-under way. Many times this is accomplished through the use of a loom, which has the primary purpose of keeping one set of threads taut and parallel. The loom also aids in selecting these threads to create a desired weave structure and pattern. The transformation of raw fiber into a piece of cloth or a work of art requires many time-consuming and labor-intensive steps before the final product can be used and enjoyed. However, once the techniques have been mastered, they can be applied to a seemingly endless variety of products. These products can include woven carriers for food or equipment, blankets, shelter, and, of course, body coverings.

SPINNING

The spinning of fibers into yarn can be done both industrially and by hand. Commercially spun yarns are readily available and used widely in different methods of production and are, in many cases, more economical than hand spinning. However, the focus here will be on the hand spinning of yarns, since it can offer several advantages over commercially spun yarn, depending on the situation. One thing is that the basics of spinning are relatively easy to learn, particularly if taught on a hand spindle (otherwise known as a drop spindle). Learning to use a spinning wheel requires a bit more persistence in coordinating the hand/foot motions of the treadle wheel, but once mastered, the rhythm of spinning becomes second nature. Another advantage hand spinning has is that the equipment required is small, light, and relatively portable, depending on the design of the spinning wheel. This allows for spinning practically anytime, anyplace.

Of course, one of the strongest arguments for hand spinning is the fact that one can create a yarn ideally suited to a particular end-use. The design of a yarn can occur at several stages during the construction process. Before spinning is even begun, certain design decisions are implied with the choice of fibers, the choice of a spinning system, and the degree of fiber preparation. Further design elements are controlled by the actual spinning—the weight of the yarn, the amount of twist, and the presence or absence of slubs or other novelty effects. And finally the yarn design continues after a single yarn has been spun, since a variety of effects can be achieved by plying two or more yarns together.

Before the actual spinning of the fibers can occur, the fibers must be prepared for spinning. The basic processes in fiber preparation are sorting, teasing, scouring, carding, combing, and forming the rolag or roving. Some of these processes are not applicable to certain fibers or in some cases may be omitted to achieve a particular effect.

Sorting the fibers is the first step and is necessary to determine their quality, length, and degree of cleanliness. This is particularly important for wool fibers purchased in fleece form, since an entire fleece will have natural variations in the length of the fibers, as well as the coarseness and fineness of the fibers. Also, parts of the fleece may be badly stained and therefore cannot be used. Certain yarns will require certain lengths and qualities. For example, a worsted yarn requires long fibers. Sorting is best done on a wire mesh or screen, which allows any coarser dirt or debris to fall through and away from the fibers.

The next step is called *teasing*, which should be done to natural fibers containing dirt or foreign material. It is accomplished through a pulling motion of the hands to open up tight fiber masses. Teasing also removes dirt particles. Ultimately teasing results in well-blended and evenly distributed fibers. Teasing should be done over a clean surface, such as paper, to allow easy notice of dirt particles. To begin teasing, take a small handful of fibers from the mass and begin to separate them using a pulling motion. Gradually transfer the fibers from one hand to the other and work the teased fibers into a blended mass, not a pile of individual handfuls of teased fibers. The end result of teasing is fibers that are cleaner, fluffier, and more airy than the original mass. While some fiber masses may require a tough pulling action, take care not to pull so hard that you damage or break the fibers.

Scouring is a cleaning process done to wool fibers to remove the natural oils as well as any other dirt still present after teasing. This step is sometimes omitted, and the wool is spun "in the grease." Scouring starts by dissolving a commercial wool-scouring solution or mild soap into warm water (up to 120°F) in a large scouring vat. The wool is placed in the bath in layers and is gently pressed to circulate it in the scouring bath. Do not agitate the fibers too much, or the fibers may start to felt together. After the oil and dirt seem to have been loosened (this may require more than one bath vat), the wool is lifted out of the bath and then rinsed thoroughly in water of the same temperature, preferably by spraying from above. The fibers are then squeezed gently to extract any excess moisture (this can also be accomplished by running them through a spin cycle of a washing machine) and then either dried in the air or in the cool cycle of an automatic dryer (but not the very long fibers that will tangle in the dryer). Before carding and spinning can occur, the fibers must be lubricated to restore some of the oil lost in the scouring process. You may either buy special oils sold for this purpose or use mineral oil diluted with water. Treat the fibers at least 24 hours before spinning so that the oil emulsion can penetrate the fibers uniformly.

After sorting, teasing, and scouring the fibers, it is time for *carding*. Carding opens up the fibers to a greater degree and partially aligns the fibers for spinning. Hand carding is accomplished using a pair of paddle-shaped wire brushes called *cards*. Cards are rated by fineness: the finer the card, the thinner the wires in the mesh and the tighter the mesh. For lighter wools, use a finer card; for heavier wools, use a coarser card. To start carding, pull the fleece across a card so that the fibers catch on the teeth, spreading the fleece evenly across the card. This is called *charging* the card. Next, pull the upper card gently across the bottom card filled with fleece. Do not let the wires touch, but let some of the fibers transfer to the top card. Repeat this several times until no more fiber will transfer easily. Now transfer the fleece from the top card to the bottom card by first reversing the top card and then pulling it across the bottom one. For this operation, the wires on the cards will actually mesh. This is called *stripping*. Stripping returns the cards to the condition of one full card and one empty card without disturbing the fiber arrangement. Continue to card until the fibers are evenly distributed and fluffy. To remove the carded fleece, first transfer it to the bottom card and then pull the top card across. If fibers are not to be combed, roll them between your palms to form a rolag (long roll of evenly distributed fibers) for spinning.

The same principles apply to machine carding. The carding machine, or drum carder, aligns the fibers by gradually transferring them from the small roll to the large drum. It is faster and more automatic than hand carding. The machine must be carefully adjusted and needs frequent lubrication. To begin carding, the fibers are spread evenly across the carding pan in such a way that the small roll can catch them. Turn the crank, and the majority of the fibers will be gradually transferred from the small roll to the large drum. Carding is finished when the fibers have been blended into a soft, evenly distributed layer across the drum. To remove the carded fiber web, insert a dowel or stiff wire in the space under the web to loosen it. The web may be torn at this point, but do not cut it with scissors. When one area has been loosened, rotate the drum backward and peel off the rest of the web from the teeth. Make the carded web into roving by tearing it into one long, narrow strip.

Combing is only done to fibers that are to be spun in the worsted system. It creates more uniform parallel fibers and helps to remove any short fibers remaining after carding. Attach one of the two combs to a post or other fixed support and charge it by stroking fibers across the tines. Draw the second comb repeatedly across the first, moving it closer and closer to the fixed comb until no more fiber will transfer easily. Remove the fiber remaining on the first comb and set it aside. Mount the full comb on the support, grasp the outermost ends of the fibers in one hand, and draw them out gradually in a thick, uniform strand.

Now the fibers are ready to form into a rolag or a roving. These are continuous, compact strands of fibers and are formed two different ways: A rolag is formed from hand carded fibers by simply rolling the fibers across a flat surface; a roving is made from machine carded fibers by drawing them through the hands under tension, resulting in a longer and narrower roving. A rolag or roving 3/4-inch thick will be a good starting size for a novice spinner.

USING A DROP SPINDLE

The drop spindle, or hand spindle, is a very simple tool. To spin with it, first tie a scrap of yarn to the base of the spindle shaft, wrap it around the base knob, then up and around the shaft tip. Fan out the ends of the leader and rolag/roving and overlap them about 2 inches. Hold the ends between thumb and forefinger of one hand as you start the spindle turning with the other hand. As the two ends start to twist together, shift the lower hand to hold the juncture and start to draw out a small amount of fiber. When the two ends are joined, move the lower hand up again to pinch the fiber just below the top hand. Now the twist can move further up the rolag but not beyond the pinch. Keep pinching until the fibers below are spun as tight as you wish. As you pinch, use the top hand to pull out more fibers. The aim is to thin out the fibers so the yarn will be as thin and even as desired. Keep repeating the process of pinching with one hand and pulling out the fibers from above. As the spindle starts to slow down, give it a twist with your fingers to keep it spinning. When the yarn spindle reaches the floor, remove the hitches at the end of the shaft, wind the yarn around the spindle, restore the hitches, and continue to spin the next bout of yarn, joining new rolag/roving as necessary.

USING A SPINNING WHEEL

There are two types of spinning wheels: a low wheel (otherwise known as a flyer wheel), in which the wheel is turned through the use of a foot treadle, and a high wheel, in which the wheel is turned using one hand. An obvious advantage of the spinning wheel over the hand spindle is convenience, since the yarn spun is directly wound onto a bobbin.

Using the low wheel takes practice to coordinate the hand and foot movements, and so it is advisable to familiarize oneself with treadling before trying to spin yarn. First, adjust the tension of the drive band so that it is almost—but not quite—tight. If it is too loose, it will slip; if it is too tight, it will turn the flyer very fast and the yarn will be kinky. Gently push and start to turn the wheel clockwise; then treadle as slowly and as smoothly as you can without letting the wheel stop or go backwards. When you are ready to begin spinning, tie a leader (approximately 2 feet long) to the bobbin, pass it over the flyer hooks, and draw the leader through the orifice with a crochet hook or a hook made with bent wire. Now fan out the ends of the rolag and leader and hold them together with one hand, while using the other hand to start the wheel clockwise. Treadle slowly and evenly and allow the rolag to twist around the leader so that they are joined. As the leader is drawn into the orifice, pinch the juncture of the leader and rolag with the right hand and use the other hand to draw out the fibers to the thickness you desire. After the rolag and leader are joined together, move your right hand to pinch the drawn-out fibers at a spot farther back on the rolag so that the twist can run farther up the fibers. As the yarn spins to the tightness you desire, continue to draw out the fibers between left and right hands while the spun yarn is drawn into the orifice. When enough fibers are drawn out, let go with right hand to let the twist travel farther up the rolag. Continue to draw out and spin fibers, adding new rolag as necessary. The flyer hooks determine where the yarn winds onto the bobbin. The yarn should be first wound at the ends of the bobbin and then moved from hook to hook to build even layers.

When using a high wheel, the spinner keeps the wheel constantly turning with one hand while drawing the fibers out with the other. As with the low wheel, the drive band must be adjusted to the proper tension and a leader attached before starting. Tie the leader at the back of the spindle, hold the yarn parallel to the wheel, turn the wheel clockwise, and wind the leader onto the spindle in an even layer. Wrap the leader around the spindle tip, pull the leader away from spindle at an approximately 120-degree angle, and hold the ends of the leader and rolag together. Turn the wheel clockwise with your free hand to start the spindle turning; the end of the rolag will twist around the end of the leader to join them. Keep turning the wheel and slide your hand farther back on the fibers. Pull out the fibers and then pinch so that the fibers in front of your hand are spun; then slide your hand back again. When a length is spun, stop the wheel and turn it counterclockwise to free the yarn. Hold the yarn parallel to the wheel and turn clockwise to wind into a cone on the inside of the spindle. Return the yarn to the spinning position and draw out fibers again. When necessary, add new rolag by holding the ends together.

FINISHING THE YARN

After the yarn is spun, you will periodically need to remove it from the spindle or bobbin. It should be wound into skeins using either a niddy-noddy or an umbrella swift. Any wool that was not previously scoured should be scoured now, unless one plans to weave in the grease. In any case, all yarn should be washed to at least remove any surface dirt and to "fix" the twist. Dip the skeins into warm water and mild soap, rinse, and hang them to dry with weights on the bottom.

WEAVING

Weaving, as stated before, is the process of interlacing threads at a right angle in an under/over manner. It can range from being a rather simple process using very primitive tools and materials to something much more complex involving computer-aided design and multimillion-dollar equipment. The most important piece of equipment required by weavers is the loom. The loom holds one set of parallel threads under tension: This set of threads is called the warp. The loom has some type of device for selecting these warp threads to be raised or lowered for weaving the weft threads across back and forth. In most looms, this device would be in the form of harnesses. The more harnesses a loom has, the more complex the weave structure of the cloth can be. For the purpose of this "weaving primer," the process of weaving will be explained using a basic floor loom. A loom is shown in Figure 3.

As with spinning, weaving requires a lot of preparation time. The first step would be choosing the yarn(s) to be used as the warp. Many factors must be considered in making this choice, including the choice of fiber, color, and thickness of the yarn. The choice will, of course, hinge on what the final product is to be. If the product is to be a light, flowing scarf, then a thin silk may be appropriate. If it is to be a rag rug, then something more substantial, such as a heavyweight cotton yarn, should be used. It should be noted that the warp threads are under tremendous tension and friction, and therefore a strong thread is required, whether it be thick or thin.

FIGURE 3
Basic floor loom.

After the warp yarns have been chosen, it is time make some calculations. First, one must determine how long to make the warp. In addition to the length of the finished piece, other factors such as shrinkage, warp take-up (the extra warp consumed as the yarns pass over and under the weft in weaving), and loom waste (the part of the warp at the end that cannot advance any farther to be woven) must be considered in the calculations. It is always better to overestimate and have a little warp left over than to run out of warp toward the end of the project. The width of the warp is calculated in a similar way, taking the width of the finished piece, shrinkage, and draw-in all into consideration. The density of the warp depends on both the thickness of the warp yarn and the density of the planned fabric. The density of the warp is often referred to as the number of ends per inch (epi). After determining the length, width, and density, one can start to wind the warp.

The purpose of winding a warp is to measure the threads in your warp to all the same length. There are several ways to do this: Using a warping board or warping reel are the most common, but some weavers use pegs or even a set of chairs to measure off their warp. Many find the warping board the most practical, unless the warp is particularly long, in which case a warping reel is more appropriate. When winding the warp, you are not only measuring out the length of the threads, you are also establishing an order to the threads. To preserve this order, a cross is made with the threads as you wind. It is basically a figure eight made between two pegs on your warping device. The cross is the most crucial element of the warp, since it prevents the threads from becoming intertwined and allows them to be arranged on the loom in the order in which they were wound. One other very important detail in winding the warp is the tension. During weaving, the tension of the warp threads should be even and consistent. If the threads are not even and consistent, some areas of the weaving will be either looser or tighter than others, which can be very problematic. Therefore, the first step to insure proper tensioning on the loom is to start with proper tensioning on the warping device.

It is sometimes helpful to use a counting thread when winding your warp. This is a contrasting thread used to divide the warp into little "bundles" or sections, each section usually being equal to the number of ends per inch. While using a counting thread slows down the warping process somewhat, it makes it up when it is time to warp the loom. Once the warp has been wound onto the warping device, it needs to be secured by tying yarns around these strategic points on the warp: the cross (or X), the warp beginning, and the warp end. It is also necessary to bind off the warp at one- to two-yard intervals to prevent the threads from shifting. After the warp has been properly secured, it may be removed from the warping device by *chaining* it to condense the length for ease of handling.

Now the loom is ready to be *dressed*—that is, attaching the warp to the loom, winding it onto the loom, and finally threading the warp through the heddles and reed. Each weaver has her or his own personal system of dressing the loom. No system is necessarily better than another as long as the final outcome is a loom warped with even tension. Basically, the chained warp is laid over the harnesses and the *lease sticks* are inserted to separate the cross. This should be done very carefully to ensure that the threads are in proper order. Then the warp is centered across the back apron bar. To help with the even distribution of the warp threads across the width of the loom, a comblike device called a *raddle* is temporarily attached to the back of the loom. The raddle is usually divided into one-inch sections. If a counting thread was used during winding the warp, then the distribution of the threads into the raddle spaces will be much less time-consuming than if one was not used. After the warp has been centered and evenly distributed across the loom, it is time to wind it onto the back roller of the loom. Depending on the size and type of loom, this step may require two people (one to hold the warp at an even tension and the other to wind it onto the loom). During the winding process, it is important to separate the layers of warp either with paper or warp sticks. This helps to ensure proper tensioning.

Once the warp has been wound onto the loom, it is time to thread the warp yarns through the heddles. How the warp is threaded through the heddles is one of the components that determines the pattern. The other components are how the harnesses are connected to the treadles and in what order the treadles are pressed. With a multiple harness loom an essentially limitless variety of patterns is possible. Once the threading plan has been devised, each warp thread is individually threaded through the eye of the corresponding heddle using a heddle hook. Next, the threads must be pulled through, or sleyed in, the reed. Threaded warp yarns, the heddle, and the harness are shown in Figure 4. Next, the warp ends are tied to the front of the loom. Finally the harnesses are connected to the treadles.

FIGURE 4
Loom with harness.

FIGURE 5
Weaving.

Weaving is done by throwing—passing—a shuttle through open sheds. The sheds are opened by depressing treadles. The sequence of sheds determines the pattern of the finished cloth. A shuttle is shown on the left in Figure 5. Inside the shuttle is a bobbin on which the weft threads have been wound. This winding is done on a quill, which can be purchased or made from a cylinder of paper. The winding needs to be firm and even.

After the shuttle has been thrown through the open shed, the weft thread is beat in or pushed against the already woven material. The beater is part of the loom. The force used in beating will determine the appearance of the finished cloth. Some experience will show the proper amount.

Eventually, so much material will be woven that the sheds will become too small to throw the shuttle conveniently. When this happens the warp is rolled forward so more warp is exposed and the finished cloth is rolled onto the cloth beam. When the bobbin is empty or the pattern calls for a new weft thread, a small tail of the old weft is wrapped around the outside warp and woven into a few warp threads. It is best to start weaving with the new bobbin from the edge of the fabric.

After the weaving has been completed and the fabric is removed from the loom, several inches of warp threads should be left to prevent unraveling. The end of the fabric should then be hemmed. It is possible to remove pieces of fabric from the loom before the entire warp is used. One should weave scrap material in for several inches past the finished piece, hemstitch this scrap material, then insert a dowel on the next shed. After the next piece has been woven another inch or so, then the first piece can be cut off. The dowel can then be sewn with an overcast stitch to the cloth stick so weaving can continue.

REFERENCE

Held, Shirley E. (1978). *Weaving a Handbook of the Fiber Arts.* Fort Worth, TX: Holt, Rinehart and Winston, Inc.

DYEING

Laura Wells

The process of dyeing has been around for centuries and even dates back to A.D. 418. Two early forms of dyeing were the batik method and the plangi (tie dye) method. Batik is a native craft of the Indian subcontinent and is a resist method of patterning cloth. During the dyeing process, something such as wax is applied to certain areas of the cloth that the dyer wants to remain its original color. Then the dye seeps into any cracks in the wax, giving batik its characteristic look.

Evidence of tie dye, or plangi, has been found in India and Java from the sixth and seventh centuries and is a method of decorating cloth by isolating areas so that they resist the dye. The most common method of tie dyeing involves tying a string or yarn around parts of the cloth.

For thousands of years colors for dyeing were obtained from natural resources such as animals, vegetables, and minerals. Other examples of natural dyes are leaves and stems, flower heads, roots, onion skins, berries, and seeds. The good thing about natural dyes is that they are free and readily available. However, the turning point in the history of the dyeing process came with the discovery of synthetic dyes by William Perkin, a student at the Royal College of Chemistry in London.

Synthetic dyes are a mixture of finely ground chemicals that come from powders, grains, granules, or liquids. There are two main categories of dyes: hot water and cold water. Cold water dyes are very stable and can withstand frequent washing, but they can only be used on natural fabrics. Cold water dyes are mainly used for tie dye and batik. Some hot water dyes can be used in washing machines. Although synthetic dyes were durable, strong, and produced wonderful colors, they were not colorfast. This led to the discovery of mordants in 1868. A mordant is something that fixes color by encouraging an attraction between the dye and the fiber and so binds the dyestuff to the fabric. Mordants can be found in natural dyes and can be produced synthetically as well. Some examples include iron, tin, salt, vinegar, and caustic soda. Probably the best way to learn the amount of mordant necessary is to experiment.

Along with synthetic dyes there are many other types of dyes. Acid dyes, which work well with wool, silk, and hairy fibers, were discovered in 1875 and were developed from experimentation with the dye elements in natural dyestuffs. In order to keep bright, acid dyes must be produced in acidic conditions with the use of vinegar as a mordant. Direct dyes are simple to use but have a low resistance to washing and therefore fade quickly. Direct dyes are most suitable for drapes and rugs.

Disperse dyes were developed in 1923 in order to dye acetate rayon. The dyes disperse into water rather than dissolve, but they have to be used with a dispersing agent so the particles won't sink to the bottom. They can be used on most synthetic fabrics such as acetate, acrylic, nylon, and polyester.

Fiber reactive dyes were developed to form stronger and more permanent bonds with fibers in fabrics such as rayon, silk, linen, cotton, and other plant fibers. Salt is used to help the dye and fibers bond.

One of the most common materials for home dyeing is cotton. There are many types of cotton that can be dyed. Calico cotton is hard wearing and does not fray. Cambric cotton is soft to handle and also will not fray. Lawn cotton is a fine fabric and is ideal for most dyeing processes. Mercerized cotton needs to be treated with caustic soda, a mordant, in order to make it more receptive to dyes, but it is easy to handle and won't fray.

The art of dyeing is a creative process that has been used for many centuries. There are various techniques of dyeing and various fabrics that can be used. For the most part, it doesn't require much skill, but it is a much-expressed form of art. A related article, "Plant Dyes," is the last one in this chapter.

TREADLE SEWING MACHINES

Based on the Web page http://www.captndick.com/

Treadle—nonelectric—sewing machines are common in Africa. On the verandahs of shops selling textiles one will often see a tailor with a treadle machine. A customer who wants a dress or a shirt will buy the material inside, then ask the tailor on the verandah to make the garment. Sewing with a treadle machine is done just as with an electric machine, except that the operator uses the treadle while sewing. Some tips on using the treadle are at the end of this article. In the article the words "Singer" and "White" refer to sewing machine manufacturers. Kerosene, used as a cleaning liquid, is called "paraffin" in some parts of the world.

SERVICING THE TREADLE

A sewing machine treadle is a very basic and simple machine—beautifully so, in its way. It is basically a framework to hold a drive wheel, a pedal, and a pitman rod that connects them. The whole is held together with large screws and bolts. When you work the pedal, the pitman rod drives the wheel through a crank or eccentric connection. That's it! Of course, a belt is connected to the wheel and in turn drives the sewing machine, but for now all we are concerned about is the treadle itself.

THINGS YOU WILL NEED

- Large flathead screwdriver
- Short but large-bladed flathead screwdriver

- Two adjustable end wrenches or a good set of open or box end wrenches, U.S. standard
- Pair of pliers
- Hammer
- Can of Liquid Wrench® or kerosene
- Tube of Singer motor grease or other light grease
- Possibly a can of really heavy grease, like water pump grease
- Tweezers
- Razor blade or knife
- Small paintbrush or two
- Wire brush

Note: This list is for a really complete overhaul. If all you are after is a cleaning and oiling, you won't need so much. If you don't have something in the list, use what you do have, except for the large screwdriver. Using too small a screwdriver is an invitation to either a ruined tool or a hole in your hand.

DISASSEMBLY

For a really thorough job, it is a good idea to remove the cabinet. First, take out the machine. To do that, loosen or remove the belt, tilt the head back, and loosen the set screws that attach the head to the hinge pins at the back. These screws are found under the machine. When they are loosened enough, you can lift the machine off the hinge pins. Set the machine aside.

You can unfasten the cabinet by crawling under it on the floor, or you can lay a cloth or old blanket on the tabletop and tip the machine upside down on the table. This allows easy access to the screws underneath. BE VERY CAREFUL TO CONTROL THE FLIPTOP LID WHEN DOING THIS! Usually, you only need to unfasten four large screws that hold the treadle legs onto the top. Have a big can or bucket to hold all screws, bolts, and small pieces. It's very embarrassing to finish the job and not have as many pieces as you started with. Lift the treadle off the top, or the top off of the treadle, as appropriate. Set the cabinet aside and put the treadle back up on the table but rightside up.

Study the treadle mechanism to see how it works. Watch what moves and what is connected to what and how. Think it out and develop a comfortable understanding of the mechanism you are working on. Basically, it just needs to be cleaned and allowed to do what it was designed for. Generally, any grease or oil in the mechanism has collected dirt and dust and dried out, making a bad situation. First, everything gets stiff and slow, but even worse, instead of acting as a lubricant, the dried, dirty oil becomes an abrasive compound and increases the wear on the parts. Figure 6 shows a Singer treadle.

A good beginning is to squirt a little Liquid Wrench® or paraffin/kerosene on all the moving joints, nuts, bolts, and screws. This will loosen things up and make disassembly easier. (*Note:* This does not apply to the connections of the wooden pitman rods. Do not put any kind of lubricant on the wood.)

For a really thorough job, it is a good idea to start by disconnecting the pitman rod at the pedal. There are two types of pitman rods: wood and metal. The wood type is usually a rod with a 1/2-inch hole in each end. There is a round rod or projection on the pedal. The rod slips over this, and there is a pin to hold it in place. (You may not need to disconnect the rod if you are not doing a really complete cleaning and repainting job.) To remove the retaining (cotter) pin, simply bend it straight and remove it. If it is a solid metal pin, tap it gently on either end. If that doesn't work, you can hit it harder. When you see it move, tap from that direction. This may be a tapered pin, so you don't want to try to drive it through the wrong way. When reassembling, I always replace a solid pin with a cotter pin.

Note: You may find actual roller ball bearings both at the pedal and at the drive wheel. Some of the better machines (White, Pfaff, others) used them. That's great. Don't let them throw you, but carefully study how they are fitted in before taking them out, and be sure to put them back in the same way after cleaning and relubricating them.

FIGURE 6
Singer treadle.

Another caution: If the pitman rod is wood, it will be very dry—which means brittle. It is easily broken or split, especially at the ends where there is a sliding wood mechanism for adjustment. Start right now and rub it with some wood restorative. (I like Howard's Feed and Wax.) When you're finished with this job, continue to rub oil on the pitman every few weeks for several months to restore moisture and strength to the wood. If disaster strikes and the pitman breaks, don't panic—you can make a new one yourself.

Generally, you don't need to take the pitman apart at the crank end. If you have a metal pitman, there will be a donut-shaped ball bearing unit at the crank end, and the pedal end will be terminated in a ball and socket nut in the corner of the pedal. You will need a pair of wrenches to disassemble the ball and socket. As you do, study how it goes together. You're going to need to get it back together after you clean everything. Also, in the future, this socket is an important adjustment for keeping your treadle operating smoothly and quietly. To disassemble, hold the top nut still with one wrench while backing off on the lock nut below the pedal. Once the lock nut is loose, you can use the stubby screwdriver to back the center screw out until the whole unit comes apart.

At this point, you should have the pitman disconnected and the pedal flopping free. Look at the ends or sides of the pedal. You will see that it is hinged on two large screws that have lock nuts on them. Use a wrench and back off on the lock nuts. Now, using your large screwdriver, back these screws all the way out until you can remove the pedal. If you study the screws, you will see that the ends are cone shaped. These are actually bearings, although they don't have rollers or balls as we usually think of bearings.

ADJUSTING THE CONE BEARINGS

The pedal is a good place to learn about cone bearings. Using a brush, a rag, and some kerosene or Liquid Wrench®, clean out the cone and clean off the cone screws. Squeeze some grease (I use the Singer motor lube) into the cone. Now put the screws back in. Note that if you screw the screws in tightly, the pedal won't move at all. If you back them off, the pedal can move freely.

If you back them out further, the pedal will have side-to-side play and be both noisy and sloppy. Here is the secret to correctly adjusting cone bearings (as taught to me by my father): Tighten the bearings until you know they are too tight and restricting movement. Now, back them out until you *think* you feel play. That should be perfect. What you are aiming for is to leave enough extra space to accommodate a film of grease, no more. If you can really feel a noticeable amount of play, that's too much. The pedal should be nice and free in its movement but not sloppy.

It would be great if this was all there is to cone bearings. Unfortunately, it's not. For the adjustment to stay in place, it is necessary to "lock" it. This is the purpose of the lock nuts on the cone bearing screws. To lock the adjustment in place, hold the screw steady with the large screwdriver while tightening the lock nut down with a wrench. You may actually find yourself putting backpressure on the screw to counter the forward pressure of the nut. The trick is get the nut locked down tight with the adjustment the same as when you were happy with it. The natural thing is to get a perfect adjustment, then make it too tight when you tighten the lock nut. Practice this—do it half a dozen times—until you get a feel for it.

You have now learned the most important single thing about treadle adjustment. However, if you are doing a complete disassembly, you will want to take the pedal out again so you can take the legs off. This is the time to remove the screws from the legs, or braces as they are sometimes called. You can lay them aside.

CLEANING THE DRIVE WHEEL

Next remove the dress guard, which is held on by one bolt at the bottom. There is no need to disassemble the little belt control lever.

At this point, you should still have the frame center with the drive wheel, crank, and pitman assembled as a unit. Study how the drive wheel is attached. It looks an awful lot like the cone bearing arrangement you saw on the pedal, doesn't it? Surprise! That's exactly what it is. If you back off on the lock nut, you can then back the cone bearing screw all the way out and remove the drive wheel. This is where a major mess may be found. People had a habit of winding two bobbins at a time—wind one, put it in the machine, wind another, ready for the next change. Can you just leave the second full bobbin in the bobbin winder? No! Sometimes the belt will pick up the loose thread end and wind miles of thread around the drive wheel, which somehow pulls it down into the drive wheel bearings. You must either disassemble everything or cut and tweeze and pull to get it all out. Before reassembling you will want to clean out the cones, clean up the screws, and grease the cones. You now have an essentially disassembled treadle frame. If you want to clean it up, this is the time to do it.

FINAL ADJUSTMENTS

Now we will make all the final adjustments that will mean a smooth working treadle. First, we need to revisit the metal pitman with its ball and socket attachment joint. This really is a variation on the cone bearings, only in a spherical shape. Clean the whole assembly, put some light grease in it, and put it back together. Be sure to do a good job of the cleaning. A tiny bit of dirt in this particular assembly can drive you nuts by making the treadling stiff.

With the whole works back together, hold the drive wheel and put pressure on the pedal. Try to rock it just a little. Does it sit firm, or is there play in it, causing it to make a clacking sound? Again, this is one of those situations where you don't want things too tight, but you don't want them so loose that the pedal has "free" travel in its joints. This causes wear and noise. If you feel there is too much play, and you can identify it as coming from the pedal joint, adjust either the ball socket or, if you have a wood pitman, the adjustment in the end of the wood pitman. (It has a little set screw in it.) A wood pitman also has an adjustment at the crank end. If you can, follow by wiping it out and then re-oiling with regular oil. WD-40 should be used on moving parts, not on the surface.

Kerosene/paraffin is a wonderful penetrant and will eventually work its way into almost anything and free it up. More than one collector has simply put a whole machine head into a pail of kerosene and left it for a while, with good, if messy, results. Liquid Wrench® is almost pure kerosene and can be bought in some hardware stores in nice little squeeze cans with spouts.

DISASSEMBLY AND CLEANING

Take off all access plates, the motor and light, if present, and the hand wheel, and study the machine. Look for old dried-up grease coating moving parts. This is usually found inside the upper pillar, behind the rear access plate and, in geared machines, wherever beveled gears meet. This point is especially critical on the underside of Singer Featherweights and Model 201s. Also, on 201s and Model 15-91, which are gear driven, remove the hand wheel and gear drive unit. You will find old grease there. Use the kerosene or Liquid Wrench® and a brush to clean all this old grease off.

Next, I squirt the lubricant liberally onto, into, and over anything that moves. Look at each part closely for little holes meant for oiling. Almost every part that has a part moving within it will have a little oil hole. Be sure to get the penetrant into these holes. After dousing with penetrant, try moving things; see if the machine starts to move more easily.

When things seem to move reasonably easily, reassemble. Lightly oil all moving parts and oil points with sewing machine oil. Grease the gears with new lubricant. (I like to keep a hand crank drive unit handy to power the machine for testing.) Once the machine is reassembled and you have either a hand crank or the treadle back on, run the heck out of it. Often, you will actually be able to hear and feel the machine start to run faster and more easily as the new lubricant works into things. Pay special attention to cleaning and lightly oiling the inside surface of the hand wheel core. This will make bobbin winding much easier. Also, there are tiny oil holes on the bobbin winder itself, and these are often missed. Finish by cleaning off all excess oil.

TOTALLY IMMOBILE MACHINES

Now a brief word about frozen machines. Sometimes a machine just won't move at all. This can be the result of general rust, dried grease, or really bad thread tangles that have been dragged into the works behind the bobbin case. Start by removing the bobbin case and checking for any bits of thread you can get out. Follow by soaking with penetrant and oiling. Some folks will immerse it in kerosene for up to a week. Look for "forks"—moving pieces that look like a musical tuning fork and have another moving piece between the tines of the fork. This structure is necessary to create certain desired movements of the machinery, but is not the easiest moving arrangement. Often it is a point where "hangups" happen. Press on or try to wiggle these points while trying to turn the hand wheel. I have had some machines just suddenly break free just by doing that. When nothing else seems to work, you will be reduced to brute force.

The machines are actually very strong. It is possible to break them but not too likely. Have a piece of leather handy. Wrap it around the clutch wheel (the small wheel inside the hand wheel). Apply a medium-sized pipe wrench with a good grip. Try turning first one way, then the other. If you can get any movement at all, even a fraction of an inch, you are winning. Apply more penetrant, then repeat the wrench movement. You should find it moving more and more, and eventually it should free up. If the very unlikely occurs and the machine breaks, well, it wasn't going to be very useful permanently frozen, was it?

USING THE TREADLE

The following notes are intended for people who have never used a treadle and want hints on how to make the treadle turn the machine.

1. Place the sewing machine on the floor. It is best if the back of machine is to the wall and the front toward open space.
2. Place a chair in front of the machine.
3. Sit in the chair and place your feet on the treadle (right one slightly ahead of left one).
4. Start the hand wheel by turning it with hand.
5. Wiggle your feet.
6. Observe what happens. If it doesn't look right, try wiggling your feet differently.
7. Curse, rethread the needle where the thread broke, and start the hand wheel in the opposite direction and try again.
8. Try it with one foot.
9. Get some kid to work the pedal while you sew.:o)

Okay, seriously, I doubt if any of us "had lessons"—unless it was from Grandma. You aren't going to hurt the machine by using it. *You* may get frustrated, but you won't ruin the machine. Treadling is a lot like roller skating or bicycling. You learn by getting the feel of it, and that can only be done by jumping in and falling a few times.

Here is something else you can do: Remove the thread from the top and the bobbin from the machine. Start the hand wheel in the direction that would move the cloth forward. Do whatever you have to do with your feet, either one or two, to keep the machine going. Sew a bunch of imaginary seams in real pieces of cloth. Stop the machine, turn the work, and so on but without thread. Once you get used to that, thread it properly, put the bobbin back in, and make sure the machine is threaded right and will make stitches. You can do that by hand, just turning the wheel by hand.

Now, sew some seams. Try back and forth, long seams, and see how fast you can go. See how slowly you can go without losing the movement. Try some patterns—large squares gradually diminishing inwards and circles. See how tiny you can get in the middle and still keep the seams evenly spaced, corners square or curves smooth. Now, try some really complex shapes where you can only go one to three stitches in a direction without lifting the foot and turning the work. *Hint:* On this type of stuff, turn the wheel by hand and let the pressure of your foot act as a brake. (Another hint: You can turn the wheel by putting your finger in a spoke, but *don't* put it in too far. Trust me—it hurts!)

SILK PRODUCTION

Silk is a fiber produced by the silkworm to make a cocoon. The fiber is unwound from the cocoon. Creating silk thread involves cultivating the worms, collecting the cocoons, unwinding the cocoons, and reeling several filaments together. Each of these processes can be done by hand, although most are mechanized.

BREEDING SILKWORMS

Silkworms eat only mulberry leaves. The entire process can be controlled by keeping the worms in a controlled environment; protecting them from ants, mice, and disease; and feeding them mulberry leaves. The silk that is produced is called "real silk." Wild silkworms will eat other kinds of leaves, as well as mulberry, and fend for themselves. The silk from wild silkworms is thicker and less lustrous, and is called "tussah" silk.

Cultivation starts with eggs collected from the previous year. These eggs are incubated until spring, when fresh mulberry leaves are available. The eggs hatch into worms, which are fed leaves for about five weeks. At the end of this time they are about two inches long, and then they spin a cocoon. A protected place—like a basket—is provided for the worms when making the cocoons. Most of the cocoons are collected for silk; a few are left so the moths will emerge and breed eggs.

Wild moths breed in a free for all. Cultivated moths are bred by putting only two in an aluminum case.

PREPARING THE SILK

Cocoons are sorted by color and then heated in an oven or with steam to kill the larvae inside. Colors can be white, yellow, or grayish. They must then be soaked to loosen the gummy substance that holds the cocoon together. A series of hot and cold immersions is used to loosen the fibers.

Unwinding, the next step, begins by finding a loose end of the fiber of a cocoon. The fiber from five or so cocoons is unwound together because a single fiber would be too fine for commercial use. The usable length of unwound fiber from a cocoon is about 1,000 feet. The fibers are then wound onto skeins of raw silk. Fibers of raw silk from two or more skeins are often twisted together to give a stronger thread. The threads are boiled to remove natural gum.

SPUN SILK

Some of the silk fibers cannot be unwound from the cocoons. The cocoons are also covered with a floss. The unused fibers and the floss can be spun to form a somewhat inferior silk thread called "spun silk." Before spinning, the fibers must be combed to straighten them out and remove impurities.

REFERENCE

http://www.rexguide.com/news_group/silk prod.htm

SOAP MANUFACTURING

Soap can be in short supply in rural areas of developing countries, but it may be made easily by mixing oils and fats with a solution of caustic soda in water. Locally available materials such as vegetables oils or animal fats are used. The soap produced in small-scale operations will not be as hard or possibly as attractive as that produced in large factories, and it will probably sell at a lower price than commercially made soaps.

INGREDIENTS

Almost any fat or nontoxic oil is suitable for soap manufacture—oil is simply liquid fat. Commonly used types of oils include animal fat, avocado oil, coconut oil, and sunflower oil. Lyes can be made from ashes, potassium hydroxide (caustic potash), or sodium hydroxide (caustic soda), or purchased from drugstores or hardware stores.

MAKING LYE

To make potash lye from wood ashes, simply fill a large (50-gallon) wooden barrel with ashes and pour boiling water in. Let the ashes soak for 12 hours or longer. As the ashes settle, more ashes can be added. Ashes from banana leaves or stems make the strongest lye. The barrel can be made of porcelain or plastic, as well as wood, but not aluminum because the lye will react with aluminum. It is convenient to use a barrel with a tap near the bottom so the water containing the lye can be drained off into another container. A cloth or straw filter should be used, either at the bottom of the tub—around the tap—or between the container that collects the water and the barrel.

The lye does not have to be a predetermined strength. A definite weight of oil reacts with a definite weight of lye. If the lye is not sufficiently strong, then more can be added until all the fat has become soap. If more lye is used than is needed, however, it will remain in the soap and produce a burning sensation on the skin.

The collected lye water is boiled down. After the water has evaporated, a dark, dry residue will remain. Continued heating will burn this residue away, leaving the grayish white potash lye. The lye can be stored but should be kept dry—and away from children.

Lye can also be made from quicklime and soda. (Soda is hydrated sodium carbonate.) Make a paste of one part quicklime and three parts water. Be careful—the mixture will get very hot. Make a solution of three parts soda in five parts boiling water. Add the lime paste, stirring vigorously. Keep the mixture at a boil until the ingredients are thoroughly mixed. The mixture is then allowed to cool and settle. The liquid is poured off and boiled down; the residue is caustic soda, a suitable lye.

SAFETY PRECAUTIONS

Lye is a highly caustic chemical. If possible, wear eye protection and rubber gloves. Clothing should consist of long sleeves, long pants, and shoes. If lye solution (or dry form) comes in contact with the skin, flush the affected area immediately with vinegar and then wash the skin well with detergent and water. The presence of lye on the skin can usually be detected as a slick feeling on the skin. When measuring out lye or stirring the solution care should be taken to avoid breathing the dust or fumes created by the lye. Always work in a well-ventilated area.

ANIMAL FAT

Beef suet, mutton tallow, or pork scraps can be used. Grease from cooking can be saved and used to make soap. The fats or grease are cleaned by heating and straining the melted fat through a coarse cloth. The melted fat is then boiled in salted water—1 tablespoon of salt per 5 kg of fat, twice as much water as fat. Boil for 10 minutes and allow to cool. When the fat is cold, it will have formed a hard cake on top of the water. Lift off the cake of fat and scrape the underside clean.

VEGETABLE OIL

Two groups of fats are used for soap making. The first group, the lauric oils, are obtained from the kernels of different types of palm. The most common oils in this category are coconut oil and palm kernel oil. They are known as lauric oils because they contain lauric acid as the major fatty acid. These fats make a hard soap that produces a fast-forming lather with a strong detergent action. The soap tends to have a harsh effect on the skin. Soaps made from lauric oils are used as saltwater soaps.

Nonlauric oils are the other group of fats. These contain virtually no lauric acid and include a large selection of liquid oils such as olive oil, corn oil, sunflower seed oil, groundnut oil, soybean oil, and cottonseed oil, as well as semisolid fats such as palm oil and tallow. Soap made from the oils in this group produces lather more slowly than lauric oil soap, but the foam is longer lasting, it has a milder detergent action, and it has a gentler effect on the skin. Palm oil and tallow have a similar foaming action, but the soaps produced are harder. Castor oil is also included in this category. It produces a hard but very soluble soap with poor foaming characteristics.

BASIC SOAP MAKING

A simple recipe for homemade soap uses 12 kg of fat, 9 kg of potash—from wood ashes, and 60 liters of water. The mixture is boiled until it becomes shining and turbid. To test the soap, put a few drops from the middle of the kettle onto a plate to cool. If the soap remains clear when it cools, it is done. When the mixture is done, it is poured into molds and allowed to solidify. Solidification takes about 24 hours, and the filled molds should be wrapped in a towel or blanket during this time so they cool evenly. Bars are the most popular shape; so the molds are shallow, wooden boxes, but gourds or coconut shells can also be used. The mold can be lined with plastic wrap or plastic garbage bags if desired, but removing the soap from the mold is usually not a problem. After 24 hours of cooling, the soap may be cut into convenient-size pieces and allowed to cure for several weeks. Curing ensure that the lye reacts fully with the fat.

Actually, it is not necessary to boil the mixture if one can be patient. If so, the fat-potash-water mixture is put into a barrel. On each of the first three days 15 kgs of boiling water are added and the mixture stirred vigorously. After that the mixture is stirred several times a day. In about a month a jelly-like soap will result.

If they are available, kitchen or bathroom scales should be used to weigh out the materials. Volume measures such as those used in kitchens or containers or bottles of known capacity, such as 1 liter, can also be used. A plastic bucket of about 8–10 liter capacity is a useful vessel for mixing the soap ingredients. Stirring rods can be made simply from suitable lengths of wood.

Soap molds can be made from wood, but cardboard boxes are also suitable. A table can be adapted to make a soap mold by attaching removable lengths of wood, about 2.5 cm high, around the edge so that the block of soap can be moved forward for cutting after it has set. Larger molds can also be made from boxes about .5 × .25 m in size, which can be taken apart to release the soap after it has set. Lining the mold with thick plastic sheeting helps to prevent the soap from sticking to the mold and makes it easier to remove. Spreading a thin layer of petroleum jelly over the plastic sheet also helps to prevent the soap from sticking to the mold.

A length of cheese-wire or wire tuna-fishing line, attached at each end to wooden handles, can be used to cut up the soap blocks. For making soap tablets, a cutting board equipped with a guide-rail to align the soap bar at right angles to the cutting wire can easily be constructed from wood.

MAKING SOAP FROM COCONUT OIL

To make soap from coconut oil start with a 1-kg pack of caustic soda. Dissolve the soda in 2 liters of water (2 liters of water weigh 2 kg) in a suitable plastic container. Measure out 6.5 liters of coconut oil (6.5 liters of coconut oil weigh 6 kg) into a plastic bucket. The oil needs to be between 30° and 40°C; normal tropical daytime temperatures are usually satisfactory.

Allow the caustic soda solution to cool to about 35°–40°C before adding it to the coconut oil. When the caustic soda solution has cooled, pour it slowly into the coconut oil, stirring the oil all the time so that the caustic soda solution is absorbed into the oil and does not separate from it. If the oil is slow in absorbing the caustic soda solution, add it in small amounts at a time and stir it well into the oil so that it is absorbed before more caustic soda solution is added. When all the caustic soda solution has been added, keep stirring the mixture until it begins to thicken and patterns can be drawn on the surface with the tip of the stirring rod. The speed at which this thickening occurs is greatly influenced by the free fatty acid content of the oil used. The higher the free fatty acid content, the quicker the oil thickens. At this stage, before it thickens any further, the soap must be poured into molds and allowed to harden for about 24 hours. During this hardening time, the soap-making reaction continues and the soap will feel warm to the touch.

After the soap has hardened, it should be removed from the mold and cut up into tablets, using cutting wire. The soap tablets need to be stored for about two weeks before being used. This is to allow time for all the caustic soda to react with the oil. About 9 kg of soap will be obtained, equivalent to 90 soap bars of 100 g each.

In this soap-making method, the 1 kg of caustic soda was dissolved in 2 liters of water. This makes a strong solution of caustic and helps to reduce the time that the soap takes to thicken when it is being mixed. However, if the oil has a high free fatty acid content, the soap may thicken too quickly and make it difficult to add all the caustic soda solution. If this happens, dissolve the 1 kg of caustic soda in 2.5 liters of water for making the next batch of soap from the same oil. This makes a weaker solution and increases the time taken for the soap to thicken but still makes a good quality soap with the added advantage of producing an additional 500 g of soap.

When oils with a very low free fatty acid content are used, the soap may take a long time to thicken. If the soap takes much longer than 45 mintues to thicken, the caustic may not be absorbed very well. If this happens, it is best to heat the oil until it is just bearable to touch, about 60°C, before mixing in the caustic soda solution. This will decrease the time taken for the soap to thicken.

To make soap from a nonlauric oil using a 1-kg pack of caustic soda, use 8.75 liters of the oil (about 8 kg) and follow the same method used to make soap from coconut oil.

FRAGRANCES AND COLORING

Fragrances and coloring may be added just as the soap mixture starts to thicken, stirring them in well just before pouring it into the molds. These fragrances are usually added in the form of essential oils at about 1 percent of the total soap weight. For the 9 kg of coconut oil soap described above, about 90 g of essence would be needed. This would be equivalent to about 20 teaspoonfuls of citronella oil.

The quantity of dye required for coloring soap is very small—usually, only .01–.03 percent is needed. For 9 kg of soap this would be about .9–2.7 g. The dye should be added as a strong solution, either in oil or water, depending on the nature of the dye used. Try dissolving about half a teaspoonful of color in about 50 ml of water or oil before adding it to the soap. Only edible dyes should be used.

REFERENCES

"Soap-Making by the Cold Process." Coconut Development Authority Advisory Leaflet No. 10. Colombo, Sri Lanka: Coconut Development Authority.

ATBRIEF, No. 6: "Soapmaking." *Appropriate Technology.* Vol. 20, No. 3, December 1993.

Donkor, P. (1986). *Small-Scale Soapmaking*. London: Intermediate Technology Publications Ltd.

Kone, Siaka. (1993). "How to Make an Improved Soap: Not Just for More Foam." Deutsches Zentrum für Entwicklungstechnologien—GATE, Deutsche Gesellschaft für Technische Zusammenarbeit (GTZ) GmbH.

Sweetman, A. A. (1992). "How to Make Soap." *Food Chain,* 5: 19. http://www.picisys.net/%7Ecosystems/soap4.htm

CANDLES

Based on the About.com Web page written by Bob Sherman //http://candle and soap.about ry/

Candles are a common source of light when electricity is not available. They are easy to make at home and can be sold at markets.

There are many factors that affect the finished candle—wick, wax, temperature, additives, type of mold, dye, scents, and so on. Every component of a candle is affected by every other component. Changing any component may require adjusting other components.

Candles may be made by dipping or in a mold. The mold technique is described first.

USING A MOLD—TOOL LIST

1. Double boiler—may be a commercial double boiler, or use a coffee can in an old pot. A seamless pot is highly recommended, though.
2. Thermometer—a candle or candy thermometer that clips to the pot works fine. Do not even consider making candles without a thermometer.
3. Potholders or pliers—depending on whether you are using a pot or a can.
4. Molds
5. Mold release—silicone spray is easiest to use, but peanut oil works well, too.
6. Cutter for wicks
7. Wooden spoon—for stirring wax
8. Baking pan at least eight inches square—numerous uses but mainly for leveling the bottom of molded candles.

WAX

There are many waxes available for candle making. A general-purpose paraffin wax that melts in the range of 135°–145°F is recommended for beginners. Other waxes that candle makers may want to experiment with are microcrystalline wax, beeswax, bayberry wax, and paraffin waxes with other melting points.

ADDITIVES

Stearine. Also called stearic acid. This has been the standard paraffin additive for a very long time. Used to make wax harder, release from mold easier, and increase opacity of the wax. Use from 5 to 30 percent (3 to 5 tablespoons per pound of paraffin).

Vybar. Available in low melting point (Vybar #260) and high melting point (Vybar #103). More economical to use than stearine. Improves color and scent retention. Difficult to find, and doesn't always release from mold easily. Use 1 to 5 percent.

Plastics. There are a variety of plastic additives (mostly polyethylenes) that will improve gloss, opacity, translucence, strength, and hardness. Marketed under a variety of names such as luster crystals, opaque crystals, and translucent crystals. These are readily attainable but are difficult to use due to their high melting point. Must be melted separately and then added to melted wax. General usage is from 1/2 to 2 percent, depending on the product. Not recommended for beginners.

WICK

There are more than 35 different wicks on the market, although only about six of these are commonly available to retail candle supply purchasers. Wicking can be broken down into three categories: flat, square, and wire core. Flat and square are used for molded and dipped candles, wire core for floating, votive, and container candles. The starting point for wick selection is to match the wick to the mold diameter. For a small mold use a small wick, and soon. If a test burn of the finished candle shows a minimal wax pool, the wick is too large for the wax formula. If the wax pool is drowning the wick by causing it to go out or have a small flame, go to a larger wick. The wick size is the easiest way to adjust to control how the candles burn, and it is important to keep in mind that changing the wax formula may require changes in wicking as well. If another size wick is not handy, adjusting the wax hardness with more or less additives may help it burn correctly.

DYE

There are two main ways to color candles: dye and pigments. Most candle making is done with dye. Pigments are very concentrated colors primarily used for overdipping and carved candles. As a general rule, never use pigments to color the core of a candle—the particles of pigment will clog the wick. Although it is common to see candle making instructions using crayons for color, this can also clog the wick. For the best results always use a dye specifically made for coloring candles. If a really deep color is needed, consider an overdip in that color. Too high a color concentration in the core of the candle may cause burning problems. Wax colors will be lighter than they appear in the melting pot. To get an idea of the finished color, place a drop of wax on a piece of white paper. An even better test is to put a half-inch of wax in a paper cup and place it in the freezer; this will give the exact finished color in a hurry. Keep in mind that wax additives affect the final color.

SCENT

Candle scent is marketed in two forms: liquid scent oil and scent blocks. Although the liquid scent is a higher outlay in cost, it works far better than scent blocks. As a general guideline follow the manufacturer's directions. Higher scent concentrations can usually be used, but too much scent can ruin a candle. Use caution with acrylic molds, since high percentages of scent may ruin the mold.

MOLDS

There are a huge variety of commercial molds on the market, as well as an almost infinite number of everyday items that make good molds. The instructions that follow will be for using a standard commercial mold—a mold that makes the candle upside down. Here is a basic rundown of mold types.

1. Metal molds—Available in a broad variety of shapes, these are simple to use and relatively durable.

2. Acrylic molds—Available in a variety of geometric shapes and sizes. They are easy to use but are easily scratched. Use caution because too much scent can damage these.

3. Two-piece plastic molds—Available in a large assortment of novelty shapes. These are more difficult to use, even though most beginners start with them.

4. Rubber molds—These are available in latex and vulcanized rubber. Both produce seamless candles, with the latex requiring a little more effort to use. Vulcanized molds tend to be expensive.

5. Top-up molds—These are molds that are used the opposite way of most candle molds—with the top of the mold being the top of the finished candle. Many floating candle and votive molds are used this way. These are easy to recognize by their lack of a wick hole.

6. Flat molds—Used to make wax appliqués and hanging ornaments. These generally do not produce good candles, but they do make nice decorations to embellish candles.

MAKING THE CANDLE WITH A MOLD

Step 1

Put enough wax in the melting pot to fill the mold. If a scale is not available, a good estimate may be made by dividing the slab into even sections. For example, divide an 11-pound slab into 11 equal sections to get one pound of wax. Add stearine at the rate of two to three tablespoons per pound of wax. Start heating in a double boiler.

Step 2

While the wax is heating, apply the mold release (gently—a little goes a long way) and then wick the mold. Follow the manufacturer's instructions for this. Prepare a water bath by submerging the empty mold in water and adding water until the level is about one-half inch below the mold top—the water bath is used for cooling in Step 4. Take care not to get any water in the mold or wax—it will ruin the candle. It is easiest to add a mold weight at this time, typically a piece of lead wrapped around the base of the mold. A more difficult alternative is placing a heavy weight on top of the filled mold once it is in the water bath, but the mold must be held down until the weight is in place.

Step 3

When the wax reaches the pouring temperature (refer to manufacturer's instructions for optimum pouring temperature), shut the heat and add the dye (optional). Stir until well dissolved. If desired add scent and stir well immediately before pouring. Set aside remaining wax for Step 5.

Step 4

Pour the wax into the mold slowly but smoothly. On taller molds it sometimes helps to tilt the mold to prevent air bubbles from excessive agitation. Always wear heavy work gloves when handling molds filled with hot wax—especially metal molds. Wetting the gloves will give even more protection if needed. Gently tap the sides of the mold, and allow 45 seconds for the air bubbles to rise. Place the mold in the water bath.

Step 5

Periodically punch one or more holes alongside the wick using a dowel or other long narrow implement. As the wax cools, it shrinks, and punching holes prevents it from shrinking away from the wick, causing air pockets. Fill the void left by shrinkage, taking care not to pour above the original level of the wax. On very large candles, it may be necessary to repeat this step more than once.

Step 6

Allow the candle to cure fully before attempting to remove from the mold. The larger the candle, the longer it takes. If the candle does not easily slide out of the mold, place it in a refrigerator for five to ten minutes. If all else fails, heat the mold with hot water until the candle will come out (this usually ruins the candle). Never pry or scrape the wax out of the mold.

Step 7

If refrigeration was used to unmold the candle, allow it to return to room temperature before proceeding. The final step is to level the base. Place the baking pan atop a pot of boiling water. Holding the candle by the wick, allow it to touch the pan until the base is flat and level.

RECORD KEEPING

One thing often overlooked by candle makers of all experience levels is the importance of keeping records. It would be a shame to develop the "ideal candle" and not be able to reproduce the results. Keeping a notebook handy in the candlemaking area is very helpful. These are some of the things you might want to record.

1. Type and quantity of wax
2. Type and quantity of additives such as stearine, vybar, luster crystals, and so on
3. Type and quantity of dye
4. Type and quantity of scent
5. Type and size of wick
6. Type and quantity of mold
7. Pouring temperature

MAKING CANDLES BY DIPPING—TOOLS AND MATERIALS

1. Container to melt the wax—this has to be as deep as the candles are tall.
2. Wire for the jig—the hanger that holds the wicking while it is being dipped into the melted wax. Each will hold four candles.
3. Rod or rope to hang the candles while they cool

FIGURE 7
The jigs.

4. Thermometer
5. Gas or kerosene stove
6. Paraffin wax
7. Stearic acid
8. Candle wicking

THE JIG

A jig made of wire coat hangers is shown in Figure 7. Each arm of the jig is about 7 cm long. The cords shown in Figure 7 represent the wicks.

The wicking is tied between the two pieces of the jig—the candles will be slightly shorter than the wicks. It is probably most convenient to prepare several jigs at a time.

PREPARING THE WAX

Cut the wax into small pieces, making sure no dirt gets into it. Melt enough wax and stearic acid to fill the container almost to the top. Use one part stearic acid to ten parts wax. Heat the wax to 70°C (158°F). Use a thermometer to check the temperature. If the wax is too hot, it won't stay on the candle, and if it is too cool, the candle will be lumpy.

The safest way to melt the wax is to set the container with the wax in a pot of water so the wax is not directly over the flame—a double boiler works this way. It is very dangerous to let the wax get too hot. Wax catches fire very easily, and a wax fire is difficult to put out. In case

FIGURE 8
Dipped candles.

of fire, cover the container and turn the stove off as quickly as possible. Be careful not to splash the hot wax. It will catch fire if it falls into flame, and it will burn skin.

DIPPING THE CANDLES

Take one of the jigs with wicking and dip it into the melted wax. Hang the jig on the rod to cool. Dip another jig with wicking into the melted wax and hang it on the rod to cool. When all the jigs have been dipped, start with the first one and dip again. Each time a little more wax will stick to the wick, and the candle will get thicker. Continue dipping until the candles are the desired size—Figure 8 shows candles between the jigs.

Do not handle the candles until they are cool and hard. Then cut them off the jigs. Trim the wicks to an even length. Store candles out of the sun and away from heat.

A wide board or plastic sheet under the rod where the jigs are hung can be used to collect wax that drips from the candles. This wax can be scraped off and reheated but must not become contaminated with dirt. Excess wax should not be disposed of in a drain, especially if it is hot (it will clog the drain).

As with molded candles, color and scent can be added to the candles, if these are locally available.

REFERENCE

Volunteers in Technical Assistance. (1988). *Village Technology Handbook*. Vita, 1815 North Lynn Street, Arlington, VA 22209.

CHALK

Everyone who has been in school knows about chalk and chalkboards. Chalk can usually be purchased locally, but it is easy to make yourself. A chalkboard can be made by painting a wooden surface or a heavy cloth with black paint. If paint is not available, a mixture of paraffin, varnish, and soot or powdered charcoal can be used. See the Peace Corps publication in the Reference section for more information.

MAKING CHALK STICKS

Chalk is used in educational facilities throughout the world. Its powder composition allows it to be applied to a chalkboard and then easily removed with a cloth. It takes little time for chalk sticks to be processed. The number of chalk sticks produced at one time can vary from one individual stick or many sticks, using large-scale production. Chalk can be manufactured in white or coloring agents can be added to make colored sticks. Molds are a good tool for larger-scale production with plaster-type material. Sticks can be hand-rolled from material with a doughlike consistency. The chalk can be dried in the sun.

These materials are needed to make one stick of chalk.

6 large, white egg shells

1/2 teaspoon cornstarch

Hot and cold water (1–2 teaspoons each: extra liquid may be needed to reach dough consistency)

1 teaspoon flour

5–8 drops dye/food coloring, if desired

3 small bowls

1 large bowl

Mortar and pestle or a rolling pin and wax paper

Rubber bands, rope, or twine

1 piece of paper, or substitute material, to hold sticks when drying

Remove the entire contents of the eggs: yolks and whites. Rinse out the shells and let them dry. Take the dry eggshells and grind them into a fine powder, using any type of tool that has a flat surface face, such as a stone, bottom of a cup, or a rolling pin. Before crushing, place the eggshells between two sheets of wax paper or any other thin paper. This will keep the powder from blowing away and also prevent any foreign objects from getting in the mixture.

Put the Eggshell Powder in a Small Bowl

In a different bowl, mix 1/2 teaspoon of cornstarch with 1/2 teaspoon of COLD water. Take another bowl and mix 1 teaspoon of flour with HOT water. The hot-water mixture should be the consistency of mashed potatoes. Start with 1/2 teaspoon and add drops from there.

Combine the contents of all three bowls in another bowl. Mix the ingredients together to make a doughlike substance, squishing it in between your fingers.

Begin to knead the dough together with your hands until it looks and handles like bread dough. It should not be wet and sticky. If it is, add a pinch of flour at a time until it is less sticky. However, if it is brittle and comes apart when handled, add a bit more HOT water. Make sure you add HOT water or you will not get the desired consistency.

Use the palms of your hands to roll the dough into a stick of the desired size. The thicker the stick, the stronger it will be when writing with it.

Take the stick and wrap a piece of paper around it. Secure it with an elastic band, rope, or twine—anything to keep it tightly wrapped.

Allow the chalk to dry in the sun for a few days. Remove the paper, and let dry for two more days.

Tip: If colored chalk is desired; add food coloring during the stage in which flour is mixed with HOT water.

Warning: This chalk is quite abrasive and can leave marks on wood, ceramics, and various wall types.

Alternative Methods

You can also make standard chalk with plaster of paris. Plaster of paris is used for large-scale chalk production. This product is a fine white gypsum plaster that is used in the making of cast molds, but is an excellent source for chalk production. Plaster of paris chalk requires the use of molds, unlike that of the hand-rolled sticks. Once the plaster is mixed (see the "Gypsum" article in the Construction chapter), it is poured into the mold. One problem with this chalk is that when it is pouring into the mold, it begins to dry immediately, even before the mold is full. One way to avoid this is to add household vinegar to the mixture. You have to experiment with the amount of vinegar—begin with a teaspoon, then add from there. The vinegar slows down the hardening time so you can fill the mold completely. It will set in just a few minutes. Push the formed sticks out of the molds and set them in sun to dry for a day or so. If colored chalk is desired, add coloring agent to the plaster.

MOLDS

Molds are preferable when it comes to allowing the chalk to dry and creating uniformly shaped sticks. Molds are necessary when the mixture is of a liquid consistency. Molds can be made out of various materials: wood, plaster, and plastic. When using molds avoid breakage when the chalk is removed. One problem when pouring the plaster into the molds is that the mixture does not reach the bottom of the mold. The bottom of the mold must have a clear exit path, allowing air to exit during the pouring process, preventing air pockets, and creating a dense chalk compound.

REFERENCE

Nonformal Education Methods. (1989). Prepared for Peace Corps by The Institute for Training and Development. Contract # PC-8 89-2016.

POTTERY

Pottery has been made in southern Africa for thousands of years. Traditional pots are used for cooking, carrying and storing water, brewing beer, and similar uses. Large pots, used for storing grain or brewing beer, are 1 meter or so high. Many of the pots produced have great aesthetic value.

Ceramic items—plates, pitchers, drinking mugs—are made in small-scale enterprises for sale in urban and export markets. Initiating such a venture could be a significant contribution to a village. If suitable clay is available in a community, most probably some local people will be proficient at making pots, although the skill may be less common now than it had been. Supporting a local pottery industry may be worthwhile simply to preserve traditional expertise. A small-scale ceramics workshop could also bring cash to the village. Of course, before starting an enterprise of this sort one should examine the market carefully—in particular, how will the output be made available to customers? (Marketing, including distribution, is discussed in the Planning and Implementation chapter.)

MANUFACTURE OF POTS

Making a traditional pot involves the following steps.

1. Collecting the clay
2. Grinding the clay to a fine powder
3. Molding the pot
4. Decorating, smoothing, and drying
5. Firing

A commercial ceramics enterprise will follow the same steps but will use different technology.

Clay is collected from layers in the subsoil, often exposed by rivers, road construction, or on hillsides. Clay is easily recognizable when wet because it is slick and shiny and has water puddles on it. Low-lying areas, especially those where water collects, probably have clay under one to four feet of peat or muck.

Two simple field tests will help to establish whether a deposit is clay or not. Moisten and knead an egg-sized piece of material. If it does not hold together or retain fingerprints, it is probably not clay. If a rolled-out pencil-sized coil cannot be wrapped around a finger without cracking, then it is probably not clay. The real test of whether the material is clay is whether it becomes hard and strong after firing.

Traditional potters grind dried clay by hand on a flat stone. In a small-scale ceramics enterprise dried clay may be grounded in a ball mill powered by hand or electricity. (A ball mill is a barrel rotated fairly rapidly around a horizontal axis with metal balls rolling loose inside.) Once the clay is reduced to a fine powder, it should be put it into water to slake. When it is soft—this can take a few hours to a few days, depending on the clay—it should be stirred vigorously with a paddle or by hand, adding water as necessary, until the clay is the consistency of thin cream. Finally, it should be strained through a window screen or a 30-mesh sieve. The resulting clay/water mixture can then be dried or used immediately, after excess water has been removed from the top of the mixture.

The steps for molding and drying a pot are illustrated in Figure 9. Molding in a small-scale factory is probably best done with a potter's wheel, which is much faster than hand molding. An automobile wheel, tire, and bearing can be the basis of a hand- or foot-powered wheel. Loading the tire with rocks will increase the momentum and keep the wheel going longer. If the wheel is propelled by kicking, then a stand that brings the clay being molded up to a convenient height is useful. Slip casting using plaster molds is an alternative molding technique.

After the pot is formed decorations can be incised. A piece of gourd can be used, as shown in Figure 10. A pointed object like a bone awl can also be used. Sometimes decorations are made by pressing beads into the neck of the pot. Ground graphite powder can also be applied at this stage—it gives a shiny, dark surface to the finished pot. Then the pot must be dried sufficiently to allow handling. After drying, a further smoothing should be done. Traditional pots

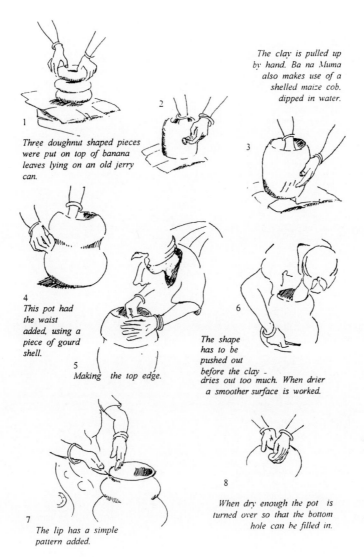

The clay is pulled up by hand. Ba na Muma also makes use of a shelled maize cob. dipped in water.

1 Three doughnut shaped pieces were put on top of banana leaves lying on an old jerry can.

2

3

4 This pot had the waist added, using a piece of gourd shell.

5 Making the top edge.

6

The shape has to be pushed out before the clay - dries out too much. When drier a smoother surface is worked.

7 The lip has a simple pattern added.

8 When dry enough the pot is turned over so that the bottom hole can he filled in.

FIGURE 9
Making a clay pot. (From *Inventors of Zambia*.)

are then fired after another 48-hour drying period. Ceramics made for sale would be glazed before firing. Generally, commercial glazes are used, although feldspar and silica can be used. The glazed ware must be dried before firing. Any moisture left in the glazed ware will cause the ware to crack and fall apart during the kiln firing process.

Figure 10 shows simple pit firing. Pit firing involves placing unfired pottery in a pit in the ground, then covering the pottery with suitable burning materials such as dried grasses and branches. A pit of the appropriate size is dug, depending on the amount of work to be fired. A bed of dry leaves and twigs and possibly coal, which will burn slowly, is placed at the bottom of the pit and the pottery placed on top of this. The work is then covered with more leaves and twigs and dung, if available, building up a mound over the pieces. Once the stacking process is finished, the pile can be lit around the edges and left to smolder for several hours or overnight. (Figure 10 indicates 15 minutes, but a longer time is better.) Traditional pots can be made

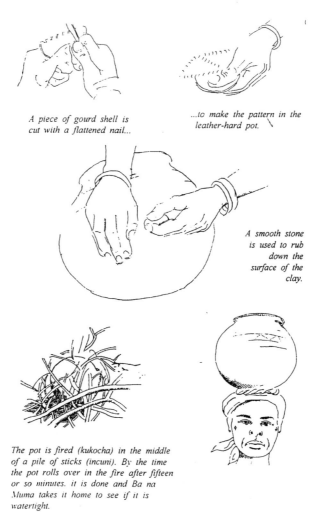

A piece of gourd shell is cut with a flattened nail...

...to make the pattern in the leather-hard pot.

A smooth stone is used to rub down the surface of the clay.

The pot is fired (kukocha) in the middle of a pile of sticks (incuni). By the time the pot rolls over in the fire after fifteen or so minutes. it is done and Ba na Muma takes it home to see if it is watertight.

FIGURE 10
Decorating a pot and pit firing. (From *Inventors of Zambia.*)

watertight by filling them with a mixture of water and maize meal or water and cow dung. The mixture releases a sealing resin.

A commercial installation uses a kiln for firing. Figure 11 shows a kiln. A kiln can be made by the potter; advice from an experienced potter would be helpful. The kiln may be heated by firewood, which may limit its cost to only labor. A steel grate and refractory bricks, which will endure intense heat, will make stacking the items easier but will add to the cost. A kiln may be heated electrically, which will require heat-resistant wiring, temperature controls, automatic switches, and, of course, electric power. The pottery is stacked in the kiln. A hot fire is created, and the pots remain in the kiln for at least several hours. The optimum length of time for firing depends on the clay and the heat generated, so experimentation is necessary.

Adequate ventilation of the kilns is necessary to help eliminate hazardous dust particles from the atmosphere, and those in continuous contact with dust should wear masks. If the products are food containers, a facility for conducting safety tests must be readily available.

After firing, the pot or other ceramic item is ready for use, although further decorations are sometimes painted on. Traditional paints came from iron ores and graphite, but commercial paints are used now.

FIGURE 11
Village kiln. (From *Inventors of Zambia.*)

REFERENCES

Crutchley, Victor. (1996). *Inventors of Zambia*. Bridport, UK: Eggardon Publications.
Ellert, H. (1984). *The Material Culture of Zimbabwe*. Zimbabwe, Harare: Longman.
Palmeri, Victor R. (1988). *Small Ceramics Plant*. VOLUNTEERS IN TECHNICAL ASSISTANCE, 1600 Wilson Boulevard, Suite 710, Arlington, Virginia 22209, USA. Tel.: 703-276-1800. vita@vita.org.
Petersham, Miska. (1984). *Understanding Clay Recognition and Processing*. Technical Paper #13. VOLUNTEERS IN TECHNICAL ASSISTANCE, 1600 Wilson Boulevard, Suite 710, Arlington, Virginia 22209, USA. Tel.: 703-276-1800. vita@vita.org.

BASKETRY

Based on Botswana Baskets, *The National Museum and Art Gallery, PO 114, Gaborone, Botswana*

Basketmaking is an ancient art that endures to the present. Baskets are both useful in everyday life and in generating cash. While it is unlikely that a basketmaking project could be successfully started in a community that did not have a tradition of making baskets, a project reviving that tradition could be beneficial to the community. Well-made baskets, especially those reflecting ancient patterns, are desired in the export market. The purpose of this article is to give an understanding of how baskets are made so the reader can work effectively with basket makers (see Figure 12).

The author of *Botswana Baskets* states the following.

An apprenticeship of approximately two years is required to be able to make an adequate basket. It takes another four to six years to master all the stitches, motifs, and designs.

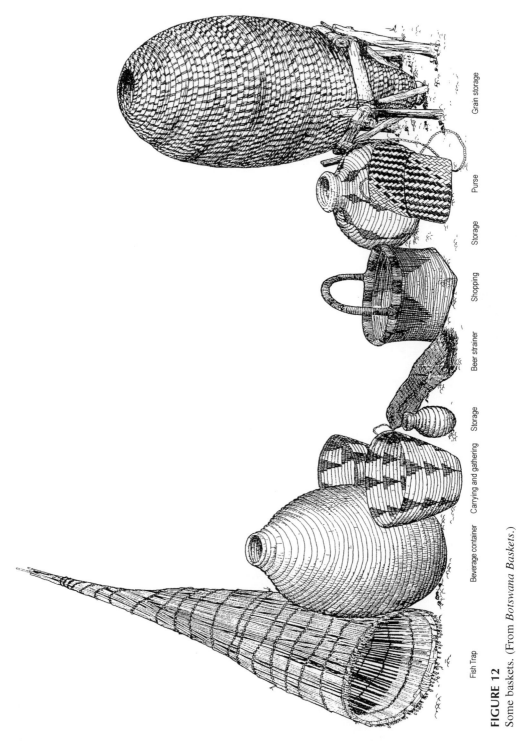

FIGURE 12
Some baskets. (From *Botswana Baskets*.)

Fish Trap   Beverage container   Carrying and gathering   Storage   Beer strainer   Shopping   Storage   Purse   Grain storage

The best baskets are usually created by women with over ten years of experience. Most of the highly skilled basketmakers are about 30 years of age and have several school-aged children, so their incentive for additional income is high. Older women, too frail to work in the lands effectively, find basketry a primary source of income. These women, with 30 to 40 years of experience, are likely to produce the masterpieces.

MATERIALS FOR MAKING BASKETS

Materials for making baskets depend on what is available—grasses, bark, vines, reeds, and the fibers of "mother-in-law's tongue" or *sansevieria*. The *sansevieria* has in some places been replaced by exotic sisal with its longer threads. The fan-shaped leaf of a fan palm (*Hyphaene ventricosa*) can be used when available. The spine of the leaf is used for the coil; the split leaf is used for the binding. These broad swordlike fronds provide a flexible and pliable fiber for basketry.

Dyes can be made from roots and bark. The palm fronds are boiled with appropriate root bark to dye the fronds various shades of brown. Commercial dyes, even ink from marking pens, can also be used to color reeds and other fibers. Purple and red patterns on some baskets made in Malawi were made using marking pens to color the fiber.

The tools required include a bowl of water to soak the leaves and keep the basket maker's hand moist while she works. Other equipment consists of a knife to cut materials and to split leaves to bind the coil; a pounding stone to flatten grass and leaves and to beat out unwanted twigs, and an awl to make the hole through which the binding material can pass. The awl can be made from any available bit of metal, flattened and pointed and then filed smooth on a stone. Some basketmakers now use large darning needles instead of the old homemade version.

COIL BASKETS

A common way of making baskets uses a coil of fiber, similar to the technique used in making pottery. Coils are prepared by one of two methods: bundles made of grass or palm-leaf spines forming a long continuous strand, or a single core coil is made from long vines or roots. To make a basket, the craftsperson starts stitching from the central coil and works progressively to the outer rim. Some stitches are shown in Figure 13. (The "plaited (purse)" is for a woven, not coil, basket.) The pattern is made from multiple stitches or embroidered stitches, as shown on the bottom of Figure 13.

The starting point for weaving a coil basket may be a simple coil. Other, more elaborate designs may be used, such as a knot or a woven square. Special weaving is also often used to bind the rim of the basket.

TWINED BASKETS

"Twined" here means woven. The basket is begun by weaving a three-inch or so square base. The fibers making up the base are then folded up to form the framework for the basket sides. Other fibers are woven around these upright fibers to create the basket. Two threads are used together: one thread being passed over and the other passed under the upright fibers—they cross between uprights. Such baskets have thinner walls than coil baskets and tend to be more malleable. To make a stronger basket, additional vertical fibers can be inserted during the weaving.

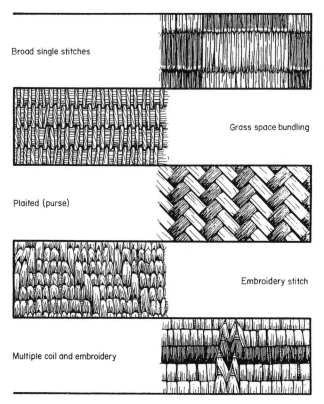

FIGURE 13
Basket stitches. (From *Botswana Baskets.*)

The rims are finished off with decorative, braidlike edging. The top of the basket may be either circular or square shaped.

MATS

Mats are used for sitting or sleeping. They are woven loosely, without a loom, using river reeds or round water grass as the warp. The weaving thread, weft, can be sisal fiber or a strip of water grass, although purchased thread is often used now. Interesting patterns can be created from the weaving thread or by crossing the reeds periodically.

Tighter mats can be made entirely of reed or of baobab bark. One traditional use for these is near grindstones, where the mat catches the ground grain.

Beer strainers are made the same way. Reeds are used, beaten to remove the fleshy part and then spun into long strands by rolling between the palm of the hand and the bare thigh. The bottom of the strainers are tightly woven and gathered to form the straining end. (This gathering is not clear in Figure 12.) A loop may be attached to the bottom to hang the strainer after washing.

REFERENCE

Marjorie Locke. (1994). *The Dove's Footprints Basketry Patterns in Matabeleland.* Harare, Zimbabwe: Baobab Books.

MAKING ARTICLES FROM WASTE PAPER

Based on Appropriate Paper-Based Technology (APT) a Manual, *Bevill Packer, Intermediate Technology Publications Ltd, London, 1995*

Appropriate paper-based technology (APT) is a way of making a wide range of strong and useful articles from waste paper. Such waste can be small scraps, strips, sheets, or pages of paper of any kind; thin cards from shoe or shirt boxes; and thick or corrugated cards from old cartons. The technology is virtually cost-free. Flour to make the paste and varnish—if it is used to finish articles—is the only purchased material needed. Articles that can be made include tables, chairs, boxes, coffins, lampstands, toy cars, dolls and dollhouses, and solar cookers.

CONSTRUCTION PROCESS

In basic terms, APT consists of applying paste to paper and, with the help of some kind of pressure, fashioning it to make an article that dries to a woodlike material. After all, paper is wood fiber, so the process makes sense.

APT can use any kind of paper, and the paper can be in four different forms.

- Odd pieces of paper and cards soaked in water, then ground to a pulp. This is called mash or papier maché.
- Paper in sheet form torn into strips
- Pieces of thin card from cereal or shirt boxes used in sheets or strips
- Pieces of thick corrugated card from old cartons

The articles are made in four stages.

1. Building the article: Several processes are possible. One is layering pasted strips of paper. Another is joining two vertical paper-layered boards to make a base and joining this to a top board. A third is to mold mash around a frame. In each case the object must be given time to dry.
2. Strengthening and drying: This includes cutting or rubbing smooth any uneven edges and binding them neatly, strapping and bandaging all joins with strong paper strips, carefully covering the whole structure to give it a smooth, tidy appearance, and allowing the article to dry.
3. Decorating: This is done in a variety of ways but always involves making the surface of the article wet with paste and, after decorating, letting it dry completely.
4. Finishing: This consists mainly of applying varnish to strengthen and protect all surfaces of the article and allowing it to dry.

Actual drying time varies from one hour to a week, depending on the size of the articles, material used, the amount of paste used, and the weather. When building some articles, especially furniture, there is a preliminary stage of making and drying components. An implication of the length of the drying time is that no article can be completed in a day, and five days is too short for a training course.

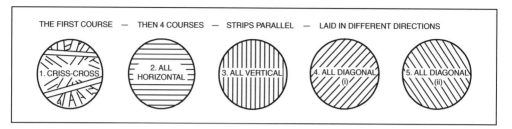

FIGURE 14

Layering. (From *Appropriate Paper-based Technology.*)

PAPER

Nearly any kind of paper can be used except paper covered with a waterproof coating. When laminating, the thicker the sheet, the quicker and easier the job. Paper should be torn or folded in the direction of the grain—the direction the fibers lie. The direction of the grain can be detected by tearing the paper—the tear will naturally go straight along the grain. The absorbency of the paper should be noted. For paste to do its work, it must penetrate the surface layer at least a little. For good layering, the paper or card should be damp right through, so it is thoroughly stretched. All paper and card swells and stretches as it is penetrated by moisture from paste, water, or very humid air, although varnishing largely prevents air penetration. As the material dries, it contracts and exerts a very strong pull. When layering, use different grain directions, as shown in Figure 14, and work rapidly so the high shrinkage in one direction is resisted by small shrinkage in another direction.

PASTE

The paste is made from some kind of flour because flour has the following properties.

- When the paste is well made, the article is very strong because it penetrates both surfaces and bonds them tightly.
- Wet flour paste softens the pieces to be joined so they can be crushed, rolled, bent, and molded to lie flat to get a good fit with joins and improve the article's contours.
- Flour paste is very cheap. If old *sadza* (the traditional maize flour porridge used in southern Africa) is used, it is free.
- Flour paste is a clean adhesive that does not harm hands and can be easily washed off skin, clothes, floors, and so on.

Paste can be made from leftover *sadza* but a better paste is made directly from flour. To make approximately 500 cc of thin paste *throughly* mix a heaped teaspoon of flour with a *little cold water* until it is about the consistency of cream and there are no lumps. *Rapidly* pour into the mixture about 500 cc of *boiling water, stirring hard. Continue stirring* for a while. The theory behind the method is that stirring in cold water floats and separates the flour's grains so that when boiling water is poured into the mixture, each grain is touched and releases gluten, which is a sticky substance.

Use hands to apply paste. Flour paste is not a contact adhesive and requires pressure to do its job.

MASH (PAPIER MACHÉ)

There are many ways of making mash. Soak pieces of paper or card—for example, newsprint—in water, preferably overnight or longer. Take handfuls of the soaked paper, and with a tool, such as a rough stone, rub, grind, or pound it into a pulp. The finer you *grind* it, the easier it is to work. Squeeze very hard to expel water. Put the mash in a strong plastic bag. Take a little thick paste or break some old *sadza* into little crumbs and work it into the mash until it has the consistency of modeling clay. Only use a little paste or *sadza*, or the finished article articles may attract weevils. The process takes only a few minutes, but the finished article is not waterproof.

Very small articles, such as jewelry or chess pieces, can be made just by building them of mash. Normally, though, an armature or frame is made over which the mash is molded. The armature is first smeared with paste. The mash is pressed onto the armature and moulded to the shape desired. The mash should be 1cm thick, not less. As the mash article dries it should be examined frequently. If any cracks develop they should be squeezed together. In doing so the shape of the article can be improved, if necessary. After drying the article should be covered with at least three layers of an absorbent paper, applied in strips and small pieces.

LAYERING

Layering is used in the production of every article. It consists of placing and pressing pasted strips of paper or card on to the surface of an article or a mould. Layering is used to strengthen, tidy, and decorate the article. Each complete covering of the object by layering is called a "course."

Layering can be used to actually make an item by applying courses of three- or four-layered strips instead of single pieces. This is called thick layering. See Figure 14. If a bowl, for example, is built using this technique the following advice applies:

- To build an average-size bowl or similar article by layering 15 to 20 layers of paper will be needed. To save time, these are applied as three- or four-layered strips. Five courses are applied to complete the job.
- Tear paper and card rather than cutting it. Tearing is quicker and leaves a beveled edge that sticks down firmly. Tear along the grain if possible.
- To neutralize the pull of drying, each course is applied in a different direction.
- The work is continually pressed, making it tight and squeezing out excess paste.
- Thick layering can be applied on the inside of a mold, for example, a bowl. The process is not difficult and small children can do it. Extra paste usually has to be squeezed out afterwards. A bowl made inside another bowl may not have a good flat base on which to rest.

A more detailed diagram showing layering of paper to make a bowl is shown in Figure 15.

A stool built by layering over a mold is shown in Figure 16. The reference to "P + P" in the Figure is to "Paste and Pressure."

TUBES

Tubes are made in different sizes and strengths according to their function. Strong hard tubes are used for lamp stands, pillars, legs, rails for furniture, for pegs, hinges, axles, bearings and so on. Hard tubes are usually made of card. However, paper an especially the strong outside paper of a corrugated card box can make excellent tubes, although they take longer to dry. Soft tubes of

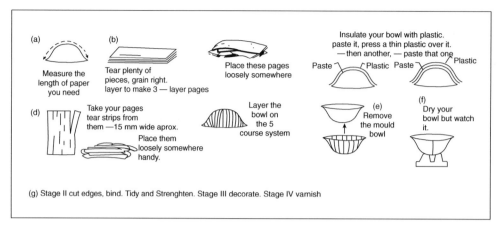

FIGURE 15

Making a bowl by layering. (From *Appropriate Paper-based Technology.*)

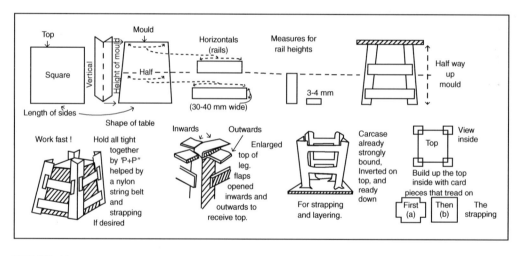

FIGURE 16

Making a square stool. (From *Appropriate Paper-based Technology.*)

paper or card are used in construction work, particularly to bond and strengthen long angle joins. They are crushed flat while still soft and folded along their lengths to make an angle piece.

The process of making a tube is shown in Figure 17. First, find or make a smooth roller. A broomstick is a good size for a roller to make a table leg or rail. Polyvinyl chloride (PVC) tubing is also excellent for the job. If reeds or sticks are used they must be of even thickness and free of ridges or bumps on the surface.

Next, prepare for rolling. Place the card on a flat, clean plastic surface with the grain lengthwise. Preferably, see that the front and back edges of the card are torn and not cut. Smear paste lightly over the card but check that it covers the card to its edges. Paste the roller well.

Then, place the roller along the grain. Roll the card tightly on to the roller and continue rolling it in your hands until the edge of the rolled card sticks down, at which point pull the tube off without delay. If necessary, stick the edge down with a pasted paper strip. Finally, dry the tube off the ground and turn occasionally to avid the risk of bending. Soft tubes are made

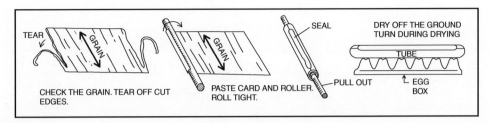

FIGURE 17

Making tubes. (From *Appropriate Paper-based Technology.*)

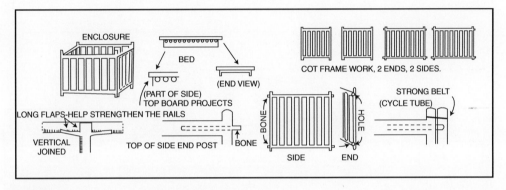

FIGURE 18

A child's cot. (From *Appropriate Paper-based Technology.*)

of soft card or thick paper and can be rolled without a roller. More details are given in the book cited.

A child's cot is shown in Figure 18. The four sides of the cot are held together by strong bands of cycle tubing placed over the tops and bottoms of each pair of corner posts. The diagram shows a thin paper tube, called a "bone," to reinforce the connection to the end posts. The diagram also shows how the vertical posts are connected to the horizontal ones—by splitting and folding out the ends of the vertical posts. It goes without saying that the posts and horizontal rails must be thick and strong.

MAKING ROPE

Laura Voorhees

Rope has been made for thousands of years. The earliest ropes were heavy vines or bands of hide tied together. Rope can be made out of just about any material that consists of long and flexible strands: yarn, binder twine, plastic twine, bailer twine, jute twine, string, and plastic bags, to name a few. Only strong and supple twines should be used to make rope that will be load bearing or used in structures.

This article gives instructions for making a three-strand laid rope with a right-hand lay, the most common type of rope. A right-hand lay has the primary strands spiraling in a clockwise

FIGURE 19
Right-hand lay.

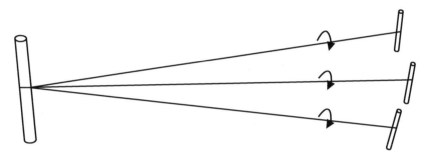

FIGURE 20
Sticks holding strands.

direction, as shown in Figure 19. Fibers are loosely twisted together to form yarn; two or three yarns are twisted together to make a strand; and three strands are twisted together to make rope. The premise of rope making is rather simple: Three strands of fiber are twisted in one direction, then the strands are then laid together and twisted in the opposite direction.

MAKING ROPE BY HAND

Four people will be needed to make rope by hand. Determine the desired length of the rope you are going to make, and choose the appropriate length of jute, sisal, or other twisted fiber twine. Keep in mind that the rope will be about 20 percent shorter than the original lengths of twine due to the twisting. Examine the already existent twist of the twine. In the finished rope, the pieces of twine will serve as the strands–see Figure 19. Cut three equal lengths of twine, and at one end of each strand, tie a small loop. Place these loops around a stick—see Figure 20. One person will hold onto the stick while the other three will each take a loose end of a strand. From the loose ends, the three people will twist the twine in the natural direction of the strands in order to tighten the twist. This twisting may be facilitated if another small loop is tied at the loose end and another stick is inserted (Figure 20). It is important for the person holding all three ropes to keep consistent tension while the other three are twisting.

Once the three strands are twisted enough to tighten, the three loose ends can be given to one person, while maintaining tension on the strands. This person will twist the three tightened strands together in the opposite direction than before (Figure 21). The two other people should monitor the lay of the rope, making sure that the strands come together evenly.

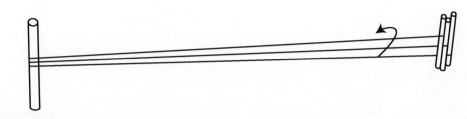

FIGURE 21
Twisting the rope.

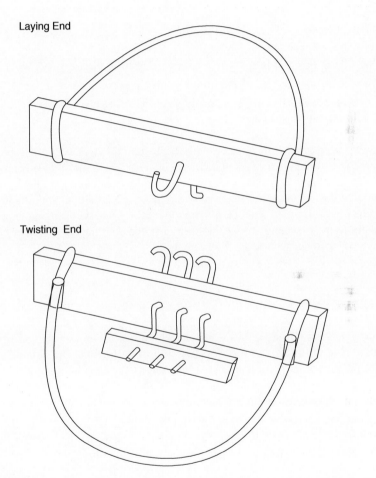

Laying End

Twisting End

FIGURE 22
Rope-making machine.

A SIMPLE ROPE-MAKING MACHINE

Making rope by hand is time consuming. If you plan on making larger quantities of rope, it may be preferable to build the simple machine shown in Figure 22. The machine is made of two devices, the twisting end and the laying end. Making a Y tool, or spreader, out of hard wood is also advisable. The twisting end always remains stationary, and the laying end must be able to move closer to the twisting end, since the rope gets shorter after twisting. After the two ends of the rope machine are correctly positioned, tie one end of the twine to the laying end hook. Then hook the twine over a hook on the twisting end and then back again over the laying end. Repeat this back and forth process until each of the hooks on the twisting end have two lengths of twine on them. It is necessary to keep equal tension on all lengths of twine from the laying end throughout the rope making process. The laying end must also remain upright at all times. You can then start turning the twisting the end clockwise. Twist the twine until it becomes firm and resists further twisting; the rope will be loose if you do not twist enough. Overtwisting will result in kinks. With practice, you will learn how much twisting is compulsory to make a strong rope.

Here are some useful tips.

- Make the twisting end out of hard wood. After long periods of winding, the holes become oversized if you use pine.
- Tie back long hair and keep loose clothing away from the rope-making activity.
- If the strands wind together before laying, they must be separated.
- If you soak natural fiber rope in soapy water, it will tighten and improve the rope, and it will help the rope last longer.

The laying of the rope will be done with the Y tool; this will keep the rope under control. The laying end is turned in a counterclockwise direction while the Y tool is moved toward the twisting end. As you are laying the rope, make sure that you cannot push the Y tool back toward the laying end; this means either the rope is too loose or you are moving the tool too quickly.

When the rope is finished, unhook it from the machine and tie overhand knots at the ends to prevent them from unraveling. To set the fibers and even out the twists, either beat the rope on the ground, roll it under your foot, or stretch it.

SOME USEFUL KNOTS

The *figure eight* knot (Figure 23) is a wonderful stopper (a knot at the end of a line). It is a good knot to hold the strands of rope together to prevent fraying. The figure eight is also used to obstruct the rope from pulling through something it has been fed through.

1. Make an underhanded loop and bring the free end around and over the working end.
2. Pass the free end under and then through the loop.
3. Pull on the free end to tighten the knot. The finished knot should look like two interlocking loops, like a figure eight with ends trailing out of each loop.

The *clove hitch* knot (Figure 24) is used when you need a simple and quick means to fasten a rope around a stationary object, commonly a post or a stake. The knot is tied around the object rather than the working end of the rope.

1. Loop the free end over the object, then pass the free end under the object.
2. Bring the free end up, and cross it over the top of the working end.
3. Pass the free end over and under the object again in the same direction as the first loop.

FIGURE 23
Figure eight knot.

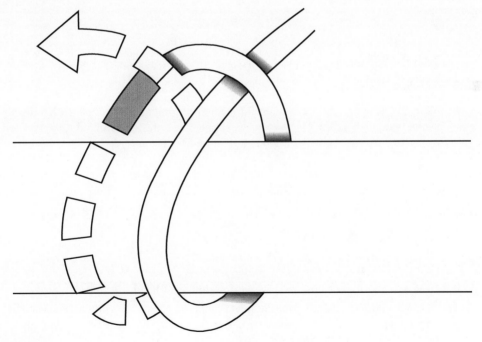

FIGURE 24
Clove hitch.

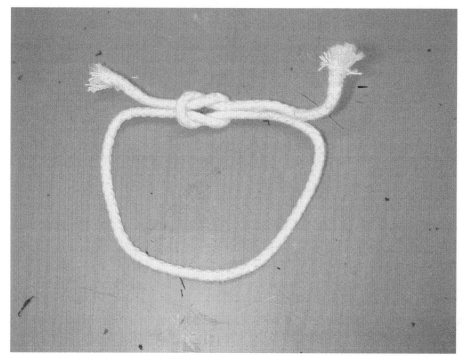

FIGURE 25
Square knot.

4. Pass the free end through the place where the rope crosses over, and pull the knot from both ends to tighten.

The *square knot* (Figure 25) is a very useful method of joining two pieces of rope together. It can be useful to tie up bundles. The drawbacks are that the knot will not hold if the two ropes are of different sizes or material, and it tends to jam under tension.

1. Pass one rope left, over, and then under the right end. Curve what is now the left end toward the right end, and cross what is now the right end over and then under the left.
2. Draw the knot tight.

The bowline (Pronounced "BO-lin") knot (Figure 26) can be used to form a loop at the end of a rope. It never slips or jams, and it is easy to untie.

1. Take the working part of the rope in your non-dominant hand.
2. Lay the free end, which is in your dominant hand, on top of the working end, making a loop (this loop should be made small because later there will be a larger loop, and it is necessary to distinguish between the two).
3. Move the free end under the loop and pass the free end up though the first loop you made.
4. Pass the free end over the working end, then around behind it, and then back down the small loop again.

FIGURE 26
Bowline.

5. Hold the free end and the side of the larger loop together and the working end in your other hand. Pull on the working end until the knot is snug. Then you can tighten the loop you have created.

REFERENCES

Graumont, Raoul, and John Hensel. (1952). *The Encyclopedia of Knots and Fancy Rope Work.* Centerville, MD: Cornell Maritime.

Ashley, Clifford W. (1990). *The Ashely Book of Knots.* New York: McGraw-Hill.

Owen, Peter. (1993). *The Book of Outdoor Knots.* New York: Lyon Press.

Findley, Gerald L. (1996). *Rope Works.* ROPE WORKS. http://www.northnet.org/ropeworks/.

RECYCLING AND REUSING RUBBER

Based on AT Brief #25 "Recycling and Re-using Rubber," Appropriate Technology, Vol. 25, No. 2, September 1998

Rubber is produced from either natural or synthetic sources. Natural rubber is made from a milky white fluid called latex that is found in many plants; synthetic rubbers are produced from unsaturated hydrocarbons. In practice, the main source of rubber in the South, that is, the Third World, is scrap tires and inner tubes, which are made from synthetic rubber. This article looks at the small-scale reclamation and reuse of rubber tires and tubes.

TABLE 1
Types of rubber in the manufacture of vehicle tires

Type of Rubber	Application
Natural rubber	Tires for commercial vehicles such as trucks, buses, and trailers
Styrene-butadiene (SBR) and butadiene rubber	Tires for small trucks, private cars, motorbikes, and bicycles
Butyl rubber (BR)	Inner tubes

SYNTHETIC RUBBER

The different types of synthetic rubber include neoprene, Buna rubbers, and butyl rubbers, and they are usually developed with specific properties for specialist applications. Styrene-butadiene rubber and butadiene rubber (both Buna rubbers) are commonly used for tire manufacture. Butyl rubber, since it is gas-impermeable, is usually used for inner tubes. Table 1 shows the typical applications of various types of rubber.

Tires are made up of both natural and synthetic rubbers, together with carbon, nylon or polyester cord, sulphur, resins, and oil. During the tire-making process, these are vulcanized into one compound product that is not easily broken down.

PRODUCTION OF RUBBER PRODUCTS

Modern rubber manufacture involves a sophisticated series of processes such as mastication, mixing, shaping, molding, and vulcanization. The following additives are included during the mixing process to give desired characteristics to the finished product.

- Fillers (like carbon black)
- Anti-degradants
- Vulcanization accelerators
- Vulcanization agents
- Colorants or pigments
- Polymers
- Activators
- Plasticizers
- Fire retardants
- Softeners

Fillers are used to stiffen or strengthen the rubber. Carbon black is a commonly used anti-abrasive. Pigments include zinc oxide, lithopone, and a number of organic dyes. Softeners, which are necessary when the mix is too stiff for proper incorporation of the various ingredients, usually consist of petroleum products, such as oils or waxes, pine tar, or fatty acids.

The molding of the compound is carried out once the desired mix has been achieved, and vulcanization is often carried out on the molded product.

VULCANIZATION

Vulcanization gives rubber its characteristic elastic quality. This process is carried out by mixing the latex with sulphur (other vulcanizing agents such as selenium and tellurium are occasionally used but sulphur is the most common) and heating it in one of two ways.

1. Pressure vulcanization. This process involves heating the rubber/sulphur under pressure at 150°C. Many articles are vulcanized in molds that are compressed by a hydraulic press (see Figure 27).

2. Free vulcanization. Used where pressure vulcanization is not possible, such as with continuous, extruded products, it is carried out by applying steam or hot air.

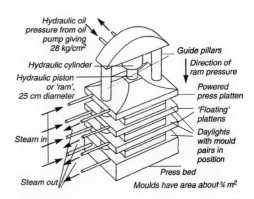

FIGURE 27
Typical small-scale rubber press.

WHY RECLAIM OR RECYCLE RUBBER?

Rubber recovery can be a difficult process. There are many reasons, however, why rubber should be reclaimed or recovered.

- Recovered rubber can cost half that of virgin natural or synthetic rubber.
- Recovered rubber has some properties that are better than those of virgin rubber.
- Producing rubber from reclaim requires less energy in the total production process than does virgin material.
- It is an excellent way to dispose of unwanted rubber products, which can otherwise be difficult and environmentally hazardous.
- It conserves nonrenewable petroleum products, which are used to produce synthetic rubbers.
- Recycling activities can generate work.
- Many useful products are derived from reused tires and other rubber products.
- If tires are incinerated to reclaim embodied energy, then they can yield substantial quantities of useful power. In Australia, some cement factories use waste tires as a fuel source.

RECOVERY ALTERNATIVES

There are many ways in which tires and inner tubes can be reused or reclaimed. The waste management hierarchy dictates that reuse, recycling, and energy recovery, in that order, are superior to disposal and waste management options. This hierarchy is outlined in Table 2 below.

PRODUCT REPAIR

Reuse results in the greatest savings of energy and resources, so damaged or worn tires are, wherever possible, repaired. Regrooving is a practice carried out in many developing countries where regulations are less stringent and standards (and speeds) lower than in the West. It is often

TABLE 2
Principal rubber recycling processing paths

Kind of Recovery		Recovery Process
Product reuse	Repair	Retreading
		Regrooving
Secondary reuse		New products
Material reuse	Physical	Tearing apart
		Cutting
		Processing to crumb
	Chemical	Reclamation
	Thermal	Pyrolysis
		Combustion
	Energy reuse	Incineration

carried out by hand and is labor intensive. Regrooving is difficult to do safely on automobile tires, but truck tires are often made with some regrooving in mind.

Another option is to retread the tire. Retreading is a well-established and acceptable (in safety terms) practice. The process involves the removal of the remaining tread (producing tire crumb—see later) and the application and vulcanization of a new tread (the "camel back") onto the remaining carcass. The tires of passenger vehicles can generally be retreaded only once, while truck and bus tires can be retreaded up to six times.

SECONDARY REUSE

The secondary reuse of whole tires is the next step in the waste management hierarchy. Tires are often used for their shape, weight, form, or volume. Whole tires are often used for erosion control, tree guards, artificial reefs, crash barriers, docking fenders, fences, or as planters. They can also be used to hold a large or heavy item in place, such as tarpaulin covering or corrugated iron sheeting. In the South there are additional uses, such as lining wells.

PHYSICAL RECOVERY

The next step in the hierarchy involves breakdown and reuse. The rubber used in tires is relatively easy to reform by hand. It behaves in a similar manner to leather and has, in fact, replaced leather for a number of applications. The few tools required for making products directly from tire rubber are not expensive. Shears, knives, tongs, hammers, and so forth are all common tools found in the recycler's workshop, along with a wide range of improvised tools for specialized applications or more sophisticated tires. The materials are recovered mechanically. The process is very labor intensive, particularly as tires become more sophisticated and new materials are used for strengthening and reinforcement. Shoes, sandals, buckets, motor vehicle parts, washers, bushes, doormats, harnesses, hinges, water containers, pots, plant pots, dustbins, and bicycle pedals are among the many products manufactured, in addition to the production of other tires, such as for carts or wheelbarrows. Old inner tubes also have many uses—swimming aids and water containers are two examples. Some of the products that result from the reprocessing of tires and tubes are particularly impressive, and the levels of skill and ingenuity are high. Recycling artisans have integrated themselves into the traditional marketplace, and both tire collection and reuse is carried out primarily by the informal sector. Making shoes from recycled tires is shown in Figure 28.

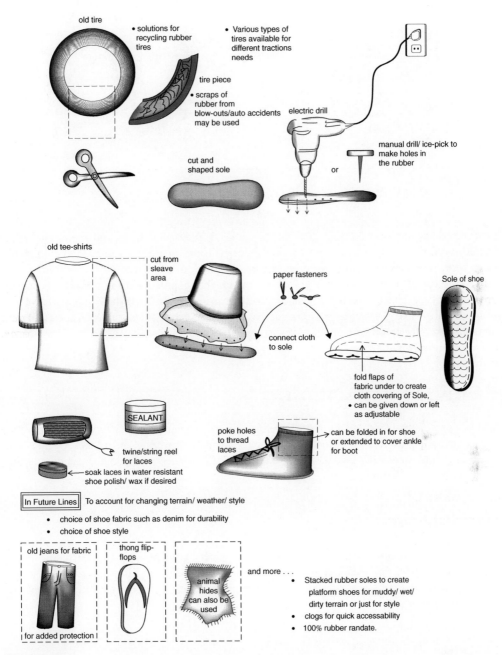

FIGURE 28
Making shoes from recycled rubber. (From Matthew Miller, Brennan Gilbane, Sasha KwesKin, Emily Roth.)

The usual process is to cut the sidewalls from the tire using a knife—see Figure 29. The beads are then cut away, and the core extracted for reuse. If the beads are nylon, the rubber is cut away; if they are steel, it is usually burned away. The nylon and steel are then sold, possibly after further processing. Next, the sidewalls can be cut away from the tread. Some products, such as containers, are made with neither or only one of the sidewalls removed. The sidewalls

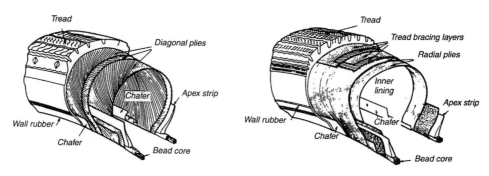

FIGURE 29
Cross section of a tire, showing components that can be recovered.

are cut into any number of shapes to make shoe soles, washers, doormats, and so on. The treads can be cut to make solid rubber tires for donkey carts. If the wear is very uneven, the tread will often be ground down to give a material of even thickness to work with.

The tire can also be reduced to a granular form and then reprocessed, but this can be a costly process, and there has to be a manufacturer willing to purchase the granules. Granulate tends to be remolded into low-grade products such as floor mats, shoe soles, rubber wheels for carts and barrows, and so on or can be added to asphalt road construction, where it improves the properties of this material.

CHEMICAL AND THERMAL RECOVERY

These types of recovery are not only lower in the waste management hierarchy but are also higher technologies requiring sophisticated equipment. Recovering the rubber in this way, however, could potentially enable southern countries to produce more sophisticated and high-value products, including new tires.

ENERGY RECOVERY

Tires consist of around 60 percent hydrocarbons, which is energy that can be recovered by incineration. Again, this technology requires sophisticated plant, and its application is limited when looking at small-scale enterprise. Tires are not as efficient a fuel as coal but are usually better than gas or oil.

LANDFILL

Landfill is the final step in the waste management hierarchy, although it is rarely needed in the South.

FURTHER READING AND USEFUL ADDRESSES

Ahmed, R., Arnold van de Klundert, and I. Lardinois. (1996). *Rubber Waste: Options for Small Scale Resource Recovery.* Gouda, The Netherlands: TOOL Publications.
Jon Vogler. (1981). *Work from Waste.* London: IT Publications and Oxfam.

ITRA Tire and Rubber Recycling Advisory Council, PO Box 37203, Louisville, Kentucky 40233-7203, USA. Tel.: +1-800-426-8835 or +1-502-968-8900. Fax: +1-502-964-7859. E-mail: itra@itra.com. Web site: http://www.itra.comlcorporate/recycling/htm.

UNDP/World Bank Integrated Resource Recovery Programme, 1818 H Street NW, Washington DC, USA. Tel.: +1-202-477-1254. Fax: +1-202-477-1052.

RA PRA Technology Ltd., Shawbury, Shrewsbury, Shropshire, 5Y4 4NR, UK. Tel.: +44-1939-250383. Fax: +44-1939-251118. Web site: http://www.rapra.net.

US Rubber Inc. Commercial Web site with an interesting range of products made from recycled rubber. http://usrubber.com/.

PAINT PRODUCTION

Paint protects as well as beautifies. Paint can be made without sophisticated equipment so can be made on a small scale although modern paints contain chemical additives that would probably have to be purchased. Offering basic paint in a village may significantly improve the quality of life there. If pigments are locally available, a village business aimed at producing inexpensive paint not intended to compete with national brands may be successful. An enterprise making paint for the commercial market, even for export in special cases, could be profitable but the economics require study. The paint produced must normally compete commercially with other paints so it must match—in uniformity and durability—paint produced by larger operations. The VITA article in the References describes a medium scale operation using imported pigments.

Paint is a suspension of finely ground pigment particles in a fluid binder, usually diluted with solvents to make application easier. After the paint is applied, it "dries" producing a thin, solid film. In some cases—shellac in alcohol, for example, the drying is simply the evaporation of the solvent. In other cases, a chemical reaction creates the film; a linseed oil binder oxidizes when exposed to air. Latex paints consist of pigment particles and tiny droplets of the binder, a resin, suspended in water. When the water evaporates, the resin particles fuse together, forming the film. Additives can be used for both oil based and latex paints to speed up drying, to resist fungus, to give special effects to the film. The word "vehicle" is sometimes used instead of "binder".

Some binders and solvents should be locally available. Linseed oil is probably the most common binder. Soybean oil is possible. Commercial paints often use a more sophisticated binder, formed by combining and heating—under controlled conditions, natural oils with various chemicals. Turpentine, distilled from the sap of pine trees, is an effective solvent for oil-based paints. Mineral spirits, or related petroleum based solvents, are more common probably because they are usually inexpensive and widely available.

Pigments can be made from a variety of naturally occurring minerals—it may not make sense to start a paint making enterprise unless appropriate minerals are available locally. Many naturally occurring pigments, such as lead oxides or chromium compounds, are poisonous to humans—especially harmful to children—and should be avoided. If they must be used the resulting paint should not be applied where children will ingest particles of paint. When planning the manufacturing care needs to be taken that poisonous pigments do not enter ground water by leaching from storage areas or through windblown dust particles. Not all pigments are mined. Lamp black is an effective black pigment. In some parts of the world a red excretion from the cochineal insect is used as a pigment. Extenders, colorless compounds, may usefully be added to paint to improve its durability at a low cost. Some extenders are clay, gypsum, and talc.

The amount of pigment needed depends on the color and the intended use of the paint. Black paint may use 8 lbs. of pigment per 100 lbs of binder; white paint may use 400 lbs of pigment per 100 lbs. of binder—the lighter the color the more pigment required. Enamel—high gloss—paints require less pigment than flat paints.

Paint is made by thoroughly mixing the pigment with the binder. The first step is washing the pigment and then grinding it as fine as possible—pigment particles in commercial paint are less then 1 micron in size. A hammer mill, described in the Agricultural Tools article in the Food chapter, can be used to pulverize the pigment. The fine pigment particles are then mixed with the binder to form a paste. The paste is then smoothed through another grinding operation.

The purpose of this second operation is not so much as to reduce the pigment particle size as to break up lumps of pigment and, more important, to ensure that pigment particles are thoroughly wetted by the binder. Grinding mills of various types can be used. One type is a ball mill—a horizontally rotating drum containing, in addition to the pigment/binder mixture, steel or porcelain balls, which shear the paste. Another type is a stone mill, similar to those used for grinding grain—an upper stone rotates against a stationary lower stone. The mixture to be ground is introduced at the center of the top stone. Both stones have radial grooves where the ground material collects and moves to the outer edge. Commercial paint operations use a set of rollers close together, rotating at different speeds. The shearing force on the paste between the rollers mixes the pigment and breaks up lumps. The paste resulting from any of these operations has the particle pigments intimately in contact with the binder.

The paste is then mixed with more binder and solvents as needed to make a workable fluid. The additives—driers and fungicides, for example—are added here. The paint should be inspected and, if suitable, packaged.

Inspection should include testing for uniformity throughout the batch. It should also verify that sufficient solids have been added to the binder. The color should be examined—laboratory tests are possible but a skilled operator can usually make a valid judgment. Long-term tests for durability and color fastness should be made from samples of several batches. Users may tend to be distrustful of products made in small-scale operations so the availability of test results is important.

WHITEWASH

Whitewash is an inexpensive paint often made by the user. The basic ingredients are lime—from limestone—and water. Chalk—actually a soft form of limestone, animal glues—from bones, hooves and skin, rice, and salt are also added to the mixture. The proportion of lime to water can best be found by experimenting, the more lime the better the whitewash covers. Pigments can added if they are available but usually the white color is accepted.

REFERENCES

Encyclopedia Britannica. (1967). "Paint." Vol. 17, William Benton, Publisher, Chicago.
Philip Heiberger. (1989). "Understanding Small Scale Paint Production." VITA Technical Paper #66, Volunteers in Technical Assistance, 1600 Wilson Boulevard, Suite 710, Arlington, Virginia 22209, USA.

SMALL-SCALE PAPERMAKING

Based on "AT Brief #27, Small Scale Paper-Making," in Appropriate Technology, *Vol. 25, No. 4, March 1999*

There is a strong link between the per capita income of a country and the amount of paper consumed. While in the industrialized countries of the world consumption can be as high as 300 kg per capita per year, in some of the world's poorest nations this figure can be as low as 1 kg. Illiteracy is also closely associated with low levels of paper consumption, since few books or newspapers are available, and schools lack basic resources. As per capita income grows and society demands higher rates of literacy, so the demand for paper grows. Only with indigenous manufacturing capacity and locally sourced raw materials can this demand be met at a reasonable cost, avoiding import taxes, high purchase prices, and loss of valuable foreign exchange. Papermaking is also an ideal example of how small industry can be developed to make of use of local resources, both in terms of raw materials and energy, while cutting transport costs and catering for a slowly growing local market. This article is an introduction to small-scale papermaking technologies for developing countries, meaning less than 30 tonnes per day.

The basic process of making paper has not changed in more than 2,000 years. It involves two stages: the breaking up of raw material (which contains cellulose fiber) in water to form a pulp (a suspension of fibers) and the formation of sheet paper by spreading this suspension on a porous surface and drying it, often under pressure.

Mechanized plants are only economically feasible if the output is more than several tonnes per day. In India, where papermaking machinery is manufactured indigenously, and hence costs are kept lower, mechanized papermaking on a small scale is very common, with plants operating at outputs of as little as five tonnes per day.

Smaller mills provide higher levels of employment, not only in the mill but also among associated industries, such as waste paper collection and machinery manufacture, and they are more flexible in their use of raw materials. With the growing concerns over deforestation and natural resource depletion in many parts of Asia, Africa, and South America, the use of agricultural residues such as bagasse, wheat, or rice straw for paper production is often a necessity. The most common types of paper are newsprint, stationery, currency, coated paper for magazines, toilet paper and cardboard or packaging paper.

RAW MATERIALS

The economic production of paper depends on a secure supply of raw materials at a reasonable price. Where wood is in short supply, other sources of fiber are used. Table 3 lists the main sources of fiber used in developing countries.

Waste paper is also an important source; compared with producing a tonne of paper from virgin wood pulp, the production of one tonne of paper from discarded paper may use half as much energy and water. It results in 74 percent less air pollution, saves 17 pulp trees, reduces solid waste going into landfill sites, and creates five times as many jobs. Thirty-four percent of the world's pulp is derived from reclaimed paper, and it is estimated that it could supply 30 percent of the needs of developing countries.

ADDITIVES

Many chemicals are used in papermaking for dying, tinting, cleaning, and improving the quality of the paper. Table 4 lists the main additives and their uses in production.

THE PAPERMAKING PROCESS

Papermaking is based on the fact that wet cellulose fibers bind together when dried under restraint (i.e., in a press or through the use of vacuum pressure). The processing of paper usually involves the initial separation of the cellulose fibers to form a wet pulp; some form of treatment while in the pulped state, such as beating and refining, to enhance the quality of the final product; the forming of the sheet paper by hand molding or by papermaking machine; and drying. Some further processing is often carried out before or during drying to acquire the desired finish.

TABLE 3
Raw materials commonly used for paper production

Raw Material	Source	Suitability
Straw (e.g., from wheat, barley or rice)	Between 5 and 10 percent of all straw which is produced is burned	Short-fibered (1.5 mm), it is often mixed with other pulp to provide a suitable pulp stock for a variety of uses.
Bagasse	from sugar cane after the sugar has been extracted	Slightly longer fibre than straw. Suitable for high-quality writing and printing paper.
Maize stalks	left over after maize harvest	The high moisture content and need for collection make maize stalks suitable only for very small-scale production. Properties similar to straw.
Bamboo	grown for this use	Fibre length of 2.7 mm, suitable for all types of papermaking without addition of other fiber. Supply is often limited.
Cotton	cuttings, lint, and fluff from cotton mills	Cotton is a high value fabric and is therefore only used for specialist papers. Has a fiber length of 25–32 mm.
Cotton rags	collected	Rags often require sorting and bleaching.
Flax	residue from the manufacture of linen	Long fibers make this material suitable for high-quality paper.
Hemp and sisal	from old ropes and tow from ropemaking factories	Fiber length is 6 mm, processing similar to that of cotton.
Jute	from old sacks and hessian	Jute does not bleach well and is therefore used for its strength rather than for high-quality grades

The papermaking process is similar whatever the raw material (or mixture of raw materials) and at whatever scale of paper production, but the complexity of the technology involved may vary considerably.

Deliveries of raw material will be made either by truck or by collectors who deliver small quantities of recycled material. It is important to ensure that there is sufficient storage capacity for the raw material, particularly where seasonally available raw materials are used, such as straw or bagasse, and a large supply will have to be stored for later use. A mill producing 10 tonnes of paper per day with a mixture of 70 percent straw pulp and 30 percent imported pulp, with three months storage capacity, will require an area of 2,000 m^2 and 16 m high.

THE PULPING PROCESS

Digestion, the first stage of the pulping process, removes lignin and other components of the wood from the cellulose fibers that will be used to make paper. Lignin is the "glue" that holds the wood together; it rapidly decomposes and discolors paper if it is left in the pulp (as in newsprint).

TABLE 4
Additives used during paper production

Chemical	Application
Caustic soda (NaOH)	Used in the cooking or digestion process in small mills.
Lime (CaOH)	Used for the cooking of low-quality materials such as jute or old rags.
Ammonia and calcium sulphate	Other chemicals used for the digestion of raw materials to form a pulp.
Chlorine	Used for bleaching paper. Chlorine is losing favour due to environmental pressures and is being replaced by other agents, such as hydrogen peroxide (H_2O_2), ozone, or enzymes.
Hypochlorite	Also used for bleaching paper.
Alum	For pH correction, which is necessary for many of the finishing processes.
Rosin	Used for sizing paper. Normally used in conjunction with alum as a "sizing system".
Alkile ketene dimer	Now used as a sizing agent in place of the alum and rosin system.
Starch	Improves the stiffness of paper and board.
China clay/chalk	A filler used to improve opacity, brightness, quality, and finish of paper. Up to 20 per cent clay is used for some grades of paper. Fillers are often cheaper than fiber and are used liberally.
Talc	Can be used instead of clay where the pH is close to neutral. Also used to reduce "stickiness" of pulp.

Mechanical pulping is an energy-intensive process that breaks down raw material into its individual fibers by grinding. The fibers are broken into smaller pieces, and relatively little lignin is released, resulting in a poor quality, "woody" paper. Mechanical pulp is used for newsprint because the paper is highly absorbent and therefore soaks up ink and dries quickly.

Chemical pulping is used by most small mills, since wood-grinding equipment is expensive and there is little suitable wood in developing countries. Forty-two percent of global pulping capacity uses a variety of chemicals as part of a high-temperature cooking, or digestion, process that breaks down the lignin, freeing the cellulose fibers. This process produces a high-quality product, and the type of chemical used determines the properties of the final product.

- Caustic soda or sodium sulphate will produce a pulp with coarse, strong fibers (known as Kraft) suitable for strong boxes.
- Ammonia or calcium sulphate will produce a finer fiber suitable for high-quality printing and writing paper.

The prepared stock is fed into a digester and mixed with the cooking chemicals, which are called "white liquor" at this point. In small mills digestion is usually carried out on a batch, rather than continuous, basis. Batch digesters can cope with various stock feeds—for example, straw, bagasse, cotton, and wood, in the same mill. As the stock and liquor move down through the digester, the lignin and other components are dissolved, and the cellulose fibers are released as pulp. After leaving the digester, the pulp is rinsed, and the spent chemicals (now known as "black liquor") are separated and recycled (see later). In a typical rotating spherical batch digester capable of handling 30 t.p.d., the complete process from filling to emptying takes approximately five to seven hours.

BLEACHING AND REFINING

The "brownstock" pulp is bleached with chlorine, chlorine dioxide, ozone, peroxide—or any of several other treatments or a combination of them.

At this point, the individual cellulose fibers are still fairly hollow and stiff, so they must be broken down somewhat to help them stick to one another in the paper web. This is accomplished by "beating" the pulp in the refiners, vessels with a series of rotating serrated metal disks. The pulp is beaten as long as necessary for the type of paper wanted. At the end of the process, the fibers will be flattened and frayed, ready to bond together in a sheet of paper.

FORMING THE SHEET

Once the pulp has been bleached and refined, it is rinsed and diluted with water, and fillers such as clay or chalk may be added. In the mechanized process, this "furnish," containing 99 percent water or more, is pumped into the flowbox of the paper machine. From the head-box, the furnish is dispensed through a long, narrow "slice" onto the "wire," a moving continuous belt of wire or plastic mesh. As it travels down the wire, much of the water drains away or is pulled away by suction from underneath. The cellulose fibers, trapped on the wire as the water drains away, adhere to one another to form the paper web. From the wire, the newly formed sheet of paper is transferred onto a cloth belt (or felt) in the press section, where rollers squeeze out much of the remaining water. In smaller paper mills the newly formed sheets may be handled manually and stacked one layer on top of another and pressed using a hydraulic press to remove the excess moisture.

DRYING, COATING, AND CALENDERING

After leaving the press section, the paper is wound up and down over many drying cylinders, large hollow metal cylinders heated internally with steam. Between dryer sections, the paper may be coated with starch to improve the printing and strength characteristics. After another round of drying, the paper sheet is passed through a series of polished, close-stacked metal rollers known as a "calender," where it is pressed smooth. Finally, the sheet is collected on a take-up roll and removed from the paper machine.

CUTTING AND PACKAGING

In many cases, the new paper roll is simply wound onto a new core, inspected, and shipped directly to the customer. Some paper grades, however, may be further smoothed by passing them through a "supercalender," where the sheet is polished between steel and hard cotton rollers, or it may be embossed with a decorative pattern. The paper may also be automatically cut, boxed, and wrapped for shipping.

SERVICES

When considering purchasing or renting premises for setting up a papermaking facility, care should be taken to ensure that there are adequate services provided at the premises or that these services can be easily accessed.

The paper industry consumes enormous quantities of water. In 1988 the average North American paper mill used 72 m^3 of water for every tonne of paper produced. This figure can be much higher where water efficiency measures have not been introduced.

Papermaking is also energy-intensive, especially when the process is mechanized. An electrical supply of sufficient capacity is required in most circumstances for powering motors, pumps, lighting, and so on. This can be supplied from the mains where the mains are accessible, from a diesel generator set, or from a renewable energy source (such as hydropower or wind power), where such power sources are available. Steam raising for the drying process can be carried out using a variety of technologies: oil-fired boiler, steam engine, and combined heat and power plants. Careful costing of the available options can bring considerable savings. The energy demand for processing of reclaimed paper is much less than that for virgin wood.

Effluent treatment and disposal is equally important. The effluent from a paper mill can contain thousands of different chemical species, which, if discharged directly into the environment, would cause untold damage. In medium and large-scale plants, specialized recovery equipment is used to reclaim chemicals for reuse or for incineration to provide energy. This is not cost effective in smaller plants, and so some form of treatment and/or disposal is required. Biological treatment plants, such as an anaerobic digester, are sometimes used to treat the effluent. This method has the added benefit of producing methane through the digestion of the organic matter in the effluent, which can be used to provide as much as 30 percent of the mill's energy requirement. The remaining sludge can then be disposed of on the land.

FURTHER READING

Small-Scale Papermaking, International Labor Organization/IT Publications, London, 1993.

A.W. Western. *Small-Scale Papermaking.* ITDG London, 1979.

Jon Vogler. (1981). *Work from Waste.* London: IT Publications.

Jorg Becker. (1988). *Paper Technology and the Third World.* GATE/GTZ, Eschborn.

Lillian Bell. (1995). *Plant Fibers For Papermaking.* McMinnville, Oregon: Liliaceae Press.

James d'A. Clark. (1985). *Pulp Technology and Treatment for Paper,* 2nd Ed., Miller Freeman, San Francisco, CA.

USEFUL ADDRESSES

Development Alternatives, B-32 TARA, Crescent, Outab Institutional Area, New Delhi- 10016, India.Tel.: +91-11-6967938. Fax: +91-116866031.

The World Resource Foundation, Heath House, 133 High Street, Tonbridge, Kent TN9 1 DH, UK. Tel.: +44-1732368333. Fax: +44-1732-368337. http://www.wrf.org.uk. E-mail: wrf@wrf.org.uk. Publishes *The Warmer Bulletin* and fact sheets on aspects of recycling.

Small Industries Research Institute (SRI), PO Box 2106,4/43 Roop Nagar, Delhi 110-007, India, Tel.: +9111291-8117. Consultancy specializing in small business and intermediate technologies.

The Two River Paper Company, Pitt Mill, Roadwater, Watchet, Somerset TA23 OQS, UK. Tel.: +44-1984-641028. Fax: +44-1984640282.

International Labor Organization, 4 Route des Morillons, CH-121 1 Geneva, Switzerland. Tel.: +41(22)-799-83-19, Fax: +41-(22)-798-86-85.

Kathryn Clark—Twinrocker Handmade Paper, 100 East Third Street, PO. Box 413, Brookston, Indiana 47923, USA. Tel.: +1-800-757-TWIN (8946) or +1-765-563-3119. Fax: +1-765-563-TWIN. E-mail: twinrock@twinrocker.com Suppliers of handmade paper products, and run courses and a bookshop.

USEFUL INTERNET SITES

http://www.ipstedu/amp/Thdex.html The Robert C. Williams American Museum of Paper Making.
http://www wrd.org.uk World Resource Foundation
http://twin rocker.com/Welcome.html/Twinrocker Handmade Paper On-Line, supplies, workshops, and tours
http://www. oneworld. org/itdg/publications.html IT Bookshop
http://wwwnbn.com/youcan/paperipaperhtml Making new paper from recycled paper
http;//wwwlearn2.com/06/0697/0697html 'Learn 2' make homemade paper

---

PLANT DYES

Helen King

INTRODUCTION

Every human culture shares the earth with plants and is dependent on them. The immediate connection is food, but think about it: Furniture, roofs, floating hollowed-out logs, and the desire to beautify the human body have all been present as long as recorded human existence. The color of various plants makes them attractive and appealing to us. In fact, as early as 3000 B.C. dye workshops were mentioned in Chinese chronology. The quest for color began.

If you find yourself in a place where simple yet economically viable projects are needed, consider dyeing with plants. As you become acquainted with the people who will carry out the project, you will surely begin to hear some memories of dyes. The older people will remember things they did as children: dyeing a weaving with stripes of colors, staining skins for garments, or using plant extracts to beautify the body on special occasions. There may be myths surrounding the dyeing process. The gathering of herbs, for example, may happen only at certain times of the year, creating a link with the environment. We have learned to use and extract the dye without destroying the plant so we can create that color again. People love using wool, which is the most common protein fiber, but one can even use dog hair.

TECHNOLOGY

The technology for a dye project is simple and requires relatively few tools. The process of dyeing will differ with fibers; one's own imagination or experience in the area should not be limited. A materials list appears in this article and can be adapted to the fibers we discuss here.

Before dyeing, thought should be given to fixing the color to the fiber. This is known as mordanting. The word *mordant* comes from the Old French (Latin) *modere*, meaning "to bite." By using a mordant (metallic salt), the chemical bond structure of the fiber is changed, and the dye can become part of the structure. This is the big difference between a dye and a stain. Stain is only on the surface of the fiber, even though it may be difficult to remove. Mordants cause the dye to become permanent. Through time, mordants have been such materials as blood, urine, ash, other plants, as well as copper and iron kettles, rusty nails, and brass or copper coins. These old mordants still work; experimenting is enjoyable but time consuming. This is a time to listen to the people in the area. Their memories and experience are invaluable. Today, however, people use common metallic salts if they are easy to procure in grocery stores or pharmacies.

By preparing or premordanting the fiber in advance, less plant product is needed for the dye bath, lessening the impact on the local environment. Several different mordanted fibers can be added to the dye pot. It is important to know that the mordants may harm the fibers, so use them with care. The mordants can also be put into the dye bath, or the fiber may be treated after dyeing.

Simmer Time	Mordant	Chemical Name	Amount Needed	Colors	Precautions
45–60 minutes	Alum	Potassium aluminum sulfate	1 tablespoon[*]	Lightest	Relatively nontoxic, probably a good mordant for children's groups. Mordant prior to dyeing.
45 minutes	Tin	Stannous chloride	Scant ½ teaspoon[*] 1½ tsp Cream of tartar, or 1 tsp Oxalic acid	Brightest	Relatively nontoxic, but too much can coarsen the wool. Can be used before or during dye bath.
45 minutes	Copper	Copper sulfate	2 teaspoons[*]	Light	Irritant to eyes and respiratory system. Brings out greens. Mordant before or during dye bath.
45–60 minutes	Chrome	Potassium dichromate	Scant ½ teaspoon	Bright	Toxic. Keep lid on the pot. Store away from light. Rich colors. Mordant before dyeing.
30–45 minutes	Iron	Ferrous sulfate	1¼ teaspoons 2 tsp Cream of tartar 1 tsp Oxalic acid	Dark colors	Ingestion fatal. Too much makes wool coarse. Sometimes called "copperus" or the "saddener."

[*]1½ teaspoons of cream of tartar will help brighten colors and soften the wool. It is generally a good idea to wash all wool after mordanting with warm, soapy water. NEVER BOIL THE WOOL! You may end up with felt.
This chart has been prepared for wool using water in my area. If you are mordanting cotton or other vegetable fiber, mordant for at least one hour at a boil. Be sure to wash cotton well before mordanting.

WOOL MORDANTING

These measurements were taken from *Dyes from Plants* by Senoid Robinson. They have been altered a bit by the author and others from experience with different quantities of water. Obtaining natural dyestuffs is limited only by the imagination and some basic plant knowledge. It would be wise to use a book that identifies local plants, as well as tap into the knowledge of the community. Remember to keep notes! If you find yourself in Mexico or South America, there are famous insects called cochineal (living on cactus) or kermes found on certain oaks. These tiny creatures produce reds and purples. Look for red-bodied insects! If you find yourself in Asia, saffron (pistils of autumn crocus) and safflower will produce wonderful yellows and reds. Wherever you are, tree bark and nuts are good bets for rich browns and blacks. My rule of thumb is one paper grocery bag of plant material per pound of wool. Other fibers such as cotton and raffia may require more. Insect dyes can use much less. Lichens are found in many places, but they should be used in small quantities. They require no mordant, and they soften the fiber.

USES AND APPLICATIONS

Plant dyes lost favor after 1856 when, by accident, William Henry Perkin produced a lavender dye using coal tar. As synthetic dyes increased in popularity, natural dyestuffs declined in commercial value. It is easy to see how economical it became to have bright, consistent colors without having to gather plant materials and create storage for natural dyestuffs. In the past 20 years or so, naturally dyed products have again come into some favor. Art galleries and specialty shops have featured apparel and handsome baskets, quilts, and hangings, all using natural dyes.

A variety of materials can be used for dyes, depending on availability. Similarly, a variety of fibers can be used. In areas where there are animals such as sheep, goats, or pack animals, people may be spinning their fibers and creating products using chemical dyes. This might be a good opportunity to introduce the natural dye process for upscale commercial markets. In some areas, such as Asia, fiber production may include cotton or silk. (Silk production is described in another article in this chapter.) Dried rushes, raffia, and cane may be the available fibers to dye for baskets and mats. Wherever you find yourself, there will be an opportunity to have a dye project. All ages may participate and contribute ideas.

The colors produced by natural dyes are so subtle and harmonious, they are works of art, set apart from the garishness of those produced with commercial dyes. Commercial opportunities will surely present themselves. In Alaska, for example, the musk ox is being developed commercially because its underbelly hair (known as kiviett) is purported to be the warmest fiber in the world. It is combed, spun, and distributed to many native villages, where it is made into the warm and light garments. The spinning is done at a mill in Woonsocket, RI, and it is then sent back for distribution to the villages. So not all of the process needs to be done at the village level.

As your dye project develops, keep detailed notes, and use all your contacts for application for your community.

CONSTRUCTION AND IMPLEMENTATION

The equipment for dyeing is the same whether used for protein fibers such as wool or animal hair or for dyeing vegetable fibers such as raffia, other grasses, or bark.

The fireplace can be outdoors or inside. Water from ponds or rivers or saved rainwater may be used. A line strung between trees is helpful for drying. The pots should be large. "Enameled" pans usually used for canning or clam boils are suitable. Brass, copper, or iron kettles may be used, but remember that metals act as mordants and will affix a color. You will also need some plastic pails as well as rods for stirring. Glass rods are ideal, but a wooden dowel will work well. A bottle of dish detergent should be handy for washing the fiber after mordanting and again after dyeing. A plastic kitchen sieve is best for straining, but a muslin cloth or a nylon stocking can be used as well. Measuring spoons and scales are helpful, especially for mordanting, but if you have none, be less generous in estimating, especially with the mordants. Be sure to have your notes and record book handy as you experiment.

The common practice is to chop the plants and gradually heat the dye bath to a simmer. Simmer for at least 20 minutes for leaves and flowers, longer for bark, woody stems, and nuts. A lid keeps the odor from escaping; stir from time to time. When the dye has been extracted, strain the plant material and return the liquid to the pot. When the dye has cooled, the clean, wetted fiber may be added and heated but not boiled. The fiber must be thoroughly wet, maybe even soaked overnight, or it will not dye evenly. If you are using wool or other protein fiber, be sure to devise some method of skeining to prevent the fibers from tangling. Simply winding and tying loosely will do.

When the desired color has been attained, remove the fiber, wash with soapy water, rinse, and dry. The wet fiber will appear darker than when dried. Cotton and other vegetable fiber will require longer mordanting and dyeing. Colors may not be as strong as those on protein fibers. Experimenting is valuable, and record keeping is essential. The people who choose to participate in this process will have ideas about plants, experimenting, and what to do with the beautifully colored fibers.

IMPLICATIONS FOR THE ENVIRONMENT AND PEOPLE

This project, whether big, small, or evolving, will have enormous satisfaction to all involved. Plants of a certain area take on special importance. Decisions have to be made regarding sustenance for the community and attention to the environment. While searching for wild plants or tending to cultivated ones, we must be careful to use and extract fruit and colors without destroying the plants.

There are many advantages to a natural dye project: It is nonpolluting, colors are harmonious, rare colors are produced, and there is the challenge of the element of chance. Some disadvantages can be the cost of the mordants, the time involved, and that some natural dyes are not completely lightfast. I do not consider this last disadvantage much of a problem because the naturally dyed materials usually change into a different color, whereas chemically dyed fibers just "fade."

If you feel that the community can sustain your project, find the links needed to make it viable. Look through your notes, take advantage of your contacts, and let the community be proud!

REFERENCES

Robertson, Senoid M. (1973). *Dyes from Plants*. New York: Van Nostrand Reinhold Co. This is a wonderful, clearly written book.

Grae, Ida. (1979). Nature's Colors. Collier Books (Division of Macmillan Publishing Co., Inc.), 1979. London: Collier Macmillan Publishers, 1974. This is a book of recipes and is fun to read.

Brown, Rachel. (1978). *The Weaving, Spinning, and Dyeing Book*. New York: Alfred A. Knopf. This book is clear and comprehensive; it could be useful in many projects.

CHAPTER 8

Water Supply

Edited by BARRETT HAZELTINE
Brown University

WATER SUPPLY

Based on Surface Water Treatment by Roughing Filters—A Design, Construction and Operation Manual, *Text Revisers: Sylvie Peter, Brian Clarke, ISBN: 3-908001-67-6, Swiss Centre for Development Cooperation in Technology and Management (SKAT), CH-9000 St. Gallen, Switzerland*

The quantity of water available to users has a direct bearing on their health. Five liters per person per day is considered the minimum consumption level, although desert dwellers exist on less. More than 50 liters per person per day, it has been estimated, gains no further health benefits. Twenty-five liters per person per day may become an acceptable goal in places where piped connections to individual houses are not feasible. An article in the Health chapter discusses water-borne diseases and is recommended as background to the present article. The "Building" article in the Construction chapter also discusses water problems—focusing on sanitation.

Here are the chief potential sources of water listed in their approximate order of preference based on cost, quality of water, need for equipment and supplies.

- *Springs.* If there are year-round springs nearby, they can usually be developed to supply clean water. This water can often be conveyed through pipes without the expense of pumps or water treatment. Springs can most often be found in hilly or mountainous regions.
- *Wells.* Because there is water at some depth almost everywhere beneath the earth's surface, a well can be sunk (using the appropriate technique) almost anywhere. The water that comes into the bottom of a well has filtered down from the surface and is, in most cases, cleaner than water that is exposed on the open ground. A separate article in this chapter deals with well construction.
- *Rainwater.* Collection and storage of rainwater may provide another source where surface and underground water supplies are limited or difficult to reach. Normally, except in the

rainiest regions, rainwater will not supply all the water needs of a locale; however, as a supplement, it can be collected from roofs or protected ground runoff areas, and stored in covered cisterns to prevent contamination. A separate article in this chapter deals with rainwater collection.

- *Surface water.* Streams, rivers, and lakes are all commonly used as sources of water. Although no construction is needed to enable them to supply water, the quality of the water is almost always poor. Only clear mountain streams flowing from protected watersheds could be considered as fit for human consumption without treatment. An article cited in the References section of this article describes a water supply project in Malawi using mountain stream water. Village people learned to plan and construct a water system to bring mountain stream water down to public standposts in the villages.

This article discusses water supplies using surface water.

From the technical point of view, the following three questions must be answered during the planning phase of a water supply scheme.

- Which raw water source should be used for the water supply scheme?
- If treatment is necessary, what type of treatment scheme should be favored?
- How much water should be distributed to the consumers, and at what service level?

Layouts of some possible water supply systems, all using surface water, are shown in Figure 1. As shown in Figure 1, surface water has to be collected, treated, and stored before it reaches the consumer.

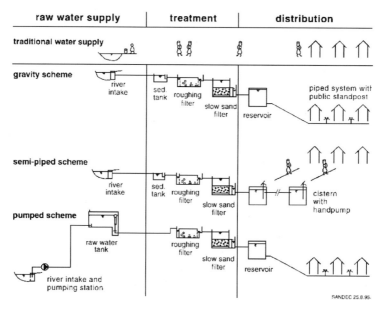

FIGURE 1
Possible water supplies using surface water.

SOURCE AND TREATMENT OF WATER

Source selection is a very basic decision entailing numerous consequences for the future water supply scheme. The different local water sources have to be evaluated with respect to their quantity, quality, and accessibility. The future water demand must be covered by the selected source with the best possible water quality and located as close as possible to the supply area.

Since water treatment is usually the most difficult element in any water supply scheme, it should be avoided whenever possible. The general statement that no treatment is the best treatment especially applies to rural water supply schemes. The use of the best available water quality sources is, therefore, an alternative that should always be taken seriously. If no other alternative is available, rural water treatment must improve the bacteriological water quality by locally sustainable treatment processes.

DISTRIBUTION

Water distribution systems depend on the type of water source used, on the topography, and on the provided supply service level. Individual water supplies, such as rainwater harvesting and shallow groundwater wells equipped with hand pumps, usually do not need piped supply systems. Treated surface water, however, is normally distributed by a piped system. A suitable topography often allows the installation of a gravity system that will improve reliability and supply continuity. Since pumped water supply schemes depend on a reliable supply of energy and spare parts, they are very susceptible to temporary standstills. Finally, the service level of water supply strongly governs water demand. Water usage increases drastically as the service improves: public stand-post, yard connection, multiple-tap house connection. The article "Saving and Reusing Water" discusses this increase in consumption with an increase in convenience. Two pumped systems are shown in Figure 2.

Water supply is always interlinked with wastewater disposal. Wastewater disposal is described in the "Sanitation" article. The health situation of a community newly supplied with treated water does not necessarily improve, especially if public health and wastewater disposal issues are neglected. The main components necessary to significantly improve the public health situation

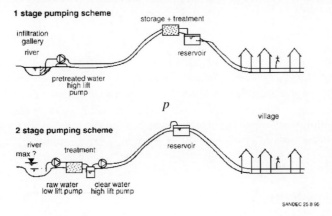

FIGURE 2
Pumped water supply schemes.

of a community are therefore a reliable and safe water supply, an adequate waste disposal system, and a comprehensive hygiene education program.

HYDRAULIC PROFILE

Selection of how to move the water is a basic criterion when planning a water supply scheme. First choice must be given to gravity supply systems, since they guarantee reliable operation at low running costs. Schemes that integrate the use of handpumps are a second choice. (Water pumps are discussed in a separate article.) The installation of mechanically driven pumps should be chosen as the last option and only applied in special cases where a reliable and affordable energy supply and the infrastructure for pump maintenance and repair work is guaranteed. Hydraulic rams—described in an article of the same name in this section—that make use of the potential energy of a large water volume to pump a small fraction of this water volume to a higher level, may be an appropriate option where abundant water is flowing.

Water treatment plants should, whenever possible, be operated by gravity and with open channels to minimize water pressure on the structures. The total head loss through the treatment plant will amount to 2 or 3 m. As noted, any type of water lifting, except through handpumps, should be avoided, since the supply of energy and sophisticated spare parts is generally unreliable. If water lifting is absolutely necessary for topographical reasons, the number of pumping steps must be limited. A one-stage pumping scheme, illustrated in Figure 2, should be chosen for raw water to be pumped to an elevated site where the treatment plant and reservoir are located. Such a one-stage pumping scheme has a great advantage over a two-stage scheme because it increases its reliability by a factor of 2. Moreover, the risk of flooding in lowland areas can often not be excluded entirely. Protecting a high-lift pumping station against floods is easier than a full-sized treatment plant. A two-stage pumping system, however, is unavoidable for a piped supply on a flat area devoid of natural elevation and in the case of serious raw water quality fluctuations, such as heavy sediment loads during the monsoon. In such a situation, installation of a low-lift raw water pump is recommended. It may consist of an irrigation unit of low efficiency but of simple design to limit high lift pumping for treated water and protect impellers and seals from damage. Hence, high-lift pumps should be used for treated water or raw water pumped from infiltration galleries or similar intake systems.

Treatment Steps

Surface water has to undergo a step-by-step treatment. Coarse solids and impurities are first removed by pretreatment, then the remaining small particles and microorganisms are separated by the ultimate treatment step. Under special local conditions, raw water collection and pretreatment may be combined in a single installation, such as intake or dynamic filters or, alternatively, by infiltration galleries. The required water treatment scheme is mainly dependent on the degree of fecal pollution, characteristics of the raw water turbidity, and the available type of surface water.

Removal of Coarse Material

Separation of coarse solids from the water is preferably carried out by a sedimentation tank (grit chamber), since sludge removal from such tanks is less troublesome than from roughing filters. Simple sedimentation tanks should be designed for a detention time between one and three hours. Such a tank is shown in Figure 3.

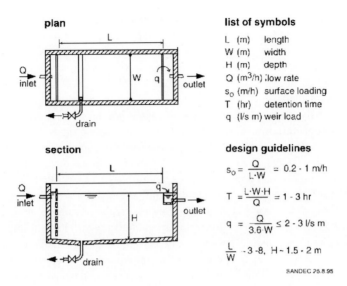

plan

list of symbols

L (m)   length
W (m)   width
H (m)   depth
Q (m³/h) flow rate
s_0 (m/h) surface loading
T (hr)   detention time
q (l/s m) weir load

section

design guidelines

$$s_0 = \frac{Q}{L \cdot W} = 0.2 - 1 \text{ m/h}$$

$$T = \frac{L \cdot W \cdot H}{Q} = 1 - 3 \text{ hr}$$

$$q = \frac{Q}{3.6 \cdot W} \leq 2 - 3 \text{ l/s m}$$

$$\frac{L}{W} \sim 3 - 8, \ H \sim 1.5 - 2 \text{ m}$$

SANDEC 25.8.95

FIGURE 3
Layout and design of a sedimentation tank.

Use of one sedimentation tank should be sufficient for a small-scale water supply scheme. The accumulated sludge can be removed during periods of low silt load. A bypass is required to maintain operation of the treatment plant during cleaning periods. In order not to interfere too much with normal operation of larger water treatment plants, two or more sedimentation tanks operating in parallel should be provided to allow cleaning, maintenance, and repair of one tank.

Aeration

The water's dissolved oxygen content plays a key role in the biology of the slow sand filtration process. Although physical processes are the main mechanisms in roughing filtration, biochemical reactions may also occur in the prefilters, especially if the raw water contains high organic loads. The activity of the aerobic biomass decreases considerably if the oxygen concentration of the water falls below 0.5 mg/l. Furthermore, nitrification of ammonia is associated with a significant consumption of oxygen—for example, 1 mg NH_4 uses 4.5 mg O_2. Hence, adequate oxygen content in the water to be filtered is of prime importance. Since turbulent surface waters are generally well oxygenated, they do not require additional aeration. Still water, however, can exhibit low oxygen contents, especially when drawn from the bottom of polluted surface water reservoirs. Multilevel draw-offs—"waterfalls"—are recommended as intake structures for stratified water bodies. Stagnant raw surface waters should be aerated.

Cascades, where water cascades from a pool at one level to a pool at a lower level, are simple but efficient aeration devices. Three or so cascades should be put in series. The cascade should preferably precede filters to meet the possible oxygen demand in the filter.

ROUGHING FILTRATION AS PRETREATMENT

In the "Filtration and Disinfection," article in this chapter the term *rapid filter* is used to mean roughing filters. Roughing filtration mainly separates the fine solids that are not retained by the preceding sedimentation tank. The effluent of roughing filters should not contain more than 2–5 mg/l solid matter to comply with the requirements of the raw water quality for slow sand filters.

Coarse gravel filters mainly improve the physical water quality as they remove suspended solids and reduce turbidity. A bacteriological water improvement can also be expected as bacteria and viruses are solids also, ranging in size between about 10 to 0.2 mm and 0.4 to 0.002 mm, respectively. Furthermore, these organisms get frequently attached by electrostatic force to the surface of other solids in the water. Hence, a removal of the solids also means a reduction of pathogens (disease-causing microorganisms). The efficiency of roughing filtration in microorganism reduction may be in the same order of magnitude as that for suspended solids—for example an inlet concentration of 10–100 mg/l can be reduced by a roughing filter to about 1–3 mg/l. The bacteriological water quality improvement could amount to about 60 to 99 percent. Larger-sized pathogens (eggs, worms) are removed to an even greater extent.

Roughing filters are used as pretreatment step prior to slow sand filters. Slow sand filtration may not be necessary if the bacteriological contamination of the water to be treated is absent or small, particularly in surface waters draining an unpopulated catchment area or where controlled sanitation prevents water contamination by human waste. Physical improvement of the water may be required because of permanent or periodic high silt loads in the surface water. Excessive amounts of solids in the water lead to the silting up of pipes and reservoirs. For technical reasons, roughing filtration may therefore be used without slow sand filtration if the raw water originates from a well-protected catchment area and if it is of bacteriologically minor contamination: in the order of less than 20–50 *E. coli*/100 ml.

For operational reasons, at least two roughing filter units are generally required in a treatment plant. Since manual cleaning and maintenance may take some time, the remaining roughing filtration unit(s) will have to operate at higher hydraulic loads. A single prefilter unit may be appropriate in small water supply schemes treating water of low turbidity.

SLOW SAND FILTRATION AS MAIN TREATMENT

The substantial reduction of bacteria, cysts, and viruses by the slow sand filters is important for public health. Slow sand filters also remove the finest impurities found in the water. For this reason they are placed at the end of the treatment line. The filters act as strainers, since the small suspended solids are retained at the top of the filter. However, the biological activities of the slow sand filter are more important than the physical processes. Dissolved and unstable solid organic matter, causing oxygen depletion or even causing fouling processes during the absence of oxygen, is oxidized by the filter biology to stable inorganic products. The biological layer on top of the filter bed, the so-called "Schmutzdecke," is responsible for oxidation of the organics and for the removal of the pathogens. A slow sand filter will produce hygienically safe water once this layer is developed.

WATER DISINFECTION

Water from a slow sand filter with a well-developed biological layer is hygienic and safe for consumption. Any further treatment, such as disinfection, is therefore not necessary. As documented by numerous examples in many developing countries, provision of a reliable chlorine disinfection system in small rural water supply schemes is often not practicable. A regular supply of mostly imported chemicals and accurate dosage of the disinfectant are the two main practical problems encountered.

However, as regards disinfection, one has to differentiate between small (rural) and large (urban) water supply schemes. Large distribution systems with often illegal connections present a risk of recontamination, especially if the supply of water is intermittent. In large urban water supply schemes, final water chlorination is recommended as a safeguard. Residual chlorine, however, will be too low and contact too short to deal with serious contamination introduced by infiltration of highly contaminated shallow groundwater in intermittently operated water supply systems. In rural water supply system, implementation of a general health education program with special emphasis on correct water handling is a more effective measure than preventive disinfection.

WATER STORAGE

To make full use of the treatment capacity and to avoid interference of the treatment process by intermittent operation, water treatment installations should preferably be operated uninter-ruptedly on a 24-hour basis. Particularly, slow sand filters should be operated continuously to provide the biological layer with a permanent supply of nutrients and oxygen. Roughing filters are less sensitive to operational interruptions, although careful restarting of filtration should be observed in order not to resuspend the solids accumulated in the filler. Water supply schemes, operated entirely by gravity, can easily handle a 24-hour operation. Pump operation, however, is often reduced to 6 to 16 hours a day in water supply systems requiring raw water lifting. In pumped schemes, construction of a raw water tank may offer an economically and technically sound option, since it enables continuous operation of the treatment plant and also acts as a presedimendation tank.

Water storage capacity must be provided to compensate for daily water demand fluctuations. In rural water supply schemes, peak water consumption occurs generally in the morning and evening hours. Therefore, a storage volume of at least 30 to 50 percent of the daily treatment capacity should be provided to compensate for the uneven daily water demand distribution.

DISTRIBUTION SYSTEM

Water accessibility rather than water quality is the most important criteria for the consumer, since her or his main concern is the walking distance between home and the water point. Consequently, treated or better quality water has to be brought nearer to the homes than the existing water sources. Treated river water is likely, for instance, to be more readily accepted if the original walking distance to the river is reduced substantially by the installation of a water supply system.

A water distribution system will therefore have to be constructed. The service level of a piped system is dependent on the economic situation; construction costs of a distribution system normally amount to 50 to 70 percent of the total investment costs of a water supply scheme, including a water treatment plant. Gravity schemes should be installed whenever possible. In many instances, however, topography is unfavorable, and differences in altitude must be over-come by water lifting. Pumps require, however, relatively high investment and operating costs, spare parts, and, particularly, energy—an aspect that will in the future gain increased importance. In rural water supply schemes, pumped systems should therefore be introduced only after careful consideration and in exceptional cases.

Figure 1 illustrated different hydraulic layout possibilities. On the raw water side, the water flows by gravity directly to the treatment plant or, if pumped, preferably first to a raw water bal-ancing tank. After passing through the treatment plant it is stored in a reservoir and later distributed to the consumers by a piped gravity scheme close to the houses. In a semipiped scheme, the water flows by gravity through the treatment plant into the reservoir equipped with handpumps, or as an extended alternative, the reservoir is connected to a system of cistern located between treatment plant and village. Treated water is now supplied by gravity to these cisterns equipped with handpumps. Each cistern acts as reservoir and water point.

Such distribution systems may increase sustainability and reliability of a water supply as the consumers keep the water supply system running at low operating costs and at village maintenance level. The system of storage tanks equipped with handpumps can best control excess water usage, prevent contamination and avoid wastewater disposal problems. The consumer, however, may require higher service levels than the "handpump option." On the one hand, higher service levels imply increased water consumption and wastewater disposal problems; on the other, collection of water charges may become easier if the distribution level is shifted from public to individual supply.

The following per capita daily water demand values are generally used:

Supply with public handpumps	15–25 l/person day
Supply with public standpipes	20–30 l/ person day
Supply with yard connections	40–80 l/ person day
Supply with multiple tap house connections	80–120 l/ person day

The effective values for the supply with public handpumps or standpipes are greatly influenced by transport distance, ranging from a few dozen to 300 and more meters. For yard and house connections, water use will be influenced by the level and manner in which the water charges are levied (such as a monthly lump sum or on an effectively used water volume basis). Furthermore, use of drinking water for backyard garden irrigation leads to an enormous water demand and should therefore be prohibited.

REFERENCES

Barrot, L. P., B. J. Lloyd, N. J. D. Graham. (1990). "Comparative Evaluation of Two Novel Disinfection Methods for Small-Community Water Treatment in Developing Countries." *Journal of Water Supply Research and Technology—Aqua,* Vol. 39, 396–404.

Basit, S. E., and D. Brown. (1986). "Slow Sand Filter for the Blue Nile Health Project." *Waterlines,* Vol. 5, No. 1.

Boller, M. (1993). "Filtermechanisms in Roughing Filters." *Journal of Water Supply Research and Technology—Aqua,* Vol. 42, 174–185.

Collins, M. R., and N. J. D. Graham. (Eds.). (1994). *Slow Sand Filtration, An International Compilation of Recent Scientific and Operational Developments.* AWWA, American Water Works Association, Denver, CO 80235, USA.

Graham, N. J. D., and M. R. Collins. (Eds.). (1996). *Advances in Slow Sand and Alternative Biological Filtration.* John Wiley & Sons.

Idelovitch, Emanuel, and Klas Ringskog. (1997). *Wastewater Treatment in Latin America: Old and New Options.* The World Bank, Washington, DC.

Liebenow, J. Gus. (1981). "Malawi: Clean Water for the Rural Poor." AUFS *Reports,* No. 40. Universities Field Staff International, PO Box 150, Hanover, NH 03755.

Lloyd B., and R. Helmer. (1991) *Surveillance of Drinking Water Quality in Rural Areas.* WHO and UNEP. John Wiley & Sons.

Wegelin, M., M. Boller, and R. Schertenleib. (1987). "Particle Removal by Horizontal-Flow Roughing Filtration." *Journal of Water Supply Research and Technology—Aqua,* Vol. 36, 80–90.

Wegelin M., R. Schertenleib, and M. Boller. (1991). "The Decade of Roughing Filters—Development of a Rural Water Treatment Process for Developing Countries." *Journal of Water Supply Research and Technology—Aqua,* Vol. 40, 304–316.

WATER HARVESTING AND SPREADING FOR CONJUNCTIVE USE OF WATER RESOURCES

Frank Simpson

Girish Sohani

INTRODUCTION

What is the best way to plan a strategy for obtaining a year-round water supply in a rural dryland area with rainfall largely restricted to a few weeks within a four-month period of monsoon? What are the most appropriate technologies for provision of a domestic water supply in a setting of rapidly dwindling surface water, where shallow aquifers are scarce and seldom found near the locations of greatest need? What is to be done about these problems when the people in need are destitute and conditioned by life in a harsh environment to think only about survival in the short term?

The first steps in such a strategy involve collecting rain (water harvesting) to satisfy the immediate water needs of village families. A more lasting solution to the problem requires significant reduction in the amount of monsoon water leaving the area as runoff. There are limits to the effectiveness of check dams and other barriers across the valleys of ephemeral streams, in this particular regard, even though these structures are essential parts of an overall strategy. The most far-reaching effects (water spreading) come from artificial reductions in the slope of the hillside, augmented by systems of ridges and trenches, configured so as to direct the surface runoff underground. The technologies employed to these ends should be cheap, small in scale, and easily replicated.

The purpose of this account is to describe the main technologies for water harvesting and spreading, employed in the research project, titled *Conjunctive Use of Water Resources in Deccan Trap, India.* The project involved Bharatiya Agro Industries Foundation (BAIF), Pune, India, and University of Windsor Earth Sciences, Windsor, Ontario, Canada, working in partnership with the tribal and rural people of Akole *Taluka*, Maharashtra. It ran from 1992 to 1996 and achieved sustainable results for the villagers of Ambevangan, Manhere, and Titvi. The authors were the Canadian (FS) and Indian (GS) project leaders.

A detailed account of the project research is given by Sohani, Simpson, et al. (1998). A summary of the main benefits, derived from the project by the three partner villages, is presented by Simpson and Sohani (2001). The main factors that set the scene for participatory management and evaluation of the project are discussed by Simpson and Sohani (in the Planning and Implementation chapter). This last mentioned article also includes a general description of the terrain and climate in Akole *Taluka* and the living conditions of the tribal and rural people, which is not repeated here.

ASSESSMENT OF TECHNOLOGIES

The tribal and rural people shared elements of local knowledge systems pertinent to the use of water and soil. This indigenous technical knowledge included information on botanical indicators of shallow ground water, such as the tree *Ficus glomerata*, known locally as *umbar*, and the associated bottomland flora. This information tended to be supported by accounts of ancient, Indian hydrology, such as the *Brahat Samhita*, written by Varaha Mihira in the sixth century (see Tagare 1992) and by modern field observations. In addition, the people provided useful information on the relationship between terrain features and groundwater discharge, local strategies of land use, and a local classification of soils.

Indigenous technical knowledge was combined with project research on hydrology and hydrogeology in the assessment of technologies for possible introduction into the area and in the selection of sites for demonstration purposes. The research involved field and laboratory study of the soils, weathered bedrock, and underlying basalt lavas making up the bedrock, as well as water from streams, springs and seepages, and shallow wells. Imagery from earth satellites in orbit was employed in the mapping of straight-line ground features (lineaments), many of which coincided with the traces of vertical fractures in the bedrock. These formed conduits for the circulation of ground water and also coincided with several of the more persistent springs.

A high priority was assigned to the selection of technologies for water harvesting and spreading that were easily understood by and acceptable to the people. In addition, it was important that the technologies under consideration were compatible with existing approaches to land use. A wide range of alternative approaches to water harvesting and spreading was considered. These were documented from other dryland regions of India and elsewhere. They included ancient technologies, such as those used in the Negev Desert by the Nabotean culture and its predecessors, more than four millennia ago (Nessler 1980).

WATER HARVESTING

Water harvesting refers to the collection and storage of water from a surface on which rain falls. Ideally, the catchment or water-collecting surface is impervious to water and may be located on natural or artificial materials. Alternatively, it may be on natural materials that have been treated so as to increase the amount of runoff associated with precipitation events. The water is contained in a storage tank, which may be connected to the catchment area by means of a pipe. A reservoir, receiving water by overland flow from the slopes of an adjacent catchment area, is also part of a water-harvesting system. Rainwater collection from buildings is described in the "Rainwater Collection" article in this chapter.

The tribal and rural people gave a high priority to a year-round water supply for domestic use. Roof water harvesting was introduced into the villages as a partial response to this need (see Figure 4). The houses in the villages are made of stone and mud and have tiled roofs, which make effective catchments. In each system, gutters, made of galvanized iron sheets, were added to the sides of the roof and connected by means of PVC pipe to a ferrocement storage tank, equipped with a tap and mounted on a stone platform. In general, a 2,500-liter tank proved adequate for the needs of a family throughout much of the year.

Check dams were constructed across the valleys of ephemeral streams. Straight stretches of the valley with low gradient on the up-slope side of the dam and shallow bedrock for its foundation provided optimum conditions. Masonry check dams, gabion structures, and gabion structures with impervious, ferrocement barriers were employed at different locations. The gabion structures were held together by galvanized iron chain link. The impervious barriers were constructed in trenches, excavated in the bedrock, and situated at the center of the dam site or on the up-slope side (see Figure 5).

The waters of selected springs were collected in stone storage tanks equipped with taps. A gravel filter was installed at the inlet of each tank. At one location, groundwater was contained for village use on the up-slope side of an underground stone dam, extending down to the bedrock. A barrier of ferrocement was included in the structure.

FIGURE 4   Roof water harvesting
Manhere Village, Akole *Taluka,* Ahmednagar District, Maharashtra State, India.

FIGURE 5   Lowermost check dam and reservoir in gravity-flow system of three reservoirs, close to completion
Middle and uppermost check dams of system also visible. Near Manhere Village, Akole *Taluka*, Ahmednagar District, Maharashtra State, India.

WATER SPREADING

Water spreading involves arresting the down-slope motion of overland flow by means of artificial reductions in the angle of slope, which may be augmented by low ridges of soil. The ridges are constructed parallel to the contours of elevation, generally along the outer margins of the areas of reduced slope. The water spread out by these means infiltrates into the soil and makes additions to soil moisture. It also may recharge shallow ground water. Infiltration may be improved through the excavation of strategically located, shallow pits and trenches, elongated parallel to contours of elevation. Infiltration at different levels on the hillside may be controlled sequentially by means of spillways.

Various approaches to slope modification were employed. The existing system of hill terraces was expanded, notably for orchard development. Contour trenches were dug close to their down-slope margins. Terrace-margin bunds (soil ridges) were built up from the excavated soil. Spill-ways were introduced at breaks in the bunds, with stone aprons to give protection against soil erosion. Farm ponds are rectangular excavations, located at the lower terrace levels, so that the removed soil forms a ridge on the down-slope side (see Figure 6).

Dry stone bunds were constructed at the upper reaches of slopes, parallel to contours of elevation, to reduce the velocity of surface runoff and minimize soil erosion. Gully plugs were positioned across incipient gullies to trap eroded soil. They serve the additional function of promoting the infiltration of trapped surface runoff (see Figure 7). Vegetative, ecological techniques were also applied, in support of these inorganic approaches to soil conservation. These included the planting of forestry and fruit trees, as well as shrubs and grasses, along terrace-margin bunds and soil ridges, associated with farm ponds.

FIGURE 6   Hillside terraces with marginal bunds
Near road between villages of Ladgaon and Titvi, Akole *Taluka*, Ahmednagar District, Maharashtra State, India.

FIGURE 7   Contour bund with contour trench and farm pond on upslope side
Near road between villages of Manhere and Ambevangan, Akole *Taluka*, Ahmednagar District, Maharashtra State, India.

Infiltration trenches and pits were excavated in the vicinity of selected dug wells, many of which also were deepened to improve their yields. Particular bore wells were given workovers for the same reason.

CAPACITY BUILDING

Villagers were involved in the introduction of the various technologies at demonstration sites. In addition to learning the techniques required, they developed basic management skills and, in some cases, demonstrated leadership qualities. In each of the three partner villages, additional training in the application and maintenance of water-harvesting and -spreading technologies was provided for interested individuals (see Figure 8).

ACKNOWLEDGMENTS

The authors gratefully acknowledge the guidance of the late Manibhai Desai, founder and first president of BAIF. They also thank his successor, Narayan Hegde, for considerable encouragement and helpful discussion.

The phenomenal success of the project was the result of teamwork and the commitment of many individuals. In particular, reference must be made to the roles of the BAIF field staff, members of the Canadian and Indian project teams, and the numerous individuals at the BAIF office in Pune and at the University of Windsor, who also made contributions.

The funding agency was the International Development Research Centre, Ottawa.

The authors gratefully acknowledge their indebtedness to all of these individuals and organizations.

FIGURE 8   Demonstration of teamwork in use of A-frame for excavation of contour bunds and trenches
Manhere Village, Akole *Taluka*, Ahmednagar District, Maharashtra State, India.

REFERENCES

Nessler, U. (1980). "Ancient Techniques and Modern Arid Zone Agriculture." *Kidma, Israel Journal of Development,* Vol. 5, 3–7.

Simpson, F., and G. Sohani. (2001). "Benefits to Villagers in Maharashtra, India, from Conjunctive Use of Water Resources." In Jeffery Roger, and Vira, Bhaskar (Eds.), *Conflict and Cooperation in Participatory Natural Resources Management: Lessons from Case Studies.* New York, Palgrave, pp.150–168.

Sohani, G. G., F. Simpson, B. K. Kakade, W. H. Blackburn, V. J. Harris, S. C. Kanekar, G. V. Jadhav, D. N. Joshi, A. Kaulagekar, S. G. Patil, P. R. Sharma, M. G. Sklash, T. E. Smith, and D. W. Steele, with assistance from S. Agarwal, and M. Macdonald. (1998). *Conjuctive Use of Water Resources in Deccan Trap, India.* Pune, BAIF Development Research Foundation, 133 p.

Tagare, G. V. (1992). "Water Exploration in Ancient India." *The Deccan Geographer,* Vol. 30, 43–48.

RAINWATER COLLECTION

Based on "Rainwater Reservoirs Above Ground Structures for Roof Catchments," Rolf Hasse, A Publication of Deutsches Zentrum für Entwicklungstechnologien–GATE in: Deutsche Gesellschaft für Technische Zusammenarbeit (GTZ) GmbH—1989

An approach to supplying clean water is to collect rainwater, usually from a roof. Systems for doing this are used presently in Thailand, New Zealand, and other places. A major design question is the size and construction of the storage tank. In Thailand, for example, large baskets are coated with cement to give tanks of 1000-gallon capacity. Figure 4 in the previous article shows a system.

ADVANTAGES OF HARVESTING RAINWATER

- Provides very high-quality water (in most areas), soft and low in minerals, so less soap is required
- Reduces mineral deposits on fixtures, pipes, and water heaters
- Offsets the need for pumping groundwater; reduces energy needed for deep well pumping and water softening
- Conserves irrigation water because plants often respond better to rainwater than groundwater, increasing yield
- Reduces erosion and flooding typically created by runoff
- Reduces silting and contamination of waterways from runoff

COMPONENTS OF A ROOFTOP RAINWATER HARVESTING SYSTEM

- Catchment Area—another potential function of a roof
- Gutters and downspouts—can include leaf screens and roof washers
- Storage tanks/cisterns—prefabricated: galvanized steel, fiberglass, polyethylene, polypropylene, PVC bladders; partially prefabricated: series of drums, cans or barrels; site built: ferrocement, stone, poured concrete, mortared block and rammed earth

A house with rainwater gutters and a cistern is shown in Figure 9.

Roof Type and Catchment

The shape of a catchment area has a considerable influence on the catchment possibilities. Of the most common roof types shown in Figure 10, the single-pitch roof is the most appropriate for rainwater harvesting, since the entire roof area can be drained into a single gutter on the lower side and one or two downpipes can be provided, depending on the area. A more difficult roof for rainwater catchment is the tent roof. It requires a gutter on each side and at least two downpipes on opposite corners. If a tent roof is large enough, it could be drained into four tanks located at each corner of the house. The main problem is always the corner. A 90-degree angle in the gutter should be avoided. It is extremely difficult to adjust gutters in such a way that water really flows easily. A gutter seldom works well when downpours occur, and it is the heavy downpours that should be caught. The hip roof is not very efficient either, since it also needs gutters all around the building. Flat roofs can be used for catchment if they are furnished with an edge, keeping the water on the slab until it has drained through the gutter or downpipe. However, using a flat roof for rainwater harvesting is not very efficient because of the extended runoff time and the evaporation losses. One way to improve the catchment is to provide the slab with a sloping cement screed. Constructing a waterproof edge on a flat roof is rather difficult because of the temperature expansion.

Roof Finish

Not all materials used for roofing finishes are equally good, but the most commonly used material, metal sheeting (corrugated galvanized iron and aluminum sheets), is very suitable for rainwater catchment. Likewise brick tiles of all variations and also thatch can be used, but these are less efficient. Lead, sometimes employed in soldering gutters or channels, should not be used.

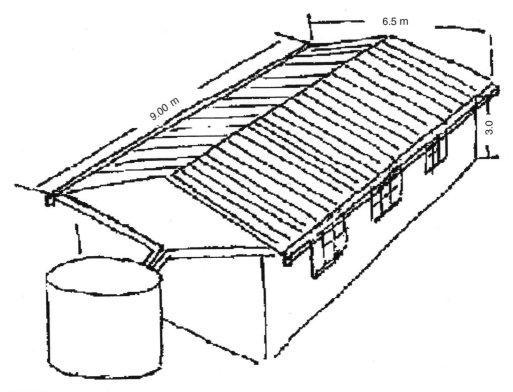

FIGURE 9
Rainwater collection.

ROOF WASHERS

The water from a roof during the first few minutes of a rainstorm will rinse dirt—often mostly bird droppings—off the roof and will be dirty. A design question, then, is how to divert that water from the storage tank if clean water is needed. One device uses a bucket to take the first flow; when it is full, it tips over, directing the rest of the flow to the storage tank (see the report by the Institute for Rural Water in the References). Another device is a vertical standpipe in the gutter between the roof and the downspout leading to the storage cistern. This standpipe extends to the ground and has a normally closed stopcock at the bottom. The first rain in a storm will flush dirty water into the standpipe. Only after the standpipe is full will rainwater go to the downspout leading to the cistern. The standpipe should hold 10 gallons of water for each 1,000 ft^2 of roof. A six- or eight-inch PVC pipe is usually suitable. The water in the standpipe can be used for irrigation between rainstorms. A screen over the gutter will keep leaves out.

DIFFERENT TYPES OF RESERVOIRS AND THEIR ADVANTAGES

In many cases in developing countries the availability of building materials outweighs the economic factor in selecting a type of reservoir.

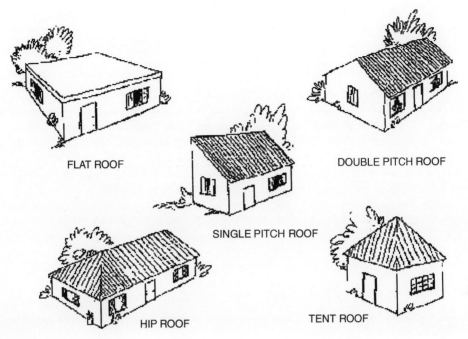

FLAT ROOF

DOUBLE PITCH ROOF

SINGLE PITCH ROOF

HIP ROOF

TENT ROOF

FIGURE 10
Types of roofs.

The Corrugated Iron Tank

This is an industrial product manufactured in many countries. Where the material for this tank is available, there are at least three capacity sizes: 2.25, 4.5, and 9.0 m^3. Although iron tanks are usually the most economical, prices have to be compared with other suitable materials; the transport aspect can also increase costs substantially. The advantage of this tank is mainly the price but certainly also the fast installation. The disadvantage is the limited lifetime due to corrosion, although this can be improved by painting. One should remember at all times that the corrugated iron tank is vulnerable to manual force. Experience has shown that this tank should not be used at public places, especially not at schools, since vandalism is likely to damage the tanks beyond repair.

The PVC Foil Tank

Several industrial producers offer tanks of PVC foil. The foil is fixed inside a reinforcement mesh framework or galvanized sheet cylinders, screwed together from sections. The tanks are available from about 5 m^3 up to 430 m^3. Their considerable advantage lies in fast assembly and low transport costs. A reservoir of 9.25 m diameter (capacity 81.0 m^3) can be transported on a small van and be assembled within a couple of hours. No foundation is needed. Dismantling and reassembly at another place can be carried out within a day or two. Apart from this advantage, which is very valuable for cases requiring immediate action—for instance, improvising a village water supply—the system has some weak points. Tanks of large capacity are uncovered, so evaporation is high and there is a danger of pollution. More important for permanent use is the problem of ultraviolet ray influence on the PVC foil. Systems in use show signs of ultraviolet light effect on the material after just a few years. Otherwise, the vulnerability to external force is great, and tanks should always be fenced in. For permanent rainwater catchment, although relatively cheap, this technique has its limitations.

The Ferro-Cement Tank Without Mold

This technique depends on the availability of welded reinforcement mesh. Since this is not to be found everywhere, other methods can be substituted. There are many examples of such reservoirs in Kenya. Ferro-cement is described in the "Boats" article in the Transportation chapter.

First, close attention must be given to the cost of the material and the transport to the site. The height of the tank will be the width of the roll of mesh or mats, about 1.80 m. This is certainly a restriction. Theoretically, it is possible to extend the height of the wall by using one and a half widths of the mesh, overlapping it a minimum of three fields, and tying it together with the bottom circle, but such is not recommended. The entire structure becomes unstable, and any vibration during the process of plastering will make the work very difficult. In addition a scaffold is needed, which might not always be available. The fixing of the scaffold requires skilled workers.

The Ferro-Cement Tank with a Factory-Made Mold

The technique has considerable advantage for rainwater storage where all tanks are of the same size. Several examples are found in Botswana.

This construction method can only be chosen if a factory or experienced workshop provides the facilities for bending corrugated sheets and welding them neatly together. The technique is highly appropriate in areas where a series of tanks are to be built, such as when new buildings, like schools, are put up and design of the buildings already includes provision for rainwater catchment.

The mold can be used 10 to 15 times, depending on the experience and careful handling of the staff. For larger projects it is advisable to have at least two molds at the site. With two molds, the work can be organized with three crews. The first crew starts preparing the ground and then casts the foundation slab. The second erects the mold and reinforces it, and the third crew does the plastering. The roof slab can be made by a fourth crew or by the first, depending on the amount of ground to be cleared. This technique will be too expensive where only four or five reservoirs have to be constructed.

The Ferro-Cement Tank with a Made-on-Site Mold

This approach should be chosen where only a few tanks are required, or even just one—in other words, where prefabricated molds do not make sense and welded reinforcement mesh is not available. All that is needed, in addition to the normal building materials for a ferro-cement structure, is some additional timber for the framework and a few corrugated iron sheets for shuttering. Fencing mesh is an additional reinforcement but could be replaced by other available mesh material.

The Reinforced Brickwork Tank

The reinforced brick tank is more expensive than the ferro-cement tank, although the cost per m^3 reduces with increased capacity. A brick tank costs about twice as much as a ferro-cement tank. For this reason this tank should be chosen only when the capacity needed is above 30 m^3 or the life of the structure is expected to be 20 years and more. The advantage of the construction method is the adaptability to the building design. Structures above 1.80 m in height can be built without problems, although plastering has to be done with great care. Especially at public buildings, which are usually higher than residential houses, it is possible to use the height between gutter and ground, avoiding large diameters and thus saving space.

HOW TO CHOOSE THE SIZE OF A RESERVOIR

The size of storage capacity is based on the mean annual rainfall, but it should be greater if funds allow. To calculate the rainwater amount that can be harvested, the mean annual rainfall figure is commonly used. Mean annual is the statistical average calculated on the basis of measured rainfall over many years. Not only the average rainfall needs to be considered but also the uniformity. If the rainfall is fairly uniform year-to-year then the average is a useful estimate. But the average is less helpful if the rainfall pattern in a given area is erratic, which is quite common in countries with drought periods. It can happen that the mean annual is not reached. It can certainly happen the other way round that considerably more rain falls than the mean annual. This makes the calculation of the storage capacity rather difficult.

As an example, let us consider the roof of the building in Figure 9 with a rainfall of 450 mm. We assume that less than 100 percent of the calculated amount of water will be collected. This is due to unavoidable small leakages in the gutter downpipe system, or rainfalls that are too light to produce sufficient runoff, or a possible overflow of gutters in the case of an extreme downpour. For these reason we can generally assume that only 90 percent of the rainwater can be collected.

For calculation we take the following formula.

Mean annual rainfall in mm $\times$ area in m^2 $\times$ runoff factor = collected rainwater in liters.

In our example this means the following

$$450 \times 6.5 \times 9 \times 0.9 = 23,700 \text{ liters}$$

The height from the ground to the gutter outlet is 3 m. A reservoir of 4-m diameter on a filling height of 1.80 m has a storage capacity of 23,000 liters. This means that one reservoir built at the gable side of the house would be sufficient for nearly all the rainwater that can be collected if average rainfall occurs. We assume that the rainfall pattern makes it unlikely that all the year's rain will occur within a short time period. Two gutters along the sides of the building should be connected with downpipes fixed to the gable wall and then bridged into the tanks.

For this storage capacity a ferro-cement tank would be less expensive than a reinforced brick tank and serves the same purpose. But if a smaller storage capacity would be sufficient, or if funds are very limited, two corrugated iron tanks, 9,000 liters each, would be cheaper and more effective.

REFERENCES

Institute for Rural Water. (1982). "Constructing, Operating, and Maintaining Roof Catchments." Water for the World Technical Note Number RWS.1.C.4 USAID. Request from Development Information Center, Agency for International Development, Washington, DC 20523.

Keller, Kent. (1982). "Rainwater Harvesting for Domestic Water Supplies in Developing Countries." WASH Working
    Paper No. 20, Water and Sanitation for Health Project, Arlington, VA.
http://www.twdb.state.tx.us/publications/reports/rainHarv.pdf
http://www.rdrop.com/users/Krishna/rainwater.htm

SAVING AND REUSING WATER

*Based on "Rainwater Reservoirs Above Ground Structures for Roof Catchment." Rolf
Hasse, A Publication of Deutsches Zentrum für Entwicklungstechnologien—GATE in:
Deutsche Gesellschaft für Technische Zusammenarbeit (GTZ) GmbH—1989*

Saving water in semidesert countries is essential and should be encouraged as much as possible.
To support saving, it is first important to understand how water is wasted. By way of example,
let us consider a self-help housing area in Botswana. This area is supplied through standpipes
on the side of streets, never more than 100 meters away from a house. Tenants are supposed to
build pit latrines before they construct dwelling rooms. People fill containers several times a
day and carry the water home. Although the supply of water is much better than in villages and
distances are much shorter, the water still has to be carried. When you have to carry water to
the point of use, you learn quickly not to waste it. It was observed that people became used to
collecting their washing water after use and watering their plants in the courtyard. Water bills
in Lobatse showed that consumption of water per standpipe, used by 7–10 families, in self-help
housing areas is lower than the consumption for one high-cost house. It appears that the consumption in residential houses rises with the number of taps and other sources connected to the
central supply. The conclusion should not be to require public standpipes. The general conclusion
should only be that the convenience of access to water raises consumption. We should have a
close look at possibilities of saving domestic water.

DOMESTIC WATER SAVING

In domestic water use the water-borne toilet system, in general, is the highest consumer of water.
Moreover, the water used is not fit for reuse and goes into the sewers. Recycling of sewer water
is possible but expensive and can only be done in special ponds. Toilets are discussed in the
"Sanitation" article in this section. At a school with water closets—12 toilets, each consuming
10 liters per flush—the consumption of these flush toilets is higher than the consumption of
1,000 pupils and their teachers for drinking, cooking one meal a day, and washing the dishes.
Consumption reduction means first reducing the consumption by the toilets. Flushing valves
consume less than flushing cisterns, but they are not appropriate, since they require a permanent
high water pressure not always available in developing countries. There are producers of toilets
consuming 4 liters of water per flush in Sweden, Great Britain, and West Germany. In the United
States all new toilets are required by federal law to use a minimum amount of water. Imports
of this highly appropriate system into developing countries, where there is a real need to save
water, should be encouraged.

Introduction of a new system—low water consumption toilets—takes time, since people have
to be convinced that the higher investment really brings returns. But there is also something that
can be done about the existing highly wasteful cisterns—toilet tanks. Some can be adjusted to
lower levels of filling by bending the bow of the cistern float downwards. This results in stopping

the filling water at a lower level. It is also possible to put stones in the cistern. The volume of the stones (blocks) will be the volume of water saved. Depending on the type of cistern, the consumption can be reduced to 7 or even 6 liters, but the cleaning effect of flushing may be reduced, since the toilet bowl is not designed for such low consumption.

BATH OR SHOWER

It is often not realized that the amount of water consumed for one bath is sufficient for three showers. In consequence, houses should be furnished with showers rather than with baths. But baths have become a status symbol in many countries, and a high-cost house must be furnished with a bath. The odd thing is that baths in Europe are rather out of fashion and much less used than showers, which produce savings in both water and time. From the hygiene point of view showers are better than baths. When a bath is installed, it should always be done that the bath can also be used for showers. At the same time the built-in bath should be chosen carefully, since the capacity varies substantially.

Several devices designed to reduce water consumption are on the market. Spray nozzles for showers, push button taps, and so on might reduce consumption but should be studied before use. When deciding on water saving equipment one has also to consider the lime content of the water. Lime precipitates at 60° C. This means that sensitive equipment in hot climates will soon clog.

REUSE OF DOMESTIC WATER

Major sources of water consumption in residential houses are the kitchen sink, the bath and/or the shower, the basin in the bathroom, and the toilet. While for obvious reasons the reuse of water from the toilet is not possible, the bathroom water, although contaminated by soap and through laundry by washing powder, can be used for cultivation, even for vegetables, if directed at the soil. Such a system is sometimes called "graywater" recycling—"blackwater" would be water from a toilet. One vegetable gardening area of 150 m at a clinic in Lobatse was only irrigated with water from sinks and hand basins for a period of one year and showed very successful results. At this clinic one sink was used for washing drug containers and equipment used for medical tests. This wastewater was drained into the sewer. All other wastewater was drained into drums dug into the ground (see Figure 11). The water was then extracted with buckets and used for gardening.

Experiments at private residential houses have shown that the reuse of water for gardening does not affect the plants. One should be careful with water running out of the kitchen sinks. Water from dishwashing usually contains much grease and is therefore not suitable for most plants or for vegetable gardening. But this water can be successfully used, for example, for cultivation of banana plants. Bananas should not be planted closer than 15 meters to a residential house because of mosquito breeding. A simple sand filter can be built to remove the grease. The cleaning material used in the household should be chosen carefully. Soaps contain fewer harmful chemicals than detergents. Boron is especially harmful to plants. It is necessary that the area of soil receiving the graywater be large enough to absorb it all—sandy soils are more absorbent than clay soils and a smaller area is necessary.

There are two ways to reuse domestic water. The first, as stated, is to disconnect the pipes of the sink outlets and fit hoses draining the water into drums dug into the ground. These drums must be provided with lids because of the danger of mosquito breeding. Water is then lifted out with buckets. The other and more convenient method is to connect long hoses direct to the

FIGURE 11
Water recycling.

outlets and draw the water straight to the place of use. See Figure 11. Attention must be paid, of course, to matching the amount of water from the drains to the needs of the garden.

Where rainwater is available and not used for the household as drinking water because of an existing centralized supply, it should be used for vegetable gardening and the waste water for cultivation of trees and other plants. It is probably best not to use graywater for irrigating root crops and not spray the water on edible leaves. Soaps and detergents are alkaline, and some plants, such as broccoli, cantaloupes, and tomatoes, thrive on alkaline soil. Beans, apricots, and peaches do not.

REFERENCE

Leckie, Jim, Gil Masters, Harry Whitehouse, and Lily Young. (1981). *More Other Homes and Garbage.* San Francisco: Sierra Club Books.

WELLS: HAND DUG AND HAND DRILLED

Based on "Wells Construction: Hand Dug and Hand Drilled," Peace Corps Information Collection & Exchange, Manual M0009, Written by Richard E. Brush, September 1982

There is water at some depth almost everywhere beneath the earth's surface. A well is a dug or drilled hole that extends deep enough into the ground to reach water. Wells are usually circular and walled with stone, concrete, or pipe to prevent the hole from caving in. They are sunk by digging or drilling through one or more layers of soil and rock to reach a layer that is at least partially full of water, called an aquifer. The top of the aquifer, or the level beneath which the ground is saturated with water, is called the water table. In some areas there is more than one aquifer beneath the water table. Deep wells, such as those sunk by large motorized equipment, can reach and pull water from more than one aquifer at the same time. However, this article will only discuss sinking wells to the first usable aquifer with hand-powered equipment.

SITE CHOICE

The well should be located at a site with the following characteristics.

- Water bearing
- Acceptable to the local community
- Suitable to the sinking methods available
- Not likely to be easily contaminated

It is not always possible for a site to meet all of these guidelines. Therefore, a site will need to be chosen that best approximates the guidelines, with particular emphasis on the likelihood of reaching water (see Figure. 12). Where there is an equal chance of reaching water at several different locations, the one closest to the users is preferable.

WHERE IS WATER LIKELY TO BE FOUND?

Choosing the site for a well can be difficult because easily available and abundant water can never be guaranteed. Even professionals, before a well is sunk, rarely know where they will reach water and how much will be available. However, there are a number of guidelines that can be very useful in providing information about possibly successful well sites. Where possible, a well can be located near a past or present water source. By doing so, you are likely to reach water at approximately the same depth as the other source.

If no other sources exist or have ever been developed nearby, you must be more cautious in choosing a well site. Unless you have the benefit of detailed geological information, it is best simply to look for the lowest spot nearby. Both surface and groundwater are likely to collect here. In some cases, plants can be indicators of the presence of groundwater—perennial plants and trees often tap into groundwater. Be careful, however, not to build in a place so low that the well would be susceptible to flooding in heavy rain.

FIGURE 12
Possible well sites.

FIgure 12 Shows four possible sites and their suitability.

1. Limited water would be available at this site because the impermeable rock layer is close to the ground surface, allowing slight fluctuations in the water table to drastically affect water availability.
2. This is the closest site to village and therefore the best site if it is possible to dig down far enough to reach water.
3. At this site, there would be a better chance of reaching more water than at site 2, but the site is farther from the village.
4. This is the site where water is most likely to be reached by a well, although it is some distance from the village. Because it is in the absolute bottom of the valley, it may be subject to flooding.

OTHER CONSIDERATIONS

Other considerations involve being acceptable to the local community, suitable to the sinking methods available, and not likely to be easily contaminated. If the most likely site would require that the well penetrate a layer of rock, and if there are no tools available to do such a difficult job, the site is not an appropriate one. The most important contamination factor is that the well not be located within 15 meters of a latrine or other sewage source. This would also include not placing the well where it might be damaged or inundated by a flood or heavy rain.

The only way to design a well to prevent water contamination is to seal it so that water can enter only through the bottom section. Dug wells need to be covered with a permanent cover through which a pump is installed to draw water. All wells should have a platform around them that is at least 1 meter wide, one that water will not penetrate. This platform ought to be sloped in such a way that any spilled water runs off away from the well.

TYPES OF WELLS

In general, there are two types of wells: dug wells and drilled wells. The obvious difference between the two is the size of the holes. Dug wells are sunk by people working down in the hole to loosen and remove the soil. The wells need to be at least 1 meter wide to give the diggers room to work. Drilled wells, on the other hand, are sunk by using special tools that are lowered into the ground and worked from the surface. These wells are normally less than 30 centimeters (cm) in diameter and usually are less than 15 cm; it is difficult to drill wide diameter holes with hand-powered tools.

WELL SECTIONS

Every well, whether drilled or dug, has three sections: top, middle, and bottom. Each of these sections varies in construction because each must function differently.

- Top section—that part of the well at or above the ground surface level. It should be designed to allow people to get water as easily as possible and, at the same time, to prevent water, dirt, and other contaminants from entering.
- Middle section—that part of the well that is between the ground surface and the water. This section is usually a circular hole. It is reinforced with some kind of lining to prevent the walls from caving in.

- Bottom section—that part of the well that extends beneath the water table into the aquifer. It should be designed to allow as much water as possible to enter and yet prevent the entrance of soil from the aquifer. Its lining will have holes, slots, or open spaces, allowing water to pass through.

Lining and casing refer to the same part of the well. Lining is used with a dug well, while casing refers to the pipe used to reinforce a drilled well. The three sections are shown in Figure 13.

A head wall should be built on all wells that will not be fitted with a permanent cover and a pump, as a inexpensive safety feature, which will prevent people and animals from accidentally falling in. It is simply a wall that extends above the surface of the ground far enough to prevent most accidental entry of people, particularly children, and animals. Its external dimension is dependent on how thick you want the head wall to be. A head wall that is unnecessarily thick will encourage people to stand on it to draw their water, creating an unsafe situation. The easiest and best way to construct the head wall is as an extension of the lining as the equipment and supplies construction will be on site. The head wall should extend 80 to 100 cm above the ground surface or apron, if there is one.

A drainage apron is most often a reinforced concrete slab 1 to 2 meters wide, which surrounds a well and, because of its slight slope, channels surface water away from the well. Wire mesh reinforcing may be used if it is available. By forcing water to flow away from the well, the apron serves two functions: It prevents contaminated surface water from flowing back down into the well before it has had a chance to be sufficiently filtered by the earth, and it prevents the formation of a mucky area immediately around the well, which can be a breeding ground for disease and a source of contaminants to the well water.

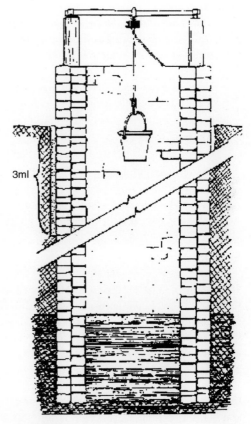

FIGURE 13
Rock-lined dug well.

All wells, except those drilled through rock, can be expected to cave in with time unless a lining is installed to support the well. The lining thus helps to keep the well open. Another function of the lining is to prevent contamination of the water. Occasionally slight ground shifts can put pressure on linings, causing them to split and separate if not strongly built. Geologists can usually predict where such shifts are likely to occur. If no such information is available, it is recommended that you build the lining strongly enough to withstand normal earth stresses.

Depending on ground conditions, you may or may not be able to dig the complete hole and then line it. In very loose, sandy soil, for example, the sand from the walls of the hole will frequently cave into the hole, seriously hampering efforts to deepen the hole. In this case the hole is lined after 1 meter or so—certainly less than 5 meters—have been dug.

Designing the lining for the middle section is largely a matter of assessing the ground conditions and materials availability to determine the lining materials and method most appropriate for the situation. The lining of a dug well can be built of reinforced concrete, concrete without reinforcement, cement alone, bricks, rocks, even wood or bamboo. Drilled wells are almost always lined with pipe. If concrete is used, the lining can be formed above ground and lowered into the well or made in place.

The purpose of the bottom section is to allow as much water as possible into the well without permitting any of the fine soil particles from the surrounding aquifer to enter the well. There are three commonly used methods of allowing water to enter the well.

- Through a porous concrete lining—lining rings sunk into the bottom section can be made of porous concrete, which acts as a filter to prevent soil particles from entering the well.
- Through angled holes in the lining—holes can be punched in a freshly poured concrete ring that, when cured, can be sunk into the bottom section. These holes are more effective at preventing soil entry if they are slanted up toward the middle of the well.
- Though the bottom—the bottom of the well should always be constructed to allow water to come up through it. Often the bottom is simply left open and uncovered, but it is preferable to prevent soil entry which will gradually fill up the well. A perforated concrete slab can be used at the bottom of the well.

MATERIALS

The two most important sections of the well are the lining (or casing) and the bottom (or intake) section. While it is not necessary that both be built of the same material, it is often cheaper and more convenient to do so. Almost all modern well linings are made of either concrete or pipe (metal or plastic). Nowadays, concrete is used most often in the lining of hand-dug wells. It can be easily mixed from cement, sand, gravel, and water, and cast in place in the well. Reinforcing bars can be added to either mortar or concrete to make a much stronger and more durable lining.

Metal pipe is normally used in the construction of drilled wells. It can easily be shaped to make the necessary tools with which to sink the well and can also serve as the permanent casing and bottom section. Plastic pipe is too soft to use during drilling but is in many situations a better casing than metal pipe because it will not rust or corrode. Large-diameter concrete pipe can be used to line dug wells. Cement and pipe are available in most countries and usually in all but the most remote regions. When both materials are available, consider such factors as transportation, type of well, depth, ease of construction, and adaptability to local practices before deciding which is more appropriate.

In emergency situations, when the best water available is immediately needed, a number of substitute materials and techniques can be used. For example, wood lining can be used instead of cement. Wells built with wood or other substitute materials and techniques will supply acceptable water for a short period of time. However, they cannot now or in the future be converted into permanent sources of clean water without rebuilding major portions of the structure.

SINKING METHOD

A sinking method is the way of sinking a well. Wells may be dug by hand, drilled with hand tools, or drilled with motorized equipment. Many methods and techniques are used. The particular choice depends on the available materials and equipment, the expected ground conditions at the well site, and the user's experience with a specific sinking technique.

DUG WELLS

Hand-dug wells are sunk by digging a hole as deep as is necessary to reach water. Once the water-bearing layer is reached, it should be penetrated as far as possible. The process is always basically the same, with only minor variations because of the particular tools and equipment available and the variety of ground conditions—see Figure 14.

The following are the advantages of dug wells.

- The procedure is a very flexible one. It can be easily adapted with a minimum of equipment to a variety of soil conditions as long as cement is available.
- Because the resulting well is wide-mouthed, it is easily adaptable to simple water-lifting techniques if pumps are not available or appropriate.
- It provides an underground reservoir, which is useful for accumulating water from ground formations that yield water slowly.

The following are the disadvantages of dug wells.

- A hand-dug well takes longer to construct than a drilled well.
- It is usually more expensive than a hand-drilled well.

FIGURE 14
Digging a well.

- It cannot easily be made into a permanent water source without the use of cement.
- Hand digging cannot easily penetrate hard ground and rock.
- It may be difficult to penetrate deeply enough into the aquifer so that the well will not dry up in the dry season.

Compared to other well-sinking methods, digging a well by hand takes a long time. An organized and experienced construction team, consisting of five workers plus enough people to lower and raise loads in the well, can dig and line 1 meter per day in relatively loose soil that does not cave in. However, the bottom section is likely to take two or three days per meter because of the difficulty in working while water continually enters the well. Depending on how the well is developed, the top section can take anywhere from a day or two to several weeks. An experienced team sinking a 20-meter well and installing pulleys on the top structure could easily take five weeks, including occasional days off (this, of course, assumes no major delays). A new or inexperienced group would be expected to take twice that time.

Hand-dug wells should be dug during the dry season when the water table is likely to be at or near its lowest point. The well can be sunk deeper with less interference from water flowing into it. The greater depth should also ensure a year-round supply of water. If the well cannot be dug during the dry season, plan to go back to it at the end of the dry season to deepen it.

DRILLED WELLS

Drilled wells are sunk by using a special tool, called a bit, that acts to loosen the soil or rock at the bottom of the hole. It is connected to a shaft or line that extends to the ground surface and above. The part of the shaft or line extending above the ground can then be rotated to operate the bit—see Figure 15.

The following are the advantages of drilled wells.

- Construction is fast.
- Where cement is not available, wells can be sunk with locally made drilling equipment and lined with local materials.
- While not easy, it is possible to penetrate hard ground and rock formations that would be very difficult to dig through.
- Drilling usually requires fewer people than hand digging.
- It is especially suitable for use in loose sand with a shallow water table.

The following are the disadvantages of drilled wells.

- Special equipment is required. There are a number of different hand-drilling techniques that are suitable for a wide range of ground conditions.
- Pumps almost always have to be used because buckets are too large to be lowered into the well.
- Limited depth can be reached with hand-powered drilling equipment.

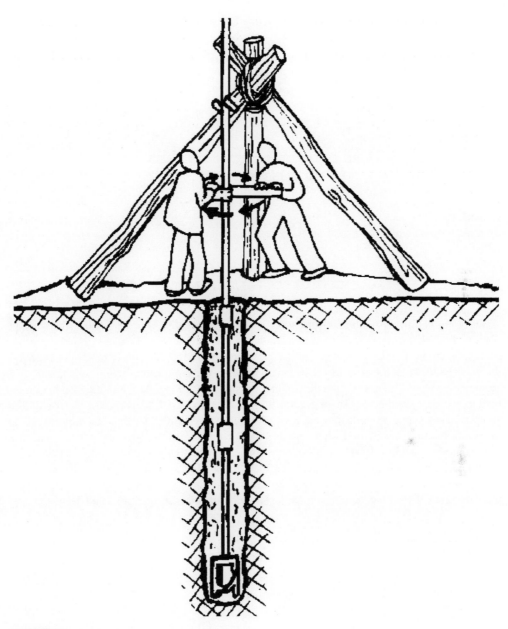

FIGURE 15
Drilling a well.

REFERENCES

Brush, Richard. (1987). *Wells Construction*. Peace Corps, 1111 20th Street, N.W., Washington, DC 20526.
Koegel, R. C. (1977). *Self-Help Wells*. Rome: FAO.
VITA. (1988). *Village Technology Handbook*. Volunteers in Technical Assistance, 1815 North Lynn Street, Washington, DC 22209, USA.

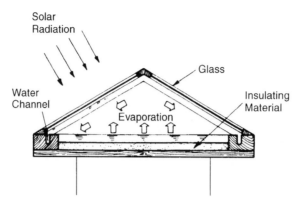

FIGURE 16
Solar still. (From *More Other Homes and Garbage.*)

SOLAR STILLS

Water is purified by distillation. If the water is very poor quality—for example, from stagnant pools or from the ocean—distillation may be the only practical way to make it drinkable. Small solar stills are fairly easy to make. One is shown in Figure 16. The sun's energy goes through the glass windows and evaporates the impure water in the pool at the bottom of the still. The evaporated water condenses on the glass windows and flows along them into the channels at the bottom, where it is collected. As long as the glass is tilted at least 20 degrees, the water will flow along it rather than drip back into the pool of impure water. Still performance improves if the temperature inside is increased, so the joints should be tight and the sides and bottom insulated. A reasonable size still—10 ft^2—will supply sufficient water for one adult, so they are useful in emergencies, but probably are not practical for a village water supply. In the Kalahari Desert village well water was much too salty to be drunk. A solar distillation facility was established that earned cash by selling the salts remaining after distillation as well as produced drinking water.

REFERENCE

Leckie, Jim, Gil Masters, Harry Whitehouse and Lily Young. (1981). *More Other Homes and Garbage.* San Francisco: Sierra Club Books.

FILTRATION AND DISINFECTION

Based on "Simple Methods for the Treatment of Drinking Water," Gabriele Heber A Publication of the Deutsches Zentrum für Entwicklungstechnologien—GATE in: Deutsche Gesellschaft für Technische Zusammenarbeit (GTZ) GmbH, 1985

Filtration is the deliberate passage of polluted water through a porous medium, utilizing the principle of natural cleansing of the soil. This widely used technique in water treatment is based on several simultaneously occurring phenomena.

- Mechanical straining of undissolved suspended particles (screening effect)
- Charge exchange, flocculation adsorption of colloidal matter (boundary layer processes)
- Bacteriological-biological processes within the filter

Filters may be divided into two principally different types

- Slow sand (or biological) filtration (filtration rates = 0.1 to 0.3 m/h)
- Rapid filtration (filtration rates = 4 to 15 m/h)

Generally, a filter consists of the following components:

- Filter medium (inert medium: quartz sand; chemically activated medium: burned material)
- Support bed (gravel) and underdrain system, influent and effluent pipes, wash and drain lines, control and monitoring appurtenances

RAPID FILTRATION

Rapid filtration is mainly based on the principle of mechanical straining of suspended matter due to the screening effect of the filter bed (sand, gravel, etc.). The particles in the water pass into the filter bed and lodge in the voids between grains of the medium.

Also operative to some degree in rapid filters are boundary layer and biological mechanisms. Their extent largely depends on the filtration rate, filter medium, depth of the filter bed, and quality of the raw water. Cleaning of the rapid filter is facilitated by backwashing—reversing the flow direction. A backwash may be conducted simply with water or a water-air mix (upward air scour). The impurities are thus dislodged and removed from the filter bed.

The performance of a rapid filter regarding the removal of suspended matter is determined by the following filtration process variables and parameters.

- Filtration rate
- Influent characteristics: particle size, distribution, and so on
- Filter medium characteristics that control the removal of the particles and their release upon backwashing.

Generally, it is true that the treatment effect can be improved by the following.

- Reduced filtration rates
- Smaller granulation size of the filter medium
- Increasing depth of the filter bed

HOUSEHOLD SIZE RAPID FILTERS

Household filters can be made from sand or gravel of different grain sizes, from ceramics, porcelain or other fine porosity materials. They basically operate on the principle of mechanical straining of the particles contained in the water. The filter performance depends on the porosity of the filter medium. Through additives in the filter material, additional effects can be obtained (adsorption, disinfection).

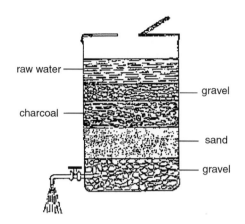

FIGURE 17
A multiple layer rapid filter.

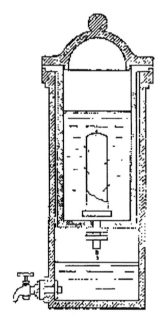

FIGURE 18
Household filter with candle.

Simple household filters can be put together using metal drums, plastic containers, or clay vessels and filling them with several 10 cm layers of sand, gravel, or charcoal. They do not perform well at removing pathogens, though. After filtration, the water therefore needs to be disinfected. A multiple-layer filter is shown in Figure 17.

Charcoal adsorbs organic substances that cause disagreeable color and taste. The effect can only be sustained, however, if the charcoal is frequently renewed. If this is not possible or if the filter (empty or filled with water) is left unused for some time, the charcoal can become a breeding ground for bacteria. The result is that the filtered water exhibits a higher bacteria count than the raw water. Monitoring of the filter condition is rendered more difficult because there is no visual indication given for the point when the charcoal should be replaced. Charcoal cannot be regenerated. It is for these reasons that the use of filters with charcoal media is not recommended.

Ceramics filters may be used for the purification of drinking water. If there are native potters, the filter can be manufactured locally. Otherwise they can be readily obtained from various commercial manufacturers. The purifying agent is a filter element, also called a candle, through which the water is passed. Suspended particles are thus mechanically retained and, depending on the size of the pores, also pathogens. Ceramics filters should only be used if the water is not too turbid, since the pores clog rather quickly. A ceramic filter is shown in Figure 18.

Ceramics filter elements can be made from various different material compositions (diatomaceous earth, porcelain); they have pore sizes of between .3 and 50 microns. If the pore size is smaller than or equal to 1.5 microns, all pathogens get removed with certainty. Posttreatment of the water prior to consumption is then unnecessary. Filters with larger pores only retain macroorganisms such as cysts and worm eggs—the filtered water must be boiled subsequently or otherwise disinfected

The candle should be cleaned because the impurities held back deposit on the candle's surface. At regular intervals—six months or so—this coating can be brushed off under running water. After the cleaning, the candle should be boiled. Ceramics filters must be handled with care. From time to time they must be checked for fissures to prevent the water from passing through the medium without being filtered.

Candles made from diatomaceous earth that contains silver have the advantage that recontamination of purified water due to infestation of the filter material with bacteria-laden washing water is avoided. Other filter inserts can be treated as follows.

Prepare a solution of 6.1 ml colloid silver in 200 ml of clean water and lay it on the filter element by means of a brush or a sponge. Let the filter dry for 24 hours. The first two filter runs should be discarded.

In this case, silver is the only component that must be imported. Though it is the most expensive part of the filter, it does achieve disinfection. Filters operating at atmospheric pressure exhibit a very slow rate of percolation. This can be increased considerably by forcing the water through the medium.

SLOW SAND FILTRATION

Slow sand filtration is accomplished by passing raw water slowly, driven by gravity, through a medium of fine sand. On the surface of the sand bed, a thin biological film develops after some time of ripening (the film makes the filter different from the rapid filter). This film consists of active microorganisms and is called "Schmutzdecke," or filter skin. It is responsible for the bacteriological purification effect.

These are the principal purification processes that take place during slow sand filtration.

- Sedimentation—the water body sitting on top of the filter bed acts as a settling reservoir. Settleable particles sink to the sand surface.
- Mechanical straining—the sand acts as a strainer. Particles too big to pass through the interstices between the sand grains are retained.
- Adsorption—the suspended particles and colloids that come in contact with the surface of the sand grains are retained by adhesion to the biological layer, by physical mass attraction (Van der Waals force), and by electrostatic and electrokinetic attractive forces (Coulomb forces).

Because of these forces, an agglomerate of oppositely charged particles forms within the top layer of sand. This process needs some ripening time to fully develop.

Several biochemical processes take place in the biological layer.

- Partial oxidation and breakdown of organic substances forming water, CO_2, and inorganic salts
- Conversion of soluble iron and manganese compounds into insoluble hydroxides, which attach themselves to the grain surfaces
- Killing of *E. coli* and of pathogens

Organic substances are deposited on the upper layer of sand, where they serve as a breeding ground and food for bacteria and other types of microorganisms (assimilation and dissimilation). These produce a slimy, sticky, gelatinous film, which consists of active bacteria, their wastes and dead cells, and partly assimilated organic materials. The dissimilation products are carried away by the water to greater depth. Similar processes occur there. The bacterial activity gradually decreases with depth. Different types of bacteria are normally found at various depths.

Algae can contribute to the breakdown of organic material and bacteria. They can improve the formation of the biological layer (filter skin). In uncovered filters, growth of algae is driven by photosynthesis. The presence of large amounts of algae in the supernatant reservoir of a filter generally impedes the functioning of the filter. Dead cell material may clog the filter. Increased consumption of oxygen due to the presence of dead cell material increases the possibility that anaerobic conditions will occur. There is always a diurnal variation in the oxygen content due to growth and decay of the algae mass. When algae growth is strong, the algae must be either removed regularly or the filter must be covered.

The conditions necessary for those biochemical processes are the following.

- Sufficient ripening of the biological layers
- Uniform and slow flow of water through the filter, approximately .1 to .3 m/hr
- Depth of the filter bed of at least 1 m (.5 m is needed solely for the biochemical process) with specific grain sizes
- Sufficient oxygen in the raw water (at least 3 mg/l) to induce biological activity.

HOUSEHOLD SLOW SAND FILTERS

Some selected slow sand filters suitable for household use are described in the following paragraphs. Because these filters are simple to build, they may be less effective biologically—necessary conditions for effectiveness are slow inflow and uniform throughflow. A pure and clean appearance of filtered water is no assurance of sufficient bacteriological quality.

A household filter can be simply made from a used metal drum, as in Figure 19. A thorough cleaning and disinfection (for example with NaOCl) is necessary prior to its use as a filter casing. A drum previously filled with oil or chemicals should not be used. The filter output is 60 l/h (as compared to up to 230 l/h for the rapid version). The filter medium is sand. The depth of supernatant is .1 to .3 m to facilitate steady flow conditions. The collection of the filtered water is in a gravel layer. Effluent discharge is through a riser pipe, which is partly perforated. The effluent pipe, mounted with a stopcock, rises just above the level of the filter bed to prevent the filter from running dry. The filtration rate is set through the effluent stopcock. The filter is cleaned whenever necessary or whenever the filtration rate is inconveniently slow. In case of high turbidity, pretreating the water is recommended by means of a rapid filter.

A two-stage coconut fiber/burnt rice husk filter is shown in Figure 20, where water flows right to left. This type of filtration plant was developed and tested in Southeast Asia, where it is widely used. Two filters are operated sequentially. The first one acts as a coarse filter, while the second one operates similarly to a slow sand filter. The filtrate is free of color, disagreeable odor, and taste. The turbidity is greatly reduced; surplus iron and manganese is removed. Since pathogen removal is not as high as using a slow sand filter, subsequent disinfection (such as chlorination in the storage tank) is recommended.

Because the plant is mostly made from locally available materials and residues, the initial capital cost and the operating cost are low. For filter vessels, clay jars or containers made of concrete, metal, or zinc-plated sheet metal can be used. Feasible operating capacities range between 1 and 15 m^3/h, depending mainly on the size of the system. The depth of filter bed is .6 m to .8 m; the depth of unfiltered water is 1 m above filter bed. The entire medium should be replaced every three to four months.

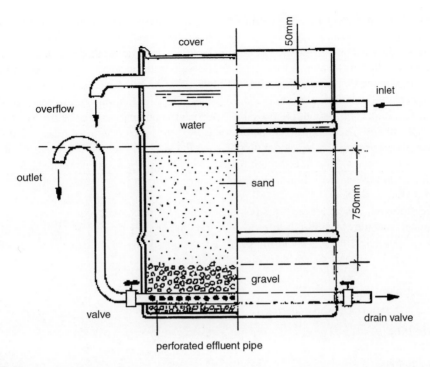

FIGURE 19
Slow sand filter in household size.

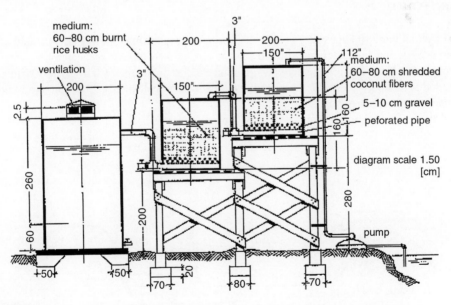

FIGURE 20
Two-stage filter.

A LARGER HORIZONTAL SAND FILTER

This type of filter, shown in Figure 21, is constructed by excavation of an earth basin, which is subsequently filled with sand. A biological skin develops at the surface of the sand around the inlet point. The inlet trough, perforated to let the water flow into the sand bed, protects the sand from disruption by the force of the incoming water. The filtration rate of the water percolating through the sand body is controlled by the filter resistance and the head differential between inflow and outflow. The filtration rate corresponds to .2 to .4 meters/hour of water—that is, .2 m^3 of water per m^2 of area. The retention time in such filters is between 36 hours and 30 days. The effect of filtration is reduction of bacteria count, turbidity, and organic content. The filter basin has a watertight lining (with plastic sheets), a depth of between .5 m and 1.0 m, a length of 5 m, and a bottom slope of 1:l0 to 1:20

When the filter starts clogging, the point of inflow is simply switched. As soon as the water has drained from the clogged inflow trough, the top sand layer is scraped off. The point of inflow can then be switched back. This technique offers the possibility of uninterrupted operation.

Figure 21 shows the following features of the filter.

1. Inlet pipe
2. Inlet trough to prevent scouring
3. Barriers
4. Gravel, 50 mm
5. Outlet trough
6. Flow direction

DISINFECTION

It is essential that drinking water be free of pathogenic organisms. Storage, sedimentation, and filtration of water, both individually and jointly, reduce the contents of bacteria in water to a certain extent. None of these methods can guarantee the complete removal of germs. Disinfection is needed at the end. Water with low turbidity may however be disinfected without any additional treatment for bacteria removal.

Groundwater abstracted from deep wells is usually free of bacteria. Surface water and water obtained from shallow wells and open, dug wells generally need to be disinfected.

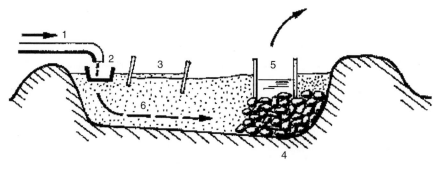

FIGURE 21
Horizontal flow sand filter.

The degree or efficiency of disinfection depends on the method employed and on the following factors influencing the process.

- Kind and concentration of microorganisms in the water
- Other constituents of the water that may impede disinfection or render it impossible
- Contact time provided (important for chemical disinfectants, since their effect is not instantaneous, a time of contact is necessary)
- Temperature of the water (higher temperatures speed up chemical reactions)

Water disinfection can be accomplished by several means.

- Physical treatment: removal of bacteria through slow sand filtration, application of heat (boiling), storage, and so on
- Irradiation, such as UV-light
- Metal ions, such as silver (and copper)
- Chemical treatment, use of oxidants (halogens and halogen compounds—chlorine, iodine, bromine, ozone, potassium permanganate, hydrogen peroxide, and so on)

A good chemical disinfectant should have the following abilities.

- Destroy all organisms present in the water within reasonable contact time, the range of water temperature encountered, and the fluctuation in composition, concentration, and condition of the water to be treated
- Accomplish disinfection without rendering the water toxic or carcinogenic
- Permit simple and quick measurement of strength and concentration in the water
- Persist in residual concentration as a safeguard against recontamination
- Allow safe and simple handling, application, and monitoring
- Ready and dependable availability at reasonable cost

Just as important as the proper choice of the disinfectant is that the agent be added to the water in a safe and controllable fashion.

CHLORINATION

The application of chlorine and its compounds for the purpose of water disinfection is the best and most tested compromise. Several chlorine compounds that have various active chlorine contents are easily used. In some form or another they are available virtually anywhere. Chlorine gas and chlorine dioxide are widely used in large-scale water treatment on account of their high efficiency and ease of application. Handling and transport, however, are too demanding and hazardous for the purposes described here. Some chlorine compounds are described next. Advice on amounts to be used will normally be given by the manufacturer.

Sodium hypochlorite (NaOCl) is commonly known as bleach or Javelle water. It is generally available in dissolved form. Its commercial strength in terms of active chlorine is between 1 and 15 percent. It is stored in dark glass or plastic bottles. The solution loses some of its strength during storage. Prior to use, the active chlorine content should be tested. Sunlight and high temperatures accelerate the deterioration of the solution. The containers therefore should be stored in cool, darkened areas. The stability of the solution decreases with increasing contents of available chlorine. A 1 percent solution is relatively stable but is not economical to store. Even though hypochlorite solutions are less hazardous than chlorine gas, every precaution should be taken to avoid skin and eye contact and to protect containers against physical damage.

Chlorinated lime, or bleaching powder, is readily available and inexpensive. It is stored in corrosion-resistant cans. When fresh, it contains 35 percent active chlorine. Exposed to air, it quickly loses its effectiveness. It is usually applied in solution form that is prepared by adding the powder to a small amount of water to form a soft cream. Stirring prevents lumping when more water is added. When the desired volume of the solution has been prepared, it is allowed to settle before decanting. Solutions should have concentrations between 5 and 1 percent of free chlorine, the latter being the most stable solution. Some 10 percent of the chlorine remains in the settled sludge. The precautions given previously pertain also to the storage of dissolved chlorinated lime.

High-test hypochlorite (HTH) is a stabilized version of calcium hypochlorite ($Ca(OCl)_2$) containing between 60 and 70 percent available chlorine. Under normal storage conditions, commercial preparations will maintain their initial strength with little loss. Even though HTH is expensive, it may be economical, thanks to its properties. It is available in tablet or granular form (commercial names: Stabo-Chlor, Caporit, or Para-Caporit).

These chemicals must be handled with great caution. They are caustic, corrosive, and sensitive to light. Chlorine corrodes metal and, to a less extent, wood and some synthetic materials. Metal parts that come in contact with the chemicals should be resistant.

REFERENCES

Rapid Filtration

"Environmental Health Directorate, Assessing the Effectiveness of Small Filtration Systems for Point-of-Use Disinfection of Drinking Water Supply." Ministry of Health and Welfare, Canada, January 1980.
ICAITI. (1981). "Water Purification Using Small Artisan Filters." Guatemala.
Thanh, N. C. (1978). "Functional Design of Water Supply for Rural Communities." Asian Institute of Technology, Bangkok.

Slow Sand Filtration

Frankel, R. J. (1977). "Manual for Design and Operation of the Coconut Fiber/Burnt Rice Husk Filter in Villages of Southeast Asia." Seatec International, Bangkok.
Frankel, R. J. (1981). Design, Construction and Operation of a New Filter Approach for Treatment of Surface Waters in S. E. Asia. *Journal of Hydrology*, 51, 319–328.
NEERI. (1976). "Slow Sand Filtration. An Interim Report." Nagpur.
IRC/WHO. (1978). "Slow Sand Filtration for Community Water Supply in D. C. a Design and Construction Manual." Technical Paper Series 11. Den Haag, Dez.
Thanh, N. C. (1978). "Functional Design of Water Supply for Rural Communities." AIT, Bangkok.

Disinfection

Burns, R. H., and J. Howard. (1974). "Safe Drinking Water. An Oxfam Technical Guide." Oxford.
Peace Corps, *Water Purification, Distribution and Sewage Disposal,* 1984, ICE/ Peace Corps, Washington, DC. http://www.peacecorps.gov
Volunteers in Technical Assistance. (1977). "Chlorination for Polluted Water and Super-Chlorination." *VITA Handbook.* Maryland.

SANITATION

Based on "Appropriate Technology for Water Supply and Sanitation," A Sanitation Field Manual, John M. Kalbermatten, DeAnne S. Julius, and Charles G. Gunnerson, The World Bank, Washington, D.C., 1980

The proper disposal of human and animal wastes is a major sanitation problem in the Third World and elsewhere. Methane digesters, described in the Energy chapter, are an approach to the animal waste problem but are less effective in killing the pathogens of human wastes, which are a bigger health problem. If they get into the water supply, pathogens from human wastes can cause cholera and hepatitis, as well as diarrhea. Flies can also spread disease by carrying germs on their feet when flying from feces to food. An article in the Health chapter, "Water and Health," discusses pathogens that affect human health. The "Building" article in the Construction chapter includes rules of thumb for sewer systems.

Most people in the world probably want flush toilets emptying into septic tanks or sewers, but the cost of providing such systems to everyone in the world makes such toilets presently out of the question. Estimates are that 2 billion people worldwide need improved sanitation facilities. The cost of providing sewers would be about $500 per person, and the annual per capita income of half of these people is less than $200. Septic tanks would not be appreciably cheaper.

The common approaches to sanitation in the Third World are as follows.

- Ventilated improved pit (VIP) latrines
- Composting toilets
- The Benjo—a toilet enclosure over a stream
- Pour-flush toilets
- Aquaprivies
- Soakaways, septic tanks, and drain fields

CULTURAL ISSUES

Float Kidha, a public health specialist in Africa, reports the following.

There are customs and beliefs about human waste disposal and even where people know the connection between human waste and diseases, they have difficulty overcoming their beliefs. Health workers have failed to reduce diarrheal diseases by forcing people to build latrines. People have built latrines for fear of the authorities but not because they understand why they must build them. The result is that nobody used them. Health education is the key. But the health educator must *know* and appreciate the culture of the people and educate them in the context of that culture. In many cultures it is believed that using a pit latrine is like being buried alive. They believe that human waste must go back to the land and fertilize it. For this reason the bush is used more often. People can be taught to bury their waste if they will not use pit latrines. Another belief is that a man would get ill and even die if he used the same latrine that has been used by a woman who is menstruating. A young woman would never use the same latrine which her mother and father-in-law use. Human waste is used for manure in many cultures. It is, therefore, not fact to assume that one's own pit latrine will ensure reduction in diseases due to poor sanitation.

People built latrines but never use them for various reasons. In one village in Kenya, health workers forced people to build latrines during a cholera outbreak. When the health workers returned to the village two weeks later everyone had a latrine, which the health workers duly noted. It turned out later that the round "huts" had holes, which were only three feet deep and were not being used as latrines but stores for illegal brew! People did not believe latrines were useful for health purposes but they did need a place to store the homemade distilled beer. Education and recognition of local customs are essential for success.

Another important cultural issue is whether people will use fertilizer processed from human waste on plants for human consumption.

POUR-FLUSH TOILETS

Two pour-flush toilets are shown in Figure 22. One or two liters of water are poured in by hand to flush the excreta into the pit. (The "soakaway" shown in the figure is discussed below.) Note the water seals in both models, which prevents odor development and mosquito breeding. An advantage of the offset pit design is that the toilet can be inside the house and the pit outside. If two pits are built, they are used alternately. When the first pit is nearly full, the second is connected to the toilet. During the period the second pit is being used, a year or so, the waste in the first pit decomposes into a humus suitable for fertilization. Such a system can be easily upgraded by attaching to a sewer line. The chief disadvantages are the amount of water that must be used—3 to 6 liters per person per day—and that the pits must be emptied annually.

The volume of the pit is approximately .05 cubic meters—or 1.75 cubic feet—times the number of people using the pit, if the pit is emptied annually. The maximum length of the connecting pipe is 8 meters—26 feet or so—and the slope should be at least 1 in 40. All the parts except the water seal can be easily made, and even the water seal can be constructed by local artisans with experience. Maintenance is minimal.

AQUAPRIVIES

An aquaprivy is shown in Figure 23. Here the waste goes into a pool of water. To avoid odors and mosquito breeding, the water level must be maintained high enough for a seal so the tank will not leak, and water must be added regularly. The waste decomposes anaerobically in the tank, but a sludge is formed that must be removed every two or three years. The volume of the tank is approximately .12 cubic meters per user but not less than 1 cubic meter. Construction of such a leakproof vault requires skill. The depth of the water is normally 1.0 to 1.5 meters. The volume of effluent flowing to the soakaway is about 6 liters per person per day, corresponding to 1.5 liters excrement and 4.5 liters for cleansing and "flushing." Aquaprivies can be upgraded to sewer systems easily. In practice, maintaining the water seal has been difficult, thus, some specialists do not recommend aquaprivies.

SOAKAWAYS, SEPTIC TANKS, AND DRAIN FIELDS

A soakaway pit is shown in Figure 24. The purpose of the soakaway is to allow waste to soak gradually into the soil. For normal soil types the infiltration of water to the soil will be 10 liters or more per day per square meter. The pit should be designed with this much area on the sides, since the bottom will probably be clogged by sludge. A pour-flush toilet used by a family of

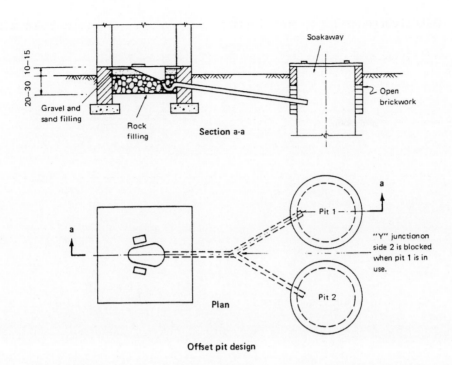

Section a-a

Plan

Offset pit design

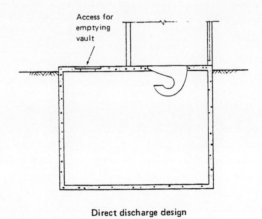

Direct discharge design

Note: In the offset pit design, the pit is placed at site of "Y" junction if only one pit is installed.

FIGURE 22
Pour-flush toilets.

eight people would be designed to receive about 48 liters a day, so the area of the sidewalls should be about 5 meters. If the depth is 1.5 meters, the corresponding diameter of the pit is about 1.1 meters. Soakaways should be located at least 15 meters—50 feet—from wells, more for sandy and gravelly soil, and 30 meters—100 feet—from streams. They should be 3 meters from buildings or large trees, whose roots can infiltrate and damage the pit. Soakaways will not

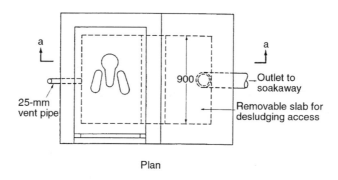

Plan

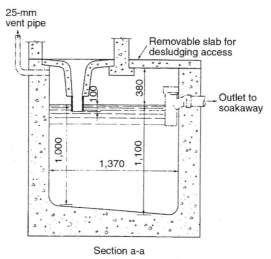

Section a-a

Source: Adapted from Wagner and Lanoix (1958).

FIGURE 23
Aquaprivy.

work with an impervious soil. Systems using soakaways require a fairly large distance between homes or a low housing density. More design guidelines are given in the "Building" article in the Construction chapter.

A septic tank is shown in Figure 25. A septic tank is used with flush toilets. It also receives wastewater from sinks, kitchens, and other household appliances. The retention time is three to five days. During this time the solids settle to the bottom of the tank, where they decompose. Most of the decomposition products—water and methane gas—flow out with the effluent, but sludge does remain, and a scum is formed on the surface of the water. Sludge should be emptied from a septic tank every five years or so. The effluent does contain harmful pathogens and should not be discharged directly to open streams or into drinking water supplies. The effluent should go to a soakaway or, alternatively, a drain field. The two chambers shown in the figure are a preferred design because they allow lighter waste particles to separate from the heavier ones. The lighter particles move into the second chamber where they can decompose more thoroughly in the absence of gas bubbles created from the heavier material. Septic tanks cost about as much as sewers and are equally attractive to users, so systems would probably not be upgraded unless the density of housing in the neighborhood increased to the point where insufficient area is available for the drain field or soakaway.

A drain field is made of two or three parallel trenches containing tile pipes with open joints. The pipe usually rests on a gravel bed. The trenches are about .5 meter deep. The length is determined by the infiltration rate of the effluent—generally 10 liters per square meter—where the area infiltrated by the effluent is the total length of the parallel trenches multiplied by the trench depth (.5 meter or so) and multiplied by 2—because the effluent infiltrates the soil on both sides of the trench. Again, the drain field should be located at least 30 meters from drinking water sources. Drain fields are effective but use much area—more than soakaways.

CHOICE OF SYSTEM

Besides the systems described in this article, VIP latrines and composting toilets, described in separate articles, should be considered. VIP latrines work well and are the least

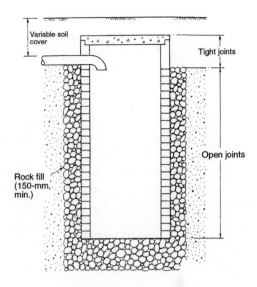

Source: Adapted from Wagner and Lanoix (1958).

FIGURE 24
Soakaway.

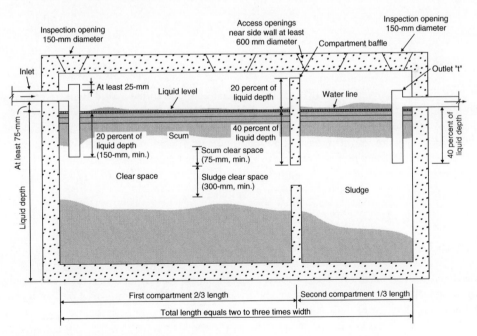

Note: If vent is not placed as shown on figures 13—2, 3, and 4, septic tank must be provided with a vent.

FIGURE 25
Septic tank.

expensive in terms of initial cost and maintenance but do need to be emptied after five years or so, and an improperly placed latrine can contaminate groundwater. Composting toilets produce a usable composted humus, but their operation requires significant user attention.

These are some questions to ask when choosing a system.

- Is it likely that the system will be upgraded to sewer-based sanitation?
- Is the decomposed excreta wanted for fertilizer?
- Are the plot sizes large enough and the soil sufficiently permeable for onsite disposal of effluent?
- Is sufficient water available for pour-flush toilets?
- Are municipal or private mechanisms available for emptying latrines?
- How do the costs compare?

REFERENCES

Leckie, Jim, Gil Masters, Harry Whitehouse, and Lily Young. (1981). *More Other Homes and Garbage*. San Francisco: Sierra Club Books.

Pickford John. (1991). *The Worth of Water, Technical Briefs on Health, Water and Sanitation,* IT Publications, 103–105 Southampton Row, London WCIB 4HH, UK. Tel.: 44 171436 9761. Fax: +44 171436 2013. orders@itpubs.org.uk. In the U.S.: Stylus Publishing, PO Box 605, Herndon, VA 20172-0605. 800-232-0223. Fax: 703-661-1501. *Stylus-pub@aol.com.*

VITA. (1988). *Village Technology Handbook*. Volunteers in Technical Assistance, 1815 North Lynn Street, Washington, DC 22209, USA.

LATRINES

The ventilated improved pit (VIP) latrine described here and in the World Bank reference was developed in Zimbabwe for rural use. It costs about $100. Even a more substantial version—for example, one useful in a settlement on the edge of a city—can be built for $150. Compared to other toilets they are low cost, easily built and maintained, and use minimal water. A principal danger is that they may pollute groundwater, so placement must be done carefully. VIP latrines are sometimes called "Blair" toilets, after their designer. An Appendix to this article is a set of plans used in a rural school in Malawi for teaching purposes.

DESIGN OF VENTILATED IMPROVED PIT LATRINE

A conventional pit latrine consists of a pit, a squatting plate, and a superstructure. The problems with a conventional pit latrine are odors and insects. The ventilated, improved pit latrine shown in Figure 26 does away with these problems. The vent pipe is the key component. The vent pipe, which is painted black, is heated by the sun so the air rises. Wind blowing across the top of the vent also pulls air out of the pipe. Air is thus pulled down the squat hole through the pit and out the vent. Inside the superstructure no odor exists because of the direction the air is moving. A way of augmenting this airflow is to locate the door opening so the prevailing wind will be caught to blow air into the pit.

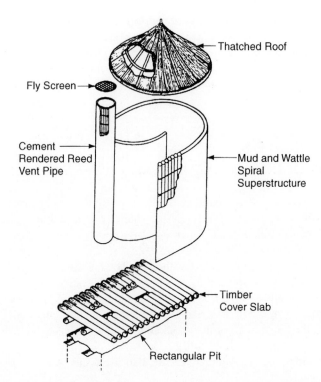

FIGURE 26
VIP latrine in Zimbabwe. (From *Ventilated Improved Pit Latrines: Recent Developments in Zimbabwe.*)

The vent pipe is screened at the top. This prevents insects from escaping and also keeps them from coming down the vent. Some flies will get into the pit through the squat hole and lay eggs. The emerging flies, though, will be attracted to the light and will fly up the vent pipe and be caught at the screen. The other troublesome insects are mosquitoes. If the pit is dry, mosquitoes are not attracted, but if the pit is wet—that is, if the level of groundwater is above the bottom of the pit—mosquitoes will breed there. Mosquitoes are not attracted as strongly to light as flies, so some will go out the hole. In this case the hole should be covered with a removable screen. It is important, of course, not to interfere with the circulation of air through the system by covering the hole with a solid cover.

It must be relatively dark inside the superstructure so flies are attracted to the vent pipe. If social custom favors an illuminated superstructure, then some type of opaque cover will be needed over the hole. This cover should be raised from the floor to allow as much ventilation as possible, using screening to keep insects in the pit. If the entrance to the latrine faces east or west, the morning or evening sun may shine into the latrine, so shading may be necessary. If wind conditions permit, north or south orientation should be used.

Lighting is an important issue. When VIP latrines were built near a health clinic in Kenya, care was taken to make them very dark inside. Snakes then found the superstructures desirable places to live. The local people in turn refused to use the latrines. One lesson is that a design must meet local conditions.

If the pit is above the groundwater level, the pit will be dry and the wastes will decompose in the pit. If the pit is below groundwater level, the pit will have water in it and some of the wastes will seep away into the ground, just as they do in septic systems not connected to sewers. In this case, wells for drinking water should be at least 150 feet from the latrine.

A dry pit 10 feet deep and 3 feet long in either direction should last a family of six for about ten years. The useful life will be twice as long if the pit is wet. (In designing one should allow for about 1.5 cubic feet of waste per person per year.) The VIP latrine is so simple to build that one might just start over somewhere else when the pit is full. Alternatively, one could build a double pit latrine and use each pit alternately for a year. At the end of a year, the material in the unused pit will have decomposed and could safely be removed and used as fertilizer, if social customs permit.

CONSTRUCTION OF THE RURAL VIP LATRINE

The basic rule is to make the construction as similar as possible to other structures in the village, so the model shown in Figure 26 would be used when the surrounding buildings are made of mud and wattle. The vent pipe is made of a reed mat, 8 feet by 3 feet, rolled up into a cylinder about 11 inches in diameter and then plastered with a mortar made with cement. Ordinary corrosion-resistant screening can be used at the top. The rest of construction is obvious. The roof thatching has to be dense to keep the interior dark. Of course, people who use thatching regularly will have no trouble with this. The walls should not have light leaks either. The people building the structure, again, will have had experience in making tight walls. If the soil crumbles easily, the top three feet or so of the pit should be lined with cement. The logs supporting the timber cover slab should extend a foot or so in each direction beyond the pit hole. They should be treated with a wood preservative. The slab itself should also be painted to protect the wood.

In areas where wood is scarce, the superstructure can be built of thatching or of locally made bricks. In more urban areas, cement over a wire frame can be used, and the squat plate can also be cement. A latrine with a cement squat plate in a refugee camp is shown in Figure 27. The latrine

FIGURE 27
Latrines under construction.

TABLE 1

Cost of a VIP latrine

Material	Amount	Cost
Cement for pit lining and vent pipe	55 lbs.	$2.00
Cement for superstructure	55 lbs.	2.00
Wire	150 ft.	0.70
Fly screen	1 ft × 1 ft.	0.20
Nails		1.00
Paint		1.00
TOTAL		$6.90

is under construction, and the squatting holes are covered with mud or bricks. The vent pipe can be asbestos cement or PVC; cast iron corrodes.

Table 1 shows estimated costs of the VIP latrine. It would take there people about one week to build the latrine

SLOPING PIPE VIP LATRINES

A variation of the VIP latrine uses a sloping pipe from the squatting plate to a pit located behind the superstructure. The advantage of this configuration is that the toilet can be inside the house and the pit outside. A removable cover can be built over the pit so it can be emptied every few years, since the toilet room will be an integral part of the house. The pipe can clog, so it may have to be cleaned regularly with a long-handled brush. Typical pipe diameters range from 150 mm to 200 mm—6 to 8 inches.

SUCCESS OF VIP LATRINES

These latrines have worked well in Zimbabwe, where 20,000 were in use in 1982. Perhaps one reason for their success is that they represent a relatively small change from what had been in use before. Another reason is probably the extensive education program that the government of Zimbabwe used, including films, instruction leaflets, and demonstration models.

REFERENCES

Morgan, Peter R., and Duncan D. Mara. (1987). *Ventilated Improved Pit Latrines: Recent Developments in Zimbabwe*. Technology Advisory Group, The World Bank, Washington, DC 20433.

VITA. (1988). *Village Technology Handbook*. Volunteers in Technical Assistance, 1815 North Lynn Street, Washington, DC 22209, USA.

APPENDIX

Plans from Evergreen Secondary School, Phulula District, Malawi. Further information from Henry Ngaiyaye, University of Malawi—The Polytechnic, Private Bag 303, Chichiri, Blantyre 3, Malawi.

HOW TO BUILD A LOW COST PIT LATRINE

By
Henry Ngaiyaye

INTRODUCTION

The Rural Housing Project has designed a ventilated improved pitlatrine that is cheap and long lasting. In this pamphlet, you can learn how to build your own pitlatrine.

The floor of the pitlatrine is held up by a brick vault. This method is not only cheaper than a reinforced concrete slab and longer lasting than the traditional slab made of timber and mud, but it is also safer.

All measurements used for the illustrations are in millimeters (mm) if not otherwise stated.

BEFORE STARTING TO BUILD, PLEASE READ THE WHOLE PAMPHLET CAREFULLY!

VENTILATION

The purpose of the ventilated pitlatrine is to; (1) keep the bad smell away and, (2) to get rid of flies by leading light into the pit. When flies see the light they fly towards that instead of the squatting hole and will be trapped against the glass or at the top of the ventpipe in the mosquito mesh fly trap.

The ventpipe should reach at least 500 mm above the highest point of the roof. Wind passing across the top of the ventpipe creates a suction within the pipe so that the air in the pit flows out through the ventpipe.

The ventpipe should ideally face north or south in order to allow the sun to heat the pipe and the air inside. The air will rise and help to ventilate the pit.

A cover must be placed over the squatting hole in order to reduce access for escaping smell, flies and minimise mosquito breeding in the pit.

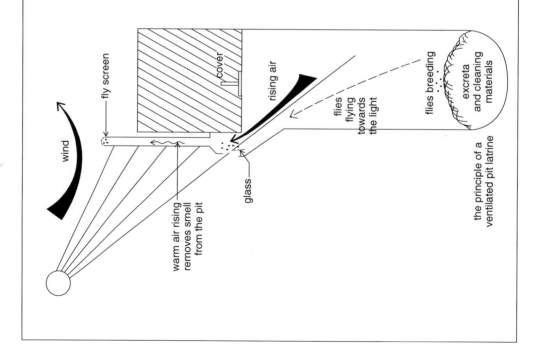

fly screen

cover

rising air

wind

warm air rising removes smell from the pit

glass

flies flying towards the light

flies breeding

excreta and cleaning materials

the principle of a ventilated pit latrine

ELEVATION AND PLAN

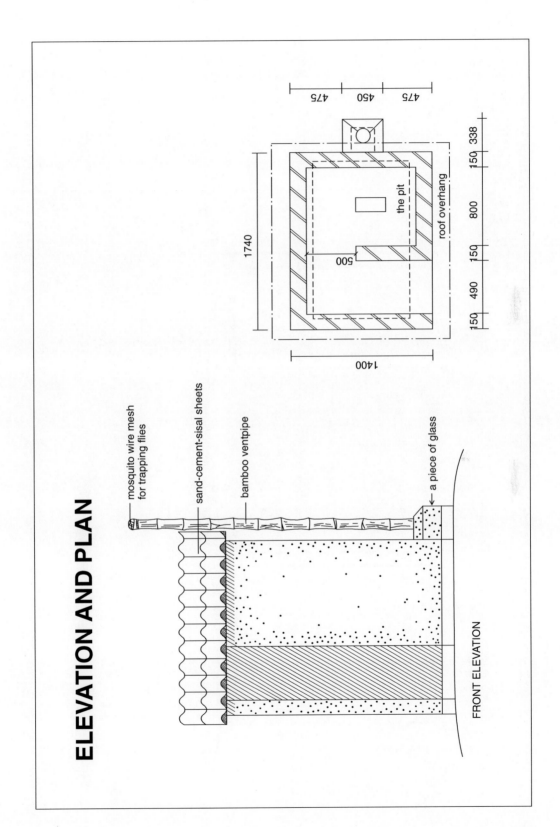

mosquito wire mesh for trapping flies

sand-cement-sisal sheets

bamboo ventpipe

a piece of glass

FRONT ELEVATION

the pit

roof overhang

1740

500

1400

475   450   475

150   338   150   800   150   490   150

781

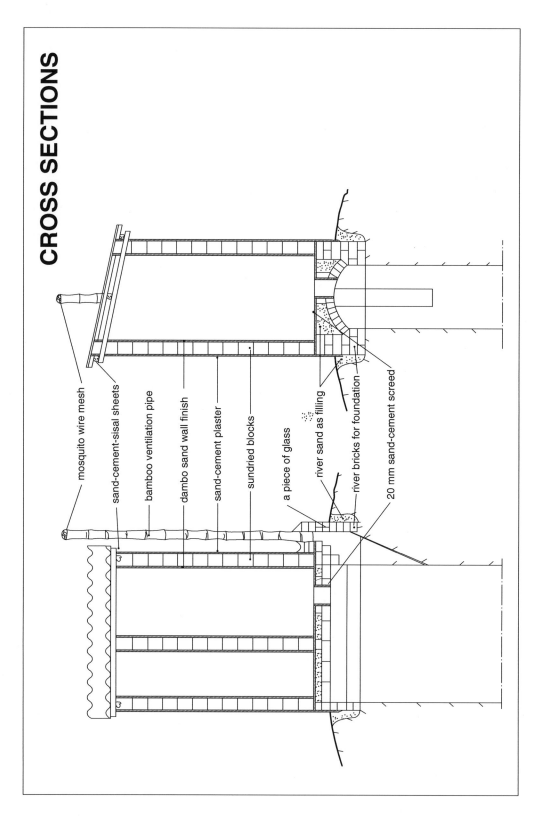

CROSS SECTIONS

mosquito wire mesh
sand-cement-sisal sheets
bamboo ventilation pipe
dambo sand wall finish
sand-cement plaster
sundried blocks
a piece of glass
river sand as filling
river bricks for foundation
20 mm sand-cement screed

MATERIALS NEEDED

500 Burnt bricks (235 mm x 112.5 mm x 85 mm)

250 Sundried blocks (300 mm x 150 mm x 150 mm)

1.5 Bags of cement

0.5 Kg Sisal

0.5 Litre bitumen paint

0.5 Litre solignum

0.5 Bag of lime

0.5 Tonne of river sand

0.5 Tonne of dambo sand

150 mm x 150 mm mosquito wire mesh

1 Big bamboo

1 Piece of glass minimum 150 mm x 150 mm

6 Sand-cement-sisal roofsheets

5 Poles (three 1.8 m and two 1.6 m long)

5 m Wire

0.3 Kg nails 4" (and 8 nails 1")

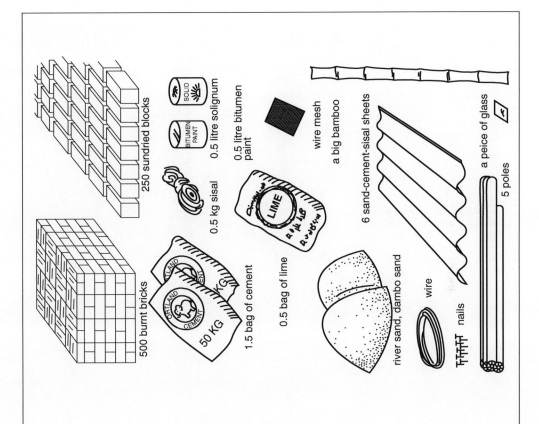

500 burnt bricks

250 sundried blocks

1.5 bag of cement

0.5 kg sisal

0.5 bag of lime

BITUMEN PAINT SOLID

0.5 litre bitumen paint

0.5 litre solignum

wire mesh

a big bamboo

6 sand-cement-sisal sheets

river sand, dambo sand

wire

nails

5 poles

a peice of glass

TOOLS NEEDED

HOE
BUCKET
ROPE - 5 m long
SAW
HAMMER
TROWEL
WOODEN FLOAT
SPIRIT LEVEL
MEASURING TAPE

Instead of a spirit level you can use a transparent plastic hosepipe 3 m long. Put two marks, 15 cm from the ends; and fill the hosepipe with water up to the marks.

When leveling, you need to have two persons, one holding each end of the hosepipe against the part that is to be leveled. If the water is above one of the marks you have to lift that end or lower the opposite end.

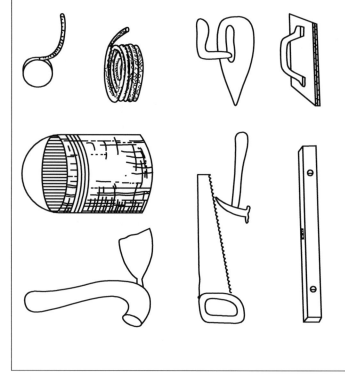

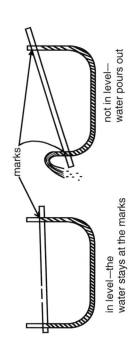

transparent hosepipe

marks

in level—the
water stays at the marks

not in level—
water pours out

WHERE TO PLACE IT

To prevent pollution, the pitlatrine should be placed down hill from a well, at a distance of not less than 30 meters.

To avoid flooding the pit by rain, the ground should slope away from the pitlatrine on all sides. Where this is not possible, a drain should be placed as shown in the drawing.

The pitlatrine must be at least 10 m from nearest house or kitchen.

There should be a clear space of not les than 3 m around the pitlatrine.

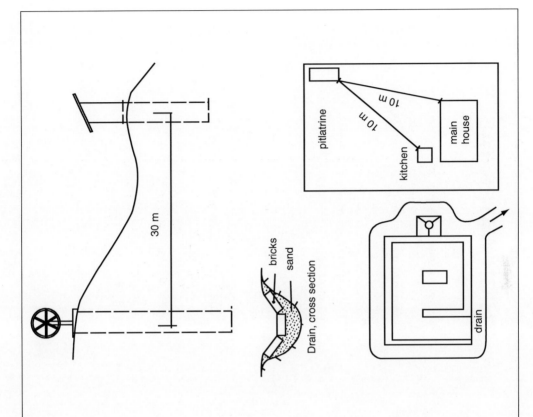

30 m

bricks
sand

Drain, cross section

pitlatrine

10 m

10 m

kitchen

main
house

drain

HIGH WATER TABLE

Where the water table is very high, the latrine must be down hill of any well at a distance of not less than 200 m.

The safe rule is that the bottom of the pit should not be less than 1 m above the highest level of the ground water level.

In some cases of high ground water, the height of the floor should be raised (700 mm) above the highest point of ground.

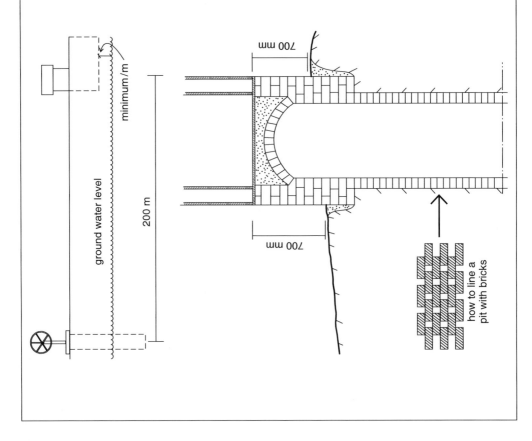

ground water level

200 m

minimum/m

700 mm

700 mm

how to line a
pit with bricks

SIZE OF THE PIT

NORMAL PIT- stable soil and low ground Water level. Width 700 mm, depth 5 m Length 1.5 m

If there is no problem with the ground-water level and the soil stability, then 5 m is a good depth.

Our normal pit is 1.5 m long and is Suitable for a pitlatrine house without a door. However, different types of houses can be used.

The life of the pit depends on the number of users. It will last for about 5 years if 6 people are using it. (0.06 m^3/per person per year plus 0.03 m^3 for cleansing material.

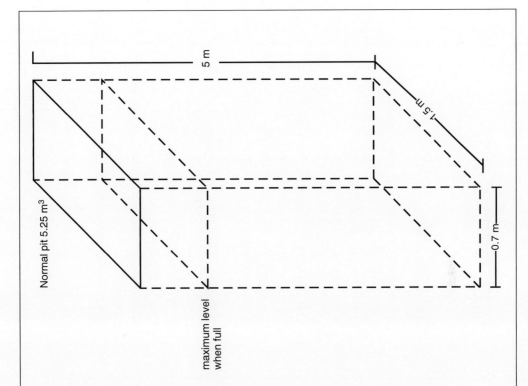

Normal pit 5.25 m^3

5 m

1.5 m

0.7 m

maximum level when full

787

SIZE OF THE PIT

High ground water

Where there is a very high water table a different size of pit should be used to avoid the pit entering the ground water and causing pollution.

A pit 2 m long, 1 m wide and 2 m deep with the floor of the latrine 600 mm above the highest ground level will give a size equal to the normal pit.

Unstable soil condition

Under unstable soil conditions, either line the pit with bricks to normal depth, or use dimensions for high ground water conditions, so as to get a size equal, or nearly equal to a normal pit.

See drawings on page 786.

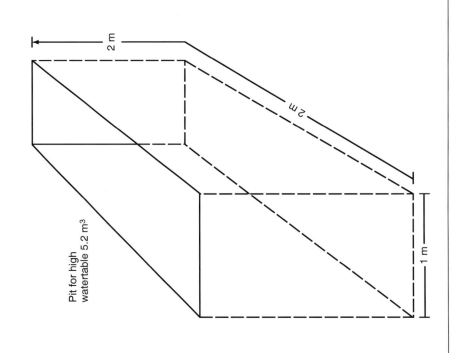

Pit for high watertable 5.2 m³

2 m

2 m

1 m

SETTING OUT

Before setting out, make sure that the ground is level.

You can use either pegs and a string or make a wooden frame. You "draw" the line on the ground. The measurements are shown on page 781.

Make sure you have right angles, check with a square and/or measure the diagonals, they should be equal.

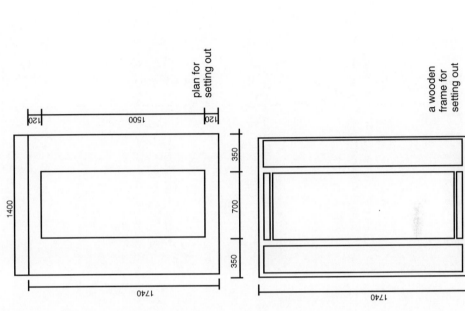

plan for setting out

a wooden frame for setting out

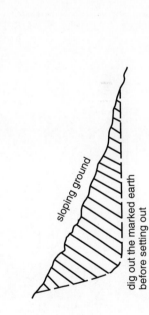

sloping ground

dig out the marked earth before setting out

DIGGING THE PIT

Start digging a pit 1400 mm wide, 1740 mm long and 400 mm deep as shown in drawing 1.

Then dig the actual latrine pit 700 x 1500 mm and 5000 mm deep as shown in drawing 2.

When you have dug about 1000 mm of the actual latrine pit, stop and start building the base and make preparations for the vault mould.

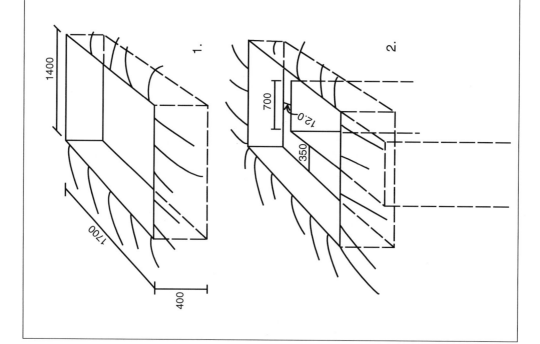

THE BASE

The base is made up of 3 courses done in English bond as shown on the drawing.

Use cement mortar 1: 5 — one part cement to 5 parts clear river sand.

All courses must be level.

Cut the bricks for the third course carefully. It must be the correct angle to support the vault fully. Different size vault need different cut angles.

Normally the arch of the vault is fixed by a radius of 350 mm. A wooden mould will help in the building of the arch.

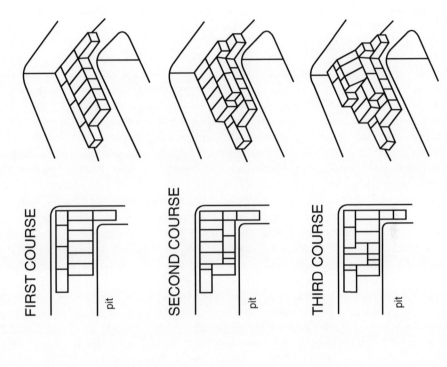

FIRST COURSE

pit

SECOND COURSE

pit

THIRD COURSE

pit

COMPOSTING TOILETS

Based on "The Composting Toilet System Book A Practical Guide to Choosing, Planning and Maintaining Composting Toilet Systems, and Alternative to Sewer and Septic Systems" David Del Porto and Carol Steinfeld, The Center for Ecological Pollution Prevention (CEPP), 1999

Composting toilets (also known as dry, waterless, and biological toilets and nonliquid saturated systems) are one of the most direct ways, among wastewater treatment technologies, to avoid pollution and conserve water and resources.

Composting toilet systems contain and control the composting of excrement, toilet paper, carbon additive, and, optionally, food wastes. Unlike a septic system, a composting toilet system relies on unsaturated conditions (material cannot be fully immersed in water), where aerobic bacteria and fungi break down wastes, just as they do in a yard waste composter. Sized and operated properly, a composting toilet breaks down waste to 10 to 30 percent of its original volume. The resulting end-product is a stable soil-like material called "humus," which is used as a soil conditioner on edible crops in many parts of the world, although in the United States such use is illegal. Humus is removed after usually a year's retention.

The main components of a composting toilet are (1) a composting reactor connected to one or more dry or micro-flush toilets; (2) a screened exhaust system (often fan-forced) to remove odors, carbon dioxide, water vapor, and the by-products of aerobic decomposition; and (3) a means of ventilation to provide oxygen (aeration) for the aerobic organisms in the composter. Other components include process controls, such as mixers, to optimize and manage the process, and an access door for removal of the end-product. The composting reactor should have a volume of 0.2 cubic meters per year per person. A composting toilet is shown in Figure 28. When a single reactor toilet like this is used, then the humus can be removed regularly from the bottom of the pile, but only material that has been in the reactor for a year or so should be removed. A double vaulted composting toilet, described later in this article, is safer because the contents of one vault can decompose during the time—often one year—that the other vault is being used.

THE AEROBIC DECOMPOSITION AGENDA

Decomposition requires aeration, a suitable moisture level, a suitable temperature, and a proper carbon/nitrogen ratio.

Aerobes require oxygen. The ventilation system in a composter should draw sufficient air across and through the decomposing material. A key factor, then, is the surface area to volume ratio of the composting substrate (which includes the microbial population) because surface area allows direct contact with oxygen. Mixing, tumbling, forced aeration, and container design are ways composters provide a good surface to volume ratio. To make the composting process work best, the materials being composted should have a loose texture to allow air to circulate freely within the pile. If the material becomes matted down, compacted, or forms too solid a mass, the air will not circulate, and the aerobic organisms will die.

Here are some ways of ensuring adequate aeration.

- Add bulking agents, such as wood chips, stale popped popcorn, and so forth to increase pore spaces that permit air to reach deep into the biomass and allow heat, water vapor, and carbon dioxide to be exhausted. Earthworms also create pores as well as help break down wastes.

- Maintain adequate airflow through the material by proper ventilation (such as pressurized air, using convection, or forced air by a fan) and/or by frequently mixing.
- Provide aerators, such as mixers, mesh, grates, air channels, and screened pipes to help increase the surface area of the composting mass that is exposed to air.

However, too much airflow can remove too much heat and moisture

The microbes in the composter need the right amount of moisture to thrive. Too much water (saturated conditions) will drown them and create conditions for the growth of odor-producing anaerobic bacteria. In optimum conditions the composting mass has the consistency of a well-wrung sponge—about 45 to 70 percent moisture,

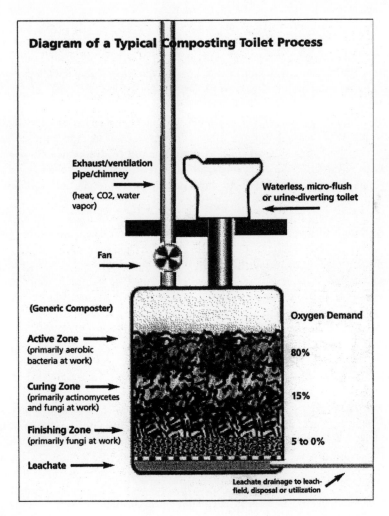

FIGURE 28
A composting toilet.

When the moisture level drops below 45 percent, it can become too dry for composting. Also, excrement, toilet paper, and additives will dry out but not decompose, thus prematurely filling the toilet (a good indicator that the mass is too dry). If the moisture level is higher than 70 percent, leachate will pool at the bottom of the composter. In this case, the leachate must be drained or evaporated; otherwise, it will drown the microbes. Urine and/or water from micro-flush toilets contributes most of the moisture in a composter and may not be distributed evenly over the mass. Fresh rainwater (which has little or no dissolved minerals) is best for moisture control, but fresh groundwater from the tap (which may contain significant dissolved minerals) will do as well. If the material is too dry, spray the compost mass with water or add a cup of water periodically.

The ambient temperature for acceptable biological decomposition is 78° to 113°F. A composter at less than 42°F will only accumulate excrement, toilet paper, and additive until the temperature rises. Generally, the rate of processing in a biochemical system is directly proportional to the increase of temperature (within certain limits, the rate doubles with every 18°F increase). The warmer the process, the more capacity in a composter. The cooler the process, the slower the rate, and more capacity needed for processing.

Microorganisms require digestible carbon as an energy source for growth and nitrogen and other nutrients, such as phosphorous and potassium, for protein synthesis to build cell walls and other structures (in the same way humans need carbohydrates and proteins). When measured on a dry weight basis, an optimum C:N ratio for aerobic bacteria is 25:1.

Human urine has a low C:N ratio (0.8:1). Therefore, oxidizing all of the nitrogen urinated into the toilet would require adding digestible carbonaceous materials on a regular basis. However, the practical fact is that urine, which contains most of the nitrogen, settles by gravity to the bottom of the composter, where it is drained away or evaporated. In either case, the nitrogen passes through the decomposing material and is lost to the process. For that reason, adding large amounts of carbon will not help process the nitrogen and will just fill up the composter faster.

Adding a small handful of dry matter per person per day or a few cups every week is a good rule of thumb to maintain a helpful C:N ratio, absorb excess moisture, and maintain pores in the composting material. The primary reason to add carbon material is to create air pockets in the composting material (that's why carbon additive is called "structure material"). Digestible carbonaceous materials include carbohydrates (sugar, starch, toilet paper, popped popcorn), vegetable or fruit scraps, finely shredded black and white newsprint, and wood chips.

LOCATING THE COMPOSTER FOR WARMTH

For solar heating, you need available solar energy, such as a clear, unobstructed south-by-southeast opening, unblocked by trees or other buildings. In most North American communities, nine square feet of solar collection area all year long will sufficiently warm a composter. There are many designs for heating small outhouses with composters in them. Some direct solar heat right onto the composting mass through a window, some on the leachate, some on the composter. (Remember that direct solar heat can dry out the surface of the mass, creating a crust that insulates the center, so you may actually need more heat. Turning the material and adding water helps.)

VENTILATION AND EXHAUST

Ensuring that air enters and exits the system in the right direction is critical for maintaining composting and preventing odors from entering the home. If your toilet room is on the side of the house opposite the prevailing winds, remember that wind pressure on the windward side of the house pulls a vacuum on the opposite, or leeward, side. So when you open your bathroom window on the leeward side, odor will be pulled from your toilet. You will know that this is the case if the window curtains blow out. If they blow into the toilet room, then the toilet is pressurized, and no odor will come into the room. Wind turbines work in areas with steady strong winds but can actually impede airflow at wind speeds of less than 10 to 15 mph. A simple plumbing T at the top of the pipe both keeps out the rain and allows the wind to suck air out of the pipe. The primary question with fan speeds is managing odor.

PULLING ODORS FROM THE COMPOSTER

Some assume that the wider the pipe from the toilet to the composter, the less chance of "skid marks" from stuff going down it. That's probably true but this connecting pipe can act as a chimney through which odors can back up into the toilet. The larger the diameter of the connecting pipe and toilet seat opening, the greater the chance of odor. To reduce this toilet-as-chimney effect, make sure that there is not a negative pressure in the toilet room so air is not pulled up the connecting pipe. Be aware that fireplaces can pull odor from the composter into the building.

Also, if the opening diameter of your toilet is large, insects have better access to the composter, and pets, toys, and infants could fall in (although we have yet to hear of a child falling into a composting toilet system). If it is too small, the pipe can get caked with excrement and may discourage ventilation/exhaust and require frequent cleaning. A good size is 8 to 12 inches.

OUTSIDE ODORS FROM THE EXHAUST PIPE

Outside, the odor from the composter will be not normally noticed on the ground if the exhaust pipe terminates at least 12 inches above the peak of the roof. Lower than the roof peak, odors from the exhaust pipe may be swept to the ground through wind downdrafting.

TOILET LOCATION

In a waterless situation, gravity is the only way to convey excrement from toilet to composter, so the composter must be almost directly under the toilet, although it can be several floors down. There should be few, if any, angles. Do not underestimate the stickiness of excrement! If this location won't work, or if you decide that aesthetic and lifestyle issues dictate a barrier between you and the composter, a micro-flush toilet or a vacuum system are alternatives.

DOUBLE VAULT SYSTEM

A double vault system—used in Mexico—is shown in Figure 29. Like most double-vault systems, one side is used at a time. When one side fills, the toilet stool is moved to the other side. The unused vault is covered and the waste therein decomposes. The decomposed excrement is removed, usually after a year or two, and occasionally used on farm fields.

USEFULNESS IN THE THIRD WORLD

Combusting toilet sanitation systems have been successful in the Third World. They do require care in operation—particularly ensuring that the waste is at an appropriate temperature and has an appropriate moisture level. Users must be educated about proper operation. Composting toilets offer particular advantages where human waste is presently recycled to gardens and where location of a latrine away from sources of drinking water is difficult.

REFERENCES

Drangert, Jan-Olof, Jennifer Bew, and Uno Winblad. (Eds.) (1997). *Ecological Alternatives in Sanitation.* Proceedings from Sida Sanitation Workshop.

Kalbermatten, John M., DeAnne S. Julius, and Charles G. Gunnerson. (1980). *Appropriate Technology for Water Supply and Sanitation, A Sanitation Field Manual.* The World Bank, Washington, DC.

Leckie, Jim, Gil Masters, Harry Whitehouse, and Lily Young. (1981). *More Other Homes and Garbage.* San Francisco: Sierra Club Books.

Steven A. Esrey, Jean Gough, Dave Rapaport, Ron Sawyer, Mayling Simpson-Hébert, Jorge Vargas, and Uno Winblad (Eds.). (1998). *Ecological Sanitation.* Sida, 105 25 Stockholm, Sweden.

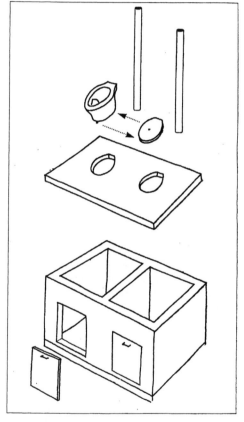

FIGURE 29
Double vault system. Adapted from graphics by César Añorvé.

WATER PUMPS

Most water supply systems rely on water pumps that raise water from ponds or streams or from underground. In addition to the pumps described in this article, a so-called ram pump, described in a separate article, can be used to lift water.

POWER SOURCES

Many types of power sources can be adapted to driving pumps. Human-powered pumps are probably the most prevalent worldwide. They can provide a moderate supply of fresh water on demand, are well understood, and are easily implemented. The simplest human-powered pump is "cranked" up and down. To power larger pumps the operator walks in a circle pushing a horizontal bar attached to a capstan. If the pump requires a reciprocating—up and down—motion gears or pulleys with a connecting rod can be used between the capstan and the pump.

If more water is required than a human-powered pump supplies, animals can be used to provide the muscle, since they do not balk at pushing a pump all day. A common configuration has the animal(s)—donkey, horse, mule, or ox—walking in a circle, pushing a horizontal rod connected to a vertical shaft. An animal-powered pump made by the Rural Industries Innovation Centre in Botswana can replace a pump driven by a small diesel engine. Through gearing the pump shaft can rotate as fast as 1,600 rpm—fast enough to power a centrifugal pump. Up to eight oxen can be used. The address of the Rural Industries Innovation Centre is given in the References section.

Wind power to drive water pumps was commonly used in the United States before electric power was available. Water pumping is an appropriate use for wind because wind's erratic nature is usually not a problem—pumped water can easily be stored. Irrigation water is effectively stored when it is applied to the field so it is not usually stored in tanks. Wind power is discussed in an article in the Energy chapter of this book.

If wind power is to be used, a design question is how big a rotor is needed. The "Wind Power" article shows how the power and the diameter of the rotor are related. Here we estimate the power required to irrigate a 1-hectare field if the water must be raised 12 meters. The power required to raise a 1 cubic meter of water 1 meter in 1 second is 9,800 watts—the explanation of this relationship uses basic mechanics and the gravitational constant in metric units. We assume 5 mm of water are required per day on the 1-hectare field. The total amount of water required, then, is 50 cubic meters per day, or 0.000579 cubic meters/second. Raising this much water requires 68 watts. Using the relationship given in the "Wind" article and making a reasonable assumption about wind speed, the diameter of the rotor can be estimated to be about 7 feet.

The familiar fan-type rotor used in the United States in water pumping applications is a well-tested technology. These "wind-mills" are reliable—they were designed for unattended operation. They normally use gears to reduce the rotor speed to a speed more suitable for displacement pumps. Scrap automobile gears may be appropriate or local machine shops may be able to make suitable gears. Other wind-powered machines are described in the "Wind" article, and the simpler machines would be suitable for water pumping. An article in the VITA handbook points out that many wind machines were installed in Third World countries in the past, and these may be available on the surplus market, since electric power was brought to commercial farms. The Rural Industries Innovation Centre makes two models of "windmills."

A seemingly attractive idea that does not seem to have caught on in the Third World is solar-powered irrigation pumps—the usual proposed configuration employs photovoltaic cells. Probably the reason photovoltaic powered systems are not widely used in the Third World is the cost of the solar panel compared to the cost of human and animal power. In the United States, where labor costs are greater and photovoltaic costs smaller, solar-powered irrigation systems have been deployed. One system in Texas uses both wind and solar—wind is erratic in the summer when the sun shines brightest, so the two systems complement each other. Another article in the VITA handbook shows how a reciprocating wire can transfer mechanical power from a water wheel to a water pump half a mile away. Amish people in Pennsylvania have used such a system for many years.

TYPES OF PUMPS

It is helpful to consider two types of pumps: lift (suction) pumps and force pumps. Lift pumps pull the water up from the well. Actually it is the pressure of the atmosphere at the bottom of the well that pushes the water up. The pump really only removes water from the top of the pipe, so more water can be pushed up. A lift pump is located at the top of the well and is limited to a lift of 6 to 7 meters, corresponding to atmospheric pressure. Force pumps actually push the water up from the bottom of the pipe.

DISPLACEMENT PUMP

The most familiar and ubiquitous pump is probably the displacement pump—a lift pump. At the beginning of the twentieth century, windmills drove millions of displacement pumps. When one thinks of early farms in the United States the image of people pushing up and down on the pump handle comes quickly to mind.

Figure 30 shows a cutaway of a typical two-cylinder displacement pump. On the upstroke the plunger valve closes and the foot valve opens. Atmospheric pressure pushes the water up the

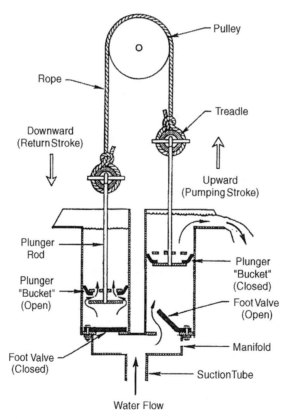

FIGURE 30
Displacement pump.

suction tube. On the downstroke, the foot valve flaps close and the plunger valve opens, allowing the piston to pass through the water already in the cylinder. On the next upstroke more water is pushed up the suction tube, and the water already in the cylinder is lifted and forced out the spout. A one-cylinder displacement pump, probably more common, is simply half of the one shown in Figure 30; water comes from the pump only half the time.

If the pump seals and valves are in good condition, this pump does not need to be primed. Perhaps an explanation of priming is in order. When a pump is not in operation, the water in the cylinder and suction tube may leak back to the well so the valves do not seal. Adding water to get the pumping started is called priming. A well-designed displacement pump is capable of pumping the air out of its system, so it does not need priming.

TWIN TREADLE PUMP

The twin treadle pump is shown in Figure 31. The prime mover is a person pedaling. The pump is actually a pair of pumps very similar to the standard displacement pump. Twin cylinders, operated alternately, provide a steady discharge, as opposed to a single cylinder that discharges only on the upstroke. A practiced treadle pump operator can work a two-hour stint; producing 2 to 3 liters per second at 2 to 4 meters lift. (The lift, or head, is the distance from the water level to the discharge spout.)

Most of the materials necessary to construct the pump will be locally available. The frame, treadles, and pipes can all be made of bamboo. The cylinders are made from sheet metal, PVC tube, cast iron, or concrete. The imported items—axles, seals, pistons, and rods—are readily available because they are replacement parts for widely used pitcher pumps. A good team of two or three people can make the bamboo pipes, strainers, and frame and sink, and install a pump in a day. The cost of a complete pump installation is around $15 to 20, which puts it within reach of all but the poorest farmers. A similar pump is made by ApproTec in Kenya—the address is given in the References.

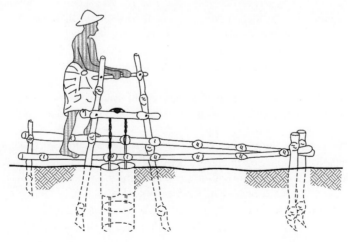

FIGURE 31
Twin treadle pump.

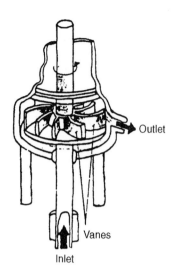

FIGURE 32
Centrifugal pump.

FORCE PUMPS

Force pumps are located at the water level—usually below ground level—and push the water up from the well. Actually most lift pumps will function as force pumps if they are submerged, but, of course, they must get power from the surface. Examples of force pumps are centrifugal pumps, rotary pumps, jet, and chain pumps.

Centrifugal pumps operate by throwing water from the center of the pump to the outlet located on the outer casing. Figure 32 shows a centrifugal pump. At the center of the impeller there is a low-pressure area, which draws more water up the suction tube. Initially water must be in this center area if low pressure is to be produced when the impeller rotates—so a dry pump must be primed. To avoid having to prime the pump before each use, a check or foot valve is installed in the suction tube. This valve prevents water from flowing back into the well—that is, water can only flow in one direction. Centrifugal pumps have a five- to ten-year lifetime, they require little maintenance, and in large installations, they have an efficiency of 80 percent, with 50 to 70 percent in smaller ones.

Axial or rotary pumps come in several variations. The best known is the Archimedes screw, which is a broadly threaded screw rotating in a snugly fitting tube. Another axial pump is made by mounting a marine propeller in a pipe. When the propeller is driven, it forces the water to the outlet. Figure 33 shows an axial pump. Axial pumps are useful in low-lift, high-volume applications. They are relatively insensitive to sediment in the water.

A jet pump, shown in Figure 34, is really two pumps. The secondary pump provides water to the primary pump. The primary pump has no moving parts; it is merely a nozzle whose outlet is a high-velocity stream inserted in a tube. The nozzle creates an area of low pressure that draws more water into the tube. Jet pumps are a good choice when the lift is more than 8 meters or the bore of the well is narrow. Their efficiency is somewhat less than that of a centrifugal pump.

A chain pump has vanes mounted on an endless chain, which push water up a channel. These are fairly easy to build but require much energy to operate. Small buckets can be substituted for the vanes.

The choice of a pump for a water supply system is based, like most appropriate design decisions, on local conditions, particularly needs and available resources. The purpose of the water (domestic, livestock, and/or irrigation), power sources, local customs, water source (shallow or deep), economic resources, climate, and local expertise all have an impact on the success of a water project. Cultural factors are significant; a pump was not accepted in the Sudan because the users thought the operators would sit in an undignified position—use of the pump violated local custom.

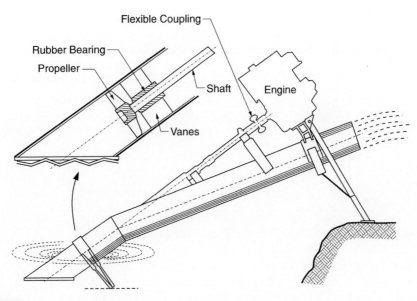

FIGURE 33
Axial pump.

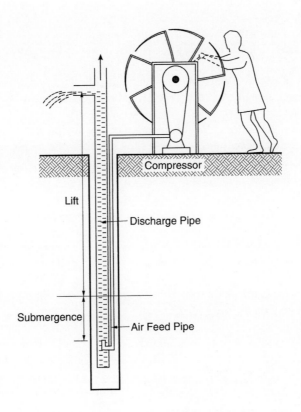

FIGURE 34
Jet pump.

REFERENCES

Approtec Catalogue. Appropriate Technologies for Enterprise Creation. PO Box 64142, Nairobi, Tel./Fax: 02-783046. approtec@nbnet.co/ke.

Een, Gillis, and S. Josde (Eds.). (1999). "The Treadle Pump." *One Hundred Innovations for Development*, IT Publication, London, also in *Appropriate Technology*. Vol. 17, No. 3.

Kristoferson, L. A., and V. Bokalders. (1986). *Renewable Energy Technologies*. Oxford: Pergamon Press.

Rural Industries Innovation Centre, P/Bag 11, Kanye, Botswana. Tel.: 340392/3 340448/9. Fax: 340642.

VITA. (1988). *Village Technology Handbook*. Volunteers in Technical Assistance, 1815 North Lynn Street, Washington, DC 22209, USA.

HYDRAULIC RAM

A hydraulic ram, shown in Figure 35, is a simple device that uses the energy in a stream with high volume but low head to lift a small volume of water to a large height. It could be used, for example, when we have a stream with a 1-foot drop carrying 100 cubic feet per minute of water and wanted to raise 2 cubic feet of water 8 feet to a storage tank on a roof to supply a kitchen and bathroom with running water. They require essentially no maintenance and so are useful in isolated installations where a fairly small amount of water is needed.

The only part of the ram in Figure 35 that is not obvious is the two valves. They are made with weights that normally push them open or closed, depending on the position of the weight. In particular the delivery valve in Figure 35 is normally closed—that is, horizontal—but can open up when water pushes from below. It closes when water pushes from above. Similarly, the waste valve in the same figure is normally open—that is, vertical—but close when water pushes up from the bottom.

It is probabbly easiest to understand how the hydraulic ram works if we go through the operation step by step. Initially, the delivery valve is closed and the waste valve is open.

1. Water from the stream flows down the input pipe, forcing the delivery valve open and the waste valve closed. Water, thus, can only flow into the compression chamber.

2. The momentum of the large amount of water from the input brings much water into the compression chamber, compressing the air on top.

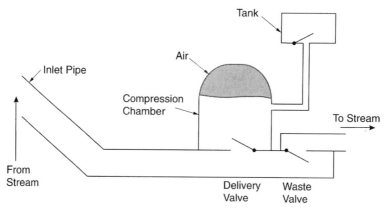

FIGURE 35
Hydraulic ram.

3. Because the water is flowing into the chamber, the pressure on the waste valve is small, so the waste valve falls open.

4. When the air in the compression chamber is sufficiently compressed, it pushes back on the water, closing the delivery valve.

5. The compressed air pushes on the water in the chamber, forcing it up to the tank.

6. Simultaneously, the water from the input, having no other place to go, flows out the waste valve, pushing it closed.

7. Meanwhile, the pressure on top of the compression chamber has decreased as water goes to the tank, so the delivery valve opens.

Now the process repeats itself, the situation being the same as it Step 1: delivery valve open, waste valve closed.

Hydraulic rams have essentially no parts rubbing against each other as many other pumps do, so they last a long time—several decades at least.

REFERENCES

Leckie, Jim, Gil Masters, Harry Whitehouse, and Lily Young. (1981). *More Other Homes and Garbage.* San Francisco: Sierra Club Books.

VITA. (1988). *Village Technology Handbook.* Volunteers in Technical Assistance, 1815 North Lynn Street, Washington, DC 22209, USA.

IRRIGATION

Based on "Small-Scale Irrigation Systems," Prepared for the United States Peace Corps by Development Planning and Research Associates, Inc., Information Collection & Exchange/Peace Corps, 1111 20th Street N.W., Washington, DC 20526, USA, September 1983

Irrigation is used for four distinct purposes.

- To enable crops to be grown where natural rainfall is too low to grow normal crops
- To provide additional water throughout the growing season, or at critical times during the crop season, when rainfall is inadequate to provide optimum crop production
- To extend the growing season
- To flood land for growing rice to prevent growth of weeds

Rice production, with some advice about irrigation, is discussed in the article "Growing Rice" in the Food chapter.

In areas where lack of natural rainfall or lack of rainfall during the cropping season limits crop production, supplemental irrigation may significantly increase yields or permit farmers to grow crops with higher yield potential or value. For example, in much of Africa sorghum and millet are traditional cereal food crops. With supplemental irrigation, yields of those crops can be increased or maize (corn) may be grown. Maize has a potential for 50 to 100 percent higher yield than sorghum or millet under optimum water availability and agronomic practices. In certain situations, the production of high value crops such as vegetables or melons may be feasible with irrigation.

Rice is probably the most valued cereal food. Supplemental irrigation may allow it to be produced in areas where it cannot be grown with natural rainfall. The capacity to keep the rice flooded during most of the growing period will increase yields and greatly reduce labor required to control weeds.

Irrigation increases a farmer's income, but developing and managing an irrigation system is expensive in labor and money and creates risks. Supplemental irrigation may allow the production of high value crops "off-season" when demand, and price, is particularly high. The system can be justified only by drastically increased crop yields or crop values. Growing crops such as fruits and vegetables might not be possible with normal rainfall but might be profitable, or socially desirable, with irrigation. If the project is only profitable if produce is sold outside of the immediate area, then the reliability of transportation is important—for one thing, will roads be passable when the crops are ready?

In general, traditional rainfed crops that suffered from lack of moisture will require, when irrigated, higher seeding rates and more fertilizer to produce optimum yields. Exact recommendations on seeding rates and fertilizer applications are "site specific," so local agronomists should be consulted for specific crop recommendations. The availability of such inputs as fertilizer must be assured. If fertilizer is not readily available when needed, the irrigation project will fail. Weed control will probably be more difficult with irrigation, so additional labor will be needed during the growing season.

Occasionally a major benefit of irrigation may be to shift or extend the growing season. For example, a short rainfall season may preclude growing maize. Also, longer season crop varieties usually have greater yield potentials. A light irrigation just to cause germination and supply the limited water requirements of young plants might bring the period when water requirements are greatest into the rainy season. One should, however, ascertain that long-season, adapted varieties or hybrids are available before building.

The economic feasibility of irrigation should be evaluated before any physical development actions are taken. If in doubt, as, for example, about the amount of water available, the optimum planting and fertilizing rates, or crops to irrigate, start with a small pilot project so the risk is not great. Economic evaluation is handled separately from the social factors, but social and cultural factors cannot be overlooked in arriving at a conclusion regarding the feasibility of an irrigation project. If it is traditional for women to weed crops, men might expect women to handle the irrigating causing a disruption of other family and household responsibilities.

AMOUNT OF WATER REQUIRED

Crops vary in the amount of water required. Typical requirements vary from 3.3 mm/day for sorghum in Pakistan to 5.3 mm/day for cotton in Hyderabad. Rice takes more, perhaps 8 mm/day. Water requirements depend on the growth stage of the plant. During early periods of plant growth, while much of the soil surface is exposed to sun and wind, the moisture loss by evaporation predominates. At later stages of crop maturity, much of the soil surface is shaded and protected from wind. Then transpiration water requirements predominate. Evaporation losses are much larger in climates where the relative humidity is low. Of course, when planning the irrigation system allow for the amount of water expected from rain.

SOURCES OF IRRIGATION WATER

Large irrigation systems usually depend on large streams or rivers for the water source. If the large streams have a constant year-round flow, water may be taken from them by building low diversion dams to raise the water level enough to allow water to be removed by gravity flow through canals. Pumps also may be installed to raise the water level to the distribution canals.

When you consider a stream as an irrigation source, first contact local residents who will know if the stream frequently "dries up." If it ceases to flow as often as one year in five, it is of questionable value for long-season crops. If stream flow is much less than desired during the dry season, high dams can be used to store water from the wet to dry season. The design of high dams requires significant expertise—such projects are not for the inexperienced.

Underground water supplies are widely used as a water source, and large projects use mechanically powered pumps to raise the water to the surface. Small projects use hand or animal powered pumps. These systems are frequently called "tube well" systems. Since the systems vary widely in size, they may be publicly or privately owned. Wells and pumps are discussed in separate articles in this chapter.

Springs that continue to flow during the dry season are ideal water sources for small private irrigation projects. The spring is at the surface and water will flow by gravity to lower elevations where it can be used. To determine the spring's potential as a water source, the quantity of flow should be measured during the season when irrigation will be required.

Small privately or community-owned ponds to store water for irrigation may be made by damming small streams. Some major advantages are the following.

- Small ponds (microdams) can be placed in almost any location.
- Gravity flow can usually be used to distribute the water.
- Local labor can be used for construction.
- The water can be used for animals or household purposes.

The following are some major disadvantages.

- Silting is frequently a problem because runoff from natural rainfall contains eroded soil. It is less of a problem where forests or good grass vegetation exists above the pond.
- The soil at the pond site should be relatively impervious (probably one of the clay types) to prevent excessive water loss through the bottom of the pond or the dam.
- Erosion around the end of the dam from excess water can be so severe that a masonry structure is required to lower excess water to the normal stream level below the dam.
- Some capital investment is required to purchase the pipe and valve required to drain water from the pond.
- Ponds should be fenced to protect them from livestock.
- Evaporation will be rapid from the pond surface during the dry season, reducing the amount of water available for other purposes.

Before making a final decision on a water source, determine if there are legal, longstanding customs or community constraints against developing the particular source. Contact the Ministry of Agriculture or Land Development officials for information about constraints that most likely apply to using water from streams. A good spring may be considered a community resource for animals and households. Using the spring water for irrigation, particularly for a private project, could cause friction in the community. Development of underground water is less likely to encounter legal or community restrictions.

WATER DISTRIBUTION

For most water sources, some sort of a distribution system is required to transfer the water from the source to the area to be irrigated. Most small irrigation systems use small ditches and canals, and the water flows by gravity.

Unlined ditches dug in ordinary soil lose a large amount of water by seepage into the surrounding earth. Part of the seepage might be recovered by growing crops along both sides of the unlined channels. Seepage losses can be reduced by using linings of masonry, concrete, or plastic sheets, but they are seldom used on small projects in developing countries. Obviously, pipes would prevent losses from seepage and evaporation, but the cost of pipes usually prevents them from being used.

Several critical requirements must be met when designing a channel to distribute water. To reduce seepage losses, the soil should not be too permeable. If the ditch must traverse an area of very permeable soil, a lining of a heavier soil type might be feasible.

The channel must have enough slope and area to convey the quantity of water required. The quantity of water depends on the cross sectional area of the channel and the velocity of the water flow. The velocity of flow in a channel is a function of the following channel characteristics.

- Cross-sectional shape
- Slope (actually the square root of the slope—to double the velocity, the slope must be quadrupled)
- Roughness

Formulas can be used to estimate the water flow through particular channels, but they are complicated and uncertain, so one should probably make estimates on the basis of observations of the flow through channels with similar soil.

For very small flow rates, channels may be of V or semicircular cross section; for larger capacities, trapezoidal cross sections are generally used. The side slope on a triangular or trapezoidal cross section will depend upon the soil type. Side slopes that are too steep will cave off into the channel. The U.S. Bureau of Reclamation recommends a side slope of 3:1 (horizontal: vertical) for sandy soil, about 2:1 for loams and clay loams, and as steep as 1:1, or vertical, on heavy clay soils. Masonry-lined channels can have vertical sides if they are properly reinforced. Designing a trapezoidal channel—determining its cross-sectional area—usually requires a trial-and-error solution.

The velocity in the channel must not be so fast that it causes excessive erosion. When slope is great enough to allow the maximum velocity to be exceeded, two solutions are possible. A wider and shallower channel can be built, or a wood or masonry drop can be installed at some point so the slope along the remainder of the channel will be reduced. The flow velocity will be very high just below the structure and will cause severe erosion on the lower side. An apron must be provided to dissipate the energy in the falling water.

Very low velocities may cause a different problem. Water from a rapidly flowing stream could be high in sand, silt, and clay carried along by the flow. Such water diverted to a slow-moving irrigation canal would settle out sediment, requiring that the canal be cleaned periodically.

Starting at the source, a survey should be made to provide a line for the ditch. It will be almost always be on a contour from the source to the field. If the contour line is quite crooked, the channel may be "smoothed" or straightened by digging deeper through high points or by filling in low areas with extra earth to build the berms higher.

FIELD IRRIGATION SYSTEMS

Basin irrigation is widely practiced where rice is irrigated. Rice (unlike most crops and most weeds) can grow when the soil is completely saturated. The basin is formed by leveling the area completely and enclosing it with berms, or levees. The side berms will run essentially on the contour. If the land is very sloping, the berms become terraces, and a large amount of earth must be moved from the upper side to the lower side. On very steep slopes, the basin will be fairly narrow to reduce the amount of leveling required. Drop structures are required to lower water from one level to another.

Border irrigation is done by laying out the land with side berms running downhill on a slight slope. The land is leveled between side berms to make the irrigation water run in a narrow sheet from the upper to the lower end of the field. When irrigation starts, the infiltration rate is high at the upper end of the border, but as the soil becomes saturated, the leading edge of the water continues to move downhill. To provide enough water at the lower end of the field without overwatering the upper end, a high berm is constructed at the lower end to hold back a pool of water. Determining the correct length and slope of a border system is by trial and error. Distributing water uniformly across the width of the border strip requires that the field be very flat (level). If the border is wide, water should be supplied at more than one point from the distribution channel. Water may be discharged from the supply channel to the border by gated pipes through the channel berm or by siphons over the berm. The border system is well adapted to watering forage crops or other crops that cover the ground entirely.

Crops normally grown in rows, such as grain or vegetable crops, are more frequently irrigated with furrow systems—a series of furrows and ridges with about 75 to 100 cm between furrows and 15 to 20 cm deep. The furrows run downhill, as with borders. Where the furrows are constructed 15 to 20 cm deep, it is possible to irrigate a field with a significant amount of side slope. As shown in Figure 36, rows of tall-growing crops like maize are planted on the ridges. Two rows of low-growing crops like onions may be planted on each ridge.

One problem that may affect row placement on the ridge is having enough soil moisture to germinate seed. In areas that normally have sufficient rain during the planting season, rainfall should provide moisture for seed germination. In drier areas, the field may be very heavily irrigated just before or after seeding so enough moisture moves sideways and up by capillary action to germinate the seed. But it is usually best to place seeds into moist soil. In severe problem cases, such as a sandy soil and low rainfall, seed may be planted on the side of the ridge so they are closer to the wetted area. Once the seedling root system develops a few inches, there should be no further problems.

As with border systems, the slope along the furrow in furrow systems must be flat enough to prevent erosion but steep enough to allow water to reach the end of the furrow so infiltration is relatively uniform the full length of the furrow. The more permeable the soil, the steeper and/or shorter the furrows must be. In general, furrow slopes should range from .1 to 2 percent. If the distribution channel is run on a very slight grade, essentially on the contour, then the furrows can be supplied and laid out on the downhill side of the channel, although they need not run perpendicular to the channel.

Because of the many variables involved, a good operating rule is that water should reach the end of the furrow within 25 percent of the total time for one irrigation. That will provide about 25 percent more irrigation at the top of the field than at the lower end. Water to irrigate a furrow can be applied at a high rate at the beginning of the period and then reduced as the soil becomes wetted. That reduces the time required for water to reach the end of the furrow and prevents excessive loss later from the end of the furrow. Also, a dam may be placed at the end of the furrow to pond water and increase infiltration rate.

Sprinkler systems are useful when evaporation is extremely high. Efficient use of irrigation water and minimum land leveling are characteristics of sprinkler systems but operating and investment costs are higher than for gravity flow systems. Pressures must be matched to sprinkler size, and manufacturers' representatives should be consulted to design the systems.

Drip irrigation is a relatively new development. Water is piped under pressure, and small outlets are located at each plant to be watered. The system is usually applied to trees, but large

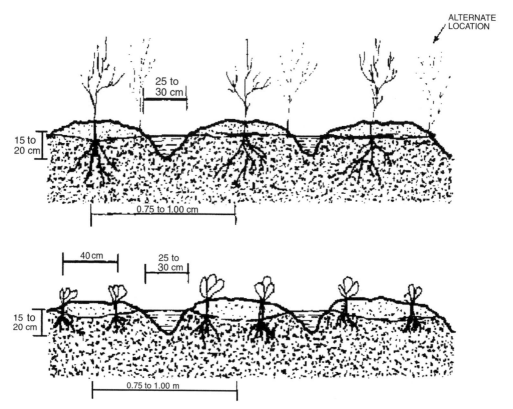

FIGURE 36
Furrow irrigation.

plants like tomatoes may be irrigated. The system is designed to apply water very slowly at a rate a specific plant needs. Other areas are not watered. Major disadvantages of the pressure system are its cost and small holes plugging up with foreign material.

In the wild flooding system, water is released from a distribution channel at the top of a field that has had little, if any, leveling. Water distribution will be very nonuniform. The system should be used only where there is a permanent ground cover such as alfalfa or grass to prevent erosion.

MAINTENANCE

Channels to and within a field require regular routine maintenance to remove weeds that reduce water velocity and cause additional evaporative losses. Some erosion will occur along channels and furrows, and some silt deposits will have to be removed to maintain channel cross-section area. With time, berms will erode and require some maintenance to maintain their height. Small leaks, particularly through or over berms, should be repaired promptly before water erodes them severely. Watch for holes made by animals through berms.

In some countries, large animals, such as water buffaloes, pose a significant threat to canal systems, since it is virtually impossible to prevent the animals from wallowing in water and thus destroying the canal bank. Consideration will need to be made for possible crossings for the animals and perhaps for policies regarding grazing alongside the canals as well.

Some leveling of basins and border systems will be required, originally and with time. High and low points should be marked when water covers the surface. Using a large plane of water is a more rapid way to locate high and low spots than using a surveying instrument.

Erosion during a rainy season can cause serious damage unless the area is well protected with drainage ditches or terraces that divert surface flood-type flow. Drainage is discussed in the next section.

DRAINAGE

Drainage is the removal of excess water from the land to prevent crop damage and salt accumulation, allow earlier planting of crops, increase the root zone, aerate the soil, favor growth of soil bacteria, and reclaim arable low-lying or swamp areas. Practically every valley where irrigation has been carried on for a considerable length of time has lands needing drainage.

These are some indications of drainage problems.

- Standing water or salt deposits on the soil's surface
- Scalding of crops by summer water ponding
- Propagation of mosquitoes in irrigated fields
- Soil compaction and resulting poor water penetration
- Difficulty in carrying on farm operations because of poor tractor footing
- Minerals accumulating in the soil
- Poor root growth due to a high water table
- Plant root diseases

A combination of field ditches and land leveling is most practical. It takes an unreasonably large network of field ditches to do a good job of moving water from most fields without land leveling. Deep channels to carry the final collection into an accepted area are often constructed on field boundaries.

To protect roads, irrigation systems, buildings, and fields, maximum rate of runoff for all drainage systems should be determined. Most structures can be flooded for a short time, but peak rainfall intensities and runoff data should be determined so that the system (bridges, culverts, etc.) can be designed to handle the runoff. It may be most economical to design the structures on a 10- to 25-year recurrence expectancy; that is, the expected runoff would be exceeded only once every 10 to 25 years.

REFERENCES

Booker, L. J. (1974). *Surface Irrigation.* FAO Agricultural Development Paper, No. 95. Rome.

Finkel, H. J. (1987). *Handbook of Irrigation Technology.* Vol. 1. Boca Raton, FL: CRC Press, Inc.

Jensen, M. E. (1980). *Design and Operation of Farm Irrigation Systems.* St. Joseph, Michigan: American Society of Agricultural Engineers.

Manzungu, Emanuel, and Pieter van der Zaag. (1996). *The Practice of Smallholder Irrigation, Case Studies from Zimbabwe.* Harare: University of Zimbabwe Publications.

Schwab, G. O., R. K. Fievert, T. W. Edminister, and K. K. Barnes. (1966). *Soil and Water Conservation Engineering.* John Wiley. New York.

Stern, Peter. (1979). *Small-Scale Irrigation.* Intermediate Technology Publications, 9 King St., London WC2 8HN, UK.

Withers, B., and S. Vipond. (1974). *Irrigation Design and Practice.* B. T. Batsford Ltd., 4 Fitzhardinge St., London W1H OAH, UK.

CHAPTER 9

Tools

Edited by ROBERT ELLIOTT
Independent Builder

THINKING ABOUT TOOLS

Robert Elliott

According to *Webster*'s the word *tool* is derived from the Old Norse *tol*, meaning "to do or to make." While we usually think of tools as objects, we would do well to keep the derivation in mind. For me, tools are extensions, enhancements, or amplifications of my body. I can pound with my fist but pound better with a rock—and better still with a hammer. I can see an amazing world using only my unaided eyes, but with tools, this world is expanded a million times beyond my natural limits.

An average person, such as myself, comes with a pretty good set of tools—a part of that wonder of evolution, the ultimate toolbox, the human body. Starting at the top we have the head, home of the brain, and the senses of sight, hearing, smell, taste, and touch. The common comparison of brain to computer is useful in some ways, and words such as *memory* and *program* are used in both systems with equal standing. Words such as *intuition, emotion*, and *judgment* are seldom used to describe computers and point to limitations of the analogy.

Senses are measuring devices, and the brain is what makes sense of sensory input. Next, from a toolbox perspective, are arms and hands. Hands and fingers are loaded with nerves able to measure temperature, texture, and pressure. But that's only half the story. Hands and fingers are also supplied with muscles, large and small, articulated by motor nerves under conscious control. The mind, having learned of the world through the senses, can now act on the world through the muscles. This input-output feedback loop yields impressive results, and it is at work in all voluntary muscles.

From a tool user's perspective, the hands are where it's at. From microsurgery to heavy machinery, hands are the operators not only of tools but also the models for many tools and the makers of most of them. Legs and feet, engineering wonders in their own right, provide locomotion, provide access, and provide power, wonderfully complementing the hands' limited mobility and relative weakness. The trunk, the body proper, is more than a point of attachment for the limbs. Much of the largely unnoticed infrastructure of life, respiration, circulation, and digestion takes place here.

Which came first, the human or the tool? Because tool making and tool use have often been judged to be *the* definitive human activity, this is very much a chicken and egg question.

To generalize (but only slightly): Without humans, there would be no tools. Equally as true (if more difficult to accept) is the reverse: Without tools, we would not be human.

One should not be afraid to use tools, but one *should* be careful. Tools are used because they can do things that the human body can't, so more damage can be done with tools than by an aided person. The next article in this chapter warns about personal safety, but one should also be careful about not ruining the thing one is building. The rule "Measure twice, cut once" makes sense.

After the safety article are articles about hand tools used in woodworking, sharpening metal tools, basic mathematics used in carpentry, and machine tools used in the United States. The final set of articles describes blacksmithing and two tools that might be used in a Third World machine shop—a machine to fold metal and a machine to make wire fences.

SAFETY

Robert Elliott

IMPORTANT! READ THIS BEFORE PROCEEDING

As one who still has the original 4 limbs, 20 digits, and 5 senses in working order, even after 40 years of working with dangerous tools, I'll admit to being lucky. But I'll also admit to learning from close calls, to getting and using safety equipment, and to developing safe practices (or habits of longevity, if you will).

Judgment is your first and best safety tool. Judgment comes with experience and reflection on that experience. Experience can be that of others (and should be when the price is a severe injury), but the reflection and learning from that experience must be your own. Just seeing my neighbor's two finger stubs was enough to impress me about table saw safety. Judgment involves anticipating danger and minimizing that danger. It means erring on the side of caution when there is uncertainty. It is weakened by fatigue, boredom, drugs, and distraction. Good judgment includes awareness of these impairments as they stealthily appear and taking corrective action, such as a nap, a snack, or a walk.

Safety equipment most often consists of a barrier between you and a hazard. Some is fairly specific to a job or operation (welding helmets), while others, such as gloves, are more generally useful. Hardhats in their simplest form protect the head from falling objects, be it raindrops or rivets. I have a logging model with ear protectors and a sawdust screen for the eyes, both of which can be swung in and out of position (see Figure 1).

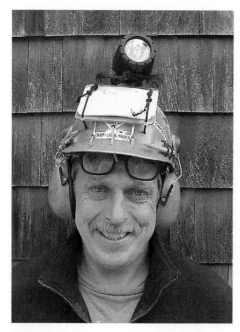

FIGURE 1
Safety hard hat.

I removed the screen but added safety glasses. I stuck foam strips on top so that I could balance loads without having them slip. Bungee cord loops at this location allow easy insertion of a flashlight whose beam points the same direction as your head. More bungee loops up front hold pencils, a note pad, and a utility knife. This is balanced by a 30-foot tape measure that slides onto a homemade bracket riveted in back. Here, too, is a small hemostat used mostly for pulling splinters. Inside, between webbing and shell, is room for a couple of dust masks and a pair of gloves. It evolved a piece at a time to make the work I do—carpentry and woodworking—easier. While thus far it has not saved my life, it has saved time and annoyance, and that's good enough. I feel safer and more competent when I'm wearing it on the job, so I always wear it. It becomes a habit, and I hardly have to think about it, which is the whole point.

Develop good safety habits. A good habit such as not walking under a ladder can even evolve into a superstition. If you put yourself at risk, you're more likely to get injured. The more aware you become of risks, the smaller the role luck plays. Listen to your gut.

BASIC HAND TOOLS

David E. Erikson

Hand tools for working wood have become ancillary to power tools in industrialized nations, but they still have an important place. Hand tools are suited to many of the details of finishing a job. Power is not always available close at hand. Most of us cannot justify buying every power tool available. Hand tools give a much closer connection to the work, and many of us appreciate the quiet and more human pace. This article introduces you to the basic hand tools of carpentry and how to use them.

Here's a good example. Local old-timers tell me that they used to pump water with a device they called a ram that captured the power of water moving down hill in little brooks to force some of the water uphill in a pipe. (Ram pumps are discussed in an article by the same name in the Water chapter.) When these rural New Hampshire farms were wired for electricity in the fifties, it became much cheaper to buy an electric pump than to clean the ram pump. Wind pumps are rusting away in all but the remotest plains location for similar reasons.

I have always enjoyed good hand tools and have lamented that they often can't compete on a job site in the United States, especially when they are dull or poorly made. When I spent a few months building in a Third World country, my interest in using and maintaining hand tools was rewarded and developed. For the past 25 years I have specialized in small repairs. Often a hand tool can do the job before the power tool is plugged in. This brief article on hand tools is written for readers who find themselves interested in using basic hand tools and fasteners, perhaps in a situation where it isn't practical to call on a repairman or plug in a power tool. It is written for folks who have little experience with hand tools.

Materials often don't arrive in nature or at the builders supply in the size and shape we want, whether we are concerned with a plank or a piece of pipe. If you stop by my workshop with a piece of pine $1'' \times 12'' \times 7'$, and I give you a piece of $1'' \times 4''$ and a handful of #7 or #8 box nails or $\#8 \times 2''$ flathead wood screws, we could build a handy saw bench in an hour or so, but we'd have to do some cutting, drilling, and planing. The completed saw bench is shown in Figure 2. A description of how to make it will make a good introductory lesson in do-it-yourselfism and allow me to explain things you may have found unfamiliar in the last sentence. Many of the tools and the techniques are transferable for building a shed or fine furniture.

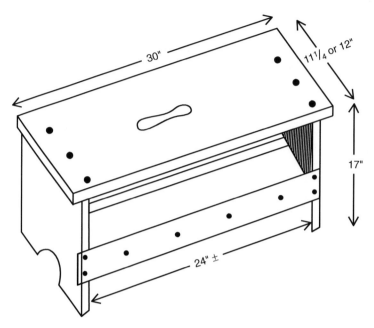

FIGURE 2
Saw bench to be constructed.

A saw bench is the right height for holding a piece of lumber with your foot while you saw it. It also makes a dandy stepstool or break-time seat, and the shelf allows you to carry the hand tools you've predicted you will need.

If you actually want to build one, you can. You'll need a measuring tape or rule, one of several kinds of squares, one or more kinds of handsaw, hammer, drill and bits, and a chisel and/or plane. The square is for accurately marking a right angle or to check that an existing cut is perpendicular to the edge of a piece of lumber. One leg or side of whatever kind of square you're using (even if it's two edges of a book or box with a square corner) is placed even with the edge of the board. You mark across the wood along the perpendicular leg. After you have checked that the ends of the lumber you have to work with are square or made them square by cutting off a half inch or so, you'd be ready to mark the lengths for the top, legs, and shelf of the bench and mark for the cuts. (Suggested dimensions are in the Figure. I make the shelf long enough for my handsaw and framing square.)

To cut across a board, you need a crosscut saw, which is the most common type. (see Figure 3). There are 7 to 12 teeth per inch, and if you look closely at the teeth, they have points resulting from sharpening files passing across them at an angle to the line of the blade. Few and bigger teeth give a faster but rougher cut. The teeth are sharpened almost straight across so they look like tiny chisels; they were designed to cut along the length of a board. Cutting parallel to the grain (fiber lines of the tree) is called ripping. Ripsaws were in every woodworker's tool chest before newer circular saws came along (the invention of a Shaker woman, I'm told). I include a ripsaw in my kit when I am working beyond the reach of extension cords. It rips many times faster than a crosscut saw (see Figure 4). A dull saw cuts slowly and can't be made to follow a line. To keep them sharp, you can slip on the plastic things that are sold to hang posters, which work just like the pieces that slide onto clear plastic termpaper covers.

I used to deliver steel-hulled boats for a living. We would make bunks and a temporary kitchen on board if the boat wasn't equipped with them. So I once asked a steelworker in the

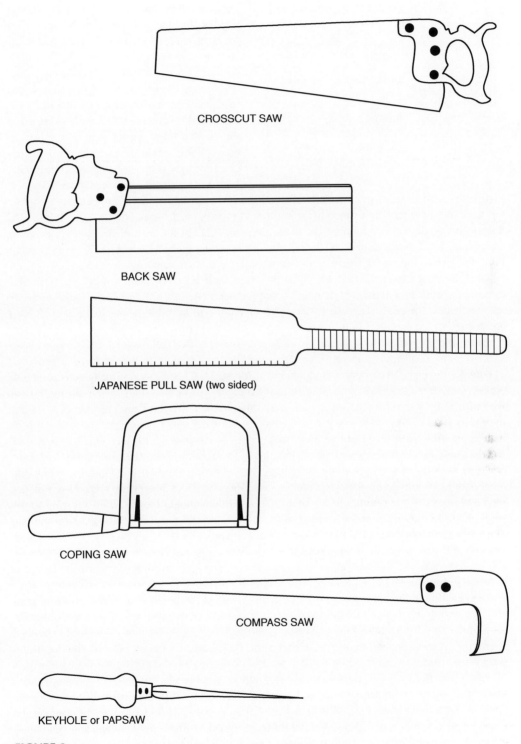

CROSSCUT SAW

BACK SAW

JAPANESE PULL SAW (two sided)

COPING SAW

COMPASS SAW

KEYHOLE or PAPSAW

FIGURE 3
Types of saws.

RIPSAW TEETH

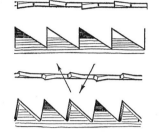

CROSSCUT SAW
TEETH

FIGURE 4
Saw teeth.

shipyard if he had a crosscut saw I could borrow. He handed me a rusty item with rounded teeth. I asked him if he had been cutting steel with it. He said, "No, it didn't cut steel good." It took forever to cut a few two by fours that a sharp saw would have cut in a few seconds. Those boat deliveries took me to several countries where I worked for a time with a relief agency helping the natives build public buildings. There was no electricity for miles, and I had to pack in my tools. Japanese saws and wooden planes are lighter.

It takes time to learn to cut accurately with handsaws, whether the European push type or the oriental pull type. Don't force them. In time, a smooth, even stroke will be natural. You also have to become aware of holding the blade perpendicular to the face of the lumber, to get a square cut. If accuracy matters, leave the pencil line on the piece of wood you'll keep, making your wobbles to the waste side. You can sand or plane the cut smooth later.

Once the top, legs, and shelf are cut, I'd rip the shelf board to be narrower than the other three by the thickness of the two sidepieces put together. In other words, when the two sidepieces are attached to the edge of the shelf, the three together will become the same width as the top and legs. Use the ripsaw if you have one.

After ripping you can smooth and straighten the edge with a jackplane or any similar plane. Like so many hand tools, the instinct of keeping the blade square to the wood takes practice. The longer the plane, the easier it is to straighten the board. Check to see if the edge is square to the two surfaces of the board with a little square. If the jackplane was sharpened in the traditional way with a slight curve, then you can adjust to cut a thicker shaving from one side or the other by which edge of the blade you cut with. As you start the cut, push on the forward knob or handle, moving to even weight on both as you move into the board, and finally move your weight to the hind handle as you go off the end of the plank to keep the blade from dipping. I could write many more pages about sharpening and adjusting planes if I had the space. I should mention one more type of plane: the block plane. Regular Block planes are about seven inches long and are good for smoothing off sharp edges. The low-angle block plane (referring to the angle made by the blade and the wood surface) is especially good for cutting across the end grain to smooth out crosscuts or to make a board a hair shorter.

Now you can fasten the four pieces together. Make squared lines for the shelf on the two legs. A few nails are fine for initial assembly, but screws will hold longer, since the bench is bumped and torqued. Box nails have a thin wire that won't split the wood as frequently. They do bend easily for someone new to driving them. Think of the hammer as an extension of your arm, and swing the tool so that the arc made by the swing of the head would go right down the shaft of the nail. If the nail bends, you can often straighten it with the claw. If one bends beyond repair, put in another and the others for that joint. Pull the bent one out later so you don't pull your whole assembly apart while pulling the nail. I use a 16-oz. curved claw hammer with a fiberglass handle for most carpentry. The straight or ripping claw is good for doing macho destructive things on construction sites, but I don't like them for pulling nails. With either one, you can put a thin piece of wood between the hammer and the project to prevent scarring the wood when you pull a nail. I like the flexibility of wood or fiberglass handles, and despite an affinity for wood, I've had to admit that the strength of fiberglass has allowed me to use one favorite hammer most of my working life (until someone filched it recently). Steel handles can cause carpal tunnel problems if you use them a lot.

Set the shelf at whatever height you want, but set it so it's centered left to right on the legs. I'd start a nail from the side without the line so that it just pokes through to hit the center of the thickness of the shelf. Then you can hold the leg piece on the upright shelf with the point of the nail poking into the shelf's end grain to hold it in position while you reach for the hammer. If you position the pieces and then start the nail, chances are the piece will vibrate away from where you set it with the first blows. All this would be much easier to show than to describe.

Once both legs are fastened to your satisfaction, nail on the top with the legs vertical, unless you want to splay them a little out at the bottom, which would make the bench a bit stronger. (Read ahead to the paragraphs on drilling if you want to cut a handhold in the top before nailing it down.) Then set the bench on its side and set a sidepiece across the two legs at the height you want to give a lip to keep tools on the shelf. Mark the length of the sidepiece in place and also mark the legs where the top and bottom edge of this side piece cross the legs. Use a sharp pencil or a knife for accuracy.

It is often better to mark from the real thing than by theoretical dimensions. A 1 × 4 is supposed to be $3\frac{1}{2}$ inches wide, but it might be $\frac{1}{16}$ bigger or smaller. Math teachers often want me to tell their students that they will need lots of math to be a carpenter. (Basic math is discussed in another article in this chapter.) Carpenters will need some, and the spatial concepts of geometry are especially pertinent, but adding together the length of your shelf board plus twice the thickness of the legs to get the theoretical length of these sidepieces doesn't work for me. First of all, it requires concentration on mundane measurements to avoid dropping a number you should have carried, and so on. I prefer to be thinking about something more interesting, carrying on a conversation, or singing, so I stick with reality whenever possible and avoid arithmetic. Second, even if you add right, it takes longer and won't match the reality of the warped or inaccurate wood. It's amazing how three or four slight variations can add up to $\frac{1}{8}$-inch error.

An aside: I'm used to inches and feet rather than metric measurement. I guess I'm a bit parochial, but I like that a foot is the length of an average man's boot and that I can divide it evenly in halves, thirds, quarters, and sixths. I haven't been convinced that a fraction of the inaccurately measured circumference of the earth at Paris divided by the number of digits on an intact carpenter's hands is the way to go. I think if Napoleon was going to torture people for using the wrong system of measurement, he should have made us all use base 12 all around, but I'll admit it's too late now. Who even heard of 12 points to the inch or a gross (144) of ball bearings in the binary age?

Back to the two marks on each side of each leg for the sidepieces. Next, use the sidepieces to mark how deep you should cut on these lines to set the sides down into the legs to touch the narrower shelf (see Figure 2). Make accurate cuts for the outer dimensions. You want the sidepieces to fit tightly to hold the bench mechanically rigid, so if the lumber isn't thoroughly dry or if it's a very humid day, make the fit especially tight to allow for shrinkage as the wood dries. You can then make several cuts down to the line in between your careful cuts. This allows you to tap the little chunks of waste wood off with a hammer. Then you can clean the joint with a chisel. Use both hands and a sharp chisel; no hammer or mallet is needed unless there is a knot in the way. If you use just two cuts and then the chisel, you could tap the chisel with a mallet, but don't start right on the finish line because the grain of the wood may cause a split below the line. Make a rough cut above the line and get a sense of how the grain is dipping.

Tap in the sidepieces. If they are too tight, adjust them with a plane. If they are a press fit, you did well. Now you can put some screws into your bench. You may wonder why I don't have you use glue, too. Glue holds well when edge grain fits tightly to edge grain and when it all is expanding and contracting in parallel with changes in humidity. There are no joints like this in your bench. (There are none in most spindle chairs either, which is why glue seldom lasts four seasons in chair joints.)

Philips screws or various other shaped insert bits have pretty much taken over from slotted screws because keeping a straight screwdriver in the slot can be frustrating. I carry a screwdriver with an insert that flips around for straight and Philips ends of several sizes. Screwdrivers wear out and work much better when they have a crisp edge. Screws can be forced into soft pine without predrilling and might not split the wood, but they will hold better and split less if the right size of pilot hole is drilled. To choose the right drill bit, I hold a bit over the screw to see if it is the thickness of the web of the screw—the thickness the shaft of the screw would be if it had no threads).

In the backcountry you could use an eggbeater drill, but get a good one with a crank $3\frac{1}{2}$ inches or more in diameter, not a little wussy one. There are dozens of two-speed models in my dad's 1940s catalog, including ones with a plate to lean your chest on so you can get all your weight into it. These are my choice for drilling steel because you want high RPMs. It makes a huge difference to keep them oiled. You'll need lightweight oil in your toolkit anyway if you sharpen with oilstones. For short forays away from the power grid, the battery driver drills are a nice luxury. It saves the wrist for putting in the screws, too.

There are other great hand tools for drilling and turning screws. The old Yankee-style push-pull screwdrivers are wonderfully speedy, and there are models for drilling small holes, which have the advantage of only requiring one hand. Again, I advise a larger model for larger jobs. If I had to choose one tool for drill bits and screwdriving, it would be the brace, which is a crank with some kind of chuck on it. It's my tool of choice for hard-to-turn screws in metal or wood, and it is ideal for use with auger bits for large holes in wood. Like the breast drill, you can lean on it.

I like to make a handhold in the center of the top of the tool bench so it can be carried one-handed. I drill two holes about four inches apart on the centerline of the bench top with the largest auger bit I have. It helps if you have the top off for this so you can avoid splintering as the bit breaks through by putting a piece of scrap wood under it or by watching for the center spur of the auger to come through and turning the piece over and starting from the other side. Auger bits have a hard time with knots even when they are sharp, so I hope you have clear wood where you want the holes.

I then connect the two holes with two comfortable curved lines (see Figure 2). A keyhole saw is the right tool for this because its thin-tapered blade can fit in the hole to start the cut, and it can cut a slight curve, especially if you use the narrow end of the blade. Don't force it or you'll bend it. You can take off the sharp edges of this handhold with a knife or chisel and clean off the edges of the top with a block plane. An artist friend of mine is always kidding me about the straight lines and angles I tend to use. Go ahead and cut some more curves on your bench if you want.

I've mentioned most of the tools I would pack if I was going to build things with lumber in the wilds of a Third World country. But don't assume that lumber is the appropriate material in other places. Spending precious time hand-sawing trees into boards for shelters nearly finished off the early Plymouth colony. Sapling and bark lodges like the ones Native Americans built would have gone up much faster. I was very impressed by the buildings that native Central Americans built with only a machete. So learn from the locals.

But I will finish with a few additions to the toolkit of western tools, which I would recommend you carry if you planned to build a stud-framed building beyond the power lines but where you could carry in 50 pounds or so of tools. I would include a couple of pry bars, perhaps a flat steel one that I like to file sharper than they come from the factory and a two-or-more-foot-long one for coaxing things and pulling spikes. I like the kind with a bent tee head so you can pry by pulling toward you or away from you. While on the subject of coaxing, I'd include what my mentor called a "persuada": a five-or-more-pound sledge with an 18- to 30-inch handle. You might want a small axe or hatchet or a bow or frame saw for some kinds of work.

Clamps can be used in numerous ways. I carry two pipe clamps of the smaller size to cut down the weight. I use a pipe coupling so I can make one long clamp out of two. I also carry a few "C" clamps and two wood and metal turn screws that are quite versatile. Some people like the new one-hand operated clamps. A piece of string line is essential for laying out straight lines. In the old days it would be rubbed with chalk to snap a straight-line mark. Now all carpenters use a chalk line filled with powdered chalk. You might also include a plumb bob to tie to your string for determining verticals when the wind isn't blowing. In a pinch you can make an old-fashioned level with a plumb bob and a straight stick, which is also a tool you will find you want on most jobs. The four-foot level with bubble vials is what most builders reach for to draw a straight line, even after the vials have gone screwy.

You might pack a couple of files. Perhaps a small triangular one that could sharpen a handsaw or auger bits if you were stuck away from a commercial sharpener, and an eight- or ten-inch flat one for rough sharpening and shaping of metal. It is also the tool to get a straight edge on your handsaw before you sharpen it. Yes, you dull it until all the teeth are the same length and then sharpen it again. Buy a good book on sharpening hand tools. (Another article in this chapter deals with sharpening tools.) Chisels and plane irons can be touched up in the field with oilstones or water stones. If you'll be away from civilization, don't forget to allow for maintaining your tools. Choose sharpenable blades instead of disposables.

I have barely touched on tools to work metals and haven't mentioned masonry, plasters, and earth, which are more practical than wood in many parts of the world. Be cautious about encouraging people to adopt our materials and tools that they then will have to import. I have written about what I know best. I have enjoyed learning from elderly people with experience in my region. I encourage you to learn from the people with experience in whatever places you find yourself.

SHARPENING TOOLS

Alan Dick

I wish I could have back the hours and days that I have wasted using dull tools. A sharp tool is truly a thing of beauty. A dull tool is the cause of frustration and discouragement. If there were one thing I wished someone had taught me as a young man, it is how to sharpen things. No one taught me, and I had to learn for myself over a period of many years. There is an unique feeling that comes from passing a hand plane over a board, producing a long, thin shaving, or easily passing a knife through a fish and, with a few strokes, have it ready to hang on the rack.

My wife's grandmother used to test her knife by holding up a hair. If she could cut the dangling hair with her knife, it was sharp enough for tanning and making rawhide. I have experimented for years, trying to learn how to sharpen to that level.

There are three considerations in sharpening a tool:

1. At what angle is the edge formed?
2. How thick or thin is the actual edge?
3. How rough or smooth are the edges and sides of the blade?

ANGLE

Many directions for sharpening say, "Sharpen the tool at 30° or 25°." I often wonder how they could possibly know what I am cutting. The material that is being cut and the toughness of the steel in the blade determine the best angle to sharpen the edge.

PICTURE THE EXTREMES

Imagine trying to chop a tree with a splitting maul. The blade is too thick, so it will smash the wood fibers before it cuts them. What if you tried to cut down the same tree with a razor blade. The blade can penetrate the wood fibers, but is so thin, it will break on first impact. Conclusion: If a blade is too thick, even if the edge is sharp, it will take too much energy to penetrate. If the edge is too thin, it will break.

DIRECTION OF FORCE

Consider the direction of the force you are using. If the blade is thick and pointed, much of the force used to cut (penetrate) is used in pushing the material away from you. Little of the force is used in pushing the material aside, which is what you want. However, it is exactly the shape of a good cold chisel used to cut thin steel. The chisel requires so much force to cut, it must be struck with a hammer. An edge that was any thinner wouldn't endure for more than a few impacts of the hammer.

If the blade is very thin and comes to a point, it will penetrate easily, but the slightest sideways motion or hard obstacle will break the thin steel. This is the shape of a razor blade. It will cut hair (and skin) well enough, but it couldn't be considered for wood or bone. What should the angle be for sharpening a blade? Once you know the quality of steel in the blade and the material you are cutting, then you know the answer to that question.

THE RULE

Follow this rule: You want an edge thin enough to penetrate easily and thick enough to last a while. If you are sharpening often, you need to thicken the edge a little. If you are pushing hard and barely penetrating, then thin the edge a little.

MIXED MATERIAL

When we cut only soft wood or meat, it is easy to figure out how to sharpen an edge. However, wood has knots, and meat and fish have bones. If we sharpen a knife to cut fish and don't think of the bones, our knife will soon be dull.

WIDTH OF THE EDGE

The actual width of the edge is very important. If you were to look at a dull edge under a magnifying glass, it would look porous, like a sponge. You can visualize that downward pressure on the blade actually pushes the material away from the operator and doesn't penetrate, pushing the material aside. If the surface area of the edge can be reduced, the pressure required to penetrate is greatly reduced.

With a sharp edge the surface area is almost zero, and all the energy can be used in separating the material. This makes a tremendous difference if you are cutting fish all day. If the edge is broken, it obviously has a large surface area. One chip in a knife or axe can make a tiring difference. New axes or other tools always come with edges that are far too thick. You can thin the edges to your needs.

DIGGING TOOLS

A hoe, pick, or shovel should be sharpened. This makes digging much easier. The thickness of the edge is determined by the kind of dirt you are working with. If it is loose soil with no rocks, the edge can be thin to cut roots. If there are hard rocks in the soil, the edge should be thicker.

SHARPENED ON ONE SIDE

Some tools are sharpened on one side only. Old-timers used to sharpen their axes on one side for chopping and shaping boat and sled parts. Shovels, hand planes, drills, circular saw blades, and so on are all sharpened on one side only, as are traditional tanning and skinning knives. The reason for this is obvious when you think of it. Think of the forces acting on each of these blades. Which way will they tend to cut as they penetrate?

ROUGH OR SMOOTH

If an edge is rough, it will have considerable friction with the surface it is penetrating. When we cut wood, a very smooth surface makes entrance of the blade much easier.

Years ago, sharpening a Swede or two-man saw was an art that everyone had to master. Since the advent of chainsaws, it is but a memory, but there are principles involved that apply in other blades. A combination blade for a circular saw has the same two kinds of teeth as a Swede saw and cuts in an identical manner.

A Swede saw or two-man saw does two things in two directions.

1. It cuts the fibers. That is the purpose of the cutting teeth.
2. It chisels out the severed fibers. It is very, very important that the chisel teeth be slightly lower than the cutting teeth.

The purpose of the chisel teeth is to remove the severed wood fibers to make room for the cutting teeth to go deeper. If there were no chisel teeth, the cutting teeth would ride back and forth in a groove, unable to go deeper. Setting a saw means bending the tips of the teeth slightly outward so the cut is wider than the thickness of the blade.

If there is no set to a blade, the friction with the sides of the cut will tire the loggers very quickly. If the set is too wide, then the loggers work too hard removing more wood than is necessary.

It is important that the sides of the blade are smooth and rust free. The friction of a rusty saw blade in the cut of a tree is tremendous, especially when it is a pitchy spruce tree. Some people lubricate the saw with a bar of soap.

STONES, FILES, AND STEELS

There are three ways to shape a blade.

1. With a file. Files do a good fast job on softer steels.
2. With a stone. Stones do well on hard steels, but aren't as fast as a file.
3. With a sharpening steel. Steels put a good finish, cutting edge on a knife to be used for meat or fish. They don't remove much material from the blade. They shape and texture the edge.

Coarse or fine? How much steel needs to be removed? If there is lots of steel that needs to come off, then a coarse file or stone is faster. However, caution must be taken with electric grinding wheels. Friction overheats the steel so that it loses its temper, turning the edge blue and soft, causing it to dull easily. If there isn't much steel to remove, use a finer file or stone to put the finishing touches on the edge.

Care of Stones

Some people put oil on the hand stone to float out the ground steel filings, so they don't plug the stone. Other people use saliva. Whichever method you use, anything is better than letting the stone get glazed and plugged. It can't cut with the abrasive particles hidden under a layer of crud.

Care of Files

When filing, it is important to put pressure only on the forward stroke. The teeth are stressed to be strong in this direction. If pressure is applied on the backstroke, the teeth are damaged. Files can also rust. They must be protected from moisture. They also become dull when they contact other harder metals. Old-timers would wrap a file in an oily cloth to protect it from rust and contact.

Use a File or a Stone?

How hard is the steel? I prefer to use a file on softer steels. If the steel of the blade is as hard or harder than the file, the file will feel slippery on the blade surface. The result is a damaged file (expensive). Files are much faster than stones. I avoid knives and axes that are too hard. Granted, the harder steels keep an edge longer, but they are far more tedious to sharpen.

Hardened Steel

If an axe has soft steel, it is easy to sharpen. However, if someone uses it to chop wood and strikes a rock, the steel at the point of impact is hardened. When the individual goes to file the axe, the hardened spot will destroy the file within a few strokes. A sharp file is a thing of beauty. Careless people don't understand the damage they do to an axe by striking it in the dirt. The only way to sharpen an axe that has been damaged this way is to file where it is soft and use a stone where it has been hardened until the hardened spot is worn away.

SHARPENING A CHAINSAW

There is nothing mystical about sharpening a chainsaw. Like a Swede saw, it is cutting simultaneously in two directions.

- The side plate severs the fibers.
- The top plate chisels the fibers out of the cut.

Cutting blocks

The side plate angle is determined by what you are cutting. If you are cutting rather dirty wood, like driftwood, you might want to sharpen the side plate angle at about 25°. If you are cutting very clean wood, you might be able to sharpen to 35°. The thinner blade is more efficient, but it dulls easier.

The top plate angle is determined by the file size. A file too small will undercut the tooth, making it very thin. This will cut amazingly fast for a little while until it dulls.

Ripping

Many Alaskans make lumber with chainsaws. It is rough lumber and puts considerable wear on a saw, but in remote locations, there is no other lumber available. There are two main considerations in filing a ripping chain. The top plate angle makes a critical difference. The long shavings are being peeled out with the grain of the wood. If the file is too big and the angle too thick, ripping will be painfully slow. (The side plate angle isn't very important for ripping because few wood fibers are being severed.)

If the log is very clean, I use an undersized file. It gives quick, clean, long shavings (but dulls quickly if any dirt is encountered).

I often peel the trees before I rip them to remove the dirt hidden in the bark from floodwaters of the past. Old-timers used to peel the trees before they cut them down with Swede saws in order to extend the life of the sharpened blade.

RAKERS

The other main consideration in ripping has to do with the rakers—the metal on the other side of the notch from the cutting tooth. The rakers determine how deeply the chainsaw tooth cuts. If the rakers are too high, the operator has to push very hard to get the saw to cut. The increased friction of this effort wears the bar and chain quickly.

If the rakers are filed to the proper length, the weight of the saw is enough to feed the saw into the wood. If the rakers are too short, the saw will bite into the wood and get stuck as a result. This produces very rough lumber and excessive clutch wear. If the rakers are of even height, the cutting is smooth. If the rakers are uneven, some teeth are biting more than the others. The cutting is therefore very erratic, putting great stress on the chain.

Getting proper raker height isn't important if you are only cutting a few boards. If you are going to rip much lumber at all, it is critical to have them filed to the proper height. For cutting blocks, standard raker height is .025"—the tooth is .025" above the raker. For ripping, I have filed them .030" to .040".

Note: Temper means hardness of the steel. Steel can lose its temper or hardness if it is heated and cooled slowly. Temper is created by heating the steel to a given color—blue, red, white—and quickly quenching (cooling) it in cold water. This hardens the steel. Some steels are easy to temper; others, because of their low carbon content, are difficult or impossible.

BASIC TECHNICAL MATH

Stephen Davis

THE EFFICACY AND BEAUTY OF RIGHT ANGLES

Squaring Up a Rectangle

It is a valuable skill to be able to lay out a rectangular shape. Suppose you are building a foundation, a large sign, or a wall. It is easy enough to measure that opposite sides are equal, but that will give you only a parallelogram, a lopsided rectangle. Making one of the angles a right angle theoretically should make all four of the angles in your figure also be right angles and, hence, should make your parallelogram a rectangle. As you will see when you first try it on a large structure, that one right angle is no guarantee.

There is a fairly easy way to do it, however; if you have the opposite sides equal, then the lengths of the diagonals become important. If you look at the parallelogram in Figure 5, you will notice that the distances between opposite corners (distance *AC* and distance *BD*) are not equal. You can also visualize that as the parallelogram becomes more lopsided, those distances will become more disparate. Also, as the parallelogram becomes a rectangle, the diagonals will become equal. In the field, then, make sure that your diagonals are equal, and if your opposite sides are equal, then you will have a rectangle.

Triangulating a Structure

If you picture the two pairs of opposite equal sides of a parallelogram, it is easy to picture how flexible the parallelogram is. There is nothing to keep the figure from flattening out. The parallelogram becomes fixed only when the diagonals are inserted, when it becomes triangulated. Four-sided structures can move relative to themselves; three-sided figures cannot. (This principle underlies the surface rigidity of geodesic domes, which are made up of triangles.) Consequently, structures require triangular bracing to maintain their rigidity. When boards are used to side and roof houses, they add little rigidity, and angled braces need to be built into the walls and roofs. With the use of plywood and other sheet materials, which are essentially fixed planes or filled-in triangles, bracing is not as crucial, since the sheet material acts as inherent triangulation.

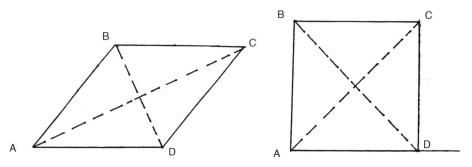

FIGURE 5
Right angles and parallelograms.

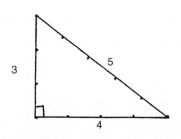
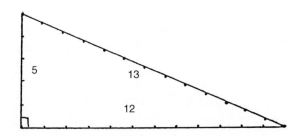

FIGURE 6
A right triangle—showing the Pythagorean theorem.

Constructing a Right Angle

Suppose that instead of making a rectangle you need to establish one accurate right angle (which would actually establish four right angles together)—For example, if you were at the four corners of Utah, Colorado, Arizona, and New Mexico, and the four governors wanted you to tell them exactly in which directions their state borders ran. Each one needs exactly a right angle. Due north you figured by the moss on the cacti. (And, shucks, you don't have a right triangle anywhere on you. Just a rope.)

This is one place where the Pythagorean theorem comes in handy. As the scarecrow in the *Wizard of Oz* said, "The square of the hypotenuse of a right triangle is equal to the sum of the squares of the other two sides." The hypotenuse is the long side, the side opposite the right angle. If you let C equal the hypotenuse and A and B be the two sides, then $C^2 = A^2 + B^2$. Also, at least in our Euclidean world, if that is true, then the triangle is a right triangle and contains a right angle.

The Egyptians apparently knew this and also knew that the simplest whole number Pythagorean triplet—that is, trio of numbers that fit the above equation—is 3-4-5. Check it out: $3^2 + 4^2 = 5^2$ (9 + 16 = 25).

If you decide on a convenient unit length for your rope (a cubit, for example) and knot it off into equal units, you can produce a right triangle with a right angle between the 3 side and the 4 side. See Figure 6.

As a variation of this process, suppose that Juan has a 5-meter rope and Lee has a 12-meter rope. They want to place them at right angles to each other. You decide that the ends of the two ropes should be a certain distance apart, if they are at right angles to each other. That distance? It would be C such that $C^2 = 5^2 + 12^2$, or $C^2 = 25 + 144 = 169$, or C = 13. If the ends of ropes are 13 meters apart, then they are forming a right angle.

TRIGONOMETRY BASICS

SOH-CAH-TOA

Trigonometry need not be intimidating. The word itself, although foreign to normal ears, means only measuring triangles. You can turn 90 percent of what you need to know about trigonometry into one easy mnemonic: SOH-CAH-TOA. (Think of banging your toe against a big construction triangle and needing to soothe it by soaking it.)

Trigonometry uses the basic measures: sine, cosine, and tangent, which are unique to particular angles, to determine something more about a given figure than is initially known. Reread this sentence when you finish this section.

Consider this mnemonic one piece at a time. For a triangle with a right angle (a right triangle), "SOH" means that the Sine of the angle being considered (here it is angle D) equals the Opposite divided by the Hypotenuse—see Figure 7.

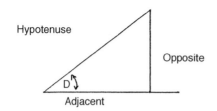

FIGURE 7   Sine, Cosine, and Tangent
• Sine is the opposite side divided by the hypotenuse.
• Cosine is the adjacent side divided by the hypotenuse.
• Tangent is the opposite side divided by the adjacent.

Realize that as a triangle gets bigger, its sides get longer, but its angles stay the same. It stays the same shape. If the angles stay the same, and each side grows or shrinks proportionately, then the ratio of corresponding sides will stay the same. The Opposite side divided by the Hypotenuse for a right triangle with a given angle will always have the same ratio. That ratio is called the sine, or sin.

Similarly, the next part of the mnemonic is "CAH." The Cosine of an angle equals the adjacent side divided by the Hypotenuse. See Figure 7 again for this relationship for angle D. Again, as the triangle grows or shrinks, the ratio will remain the same.

"TOA" means that the Tangent of angle D is the Opposite divided by the Adjacent. Once more, see Figure 7.

If you can use these relationships, then you will have the basics of trigonometry. Here is an example.

Suppose you are standing next to a river, and you need to know how far it is across the river (may be you have to design a bridge for the river). You are standing directly across from a tree on the edge of the river—Figure 8. You walk 100 feet down the river (conveniently straight) and look back at the tree. The angle to the tree from where you are is 30 degrees. You correctly deduce that the width of the river is the side opposite the 30-degree angle. The 10 meters that you walked is the adjacent side to the 30-degree angle. The ratio of opposite to adjacent is "TOA," or the tangent. The tangent of 30 degrees can be determined by a calculator or table of tangents to be equal to 0.58. You substitute in for the ratio tangent = opp/adj, which you conveniently rearrange to give opp = tangent × adj. and get opposite = .58 × 10, which equals 5.8 meters, the width of the river.

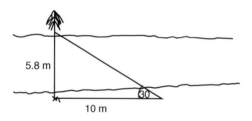

FIGURE 8
Finding the width of a river.

(*Note:* A 30-degree angle makes for a special 30-60-90 right triangle. The ratio of the sides of a 30-60-90 triangle are "1:2:square-root 3," a mnemonic of its own. The longest side is opposite the largest angle.)

Here is another example: You are standing 10 meters from a large tree—Figure 9. You measure the angle relative to the ground from your feet to the top of the tree. That measure is 70 degrees. The tangent of 70 degrees is 2.75. The height of the tree, then, is 10 meters × 2.75, or 27.5 meters.

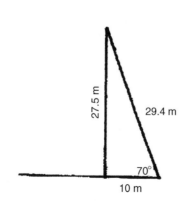

FIGURE 9
Finding the height of a tree.

TABLE 1
Basic trigonometric functions

Angle	Sine	Cosine	Tangent
5	0.09	0.99	0.09
10	0.17	0.98	0.18
15	0.26	0.97	0.27
20	0.34	0.94	0.36
25	0.42	0.91	0.47
30	0.50	0.87	0.58
35	0.57	0.83	0.70
40	0.64	0.77	0.84
45	0.71	0.71	1.00

You can also determine how far it is from your feet to the top of the tree by using the cosine. The cosine is the ratio of the adjacent leg of the triangle to the hypotenuse use. The cosine of 70 degrees is 0.34, so the ratio of the 10-meter distance that you are away from the tree to the distance from your feet to the top of the tree is 0.34. Using the equation

$$\frac{\text{Adjacent side}}{\text{Hypotenuse}} = \frac{\text{Distance to base of tree}}{\text{Distance from feet to top}} = \frac{\text{Meters}}{X} = 0.34$$

the distance to the top of the tree from your feet is 10 m/0.34 = 29.4 m.

Similarly, the trigonometric functions can be used in reverse to determine angles. If you can measure that the rise over the run of a stairway is 4.5 meters over 6 meters, then the tangent of the angle at the bottom of the stairs is 4.5/6.0, or 0.75. The angle whose tangent is 0.75 is just about 37 degrees, so you know that the stairs go up at a 37-degree angle.

As mentioned previously, primary trigonometric functions of the sine, cosine, and tangent can determine unknown components of triangles. The following table, Table 1, of trig functions (with interpolation where necessary) should serve for most purposes. Remember also that the $\sin A = \cos (90° - A)$—for example, $\sin 30° = \cos 60°$ and $\cos A = \sin (90° - a)$. Also $\sin A/\cos A = \tan A$—$\tan 45° = .71/.71 = 1$, and that $\tan (90° - A) = 1/\tan A$—$\tan 60° = 1/\tan 30° = 1/.58 = 1.72$.

CARPENTRY SQUARES

There are two basic types of carpenter squares that can help with framing measurements: the framing square and the "speed" square. Each has measurements embossed on the surface to make specific values easier to determine.

Framing Square

The framing square shown in Figure 10 is usually 16 inches by 24 inches. The blades of the square themselves are $1\frac{1}{2}$ inches and 2 inches wide to assist with those measurements. The 16-inch measure is chosen presumably because that is a standard distance between studs.

Framing squares have six rows of numbers along the blade, which can be used to determine the appropriate lengths of rafters, shown enlarged in Figure 11. The top row of numbers gives the length of the hypotenuse, or common rafter length, for a given rise or pitch of a roof. The pitch per foot is the inch measure at the edge of the square, and the rafter length is the first number under that value. For example, a pitch of 8 inches would give a rafter length of 14.42 inches for each foot of run.

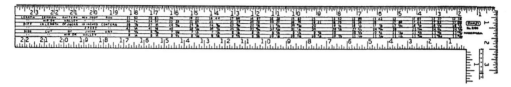

FIGURE 10
Framing square.

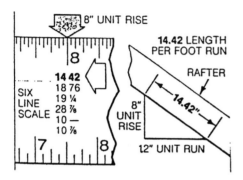

FIGURE 11
Rafter lengths shown on framing square.

The second row of numbers, similarly, produces the length of a hip or valley rafter length. A hip rafter is at the corner of a hip roof—see Figure 10 in Chapter 8. A roof that rises 8 inches in 12 will have a hip rafter length of 18.76 inches per foot of run; a roof that rises 10 inches in 12 will have a hip rafter length of 19.70 inches per foot of run.

The math is suitable for the metric system as well. If a roof rises 80 cm for each 120 cm, then the common rafter length will be 144.2 cm. If a roof rises 100 cm in 120 cm, then the hip rafter measure will be 197 cm.

The other rows of numbers on the framing square help with determining the lengths of other roofing members: jack lengths and side cuts of jacks and hips and valleys.

"Speed" Square

Although less precise, the smaller speed square can be used to determine the lengths of common rafters as well. Used in reverse, it can also be used to measure angles, if needed.

REFERENCES

Amary, A.B. (1976). *Carpentry for Builders.* New York: Drake Publishers.
Feirer, John, and Gilbert Hutchins. (1981). *Carpentry and Building Construction,* 2nd Ed. Peoria, IL: Chas. A. Bennett Co. Inc.
Koel, Leonard. (1985). *Carpentry.* Alsip, IL: American Technical Publishers.

MACHINE TOOLS

Based on "Introduction to Machine Tools,"
http://me.mit.edu/2.670/Tutorial/Machining/Description.html

Machine tools are the basis for manufacturing in the United States and Europe. They are essential for making replacement parts for most imported machines and other equipment operating at high speeds or to close tolerances. Useable but out-of-date machines are sometimes available to village artisans. Their operation can be learned fairly easily by observing, closely, an experienced machinist, but using a machine tool without training is risky. The tools tend to operate at high speeds; they are powerful; and they can do much damage to the operator, the work, and the machine itself if a mistake is made. This article describes some basic machines and what they do. Before doing things, learn more.

MEASUREMENT

A caliper measures lengths from 0 to 7.5 inches to the precision of thousandths of an inch. One can measure the outside of a piece, the inside of a hole or slot, or the depth of a hole or slot. Some calipers have a vernier scale; some have digital readouts. A micrometer provides greater precision but can measure a smaller range of lengths and is basically intended to measure outside dimensions. Micrometers have vernier scales.

BAND SAW

A band saw cuts metal or wood to a convenient size and to a desired shape. A typical band saw is about six feet high. The working part is a blade with sharp teeth in the form of a continuous band. The rest of a band saw is a metal table to hold the work and a motor to drive the blade. Blades do not always come in proper lengths, so many band saws are equipped with a welder to join ends of a new blade. The blade fits over pulleys, and the upper pulley can be lowered to put a new blade in easily. Band saws are equipped with a blade guard, which should be adjusted to cover as much of the blade as possible and still give access to the material being cut. Both the blade and the cutting speed should be chosen appropriately—depending on what is being cut. Many band saws have guidelines mounted on the case.

BELT SANDER

A belt sander removes burrs and rough edges. Wood, most metals, and some plastics can be smoothed and shaped by a belt sander. The working part is a belt—usually cloth—coated with abrasive. The rest of the belt sander again is a motor and a table to hold the work. Some belt sanders also have an abrasive disk powered by the same motor. The "sawdust" created by the belt sander can be toxic, so it is a good idea to wear a mask when using the sander. It is best if a way of collecting the dust is employed. The table should be adjusted so the gap between the table and the moving belt is less than the size of the work being sanded.

DRILL PRESS

A drill press is built around a rigid column that holds a table for the work being drilled and a motor driving a drill. A drill press might be five feet tall with a footprint of nearly a square yard or meter. The drill moves up and down, either manually or controlled by the press. A drill press is preferable to a hand drill when the location and direction of the hole must be controlled accurately. As it is sometimes difficult locating the hole accurately because conventional drill bits will bend slightly under pressure, a center drill with a thick shaft is used in starting a hole.

A tap is used to cut screw threads inside a hole. It can be used manually or with a drill press. Tapping is a delicate operation—if done manually, one should go slowly. Make sure the hole is the proper size for the tap. Taps are hard and, therefore, brittle—treat them carefully.

GRINDING AND BUFFING

A grinding wheel is needed in a machine shop to sharpen tools—drill bits, for example. They are meant to be used with hard steel. Soft material such as aluminum and plastic will coat the wheel and the abrasive will not be effective. If this does happen, the wheel must be "dressed"— see a machine shop handbook. A deburring wheel is a mesh of abrasive fibers held together by adhesive. It is used to polish metal, even remove tool steel stuck on the piece.

Grinding wheels often have a cloth buffing wheel on the same shaft. Buffing wheels put a high polish on a metal piece. An abrasive must be applied to the cloth before use and periodically during use.

LATHE

A lathe fabricates parts that have a circular cross section—such as screw threads on a shaft. A modern metal lathe is approximately one yard (meter) wide by two yards (meters) long. The part being fabricated is rotated against a stationary cutting tool. A handmade lathe is shown in the Household Technologies chapter. The working parts of a lathe are a spindle that holds the material being cut and a carriage that holds the cutting tool. The carriage can move along the length of the material, as when screw threads are being made. Often the operator of a wood lathe cutting legs for a chair, for example, simply moves a chisel—actually a gouge with a semicircular cross section—by hand. Metal working lathes usually have power feeds. In order to do precision work, metal lathes are built very solidly to minimize vibration or bending of the tools or the work. Various kinds of spindles are available. Several types of cutting tools are also available—the choice depends on the metal being cut and the smoothness of the finish needed.

A lathe can also be used for drilling—the drill is stationary and the piece rotates. The hole can only be at the center of the rotation. Lubrication should definitely be used.

MILLING MACHINE

A milling machine removes metal by rotating a multitoothed cutter that is fed into a moving work piece. Different configurations are possible: a common one has a vertical spindle, like a drill press. Such a milling machine differs from a drill press in that different-shaped cutters are used and the table holding the work moves under power. The cutters look like thick circular saw blades, and the cutting edges are on the side rather than the end, as on a drill. Both the rotating speed of the cutter and the feed speed—how fast the work is moved past the cutter—must be chosen appropriately. A machinist's handbook or the manufacturer's instructions will give advice. Good milling requires accurate placement of the piece on the table.

SHEET METAL TOOLS

Thin sheet metal can be cut or bent with a shear or brake. (A brake that can be built and used in a small workshop is described in another article in this chapter.) The working part of a shear is a heavy blade that slices the sheet metal. The shear has a heavy frame to hold the material. Usually the shear is operated by pressing down with the foot. A notcher cuts notches in sheet metal so the metal can be folded to create boxes or bread pans. It can be smaller than a shear because not as much metal is cut at a time. A brake bends sheet metal. A brake might occupy an area of five feet by four feet. It is a strong frame to hold the sheet metal and a moving piece, which pushes down and bends the metal. It is necessary to bend the work slightly beyond the desired angle, since the metal will "spring back" after bending.

BLACKSMITHING

Based on "Farm Shop Practice," War Department Education Manual, EM 862

Blacksmithing is an important art in much of the Third World. Figure 12 shows farm tool blades being forged by village craftspeople. Blacksmith shops make other farm implements. They also make products sold in local and urban markets, such as door and window frames, decorative lattices, and burglar bars. Welding is a common and essential business in many places. Where electricity is available, electric welders—see the reference to Silveira House—make sense. Otherwise, gas welders or forges are used.

TOOLS

Blacksmiths need a forge, where the fire burns. Coal is the most common fuel. Bellows are needed to obtain a suitably high temperature. Mechanical fans can be used. Manually operated water-sealed bellows are an alternative. Bellows can also be made of leather and wood.

Anvils, like the one shown in Figure 13, are common in the United States. If such is not available, a piece of rail from a railroad can substitute.

At least one or two pairs of tongs will be needed. Various types are available, but the hollow-bit, curved-lip bolt tongs are probably the most useful. Tongs are shown in Figure 14. The styles in Figure 14 are very good for the farm shop. Flat bars as well as round rods and bolts can be held in them. If tongs cannot be found to fit the work, a pair should be reshaped by heating and hammering the jaws over the piece to be held. Poorly fitting tongs are a source of continual trouble and should not be used. A blacksmith's hand hammer weighing $1\frac{1}{2}$ or 2 pounds and another weighing 3 or $3\frac{1}{2}$ pounds will handle all ordinary work very satisfactorily. An assortment of hand cold chisels and punches should also be available. A large vise is useful, as well as a metal rule.

THE FORGE FIRE

A good fire is the first requirement for good blacksmithing. Many beginners do poor work simply because they do not recognize the importance of a good fire. A good forge fire has three characteristics: It is clean—that is, free from clinkers, cinders, and so on; it is deep, with a big center of live burning coke; and it is compact, being well banked with dampened coal. From time to time—usually every half hour when welding—the clinkers and cinders should be removed. This can be done by passing the shovel along the bottom of the hearth

HEATING THE IRONS

To heat irons in a forge, they should be placed in the fire in a horizontal position, not pointing down. There should be burning coke below the irons, on both sides of them, and on top of them. Irons heated in a deep, compact fire heat much more rapidly and oxidize or scale off less than when heated in a shallow, burned-out fire. Some scale will form in spite of a good fire, but it

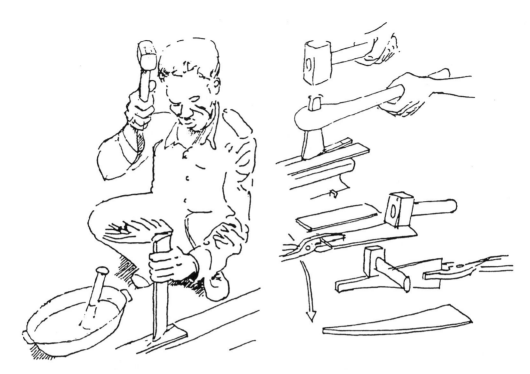

When cutting hot metal, the cutter blades must be dipped in water to keep the temper. Old shafts, heavy coil or leaf springs make good cutters. On the left, Ba Kabuswe is cutting with a leaf spring cutter. On the top right, an old axe is used.

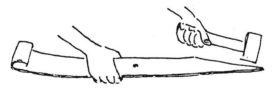

It is easier to wrok on the blade if it is still part of the whole leaf spring.

The profile of the blade can be adapted for different jobs. The rounded blade is more ideal for cutting branches and trees and flat surfaces.

The flatter head is good for trimming tool handles...

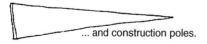

... and construction poles.

Once cut off the blade can be held in a short piece of stick, an alternative to using tongs or pliers. When blunt, such tools are drawn out by a blacksmith several times in their life, which also allows them to be retempered.

The blade of an axe is straight...

... whereas adzes have a slightly curved blade.

FIGURE 12
Forging farming tools. (From *Inventors of Zambia*.)

should be kept to a minimum. A good black-smith keeps the scale brushed from the face of the anvil with her or his hands.

Small, thin parts heat much more rapidly than heavier and thicker parts. To prevent burning the thinner parts, they may be pushed on through the fire to a cooler place or the position of the irons otherwise changed to make all parts heat uniformly. Mild steel should be heated to a good, bright-red heat for forging. It should not be allowed to get white hot and sparkle, since it is then burning.

A considerable amount of work can be done without tongs. An eyebolt, for instance, can be made on the end of a rod 20 or 30 inches long and then cut off when finished.

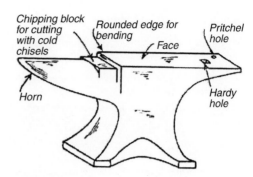

FIGURE 13
Parts of an anvil.

FUNDAMENTAL FORGING OPERATIONS

Forging may be defined as changing the shape of a piece of metal by heating and hammering. All the various operations that a blacksmith performs in forging iron may be classified into a surprisingly small number of fundamental or basic processes. Once these are mastered, the beginner is well on her or his way to success, and he or she can do practically any ordinary piece of forge work. These fundamental operations are the following.

1. Bending and straightening
2. Drawing, or making a piece longer and thinner
3. Upsetting, the opposite of drawing, or making a piece shorter and thicker
4. Twisting
5. Punching

Other operations commonly done by a blacksmith, but that are not strictly forging, are welding, tempering, drilling, threading, and filing.

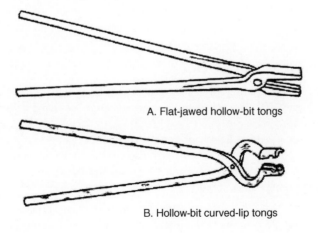

A. Flat-jawed hollow-bit tongs

B. Hollow-bit curved-lip tongs

FIGURE 14
Types of tongs.

Bending and Straightening

In bending at the anvil (see Figure 15) two things are most important.

1. Heat the iron to a good, bright-red heat, almost but not quite white hot, throughout the section to be bent.
2. Use bending or leverage blows—not mashing blows.

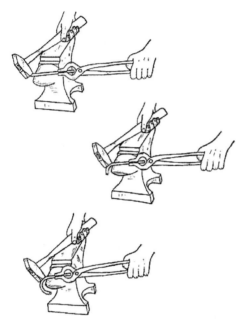

FIGURE 15
Bending.

The iron should be so placed on the anvil and so struck that it can bend down under the hammer blow without being forced against the anvil and mashed. If the iron is struck at a place where it is resting firmly on the anvil, it will be mashed instead of bent. A few moderately sharp blows are better than several lighter blows.

Abrupt square bends can be made over the face of the anvil where the corner of the anvil is rounded to prevent marring or galling the iron. To make a uniform bend in the end of a rod, strike the part that projects beyond the horn, and keep feeding the rod forward with the tongs as the bending progresses. Keep the iron at a good working heat and do not strike the rod where it rests on the horn.

Care should be taken to keep the iron at the proper bending heat. If it gets below a red heat, it should be put back in the fire and heated again. To bend a piece at a certain point without bending the adjacent section, the piece may be heated to a high red heat and then quickly cooled up to the point of bending by dipping in water. Bending is then done quickly by hammering or other suitable methods.

Straightening can usually best be done on the face of the anvil. The stock should always be firmly held and then struck with the hammer at points where it does not touch the face. Sighting is the best way to test for straightness and to locate the high points that need striking.

Bending and Forming an Eye

One of the most common bending jobs in the blacksmith shop is that of forming an eye on the end of a rod (see Figure 16). The following is a good method of making such an eye.

1. Heat the rod to a good red heat back for a distance of about 5 to 8 inches, depending on the size of the eye. Place the well-heated iron across the anvil with enough stock projecting over to form the eye. Where the eye must be made accurately to size, use a metal rule or square for measuring. Work rapidly.
2. Bend the projecting portion down, forming a right angle.
3. Finish the right angle bend by striking alternately on top and on the side, keeping the iron at a good working heat all the while. If the square bend at the juncture of the stem and

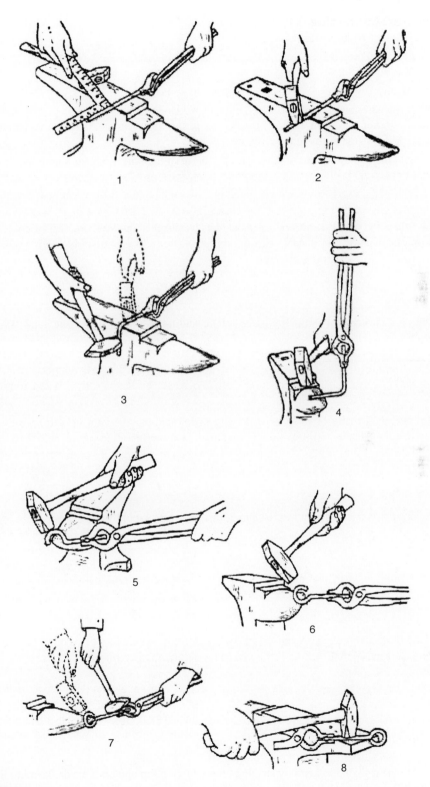

FIGURE 16
Steps in making an eye.

eye tends to straighten out, it is an indication that the end of the stock is not being kept hot enough while being bent.

4. Start bending the tip end around the horn, being careful to strike "overhanging" or bending blows.

5. Gradually work back from the end to the square bend.

6. Turn the eye over and close it up. Exert considerable backpull on the tongs to keep the upper part of the eye up off the horn. In this position the hammer can strike bending blows instead of flattening or mashing blows.

7. Round the eye by driving it back over the point of the horn. Carefully note where the eye does not touch the horn, and strike down lightly in these places. Keep the iron well heated.

8. To straighten the stem of an eye, place it across the corner of the anvil face and strike the high points while the iron is at a good working heat. The stock should be well heated at the juncture of the stem and eye, but the eye itself should be practically cold. Such a condition can be produced by heating the whole eye and then quickly cooling most of the rounded part by dipping in water.

Drawing

Drawing is the process of making a piece longer and thinner. Two important points should be kept in mind while drawing.

1. The iron must be kept at a good forging heat, a high red or nearly white.
2. Heavy, straight down, square blows should be struck.

Many beginners make the mistake of striking a combination down-and-forward pushing blow, thinking that the pushing helps to stretch the metal. Hammering after the red heat leaves is hard work and accomplishes little. Also, the iron is apt to split or crack if hammered too cold.

Pointing a Rod

If a round point is desired on a rod, a square, tapered point should first be made. It is then easy to make it eight-sided and finally round. In pointing a rod or bar, raise the back end, tilt the toe of the hammer down, and work on the far edge of the anvil. Round points should be made square first, then eight-sided, and finally round (see Figure 17).

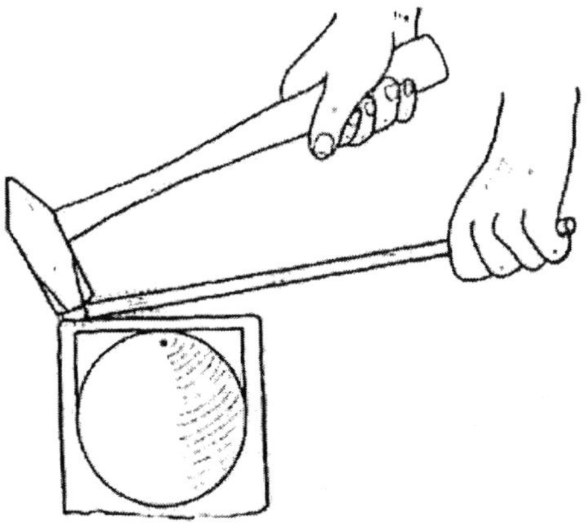

FIGURE 17
Pointing a rod.

Upsetting

Upsetting is simply the reverse of drawing, or the process of making a piece shorter and thicker (see Figure 18). It is done when more metal is needed to give extra strength, as when a hole is to be punched for an eye. There are two main points to be observed in upsetting.

1. Heat the bar or rod to a high red or nearly white heat throughout the section to be upset.
2. Strike extremely heavy, well-directed blows.

To insure success in upsetting, work the iron just under a white heat and strike tremendously heavy blows. Light blows simply flare the end without upsetting very far back from the end.

Probably the best way to upset a short piece is to place the hot end down on the anvil and strike the cold end. The hot end, of course, may be up, but it is usually easier to upset without bending if the hot end is down. If the bar starts to bend, it should be straightened at once. Further hammering will simply bend it more instead of upsetting it.

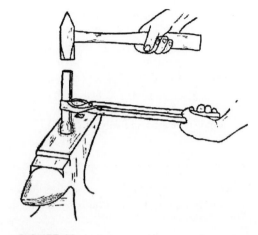

FIGURE 18
Upsetting.

In order to heat thoroughly the part to be upset and yet confine the heat to this part, it is sometimes better to heat the work somewhat further than the upsetting is to go and then cool it quickly back to the line of upsetting by dipping in the water.

The end of a long bar may be upset by laying it on the anvil face with the hot end projecting beyond the edge and striking heavy blows endways with the hammer. If the bar is long and heavy enough, it may be upset easily by ramming the hot end against the face or the side of the anvil.

Striking with the Hammer

Success in blacksmithing depends largely on the ability to strike effectively with the hammer. Most blacksmithing requires heavy, well-directed blows. Where light blows are better, however, they should be used. Light blows are struck mostly with motion from the wrist, while heavier blows require both wrist and elbow action. Very heavy blows require action from the shoulder in addition to wrist and elbow motion. To direct hammer blows accurately, strike one or two light taps first to get the proper direction and feel of the hammer and then follow with quick, sharp blows of appropriate force or strength. It is also important to use a hammer of appropriate size. A heavy hammer on light work is awkward, and blows cannot be accurately placed. Using a light hammer on heavy work is very slow and tedious.

FORGING AND TEMPERING TOOL STEEL

Tool steel for making cold chisels and punches and similar tools may be bought from a blacksmith or ordered through a hardware store, or it may be secured from parts of old machines, such as hay rake teeth, pitchfork tines, and axles and drive shafts from old automobiles.

Nature of Tool Steel

The smaller the size of the grains or particles in tool steel, the tougher and stronger it is. When tool steel is heated above a certain temperature, called the critical temperature, the grain size increases. (The critical temperature is usually between 1,300° and 1,600°F, depending on the carbon content and for practical purposes is indicated by a dark red color.) If the steel is heated only slightly above the critical temperature, the fine-grain size may be restored by allowing it to cool slowly and then reheating it to just the critical temperature. If the steel is heated to a white heat, however, the grain size will be permanently enlarged and the steel damaged or possibly ruined. If tool steel is hammered with heavy blows while it is just above the critical temperature, the grain size will be made smaller and the steel thereby refined and improved. It is evident, therefore, that a piece of steel may be improved or damaged or even ruined, depending on how it is heated and forged.

Forging Tool Steel

Since the making of a satisfactory tool depends so largely on the proper heating and handling of the steel, the following points should be kept in mind when forging with it.

1. Tool steel has a much narrower range of forging temperatures than mild steel. Hammering below a red heat may cause cracking or splitting, while temperatures above a bright red or dark orange may damage the grain structure.
2. Tool steel should always be uniformly heated throughout before it is hammered. Otherwise the outside parts, which are hotter, may stretch away from the inside parts, which are colder and thus cause internal flaws.
3. Very light hammering should be avoided, even when the steel is well heated, because this may likewise draw the outer surface without affecting the inner parts.
4. As much of the forging as possible should be done by heavy hammering at a bright red or dark orange heat—slightly above the critical temperature—because this will make the grain size smaller and thus refine and improve the steel.
5. When a piece is being finished and smoothed by moderate blows, it should not be above a dark red heat.

Annealing Tool Steel

After a tool has been forged, it is best to anneal it, or soften it, before hardening and tempering. This is to relieve any strains that may have been set up by alternate heating and cooling and by hammering. Annealing is done by heating the tool to a uniform dark red heat and placing it somewhere out of drafts, as in dry ashes or lime, and allowing it to cool very slowly. (Copper and brass may be softened by heating to a red heat and plunging quickly into water.)

Hardening and Tempering Tool Steel

If tool steel is heated to a dark red or the critical temperature and then quenched (cooled quickly by dipping in water or other solution), it will be made very hard, the degree of hardness depending on the carbon content of the steel and the rapidity of cooling. The higher the carbon content, the harder it will be, and the more rapid the cooling, the harder it will be.

A tool thus hardened is too hard and brittle and must be tempered, or softened somewhat. This is done by reheating the tool to a certain temperature (always below the hardening temperature) and quickly cooling it again. The amount of softening accomplished will depend on the temperature to which the tool is reheated. For practical purposes, these temperatures are judged by the color of the oxide or scale on the steel as it is being reheated. A straw color, for example, indicates that the tool has been reheated to a comparatively low temperature, and if

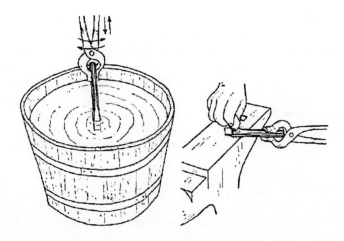

FIGURE 19
Tempering a cold chisel.

quenched on a straw color, it will be rather hard. A blue color, on the other hand, indicates that the tool has been reheated considerably higher and, if quenched, will be softer.

Hardening and Tempering a Cold Chisel

After a cold chisel is forged and annealed, it may be hardened and tempered (see Figure 19) as follows.

1. Heat the end to a dark red, back two or three inches from the cutting edge.
2. Cool about half of this heated part by dipping in clean water and moving it about quickly up and down and sideways until the end is cold enough to hold in the hands.
3. Quickly polish one side of the cutting end by rubbing with emery cloth, a piece of an old grinding wheel, a piece of brick, or an old file.
4. Carefully watch the colors pass toward the cutting end. The first color to pass down will be yellow, followed in turn by straw, brown, purple, dark blue, and light blue.
5. When the dark blue reaches the cutting edge, dip the end quickly into water and move it about rapidly. If much heat is left in the shank above the cutting edge, cool this part slowly so as not to harden the shank and make it brittle. This is done by simply dipping only the cutting end and keeping it cool—while the heat in the shank above slowly dissipates into the air.
6. When all redness has left the shank, drop the tool into the bucket or tub until it is entirely cool.

WELDING

A good fire is the first requirement for welding. It is important for any blacksmithing work, but for welding it is indispensable. Ends to be welded together should first be properly shaped or scarfed. Scarfed ends should be short, usually not over $1\frac{1}{2}$ times the thickness of the stock. They should have rounded or convex surfaces, so that when they come together any slag or impurities will be squeezed out rather than trapped in the weld. Long, thin, tapering scarfs are to be avoided because they are easily burnt in the fire and because they cool and lose their welding heat very rapidly when removed from the fire, thus making welding exceedingly difficult.

In order to counteract the wasting away of the irons by scaling and the tendency to draw out from hammering when they are welded, the ends are commonly upset before scarfing. Scarfs on the ends of bars are made by working on the far edge of the anvil, striking backing-up or semi-upsetting blows with the toe of the hammer lower than the heel.

Welding Flux

Borax or clean sand, or a mixture of the two, may be used as a welding flux. Commercial welding flux, however, such as may be bought from hardware stores, is usually more satisfactory, and since but a little is needed, it is probably best to buy a small package. Flux is applied to the pieces to be welded after they are at a red or white heat and just before the welding heat is to be taken. It covers the irons and causes the oxide to melt at a lower temperature. The oxide must be melted before the irons can be welded.

Heating the Irons

The irons should be heated slowly at first, so they will heat thoroughly and uniformly throughout. After the irons reach a bright red heat, remove them and dip the scarfed ends into flux, or sprinkle the flux on them with your fingers. Replace the irons in the fire and continue to heat, being careful not to brush the flux off the irons before it melts. Pull a few lumps of coke on top of the irons and raise the coke occasionally with the poker to see how the heating is progressing.

Care must be taken to see that both irons reach the welding heat at the same time. If one heats faster, pull it back into the edge of the fire for a few seconds. During the last part of the heating, have the scarfed sides of the irons down so they will be fully as hot as the other parts of the pieces. When the irons reach the welding temperature, they will be a brilliant, dazzling white. Their surfaces will appear molten, much like a melting snowball, and a few explosive sparks will be given off.

Welding a Link or Ring

To make a link or ring, the stock is first heated and bent into a horseshoe or U shape. The ends are then scarfed by placing on the anvil, with one end diagonally across the shoulder between the anvil face and the chipping block, and with the other end against the vertical side of the anvil. A series of three or four medium or light blows are struck on the end on the shoulder, swinging the tongs a little between each blow. In this manner the end of the U is given a short, blunt, angling taper with a slightly roughened surface. The piece is then turned over and the other end scarfed in the same manner. The scarfs may be finished by striking lightly with the cross peen of the hammer (see Figure 20).

The legs of the U are next bent over the horn, lapped together, and hammered shut. It is important that the link or ring be somewhat egg-shaped at this stage—not round. The ends

FIGURE 20
Steps in making a link.

should cross each other at an angle of about 90 degrees. This insures plenty of material at the joint for finishing the link and prevents a thin, weak section at the weld.

The link is then placed in a good welding fire and heated, flux being applied after a red heat is reached. The link may need to be turned over in the fire a time or two in order to ensure even heating. When the welding heat is reached, the work is quickly removed from the fire, given a quick rap over the anvil to shake off any slag or impurities, and then put in place on the face of the anvil and the ends hammered together. The link is struck two or three quick, medium blows on one side, then turned over and struck on the other side.

Medium blows are used because the iron at welding heat is soft, and heavy blows would mash it out of shape. Forcing the parts firmly together is all that is required. It is essential to work fast before the iron loses the welding heat. A second or even a third welding heat may be taken if necessary to completely weld the ends down.

After the ends are welded together, the link is finished by rolling it slowly on the horn (by twisting or swinging the tongs back and forth) while hammering rapidly with light blows.

Welding Rods or Bars

To weld rods or bars, it is best to upset the ends somewhat before scarfing. The scarfs should be short and thick with rounded convex surfaces (see Figure 21). The irons are fluxed and brought up to the welding heat in the usual manner. When they reach the welding heat, they are removed from the fire, struck quickly over the edge of the anvil to shake off any slag or impurities, and put in place on the anvil and hammered together first on one side and then the other with light or medium blows, followed by heavier ones. After the first blow or two to stick the irons, the ends of the scarfs should be welded down immediately because they are thin and lose their welding temperature rapidly.

Figure 22 shows how to weld irons together. Steady the irons over the edges of the anvil, the one in the left hand being on top. Gradually raise the hands until the iron in the left hand holds the other one against the anvil while the right hand strikes with the hammer.

Finishing the Weld

If it is not possible to get all parts welded down at the first heat, then flux is reapplied and another heat taken. Once the pieces are stuck well enough to hold together, however,

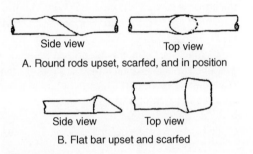

Side view   Top view

A. Round rods upset, scarfed, and in position

Side view   Top view

B. Flat bar upset and scarfed

FIGURE 21
Welding rods and bars.

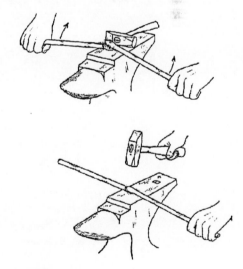

FIGURE 22
Welding irons.

they are much more easily handled. In taking an extra heat to weld down a lap, the lap should be on the underside in the fire just before removing. This ensures thorough heating.

If the irons do not stick at the first attempt, do not continue hammering but reshape the scarfs and try again, being sure that the fire is clean and that it is deep and compact. Irons will not stick if there is clinker in the fire or if it has burned low and hollow. Be sure, also, that the irons are brought well up to the welding temperature. It is generally not possible to make irons stick after two or three unsuccessful attempts because they will most likely be burned somewhat, and burned irons are difficult or impossible to weld. In such cases the ends should be cut off and rescarfed.

REFERENCE

Murewa Silveira Blacksmiths Shop, PO Box 545, Harare. Tel.: 491066/7. silveira@samara.co.zw.

FOLDING MACHINE FOR SHEET METAL WORK

Based on How to Make a Folding Machine for Sheet Metal Work, *Rob Hitching, Intermediate Technology Publications, London, 1999*

A machine that folds metal and can be made in a village workshop is shown in Figure 23. This machine can increase the range of products that a village metal worker can produce. The machine can be made in a small-scale shop by the person who will use it. Machines like this can improve greatly the capability of small shops.

FIGURE 23
Sheet metal folding machine.

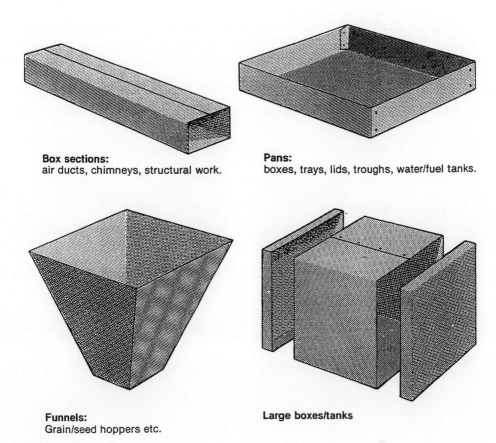

Box sections:
air ducts, chimneys, structural work.

Pans:
boxes, trays, lids, troughs, water/fuel tanks.

Funnels:
Grain/seed hoppers etc.

Large boxes/tanks

FIGURE 24
Items made by the folding machine.

The machine can bend sheet metal to produce the shapes shown in Figure 24. It will bend metal of 16 gauge or less, corresponding to 1.5 mm thick, and 1 meter wide.

The machine can be built in a small workshop using locally obtained materials. The only tools necessary are a drill, an electric welder, clamps, and basic hand tools. The parts of the machine are shown in Figure 25. Complete plans are in the booklet *How to Make a Folding Machine for Sheet Metal Work*, from ITDG in London.

WIRE FENCE–MAKING MACHINE

Based on Rural Industries Innovation Center, Catalogue of Goods and Services, *3rd Edition, RIIC, Private Bag 11, Kanye, Botswana*

Chain-link fences are very common in the United States—used usually for security purposes, to keep out intruders. Such fences are less common in the Third World but could be a worthwhile substitute for pole and brushwood fences used to keep animals out of gardens, especially where local vegetation needs protection. Motorized machines to make such fences are available from several organizations. The machine described here and shown in Figure 26 is produced in Botswana. It, or one like it, might be the basis of a small-scale business.

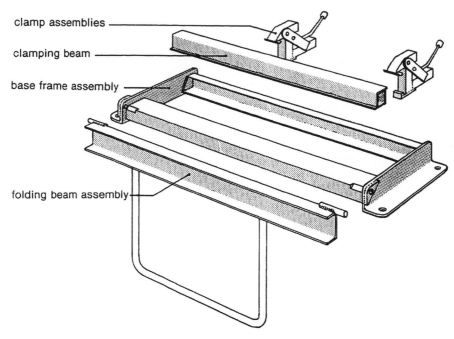

clamp assemblies

clamping beam

base frame assembly

folding beam assembly

FIGURE 25
Disassembled machine.

FIGURE 26
Wire fencing making machine made at RIIC in Botswana. (From RIIC *Catalogue of Goods and Services*, 3rd Edition.)

MOTORIZED DIAMOND MESH WIRE MACHINE

The machine produces wire mesh for fencing in either 1.2 m or 1.8 m widths—the corresponding maximum fence height is 1.8 meters. The nominal mesh size is 75 mm. The input to the machine is soft 2 mm wire. From a 50 kg wire roll approximately 60 m of 1.8 m fencing is produced. The electric input power required is 220 V, single phase. The machine is approximately 2.5 meters long, 1 meter wide, and 1.5 meters high. It is easily transportable and requires low maintenance. Only minimum training is needed for the single operator.

The actual mesh-weaving device is surrounded by panels for safety. The weaving mechanism features a cooling/lubrication system, which minimizes wear of the blade and coil. The electric motor connects to the weaving mechanism through a foot-operated clutch. The main drive shaft is supported by two ball bearings for quiet operation and reliable service.

CHAPTER 10

Other Technologies

Edited by BARRETT HAZELTINE
Brown University

OTHER TECHNOLOGIES

Three technologies are discussed in this chapter, together with some notes about the use of computers in the Third World. The technologies are: provision of telephone service to villages, land mine detection, and small-scale mining.

In Europe and the United States, cell phones are inexpensive and the coverage extends nearly everywhere in the country. In the Third World, cell phones are moderately expensive and coverage is spotty. Where coverage exists, cell phones can make a significant improvement to the lives of village people. One person in a village owning a cell phone can serve as the equivalent of a telephone company, answering the telephone, finding the called party, or taking a message, connecting wage earners in cities with their families in villages. However, because the density of traffic in rural areas in the Third World is low, so it will be a long time before either cell phones or land lines are generally available there. Two problems with landlines are that copper wire is prized for jewelry and that large animals—like elephants—knock over the poles over. The article describes alternatives less expensive than either cell phones or landlines.

Unfortunately, many buried land mines remain in the Third World and serious accidents continue to occur. In some places large areas of arable land are not usable because they contain land mines. One of the articles in this chapter reviews the technology of land mines and another discusses a way of detecting their presence.

Many small-scale mines are functioning in the Third World. They are often dangerous and polluting but earn cash for people who have only worse alternatives. The article gives the rudiments of the technologies used and suggests ways of making the process more benign.

Computers are potentially of significant value in the Third World, and it is recommended that volunteers bring laptops with them. Volunteers should know word processing and use of spreadsheets. These can be learned quickly; while an instruction book can be helpful, many people have learned to use both word processors and spreadsheets by experimenting. Basic modern software is nearly always user-friendly. It is advisable to bring backup disks of all software, including the operating system, and any manuals supplied with hardware or software. The number of people one can turn to for advice about computer problems will be much fewer than in the United States and Europe; the author, however, has found some very capable computer experts in the Third World who are often associated with universities.

Connecting a computer or other electrical device to power is not difficult overseas, but does require attention. The plugs going into electrical outlets will usually be a different shape from those in the United States. One can buy plug adaptors in the United States or in cities worldwide. It is probably cheaper, though, to purchase plugs and replace the ones on the equipment—a knife and screwdriver will be required. Of course, replacing a plug does not affect the amplitude of the voltage, which is 240 volts in much of the world but 120 volts in the United States. Generally, electrical appliances will be damaged if plugged into the wrong size voltage. Most laptops, however, can be powered with either 120 volts or 240 volts, so one can usually use the voltage available. It goes without saying, however, that one should be prudent and check the label on the power supply. If it does not indicate "INPUT 100–240 volts," or similar, be careful. Printers, however, are normally designed for either 120 or 240 volts so the input voltage will probably have to be changed to use a printer designed for the U.S. in the Third World. Big cities in the Third World usually have electrical supply stores where suitable voltage adapters can be purchased. Transformers are used to convert from one voltage to another and are the basis of the best adaptors. The current rating of the adaptor should be higher than the nominal current specified on the printer because peak current demand can be several times the nominal current. Lightly loaded electrical power systems—as many are in the Third World—are likely to have large voltage surges that can damage a computer. A surge protector is worth having. It is also wise to unplug a computer during a lightning storm.

The Internet is a valuable resource both in the Third World and the U.S., but not available in much of the world, and when nominally available may be too slow to be really useful. The bandwidth of the telephone lines is often too small. E-mail is more likely to be available if one has access to the local telephone service. A volunteer may have to do some searching to find an e-mail provider but most volunteers will agree the search is worth it. The details of making a connection will depend on local conditions. It is safest to have a widely used modem and computer, that is, one made by a well-known company, so the appropriate software can be easily installed.

TELECOMMUNICATION SERVICES IN RURAL DEVELOPMENT CENTERS

Harry S. H. Gombachika

ABSTRACT

Provision of telecommunication services has long been identified as one of the factors that contribute toward development. In developing countries, telecommunication development programs for rural areas have been highly controversial in terms of whether they should receive high priority or not. Financial return on investments is still the important determining factor to justify telecommunication programs. Rural areas are characterized by low financial return on investments; as such they receive low priority in telecommunication development programs. Studies have indicated that telecommunication services contribute toward development of rural areas, although the contribution cannot easily be quantified in financial terms. It is time for rural areas to receive some priority in telecommunication infrastructure development. This can be realized by considering technology opportunities that can be adopted to develop economically optimum telecommunication services. This paper describes and discusses technologies that can be adopted for telecommunications services appropriate for rural development centres. It conceptually examines implementation methodologies, highlighting technical challenges that offer areas of research.

INTRODUCTION

The need for telecommunication facilities in rural areas of developing countries has been justified in many studies. Development of telecommunication systems acts as a catalyst for economic development, an effect that is more pronounced in rural areas of poor countries than urban areas. Telecommunication facilities act as a linkage in international trade through development of exports from rural areas, besides the export from industrial zones, and conceptually enable equal distribution of national income. These facilities bring together people in rural and those in urban areas—for example, sharing information through the telephone. This is one of the factors that reduce the need for people to migrate to urban centers. Telecommunication facilities have been used in medicine and distant education through telemedicine and tele-education. Some studies indicate that telephone facilities evidently reduce business enterprise costs in the following areas: business expansion, managerial and labor time, inventory level, production stoppage, vehicle fleet scheduling, purchasing, selling, and supply. Transport needs and energy consumption are reduced through technologies like teleconferencing and telecommuting (Yusuf 1991; Hudson 1989; Bulter 1983).

These studies clearly demonstrate that telecommunication development plays a major role in the economic and social development of rural areas. However, provision of telecommunication facilities in rural areas of developing countries is still controversial. Many have argued that the benefits accrued in rural areas do not economically justify the cost of installing and maintaining telecommunication facilities due to long distance and small population density. Investment in telecommunication is considered to benefit only the affluent urban segment of society and hence is less important among public utilities. There is an apparently low demand for telephone services. Although the need for telephone services is high, this is not translated into demand because many people in rural areas are poor with low or no income. The development of telecommunication services has also been retarded because of some criminal acts in other rural areas as telephone lines run through vast areas unguarded. These unguarded lines are vulnerable to the acts of vandalism—for example, telephone poles are cut down (Ng'-eny 1989).

It is a technical challenge to provide efficient and reliable telecommunication facilities to rural areas cost effectively. Tremendous efforts have been made to develop telecommunication facilities to be deployed in rural areas. These efforts have been concentrated on the use of satellite technology (FWTF 1983). This technology requires the use of space satellites. Poor developing countries do not have the technical and financial capacity to launch and manage these satellites. They have to buy access time from organizations operating from industrialized countries, using foreign reserves. This may prove difficult for such countries whose foreign reserves are meager. Technologies must be explored that comply with the requirements of rural areas using mostly the available resources. This paper describes such technologies that can be adopted and adapted to appropriately provide telecommunication services to rural areas of developing countries.

TECHNOLOGIES FOR RURAL AREAS OF DEVELOPING COUNTRIES

Technologies presented in this article are not necessarily new. They have been used in other applications. We intend to introduce some modifications to the technologies so that they can be used in rural areas. We are interested in the technologies that can be used to link centers of development in rural areas such as trading centers and central switching offices. Subscribers within the center of development could be linked using the traditional local wire-loop because within the center the distances involved are small and lines do not run through isolated areas—hence they are well protected from acts of vandalism. Three technologies have been identified: cellular; point to multipoint using radio links; and use of electricity power lines.

Cellular Technology

A cellular technology is based on the concept that any service area is divided into small areas called cells, as shown in Figure 1. Each cell is served by a radio site having a combination transmitter and receiver (transceiver) called a base station. Each base station is linked to a mobile unit using radio signals. A mobile unit acts as an interface between subscriber and the base station. It consists of a transceiver, a handset, and control and interface facilities.

GSM is a cellular technology standard for Europe. Each base station can accommodate 123 frequency bands, each 200 kHz wide, through Frequency Division Multiple Access (FDMA). Each frequency band has the capacity to handle 8 speech channels, which are multiplexed in time resulting in Time Division Multiple Access (TDMA). This results in 984 channels. Not all of these channels are used for voice communication; some are used for signaling and control. It may appear an oxymoron to suggest that cellular technology be used in rural areas, since it has demonstrated that it is an expensive technology because it deploys mobile terminals. These terminals require sophisticated facilities to deal with phenomenon resulting from mobility such as paging, roaming, hand-over, and fading. The terminals have to be powered using small batteries, resulting in small coverage areas and many base stations, which may not be justified in rural areas where traffic is small. The batteries must also be frequently charged.

It has been suggested that fixed terminals be used. A fixed terminal system will not need all the facilities to deal with issues that are associated with mobility. The terminal will be a simple radio transceiver, which can be located at an optimal position for maximum reception. The transceiver can easily be powered by mains or solar. This will enable the use of high-powered systems that can cover large areas resulting in a small number of base stations to cater for the

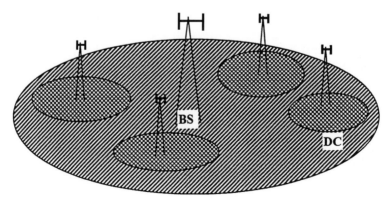

FIGURE 1
Cellular technology architecture.

service area. The author envisages one base station to cater for one government district. This may tremendously reduce the cost of providing telecommunication service to rural areas.

A high-powered base station transceiver, using omni-directional antenna, could be installed at the center of a district with the possibility of covering an area of 50 Km radius. A fixed terminal could be installed at each development center and tuned to one or more of the 123 frequency bands. This means that there will be at least 8 communication channels at each development center. These 8 channels offer enough capacity to cater for each center. These channels could be distributed among the subscribers using the traditional local wire loop through either multiplexer/demultiplexer or switching systems. If the number of subscribers is small, a simple multiplexer/demultiplexer arrangement could be used, leaving all the switching and billing to be done at the central exchange. If the number of subscribers in the center is large, demanding sharing of the channels, a switching system can be used. However, to make the system simple, we suggest that the billing be done at the central exchange. Information about start and termination of a call could then be sent to the central exchange for this purpose through the signaling and control channels.

Point to Multipoint Radio Link

A composite of radio transmitters/receiver arrangement deploying directional antennas are used. This concept has been used in telephone systems for transmission trunks at microwave. In this system the author suggests deploying simple radio links operating in the very high frequency (VHF) band.

A station consisting of a number of transceivers could be installed at a central position of a district or service area, as shown in Figure 2. The number of transceivers will depend on the number of development centers. The author suggests that for every center there must be at least one transceiver. The antenna of each transceiver could be directed to the center where a similar transceiver system is installed. This forms a radio link whose capacity may range from 8 up to 1,024 channels and length going up to 50 Km without deploying repeaters depending on the terrain. Subscribers could share these channels using a wire-line local loop through either simple multiplexer/demultiplexer arrangement or simple switching network. The billing could be done at the central exchange.

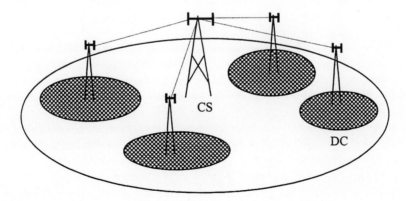

FIGURE 2
Point to multipoint technology architecture.

Use of Electrical Power Lines

Electrical power lines are used to transport electrical power from the generation plant to the consumer at very high voltage and current levels. These lines operate at 33 KV, 66 KV, and 132 KV. As the campaign against the use of trees for energy intensities, more of these lines will be installed in rural areas under Rural Electrification programs.

A network of such lines spanning the whole country offers an opportunity for provision of telephone services. The electrical power lines can also be used as telephone lines. Using transformers, telephone signals can be coupled to the lines. As long as these signals operate at appreciable higher frequency than the electric power voltage frequency (50 Hz), there will be no mutual interference. This will require some modulation and simple amplitude or frequency modulation can be used. At the receiver the signal could be de-coupled using a similar transformer before being demultiplexed and distributed among subscribers using local wire loop network.

This concept is used by ESCOM—the Malawi Power Company—for their voice and data communication. It could be adopted for voice communication on a public network.

TECHNICAL CHALLENGES

Implementation of systems based on these technologies may pose some problems. Perhaps the greatest challenge that may be encountered is the development of models to be used in the performance analysis. Before these technologies are adopted for rural areas, they have to be analyzed to establish their performance based on the constraints and conditions prevailing in rural areas. The performance of each system can be measured using several metrics, which may include capacity, signal to noise ratio (SNR) cost (both capital and running), link budget, congestion, and maintainability. These metrics in turn are functions of many parameters, such as distance covered, terrain, available technical expertise, telecommunication regulations, and power requirements. All these are interrelated. It is a technical challenge to establish either a mathematical or simulation model that captures the relationships among all the parameters, taking into account assumptions and constraints. Without such comprehensive models it may be difficult to establish facts to back their adoption for rural areas.

It has been suggested that the equipment to be used must be simple, manufactured using local expertise. This may be an attractive suggestion, since it may promote local industry and generate employment for the local population. However, the volumes involved may be low, so the industries involved may not break even. Regional collaboration such as SADC should be involved to benefit from economies of scale.

The use of electric power lines poses another challenge. Attenuation of signals increases with frequency. At high frequencies, the attenuation is prohibitively high. This means that low frequencies must be used, resulting in a relatively low bandwidth available, translating into a small number of communication channels. It is a challenge to develop digital coding systems that use the available channels effectively. In voice communication, predictive coding schemes could be employed, for example, in cellular telephone systems, with the use of Codebook Excited Linear Predictive (CELP) coding, a bandwidth of 32 kHz has been reduced to 6.5 kHz, (fivefold reduction in bandwidth requirements).

This decade has seen the proliferation of the concept that state-controlled telecommunication services must be deregulated and providers must be privatized. This means that telecommunication has become a commodity driven by demand. There is also a tendency for private service providers to invest where return on investment is high and mostly in monetary terms. Most of the benefits in rural areas realized from telecommunications are not in monetary form; hence, it will be difficult to attract companies to invest in rural telecommunications. However, governments could offer incentives such as tax deductions to service providers operating in rural areas.

CONCLUSIONS

The benefits realized from telecommunications service indicate that rural areas must be given some priority by governments when setting up telecommunication policies. Implementation of such policies must use local resources considering local conditions and constraints. Cellular technology, point to multipoint using radio link, and the use of power lines offer opportunities for provision of telecommunication services in rural areas of poor, developing countries. However, there is need to conduct analyses of their performance under prevailing local conditions, such as traffic distribution. There may be also the need to revisit government policies to cater for regional collaboration in developing telecommunication systems and effects of deregulation and privatization.

REFERENCES

Bulter, R.E. (1983). "Telecommunication for Development." ITU, International Telecommunications Union, Place des Nations CH-1211, Geneva, ZO, Switzerland.

Hudson, H.E. (1989). "Telecommunication and the Developing World." *IEEE Communication Magazine,* Vol. 25, No. 10. Fourth World Telecommunication Forum, Geneva, 1983.

Ng'-eny, K. Arap. (1989). "Challenges of Rural Communications; African Experience." *Telecommunication Journal,* Vol. 56, No. 9.

Yusuf, R.M. (1991). "Using Telecommunications as a Catalyst for Growth." *Telecommunication Journal,* Vol. 58, No. 7.

LAND MINES

Kevin J. Weddle
Colonel, U.S. Army Corps of Engineers

The land mine is one of the most capable weapons systems available today. It is cheap, easily mass-produced, and deadly. It is also controversial: It maims and kills soldiers and civilians alike. For much of this century, mine and countermine technology have gone round for round in innovation and counterinnovation. Here are some of the salient events of that fight.

In the 1920s Germany began producing antitank pressure mines, with fuses that detonated several pounds of explosives (up to 20 pounds or more in later models) when a vehicle ran over them. Other countries, including the United States, followed, and by World War II the devices were being put to use in Europe, Africa, and the Pacific. Since they were pressure-detonated, these early antitank mines typically did most of their damage to a tank's treads, leaving its crew unharmed and its guns still operational but immobilized and vulnerable to aircraft and enemy antitank weapons. Manufacturers soon began to produce both blast resistant tank treads and mines aimed at piercing a tank's thin belly armor, causing a catastrophic kill. In the latter, the Germans once again took the lead. During World War II they began using a mine with a "tilt rod" fuse, a thin rod standing approximately two feet up from the center of the charge and nearly impossible to see after the mine had been buried. As a tank passed over the mine, the rod was pushed forward, causing the charge to detonate directly beneath it. The blast often killed the crew and sometimes exploded onboard ammunition. Now that tank crews were directly at risk, they were less likely to plow through a minefield.

These antitank mines, particularly when coupled with smaller antipersonnel mines (to keep soldiers from removing the antitank mines), were so effective that by the early 1950s most armed forces considered them a standard part of their arsenal. In fact, no one saw a need for further major innovations in mine technology until the era of the Vietnam War, when close fighting and ambushes became the norm. Tripwires were then used more frequently, as were remotely detonated mines, such as the dreaded claymore antipersonnel mine. The claymore, which entered service in 1961, consists of a plate filled with explosives and with hundreds of small steel balls that fly out over a 60-degree arc when detonated, causing massive damage to a distance of 50 feet or more.

During the Korean and Vietnam Wars two increasingly evident problems caused a technological shift in mine development. The first was that people had figured out how to detect and move mines without detonating them. American mines were frequently replanted and used against American troops. The second was that they posed a lingering threat to civilians for years after the fighting ended. To prevent enemy soldiers from picking up a mine and moving it, the United States added antihandling devices, secondary fuses designed to detonate when a mine is moved. Next came "smart" or self-destructing mines, which helped decrease, if not eliminate, the threat to civilians. One version, for example, became inert once water seeped in and dissolved the explosive chemicals.

The United States tried a variety of smart mines during Vietnam, but the safety mechanisms were unreliable and a number of U.S. soldiers were injured by mines that were supposed to be dead. Then in the early 1970s a formal research and development program was begun. The results today are both highly reliable and capable of being rapidly disabled by hand, by armored vehicle, or even by aircraft. Exhaustive tests of the U.S. Army stockpile have proved the latest smart mines to be 99.999 percent reliable. These have limited lives of 4 hours to 15 days, after which they either self-destruct or self-neutralize. The antitank version even contains a fuse that can detect the magnetic signature of an armored vehicle, ensuring that a pedestrian won't cause an explosion.

As mine technology has progressed, countermining has had to keep pace. There are three main branches of countermining: detection, breaching, and clearing. Detection is obviously the most important because it is the key to preventing civilian casualties and ensuring maximum military flexibility. The earliest countermine efforts consisted of probing the ground with long poles or bayonets, being alert for such visual cues as disturbed earth and tripwires and, all too frequently, detonation. Unfortunately, the only real improvements, in active use, in mine detection since World War II are fiberglass probes (bayonets can activate magnetic fuses) and metal detectors.

The earliest metal detectors were developed during World War II; in response, manufacturers soon began to develop mines that contained almost no metal. The small amount of metal in most modern mines is often lost in background noise in the detector's earphones. At one point the Army tried a detector that measured the relative density of soil versus that of objects buried in it, but the device couldn't distinguish between a rock and a mine. So American soldiers in Desert Storm had virtually the same detection capabilities as their predecessors in World War II. Today the Army is working on a full suite of mine detection equipment using lasers, infrared and multispectral imaging, and ground-penetrating radar—see following article.

Of course, once mines have been spotted, they must be removed. This task is often more difficult than detection. In wartime it is usually done by breaching, cutting a lane through a minefield so that forces can pass and maintain the momentum of attack. Breaching places a premium on speed rather than safety. Since World War II it has usually been accomplished either by destroying individual mines or by using a special vehicle to plow them out of the way. The first method, in which engineers place charges next to each mine, is time-consuming, leaves soldiers exposed to enemy fire, and is ineffective when mines are buried. Frequently an explosive line charge is used instead: An armored vehicle fires a rocket-propelled flexible tube filled with plastic from the edge of the minefield to destroy or detonate mines and create a pathway. Unfortunately, modern blast-resistant mines can often survive even the largest line charges.

In the second method, mechanical breaching, a plow designed to withstand the blast of several mines at once is attached to the front of a tank or armored vehicle. It lifts mines out and off to the side. Mechanical breaching can conquer most mines, provided the ground is not too frozen or rocky, but mine technology has been keeping pace, and a number of countries are developing side attack or "off route" mines that can defeat breaching vehicles as they proceed through a minefield. These devices detect the presence of vehicles traveling along a path (through acoustical, seismic, or infrared means or with a tripwire or pressure tape) and fire armor-piercing projectiles at them from the side.

Military breaching operations remove or destroy only enough mines to enable a force to pass. Clearing all mines in peacetime, with a premium on safety and thoroughness, is much more difficult. However, clearing personnel can use wide variety of detection and removal tools that are too deliberate and time-consuming for soldiers. Blast shields, mini-mine detectors, encasing foam that allows mines to be moved without detonations, and robotic mine clearing machines are just a few.

For every invention that has rendered mines harmless or, at the very least, less harmful, a new innovation has made mines all the more unstoppable. And vice versa.

Reprinted by permission of copyright owner from *Invention & Technology*, Summer 1999.

LAND MINE DETECTION

Dimitri Donskoy

ABSTRACT

Stevens Institute of Technology has develop a land mine detection technology using a nonlinear seismo-acoustic technique. The detection system utilizes a combination of seismic and/or acoustic sources, noncontact vibration sensors, and detection algorithms. The system detects buried mines analyzing ground vibrations induced by the seismic and/or acoustic sources. The presence of a mine manifests itself with specific frequency response. The system and underlying physical mechanism have been experimentally validated under a wide variety of conditions. The field system under development will have high probability of detection with very low false alarm rate. It could be deployed on vehicles, unmanned robotic platforms, or as a handheld device. The system could be integrated with other detection sensors, such as linear seismo-acoustics and GPR. It can also be a part of a mine detection/removal system.

The last few years brought significant progress in landmine detection using acoustic energy. The work by the University of Mississippi (Sabatier & Xiang 1999), Georgia Institute of Technology (Scott & Martin 1999), Stevens Institute of Technology (Donskoy 1998, 1999), and other research groups (McKnight et al. 1999; Rechtien & Mitchell 1999) demonstrated that acoustic/seismic techniques have significant potential for reliable low false alarm detection of buried landmines. There are three principal approaches in acoustic detection.

1. Linear approach, based on measurements of local acoustic impedance above buried mine (Sabatier & Xiang 1999; Scott & Martin 1999)

2. Nonlinear approach, based on the measurements of nonlinear distortion of probing acoustic signal on mine-soil interface

3. Acoustic imaging approach (McKnight et al. 1999)

Some of these approaches, such as impedance measurements and imaging techniques, have been tested before but did not produce appreciable results. The recent successes of these techniques are owed to a great extent to the application of advanced sensors and signal processing algorithms. Thus, remote laser and electromagnetic sensors (vibrometers) allow for fast scan of large areas, providing spatial averaging, which is critical for detection in highly inhomogeneous mediums such as soil.

All of the acoustic techniques, including nonlinear ones, have their advantages and limitations. They all utilize similar hardware (acoustic/seismic sources and remote sensors) but different underlying physical mechanisms and, respectively, different signal processing algorithms. This opens the unique possibility of combining all of these techniques into one system. All of the techniques could complement each other, increasing overall capabilities of the resulting detection system.

To date, Stevens's contribution to acoustic detection techniques consists of the invention of the nonlinear seismo-acoustic detection method; the development of the practical physical model of seismo-acoustic detection, which explains and predicts results of linear impedance measurements as well as nonlinear ones; and the development of advanced remote sensors, which enables practical implementation of acoustic detection.

Stevens's Nonlinear Land Mine Detection technique is based on discovered phenomenon of nonlinear interaction of acoustic energy on the mine/soil interface. The technique does not depend on the material from which the mine is fabricated, whether it is metal, plastic, wood, or any other material. It depends on the fact that a mine is a "container" whose purpose is to contain explosive materials and associated detonation apparatus. The mine container is in contact with the soil in which it is buried. The container is an acoustically compliant article, whose compliance is notably different from the compliance of the surrounding soil. This difference is responsible for the mechanically nonlinear behavior of the soil/container interface, making it the detectable entity. Thus, for this new technology, the fact that the mine is buried is turned to a detection advantage. Because the technique intrinsically detects buried "containers," it is insensitive to rocks, tree roots, chunks of metal, bricks, and so forth.

To date this phenomenon has been experimentally and theoretically examined. The experimental studies were performed with real inert antipersonnel plastic and wooden mines under various soil conditions in the laboratory and outdoors. The experimental verification of the nonlinear behavior of the soil/mine interface was performed using two sound frequencies, f_1 and f_2, broadcasted from sound and/or seismic sources toward the buried object, as shown in Figure 3. The soil vibration is measured with an accelerometer and/or a Laser-Doppler Vibrometer (LDV). The presence of a mine manifests itself at a difference frequency $f_2 - f_1$.

Figures 4 and 5 demonstrate result of a nonlinear (different frequency $f_2 - f_1$) acoustic scan of a "mine field" with buried mines (simulated and real inert mines) and false targets (solid steel disk, solid wood, and brick). This plot clearly indicates the detection and discrimination capabilities of the developed technique.

The developed nonlinear approach detects features of a mine that are unique to mines, being insensitive to the false targets. This promises to deliver a high probability of detection with a very low false alarm rate, which is difficult to achieve, especially for AP mines, using competing technologies.

One of the most important steps in enabling and advancing the developed technique is to use remote sensors to measure ground vibration. Commercially available laser-doppler vibrometer proved to be applicable for the ground vibration measurements. However, it has some limitations: LDV is quite sensitive to environmental conditions and may not work for surfaces covered by

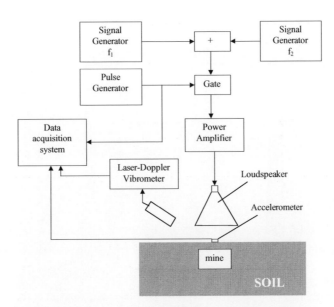

FIGURE 3
Experimental setup.

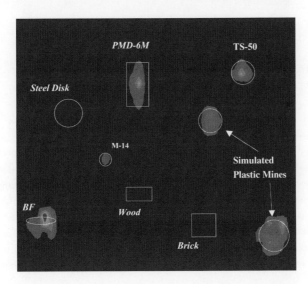

FIGURE 4
Mines and false targets. Nonlinear scan of mines and false targets.

grass, foliage, and so on. To address this problem, we developed a new remote sensing device—microwave vibrometer.

Two prototype vibrometers have been designed and fabricated. One operates on a 38-GHz sensor and is intended to receive surface vibrations, while the other, low-frequency 2.5 GHz vibrometer is capable to received vibration directly from the mine-soil interface. The last version of the 2.5-GHz vibrometer, developed in collaboration with Tegam, Inc., is a small, robust, and relatively inexpensive device suitable for field applications. A number of vibrometers can be deployed as an array for faster scanning of the investigated areas.

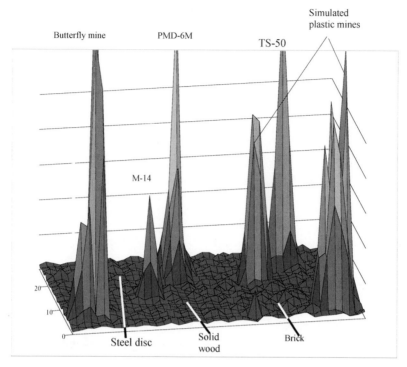

FIGURE 5

Result of scan. 3D representation of the nonlinear scan of the surface of the laboratory soil tank with buried mines and false targets.

CONCLUSION

A new nonlinear vibro-acoustic technique for land mine detection and discrimination has been developed and tested under laboratory and field conditions. The theoretical model and experimental results demonstrate the applicability of the developed technique to detect mines with a significantly reduced false alarm rate. The developed method utilizes the effect of the nonlinear transformation of probing bi-harmonic acoustic signals at the soil-mine interface. The resulting signal with a different frequency indicates the presence of a dynamically compliant object, such as a mine. The preliminary study, performed at Stevens, indicates that these signals can be detected with remote sensors such as laser-doppler and microwave vibrometers.

REFERENCES

Donskoy, D. M. (1998). "Nonlinear Seismo-Acoustic Technique for Land Mine Detection and Discrimination." 2nd Int. Conf. on Detection of Abandoned Land Mines, Edinburg, UK.

Donskoy, D. M. (1999). "Detection and Discrimination of Nonmetallic Mines." Proceedings of SPIE, Vol. 3710, 239–246.

Donskoy, D. M., N. Sedunov, and E. Whittaker, "Method and Apparatus for Remote Sensing of Vibration and Properties of Objects." U.S. Patent Pending.

McKnight, S. W., W. Li, and C. A. DiMarzio. (1999). "Imaging of Buried Objects by Laser-Induced Acoustic Detection." Proceedings of SPIE, Vol. 3710, 231–238.

Rechtien, R. D., and O. R. Mitchell. (1999). "Airborne Acoustic Focusing System for Landmine Detection." Proceedings of SPIE, Vol. 3710, 223–230.

Sabatier, J. M., and N. Xiang. (1999). "Laser-Doppler Based Acoustic-to-Seismic Detection of Buried Mines." Proceedings of SPIE, Vol. 3710, 215–222.

Scott, W. R., and J. S. Martin. (1999). "Experimental Investigation of the Acousto-Electromagnetic Sensor for Locating Land Mines." Proceedings of SPIE, Vol. 3710, 204–214.

U.S. Patent #5,974,881. "Method and Apparatus for Acoustic Detection of Mines and Other Buried Man-Made Objects."

MINING

Small-scale mining can be an important source of cash to the people who do it. It is widely done in the Third World. It is difficult work. It can be unsafe because of both accidents and airborne fumes and dust. It can be harmful to the environment because of poisoned runoff water and because of the destruction of vegetation, leading to erosion and landslides. Mining enterprises can vary in size from one or two people, usually using hand tools, through small companies with some task specialization, to very large companies. A few large companies are much easier for governments to regulate than many small activities, so in most cases small-scale mining is not supported or even closely monitored by governments. Large mining operations may be made less profitable by small operations, which remove high-grade ores, reducing the quality of the ore remaining, so later mining becomes technically difficult or impossible. Technical support to artisanal miners could make a significant difference in improving earnings, reducing hardships, and protecting the natural environment.

Gold, diamonds, and, sometimes copper and tin are the minerals commonly mined in low-technology operations. Another product fruitfully mined without mechanization is phosphate rock to be used as fertilizer. Coal and talc can be mined on the small scale. Rock quarries—yielding stones for building, for roads, and for sculpture—are also possible successful ventures.

Mining involves five operations: prospecting, digging, transporting, crushing and separating, and refining. The technology of each of these can be improved. A major question for small scale operating is whether shared facilities should be promoted—for example, a central rock crusher and separating apparatus used by several miners or a single professional geologist who serves miners in a particular area rather than improved facilities for individual miners.

Prospecting

Most small-scale mines extract only high-grade ores because only such are profitable for labor-intensive operations. Local people probably know where such deposits are or how to detect their presence—in southern Africa, for example, particular vegetation indicates copper. A professional geologist can suggest less apparent areas that will be susceptible to small-scale mining. A professional can also assay the quality of the ore, both for its mineral content and for the presence of impurities that will be hard to remove. The information obtained can be used to decide if a site is worth pursuing. Useful assaying results can be obtained from simple equipment, but much training in chemical techniques is essential. A shared central testing facility suitably staffed would make a useful contribution to the success of small-scale miners. The government usually does geological surveys, but these may be hard to interpret by informally trained miners. A trained person who will be trusted by miners could assist in suggesting promising places to dig.

Digging

Mining is done successfully, but with much effort, with shovels and double-pointed picks. A slightly more mechanized approach utilizes drilling and blasting performed by hand. Mining can be done through underground shafts or in an open pit—called "surface mining." Underground mining usually is done by at least moderate-size organizations and very seldom by informal groups of three or four artisans. The problems associated with underground mining include drainage of groundwater, ventilation (especially after blasting), lighting, and roof support. Groundwater is pumped out, but the effort can be considerable. Ventilation can be improved by arranging the entrances so only one collects prevailing winds. Problems common to both underground and surface mining include digging the ore out of the vein, collecting the ore, and haulage—carrying the ore out of the mine. The ore can be blasted from the vein by first drilling a hole, loading the hole with explosive, and detonating. Traditional miners heated the rock with fires and sometimes quenched the fire with cold water to crack the rock. Manual drills are available to make blasting holes—one such drill is called a Lisbeths hand rotary drill. Pneumatic drills are now commonly used but are relatively expensive—the expense is chiefly in the compressed air supply. A possibility is to have mobile drilling units shared among miners so blasting is done in each mine every week or so. In between blasts, the miners clear the blasted rock and may do sorting onsite.

Transporting

In many cases baskets are used to carry ore by hand, but more efficient means can be devised. Pieces of ore can be collected on a sheet of heavy cloth to be loaded onto wheelbarrows. A pulley arrangement, a windlass, can lift ore out of the shaft or open pit more easily than it can be carried. The mine should be planned with the lifting apparatus loaded at the lowest part of the mine so gravity can help move the ore to the lift. If some sorting of the ore is done in the mine, the amount of material hauled will be reduced. The depth of small mines is normally limited by limits on haulage, as well as by the water table and ventilation of an underground mine.

Crushing and Separating

Ore is crushed to facilitate separating the metal or metallic compounds from the base rock. Traditional metal workers used a mortar and pestle. The process was very laborious and dangerous because of dust and flying splinters. Mechanized crushers are large pieces of machinery, which is why a central shared facility may make sense. One problem with a crusher is that it can create very fine particles, making extraction of the metal difficult. Careful design of the equipment and skillful operation is needed. The process of crushing and separating is part of "beneficiation." Beneficiation is necessary to make the ore rich enough to be worth transporting and refining. In some cases, including small-scale mining, hand inspection and sorting is an efficient approach rather than more mechanized techniques.

Refining

Refining is usually done commercially—small-scale miners sell their enriched ore to large refineries. Gold is an exception. It can be separated using hydraulic process or can be purified by amalgamation with mercury.

Gold grains and nuggets are commonly separated from ore using water in pans and sluices. Panning is a manual process done by adding the ore to water in a pan and swirling the mixture; the heavier gold sinks to the bottom of the pan, where the miner can pick it out. A sluice is a long trough with low obstacles—riffles—on the bottom. Water containing the ore passes down the sluice, and the heavier gold is trapped by the riffles.

In the amalgamation process mercury is mixed with the oar. The gold and the mercury form an amalgam—analogous to a solution. The amalgam is heavy and can be trapped easily, in either a dry or wet process. Amalgamation can be done during crushing or after. The dry process is similar to the winnowing of grain. In the wet process the amalgam and ore mixture is washed with water and passed down a sluice—the added weight of the amalgam increases recovery compared to simply using water. The heavier amalgam-mercury mixture sinks below the water and can be recovered without difficulty. The amalgam is separated from the mercury by wringing through a cloth—often the miner's shirt. Gold is separated from the amalgam by burning—one technique uses a butane torch. The mercury evaporates, leaving the gold.

The use of mercury just described is common but unhealthy. Mercury, both in the liquid and gaseous state, is highly poisonous. The vapor affects not only the miner but also everyone else in the immediate area. A technique for collecting and condensing the vapor would improve the environment and also reduce costs for the miner. A condensing apparatus made of iron, according to one report, introduces compounds into the mercury that discolor the gold. In large-scale mining, cyanide leaching has replaced amalgamation, but cyanide is also poisonous and the technique has not found favor among small-scale miners.

Traditional African miners refined both copper and iron by smelting, within the memory of living people. Smelting took place in a large clay furnace heated with a wood fire augmented by air pumped from ingenious bellows. Only a small amount of metal could be obtained from a single smelt. Copper in Africa was used for jewelry, household implements, and currency. Iron was chiefly used for tools. The small amount of metal obtained from a single smelting operation was enough to be useful.

DIAMONDS

An appreciable amount—at least 10 percent—of the world diamond production comes from small-scale miners. At least some of this enters the market outside of regulated (legal) channels. Diamonds are found in volcanic "chimney vents" of hard rock or in weathered, decomposed products of diamond-containing rocks, sometimes as hardened sediments—conglomerate rocks—or as loose sediments in riverbeds. Diamonds are susceptible to splitting along cleavage planes, so crushing must be done carefully. In Southern Africa the feed material is thinly distributed over the ground surface, where it is exposed to natural weathering, sufficient to expose gems. Small and medium-scale operations consist basically of separating the diamonds from the base material.

Diamonds in small-scale operations are usually separated from the base material by hand sorting. The feed material is spread out over a table and the diamonds picked out because of their brilliance. Another low-technology approach, a water-borne system, makes use of the high weight of the diamonds. Greasing the interior surface of the sluice will assist in holding the diamonds as the water flows away.

PHOSPHATE FERTILIZER

About 10 percent of the world production of phosphate fertilizer comes from small-scale mines. Ground phosphate rock can be directly applied to fields. Although it is less soluble than commercial phosphate fertilizer made by processing the same rock, it may be less expensive than the commercial variety because of transportation costs. An advantage of an only slightly soluble fertilizer is that it is "slow release." Mixtures of ground phosphate rocks and sulphur-bearing minerals have greater solubility—the sulphur reacts to produce sulfuric acid, which reacts with the phosphate. If sulphuric acid is available for other reasons, it can be used directly on the phosphate. Nitric acid, produced by electrolysis, combines with phosphate to produce an effective fertilizer.

Small-scale limestone quarries can be worthwhile. Crushed limestone is useful in reducing the acidity of soils, and many soils in the humid tropics are very acidic. Lime is a low-value material, so a local source is necessary to make the enterprise pay, but for the same reason, a local source has a competitive advantage.

REFERENCES

Appleton, Don. (1990) "Rock and Mineral Fertilizer." *Appropriate Technology,* Vol. 17, No. 3, 25–27.

Bugnosen, Edmund. (1990). "Small-Scale Mining and Resource Development." *Appropriate Technology,* Vol. 17, No. 3, 1–3.

Cleary, David. (1990). "Gold Mining and Mercury Use in the Amazon Basin." *Appropriate Technology,* Vol. 17, No. 3, 17–19.

Legge, Christopher. (1990). "Small-Scale Mining—Making It Work." *Appropriate Technology,* Vol. 17, No. 3, 10–13

Priester, Michael, Thomas Hentschel, and Bernd Benthin. (1993). *Tools for Mining: Techniques and Processes for Small-Scale Mining.* A Publication of the Deutsches Zentrum für Entwicklungstechnologien—GATE. A Division of the Deutsche Gesellschaft für Technische Zusammenarbeit (GTZ) GmbH.

Spiropoulos, John. (1990). "Shared Mining Services—Making Technology Available." *Appropriate Technology,* Vol. 17, No. 3, 14–16.

INDEX

Printed and bound by CPI Group (UK) Ltd, Croydon, CR0 4YY

10/05/2025

01866494-0002